Sparse Arrays for Radar, Sonar, and Communications

IEEE Press
445 Hoes Lane
Piscataway, NJ 08854

Sparse Arrays for Radar, Sonar, and Communications

Edited by

Moeness G. Amin
Villanova University
Villanova, PA, USA

IEEE PRESS
WILEY

Published by John Wiley & Sons, Inc., Hoboken, New Jersey.
Published simultaneously in Canada.

Library of Congress Cataloging-in-Publication Data:

Names: Amin, Moeness G., editor. | John Wiley & Sons, publisher.
Title: Sparse arrays for radar, sonar, and communications / [edited by] Moeness G. Amin.
Description: Hoboken, New Jersey : Wiley-IEEE Press, [2024] | Includes index.
Identifiers: LCCN 2023040085 (print) | LCCN 2023040086 (ebook) | ISBN 9781394191017 (cloth) | ISBN 9781394191024 (adobe pdf) | ISBN 9781394191031 (epub)
Subjects: LCSH: Antenna arrays. | Antenna arrays–Design and construction. | Signal processing.
Classification: LCC TK7871.67.A77 S63 2024 (print) | LCC TK7871.67.A77 (ebook) | DDC 621.382/2–dc23/eng/20230907
LC record available at https://lccn.loc.gov/2023040085
LC ebook record available at https://lccn.loc.gov/2023040086

Cover Design: Wiley
Cover Image: © Lars Poyansky/Shutterstock; Yuichiro Chino/Getty Images; enot-poloskun/Getty Images

Set in 9.5/12.5pt STIXTwoText by Straive, Chennai, India

I dedicate this book to my family whose love and support have been instrumental for all accomplishments in my academic career.

Contents

About the Editor

Moeness G. Amin received his PhD degree from the University of Colorado, Boulder, in 1984. Since 1985, he has been with the Faculty of the Department of Electrical and Computer Engineering, Villanova University, Villanova, PA, USA, where he became the Director of the Center for Advanced Communications, College of Engineering, in 2002.

Dr. Amin is a Life Fellow of the Institute of Electrical and Electronics Engineers (IEEE); Fellow of the International Society of Optical Engineering (SPIE); Fellow of the Institute of Engineering and Technology (IET); and a Fellow of the European Association for Signal Processing (EURASIP). He is the recipient of the 2022 IEEE Dennis Picard Gold Medal in Radar Technologies and Applications; the 2017 Fulbright Distinguished Chair in Advanced Science and Technology; the 2016 Alexander von Humboldt Research Award; the 2016 IET Achievement Medal; the 2014 IEEE Signal Processing Society Technical Achievement Award; the 2009 Technical Achievement Award from the European Association for Signal Processing; the 2015 IEEE Aerospace and Electronic Systems Society Warren D. White Award for Excellence in Radar Engineering. He is also the recipient of the 2010 Chief of Naval Research Challenge Award; the 1997 IEEE Philadelphia Section Award, the 1997 Villanova University Outstanding Faculty Research Award. He is the recipient of the IEEE Third Millennium Medal.

Dr. Amin served as Chair/Member of the Electrical Cluster of the Franklin Institute Committee on Science and the Arts (2001–2016). He served on the Editorial Board of the *Proceedings of the IEEE* and is currently a member of the Editorial Board of the IEEE Press. Dr. Amin has over 900 journal and conference publications in signal processing theory and applications, covering the areas of wireless communications, radar, satellite navigations, ultrasound, healthcare, and RFID. He is the editor of four books on radar.

List of Contributors

Elias Aboutanios
School of Electrical Engineering and Telecommunications
University of New South Wales
Kensington
NSW
Australia

Kaushallya Adhikari
Electrical, Computer, and Biomedical Engineering Department
University of Rhode Island
Kingston
RI
USA

Fauzia Ahmad
Department of Electrical and Computer Engineering
Temple University
Philadelphia
PA
USA

Saleh A. Alawsh
Center for Communication Systems and Sensing
King Fahd University of Petroleum and Minerals
Dhahran
Saudi Arabia

Ahmed Alkhateeb
School of Electrical
Computer and Energy Engineering
Arizona State University
Tempe
AZ
USA

M. Burak Alver
Faculty of Engineering & Natural Sciences
Maltepe University
Istanbul
Turkey

Moeness G. Amin
Center for Advanced Communications (CAC)
College of Engineering
Villanova University
Villanova
PA
USA

Nicola Anselmi
DICAM - Department of Civil, Environmental, and Mechanical Engineering
ELEDIA Research Center
University of Trento
Trento
Italy

Müjdat Çetin
Department of Electrical and Computer Eng. &
Georgen Institute for Data Science
University of Rochester
Rochester
NY
USA

Sundeep Prabhakar Chepuri
Department of Electrical Communication
Engineering (ECE)
Indian Institute of Science
Bengaluru
India

Mario Coutino
Netherlands Organisation for Applied
Scientific Research (TNO)
The Hague
The Netherlands

Ahmet M. Elbir
Interdisciplinary Centre for Security
Reliability and Trust (SnT)
University of Luxembourg
Esch-sur-Alzette
Luxembourg

Xiang Gao
National Key Laboratory of Science and
Technology on Communications
University of Electronic Science and
Technology of China
Sichuan
China

Peter Gerstoft
NoiseLab
University of California San Diego
La Jolla
CA
USA

Sotirios Goudos
ELEDIA Research Center
Aristotle University of Thessaloniki
Thessaloniki
Greece

Yujie Gu
Advanced Safety & User Experience
Aptiv
Agoura Hills
CA
USA

Alper Güngör
Aselsan Research Center
Aselsan
Ankara
Turkey

Emre Güven
Cognizen Arge Mühendislik ve Yazılım A.Ş.
Ankara
Turkey

Syed A. Hamza
Electrical Engineering Department
School of Engineering
Widener University
Chester
PA
USA

Koichi Ichige
Department of Electrical and Computer
Engineering
Yokohama National University
Yokohama
Japan

Kyle Juretus
ECE Department
College of Engineering
Villanova University
Villanova
PA
USA

Visa Koivunen
Department of Information and Communications Engineering
Aalto University
Espoo
Finland

Pranav Kulkarni
Department of Electrical Engineering
California Institute of Technology
Pasadena
CA
USA

Geert Leus
Faculty of Electrical Engineering, Mathematics and Computer Science (EEMCS)
Delft University of Technology
Delft
The Netherlands

Wei Liu
School of Electronic Engineering and Computer Science
Queen Mary University of London
London
UK

Andrea Massa
DICAM - Department of Civil, Environmental, and Mechanical Engineering
ELEDIA Research Center
University of Trento
Trento
Italy

School of Electronic Science and Engineering
ELEDIA Research Center
University of Electronic Science and Technology of China
Chengdu
China

ELEDIA Research Center
Tsinghua University
Beijing
Haidian
China

School of Electrical Engineering
Tel Aviv University
Tel Aviv
Israel

and

ELEDIA Research Center
University of Illinois Chicago
Chicago
IL
USA

Kumar Vijay Mishra
Computational and Information Sciences Directorate (CISD)
United States DEVCOM Army Research Laboratory
Adelphi
MD
USA

Ali H. Muqaibel
Electrical Engineering Department
King Fahd University of Petroleum and Minerals
Dhahran
Saudi Arabia

and

Center for Communication Systems and Sensing
King Fahd University of Petroleum and Minerals
Dhahran
Saudi Arabia

Giacomo Oliveri
DICAM - Department of Civil, Environmental, and Mechanical Engineering
ELEDIA Research Center
University of Trento
Trento
Italy

Piya Pal
Department of Electrical and Computer Engineering
University of California San Diego
La Jolla
CA
USA

Lorenzo Poli
DICAM - Department of Civil, Environmental, and Mechanical Engineering
ELEDIA Research Center
University of Trento
Trento
Italy

Heng Qiao
University of Michigan-Shanghai Jiao Tong University Joint Institute
Shanghai Jiao Tong University
Shanghai
China

Guodong Qin
School of Electronic Engineering
Xidian University
Xi'an
Shaanxi
China

Si Qin
Microsoft China
Beijing
China

Robin Rajamäki
Electrical and Computer Engineering Department
University of California San Diego
La Jolla
CA
USA

and

Department of Information and Communications Engineering
Aalto University
Espoo
Finland

Paolo Rocca
DICAM - Department of Civil, Environmental, and Mechanical Engineering
ELEDIA Research Center
University of Trento
Trento
Italy

and

ELEDIA Research Center
Xidian University
Xi'an
Shaanxi Province
China

Ammar Saleem
Faculty of Engineering & Natural Sciences
Sabanci University
Istanbul
Turkey

Marco Salucci
DICAM - Department of Civil, Environmental, and Mechanical Engineering
ELEDIA Research Center
University of Trento
Trento
Italy

Zhiguo Shi
Key Laboratory of Collaborative Sensing and Autonomous Unmanned Systems of Zhejiang Province
Zhejiang University
Hangzhou
China

and

College of Information Science and Electronic Engineering
Zhejiang University
Hangzhou
China

Palghat P. Vaidyanathan
Department of Electrical Engineering
California Institute of Technology
Pasadena
CA
USA

Kathleen E. Wage
Electrical and Computer Engineering Department
George Mason University
Fairfax
VA
USA

Xiangrong Wang
School of Electronic and Information Engineering
Beihang University
Beijing
China

Xianghua Wang
School of Artificial Intelligence
Beijing University of Posts and Telecommunications
Beijing
China

Weitong Zhai
School of Electronic and Information Engineering
Beihang University
Beijing
China

Yimin D. Zhang
Department of Electrical and Computer Engineering
Temple University
Philadelphia
PA
USA

Chengwei Zhou
State Key Laboratory of Industrial Control Technology
Zhejiang University
Hangzhou
China

and

College of Information Science and Electronic Engineering
Zhejiang University
Hangzhou
China

Preface

The timing of this book is motivated by recent advances in sparse arrays design. The book builds on strides made over the last decade in convex optimizations and machine learning, and covers various emerging applications in sensing and communications. It is intended as a reference book for students, faculty, and scientists pursuing research in the broad areas of active and passive sensing and wireless technologies. Also, benefiting from this book are the engineers who engage in system developments design, and manufacturing using multi-sensor platforms.

Sensor arrays have endowed transmitters and receivers with far more capabilities in radar, sonar, and communications to achieve system objects compared to a single sensor deployment. Sparse arrays are a type of sensor arrays which have been around for many years. It was recognized early on that sensors can be arranged in random or nonuniform ways, different from the conventional uniform array. Different array structures, in essence, can provide different tradeoffs comprising performance, aperture size, mutual coupling, cost, and implementation.

Recent research findings in sparse arrays has shown increased capacity in wireless communications and improved sensing in radar and sonar using both narrowband and wideband signal models. Sparse arrays, in lieu of conventional uniform arrays, allow dealing with more targets and radio frequency (RF) emitters than the number of physical sensors. Optimum Sparse arrays for direction finding and beamforming have been developed which provide superior performance in target detection and tracking compared to uniform arrays of the same aperture or number of sensors. Sparse array applications have broadened to include synthetic aperture radar (SAR), multiple-input multiple-output (MIMO) radar, Massive MIMO communications, 5G and Beyond, RF surveillance and sensing, target localization, wideband urban radar processing, and space–time adaptive processing.

This book captures new important applications of sparse arrays and delineates their utilities. It describes various array configurations for stationary and moving sensor platforms and examines transceiver array design under fixed and dynamic environments. The book includes array designs that seek to increase the corresponding one-dimensional and two-dimensional virtual array apertures and others that configure and synthesize the sparse arrays based on solutions of constrained minimization problems. Optimality is also considered for combined sparse array-waveform design. The book presents a view of the Perception–Action Cycle in cognitive sensing through the lens of sparse arrays and antenna switching technology for direction finding and beamforming.

I am grateful to the authors of all chapters. They have provided the state-of-the-art coverage of sparse arrays supported by analytical rigor and validated by examples involving real and simulated data.

The book consists of 15 chapters. Each chapter addresses a different aspect of sparse array design and application. *Chapter 1 introduces recent and conventional sparse arrays and explains their advantages*. Using the notion of a difference coarray, the chapter delineates how sparse sensor arrays can identify more sources than the number of sensors. *Chapter 2 considers sparse array interpolation*. It addresses the interpolation problem of obtaining a fully augmented sparse array where the co-array (virtual array) has consecutive elements, that is, all spatial autocorrelation lags are available. *Chapter 3 considers wideband and multi-frequency operations of sparse arrays for both beamforming and direction-of-arrival (DOA) estimation*. It demonstrates enhanced sparse array degrees of freedom (DOF) beyond those obtained using narrowband signals.

Chapter 4 deals with sparse arrays in view of limited data. It characterizes array performance by deriving the interplay between the array geometry, number of snapshots and signal-to-noise ratio (SNR). *Chapter 5 deals with two-dimensional (2D) sparse array design which is fundamental to source localization applications*. The main objective is to increase the DOF which is a measure of the number of sources that can be correctly localized.

Chapters 6 and 7 apply deep learning (DL) for sparse array design. The two chapters, respectively, present an overview of DL applications of sparse array design for high resolution direction finding and maximum signal-to-interference and noise ratio (MaxSINR) beamforming. *Chapter 8 deals with optimum placements of sensors in sensor networks*. It considers both detection and estimation problems utilizing a limited number of network sensors. *Chapter 9 focuses on active sensing based on sparse arrays*. Using the notion of sum-coarray, it illustrates several advantages of sparse arrays over conventional uniform arrays, including improved resolution, fewer physical sensors, and the capability to identify a large number of scatterers.

Chapter 10 covers sparse arrays for MIMO array radar. In this regard, it deals with active sensing and complements Chapter 9. It examines joint sparse optimizations of MIMO radar transceiver under different task requirements. *Chapter 11 introduces and analyzes generalized structured sparse arrays for fixed and moving platforms*. It shows that both platforms yield increased DOF of the associated virtual arrays, thus improving DOA estimation performance. *Chapter 12 deals with sparse observation synthetic aperture radar (SAR)* and presents optimization and learning-based algorithms for SAR imaging. It applies compressed sensing-based imaging and deep learning-based regularizers within a physics-based observation model, and provides a supportive real SAR data example.

Chapter 13 tackles sparse arrays for sonar systems. It introduces fundamental concepts in passive and active sonar using sparse arrays and presents a unifying framework for understanding standard processing techniques and performance metrics. *Chapter 14 discusses sparse arrays in the context of phased arrays and antenna technology*. It reviews recent advances and trends in the synthesis of unconventional array architectures and presents the guidelines for the state-of-the-art strategies in phased array synthesis for next generation wireless communications. *Chapter 15 provides an overview of the different sparse array architectures for MIMO communication systems and their design approaches*. It draws insights into the potential gains achieved under fully digital and hybrid analog–digital architectures in various communication scenarios.

Moeness G. Amin
Villanova University
Villanova
Pennsylvania, USA

1

Sparse Arrays: Fundamentals

Palghat P. Vaidyanathan and Pranav Kulkarni

Department of Electrical Engineering, California Institute of Technology, Pasadena CA, USA

1.1 Introduction

A sensor array or just "array" is a collection of sensors that performs spatial sampling of signals arriving from various directions. In general, the signals can be EM waves, sound waves, or other types of signals. In this chapter, we concentrate on plane monochromatic EM waves arriving from different directions and impinging on a sensor array. Arrays typically processes the signals received at the sensors to estimate the *directions of arrival* or **DOA**s. In some cases the arrays operate in the beamforming mode, by providing large gain in certain directions of arrival, compared to other directions.

An array is said to be linear if the sensors are on a straight line, and this is what we consider in this introductory chapter. More general arrays are discussed in Chapter 5 in the book. The most commonly used arrays are uniform linear arrays or **ULA**s, which have adjacent sensors separated by $\lambda/2$ where λ is the wavelength of the impinging monochromatic waves. But there are also other types of arrays where a number of adjacent sensor pairs have larger separation. These are called **sparse arrays**, and they have a number of advantages.

In this chapter, we will introduce sparse arrays and discuss their advantages. In the world of array signal processing, such arrays have been known for many decades, especially in the context of imaging and beamforming. Sparse arrays received more attention in the last decade because of the introduction of **nested** and **coprime** arrays, which have certain advantages compared to theoretically optimal arrays that have been well-known for much longer.

We begin by reviewing the fundamentals of Array Signal Processing in Section 1.2, including DOA estimation and beamforming. Sparse arrays are introduced in Section 1.3. A unique property of some sparse arrays is that the number D of DOAs that can be identified can exceed the number N of sensors in the array. The reason for this counterintuitive property is explained in Section 1.4, and in greater detail in Section 1.5. The **difference coarray**, which is central to this property, is also introduced. Specific algorithms for DOA estimation with sparse arrays, such as coarray MUSIC, are introduced in Section 1.6. Well-known classes of sparse arrays such as nested arrays and coprime arrays are introduced in Section 1.7, and their properties are discussed. Then, in Section 1.8, we review classical optimal sparse arrays such as the minimum redundancy arrays or **MRA**s, minimum hole arrays or **MHA**s (also called Golomb rulers), and **perfect arrays**.

Section 1.9 explains how coprime arrays can be used to develop a novel type of beamformer, where the number of nonoverlapping spatial beams can be much larger than the number of sensors.

Sparse Arrays for Radar, Sonar, and Communications, First Edition. Edited by Moeness G. Amin.

This is another unique advantage of some sparse arrays. In Section 1.10, we conclude the chapter by providing directions for further reading.

Notations We use bold face letters for matrices and vectors, as in $\mathbf{A}$ and $\mathbf{v}$, where usually uppercase letters denote matrices and lowercase letters denote column vectors. The notations $\mathbf{A}^T, \mathbf{A}^*$, and $\mathbf{A}^H$ denote, respectively, transpose, conjugate, and transpose-conjugate. The notation $(n_1, n_2, \ldots, n_{N-1})$ stands for the greatest common divisor (gcd) of the integers $n_1, n_2, \ldots, n_{N-1}$.

1.2 Basics of Array Processing

Figure 1.1 shows a linear array [1], that is, an array with sensors located on a straight line. There are N sensors located at $z_0, z_1, \ldots, z_{N-1}$. The figure also schematically shows a plane EM wave impinging on the array. The **direction of arrival** (DOA) is indicated by the angle θ shown in the figure. The "plane wave" property arises from the assumption that the source which generates the wave is far away. With λ denoting the wavelength, we assume in this chapter that z_i are at integer multiples of $\lambda/2$, that is,

$$z_i = n_i \frac{\lambda}{2}, \tag{1.1}$$

where $n_i < n_{i+1}$ are integers. Such arrays are called **integer arrays** or integer linear arrays. Either one of the following sets

$$\mathcal{Z} = \{z_0, z_1, \ldots, z_{N-1}\}, \quad \mathcal{N} = \{n_0, n_1, \ldots, n_{N-1}\} \tag{1.2}$$

specifies such an array. When we say that the integer array is $\{n_0, n_1, \ldots, n_{N-1}\}$, we actually mean that the sensors are at (1.1). We will refer to the quantity $A = n_{N-1} - n_0$ as the **aperture** of the array. The array

$$\mathcal{N} = \{0, 1, 2, \ldots, N-1\} \tag{1.3}$$

is called a uniform linear array or **ULA** and has been widely used for decades. Arrays of the form $\mathcal{N} = \{0, M, 2M, \ldots, (N-1)M\}$, $M > 1$ are also called ULAs, but they are "sparse" ULAs because $M > 1$. To distinguish, we sometimes refer to (1.3) as the standard ULA.

1.2.1 Expression for the Array Output

A plane monochromatic propagating wave can be expressed as

$$f(t, x) = Ae^{j(\omega_c t - \kappa x)}, \tag{1.4}$$

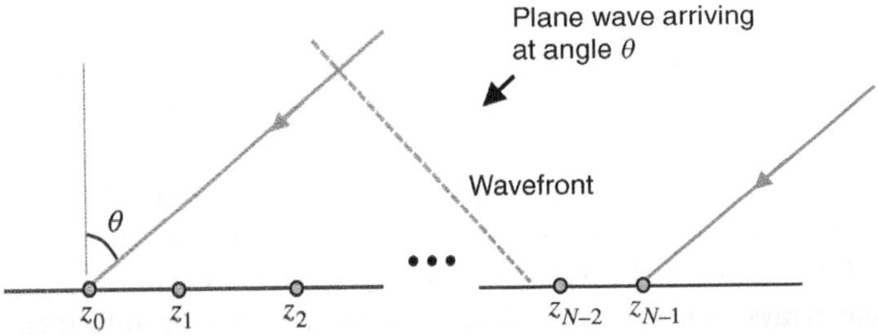

Figure 1.1 A plane wave arriving at angle θ and impinging on a N-sensor linear array.

where A is its amplitude and ω_c is the frequency. Here, t is time, x is the spatial coordinate along the line of propagation, and κ is the wavenumber, so that $\lambda = 2\pi/\kappa$. The velocity of propagation is given by

$$c = \frac{\omega_c}{\kappa} = \frac{\omega_c}{2\pi}\lambda. \tag{1.5}$$

In practice, the signal received is not monochromatic. A more reasonable assumption would be that it has a narrow bandwidth around the frequency ω_c. In this case, the signal received by the ith sensor has the general form

$$\alpha_i s(t) e^{j\omega_c t}, \tag{1.6}$$

where α_i is a constant (i.e., independent of time) and $s(t)$ is a narrowband lowpass signal.[1] Its bandwidth is small compared to ω_c, or equivalently $s(t)$ changes very slowly compared to $e^{j\omega_c t}$. Note that (1.6) has Fourier transform $\alpha_i S(j(\Omega - \omega_c))$, where $S(j\Omega)$ is the Fourier transform of $s(t)$.

The EM wave has the form $e^{j(\omega_c t - \kappa x)}$, where x is the distance coordinate along the direction of propagation. Let us say at time t the signal at sensor z_{N-1} is exactly $e^{j\omega_c t}$. Then the signal at sensor i is $e^{j(\omega_c t - \kappa\Delta x)}$ where Δx is the extra distance traveled by the wave to arrive at z_i (as compared to z_{N-1}). It is clear that $\Delta x = (z_{N-1} - z_i)\sin\theta$. So the signal at sensor i at time t is

$$e^{j(\omega_c t - \kappa(z_{N-1} - z_i)\sin\theta)} = e^{j\omega_c t} e^{-j\kappa z_{N-1}\sin\theta} e^{j\kappa z_i \sin\theta} = \alpha e^{j\kappa z_i \sin\theta}, \tag{1.7}$$

where α is independent of i. Since $z_i = n_i\lambda/2$ and $\kappa\lambda = 2\pi$, the above can be written as $\alpha e^{jn_i\omega}$, where $\omega = \pi\sin\theta$. It therefore follows that the set of N received signals has the form

$$\begin{bmatrix} r_0(t) \\ r_1(t) \\ \vdots \\ r_{N-1}(t) \end{bmatrix} = \alpha\, s(t) e^{j\omega_c t} \times \begin{bmatrix} e^{jn_0\omega} \\ e^{jn_1\omega} \\ \vdots \\ e^{jn_{N-1}\omega} \end{bmatrix} \tag{1.8}$$

where ω is the dimensionless quantity

$$\omega = \pi\sin\theta, \tag{1.9}$$

called the **directional frequency**, and is a different way to represent the DOA θ. For simplicity, we often refer to ω also as DOA. We always assume θ to be in the range

$$-\frac{\pi}{2} \le \theta < \frac{\pi}{2}, \tag{1.10}$$

so that $-\pi \le \omega < \pi$. The high frequency signal $e^{j\omega_c t}$ can be removed by demodulation at the output of every sensor. Assuming that all sensors are omnidirectional and have identical physical properties, the constant α in (1.8) is independent of sensor number i. So we shall ignore α and $e^{j\omega_c t}$ from now on, and write the output of the array in response to the plane wave as

$$\underbrace{\begin{bmatrix} x_0(t) \\ x_1(t) \\ \vdots \\ x_{N-1}(t) \end{bmatrix}}_{\text{call this } \mathbf{x}(t)} = s(t) \times \begin{bmatrix} e^{jn_0\omega} \\ e^{jn_1\omega} \\ \vdots \\ e^{jn_{N-1}\omega} \end{bmatrix}, \tag{1.11}$$

1 For generality, we assume $s(t)$ is complex. We can regard $s(t)$ as the complex envelope of some real lowpass signal [2].

where $s(t)$ is the amplitude of the signal arriving from DOA ω (or more precisely DOA θ). More generally, if we have D sources with the same wavelength λ (frequency ω_c) with DOAs

$$\theta_0, \theta_1, \dots, \theta_{D-1}, \tag{1.12}$$

the array output is the superposition

$$\begin{bmatrix} x_0(t) \\ x_1(t) \\ \vdots \\ x_{N-1}(t) \end{bmatrix} = \begin{bmatrix} e^{jn_0\omega_0} & e^{jn_0\omega_1} & \dots & e^{jn_0\omega_{D-1}} \\ e^{jn_1\omega_0} & e^{jn_1\omega_1} & \dots & e^{jn_1\omega_{D-1}} \\ \vdots & \vdots & \ddots & \vdots \\ e^{jn_{N-1}\omega_0} & e^{jn_{N-1}\omega_1} & \dots & e^{jn_{N-1}\omega_{D-1}} \end{bmatrix} \begin{bmatrix} s_0(t) \\ s_1(t) \\ \vdots \\ s_{D-1}(t) \end{bmatrix} \tag{1.13}$$

where $\omega_m = \pi \sin \theta_m$, and $s_m(t)$ is the amplitude of the mth source, assumed to be "narrowband" (i.e., bandwidth $<< \omega_c$).

1.2.2 Sampling the Array Outputs

In practice, the array outputs are uniformly sampled to obtain $x_i[k] = x_i(kT)$, where T is the sample spacing. So the set of sampled outputs is

$$\underbrace{\begin{bmatrix} x_0[k] \\ x_1[k] \\ \vdots \\ x_{N-1}[k] \end{bmatrix}}_{\text{call this } \mathbf{x}[k]} = \underbrace{\begin{bmatrix} e^{jn_0\omega_0} & e^{jn_0\omega_1} & \dots & e^{jn_0\omega_{D-1}} \\ e^{jn_1\omega_0} & e^{jn_1\omega_1} & \dots & e^{jn_1\omega_{D-1}} \\ \vdots & \vdots & \ddots & \vdots \\ e^{jn_{N-1}\omega_0} & e^{jn_{N-1}\omega_1} & \dots & e^{jn_{N-1}\omega_{D-1}} \end{bmatrix}}_{\text{call this } \mathbf{A}} \underbrace{\begin{bmatrix} s_0[k] \\ s_1[k] \\ \vdots \\ s_{D-1}[k] \end{bmatrix}}_{\text{call this } \mathbf{s}[k]}, \tag{1.14}$$

where $s_i[k] = s_i(kT)$. Since there is always noise in the received signal, a more accurate model for the array output samples has the form

$$\mathbf{x}[k] = \mathbf{A}\mathbf{s}[k] + \mathbf{e}[k], \quad 0 \le k \le K-1, \tag{1.15}$$

where $\mathbf{e}[k]$ is the additive noise term. Here K is the number of samples or **snapshots**. The matrix $\mathbf{A}$ in (1.14) can be written as

$$\mathbf{A} = \left[\mathbf{a}(\omega_0) \ \mathbf{a}(\omega_1) \ \cdots \ \mathbf{a}(\omega_{D-1})\right], \tag{1.16}$$

where $\mathbf{a}(\omega_i)$ are column vectors:

$$\mathbf{a}(\omega) = \left[e^{jn_0\omega} \ e^{jn_1\omega} \ \dots \ e^{jn_{N-1}\omega}\right]^T. \tag{1.17}$$

$\mathbf{A}$ is called the **array manifold matrix**, and $\mathbf{a}(\omega)$ is called the **array manifold vector** or the **steering vector**.

1.2.3 Covariance of the Array Output

We assume that the signal $\mathbf{s}[k]$ and noise $\mathbf{e}[k]$ are jointly wide sense stationary (WSS) random processes with the following properties:

- A1. Signal and noise have zero mean: $E(\mathbf{s}[k]) = \mathbf{0}$, $E(\mathbf{e}[k]) = \mathbf{0}$.
- A2. Signal and noise are uncorrelated: $E\left(\mathbf{s}[k]\mathbf{e}^H[k]\right) = \mathbf{0}$.

- A3. $\mathbf{R_{ee}} = E\left(\mathbf{e}[k]\mathbf{e}^H[k]\right) = \sigma_e^2\mathbf{I}$. In words, noise components $e_{i_1}[k]$ and $e_{i_2}[k]$ at any two sensors are uncorrelated, with *equal variance* $\sigma_e^2 = E[|e_i[k]|^2]$ at all sensors.
- A4. $\mathbf{R_{ss}} = E\left(\mathbf{s}[k]\mathbf{s}^H[k]\right) > \mathbf{0}$, that is, the $D \times D$ signal covariance is positive definite. In particular, $\mathbf{R_{ss}}$ has rank D.

These assumptions typically hold over short durations of time (a few hundred samples). Under these assumptions, the array output $\mathbf{x}[k]$ is a zero-mean WSS process with a $N \times N$ covariance matrix

$$\underbrace{\mathbf{R_{xx}}}_{N\times N} = E\left(\mathbf{x}[k]\mathbf{x}^H[k]\right) = \underbrace{\mathbf{A}}_{N\times D}\underbrace{\mathbf{R_{ss}}}_{D\times D}\mathbf{A}^H + \sigma_e^2\mathbf{I}. \tag{1.18}$$

In expressions involving expectation operators, we often drop the snapshot index k for simplicity. For example, we write $E(\mathbf{s})$, $E(\mathbf{x}\mathbf{x}^H)$, etc., instead of $E(\mathbf{s}[k])$, $E(\mathbf{x}[k]\mathbf{x}^H[k])$. This is because these expectations do not depend on k, in view of the WSS assumption. The $N \times N$ covariance (1.18) is usually estimated by computing the **sample covariance matrix**

$$\hat{\mathbf{R}}_{\mathbf{xx}} = \frac{1}{K}\sum_{k=0}^{K-1}\mathbf{x}[k]\mathbf{x}^H[k]. \tag{1.19}$$

Here, K is the number of shapshots. Since k is the time index, the WSS property mentioned above is called temporal WSS property, to distinguish it from something called spatial WSS property, which we shall encounter later.

1.2.4 The MUSIC Algorithm

For the case $D < N$, the DOAs can be found using algorithms such as MUSIC [3] and ESPRIT [4] by starting from (an estimate of) the eigenvectors of $\mathbf{R_{xx}}$. A *sufficient* condition for these algorithms to identify $D < N$ DOAs without ambiguity is that the array be a standard ULA as in (1.3). Assuming this is the case, we briefly mention how MUSIC works, as it is essential for an understanding of Section 1.6. With $\sigma_e^2 > 0$, the autocorrelation matrix $\mathbf{R_{xx}}$ in (1.18) is positive definite, and its eigenvalues can be ordered as

$$\lambda_0 \geq \lambda_1 \geq \cdots \geq \lambda_{D-1} > \lambda_D = \cdots = \lambda_{N-1} = \sigma_e^2 > 0. \tag{1.20}$$

Since $\mathbf{R_{xx}}$ is Hermitian, we can always find an orthonormal set of eigenvectors $\mathbf{e}_i$, so that the $N \times N$ eigenvector matrix

$$\mathbf{E} = \left[\mathbf{e}_0 \; \mathbf{e}_1 \; \cdots \; \mathbf{e}_{D-1} \mid \mathbf{e}_D \; \mathbf{e}_{D+1} \; \cdots \; \mathbf{e}_{N-1}\right], \tag{1.21}$$

is unitary. Now, it can be shown that the subspace spanned by the first D eigenvectors above is the same as the subspace spanned by the D columns $\mathbf{a}(\omega_i)$ of $\mathbf{A}$. This is called the **signal subspace** because $\omega_i = \pi \sin\theta_i$, where θ_i are the DOAs of the signals. The vectors $\mathbf{a}(\omega_i)$ are therefore orthogonal to the last $N - D$ eigenvectors in (1.21). Defining the **noise subspace** matrix

$$\mathbf{E}_n = \left[\mathbf{e}_D \; \mathbf{e}_{D+1} \; \cdots \; \mathbf{e}_{N-1}\right], \tag{1.22}$$

it then follows that $\mathbf{a}^H(\omega_i)\mathbf{E}_n = \mathbf{0}$ for all DOAs ω_i. Under some conditions (which are surely satisfied for the ULA when $D < N$) it can also be shown that $\mathbf{a}^H(\omega)\mathbf{E}_n = \mathbf{0}$, *only if* ω is a DOA. We can

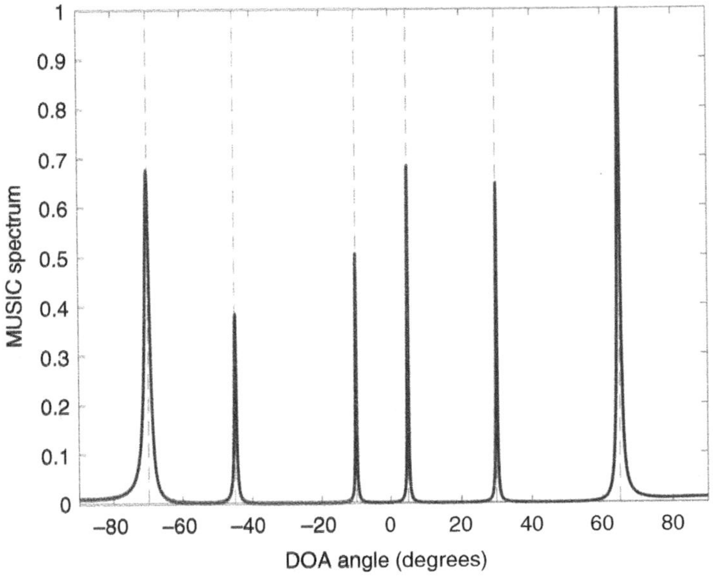

Figure 1.2 Example of a MUSIC spectrum. See the text for details.

therefore identify the $D < N$ DOAs by finding the D values of $\omega \in [-\pi, \pi)$ such that $\mathbf{a}^H(\omega)\mathbf{E}_n = \mathbf{0}$.[2] In practice, since $\mathbf{E}_n$ can only be estimated from the estimate $\hat{\mathbf{R}}_{xx}$, one plots the quantity

$$P(\omega) = \frac{1}{\mathbf{a}^H(\omega)\mathbf{E}_n\mathbf{E}_n^H\mathbf{a}(\omega)} \tag{1.23}$$

on a dense grid of ω and locates its D peaks to find ω_i. $P(\omega)$ is called the **MUSIC spectrum**. An example of the MUSIC spectrum is shown in Fig. 1.2. In this example, we have a 10-sensor ULA ($N = 10$) and there are six signals ($D = 6$) arriving from various angles, which are indicated by the vertical dashed lines. The number of snapshots used is $K = 200$, and the SNR is 5 dB.[3] Indeed, the peaks of the MUSIC spectrum are very close to the actual source angles. It should be mentioned here that the peak heights are not necessarily proportional to the signal powers p_i.

Variations of the above method lead to other algorithms such as root-MUSIC [5]. Since these algorithms are based on the properties of the signal and noise subspaces, they are called **subspace algorithms**. Even though the above description of MUSIC is for the ULA, it is possible to extend it to non-ULAs under certain conditions. One such example is described in Section 1.10.3.

1.2.5 Invertibility of the Array Manifold

We know that the output of an array in response to a single source coming from direction ω is proportional to $\mathbf{a}(\omega)$, as defined in (1.17). Let the integer array be $\{n_0, n_1, \ldots, n_{N-1}\}$ with $n_i < n_{i+1}$ as usual. Without loss of generality, let us assume $n_0 = 0$, so that $\mathbf{a}(\omega) = \begin{bmatrix}1 & e^{jn_1\omega} & \cdots & e^{jn_{N-1}\omega}\end{bmatrix}^T$. It is clear that we can identify ω uniquely from measurements of the array output only if $\omega \in [-\pi, \pi)$ is uniquely determined by $\mathbf{a}(\omega)$. This property is called the array manifold invertibility.

2 This assumes that we can find the value of ω by knowing $\mathbf{a}(\omega)$. This is certainly the case for the ULA. See Section 1.2.5 for more discussion.
3 In all examples, the SNR is computed as $\sum_i p_i/(D\sigma_e^2)$, and p_i is the same for all i unless mentioned otherwise.

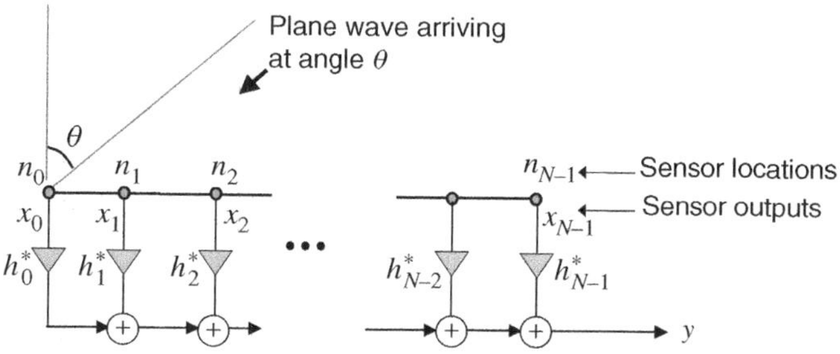

Figure 1.3 A digital beamformer with weights or taps h_m^*, working on a linear array of sensors.

Thus, a necessary condition for unambiguous identifiability of DOAs using a sparse array is that the array manifold vector be "invertible." It can be shown [6, 7] that the array satisfies this property if and only if the greatest common divisor of the integers $n_1, n_2, \ldots, n_{N-1}$ is unity, that is, $(n_1, n_2, \ldots n_{N-1}) = 1$. For example, the sparse ULA $\{0, 2, 4, 6\}$ has $(2, 4, 6) = 2$ and does not satisfy invertibility whereas the sparse array $\{0, 2, 5\}$ satisfies it and so does $\{0, 2, 4, 9\}$. The array $\{0, 1, n_2, \ldots\}$ satisfies invertibility no matter what the elements $n_2, n_3, \ldots$ are. Needless to say, the standard ULA satisfies invertibility. More extensive properties of the geometry of array manifolds can be found in [8].

1.2.6 Beamforming

In some scenarios, the array output covariance is analyzed using subspace methods in order to estimate the DOAs as explained above. In some other situations, the array is used as a beamformer. Figure 1.3 shows the schematics of a beamformer. Here, the outputs of the sensors $x_m[k]$ are linearly combined using weights (or taps) h_m^* to obtain the output $y[k]$. (The snapshot index k is omitted in the figure for simplicity.) If a plane monochromatic wave with amplitude $s[k]$ arrives at an angle θ, then the array output (at the sensor locations) is $\left[x_0[k]\ x_1[k]\ \cdots\ x_{N-1}[k]\right]^T = s[k]\left[e^{jn_0\omega}\ e^{jn_1\omega}\ \cdots\ e^{jn_{N-1}\omega}\right]^T$. So the beamformer output is

$$y[k] = s[k]\sum_{m=0}^{N-1} h_m^* e^{jn_m\omega} = H^*(e^{j\omega})s[k], \tag{1.24}$$

where $\omega = \pi \sin\theta$ as before, and

$$H^*(e^{j\omega}) = \sum_{m=0}^{N-1} h_m^* e^{jn_m\omega} \tag{1.25}$$

is called the beamformer gain or transfer function. If $\{h_m\}$ are designed such that $|H(e^{j\omega})|$ is large in the neighborhood of some $\omega = \omega_0$ and small elsewhere, then the signals received at or near the DOA ω_0 are admitted by the beamformer and other DOAs are attenuated. So the beamformer can receive signals in one direction ω_0 in preference to others.

1.3 What Are Sparse Arrays?

The most commonly used integer linear array is the standard ULA, which satisfies Eq. (1.3). This is shown in Fig. 1.4a. In contrast if $n_{i+1} - n_i > 1$ for some (or all) i, the array is said to be **sparse**.

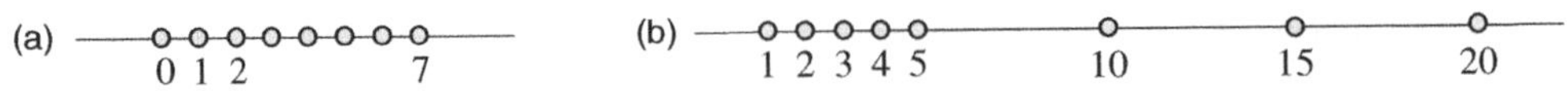

Figure 1.4 (a) A standard uniform linear array (ULA) and (b) a sparse array called the nested array (Section 1.7.1).

Figure 1.4b shows an example of a sparse array, called the nested array (to be defined formally in Section 1.7.1). Sparse arrays have been known for a long time although they received significant attention only in the last decade or so. We will discuss the details later, but here are some basic points about sparse arrays:

1. *More sources than sensors*. A well-designed sparse array with N sensors can identify $D > N$ DOAs unambiguously, in fact $D = \mathcal{O}(N^2)$ DOAs. This is possible if the sources are uncorrelated and the number of snapshots K is large enough. This is unlike the ULA which can identify at most $N - 1$ DOAs.[4]
2. *Resolution advantage*. For fixed N (number of sensors), a sparse array has a larger aperture $A = n_{N-1} - n_0$ than the standard ULA. A larger aperture typically implies that the mean square error (MSE) in DOA estimation is smaller, and that the array can resolve closely spaced DOAs more accurately than a standard ULA.[5]
3. *Ambiguity*. However, the sparse array can also create ambiguity in DOA estimation because of "sparse spatial sampling." For example, consider the array $\{0, 2, 4, \ldots, 2(N-1)\}$ which is a **sparse ULA**. It has output of the form $\alpha \cdot \begin{bmatrix} 1 & e^{j2\omega} & e^{j4\omega} & \cdots & e^{j2(N-1)\omega} \end{bmatrix}^T$ in response to a single source with DOA ω. Thus, a source at ω_1 and a source at $\omega_1 + \pi$ produce the same output, and these DOAs cannot be distinguished. However, as we shall see in this chapter, for some of the "appropriately designed" sparse arrays it is possible to resolve such ambiguities by using the so-called co-array domain processing.

One of the earliest known sparse arrays is the minimum redundancy array or **MRA**, which we shall discuss in Section 1.8.1. More recently studied sparse arrays are **nested** arrays, **coprime** arrays, and generalizations of these (Section 1.7). All these arrays can identify $\mathcal{O}(N^2)$ uncorrelated DOAs unambiguously using N sensors. For MRAs, it is cumbersome to find the sensor locations for large N, so they are usually tabulated. But nested and coprime arrays have closed form expressions for sensor locations, which makes them attractive. Many variations and improvements of these arrays have been proposed, which also have readily calculable sensor locations.

There are many algorithms to identify $D > N$ DOAs using the above types of sparse arrays. One of these is called **coarray MUSIC** and will be introduced in Section 1.6. Figure 1.5 shows the coarray MUSIC spectrum for an example where a 10-sensor nested array ($N = 10$) receives 25 sources ($D = 25$) from various angles. The peaks in the coarry MUSIC spectrum are quite close to the source locations indicated by the dashed vertical lines. In this example, $K = 500$ snapshots are used and the SNR is 10 dB.

---

4 This chapter assumes that only second-order statistics (i.e., correlations) of array outputs are used to estimate DOAs. If higher order statistics (HOS) are available, one can estimate more than N sources even with ULAs [9]. We do not go into details here.

5 More precisely though, the second central moment and fourth central moment of the sensor positions, rather than the aperture, determine resolution and accuracy of estimation [10]. However, in most situations a larger aperture has the said benefits.

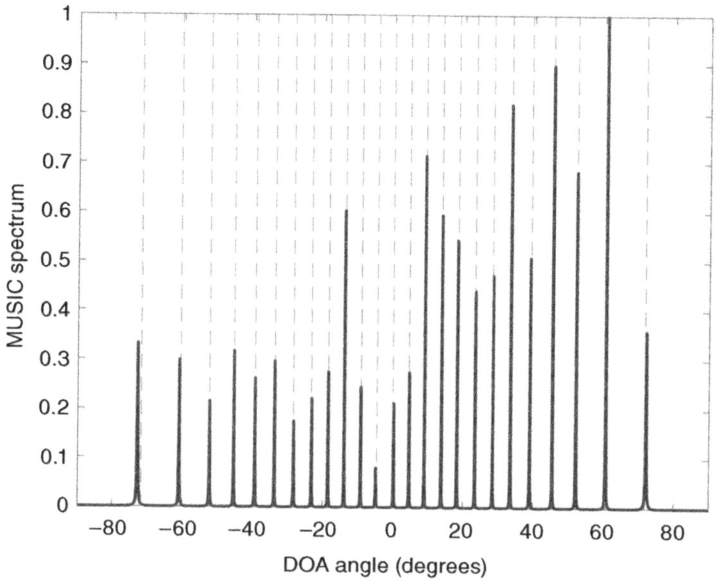

Figure 1.5 Example of a coarray MUSIC spectrum (Section 1.6) for the nested array (a sparse array). Here, $N = 10$ and $D = 25$, that is, the 10-sensor nested array identifies 25 DOAs.

1.4 How Sparse Arrays Identify $\mathcal{O}(N^2)$ Sources

In addition to the standard statistical assumptions A1–A4 (Section 1.2.3), suppose the D sources impinging on the array are also **uncorrelated**, that is, $E[s_k s_m^*] = 0$ for $k \neq m$ or equivalently,

$$\text{Assumption A5:} \qquad \mathbf{R}_{\mathbf{ss}} = \begin{bmatrix} p_0 & 0 & \cdots & 0 \\ 0 & p_1 & \cdots & 0 \\ \vdots & \vdots & \ddots & \vdots \\ 0 & 0 & \cdots & p_{D-1} \end{bmatrix}, \tag{1.26}$$

where $p_i > 0$ are the source powers. Then, $\mathbf{R}_{\mathbf{xx}}$ in Eq. (1.18) simplifies to

$$\begin{aligned} \mathbf{R}_{\mathbf{xx}} &= \mathbf{A}\mathbf{R}_{\mathbf{ss}}\mathbf{A}^H + \sigma_e^2\mathbf{I} = \sum_{i=0}^{D-1} p_i \mathbf{a}(\omega_i)\mathbf{a}^H(\omega_i) + \sigma_e^2\mathbf{I} \\ &= \sum_{i=0}^{D-1} p_i \begin{bmatrix} 1 & e^{j\omega_i(n_0-n_1)} & \cdots & e^{j\omega_i(n_0-n_{N-1})} \\ e^{j\omega_i(n_1-n_0)} & 1 & \cdots & e^{j\omega_i(n_1-n_{N-1})} \\ \vdots & \vdots & \ddots & \vdots \\ e^{j\omega_i(n_{N-1}-n_0)} & e^{j\omega_i(n_{N-1}-n_1)} & \cdots & 1 \end{bmatrix} + \sigma_e^2\mathbf{I}, \end{aligned} \tag{1.27}$$

so that the cross-correlation between outputs of the kth and mth sensors is

$$[\mathbf{R}_{\mathbf{xx}}]_{km} = E\left(x_k x_m^*\right) = \sum_{i=0}^{D-1} p_i e^{j\omega_i(n_k-n_m)} + \sigma_e^2\delta[k-m], \tag{1.28}$$

where $p_i > 0$ and $\sigma_e^2 > 0$. Thus, $[\mathbf{R}_{\mathbf{xx}}]_{km}$ depends only on the difference $n_k - n_m$ between sensor positions and not on individual values of n_k and n_m. We refer to this as the spatial wide sense stationarity, or the **spatial WSS** property.[6] Letting $l = n_k - n_m$, we can therefore define a sequence

$$R[l] = E\left(x_k x_m^*\right) = \sum_{i=0}^{D-1} p_i e^{j\omega_i l} + \sigma_e^2 \delta[l], \quad l = n_k - n_m. \tag{1.29}$$

In particular, $R[0] = [\mathbf{R_{xx}}]_{kk} = E(|x_k|^2) = \sum_{i=0}^{D-1} p_i + \sigma_e^2$ for all k. Notice also that $R[l]$ is Hermitian symmetric, that is, $R[l] = R^*[-l]$. So, if $R[l_0]$ is known, then so is $R[-l_0]$. An important result to be proved later (Section 1.5) is the following:

Theorem 1.1 ***Finding DOAs from R[l].*** In (1.29), if $R[l]$ is known for $0 \leq l \leq M-1$ (i.e., for M consecutive values of l) for some integer M, the D DOAs ω_i satisfying (1.29) can be identified, provided $D \leq M-1$.

With this result in mind, let us compare two scenarios:

1. *N-element ULA*. This can only estimate $R[0], R[1], \ldots, R[N-1]$, so $M = N$, and we can identify at most $N-1$ DOAs.
2. *N-element sparse arrays*. These can be designed such that $M > N$; In fact, $M = \mathcal{O}(N^2)$, as we shall see. So, a well-designed sparse array can identify $\mathcal{O}(N^2)$ DOAs. Algorithms that achieve this will be described later (e.g., Section 1.6).

1.4.1 The Difference Coarray

At this point, it is useful to introduce the notion of difference coarrays [11, 12]. Given a physical array $\mathcal{N} = \{n_0, n_1, \ldots, n_{N-1}\}$ with N sensors, the set C of all *distinct* differences of the form

$$n_k - n_m, \quad 0 \leq k, m \leq N-1, \tag{1.30}$$

is said to be the **difference coarray** or difference set of the array $\mathcal{N}$. For example, Fig. 1.6a,b show the difference coarrays for the eight-element arrays in Fig. 1.4. The number of elements N_c in the difference coarray of an N-sensor array is bounded as

$$N_c \leq N^2 - N + 1, \tag{1.31}$$

since $n_i - n_i = 0$ repeats N times. The coarray is **symmetric** by definition, that is, if $l \in C$, then $-l \in C$. With $C = \{0, \pm l_1, \pm l_2, \ldots, \pm l_{L-1}\}$, we have

$$N_c = 2L - 1. \tag{1.32}$$

Now, we know that $R[l]$ in (1.29) can be estimated from snapshot averages of the products of sensor outputs. Thus, the *difference coarray is precisely the set of integers l for which R[l] can be estimated.* For a ULA, the coarray C is the set of integers $-(N-1) \leq l \leq N-1$ and has only $\mathcal{O}(N)$ elements, as demonstrated in Fig. 1.6a. But it is possible to design N-sensor sparse arrays such that the coarray C has $\mathcal{O}(N^2)$ *consecutive* elements, as demonstrated in Fig. 1.6b. So $R[l]$ can be estimated for $\mathcal{O}(N^2)$ consecutive values of l, using such sparse arrays. Because of Theorem 1.1, this makes it possible for sparse arrays to identify $\mathcal{O}(N^2)$ uncorrelated DOAs from sensor output correlations. The notion of

6 The spatial WSS property should not be confused with the temporal WSS property, which says that $E(s[k])$ and $E(s[k]s^*[k-m])$ are independent of time index (snapshot index) k. Spatial WSS is a consequence of the uncorrelated assumption (1.26).

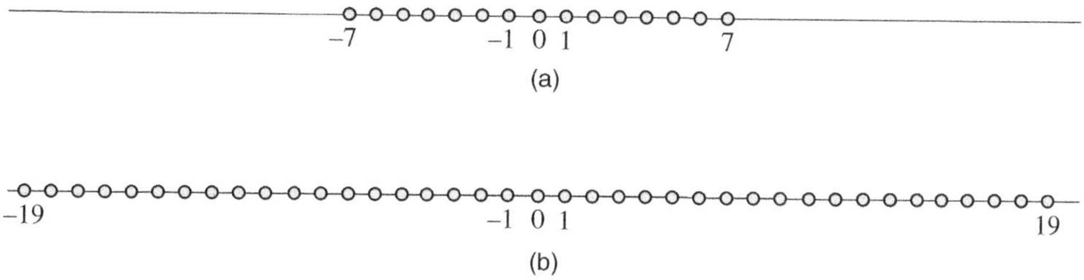

Figure 1.6 Difference coarrays for the eight-element arrays in Fig. 1.4. (a) Coarray for ULA and (b) coarray for nested array.

Khatri–Rao products [1, 13] is useful to understand difference coarrays as done in [12], although we have not used it in this chapter.

Another type of coarray called the sum-coarray arises in the context of MIMO radar and imaging applications [11, 14–16]. In this chapter, "coarray" always refers to the difference coarray. The sum coarray is also very important and is discussed in Chapter 6.

1.4.2 The Weight Function and the Estimation of $R[l]$

Given a linear array $\mathcal{N} = \{n_0, n_1, \dots, n_{N-1}\}$, we define the weight function $w[l]$ as follows:

$$w[l] = \text{number of sensor pairs } \{n_k, n_m\} \text{ such that } l = n_k - n_m. \tag{1.33}$$

For integer arrays since n_i are integers, l is an integer as well. Clearly

$$\sum_{l \in C} w[l] = N^2. \tag{1.34}$$

For a fixed l, each of the sensor pairs $\{n_k, n_m\}$ satisfying $l = n_k - n_m$ can be used to estimate $R[l]$. Thus, we can estimate $R[l]$ as follows:

$$\hat{R}[l] = \frac{1}{w[l]} \sum_{\substack{k,m \text{ s.t.} \\ n_k - n_m = l}} \frac{1}{K} \sum_{n=0}^{K-1} x_k[n] x_m^*[n], \tag{1.35}$$

where K is the number of snapshots. There is a simple way to calculate $w[i]$ for any array. Thus, let $c[n]$ be the *indicator function* for the array, that is,

$$c[n] = \begin{cases} 1 & \text{if there is a sensor at } n, \\ 0 & \text{otherwise.} \end{cases} \tag{1.36}$$

Then, $w[i]$ is the deterministic autocorrelation of $c[n]$, that is,

$$w[i] = \sum_{n \in \mathcal{N}} c[n] c[n-i]. \tag{1.37}$$

This can be computed readily indeed.

1.4.3 Central ULA

In general, the coarray of a sparse array may or may not be a ULA. The sparse array in Fig. 1.4b had the ULA coarray shown in Fig. 1.6b. But the sparse array shown in Fig. 1.7a has a coarray that is not

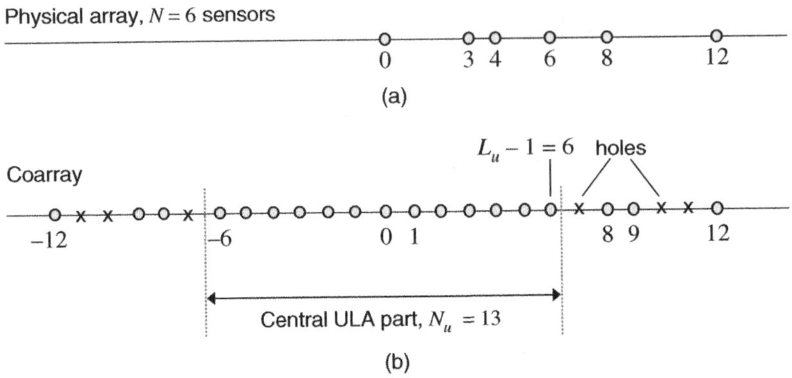

Figure 1.7 (a) A sparse linear array. (b) Its coarray, with the central ULA segment indicated. Coarray has $N_c = 19$ elements. Central ULA has $N_u = 13$ elements.

a ULA (see Fig. 1.7b). The missing coarray elements, called **holes**, are indicated in the figure with a cross. For the case where the coarray is a ULA (i.e., no holes), we have $C = \{0, \pm1, \pm2, \dots, \pm(L-1)\}$. That is, $R[0], R[1], \dots, R[L-1]$ can be estimated. In this case, we can theoretically prove (using Theorem 1.1) that $L-1$ uncorrelated DOAs can be identified. But when the coarray is not a ULA, we can only estimate $R[0], R[l_1], \dots, R[l_{L-1}]$, and there is no guarantee that $L-1$ uncorrelated DOAs can be identified from these.

There are some sparse arrays for which even though the coarray is not a ULA, there is a large **central ULA segment**, as demonstrated in Fig. 1.7a,b. Say, the central ULA segment is $-(L_u-1) \leq l \leq (L_u-1)$, which has

$$N_u = 2L_u - 1 \leq N_c \tag{1.38}$$

elements. In the example shown in Fig. 1.7b, we have $N_c = 19, N_u = 13$, and $L_u = 7$. Then $R[l]$ can be estimated for consecutive lags: $R[0], R[1], \dots, R[L_u-1]$ (and for some nonconsecutive lags beyond that). In this case, it again follows (Theorem 1.1) that if there are

$$D \leq L_u - 1, \tag{1.39}$$

uncorrelated sources, then all D DOAs can be identified. We will see that for sparse arrays, L_u can be much larger than N, enabling the identification of many more DOAs than the number of sensors N. Two remarks are here in order:

1. There is no implication in the above discussion that more than $L_u - 1$ DOAs *cannot* be identified. It is just that there exists a theoretical guarantee of identifiability when $D \leq L_u - 1$.
2. In practice, as the number of DOAs is increased to approach $L_u - 1$, the accuracy of estimation typically becomes poorer unless we have a very large number of snapshots.

Notice that $L_u - 1 = 0.5(N_u - 1) \leq 0.5(N_c - 1) \leq 0.5N(N-1)$, where we have used (1.31) to get the last inequality.

1.4.3.1 Degrees of Freedom

With a difference coarray having $N_c = 2L - 1$ elements, we can estimate the correlations $R[0], R[\pm l_1], \dots, R[\pm l_{L-1}]$ between sensor pairs. Since $R[-l] = R^*[l]$ and a scale factor in the sequence $R[l]$ is immaterial, we only have $L-1$ pieces of information here. In this sense, we say that the sparse array offers $L - 1 = (N_c - 1)/2$ degrees of freedom (DOF). We can identify the

DOAs of *at most* $L - 1$ uncorrelated sources based on these, although this is not guaranteed. On the other hand, if there exists a central ULA segment of size $N_u = 2L_u - 1$ in the difference coarray, the DOAs of $D \leq L_u - 1$ uncorrelated sources can in principle be identified, although the accuracy depends on the signal-to-noise ratio, number of snapshots, and so on. Clearly DOF $\geq L_u - 1$. A more fundamental approach to understanding the notion of DOF can be found in [17].

1.4.4 How Coarrays Arise in Other Contexts

Difference sets arise in other ways, both in array processing and in various fields of science and engineering. We now remark on some of these.

1. *Beamforming*. Recall that when an array is used as a beamformer, the beamformer gain is given by (1.25). We can readily verify that

$$|H(e^{j\omega})|^2 = \sum_{k=0}^{N-1} \sum_{m=0}^{N-1} h_k h_m^* e^{-j(n_k - n_m)\omega}. \tag{1.40}$$

So, all the differences $n_k - n_m$ which appear in the difference coarray C appear in this expression. Moreover, if the taps are chosen to be $h_m = 1$ for all m, then

$$|H(e^{j\omega})|^2 = \sum_{l=-A}^{A} w[l]e^{-j\omega l} = \sum_{l\in C} w[l]e^{-j\omega l}, \tag{1.41}$$

where $w[l]$ is the weight function (i.e., the difference $l = n_k - n_m$ occurs $w[l]$ times) and $A = n_{N-1} - n_0$ is the array aperture. Equation (1.41) is nothing but the Fourier transform of the weight function $w[l]$. The range C of the summation in (1.41) is precisely the difference coarray.
2. *Mathematics*. In discrete mathematics, the concept of a difference basis has been studied extensively [18–21]. In our notation, $\{n_0, n_1, \ldots, n_{N-1}\}$ is a **difference basis** with respect to an integer B if the difference coarray C contains a ULA segment $-B \leq l \leq B$ [21]. Several properties of difference bases, especially asymptotic properties (as $N \to \infty$) were studied by these authors. In mathematics, the sum and difference of sets have been referred to as the Minkowski set addition and set dilation, respectively [11, 22].
3. *Physics*. Difference coarrays have existed in Nature even before humans thought of studying them. Thus, consider the **energy levels** $\{E_k\}$ of the electron orbiting the nucleus in a hydrogen atom. It is well known that the emission and absorption spectra of the atom are determined by the differences $E_k - E_m$ of these energy levels [23]. Photons with energies $h\nu = E_k - E_m$ are emitted or absorbed when electrons jump between quantum orbits. Thus, the line spectra of atoms are nothing but the difference coarrays of the quantum energy levels.

The importance of coarrays in imaging and array processing was brought to the limelight by the landmark paper of Hoctor and Kassam [11], which discussed the sum coarray as well as the difference coarray in a general setting and also placed in evidence their usefulness and importance.

1.5 Identifying DOAs from Correlations

It remains to prove Theorem 1.1. Thus, referring to Eq. (1.29), where $p_i > 0$ and $\sigma_e^2 > 0$, we now prove that if $R[l]$ is known for $0 \leq l \leq M - 1$ for some integer M, we can identify $D \leq M - 1$ DOAs

ω_i uniquely. For this we define an $M \times M$ matrix $\mathbf{R}$ of correlations $R[l]$, demonstrated below for $M = 4$:

$$\mathbf{R} = \begin{bmatrix} R[0] & R^*[1] & R^*[2] & R^*[3] \\ R[1] & R[0] & R^*[1] & R^*[2] \\ R[2] & R[1] & R[0] & R^*[1] \\ R[3] & R[2] & R[1] & R[0] \end{bmatrix}. \tag{1.42}$$

Since $R^*[l] = R[-l]$, the entries are nothing but $R[l]$ in $-(M-1) \leq l \leq (M-1)$.

1.5.1 Factorization of the Matrix R

Using Eq. (1.29), the left most column of $\mathbf{R}$ can be written as

$$\begin{bmatrix} R[0] \\ R[1] \\ R[2] \\ R[3] \end{bmatrix} = \underbrace{\begin{bmatrix} 1 & 1 & 1 & 1 \\ e^{j\omega_0} & e^{j\omega_1} & e^{j\omega_2} & e^{j\omega_3} \\ e^{j2\omega_0} & e^{j2\omega_1} & e^{j2\omega_2} & e^{j2\omega_3} \\ e^{j3\omega_0} & e^{j3\omega_1} & e^{j3\omega_2} & e^{j3\omega_3} \end{bmatrix}}_{\text{call this } \mathbf{V}} \underbrace{\begin{bmatrix} p_0 & 0 & 0 & 0 \\ 0 & p_1 & 0 & 0 \\ 0 & 0 & p_2 & 0 \\ 0 & 0 & 0 & p_3 \end{bmatrix}}_{\text{call this } \Lambda_p} \begin{bmatrix} 1 \\ 1 \\ 1 \\ 1 \end{bmatrix} + \begin{bmatrix} \sigma_e^2 \\ 0 \\ 0 \\ 0 \end{bmatrix}. \tag{1.43}$$

Similarly, the next column can be written as

$$\begin{bmatrix} R^*[1] \\ R[0] \\ R[1] \\ R[2] \end{bmatrix} = \begin{bmatrix} e^{-j\omega_0} & e^{-j\omega_1} & e^{-j\omega_2} & e^{-j\omega_3} \\ 1 & 1 & 1 & 1 \\ e^{j\omega_0} & e^{j\omega_1} & e^{j\omega_2} & e^{j\omega_3} \\ e^{j2\omega_0} & e^{j2\omega_1} & e^{j2\omega_2} & e^{j2\omega_3} \end{bmatrix} \begin{bmatrix} p_0 & 0 & 0 & 0 \\ 0 & p_1 & 0 & 0 \\ 0 & 0 & p_2 & 0 \\ 0 & 0 & 0 & p_3 \end{bmatrix} \begin{bmatrix} 1 \\ 1 \\ 1 \\ 1 \end{bmatrix} + \begin{bmatrix} 0 \\ \sigma_e^2 \\ 0 \\ 0 \end{bmatrix}$$

$$= \begin{bmatrix} 1 & 1 & 1 & 1 \\ e^{j\omega_0} & e^{j\omega_1} & e^{j\omega_2} & e^{j\omega_3} \\ e^{j2\omega_0} & e^{j2\omega_1} & e^{j2\omega_2} & e^{j2\omega_3} \\ e^{j3\omega_0} & e^{j3\omega_1} & e^{j3\omega_2} & e^{j3\omega_3} \end{bmatrix} \begin{bmatrix} p_0 & 0 & 0 & 0 \\ 0 & p_1 & 0 & 0 \\ 0 & 0 & p_2 & 0 \\ 0 & 0 & 0 & p_3 \end{bmatrix} \begin{bmatrix} e^{-j\omega_0} \\ e^{-j\omega_1} \\ e^{-j\omega_2} \\ e^{-j\omega_3} \end{bmatrix} + \begin{bmatrix} 0 \\ \sigma_e^2 \\ 0 \\ 0 \end{bmatrix}.$$

By repeating this for all columns of $\mathbf{R}$, we therefore obtain the identity

$$\mathbf{R} = \underbrace{\mathbf{V}}_{M \times D} \underbrace{\Lambda_p}_{D \times D} \underbrace{\mathbf{V}^H}_{D \times M} + \sigma_e^2 \mathbf{I}_M. \tag{1.44}$$

The matrix $\mathbf{V}$ is a **Vandermonde**[7] matrix, and therefore, assuming ω_i are distinct in $-\pi \leq \omega < \pi$, it has rank D (since $D < M$). Since $p_i > 0$ for all i, the first term of (1.44) is therefore (Hermitian and) positive semidefinite with rank D. Assuming $\sigma_e \neq 0$, the matrix $\mathbf{R}$ is therefore positive definite. It is also a **Toeplitz** matrix[8] as we can see directly from (1.42). Thus, $\mathbf{R}$ has all the algebraic properties of the autocorrelation matrix of a *temporally* WSS random process (even though we obtained it from $R[l]$ whose definition depended on the *spatially* WSS property; see Section 1.4). The factorization (1.44) is called the *Vandermonde decomposition* of the Toeplitz matrix $\mathbf{R}$. At this point, the difference between $\mathbf{R}$ and $\mathbf{R}_{\mathbf{xx}}$ must be noted:

1. $\mathbf{R}_{\mathbf{xx}}$ in Eq. (1.27) is the $N \times N$ covariance $E(\mathbf{x}\mathbf{x}^H)$ of the array output vector. Clearly it is positive definite if $\sigma_e \neq 0$. But $\mathbf{R}_{\mathbf{xx}}$ is Toeplitz only when the array is a ULA in which case $n_k - n_m = k - m$. But for arbitrary arrays, $\mathbf{R}_{\mathbf{xx}}$ is not Toeplitz because $n_k - n_m$ is not determined by $k - m$.

7 That is, each column has the form $\begin{bmatrix} 1 & \alpha & \alpha^2 & \cdots & \alpha^{M-1} \end{bmatrix}^T$.
8 That is, $[\mathbf{R}]_{km}$ depends only on the difference $k - m$ and not on k and m individually.

2. The matrix $\mathbf{R}$, on the other hand, is Toeplitz by construction. It is well defined as long as the coarray has the consecutive elements $0, 1, \ldots, M-1$, so that $R[0], R[1], \ldots, R[M-1]$ are meaningful. The fact that $\mathbf{R}$ is positive definite is clear from (1.44), although not obvious from (1.42), which does not have the general form $E(\mathbf{y}\mathbf{y}^H)$.

1.5.2 Proof of Theorem 1.1

First assume $D = M-1$. The first term of (1.44) has rank $M-1$ so that there is a vector $\mathbf{u} = \begin{bmatrix} u_0 & u_1 & \cdots & u_{M-1} \end{bmatrix}^T \neq \mathbf{0}$, unique up to scale, satisfying $\mathbf{u}^H\mathbf{V}\Lambda_p\mathbf{V}^H\mathbf{u} = 0$. Since Λ_p is itself positive definite, this implies $\mathbf{V}^H\mathbf{u} = \mathbf{0}$. That is, $\sum_{n=0}^{M-1} u_n e^{-j\omega_i n} = 0$ for $0 \leq i \leq D-1$. In other words, the FIR filter or polynomial $U(z) = \sum_{n=0}^{M-1} u_n z^{-n}$ has the $M-1$ **unit-circle zeros** $e^{j\omega_i}$, $0 \leq i \leq M-2$. Thus, knowing $\mathbf{u}$, we can find all the $M-1$ DOAs ω_i. In practice, we do not know the first term, we only know $\mathbf{R}$. But we can still find $\mathbf{u}$ because $\mathbf{u}$ is the unique eigenvector of $\mathbf{R}$ corresponding to its smallest eigenvalue $\sigma_e^2 > 0$. Summarizing, here is how we can find the $M-1$ DOAs:

1. Given the correlations $R[l], 0 \leq l \leq M-1$, form the matrix $\mathbf{R}$.
2. Compute the unique eigenvector $\mathbf{u}$ corresponding to the smallest eigenvalue of $\mathbf{R}$.
3. Form the degree-$(M-1)$ polynomial $U(z)$ from $\mathbf{u}$, and find its $M-1$ roots. These are guaranteed to be on the unit circle, that is, they have the form $z_i = e^{j\omega_i}$ for real ω_i. This reveals the $M-1$ DOAs ω_i.

More generally, assume $D \leq M-1$, and that D is possibly unknown as well. In principle, we can now do this: find the eigenvalues $\lambda_i > 0$ of $\mathbf{R}$. These can be ordered as $\lambda_0 \geq \lambda_1 \geq \cdots \geq \lambda_{D-1} > \lambda_D = \cdots = \lambda_{M-1} = \sigma_e^2$. Thus,

$$\lambda_i = \begin{cases} \mu_i + \sigma_e^2 & 0 \leq i \leq D-1, \\ \sigma_e^2 & D \leq i \leq M-1, \end{cases} \tag{1.45}$$

where $\mu_i > 0$ are the D nonzero eigenvalues of the first term of (1.44), which is a rank-D positive semidefinite matrix. Thus, D can be estimated by plotting the eigenvalue pattern. Once this is done, we simply retain $R[l], 0 \leq l \leq D$ and discard the rest. That is, we define a new smaller M, namely $M' = D+1$. Then we can find the D DOAs as described above.

Summarizing, given $R[l], 0 \leq l \leq M-1$, we can always find the number of sources $D \leq M-1$, and then the DOAs $\omega_i, 0 \leq i \leq D-1$, as explained above. A warning is here in order. While the above serves as a proof that $D \leq M-1$ DOAs can be estimated, the procedure described requires modification. This is partly because in practice $\mathbf{R}$ can only be estimated from data and the estimate $\hat{\mathbf{R}}$ may not be positive semidefinite. Additionally, one should not just discard extra samples of $R[l]$ as mentioned above.

There is a more practical way to estimate DOAs from $\hat{\mathbf{R}}$ using the MUSIC algorithm. This is called coarray MUSIC [12, 24, 25], and will be described next. Similarly, algorithms such as root-MUSIC, ESPRIT, and their variations can be used in the coarray domain.

1.6 Coarray MUSIC

Given a sparse array and its coarray $C = \{0, \pm l_1, \pm l_2, \ldots, \pm l_{L-1}\}$, we know we can estimate $R[l]$ for all $l \in C$. In particular, we can find $R[l]$ in the central ULA part $0 \leq l \leq L_u - 1$. We can therefore use this to estimate $D \leq L_u - 1$ DOAs for uncorrelated sources (Theorem 1.1). For sparse arrays which have $L_u = \mathcal{O}(N^2)$, this shows that we can identify $\mathcal{O}(N^2)$ uncorrelated DOAs using N-sensor arrays. Procedurally, this can be accomplished by applying the MUSIC algorithm (Section 1.2.4) to the $M \times M$ matrix $\mathbf{R}$ in (1.42), with M taken to be $M = L_u$. Application of MUSIC to $\mathbf{R}$, instead of to the array output covariance $\mathbf{R}_{\mathbf{xx}}$, is called **coarray MUSIC**. For sparse arrays, coarray MUSIC can unambiguously identify more (uncorrelated) sources than sensors, but direct MUSIC on $\mathbf{R}_{\mathbf{xx}}$ cannot.

With $M = L_u$ and the eigenvalues of $\mathbf{R}$ as in (1.45), let the L_u orthogonal eigengectors $\mathbf{e}_i$ be arranged as follows:

$$\mathbf{E} = \begin{bmatrix} \mathbf{e}_0 & \mathbf{e}_1 & \cdots & \mathbf{e}_{D-1} & | & \mathbf{e}_D & \mathbf{e}_{D+1} & \cdots & \mathbf{e}_{L_u-1} \end{bmatrix}. \tag{1.46}$$

The vectors $\mathbf{e}_i$ have size $M \times 1$, and should not be confused with the $N \times 1$ vectors $\mathbf{e}_i$ in (1.21). As in the derivation of the standard MUSIC algorithm, it can be shown from (1.44) that the column space of $\mathbf{V}$ is the same as the space spanned by the first D eigenvectors in (1.46). This is the D-dimensional signal subspace $\mathcal{S}$. The noise subspace has dimension $L_u - D$ and is spanned by the last $L_u - D$ vectors in (1.46). Thus, each of the D vectors in $\mathbf{V}$ is orthogonal to the columns of the noise subspace matrix

$$\mathbf{E}_n = \begin{bmatrix} \mathbf{e}_D & \mathbf{e}_{D+1} & \cdots & \mathbf{e}_{L_u-1} \end{bmatrix}. \tag{1.47}$$

That is, $\mathbf{a}^H(\omega_i)\mathbf{E}_n = \mathbf{0}$, for each DOA ω_i. Morever, as explained in Section 1.6.1, the *only* vectors satisfying $\mathbf{a}^H(\omega)\mathbf{E}_n = \mathbf{0}$ are such that $\omega = \omega_i$. We can therefore uniquely identify the D DOAs ω_i by finding the unique set of D Vandermonde vectors $\mathbf{a}(\omega_i) = \begin{bmatrix} 1 & e^{j\omega_i} & e^{j2\omega_i} & \ldots & e^{j(L_u-1)\omega_i} \end{bmatrix}^T$ satisfying $\mathbf{a}^H(\omega_i)\mathbf{E}_n = \mathbf{0}$. In practice, we can plot the **coarray MUSIC spectrum**

$$P(\omega) = \frac{1}{\mathbf{a}^H(\omega)\mathbf{E}_n\mathbf{E}_n^H\mathbf{a}(\omega)}, \tag{1.48}$$

on a dense grid of ω and locate its peaks to find ω_i. We can also modify the above procedure appropriately to use root-MUSIC for finding ω_i from $\mathbf{E}_n$.

1.6.1 Unique Identifiability

It can be shown that for any set of D DOAs $\{\omega_i\}$, the *only* vectors $\mathbf{a}(\omega)$ satisfying $\mathbf{a}^H(\omega)\mathbf{E}_n = \mathbf{0}$ are the D vectors $\mathbf{a}(\omega_i)$, if and only if the following $M \times (D+1)$ matrix

$$\begin{bmatrix} \mathbf{a}_M(\beta_0) & \mathbf{a}_M(\beta_1) & \cdots & \mathbf{a}_M(\beta_{D-1}) & \mathbf{a}_M(\beta_D) \end{bmatrix} \tag{1.49}$$

has rank $D+1$ for all possible choices of the $D+1$ real constants $\beta_0, \beta_1, \ldots, \beta_D$, where $\beta_i \neq \beta_k$ mod 2π for $i \neq k$, and

$$\mathbf{a}_M(\beta) = \begin{bmatrix} 1 & e^{j\beta} & e^{j2\beta} & \ldots & e^{j(M-1)\beta} \end{bmatrix}^T. \tag{1.50}$$

Since the matrix (1.49) is Vandermonde, this rank condition is satisfied if and only if $D + 1 \leq M$. Thus as long as $D \leq M - 1$, the coarray MUSIC algorithm will identify the D DOAs uniquely without ambiguity or false peaks. For sparse arrays, since we can make $M = L_u = \mathcal{O}(N^2)$, we can therefore identify $\mathcal{O}(N^2)$ uncorrelated DOAs.

1.6.2 Estimating the Signal Powers

Let $\mathbf{r}_0$ be the leftmost column of $\mathbf{R}$. Then from (1.43) we see that

$$\mathbf{r}_0 = \mathbf{V}\mathbf{p} + \sigma_e^2\mathbf{1}, \tag{1.51}$$

where $\mathbf{p} = \begin{bmatrix} p_0 & p_1 & \cdots & p_{D-1} \end{bmatrix}^T$ and $\mathbf{1} = \begin{bmatrix} 1 & 0 & \cdots & 0 \end{bmatrix}^T$. If $D \leq M - 1$, the Vandermonde matrix $\mathbf{V}$ has rank D so that $\mathbf{V}^H\mathbf{V}$ is nonsingular. So we can rewrite the above as

$$\mathbf{p} = \left(\mathbf{V}^H\mathbf{V}\right)^{-1}\mathbf{V}^H\left(\mathbf{r}_0 - \sigma_e{}^2\mathbf{1}\right). \tag{1.52}$$

This is useful to estimate the signal powers p_i. To see this, recall that the eigenvalues of the Toeplitz matrix $\mathbf{R}$ have the form (1.45). Thus, by estimating and plotting these, we can obtain estimates of D and the noise variance σ_e^2. Assume that the DOAs ω_i have also been estimated using coarray MUSIC, as explained above. By substituting these estimates into the right-hand side of (1.52), we can estimate the power vector $\mathbf{p}$. Thus, using an N-sensor sparse array with an appropriate difference coarray, we are able to estimate $D > N$ DOAs (for uncorrelated signals), and also the powers of these D signals.

1.6.3 Subtleties Which Arise in Practice

Coarray MUSIC is applied on an estimate of the Toeplitz matrix $\mathbf{R}$, whereas traditional MUSIC is done on an estimate of the positive definite matrix $\mathbf{R}_{\mathbf{xx}}$ (array output covariance). To show the relation between these matrices, consider the sparse array $\{0, 1, 3\}$ whose coarray is $\{-3, -2, -1, 0, 1, 2, 3\}$. In this case,

$$\mathbf{R}_{\mathbf{xx}} = \begin{bmatrix} R[0] & R^*[1] & R^*[3] \\ R[1] & R[0] & R^*[2] \\ R[3] & R[2] & R[0] \end{bmatrix}, \quad \mathbf{R} = \begin{bmatrix} R[0] & R^*[1] & R^*[2] & R^*[3] \\ R[1] & R[0] & R^*[1] & R^*[2] \\ R[2] & R[1] & R[0] & R^*[1] \\ R[3] & R[2] & R[1] & R[0] \end{bmatrix} \tag{1.53}$$

In practice, MUSIC can only be applied to an *estimate* of $\mathbf{R}_{\mathbf{xx}}$ or $\mathbf{R}$. These estimates have to be positive semidefinite in order for MUSIC to be successful. In this context, the difference between direct MUSIC on $\mathbf{R}_{\mathbf{xx}}$ and coarray MUSIC on $\mathbf{R}$ is the following.

1. Since each term in (1.19) is positive semidefinite, the estimate (1.19) is also positive semidefinite, and for large enough K, $\hat{\mathbf{R}}_{\mathbf{xx}}$ turns out to be positive definite in practice, due to noise. So there are no difficulties in applying MUSIC to $\hat{\mathbf{R}}_{\mathbf{xx}}$.
2. However, the estimate $\hat{\mathbf{R}}$ of $\mathbf{R}$ based on array output data is not guaranteed to be positive definite. This is because $\mathbf{R}$ does not have a definition of the form $E(\mathbf{y}\mathbf{y}^H)$ and cannot be estimated by snapshot averaging of positive semidefinite quantities. Each $R[l]$ in $\mathbf{R}$ is estimated separately using (1.35) to obtain the estimate $\hat{\mathbf{R}}$. Thus, the method of separating the signal subspace from the noise subspace by ordering of the eigenvalues of $\hat{\mathbf{R}}$ needs to be examined more closely.

In this context, it is helpful to introduce the notion of spatial smoothing next.

1.6.4 Spatial Smoothing

The $N_u(= 2L_u - 1)$ estimates $\hat{R}[l]$ in the central ULA region of the coarray can be written as

$$\hat{R}[l] = \sum_{i=0}^{D-1} p_i e^{j\omega_i l} + \epsilon[l], \qquad 0 \leq |l| \leq L_u - 1, \tag{1.54}$$

where $\epsilon[l]$ includes the effect of estimation errors and the noise term $\approx \sigma_e^2\delta[l]$, both of which are presumably small. The estimates $\hat{R}[l]$ are still random variables. One way to estimate the quantities

ω_i would be to first estimate the $L_u \times L_u$ autocorrelation matrix of the random sequence $\hat{R}[l]$ using sample averaging as follows:

$$\mathbf{Z} = \frac{1}{L_u} \sum_{i=0}^{L_u-1} \mathbf{r}_i \mathbf{r}_i^H \tag{1.55}$$

Here $\mathbf{r}_i$ is the ith block of samples from the sequence $\hat{R}[l]$ defined as

$$\mathbf{r}_i = \left[\hat{R}[-L_u + 1 + i] \ \hat{R}[-L_u + 2 + i] \ \cdots \ \hat{R}[i]\right]^T. \tag{1.56}$$

The sample locations used by the blocks are demonstrated in Fig. 1.8 for the case where $L_u = 7$. Thus, there are L_u blocks, each of size L_u. The averaging process (1.55) is called **spatial smoothing**. Note that $\hat{\mathbf{R}}$ can be written in terms of the vectors $\mathbf{r}_i$ as follows:

$$\hat{\mathbf{R}} = \left[\mathbf{r}_{L_u-1} \ \mathbf{r}_{L_u-2} \ \cdots \ \mathbf{r}_0\right]. \tag{1.57}$$

Using this and the fact that $\hat{\mathbf{R}}$ is Hermitian (which you can verify by writing out the entries of (1.57)), it readily follows that

$$\mathbf{Z} = \frac{1}{L_u} \hat{\mathbf{R}}^2. \tag{1.58}$$

Since $\hat{\mathbf{R}}$ is Hermitian, we have the eigendecomposition $\hat{\mathbf{R}} = \hat{\mathbf{E}}\hat{\Lambda}\hat{\mathbf{E}}^H$ where $\hat{\mathbf{E}}$ is unitary ($\hat{\mathbf{E}}^H\hat{\mathbf{E}} = \mathbf{I}$). It then follows that

$$\mathbf{Z} = \frac{1}{L_u} \hat{\mathbf{E}}\hat{\Lambda}^2\hat{\mathbf{E}}^H \tag{1.59}$$

Clearly $\mathbf{Z}$ is positive semidefinite even though the estimate $\hat{\mathbf{R}}$ may not be. Both matrices have the same set of eigenvectors (columns of $\hat{\mathbf{E}}$), and the eigenvalues of $\mathbf{Z}$ are λ_k^2 where λ_k are the eigenvalues of $\hat{\mathbf{R}}$ (which are real because $\hat{\mathbf{R}}$ is Hermitian). Thus, we can estimate the noise subspace in one of two ways:

1. *Spatial smoothing method.* Compute the $L_u \times L_u$ matrix $\mathbf{Z}$, find its $L_u - D$ eigenvectors $\hat{\mathbf{e}}_k$ corresponding to its smallest $L_u - D$ eignevalues λ_k^2 (which are clearly nonnegative), and take $\hat{\mathbf{E}}_n = \left[\hat{\mathbf{e}}_D \ \hat{\mathbf{e}}_{D+1} \ \cdots \ \hat{\mathbf{e}}_{L_u-1}\right]$ to be the estimated noise subspace matrix.

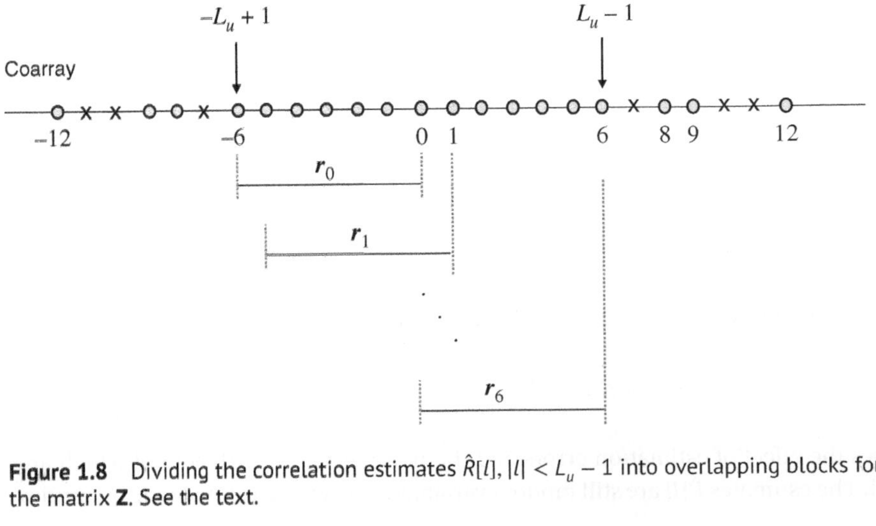

Figure 1.8 Dividing the correlation estimates $\hat{R}[l]$, $|l| < L_u - 1$ into overlapping blocks for computation of the matrix **Z**. See the text.

2. *Direct method*. Compute the eigenvalues and eigenvectors of the $L_u \times L_u$ matrix $\hat{\mathbf{R}}$, arrange them according to decreasing magnitudes $|\hat{\lambda}_k|$, and use the $L_u - D$ eigenvectors $\hat{\mathbf{e}}_k$ corresponding to the smallest $L_u - D$ magnitudes $|\hat{\lambda}_k|$ to define noise subspace matrix $\hat{\mathbf{E}}_n = [\hat{\mathbf{e}}_D \; \hat{\mathbf{e}}_{D+1} \; \cdots \; \hat{\mathbf{e}}_{L_u-1}]$.

Both of these yield exactly the same estimated matrix $\hat{\mathbf{E}}_n$. The estimated DOAs $\hat{\omega}_i$ are therefore the same using either method. The first method was used in [12], and the second method, which is computationally simpler, was used in [25]. Early work on the use of MUSIC for sparse arrays such as MRAs can be found in [26].

1.6.4.1 Steps in the Computation of Coarray-MUSIC for Sparse Arrays

In Fig. 1.5, we showed an example of coarray MUSIC spectrum but did not explain how it was obtained. It was obtained just by plotting an estimate of (1.48). For this, we proceed as follows:

1. First estimate $R[l]$ using snapshot averaging as in Eq. (1.35), and form an estimate $\hat{\mathbf{R}}$ of the $L_u \times L_u$ Toeplitz matrix $\mathbf{R}$ (demonstrated in (1.42)).
2. Estimate the eigenvectors $\hat{\mathbf{e}}_i$ of $\hat{\mathbf{R}}$.
3. Using the "Direct method" above, identify the noise subspace matrix $\hat{\mathbf{E}}_n$. Then (1.48) can be computed and plotted.

In the example of Fig. 1.5, a 10-sensor nested array ($N = 10$) receives 25 sources ($D = 25$) from various angles, and all the DOAs can be identified from the coarray MUSIC spectrum. $K = 500$ snapshots were used, and the SNR is 10 dB.

1.7 Examples of Sparse Arrays

In this section and in Section 1.8, we will discuss some of the commonly encountered sparse arrays such as nested arrays, coprime arrays, minimum redundancy arrays (MRAs), and Golomb rulers or minimum hole arrays (MHAs).

1.7.1 Nested Arrays

A nested array is defined in terms of two integers N_1 and N_2, as shown in Fig. 1.9a. Thus, the array is the union of a standard ULA and a sparse ULA:

$$\begin{aligned} \text{Standard ULA } \mathcal{N}_1: &\quad \{1, 2, \ldots, N_1\}, \\ \text{Sparse ULA } \mathcal{N}_2: &\quad \{N_1 + 1,\; 2(N_1 + 1), \cdots, N_2(N_1 + 1)\}. \end{aligned} \tag{1.60}$$

It can be shown [12] that the difference coarray of this array is a ULA, and is given by

$$\{-L_u + 1, -L_u + 2, \ldots - 1, 0, 1, \ldots, L_u - 2, L_u - 1\}. \tag{1.61}$$

where

$$L_u = N_2(N_1 + 1). \tag{1.62}$$

The coarray is shown in Fig. 1.9b. For example, the eight-sensor array in Fig. 1.4b is a nested array with $N_1 = N_2 = 4$, and its difference coarray is as in Fig. 1.6b. For a fixed number of sensors, that is, $N = N_1 + N_2$, it can be shown [12] that the values of N_1 and N_2 which maximize L_u are as follows:

$$\begin{aligned} \text{Odd } N: &\quad N_1 = \frac{N-1}{2}, \quad N_2 = \frac{N+1}{2}, \\ \text{Even } N: &\quad N_1 = N_2 = \frac{N}{2}. \end{aligned} \tag{1.63}$$

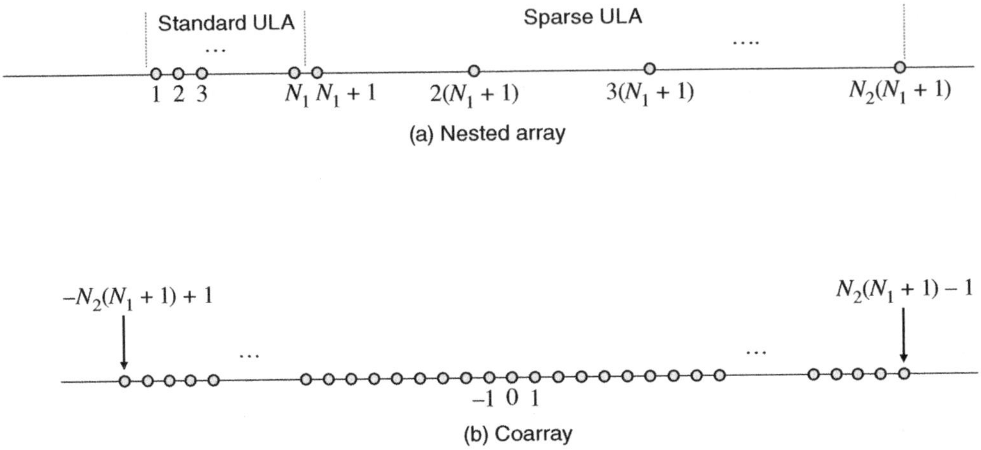

Figure 1.9 (a) The nested array with $N = N_1 + N_2$ sensors. The standard ULA part has N_1 sensors and the sparse ULA part has N_2 sensors. (b) Difference coarray of the nested array.

Thus the maximum number of DOAs that can be identified by the nested array based on correlations is $D_{\text{max}} = L_u - 1 = N_2(N_1 + 1) - 1$ and simplifies to

$$D_{\text{max}} = \begin{cases} \dfrac{N^2 + 2N - 3}{4} & \text{for odd } N, \\ \dfrac{N^2 + 2N - 4}{4} & \text{for even } N. \end{cases} \tag{1.64}$$

Thus, $D_{\text{max}} \approx N^2/4$ for the nested array. For example,

$$D_{\text{max}} = \begin{cases} 99 & \text{for } N = 19, \\ 109 & \text{for } N = 20. \end{cases} \tag{1.65}$$

Thus, a nested array with 20 sensors can identify as many as 109 uncorrelated sources!

1.7.2 Coprime Arrays

A **basic coprime array** is a union of two sparse ULAs as shown in Fig. 1.10 [27]. The first array has adjacent elements spaced by the integer N_1 and has N_2 sensors. The second array has adjacent elements spaced by the integer N_2 and has N_1 sensors. Furthemore, the integers are chosen to be coprime, that is,

$$(N_1, N_2) = 1, \tag{1.66}$$

where (N_1, N_2) stands for the greatest common divisor (gcd) of N_1 and N_2. Thus the array is

$$\{0, N_1, 2N_1, \ldots, (N_2 - 1)N_1, \;\; | \;\; N_2, 2N_2, \ldots, (N_1 - 1)N_2\} \;\; \text{(basic coprime array)}.$$

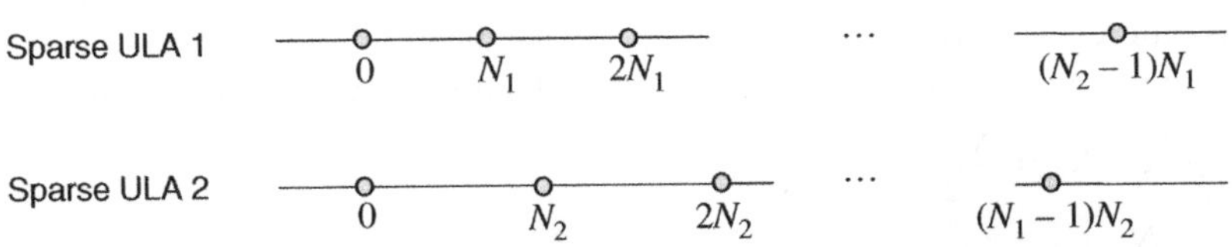

Figure 1.10 The two sparse ULAs making up a basic coprime array with $N = N_1 + N_2 - 1$ sensors.

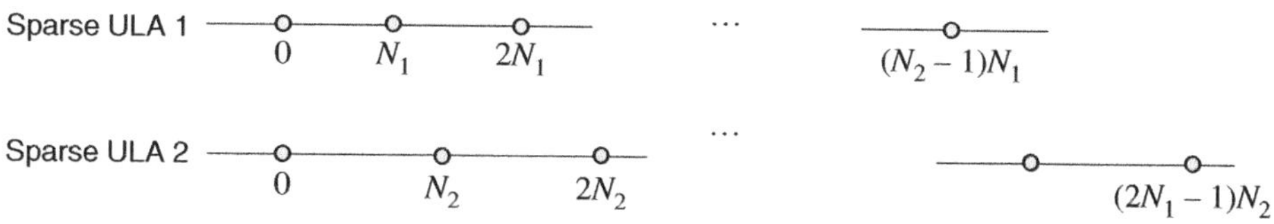

Figure 1.11 The two sparse ULAs making up the extended coprime array. There are $N = 2N_1 + N_2 - 1$ sensors.

Since the sensor at zero need not be counted twice, the array has

$$N = N_1 + N_2 - 1 \tag{1.67}$$

sensors. Before we study the properties of this array, we also define a generalization called the **extended coprime array**:

$$\{0, N_1, 2N_1, \ldots, (N_2-1)N_1, \;\; | \;\; N_2, 2N_2, \ldots, (2N_1-1)N_2\}. \tag{1.68}$$

This is shown in Fig. 1.11, and the only difference is that the second sparse ULA has N_1 extra sensors ($N_1 < N_2$), so the total number of sensors is

$$N = 2N_1 + N_2 - 1. \tag{1.69}$$

Figure 1.12a,b show details of the basic coprime array with $N = 6$ sensors ($N_1 = 3, N_2 = 4$). The corresponding difference coarray is shown in part (c). The coarray is not a ULA, as it has a hole at 7. However, it does have a central ULA part $-6 \leq n \leq 6$ with 13 sensors. Figure 1.13 shows the extended coprime array with $N = 9$ sensors ($N_1 = 3, N_2 = 4$ again). Notice that the difference coarray C now has a large ULA segment $-14 \leq n \leq 14$ with 29 sensors and some holes after that. While C is symmetric, the negative portion is not fully shown due to lack of space.

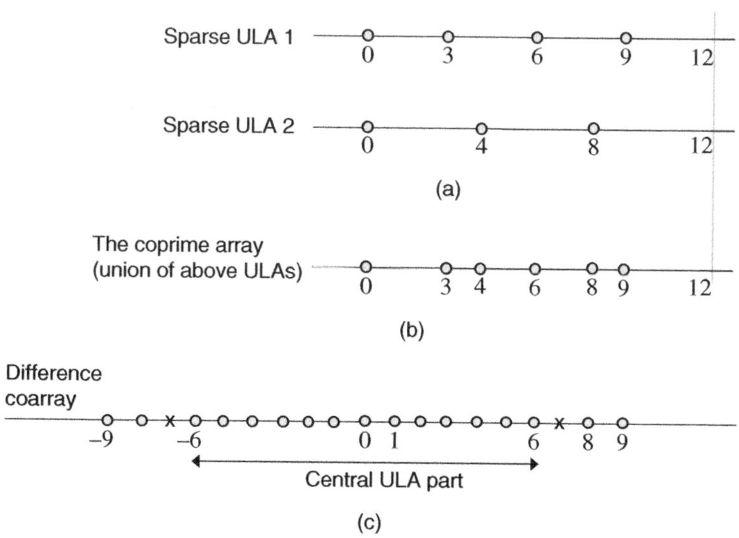

Figure 1.12 Basic coprime array with $N = 6$ sensors ($N_1 = 3, N_2 = 4$). (a) The two sparse ULAs that make up the coprime array; (b) the coprime array, which is the union of the two sparse ULAs; and (c) the difference coarray of the coprime array, which has a central ULA part with 13 sensors.

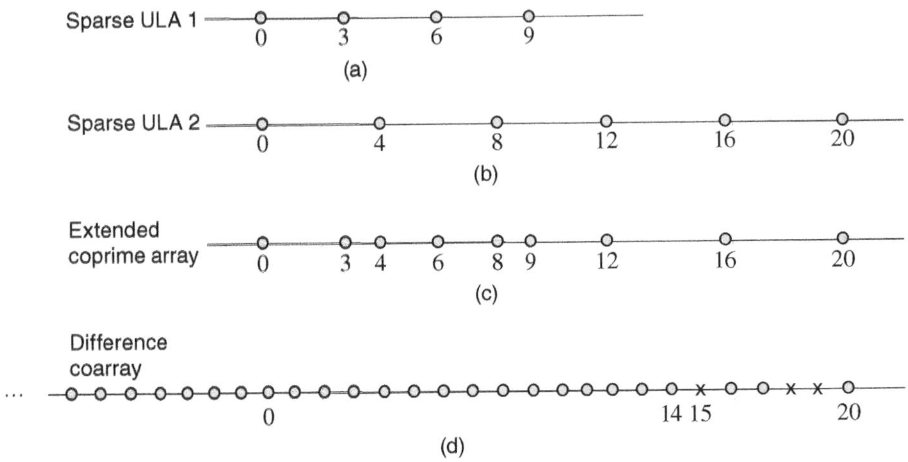

Figure 1.13 The extended coprime array, with $N_1 = 3, N_2 = 4$. (a), (b) The sparse ULAs, (c) the extended copime array, and (d) the difference coarray $\mathcal{C}$. Note that $\mathcal{C}$ is symmetric, but only the nonnegative part $\mathcal{C}^+$ is fully shown due to lack of space.

In view of Theorem 1.1, we can conclude the following:

1. The basic coprime array $\{0, 3, 6, 9, \mid 4, 8\}$ with six sensors can identify six DOAs, which is one more that what a six-sensor ULA can do. While this is only a small example, for larger number of sensors, the basic coprime array can do significantly better than the ULA.
2. The extended coprime array $\{0, 3, 6, 9, \mid 4, 8, 12, 16, 20\}$ with nine sensors can identify 14 DOAs, which is significantly more than what a nine-sensor ULA can do.

1.7.2.1 Coarray of the Coprime Array

As seen from the above examples, the difference coarray of a coprime array has holes, although there is a large central ULA portion in the coarray if N_1 and N_2 are properly chosen. Two results on the coarray are summarized next. All proofs can be found in [27]:

1. *Basic coprime array*. The difference coarray of the basic coprime array is the union $\mathcal{C} = -\mathcal{C}_1 \cup \mathcal{C}_1$, where $\mathcal{C}_1$ is the set of N_1N_2 differences

$$N_1m_1 - N_2m_2, \quad 0 \le m_1 \le N_2 - 1, \quad 0 \le m_2 \le N_1 - 1, \tag{1.70}$$

which can be shown to be distinct when N_1 and N_2 are coprime.[9] Thus there are significantly more than $N_1 + N_2$ elements in the coarray, but the central ULA segment can be smaller because there are holes.

2. *Extended coprime array*. The difference coarray of the extended coprime array is guaranteed to contain a central ULA segment which includes

$$-N_1N_2 \le k \le N_1N_2. \tag{1.71}$$

Thus, we can identify at least N_1N_2 DOAs for uncorrelated sources, using the extended coprime array which has $N = 2N_1 + N_2 - 1$ sensors. Thus, $\mathcal{O}(N^2)$ sources can be identified. In fact, (1.71) has been slightly strengthened to $|k| \le N_1(N_2 + 1) - 1$ in [28].

9 It turns out that all differences of the form $N_1(m_1 - m_1')$ and $N_2(m_2 - m_2')$ are automatically included in the union $\mathcal{C} = -\mathcal{C}_1 \cup \mathcal{C}_1$.

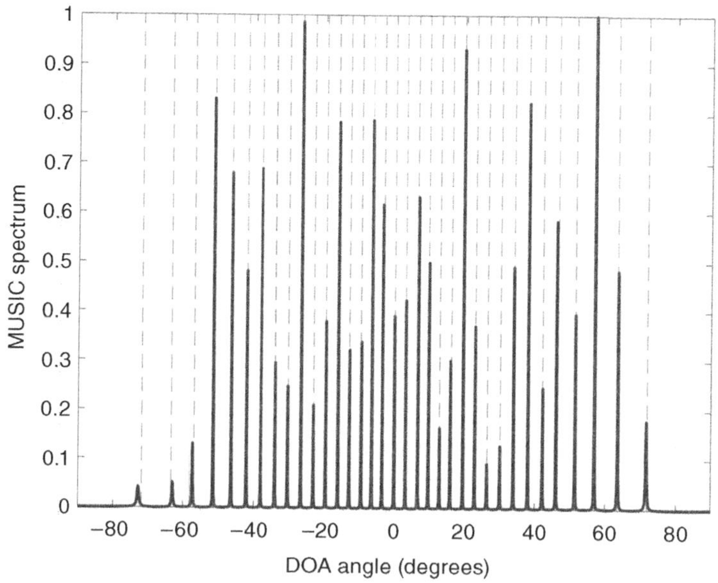

Figure 1.14 Coarray MUSIC spectrum for an extended coprime array with $N = 17$ sensors ($N_1 = 5, N_2 = 8$) receiving $D = 35$ DOAs.

We usually choose $N_1 < N_2$ in the extended coprime array because it reduces the number of sensors for fixed N_1N_2-product.

Figure 1.14 shows the coarray MUSIC spectrum for an extended coprime array with $N = 17$ sensors ($N_1 = 5, N_2 = 8$) receiving $D = 35$ DOAs. The difference coarray contains the ULA segment $-44 \leq k \leq 44$, which guarantees that up to 44 DOAs can be identified. In this example, number of snapshots $K = 300$ and SNR $= 0$ dB. As usual, the dashed vertical lines show the actual DOAs and the peak locations in the MUSIC spectrum show the estimated DOAs.

1.8 Examples of Optimal Sparse Arrays

In this section, we consider some arrays which can be regarded as optimal according to some criteria. We begin with what is called a **zero-redundancy** array. An array $\mathcal{N} = \{n_0, n_1, \ldots, n_{N-1}\}$ is said to have zero-redundancy if the weight function (1.33) satisfies $w[l] \in \{0, 1\}$ for $l \neq 0$. That is, for any l there is at most one pair of sensors $\{n_k, n_m\}$ such that $l = n_k - n_m$. An array with zero-redundancy is also called a **Golomb ruler**[10] [29]:

$$w[l] \in \{0, 1\}, \ l \neq 0 \qquad \text{(Golomb ruler).} \tag{1.72}$$

A subclass of Golomb rulers called minimum hole arrays (MHA) will be discussed in Section 1.8.2. If the aperture is large compared to N, then a random assignment of sensor positions $\{n_i\}$ yields a zero-redundancy array with high probability. As another example, an array with exponential sensor positions $n_i = 2^i, 0 \leq i \leq N - 1$ achieves zero-redundancy; it was introduced in [12] as a limiting

10 A common *ruler* has uniform markings $\{0, 1, \ldots, N-1\}$ like a ULA and can be used to measure any distance equal to the difference between any pair of markings. So, every distance in the set $\{0, 1, \ldots, N-1\}$ can be measured. A *sparse ruler* $\{n_0, n_1, \ldots, n_{N-1}\}$ has markings at nonconsecutive integers n_i and the distances measurable are those in the difference set.

case of multi-level nested arrays. More interesting is the problem of finding zero-redundancy arrays whose coarrays are ULAs. It turns out that there are only four such integer arrays (except for minor variations such as shifts and reversals) [30]. *Zero-redundancy arrays $\mathcal{N}$ with ULA coarrays C are also called* **perfect arrays** and are given below:

1. $\mathcal{N} = \{0\}$ and $C = \{0\}$.
2. $\mathcal{N} = \{0, 1\}$ and $C = \{-1, 0, 1\}$
3. $\mathcal{N} = \{0, 1, 3\}$ and $C = \{-3, -2, -1, 0, 1, 2, 3\}$
4. $\mathcal{N} = \{0, 1, 4, 6\}$ and $C = \{-6, -5, -4 - 3, -2, -1, 0, 1, 2, 3, 4, 5, 6\}$

DOA estimation based on sparse arrays viewed as rulers is discussed in [31].

1.8.1 Minimum Redundancy Arrays

For $N > 4$ there are no examples of perfect arrays. However, there are arrays with very small amount of redundancy. The family of **minimum redundancy arrays** or **MRAs** is one such family[11] [30, 32]. The name "minimum redundancy arrays" could be misleading, but this is how the sensor locations are optimized: for fixed number of sensors N, the MRA has the sensors $\{n_0, n_1, \dots, n_{N-1}\}$ chosen such that

$$|C| \text{ is maximized under the constraint } C = \text{ULA} \quad \text{(MRA)}. \tag{1.73}$$

Thus, the coarray size $|C|$ is the largest, subject to being a ULA. With MRAs, $R[l]$ can be estimated for a large number (typically $O(N^2)$) of consecutive lags l as we shall see.

These arrays have $|C| = \mathcal{O}(N^2)$, with $|C| < N^2 - N + 1$ for $N > 4$. MRAs do not have a simple closed form formula for the sensor locations n_i (unlike nested and coprime arrays). The optimal sensor locations are usually found by numerical search and tabulated, and the solution is not unique [32]. Table 1.1 shows some examples of MRAs [30], along with the size of the coarray which is a ULA. For example, the MRA with $N = 8$ sensors has the coarray $-23 \leq l \leq 23$ which has 47 elements. The coarray size for an ULA with the same number of sensors is also shown.

Let us now compare the MRA with the nested array and ULA for $N = 8$ sensors. From Section 1.7.1 we can find the coarray of the eight-sensor nested array with $N_1 = N_2 = 4$. Thus, for

Table 1.1 Examples of minimum redundancy arrays.

N	Physical array	Coarray size for MRA	Coarray size for ULA
5	0, 1, 4, 7, 9	19	9
6	0, 1, 6, 9, 11, 13	27	11
7	0, 1, 4, 10, 12, 15, 17	35	13
8	0, 1, 4, 10, 16, 18, 21, 23	47	15
9	0, 1, 4, 10, 16, 22, 24, 27, 29	59	17
10	0, 1, 3, 6, 13, 20, 27, 31, 35, 36	73	19
11	0, 1, 3, 6, 13, 20, 27, 34, 38, 42, 43	87	21

Source: Adapted from [30].

11 In [30], the term "restricted array" is used to indicate an array whose coarray is a ULA, and the term "general array" is used for arrays without this restriction.

eight-sensor arrays:

$$\text{Coarray size} = \begin{cases} 15 & \text{ULA,} \\ 39 & \text{nested array,} \\ 47 & \text{MRA.} \end{cases} \tag{1.74}$$

Thus the number of DOAs that can be identified for uncorrelated sources are:

$$\text{Number of DOAs} = \begin{cases} 7 & \text{ULA,} \\ 19 & \text{nested array,} \\ 23 & \text{MRA.} \end{cases} \tag{1.75}$$

While the nested array is significantly better than the ULA, the MRA is even better than the nested array. Let us also compare these with coprime arrays. The extended coprime array with $N_1 = 2$, $N_2 = 5$ has $2N_1 + N_2 - 1 = 8$ sensors. How many sources can it identify? As mentioned in Section 1.7.2, the coarray has a ULA segment at least as large as $-N_1N_2 \leq l \leq N_1N_2$. It can be verified that the ULA segment is even larger, namely $-11 \leq l \leq 11$, showing that it can identify 11 DOAs with only eight sensors. Even though nested arrays can identify more sources than coprime arrays with the same number of elements, most adjacent sensor pairs in coprime arrays are widely separated, which helps to reduce mutual coupling effect between sensors [33]. Thus, comparing these 8-sensor arrays we can verify the following:

$$\text{Number of sensor pairs with unit separation} = \begin{cases} 7 & \text{ULA,} \\ 4 & \text{nested array,} \\ 2 & \text{extended coprime array,} \\ 1 & \text{MRA.} \end{cases}$$

Thus, coprime arrays are better than nested arrays (and of course ULAs) from the view point of mutual coupling. However, MRAs are even better. The only disadvantage of MRAs is that for arbitrary large N we cannot easily find the sensor locations, as there is no closed form.

Figure 1.15 shows the weight functions for these four arrays. As we know, $w[0] = N = 8$ for all arrays. Here are the other points to be noticed:

1. The ULA has a triangular weighting function because its indicator function $c[i]$ is a rectangle, so the correlation (1.37) produces a triangle.
2. As expected, only the coprime array has holes (indicated by the points where $w[k] = 0$).
3. The MRA has $w[k] = 1$ at most places, and $w[k] = 2$ at a few places, demonstrating low redundancy.
4. The nested array has $w[k] = 1, 2, 3$, or 4 at various places, which shows that it has more redundancy than the MRA.
5. None of the arrays is a zero-redundancy array, since $w[k] > 1$ for some $k \neq 0$ for all of them.

1.8.2 Minimum Hole Arrays

Recall here that a perfect array has zero-redundancy *and* a ULA coarray. The MRA retains the ULA-coarray (or "no hole") property, but minimizes redundancy. Instead, it is possible to retain the zero-redundancy (or Golomb-ruler) property (1.72) and minimize the number of holes in the difference coarray C. Such an optimum Golomb ruler is called a **minimum hole array** or **MHA**:

$$\text{Number of holes in } C \text{ is minimized subject to } w[k] \in \{0, 1\}, k \neq 0 \text{ (MHA).} \tag{1.76}$$

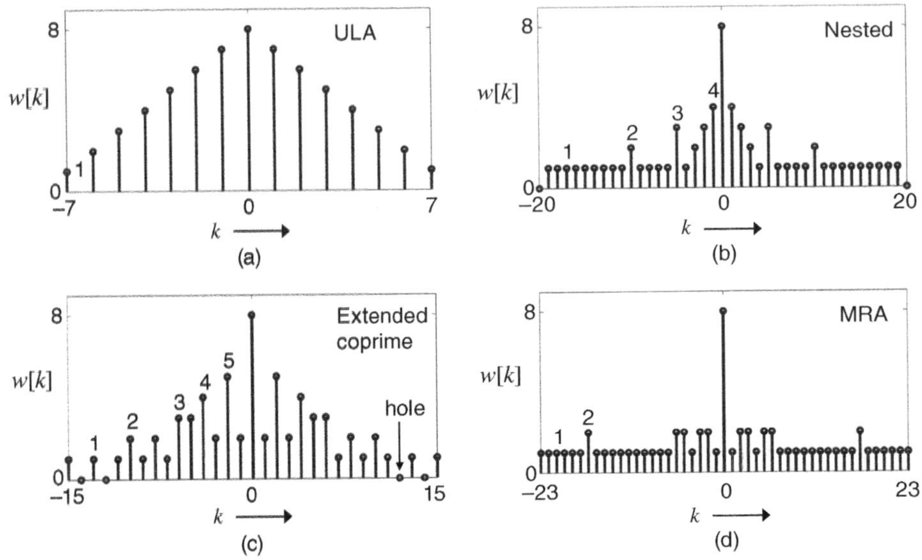

Figure 1.15 Weight functions for several arrays with $N = 8$ sensors. (a) ULA, (b) nested array with $N_1 = N_2 = 4$, (c) extended coprime array with $N_1 = 2, N_2 = 5$, and (d) MRA.

It can be shown that this is equivalent to minimizing the *aperture* $n_{N-1} - n_0$ subject to $w[k] \in \{0, 1\}, k \neq 0$. We shall not discuss these arrays but details can be found in [34–37]. Variations of these arrays such as "linear minimum-holes-plus-redundancies arrays" are also mentioned in [11] and references therein.

Besides MRA and MHA, sparse arrays have also been optimized based on other criteria. Examples include minimizing the Cramér–Rao bound or CRB for the DOA estimation errors, or maximizing the signal-to-noise-plus-interference ratio (SINR) at the output of an array. These results are discussed in Chapter 10.

1.9 Coprime DFT Beamformers

We now explain how the basic coprime array with $N_1 + N_2 - 1$ sensors can be used as a beamformer that can, under some assumptions, produce N_1N_2 narrow beams with beamwidths $2\pi/N_1N_2$ in the ω-domain [27]. This is comparable to what can be achieved using a ULA with as many as N_1N_2 physical sensors. The construction involves the definition of N_1N_2 product filters and the practical realization of the product filters.

1.9.1 Definition of a Set of N_1N_2 Product Filters

Consider Fig. 1.16, which shows one of the two sparse ULAs that constitute the basic coprime array. We have shown it in its beamforming mode, where taps h_m^* are used at the sensor outputs to obtain a linear combination y_1. Since adjacent sensors are spaced apart by N_1 the beamformer gain (1.25) is replaced with

$$H^*(e^{jN_1\omega}) = \sum_{m=0}^{N_2-1} h_m^* e^{jmN_1\omega}. \tag{1.77}$$

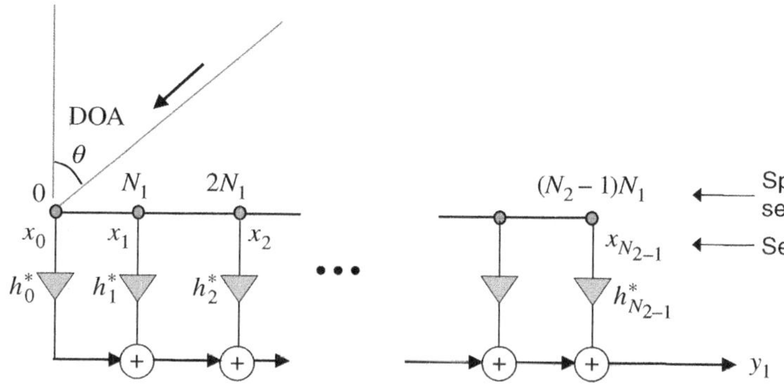

Figure 1.16 Beamformer based on the sparse ULA $\{0, N_1, 2N_1, \ldots, (N_2-1)N_1\}$.

Similarly, we can use taps g_m^* on the other sparse ULA in the coprime array to obtain the beamformer gain

$$G^*(e^{jN_2\omega}) = \sum_{m=0}^{N_1-1} g_m^* e^{jmN_2\omega}. \tag{1.78}$$

Now assume $H(e^{j\omega})$ is a lowpass filter[12], as shown in Fig. 1.17a, with passband edge π/N_2. (In all figures we use $N_1 = 3, N_2 = 4$ for demonstration.) Then, $H(e^{jN_1\omega})$ has N_1 passbands in $[0, 2\pi)$, each with width $2\pi/N_1N_2$, as demonstrated in Fig. 1.17b.

Next, consider a filter bank with N_2 filters

$$H_k(e^{j\omega}) = H(e^{jN_1\omega}W_{N_2}^k), \quad 0 \leq k \leq N_2 - 1, \tag{1.79}$$

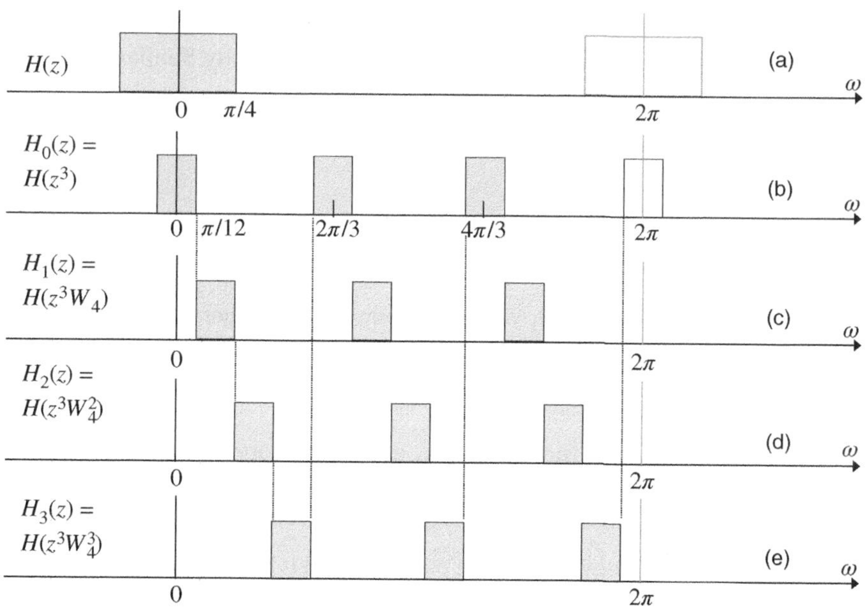

Figure 1.17 (a)–(e) Magnitude responses of the filter $H(z)$ and the N_2 filters $H_k(z)$. Here $N_1 = 3, N_2 = 4$.

12 We can get good approximations of this in practice.

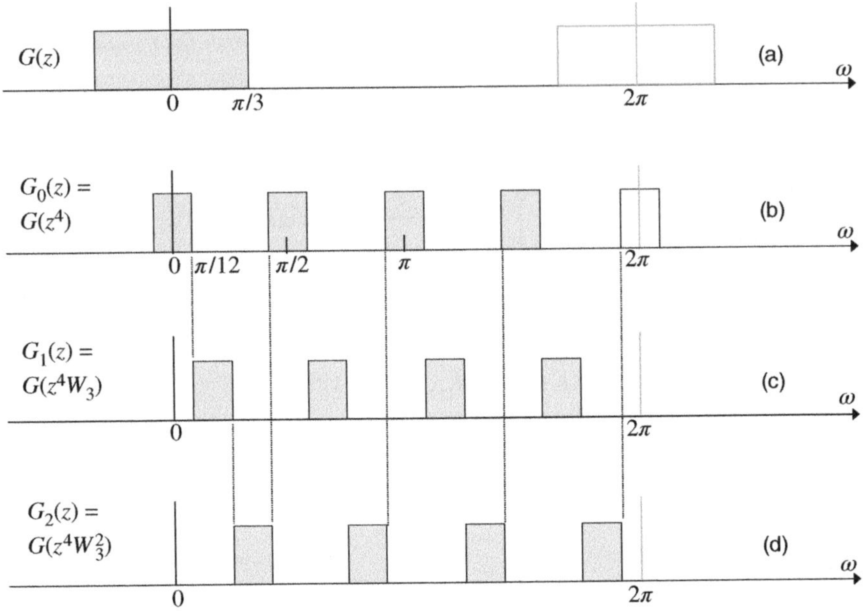

Figure 1.18 (a)–(d) Magnitude responses of the filter $G(z)$ and the N_1 filters $G_l(z)$. Here $N_1 = 3, N_2 = 4$.

where $W_M = e^{-j2\pi/M}$. These can be rewritten as

$$H_k(e^{j\omega}) = H(e^{jN_1\omega}e^{-j\frac{2\pi k}{N_2}}) = H(e^{jN_1(\omega-\frac{2\pi k}{N_1N_2})}). \tag{1.80}$$

Thus, $H_k(e^{j\omega})$ are uniformly shifted versions of $H(e^{jN_1\omega})$, with shifts in multiples of $2\pi/N_1N_2$. These are demonstrated in Fig. 1.17. Each $H_k(e^{j\omega})$ has N_1 passbands of width $2\pi/N_1N_2$ each. These N_2 filters are nonoverlapping, and the union of their passbands tiles $[0, 2\pi)$ perfectly. Similarly, we can define

$$G_l(e^{j\omega}) = G(e^{jN_2\omega}W_{N_1}^l), \quad 0 \le l \le N_1 - 1, \tag{1.81}$$

so that

$$G_l(e^{j\omega}) = G(e^{jN_2(\omega-\frac{2\pi l}{N_1N_2})}), \tag{1.82}$$

which are demonstrated in Fig. 1.18.

Each $G_l(e^{j\omega})$ has N_2 passbands of width $2\pi/N_1N_2$ each. Again, these N_1 filters are nonoverlapping, and the union of their passbands tiles $[0, 2\pi)$. Now define the N_1N_2 **product filters**

$$F_{k,l}(e^{j\omega}) = H_k^*(e^{j\omega})G_l(e^{j\omega}) = H^*(e^{jN_1(\omega-\frac{2\pi k}{N_1N_2})})G(e^{jN_2(\omega-\frac{2\pi l}{N_1N_2})}), \tag{1.83}$$

where $0 \le k \le N_2 - 1, 0 \le l \le N_1 - 1$. Using the fact that N_1 and N_2 are coprime, it can be shown [27] that out of the N_1 passbands of $H_k(e^{j\omega})$ and N_2 passbands of $G_l(e^{j\omega})$, one pair of passbands overlap completely and the other passbands have no overlap. Thus, two conclusions can be drawn:

1. The product $F_{k,l}(e^{j\omega})$ is a bandpass filter with a single passband and it has width $2\pi/N_1N_2$.
2. Furthermore, coprimality of N_1 and N_2 also ensures that any two filters $F_{k_1,l_1}(e^{j\omega})$ and $F_{k_2,l_2}(e^{j\omega})$ with $\{k_1, l_1\} \ne \{k_2, l_2\}$ have nonoverlapping passbands.

Thus, the N_1N_2 filters (1.83) are bandpass with one passband each, and their passbands tile the region $\omega \in [0, 2\pi)$ perfectly, as shown in Fig. 1.19.

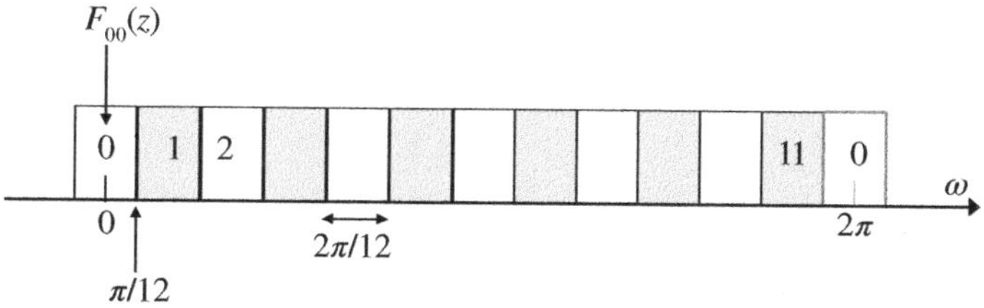

Figure 1.19 The filter bank of N_1N_2 filters $F_{k,l}(z)$ defined as in (1.83). Magnitude responses are shown. Here $N_1 = 3, N_2 = 4$. Adjacent filters are shown in alternating shades for clarity.

1.9.2 Realization of the Set of N_1N_2 Beamformers

Next, we explain how the N_2 beamformers $\{H_k^*(e^{j\omega})\}$ and N_1 beamformers $\{G_l^*(e^{j\omega})\}$ can be realized, along with the product beamformers $H_k^*(e^{j\omega})G_l(e^{j\omega})$. We will see that merely multipling the outputs of the beamformers $H_k^*(e^{j\omega})$ and $G_l(e^{j\omega})$ will not produce the product beamformer $H_k^*(e^{j\omega})G_l(e^{j\omega})$. Some further steps are necessary as explained next. Consider Fig. 1.20a.

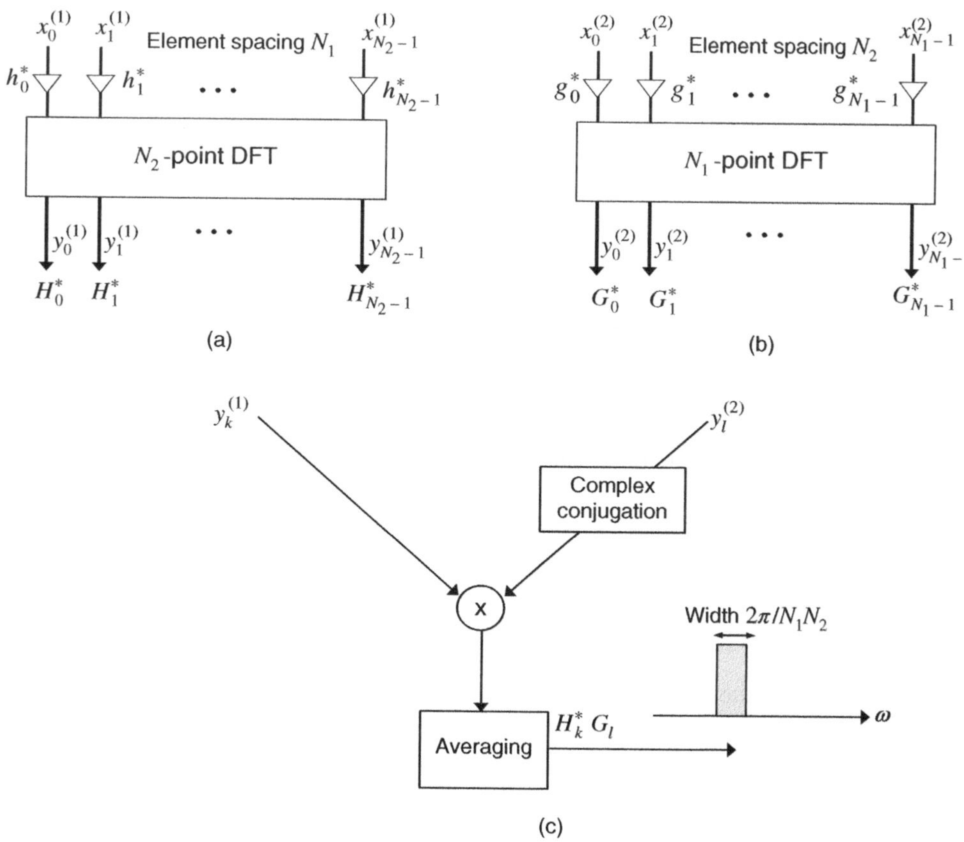

Figure 1.20 The coprime DFT filter bank. (a) The N_2 beamformer gains H_k^* from the first sparse ULA, (b) the N_1 beamformer gains G_l^* from the second sparse ULA, and (c) combining pairs of beamformer outputs $y_k^{(1)}$ and $y_l^{(2)}$ to produce N_1N_2 sharp (narrow-band) beams. One of these beams is shown in part (c).

Here the N_2-point DFT matrix has the form

$$\mathbf{W}_{N_2} = \begin{bmatrix} 1 & 1 & \cdots & 1 \\ 1 & W_{N_2} & \cdots & W_{N_2}^{N_2-1} \\ \vdots & \vdots & \ddots & \vdots \\ 1 & W_{N_2}^{N_2-1} & \cdots & W_{N_2}^{(N_2-1)(N_2-1)} \end{bmatrix} \tag{1.84}$$

In view of the definition (1.80), it can be verified [38] that the beamformer transfer functions at the output of the DFT block are precisely $H_k^*(e^{j\omega})$. Similarly, the beamformer transfer functions at the output of the DFT block in Fig. 1.20b are $G_l^*(e^{j\omega})$. Thus, if there are D sources arriving at angles ω_i with amplitudes s_i, then the outputs $y_k^{(1)}$ and $y_l^{(2)}$ are given by

$$y_k^{(1)} = \sum_{i=0}^{D-1} s_i H_k^*(e^{j\omega_i}) \quad \text{and} \quad y_l^{(2)} = \sum_{i=0}^{D-1} s_i G_l^*(e^{j\omega_i}). \tag{1.85}$$

The products $y_k^{(1)}(y_l^{(2)})^*$ therefore take the form

$$y_k^{(1)}(y_l^{(2)})^* = \sum_{i=0}^{D-1}\sum_{m=0}^{D-1} s_i s_m^* H_k^*(e^{j\omega_i})G_l(e^{j\omega_m}), \tag{1.86}$$

and contain cross-products $s_i s_m^*$. We now make use of the important assumption that the sources are zero mean and uncorrelated:

$$E[s_i] = 0, \quad E[s_i s_m^*] = p_i\delta[i-m]. \tag{1.87}$$

Then (1.86) implies $E[y_k^{(1)}(y_l^{(2)})^*] = \sum_{i=0}^{D-1} p_i H_k^*(e^{j\omega_i})G_l(e^{j\omega_i})$. Thus, in practice, we can use snapshot averaging to obtain

$$\frac{1}{K}\sum_{n=0}^{K-1} y_k^{(1)}[n](y_l^{(2)}[n])^* \approx \sum_{i=0}^{D-1} p_i H_k^*(e^{j\omega_i})G_l(e^{j\omega_i}). \tag{1.88}$$

The operation on the left is schematically shown in Fig. 1.20c. The system in Fig. 1.20 is therefore a beamformer with response $H_k^*(e^{j\omega})G_l(e^{j\omega})$. This is a bandpass response as shown in the figure. All the N_1N_2 beams $H_k^*(e^{j\omega})G_l(e^{j\omega})$ can thus be realized by using N_1N_2 multiplication-and-averaging operations.

It should be mentioned that we can add extra sensors to each of the sparse ULAs to increase the filter lengths. That is, we can replace (1.77) and (1.78) with

$$H^*(e^{jN_1\omega}) = \sum_{m=0}^{N_2'-1} h_m^* e^{jmN_1\omega}, \quad G^*(e^{jN_2\omega}) = \sum_{m=0}^{N_1'-1} g_m^* e^{jmN_2\omega}, \tag{1.89}$$

where $N_i' > N_i$. Figure 1.21 shows an example with $N_1 = 5$ and $N_2 = 6$ which are coprime. There are $N_1N_2 = 30$ filters and their magnitudes $|F_{k,l}(e^{j\omega})|$ are shown in Fig. 1.21, where the filter lengths are $N_2' = N_2 + 4$ and $N_1' = N_1 + 4$. The filter coefficients h_m and g_m are designed using the Parks–McClellan (or Remez exchange) method [39].

1.9.3 Summary: Coprime DFT Beamformer

Summarizing, this is what we have shown: under the assumptions that (a) the sources are uncorrelated and (b) N_1 and N_2 are coprime, the product beamformer system in Fig. 1.20 can be used to

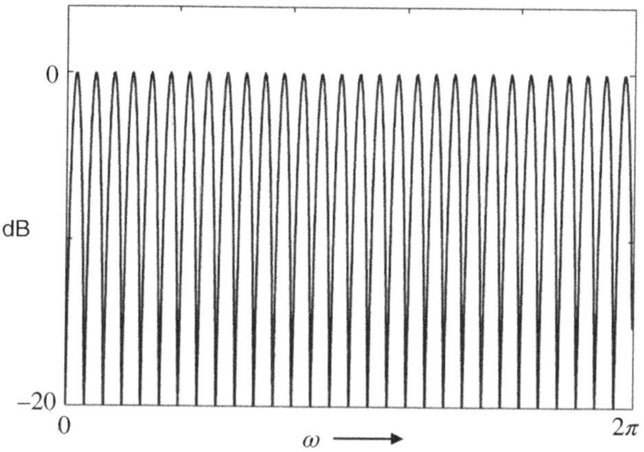

Figure 1.21 © Adopted from [27]. Magnitude responses of the filters $F_{k,l}(e^{j\omega})$ for the coprime DFT filter bank with $N_1 = 5$ and $N_2 = 6$. The filters are Parks–McClellan filters with lengths $N_2' = N_2 + 4$ and $N_1' = N_1 + 4$.

generate $N_1 N_2$ nonoverlapping beams in parallel. More examples of such beams can be found in [27]. Further developments on coprime DFT beamformers along the above lines can also be found in [40, 41].

Another interesting idea was developed by Liu and Buck [42] where the outputs of the DFT beamformers are combined using the so-called min-processor (instead of using multiplication as above). This has some advantages as explained in detail [42, 43]. Further interesting work on beamforming based on coprime arrays can be found in [44].

1.10 Directions for Further Reading

There has been significant amount of research on sparse arrays in the last decade and a half. In this concluding section, we give a flavor for a few areas we did not elaborate upon and provide references for further reading. The other chapters in this book will address many other research directions in the field.

1.10.1 Sparse Reconstruction Methods for DOAs

It turns out that the DOA estimation problem can be formulated as a "sparse recovery problem" [45], and this serves as an alternative to subspace methods such as MUSIC and ESPRIT. To explain further, let us go back to the array output equation (1.15). To estimate the unknown DOAs ω_i (which are contained in the columns $\mathbf{a}(\omega_i)$ of $\mathbf{A}$) we can reformulate the problem as follows. Define a dense grid of potential DOAs $\Omega_i \in [-\pi, \pi)$ and form a dictionary $\mathbf{D}$ of steering vectors $\mathbf{a}(\Omega_i)$ as follows:

$$\mathbf{D} = \left[\mathbf{a}(\Omega_0)\ \mathbf{a}(\Omega_1)\ \cdots\ \mathbf{a}(\Omega_{J-1})\right]. \tag{1.90}$$

This is a matrix of size $N \times J$ where N is the number of sensors and J is the number of "atoms" (potential DOAs) with $J >> D$ (D being the actual number of DOAs) and $J >> N$. So $\mathbf{D}$ is a "fat" matrix. If the grid $\{\Omega_i\}$ is dense enough, then each actual DOA ω_i is close to one of the grid points Ω_m. Thus the measured array output $\mathbf{x}[k]$ in (1.15) is approximately expressible in the form

$$\mathbf{x}[k] \approx \mathbf{D}\mathbf{u}[k] + \mathbf{e}[k], \tag{1.91}$$

where $\mathbf{u}[k]$ is a sparse vector (ideally with D nonzero elements). Since the noise $\mathbf{e}[k]$ is not known, the problem of finding a sparse $\mathbf{u}[k]$ to satisfy the above equation is typically solved by finding $\mathbf{u}$ such that

$$\phi = \|\mathbf{x} - \mathbf{D}\mathbf{u}\|_2^2 + \lambda\|\mathbf{u}\|_1 \tag{1.92}$$

is minimized, where $\|\mathbf{u}\|_p$ is the l_p norm of $\mathbf{u}$ and λ is a hyper parameter. The use of the l_1-norm term $\lambda\|\mathbf{u}\|_1$ promotes sparsity of the solution $\mathbf{u}$. There are many other ways to formulate the problem of finding a sparse $\mathbf{u}$, and well-known algorithms include the Lasso, the Orthogonal Matching Pursuit (OMP), and so forth [46–48]. We do not of course need to solve (1.92) for every snapshot index k. There is a way to reformulate the problem in the multiple measurement vector (MMV) mode and solve one optimization problem that takes advantage of all snapshots of the array output.

Assume then that a sparse solution $\mathbf{u}$ with $D << J$ nonzero components has been found. From the locations of these nonzero components in $\mathbf{u}$, the columns of $\mathbf{D}$ which are required in the approximation (1.91) can be identified, and the Ω_i corresponding to these columns are regarded as estimates of the actual DOAs ω_i. Detailed discussions and examples of DOA estimation based on these ideas can be found in [45]. When the array is a ULA, the matrix $\mathbf{D}$ is Vandermonde, and this helps to develop some theoretical guarantees on identifiability when $D < N$.

Can we generalize the above idea to sparse arrays, and thereby estimate $D > N$ DOAs? Indeed, this is possible as well. For this recall that if the sources are uncorrelated, we can define $R[k]$ in (1.29) and obtain an equation of the form (1.51) where $\mathbf{V}$ is $L \times D$ (assuming for simplicity that the coarray is the ULA $-L < l < L$). The information about the D DOAs is in the D columns of $\mathbf{V}$. To estimate these DOAs, we can again formulate this estimation problem as a sparse reconstruction problem [49–53]. For this we only have to replace the objective function (1.92) with

$$\phi = \|\hat{\mathbf{r}} - \mathbf{D}_1\mathbf{u}\|_2^2 + \lambda\|\mathbf{u}\|_1, \tag{1.93}$$

where $\mathbf{D}_1$ is an $L \times J$ dictionary similar to (1.90), and $\hat{\mathbf{r}}$ is an estimate of $\mathbf{r}_0$ in (1.51). Once we optimize $\mathbf{u}$ to minimize (1.93) we can estimate the DOAs ω_i from locations of the nonzero elements in $\mathbf{u}$ as before. Even though the physical array is a sparse array, $\mathbf{D}_1$ will be a Vandermonde matrix when the difference coarray has a ULA segment $-L < l < L$, and this helps to develop some theoretical guarantees on identifiability when $D < L$. In particular, when the coarray has a large ULA segment, we will have $L = \mathcal{O}(N^2)$, so that $D = \mathcal{O}(N^2)$ DOAs can be estimated. Further results and generalizations can be found in a number of papers [54, 55]. For an overview of sparse recovery methods for DOA estimation see [56].

1.10.2 Cramér–Rao Bounds for Sparse Arrays

The Cramér–Rao bound or **CRB** can be used to obtain a lower bound on the variances of unbiased estimates of the DOAs [1]. For an N sensor array with $D < N$ sources, expressions for CRB have been derived many years ago in the seminal papers by Stoica and Nehorai [57, 58]. The expressions involve the matrix inverse $(\mathbf{A}^H\mathbf{A})^{-1}$ where $\mathbf{A}$ is the $N \times D$ array manifold matrix. This inverse does not exist if $D > N$. For $D = N$, the matrix $(\mathbf{A}^H\mathbf{A})^{-1}$ can exist, but in that case, the projection matrix $\mathbf{I} - \mathbf{A}(\mathbf{A}^H\mathbf{A})^{-1}\mathbf{A}^H$ (which appears in the CRB expression) becomes identically zero and the CRB expression becomes meaningless. Furthermore, even for $D < N$, the above CRB expression does not use the apriori assumption that the sources are uncorrelated (which is a common assumption in coarray-based methods). Jansson, et al. [59] derived an expression for CRB which included this assumption. Not surprisingly this assumption makes the CRB smaller. Unfortunately, the expression due to Jansson, et al. is not valid when $D = N$, and for $D > N$, certain projection matrices in the expression become meaningless.

However, even when $D \geq N$ we know it is possible to identify DOAs unambiguously using sparse arrays if we have the apriori knowledge that the sources are uncorrelated. So it is of great interest to study the CRB for this case. In fact, the authors in [60] numerically computed and plotted the CRB for this case and made some interesting observations. For example, while the CRB decreases monotonically with SNR when $D < N$, the situation when $D > N$ is different. Namely, with a fixed number of snapshots the CRB does not keep decreasing as SNR increases; it decreases initially but saturates for large SNR. The above observations were not based on closed form expressions but on numerical computations. So, no theoretical justifications were provided.

An expression for the CRB for the DOA estimation problem has been derived in [61], which works even when $D \geq N$. Thus it is valid for sparse arrays receiving more sources than sensors. Among other things, this expression explains the experimentally observed behavior that the CRB can saturate with SNR when $D \geq N$. The expression is valid as long as the Fisher information matrix (FIM) is nonsingular, and the precise condition for this nonsingularity is also given in that work. A number of corollaries of this are also pointed out. Thus, referring to the difference coarray where the central ULA is $-(L_u - 1) \leq l \leq L_u - 1$, it has been shown that as long as $D \leq L_u - 1$, the FIM is invertible and the CRB expression remains valid. Furthermore, if $D > L - 1$ (where the coarray has $2L - 1$ elements), the FIM is guaranteed to be singular. Thus the CRB results in [61] answer many questions for sparse arrays satisfactorily. Examples of CRB plots based on this are shown in Fig. 1.22 for some sparse arrays. In this example, $N = 16$ for all arrays. The nested array has $N_1 = N_2 = 8$ and the extended coprime array has $N_1 = 4, N_2 = 9$. There are 30 DOAs uniformly spaced in $-0.98\pi \leq \omega \leq 0.98\pi$, and $K = 500$ snapshots are used. Observe that the CRB saturates for large SNR, as predicted by theory. The MRA has the smallest CRB. In the plot, the CRB is for DOAs specified as ω_i (and not θ_i).

Similar results for sparse arrays were also simultaneously reported in [62], which also included expressions for the MSE in DOA estimation for popular algorithms such as coarray MUSIC. Important results on the topic were also reported in [63].

When $D < N$ (fewer sources than sensors) we can compare the CRB of the ULA with those of sparse arrays. This has been done in [61] (see Fig. 4 therein), and it has been demonstrated that for

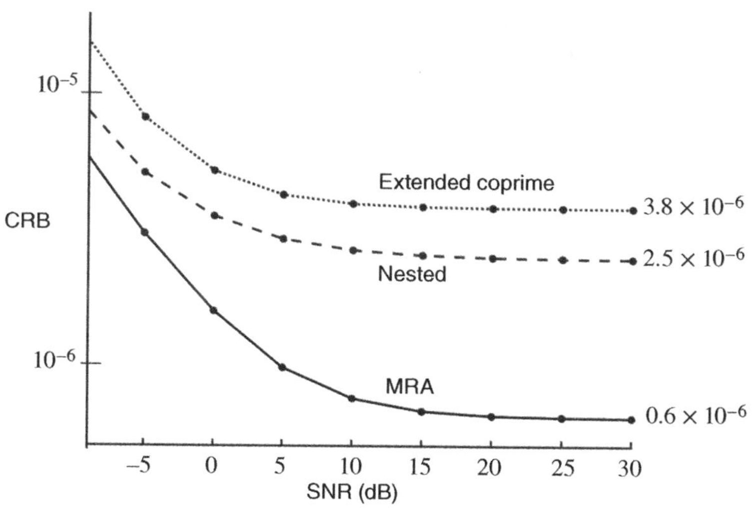

Figure 1.22 Plots of the CRB for sparse arrays identifying more DOAs than sensors ($D = 30, N = 16$). See the text for details.

nested, coprime, and minimum-redundancy arrays, the CRB is typically much smaller than that of the ULA. Thus, even when $D < N$, sparse arrays offer some significant advantages.

1.10.2.1 CRB Versus MSE for Coarray Methods

As mentioned above, when $D \geq N$, the CRB does not keep decreasing as the SNR is increased, but rather saturates [61]. Consequently, the MSE of DOA estimation using coarray domain algorithms (Section 1.6) also saturates. Wang and Nehorai [62] further observed that the DOA estimation using coarray MUSIC is not *efficient*, that is, the ratio CRB/MSE < 1 even as the SNR goes to infinity.

The efficiency of coarray MUSIC is even worse for the case $D < N$ (except when $D = 1$). In this regime, the CRB monotonically decreases without bound as SNR is increased [61]. However, MSE of coarray MUSIC still saturates for larger SNRs [62]. That is, the CRB/MSE ratio goes to 0 as the SNR increases!

To alleviate these drawbacks of coarray MUSIC, algorithms based on weighted least squares [64] and low-rank Toeplitz matrix reconstruction [65] have been developed. The weighted least squares approach provably achieves asymptotic efficiency for both $D \geq N$ and $D < N$ cases, and the gap between CRB and MSE for coarray DOA estimation becomes negligible as the number of snapshots increases.

1.10.3 Direct MUSIC on Sparse Arrays

The main advantage of N-element sparse arrays is that they can identify $D > N$ sources. For this we use algorithms such as coarray MUSIC (Section 1.6) or other coarray methods (Section 1.10.1). When $D < N$, we can use a ULA instead of a sparse array, but there are still some advantages for sparse arrays as mentioned above, such as better resolution, smaller MSE, and so forth. Now the question is, when $D < N$ can we directly use subspace algorithms such as MUSIC on the original array, instead of on the coarray? That is, can we do MUSIC directly based on the eigendecomposition of the $N \times N$ covariance matrix $\mathbf{R}_{\mathbf{xx}}$ (instead of doing it on the larger matrix $\mathbf{R}$ of Section 1.5)? This is referred to as "Direct MUSIC." Usually direct MUSIC on sparse arrays leads to ambiguity but under some special assumptions it can be successful. We will mention one such result here and refer the reader to [6, 7] for details. For this recall coprime arrays which were introduced in Section 1.7.2. Consider a variation of this, namely

$$\{0, N_1, 2N_1, \ldots, (M-1)N_1, \mid N_2\}, \tag{1.94}$$

where N_1 and N_2 are coprime. This array has $N = M + 1$ elements. The claim is the following: if we apply direct MUSIC to this array, then we can identify $D \leq N - 2$ DOAs unambiguously, as long as the DOAs are distinct modulo $2\pi/N_1$, that is, for any two DOAs $\omega_k, \omega_m \in [-\pi, \pi)$,

$$\omega_k - \omega_m \neq \frac{2\pi i}{N_1}, \tag{1.95}$$

for any integer i. It is demonstrated in the above work that the sparse array (1.94) (using direct MUSIC) has significantly better performance than the N element ULA, in terms of DOA resolution as well as MSE of estimation. Notice that the array manifold invertibility (Section 1.2.5), which is a necessary condition for unambiguous identifiability, is satisfied by (1.94) because N_1 and N_2 are coprime.

To demonstrate, let $N_1 = 4$, $N_2 = 5$, and $M = 7$, so that there are $N = 8$ sensors. Then we can identify $D \leq N - 2 = 6$ sources unambiguously using (1.94). For a specific example where there are $D = 5$ DOAs and the condition (1.95) is satisfied, Fig. 1.23a shows the direct MUSIC spectrum

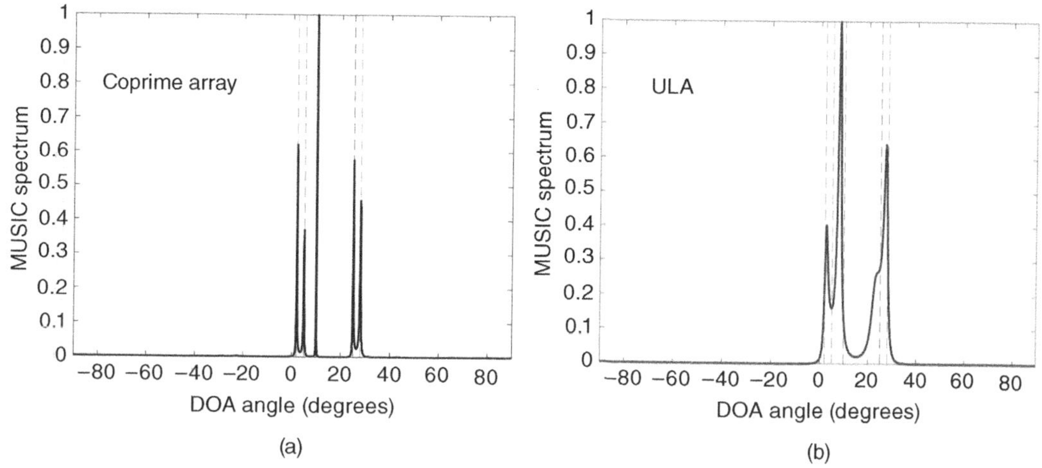

Figure 1.23 Direct MUSIC on the coprime array (1.94) and a ULA both with $N = 8$ sensors and $D = 5$ sources. The coprime array can resolve the sources but the ULA cannot.

using this coprime array. The dashed vertical lines show the ideal DOA locations. The MUSIC spectrum for an eight-sensor ULA is also shown in Fig. 1.23b. In this example, the SNR $= 5\,\text{dB}$ and the number of snapshots $K = 500$. It is clear that even though $D < N$, the ULA cannot resolve such closely spaced sources with this SNR and K. But the coprime array with its larger aperture can clearly resolve the sources.

1.10.4 Further Developments on Sparse Array Geometry

The nested array has a hole-free coarray, while the coprime array has holes. For a fixed number of sensors, the ULA coarray of the nested array is longer than the central ULA segment in the coprime coarray. This is an advantage of the nested array over the coprime array. But the (basic) coprime array has the advantage that there is only one pair of elements with spacing $\lambda/2$, while the nested array has a dense ULA component with many sensor pairs separated by $\lambda/2$ (Section 1.8.1). Thus, the nested array suffers more from the effect of mutual electromagnetic coupling between sensors.

A number of generalizations of nested and coprime arrays have been reported by many authors in recent years, to combine the best features of nested and coprime arrays. A generalization called the super nested array [66] was developed by starting from the nested array and systematically moving the sensors in such a way that the difference coarray remains unchanged, while the weights $w[1]$, $w[2]$, and $w[3]$ are systematically reduced. This results in an array with the same ULA coarray like the nested array, but with reduced mutual coupling. This principle has also been generalized to two-dimensional arrays, resulting in the so-called hourglass array [67]. Earlier extensions of sparse arrays to two dimensions include [68–70]. Extensions of nested arrays to identify DOAs from higher order statistics (HOS) can be found in [71].

Qin et al. [28] reported some generalizations of the coprime array based on the so-called compression and displacement operations, resulting in important advantages. Another novel variation of coprime arrays is frequency-diverse coprime arrays with coprime frequency offsets [53]. Yet another interesting line of work is coarray interpolation to fill the holes in the coarray of a coprime array [72, 73], and [74]. Other recent generalizations include the dilated nested array (DNA) [75] and more extensive variations of coprime arrays reported in [76] and references therein.

Another novel advancement is nested vector-sensor array processing [77]. Extension of the use of coprime arrays to moving platform scenarios can be found in [78] and [79]. Some authors have reported novel geometries for sparse arrays with a focus on the properties of sum coarrays [14–16, 80]. Sum coarrays are important in active sensing [11]. Although not discussed in this chapter, they can be found elsewhere in this book.

In recent years, convolutional neural networks (CNN) have been shown to be attractive for DOA estimation especially because they are robust to calibration errors and other model imperfections [81, 82]. The use of CNN for sparse arrays, to find more DOAs than sensors, is described in [83] which also contains more references for this line of research.

A broad class of non-integer arrays (as opposed to the widely used integers arrays (1.1)) was recently studied systematically in [84]. In particular, sparse and coprime rational arrays introduced in this work can estimate the DOAs more accurately than integer arrays when a fixed number of sensors are available, and the aperture is constrained. Furthermore, rational arrays are also advantageous when the DOAs are apriori known to be in a proper subset of $[-\pi/2, \pi/2)$.

Sparse spatial sampling and difference coarrays also find application in nonuniform decimation of the outputs of convolutional beamspace processors [85]. This has recently been used for DOA estimation of mmWave sources using hybrid (analog/digital) architectures, which are more suitable for large arrays [86]. The method allows the identification of more number of DOAs in the beamspace band than the number of RF chains in the system.

Finally, sparse array design and reconfigurability for optimum direction finding and beamforming under time-varying environment is discussed in Chapters 6, 7 and 10 of this book [87, 88].

Acknowledgment

This work was supported by the ONR grant N00014-21-1-2521 and the California Institute of Technology.

References

1 Van Trees, H.L. (2002). *Optimum Array Processing*. New York: Wiley.

2 Proakis, J.G. (1989). *Digital Communications*. New York: McGraw Hill, Inc.

3 Schmidt, R. (1986). Multiple emitter location and signal parameter estimation. *IEEE Transactions on Antennas and Propagation* 34: 276–280.

4 Roy, R. and Kailath, T. (1989). ESPRIT –estimation of signal parameters via rotational invariance techniques. *IEEE Transactions on Acoustics, Speech, and Signal Processing* 37: 984–995.

5 Barabell, A.J. (1983). Improving the resolution performance of eigenstructure-based direction-finding algorithms. *Proceedings of the IEEE International Conference on Acoustics, Speech, and Signal Processing*, 336–339.

6 Vaidyanathan, P.P. and Pal, P. (2012). Direct-MUSIC on sparse arrays. *IEEE International Conference on Signal Processing and Communications.*

7 Vaidyanathan, P.P. and Pal, P. (2013). Why does direct-MUSIC on sparse-arrays work. *47th Proceedings of Asilomar Conference on Signals, Systems and Computers.*

8 Manikas, A. (2004). *Differential Geometry in Array Processing*. Imperial College Press.

9 Shamsunder, S. and Giannakis, G.B. (1993). Modeling of non-Gaussian array data using cumulants: DOA estimation of more sources with less sensors. *Signal Processing* 30: 279–297.

10 Chambers, C., Tozer, T.C., Sharman, K.C., and Durrani, T.S. (1996). Temporal and spatial sampling influence on the estimates of superimposed narrowband signals: when less can mean more. *IEEE Transactions on Signal Processing* 44: 3085–3098.

11 Hoctor, R.T. and Kassam, S.A. (1990). The unifying role of the coarray in aperture synthesis for coherent and incoherent imaging. *Proceedings of the IEEE* 78: 735–752.

12 Pal, P. and Vaidyanathan, P.P. (2010). Nested arrays: a novel approach to array processing with enhanced degrees of freedom. *IEEE Transactions on Signal Processing* 58: 4167–4181.

13 Ma, W.-K., Hsieh, T.-H., and Chi, C.-Y. (2010). DOA estimation of quasi-stationary signals with less sensors than sources and unknown spatial noise covariance: a Khatri–Rao subspace approach. *IEEE Transactions on Signal Processing* 58: 2168–2180.

14 Bliss, D.W. and Forsythe, K.W. (2003). Multiple-input multiple-output (MIMO) radar and imaging: degrees of freedom and resolution. *37th Asilomar Conference on Signals, Systems and Computers*, 54–59.

15 Chen, C.-Y. and Vaidyanathan, P.P. (2008). MIMO radar space-time adaptive processing using prolate spheroidal wave functions. *IEEE Transactions on Signal Processing* 56: 623–635.

16 Rajamäki, R. and Koivunen, V. (2021). Sparse symmetric linear arrays with low redundancy and a contiguous sum co-array. *IEEE Transactions on Signal Processing* 69: 1697–1712.

17 Friedlander, B. (2022). Degrees-of-freedom and the covariance manifold of an array. *56th Asilomar Conference on Signals, Systems and Computers.*

18 Singer, J. (1938). A theorem in finite projective geometry and some applications to number theory. *Transactions of the American Mathematical Society* 43: 377–385.

19 Brauer, A. (1945). A problem of additive number theory and its application in electrical engineering. *Journal of the Elisha Mitchell Scientific Society* 61: 55–66.

20 Erdös, P. and Gál, I.S. (1948). On the representation of 1, 2, …, N by differences. *Indagationes Mathematicae* 10: 379–382.

21 Leech, J. (1956). On the representation of 1, 2,…, N by differences. *Journal of the London Mathematical Society* s1-31: 160–169.

22 Serra, J. (1982). *Image Analysis and Mathematical Morphology*. Orlando, FL: Academic Press.

23 Feynman, R.P., Leighton, R.B., and Sands, M. (1963). *The Feynman Lectures on Physics.* Reading, MA: Addison-Wesley.

24 Pal, P. and Vaidyanathan, P.P. (2011). Coprime sampling and the MUSIC algorithm. *Digital Signal Processing and Signal Processing Education Meeting (DSP/SPE)*, 289–294.

25 Liu, C.-L. and Vaidyanathan, P.P. (2015). Remarks on the spatial smoothing step in coarray MUSIC. *IEEE Signal Processing Letters* 22: 1438–1442.

26 Pillai, S.U., Bar-Ness, Y., and Haber, F. (1985). A new approach to array geometry for improved spatial spectrum estimation. *Proceedings of the IEEE* 73: 1522–1524.

27 Vaidyanathan, P.P. and Pal, P. (2011). Sparse sensing with co-prime samplers and arrays. *IEEE Transactions on Signal Processing* 59: 573–586.

28 Qin, S., Zhang, Y.D., and Amin, M.G. (2015). Generalized coprime array configurations for direction of arrival estimation. *IEEE Transactions on Signal Processing* 63: 1377–1390.

29 Bloom, G.S. and Golomb, S.W. (1977). Applications of numbered undirected graphs. *Proceedings of the IEEE* 65: 562–570.

30 Moffet, A.T. (1968). Minimum-redundancy linear arrays. *IEEE Transactions on Antennas and Propagation* 16: 172–175.

31 Shakeri, S., Ariananda, D.D., and Leus, G. (2012). Direction of arrival estimation using sparse ruler array design. *IEEE 13th International Workshop on Signal Processing Advances in Wireless Communications*, 525–529.

32 Ishiguro, M. (1980). Minimum redundancy linear arrays for a large number of antennas. *Radio Science* 15: 1163–1170.

33 BouDaher, E., Ahmad, F., Amin, M.G., and Hoorfar, A. (2015). DOA estimation with co-prime arrays in the presence of mutual coupling. *Proceedings of the 23rd European Signal Processing Conference*, 2830–2834.

34 Vertatschitsch, E. and Haykin, S. (1986). Nonredundant arrays. *Proceedings of the IEEE* 74: 217–217.

35 Robinson, J. (1979). Optimum Golomb rulers. *IEEE Transactions on Computers* 28: 943–944.

36 Shearer, J.B. (1990). Some new optimum Golomb rulers. *IEEE Transactions on Information Theory* 36: 183–184.

37 Drakakis, K. (2009). A review of the available construction methods for Golomb rulers. *Advances in Mathematics of Communications* 3: 235–250.

38 Vaidyanathan, P.P. (1993). *Multirate Systems and Filter Banks*. Englewood Cliffs, NJ: Prentice Hall.

39 Oppenheim, A.V. and Schafer, R.W. (2010). *Discrete-Time Signal Processing*. Upper Saddle River, NJ: Prentice Hall, Inc.

40 Liu, C.-L. and Vaidyanathan, P.P. (2014). Design of coprime DFT arrays and filter banks. *48th Asilomar Conference on Signals, Systems and Computers*.

41 Liu, C.-L. and Vaidyanathan, P.P. (2015). Coprime DFT filter bank design: theoretical bounds and guarantees. *IEEE International Conference on Acoustics, Speech and Signal Processing*, 3861–3865.

42 Liu, Y. and Buck, J.R. (2015). Detecting Gaussian signals in the presence of interferers using the coprime sensor arrays with the min processor. *49th Asilomar Conference on Signals, Systems and Computers*, 370–374.

43 Liu, Y. and Buck, J.R. (2018). Gaussian source detection and spatial spectral estimation using a coprime sensor array with the min processor. *IEEE Transactions on Signal Processing* 66: 186–199.

44 Adhikari, K., Buck, J.R., and Wage, K.E. (2014). Extending coprime sensor arrays to achieve the peak side lobe height of a full uniform linear array. *EURASIP Journal on Advances in Signal Processing* 2014: 148.

45 Malioutov, D., Çetin, M., and Willsky, A.S. (2005). A sparse signal reconstruction perspective for source localization with sensor arrays. *IEEE Transactions on Signal Processing* 53: 3010–3022.

46 Tropp, J.A. and Wright, S.J. (2010). Computational methods for sparse solution of linear inverse problems. *Proceedings of the IEEE* 98: 948–958.

47 Elad, M. (2013). *Sparse and Redundant Representations*. New York: Springer.

48 Eldar, Y.C. (2015). *Sampling Theory: Beyond Bandlimited Systems*. Cambridge: Cambridge University Press.

49 Pal, P. and Vaidyanathan, P.P. (2012). Correlation-aware techniques for sparse support recovery. *IEEE Statistical Signal Processing Workshop (SSP)*, 53–56.

50 Pal, P. and Vaidyanathan, P.P. (2012). On application of LASSO for sparse support recovery with imperfect correlation awareness. *46th Asilomar Conference on Signals, Systems and Computers*, 958–962.

51 Pal, P. and Vaidyanathan, P.P. (2015). Pushing the limits of sparse support recovery using correlation information. *IEEE Transactions on Signal Processing* 63: 711–726.

52 Zhang, Y.D., Amin, M.G., and Himed, B. (2013). Sparsity-based DOA estimation using co-prime arrays. *Proceedings of the IEEE International Conference on Acoustics, Speech, and Signal Processing*, 3967–3971.

53 Qin, S., Zhang, Y.D., Amin, M.G., and Gini, F. (2017). Frequency diverse coprime arrays with coprime frequency offsets for multitarget localization. *IEEE Journal of Selected Topics in Signal Processing* 11: 321–335.

54 Pal, P. and Vaidyanathan, P.P. (2014). A grid-less approach to underdetermined direction of arrival estimation via low rank matrix denoising. *IEEE Signal Processing Letters* 21: 737–741.

55 Qiao, H. and Pal, P. (2017). Gridless line spectrum estimation and low-rank Toeplitz matrix compression using structured samplers: a regularization-free approach. *IEEE Transactions on Signal Processing* 65: 2221–2236.

56 Yang, Z., Li, J., Stoica, P., and Xie, L. (2018). Sparse methods for direction-of-arrival estimation. In: *Academic Press Library in Signal Processing*, vol. 7 (ed. R. Chellappa and S. Theodoridis), 509–581. Academic Press.

57 Stoica, P. and Nehorai, A. (1989). MUSIC, maximum likelihood, and Cramér–Rao bound. *IEEE Transactions on Acoustics, Speech, and Signal Processing* 37: 720–741.

58 Stoica, P. and Nehorai, A. (1990). Performance study of conditional and unconditional direction-of-arrival estimation. *IEEE Transactions on Acoustics, Speech, and Signal Processing* 38: 1783–1795.

59 Jansson, M., Goransson, B., and Ottersten, B. (1999). A subspace method for direction of arrival estimation of uncorrelated emitter signals. *IEEE Transactions on Signal Processing* 47: 945–956.

60 Abramovich, Y.I., Gray, D.A., Gorokhov, A.Y., and Spencer, N.K. (1998). Positive-definite Toeplitz completion in DOA estimation for nonuniform linear antenna arrays –Part I: Fully augmentable arrays. *IEEE Transactions on Signal Processing* 46: 2458–2471.

61 Liu, C.-L. and Vaidyanathan, P.P. (2017). Cramér–Rao bounds for coprime and other sparse arrays, which find more sources than sensors. *Digital Signal Processing* 61: 43–61.

62 Wang, M. and Nehorai, A. (2017). Coarrays, MUSIC, and the Cramér–Rao bound. *IEEE Transactions on Signal Processing* 65: 933–946.

63 Koochakzadeh, A. and Pal, P. (2016). Cramér–Rao bounds for underdetermined source localization. *IEEE Signal Processing Letters* 23: 919–923.

64 Sedighi, S., Rao, B.S.M.R., and Ottersten, B. (2020). An asymptotically efficient weighted least squares estimator for co-array-based DoA estimation. *IEEE Transactions on Signal Processing* 68: 589–604.

65 Zheng, Z., Huang, Y., Wang, W.-Q., and So, H.C. (2021). Augmented covariance matrix reconstruction for DOA estimation using difference coarray. *IEEE Transactions on Signal Processing* 69: 5345–5358.

66 Liu, C.-L. and Vaidyanathan, P.P. (2016). Super nested arrays: linear sparse arrays with reduced mutual coupling –Part I: Fundamentals. *IEEE Transactions on Signal Processing* 64: 3997–4012.

67 Liu, C.-L. and Vaidyanathan, P.P. (2017). Hourglass arrays and other novel 2-D sparse arrays with reduced mutual coupling. *IEEE Transactions on Signal Processing* 65: 3369–3383.

68 Pal, P. and Vaidyanathan, P.P. (2012). Nested arrays in two dimensions, Part I: Geometrical considerations. *IEEE Transactions on Signal Processing* 60: 4694–4705.

69 Vaidyanathan, P.P. and Pal, P. (2011). Theory of sparse coprime sensing in multiple dimensions. *IEEE Transactions on Signal Processing* 59: 3592–3608.

70 Pal, P. and Vaidyanathan, P.P. (2011). Coprimality of certain families of integer matrices. *IEEE Transactions on Signal Processing* 59: 1481–1490.

71 Pal, P. and Vaidyanathan, P.P. (2012). Multiple level nested array: an efficient geometry for $2q$th order cumulant based array processing. *IEEE Transactions on Signal Processing* 60: 1253–1269.

72 Liu, C.-L., Vaidyanathan, P.P., and Pal, P. (2016). Coprime coarray interpolation for DOA estimation via nuclear norm minimization. *IEEE International Symposium on Circuits and Systems*, 2639–2642.

73 Zhou, C., Gu, Y., Shi, Z., and Zhang, Y.D. (2018). Off-grid direction-of-arrival estimation using coprime array interpolation. *IEEE Signal Processing Letters* 25: 1710–1714.

74 Fu, M., Zheng, Z., Wang, W.-Q., and So, H.C. (2021). Coarray interpolation for DOA estimation using coprime EMVS array. *IEEE Signal Processing Letters* 28: 548–552.

75 Shaalan, A.M.A., Du, J., and Tu, Y.H. (2022). Dilated nested arrays with more degrees of freedom (DOFs) and less mutual coupling–Part I: The fundamental geometry. *IEEE Transactions on Signal Processing* 70: 2518–2531.

76 Mohsen, N., Hawbani, A., Agrawal, M. et al. (2022). A new coprime-array-based configuration with augmented degrees of freedom and reduced mutual coupling. *Proceedings of the IEEE International Conference on Acoustics, Speech and Signal Processing*, 5033–5037.

77 Han, K. and Nehorai, A. (2014). Nested vector-sensor array processing via tensor modeling. *IEEE Transactions on Signal Processing* 62: 2542–2553.

78 Ramirez, J. and Krolik, J. (2015). Multiple source localization with moving co-prime arrays. *Proceedings of the IEEE International Conference on Acoustics, Speech and Signal Processing*, 2374–2378.

79 Qin, G., Amin, M.G., and Zhang, Y.D. (2019). DOA estimation expoliting sparse array motions. *IEEE Transactions on Signal Processing* 67: 3013–3027.

80 Rajamäki, R. and Koivunen, V. (2017). Sparse linear nested array for active sensing. *Proceedings of the 25th European Signal Processing Conference*, 1976–1980.

81 Liu, Z.-M., Zhang, C., and Yu, P.S. (2018). Direction-of-arrival estimation based on deep neural networks with robustness to array imperfections. *IEEE Transactions on Antennas and Propagation* 66: 7315–7327.

82 Wu, L., Liu, Z.-M., and Huang, Z.-T. (2019). Deep convolution network for direction of arrival estimation with sparse prior. *IEEE Signal Processing Letters* 26: 1688–1692.

83 Kulkarni, P. and Vaidyanathan, P.P. (2021). Feature engineering for DOA estimation using a convolutional neural network, for sparse arrays. *55th Asilomar Conference on Signals, Systems and Computers*.

84 Kulkarni, P. and Vaidyanathan, P.P. (2022). Non-integer arrays for array signal processing. *IEEE Transactions on Signal Processing* 70: 5457–5472.

85 Chen, P.-C. and Vaidyanathan, P.P. (2020). Convolutional beamspace for linear arrays. *IEEE Transactions on Signal Processing* 68: 5395–5410.

86 Chen, P.-C. and Vaidyanathan, P.P. (2022). Hybrid convolutional beamspace for DOA estimation of millimeter wave sources. *56th Asilomar Conference on Signals, Systems and Computers*.

87 Wang, X., Amin, M., and Cao, X. (2018). Analysis and design of optimum sparse array configurations for adaptive beamforming. *IEEE Transactions on Signal Processing* 66 (2): 340–351.

88 Hamza, S.A. and Amin, M.G. (2021). Sparse array beamforming design for wideband signal models. *IEEE Transactions on Aerospace and Electronic Systems* 57 (2): 1211–1226.

2

Sparse Array Interpolation for Direction-of-Arrival Estimation

Chengwei Zhou[1,3], Yujie Gu[4], Yimin D. Zhang[5], and Zhiguo Shi[2,3]

[1]*State Key Laboratory of Industrial Control Technology, Zhejiang University, Hangzhou, China*
[2]*Key Laboratory of Collaborative Sensing and Autonomous Unmanned Systems of Zhejiang Province, Zhejiang University, Hangzhou, China*
[3]*College of Information Science and Electronic Engineering, Zhejiang University, Hangzhou, China*
[4]*Advanced Safety & User Experience, Aptiv, Agoura Hills, CA, USA*
[5]*Department of Electrical and Computer Engineering, Temple University, Philadelphia, PA, USA*

2.1 Introduction

As discussed in Chapter 1, sparse arrays can offer an enlarged array aperture with less sensors, and the augmented virtual arrays enable direction-of-arrival (DOA) estimation with enhanced source identifiability and improved resolution [1–6]. However, for partially augmentable sparse arrays, such as the prototype coprime array and its variants, the corresponding virtual arrays are non-uniform, that is, they contain holes at certain positions [7]. Therefore, it is difficult to directly utilize the entire derived signal statistics corresponding to the nonuniform virtual array. To address this issue, a common solution is to extract the maximum contiguous part from the nonuniform virtual array to form a virtual uniform linear array (ULA) [8]. However, the virtual ULA generated in such a way is at the expense of reducing the number of the achievable degrees of freedom (DOFs) and virtual array aperture, since the discontiguous virtual sensors are discarded and, as a result, the information contained in the virtual array is not fully utilized.

The nonuniformity of the virtual array is inherently caused by the sub-Nyquist spatial sampling of sparse arrays. To directly process sparse array signals in the physical domain for DOA estimation, the DECOM method [9] is proposed to eliminate the phase ambiguity exploiting a coprimality-based sparse sensor deployment. In this approach, a coprime array is decomposed into two subarrays, and subspace processing is separately implemented on the subarray signals. Based on the prime factorization theorem, the property of nonoverlapped spatial aliasing of coprime subarrays is analyzed. Subsequently, the source DOAs can be accurately estimated by searching for the closest spectral peaks corresponding to the subarrays. However, this method cannot exploit the entire aperture of the sparse array, and it also reduces at least half of the DOFs.

On the other hand, from the perspective of Nyquist matching in the virtual domain, sparse array configurations have been optimized with specific structural constraints to yield contiguous virtual arrays [10–12]. In particular, the length of contiguous coarray lags is often considered as the principal objective of sparse array optimization problems. Nevertheless, these methods dynamically select different sparse array configurations, thus resulting in a high computational complexity and hindering their implementation in practical applications. Therefore, it is of significant

Sparse Arrays for Radar, Sonar, and Communications, First Edition. Edited by Moeness G. Amin.

interest to make full use of the correlation information of structured sparse arrays with systematic configurations.

To this end, the theory of sparse array interpolation has been widely studied. To utilize the entire virtual array signals, a high-resolution DOA estimation algorithm is proposed in [13] by utilizing multiple frequencies to fill the missing elements in the difference coarray. Moreover, a gridless DOA estimation algorithm is proposed by interpolating the coprime coarray through nuclear norm minimization [14], based on which the minimum number of virtual sensors required to complete the interpolated virtual array covariance matrix is derived [15]. However, these algorithms are based on the matrix completion principle, indicating that the virtual signals corresponding to a discontiguous virtual array remain unchanged in the optimized covariance matrix. Since the virtual signals are obtained from the sample covariance matrix, the deviation caused by the finite number of snapshots may affect the estimation accuracy of the completed covariance matrix. In this regard, a covariance matrix reconstruction-based algorithm is proposed for virtual array interpolation [16], where gridless DOA estimation with full utilization of the correlation information is guaranteed. On the other hand, since the precise coarray information relies on the intact sparse array structure, the derived virtual array geometry is quite sensitive to the original sparse array configurations, and the array interpolation in the physical domain provides effective remedy for such scenario. In particular, an off-grid coprime array interpolation algorithm is proposed in [17], which realizes enhanced source identifiability beyond that of the virtual array interpolation-based methods. This approach has also been extended to deal with the DOA estimation of coherent sources [18].

In this chapter, we elaborate two sparse array interpolation algorithms. A gridless DOA estimation algorithm based on virtual array interpolation is presented in Section 2.2, whereas an off-grid DOA estimation algorithm based on physical array interpolation is described in Section 2.3. In Section 2.4, some future directions are briefly outlined.

The notations listed in Table 2.1 are used throughout this chapter.

Table 2.1 List of notations.

Symbol	Description
$a, \boldsymbol{a}, \boldsymbol{A}$	Scalar, vector, and matrix
j	Imaginary unit
$\mathbb{C}, \mathbb{R}, \mathbb{Z}$	Complex number, real number, and natural number sets
$(\cdot)^{\mathrm{T}}, (\cdot)^{H}, (\cdot)^{*}$	Transpose, Hermitian transpose, and conjugation
$\mathrm{E}[\cdot]$	Statistical expectation
$\mathrm{vec}(\cdot)$	Vectorization
$\mathrm{diag}(\cdot)$	Diagonal matrix
$\otimes, \circ$	Kronecker product and Hadamard product
$\succeq$	Matrix inequality
$\lVert\cdot\rVert_2, \lVert\cdot\rVert_F, \lVert\cdot\rVert_*$	Euclidean norm, Frobenius norm, and nuclear norm
$(\cdot)^{-1}, \mathrm{Tr}(\cdot), \mathrm{rank}(\cdot)$	Inverse, trace, and rank of a matrix
$\lvert\cdot\rvert$	Cardinality of a set
$\langle \boldsymbol{a} \rangle_{\ell}$	ℓth element of vector $\boldsymbol{a}$
$\boldsymbol{I}$	Identity matrix
$\boldsymbol{0}$	Zero matrix

2.2 Virtual Array Interpolation for Gridless DOA Estimation

In this section, we elaborate the virtual array interpolation-based algorithm for coprime array DOA estimation, where the virtual array interpolation is implemented to maintain the full utilization of the coarray statistics, and the atomic norm tailored for the second-order coarray statistics is formulated to avoid the basis mismatch problem in DOA estimation.

2.2.1 Discontiguous Coarray Model

Consider a pair of sparse ULAs as shown in Fig. 2.1a, where the top sparse ULA consists of M sensors spaced Nd apart, whereas the bottom one consists of N sensors spaced Md apart. Here, M and N are coprime integers, and d is assumed to be a half-wavelength, i.e., $d = \lambda/2$. Aligning these two sparse ULAs with the first sensor as the reference yields a coprime array configuration as shown in Fig. 2.1b. Due to the coprimality, the sensors do not overlap except at the reference position; hence, the coprime array consists of $M + N - 1$ sensors in total.

Assuming K far-field, narrowband, and uncorrelated sources impinging from directions $\boldsymbol{\theta} = [\theta_1, \theta_2, \dots, \theta_K]^{\mathrm{T}}$, the received signals of the coprime array can be modeled as

$$\boldsymbol{x}(t) = \sum_{k=1}^{K} \boldsymbol{a}(\theta_k)s_k(t) + \boldsymbol{n}(t) = \boldsymbol{A}\boldsymbol{s}(t) + \boldsymbol{n}(t), \tag{2.1}$$

where $\boldsymbol{A} = \left[\boldsymbol{a}(\theta_1), \boldsymbol{a}(\theta_2), \dots, \boldsymbol{a}(\theta_K)\right] \in \mathbb{C}^{(M+N-1)\times K}$ denotes the coprime array steering matrix, $\boldsymbol{s}(t) = [s_1(t), s_2(t), \dots, s_K(t)]^{\mathrm{T}}$ denotes the signal waveform vector, and $\boldsymbol{n}(t) \sim \mathcal{CN}(\boldsymbol{0}, \sigma_n^2\boldsymbol{I})$ represents the independent and identically distributed zero-mean additive white Gaussian noise vector with σ_n^2 denoting the noise power. The kth column of $\boldsymbol{A}$ represents the steering vector of the kth source signal, expressed as

$$\boldsymbol{a}(\theta_k) = \left[1, e^{-j\frac{2\pi}{\lambda}u_2 d\sin(\theta_k)}, \dots, e^{-j\frac{2\pi}{\lambda}u_{M+N-1} d\sin(\theta_k)}\right]^{\mathrm{T}}, \tag{2.2}$$

where $u_l d, l \in \{2, 3, \dots, M+N-1\}$, denotes the position of the lth sensor in the coprime array.

The covariance matrix of the received signal vector $\boldsymbol{x}(t)$ of the coprime array can be expressed as

$$\boldsymbol{R}_x = \mathrm{E}\left[\boldsymbol{x}(t)\boldsymbol{x}^H(t)\right] = \sum_{k=1}^{K} p_k \boldsymbol{a}(\theta_k)\boldsymbol{a}^H(\theta_k) + \sigma_n^2\boldsymbol{I}, \tag{2.3}$$

Figure 2.1 Illustration of a coprime array. (a) A coprime pair of sparse ULAs; (b) Coprime array configuration.

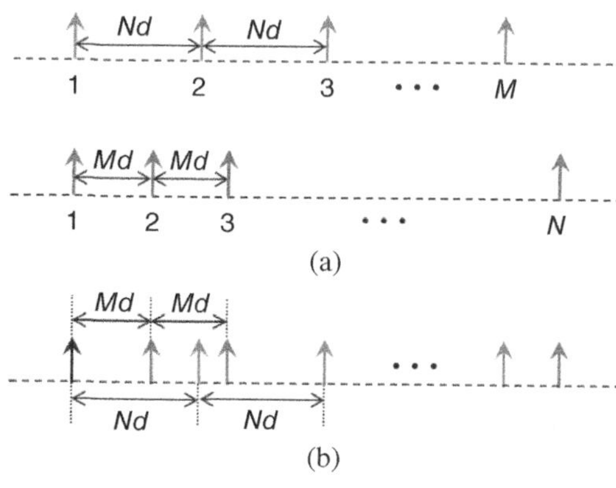

where p_k denotes the power of the kth source signal. The exact covariance matrix $\boldsymbol{R}_x$ is unavailable in practice and is approximated by its sample estimate

$$\hat{\boldsymbol{R}}_x = \frac{1}{T}\sum_{t=1}^{T}\boldsymbol{x}(t)\boldsymbol{x}^H(t), \tag{2.4}$$

where T is the number of snapshots. The sample covariance matrix $\hat{\boldsymbol{R}}_x$ is the maximum-likelihood estimator of $\boldsymbol{R}_x$ and converges to $\boldsymbol{R}_x$ when T tends to infinity under the stationarity and ergodicity assumptions [19].

By vectorizing the covariance matrix $\boldsymbol{R}_x$, we obtain the equivalent virtual array signals as

$$\boldsymbol{y}_v = \text{vec}(\boldsymbol{R}_x) = \boldsymbol{A}_v\boldsymbol{p} + \sigma_n^2\boldsymbol{i}, \tag{2.5}$$

where $\boldsymbol{A}_v = [\boldsymbol{a}^*(\theta_1)\otimes\boldsymbol{a}(\theta_1), \boldsymbol{a}^*(\theta_2)\otimes\boldsymbol{a}(\theta_2), \ldots, \boldsymbol{a}^*(\theta_K)\otimes\boldsymbol{a}(\theta_K)] \in \mathbb{C}^{(M+N-1)^2\times K}$, $\boldsymbol{p} = [p_1, p_2, \ldots, p_K]^{\mathrm{T}}$, and $\boldsymbol{i} = \text{vec}(\boldsymbol{I})$. Here, the steering matrix $\boldsymbol{A}_v$ corresponds to an augmented array with the virtual sensors located at $\mathcal{S}d$, where $\mathcal{S} = \{u_m - u_n | \, m, n = 0, 1, \ldots, M+N-1\}$. By removing the repeated elements in $\mathcal{S}$, a proper subset $\mathcal{S}_V \subsetneq \mathcal{S}$ is obtained, and the virtual sensors located at $\mathcal{S}_V d$ constitute a virtual array. For the systematic coprime array configuration shown in Fig. 2.1b, the elements in $\mathcal{S}_V$ can be obtained by picking all the unique values from its difference coarray as

$$\mathcal{S}_V \subsetneq \{\pm(Mn - Nm) | \, m = 0, 1, \ldots, M-1, n = 0, 1, \ldots, N-1\}. \tag{2.6}$$

Accordingly, the equivalent virtual signals of the derived virtual array $\mathcal{S}_V$ can be obtained by selecting the corresponding elements from $\boldsymbol{y}_v$ as

$$\bar{\boldsymbol{y}}_v = \bar{\boldsymbol{A}}_v\boldsymbol{p} + \sigma_n^2\bar{\boldsymbol{i}}, \tag{2.7}$$

where $\bar{\boldsymbol{A}}_v \in \mathbb{C}^{|\mathcal{S}_V|\times K}$ denotes the steering matrix of the derived virtual array $\mathcal{S}_V$, and $\bar{\boldsymbol{i}}$ denotes a sub-vector of $\boldsymbol{i}$ corresponding to the selected positions of the virtual sensors of $\mathcal{S}_V$ in $\mathcal{S}$.

Since the coprime array is partially augmentable, its difference coarray contains missing elements, which are referred to as holes [7, 20], leading to a nonuniform virtual array geometry. To have an intuitive understanding, a coprime array configuration with $M = 3$ and $N = 5$ is illustrated in Fig. 2.2a, and the derived nonuniform virtual array $\mathcal{S}_V = \{-12, -10, -9, -7, -6, -5, \ldots, 5, 6, 7, 9, 10, 12\}$ is depicted in Fig. 2.2b, where the missing elements $\{-11, -8, 8, 11\}$ are holes. Since the nonuniformity results in difficulties in subsequent statistical signal processing, a common solution is to pick the maximum contiguous part $\mathcal{S}_C$ from $\mathcal{S}_V$ while discarding the discontiguous part $\mathcal{S}_V - \mathcal{S}_C$ [8, 21]. Although the obtained virtual ULA $\mathcal{S}_C$ as shown in Fig. 2.2c is easy to operate, part of the information received by the coprime array is apparently ignored due to the discarded discontiguous virtual sensors, resulting in performance degradation.

2.2.2 Virtual Array Interpolation and Its Atomic Norm

This subsection introduces the virtual array interpolation method and its corresponding atomic norm. First, the derived nonuniform virtual array is transformed to a virtual ULA through array interpolation. The atomic norm of multiple virtual measurements is then considered based on an ideal assumption, from which the relationship between the second-order equivalent virtual array signals and the covariance matrix of the interpolated virtual array is established. Furthermore, the properties of virtual domain atomic norm are analyzed, such that gridless Toeplitz matrix reconstruction can be performed for underdetermined DOA estimation.

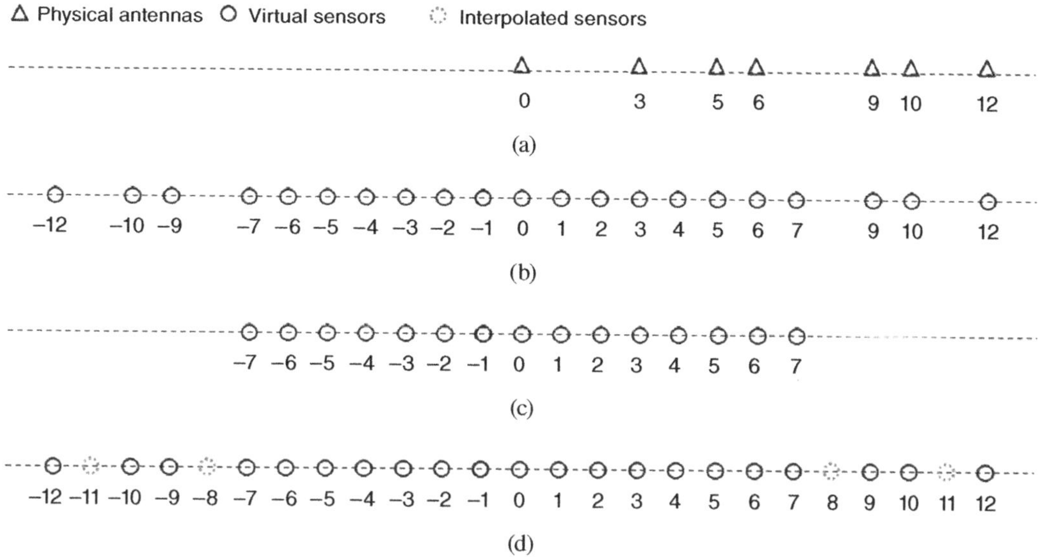

Figure 2.2 Illustration of various array representations with an example of $M = 3$ and $N = 5$. (a) Coprime array; (b) $\mathbb{S}_V$, virtual array derived from the difference coarray of the coprime array; (c) $\mathbb{S}_C$, contiguous part of the virtual array; and (d) $\mathbb{S}_I$, interpolated virtual array.

2.2.2.1 Array Interpolation for Virtual ULA

In order to make full use of the information available in the nonuniform virtual array $\mathbb{S}_V$, the concept of array interpolation is introduced to fill in the holes in $\mathbb{S}_V$ with nominal sensors. As such, a virtual ULA $\mathbb{S}_I$ with $2M(N-1)+1$ sensors is constructed as shown in Fig. 2.2d, where none of the virtual sensors in $\mathbb{S}_V$ are discarded. Accordingly, the $|\mathbb{S}_I|$-dimensional interpolated virtual array signals $\boldsymbol{y}_I$ is initialized as

$$[\boldsymbol{y}_I]_i = \begin{cases} [\bar{\boldsymbol{y}}_v]_i, & i \in \mathbb{S}_V, \\ 0, & i \in \mathbb{S}_I - \mathbb{S}_V, \end{cases} \tag{2.8}$$

where $[\,\cdot\,]_i$ denotes the virtual signal of the virtual sensor at position *id*. Clearly, there are two types of sensors in the interpolated virtual array $\mathbb{S}_I$, namely, virtual sensors and nominal sensors. The virtual signals in $\boldsymbol{y}_I$ corresponding to the derived virtual sensors in the nonuniform virtual array capture those in $\bar{\boldsymbol{y}}_v$, whereas the remaining elements corresponding to the nominal sensors are zero. The *de facto* ULA-based virtual signals $\boldsymbol{y}_I$ enable the potential utilization of the mature DOA estimation techniques designed for ULAs with all the information contained in $\bar{\boldsymbol{y}}_v$. Therefore, in lieu of using the contiguous virtual array $\mathbb{S}_C$, we perform DOA estimation using the equivalent signals of the interpolated virtual array $\mathbb{S}_I$ to resolve more sources with a higher resolution.

2.2.2.2 Atomic Norm of Multiple Virtual Measurements

Since the sparse signal reconstruction is formulated as a basis pursuit denoising problem, the pre-defined sampling grid with a specific sampling interval results in an inherent basis mismatch for DOA estimation. In order to overcome this basis mismatch problem, gridless methods provide a novel insight for statistical signal processing. The atomic norm is one of the most important mathematical tools to utilize the signal characteristics without pre-defined sampling grid. The analysis for the atomic norm in the virtual domain begins with an ideal scenario assumption, where the interpolated virtual array signals are assumed to be accurate, i.e., the number of available snapshots for

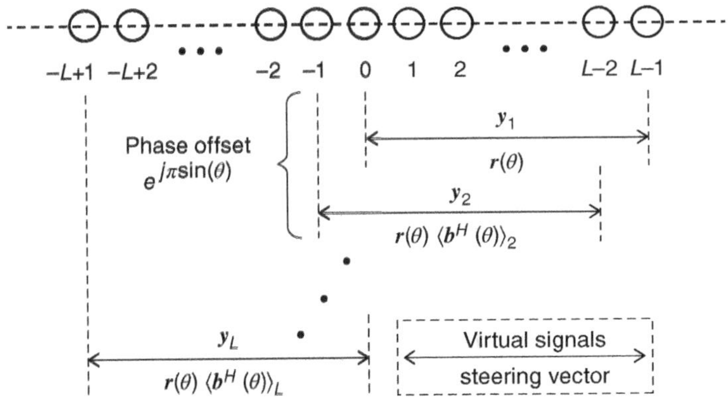

Figure 2.3 Phase offsets among the virtual measurements of each subarray.

calculating the sample covariance matrix tends to infinity, and the coprime array received signals are noise-free. Similar to (2.5), the ideal virtual signals of the interpolated virtual array $\mathcal{S}_\mathcal{I}$ can be modeled as

$$\boldsymbol{y} = \sum_{k=1}^{K} \boldsymbol{v}(\theta_k)p_k = \boldsymbol{V}\boldsymbol{p}, \tag{2.9}$$

where $\boldsymbol{V} = [\boldsymbol{v}(\theta_1), \boldsymbol{v}(\theta_2), \dots, \boldsymbol{v}(\theta_K)] \in \mathbb{C}^{|\mathcal{S}_\mathcal{I}|\times K}$ denotes the steering matrix of the interpolated virtual array $\mathcal{S}_\mathcal{I}$. Although the interpolated virtual array signal $\boldsymbol{y}$ modeled in (2.9) has a similar structure as the coprime array received signal $\boldsymbol{x}(t)$ in a noise-free case, $\boldsymbol{y}$ is actually the *second-order* statistics derived from the correlation of the *first-order* coprime array received signals. Because $\boldsymbol{y}$ behaves like a single-snapshot data, the corresponding correlation matrix is rank-deficient and thus makes multi-source identification difficult.

To overcome this problem, we divide the interpolated virtual array $\mathcal{S}_\mathcal{I}$ into $L = (|\mathcal{S}_\mathcal{I}| + 1)/2$ overlapping subarrays with L contiguous virtual sensors in each subarray, as shown in Fig. 2.3. Since $\mathcal{S}_\mathcal{I}$ is symmetric about the zeroth position, $|\mathcal{S}_\mathcal{I}|$ is odd and the value of L is always an integer. Accordingly, the equivalent virtual signals of each subarray can be obtained by dividing vector $\boldsymbol{y}$ into L sub-vectors $\{\boldsymbol{y}_1, \boldsymbol{y}_2, \dots, \boldsymbol{y}_L\}$ as

$$\boldsymbol{y}_\ell = \sum_{k=1}^{K} \boldsymbol{v}_\ell(\theta_k)p_k = \boldsymbol{V}_\ell\boldsymbol{p}, \quad \ell = 1, 2, \dots, L, \tag{2.10}$$

where $\boldsymbol{V}_\ell = [\boldsymbol{v}_\ell(\theta_1), \boldsymbol{v}_\ell(\theta_2), \dots, \boldsymbol{v}_\ell(\theta_K)] \in \mathbb{C}^{L\times K}$. Here,

$$\boldsymbol{v}_\ell(\theta_k) = \left[e^{-j\pi v_{L-\ell+1}\sin(\theta_k)}, e^{-j\pi v_{L-\ell+2}\sin(\theta_k)}, \dots, e^{-j\pi v_{2L-\ell}\sin(\theta_k)}\right]^{\mathrm{T}} \tag{2.11}$$

denotes the steering vector of the ℓth virtual subarray corresponding to the kth signal, where v_i denotes the ith element in $\mathcal{S}_\mathcal{I}$. Collecting these L sub-vectors, $\boldsymbol{Y} = [\boldsymbol{y}_1, \boldsymbol{y}_2, \dots, \boldsymbol{y}_L] \in \mathbb{C}^{L\times L}$ yields the equivalent virtual signals of the interpolated virtual subarray as L virtual measurements and is referred to as the virtual measurements in the sequel.

To analyze the atomic norm of $\boldsymbol{Y}$, an atom with the same dimension is required to represent $\boldsymbol{Y}$. It should be emphasized that there are striking differences between atoms in the virtual domain and those in the physical domain [22–24]. On one hand, for each source θ_k, the virtual signals $\boldsymbol{y}_\ell$ formulated in (2.10) belong to the second-order statistics, containing the real-valued source power p_k, unlike the first-order received signals $\boldsymbol{x}(t)$ containing complex-valued signal waveforms $s_k(t)$, for $k = 1, 2, \ldots, K$. On the other hand, for each virtual measurement, the displacement among the L subarrays of the interpolated virtual array $\mathbb{S}_{\mathcal{I}}$ creates phase offsets, which can be utilized to characterize the difference among the L subarrays in the virtual domain. While $\boldsymbol{Y}$ contains the second-order virtual measurements received by the subarrays as illustrated in Fig. 2.3, a series of atoms can be formulated to represent $\boldsymbol{Y}$ based on the steering vector of a certain subarray and the phase offsets among the L subarrays. In particular, we set the first subarray of $\mathbb{S}_{\mathcal{I}}$ as the reference virtual array, whose steering vector can be calculated by setting $\ell = 1$ in (2.11) as

$$\boldsymbol{r}(\theta) = \boldsymbol{v}_1(\theta) = \left[e^{-j\pi v_L \sin(\theta)}, e^{-j\pi v_{L+1} \sin(\theta)}, \ldots, e^{-j\pi v_{2L-1} \sin(\theta)}\right]^{\mathrm{T}}. \tag{2.12}$$

Accordingly, the phase offsets between the L subarrays and the reference virtual array become

$$\boldsymbol{b}(\theta) = \left[1, e^{-j\pi \sin(\theta)}, \ldots, e^{-j\pi(L-1)\sin(\theta)}\right]^{\mathrm{T}}, \tag{2.13}$$

and the steering vector of the ℓth subarray can be represented as

$$\boldsymbol{v}_\ell(\theta) = \boldsymbol{r}(\theta)\langle \boldsymbol{b}^H(\theta)\rangle_\ell, \quad \ell = 1, 2, \ldots, L. \tag{2.14}$$

As a result, the virtual measurements $\boldsymbol{Y}$ can now be regarded as L second-order virtual signal snapshots received by the reference virtual array, behaving like the snapshots of the first-order signal waveforms. The phase offsets contained in $\boldsymbol{b}(\theta)$ characterize the difference among different virtual measurements.

Based on $\boldsymbol{r}(\theta)$ and $\boldsymbol{b}(\theta)$, an atom for representing $\boldsymbol{Y}$ can be defined as

$$\boldsymbol{B}(\theta) = \boldsymbol{r}(\theta)\boldsymbol{b}^H(\theta), \tag{2.15}$$

where $\boldsymbol{B}(\theta) \in \mathbb{C}^{L \times L}$ with $\theta \in [-90^\circ, 90^\circ]$, and the corresponding atom set is

$$\mathcal{A} = \left\{\boldsymbol{B}(\theta) | \, \theta \in [-90^\circ, 90^\circ]\right\}. \tag{2.16}$$

With the specifically defined atom set $\mathcal{A}$, the smallest number of atoms for the representation of the virtual measurements $\boldsymbol{Y}$ can be defined as

$$\|\boldsymbol{Y}\|_{\mathcal{A},0} = \inf_K \left\{ \boldsymbol{Y} = \sum_{k=1}^{K} p_k \boldsymbol{B}(\theta_k), \; p_k \geq 0 \right\}, \tag{2.17}$$

where inf $\cdot$ denotes the infimum. While performing the atomic decomposition of $\boldsymbol{Y}$ with the minimal number of atoms, i.e., minimizing (2.17), is an NP-hard problem, we can introduce the atomic norm convex relaxation as[1]

$$\|\boldsymbol{Y}\|_{\mathcal{A}} = \inf\left\{h > 0 : \; \boldsymbol{Y} \in h \operatorname{conv}(\mathcal{A})\right\} = \inf \left\{ \sum_k p_k \,\middle|\, \boldsymbol{Y} = \sum_k p_k \boldsymbol{B}(\theta_k), \; p_k \geq 0 \right\}, \tag{2.18}$$

where conv($\mathcal{A}$) denotes the convex hull of the atom set $\mathcal{A}$.

2.2.2.3 Properties of Virtual Domain Atomic Norm

In this subsection, we analyze the properties of the virtual domain atomic norm defined in (2.18) that can be used for the design of subsequent DOA estimation algorithm. Specifically, we have the following theorem for the equivalent representation of $\|\boldsymbol{Y}\|_{\mathcal{A}}$.

1 Note that the atomic norm is formally defined as the gauge function of $\mathcal{A}$ [25]. Since the atom set $\mathcal{A}$ for representing the multiple virtual measurements $\boldsymbol{Y}$ is not centrally symmetric, our analysis is based on the underlying convex geometry of conv($\mathcal{A}$), and $\|\cdot\|_{\mathcal{A}}$ is referred to as the atomic norm of set $\mathcal{A}$ with an abuse of terminology [26].

Theorem 2.1 The atomic norm of the virtual measurements $\boldsymbol{Y}$ defined in (2.18) can be represented in an equivalent semi-definite programming (SDP) form as

$$\|\boldsymbol{Y}\|_{\mathcal{A}} = \inf_{\boldsymbol{z}\in\mathbb{C}^{L},\boldsymbol{W}\in\mathbb{C}^{L\times L}} \left\{ \frac{1}{2L}\mathrm{Tr}\,(\mathcal{T}(\boldsymbol{z})) + \frac{1}{2L}\mathrm{Tr}(\boldsymbol{W}) \middle| \begin{bmatrix} \mathcal{T}(\boldsymbol{z}) & \boldsymbol{Y} \\ \boldsymbol{Y}^{H} & \boldsymbol{W} \end{bmatrix} \succeq \boldsymbol{0} \right\}, \tag{2.19}$$

where $\mathcal{T}(\boldsymbol{z})$ denotes a Hermitian Toeplitz matrix with vector $\boldsymbol{z}$ as its first column.

Proof: Theorem 2.1 is proved from two aspects, namely, the lower bound of $\|\boldsymbol{Y}\|_{\mathcal{A}}$ and the upper bound of $\|\boldsymbol{Y}\|_{\mathcal{A}}$.

Part A: The lower bound of $\|\boldsymbol{Y}\|_{\mathcal{A}}$
Assume that the atomic decomposition of $\boldsymbol{Y}$ is

$$\boldsymbol{Y} = \sum_{k=1}^{K} p_k \boldsymbol{B}(\theta_k). \tag{2.20}$$

According to the Vandermonde decomposition lemma [27], there exists a variable vector $\boldsymbol{z}$ satisfying

$$\mathcal{T}(\boldsymbol{z}) = \sum_{k=1}^{K} p_k \boldsymbol{r}(\theta_k)\boldsymbol{r}^{H}(\theta_k). \tag{2.21}$$

Accordingly, we have

$$\begin{aligned} \begin{bmatrix} \mathcal{T}(\boldsymbol{z}) & \boldsymbol{Y} \\ \boldsymbol{Y}^{H} & \boldsymbol{W} \end{bmatrix} &= \begin{bmatrix} \sum_{k=1}^{K} p_k \boldsymbol{r}(\theta_k)\boldsymbol{r}^{H}(\theta_k) & \sum_{k=1}^{K} p_k \boldsymbol{B}(\theta_k) \\ \sum_{k=1}^{K} p_k \boldsymbol{B}^{H}(\theta_k) & \sum_{k=1}^{K} p_k \boldsymbol{b}(\theta_k)\boldsymbol{b}^{H}(\theta_k) \end{bmatrix} \\ &= \sum_{k=1}^{K} p_k \begin{bmatrix} \boldsymbol{r}(\theta_k) \\ \boldsymbol{b}(\theta_k) \end{bmatrix} \begin{bmatrix} \boldsymbol{r}^{H}(\theta_k) & \boldsymbol{b}^{H}(\theta_k) \end{bmatrix} \succeq \boldsymbol{0}. \end{aligned} \tag{2.22}$$

Based on the definitions of $\boldsymbol{r}(\theta)$ in (2.12) and $\boldsymbol{b}(\theta)$ in (2.13), we have

$$\mathrm{Tr}\left(\boldsymbol{r}(\theta_k)\boldsymbol{r}^{H}(\theta_k)\right) = \mathrm{Tr}\left(\boldsymbol{b}(\theta_k)\boldsymbol{b}^{H}(\theta_k)\right) = L, \tag{2.23}$$

thus the subsequent equation can be obtained as

$$\frac{1}{2L}\mathrm{Tr}\,(\mathcal{T}(\boldsymbol{z})) + \frac{1}{2L}\mathrm{Tr}\,(\boldsymbol{W}) = \sum_{k=1}^{K} p_k = \|\boldsymbol{Y}\|_{\mathcal{A}}. \tag{2.24}$$

Therefore,

$$\|\boldsymbol{Y}\|_{\mathcal{A}} \geq \inf_{\boldsymbol{z}\in\mathbb{C}^{L},\boldsymbol{W}\in\mathbb{C}^{L\times L}} \left\{ \frac{1}{2L}\mathrm{Tr}\,(\mathcal{T}(\boldsymbol{z})) + \frac{1}{2L}\mathrm{Tr}(\boldsymbol{W}) \middle| \begin{bmatrix} \mathcal{T}(\boldsymbol{z}) & \boldsymbol{Y} \\ \boldsymbol{Y}^{H} & \boldsymbol{W} \end{bmatrix} \succeq \boldsymbol{0} \right\}. \tag{2.25}$$

□

Part B: The upper bound of $\|\boldsymbol{Y}\|_{\mathcal{A}}$
Assume that the positive semi-definite (PSD) constraint in (2.19) holds. With the Vandermonde decomposition, the PSD Toeplitz matrix $\mathcal{T}(\boldsymbol{z})$ can be represented as $\mathcal{T}(\boldsymbol{z}) = \boldsymbol{D}\boldsymbol{C}\boldsymbol{D}^{H}$, where $\boldsymbol{D} \in \mathbb{C}^{L\times K}$ is a Vandermonde matrix, and $\boldsymbol{C} = \mathrm{diag}(c_k)$ with $c_k \geq 0$ [22]. Since $\|\boldsymbol{r}(\theta_k)\|_2 = \sqrt{L}$, we have

$$\mathrm{Tr}\,(\mathcal{T}(\boldsymbol{z})) = \mathrm{Tr}\left(\sum_{k=1}^{K} p_k \boldsymbol{r}(\theta_k)\boldsymbol{r}^{H}(\theta_k) \right) = L\ \mathrm{Tr}(\boldsymbol{C}). \tag{2.26}$$

Consequently, each $\boldsymbol{r}(\theta_k)$, $k = 1, 2, \ldots, K$, lies in the range spanned by $\boldsymbol{D}$, and $\boldsymbol{Y}$ can be rewritten as $\boldsymbol{Y} = \boldsymbol{D}\boldsymbol{E}$, where matrix $\boldsymbol{E} \in \mathbb{C}^{K\times L}$ contains the information of both the power and the phase offsets among the virtual measurements. Denoting $\boldsymbol{e} = [e_1, e_2, \ldots, e_K]^{\mathrm{T}}$ as the first column of $\boldsymbol{E}$, the PSD matrix $\boldsymbol{W}$ can then be represented as $\boldsymbol{W} = \boldsymbol{E}^H\boldsymbol{Q}\boldsymbol{E}$, where $\boldsymbol{Q}$ is also a PSD matrix. Since

$$\begin{bmatrix} \mathcal{T}(\boldsymbol{z}) & \boldsymbol{Y} \\ \boldsymbol{Y}^H & \boldsymbol{W} \end{bmatrix} = \begin{bmatrix} \boldsymbol{D} & \boldsymbol{0} \\ \boldsymbol{0} & \boldsymbol{E}^H \end{bmatrix} \begin{bmatrix} \boldsymbol{C} & \boldsymbol{I} \\ \boldsymbol{I} & \boldsymbol{Q} \end{bmatrix} \begin{bmatrix} \boldsymbol{D}^H & \boldsymbol{0} \\ \boldsymbol{0} & \boldsymbol{E} \end{bmatrix} \succeq \boldsymbol{0}, \tag{2.27}$$

according to the Schur complement lemma [28],

$$\begin{bmatrix} \boldsymbol{C} & \boldsymbol{I} \\ \boldsymbol{I} & \boldsymbol{Q} \end{bmatrix} \succeq \boldsymbol{0}, \quad \boldsymbol{Q} \succeq \boldsymbol{C}^{-1}. \tag{2.28}$$

Therefore, we can naturally obtain that

$$\mathrm{Tr}\,(\boldsymbol{W}) = \mathrm{Tr}\left(\boldsymbol{E}^H\boldsymbol{Q}\boldsymbol{E}\right) \geq \mathrm{Tr}\left(\boldsymbol{E}^H\boldsymbol{C}^{-1}\boldsymbol{E}\right). \tag{2.29}$$

Furthermore, the following equation holds

$$\mathrm{Tr}\left(\boldsymbol{E}^H\boldsymbol{C}^{-1}\boldsymbol{E}\right) = \mathrm{Tr}\left(\boldsymbol{C}^{-1}\boldsymbol{E}\boldsymbol{E}^H\right) = L\sum_k c_k^{-1}|e_k|^2. \tag{2.30}$$

Substituting (2.30) into (2.29) yields

$$\mathrm{Tr}(\boldsymbol{W}) \geq L\sum_k c_k^{-1}|e_k|^2. \tag{2.31}$$

With (2.26) and (2.31), we have

$$\begin{aligned} \frac{1}{2L}\mathrm{Tr}\,(\mathcal{T}(\boldsymbol{z})) + \frac{1}{2L}\mathrm{Tr}\,(\boldsymbol{W}) &\geq \frac{1}{2}\mathrm{Tr}\,(\boldsymbol{C}) + \frac{1}{2}\sum_k c_k^{-1}|e_k|^2 \\ &= \frac{1}{2}\sum_k c_k + \frac{1}{2}\sum_k c_k^{-1}|e_k|^2 \\ &\geq \sqrt{\sum_k c_k \sum_k c_k^{-1}|e_k|^2} \\ &\geq \sum_k |e_k| \geq \|\boldsymbol{Y}\|_{\mathcal{A}}. \end{aligned} \tag{2.32}$$

Therefore,

$$\|\boldsymbol{Y}\|_{\mathcal{A}} \leq \inf_{\boldsymbol{z}\in\mathbb{C}^L, \boldsymbol{W}\in\mathbb{C}^{L\times L}} \left\{ \frac{1}{2L}\mathrm{Tr}\,(\mathcal{T}(\boldsymbol{z})) + \frac{1}{2L}\mathrm{Tr}(\boldsymbol{W}) \middle| \begin{bmatrix} \mathcal{T}(\boldsymbol{z}) & \boldsymbol{Y} \\ \boldsymbol{Y}^H & \boldsymbol{W} \end{bmatrix} \succeq \boldsymbol{0} \right\}. \tag{2.33}$$

□

Combining (2.25) and (2.33), we can draw the conclusion that

$$\|\boldsymbol{Y}\|_{\mathcal{A}} = \inf_{\boldsymbol{z}\in\mathbb{C}^L, \boldsymbol{W}\in\mathbb{C}^{L\times L}} \left\{ \frac{1}{2L}\mathrm{Tr}\,(\mathcal{T}(\boldsymbol{z})) + \frac{1}{2L}\mathrm{Tr}(\boldsymbol{W}) \middle| \begin{bmatrix} \mathcal{T}(\boldsymbol{z}) & \boldsymbol{Y} \\ \boldsymbol{Y}^H & \boldsymbol{W} \end{bmatrix} \succeq \boldsymbol{0} \right\}, \tag{2.34}$$

which can be viewed as an equivalent expression of $\|\boldsymbol{Y}\|_{\mathcal{A}}$ defined in (2.18). ■

Based on the atomic norm of the virtual measurements $\boldsymbol{Y}$ defined in (2.18) and its equivalent SDP form in (2.19), we have the following two corollaries regarding the properties of $\mathcal{T}(\boldsymbol{z})$.

Corollary 2.1 Denoting $\boldsymbol{z}^\star$ as the optimum solution to (2.19), the Hermitian PSD Toeplitz matrix $\mathcal{T}(\boldsymbol{z}^\star)$ is the covariance matrix of signals received by the reference virtual array, and can be obtained from the principal square root of $\boldsymbol{Y}\boldsymbol{Y}^H$.

Proof: According to (2.21), the Hermitian PSD Toeplitz matrix $\mathcal{T}(z^\star)$ can be represented as

$$[\mathcal{T}(z^\star)]^2 = \left[\sum_{k=1}^{K} p_k \mathbf{r}(\theta_k)\mathbf{r}^H(\theta_k)\right]^2. \tag{2.35}$$

Since the reference virtual array consists of the virtual sensors located at $\{0, d, 2d, \dots, (L-1)d\}$ as shown in Fig. 2.3, its corresponding steering vector can be described as $\mathbf{r}(\theta_k) = [1, e^{-j\pi\sin(\theta_k)}, \dots, e^{-j\pi(L-1)\sin(\theta_k)}]^T$, which coincides with the phase offset vector $\mathbf{b}(\theta_k)$ in (2.13), i.e., $\mathbf{r}(\theta_k) = \mathbf{b}(\theta_k)$. Thus, (2.35) can be rewritten as

$$\begin{aligned}[\mathcal{T}(z^\star)]^2 &= \left(\sum_{k=1}^{K} p_k \mathbf{r}(\theta_k)\mathbf{r}^H(\theta_k)\right)\left(\sum_{k=1}^{K} p_k \mathbf{r}(\theta_k)\mathbf{r}^H(\theta_k)\right)^H \\ &= \underbrace{\sum_{k=1}^{K} p_k \mathbf{r}(\theta_k)\mathbf{b}^H(\theta_k)}_{Y} \underbrace{\sum_{k=1}^{K} p_k \mathbf{b}(\theta_k)\mathbf{r}^H(\theta_k)}_{Y^H}.\end{aligned} \tag{2.36}$$

Therefore, the Hermitian PSD Toeplitz matrix $\mathcal{T}(z^\star)$ is the covariance matrix of the signals received by the reference virtual array, containing all the information of the interpolated virtual array. In addition, it is clear from (2.36) that the interpolated virtual array covariance matrix $\mathcal{T}(z^\star)$ can be obtained from the principal square root of $\boldsymbol{Y}\boldsymbol{Y}^H$. □

Corollary 2.2 The first column of the Hermitian PSD Toeplitz matrix $\mathcal{T}(z^\star)$ is equivalent to the virtual signals (second-order statistics) of the reference virtual array.

Proof: Since $\mathcal{T}(z^\star)$ satisfies a Hermitian Toeplitz structure, its first column $z^\star$ can be calculated as

$$z^\star = \sum_{k=1}^{K} p_k \mathbf{r}(\theta_k)\langle \mathbf{b}(\theta_k)\rangle_1 = \sum_{k=1}^{K} p_k \mathbf{r}(\theta_k). \tag{2.37}$$

Based on the equivalence relationship between $\mathbf{r}(\theta_k)$ and $\boldsymbol{v}_1(\theta_k)$ as established in (2.12), it is clear that $z^\star$ is equal to the virtual signals of the reference virtual array $\boldsymbol{y}_1$ as defined in (2.10), thus verifying the corollary. □

The atomic norm defined in (2.18) provides a gridless model tailored for the second-order virtual measurements, and the associated corollaries and the equivalent SDP form reveal the important relationship between the interpolated virtual array covariance matrix and the second-order virtual signals. Since the involved directional parameters are continuously parameterized with an atomic norm, the incorporation of the above-mentioned theories enables to estimate off-grid DOAs in the virtual domain.

2.2.3 Toeplitz Matrix Reconstruction for DOA Estimation with Interpolated Virtual Array

Corollary 2.1 provides a relationship between the virtual measurements $\boldsymbol{Y}$ and the Hermitian Toeplitz matrix $\mathcal{T}(z^\star)$ containing the signal information of the interpolated virtual array, while Corollary 2.2 relates the reference virtual array signals to the interpolated virtual array covariance matrix $\mathcal{T}(z^\star)$, such that $\mathcal{T}(z^\star)$ can be directly constructed from $\boldsymbol{y}_1$. However, both $\boldsymbol{Y}$ and $\mathcal{T}(z^\star)$ are derived based on an ideal assumption, i.e., infinite number of snapshots, noise-free signal model, and accurate virtual signals for the interpolated nominal sensors. According to the initial array interpolation process (2.8), the virtual signals corresponding to the interpolated nominal sensors in $\mathbb{S}_\mathcal{I}$ are initialized to be zeros in $\boldsymbol{y}_\mathcal{I}$. Also, the number of snapshots is finite, and the noise term in (2.5) is non-negligible. In this subsection, we focus on the reconstruction of the interpolated virtual array covariance matrix $\mathcal{T}(\boldsymbol{z}^\star)$ based on the initialized interpolated virtual array signals $\boldsymbol{y}_\mathcal{I}$ under the gridless framework of the defined atomic norm for virtual measurements (2.19).

In practice, the virtual signals of the L subarrays in Fig. 2.3 are obtained by selecting corresponding elements of $\boldsymbol{y}_\mathcal{I}$ (2.8). Similar to (2.10), by dividing the initialized interpolated virtual array signals $\boldsymbol{y}_\mathcal{I}$ into L sub-vectors as $\tilde{\boldsymbol{y}}_\ell, \ell = 1, 2, \dots, L$, we have multiple virtual measurements as $\tilde{\boldsymbol{Y}} = [\tilde{\boldsymbol{y}}_1, \tilde{\boldsymbol{y}}_2, \dots, \tilde{\boldsymbol{y}}_L]$. According to Corollary 2.1, the following equation holds

$$\tilde{\boldsymbol{R}}_v^2 = \sum_{\ell=1}^{L} \tilde{\boldsymbol{y}}_\ell \tilde{\boldsymbol{y}}_\ell^H = \tilde{\boldsymbol{Y}}\tilde{\boldsymbol{Y}}^H, \tag{2.38}$$

where $\tilde{\boldsymbol{R}}_v \in \mathbb{C}^{L\times L}$ denotes the reference virtual array covariance matrix, containing all information in $\boldsymbol{y}_\mathcal{I}$. Since there exist several zero elements in $\tilde{\boldsymbol{Y}}$ resulting from the interpolated nominal sensors, the deviation accumulates for each element in $\tilde{\boldsymbol{R}}_v^2$ during the summation process in (2.38). Hence, $\tilde{\boldsymbol{R}}_v$ cannot be directly calculated from the principal square root of $\tilde{\boldsymbol{Y}}\tilde{\boldsymbol{Y}}^H$. Encouragingly, the relationship established in Corollary 2.2 enables to directly construct $\tilde{\boldsymbol{R}}_v$ according to the initialized virtual signals of the reference virtual array as

$$\tilde{\boldsymbol{R}}_v = \begin{bmatrix} \langle \boldsymbol{y}_\mathcal{I} \rangle_L & \langle \boldsymbol{y}_\mathcal{I} \rangle_{L+1}^* & \cdots & \langle \boldsymbol{y}_\mathcal{I} \rangle_{2L-1}^* \\ \langle \boldsymbol{y}_\mathcal{I} \rangle_{L+1} & \langle \boldsymbol{y}_\mathcal{I} \rangle_L & \cdots & \langle \boldsymbol{y}_\mathcal{I} \rangle_{2L-2}^* \\ \vdots & \vdots & \ddots & \vdots \\ \langle \boldsymbol{y}_\mathcal{I} \rangle_{2L-1} & \langle \boldsymbol{y}_\mathcal{I} \rangle_{2L-2} & \cdots & \langle \boldsymbol{y}_\mathcal{I} \rangle_L \end{bmatrix}, \tag{2.39}$$

i.e., $\tilde{\boldsymbol{R}}_v = \mathcal{T}(\tilde{\boldsymbol{y}}_1)$. Due to the fact that $\boldsymbol{y}_\mathcal{I}$ contains several zero elements, the diagonals in $\tilde{\boldsymbol{R}}_v$ corresponding to the positions of holes in the reference virtual array are zeros. It is clear that the Hermitian Toeplitz matrix $\tilde{\boldsymbol{R}}_v$ formulated in (2.39) is the reference virtual array covariance matrix because it satisfies the equivalent condition in (2.38). Considering that the interpolated virtual array $\mathbb{S}_\mathcal{I}$ is symmetric to the zeroth position, the complex-valued virtual signals of the symmetrical pair are mutually conjugate. Therefore, on the premise of the Hermitian Toeplitz structure, the conclusion in Corollary 2.2 is consistent with the relationship between the coprime coarray covariance matrix and the second-order virtual signals established in [29].

On the other hand, since the additive noise in (2.1) is independent of the source signals, its correlation term $\sigma_n^2\boldsymbol{I}$ does not destruct the Hermitian Toeplitz structure of the interpolated virtual array covariance matrix. As such, this special covariance matrix structure can be exploited as *a priori* information for DOA estimation, permitting us focusing on the reference virtual array to investigate the covariance matrix corresponding to the interpolated virtual array. Accordingly, we define a binary vector $\boldsymbol{g} \in \mathbb{R}^L$ to distinguish nominal sensors and virtual sensors in the reference virtual array, whose element is 0 for the nominal sensor positions and 1 otherwise.

The reconstruction of the interpolated virtual array covariance matrix $\mathcal{T}(\boldsymbol{z}^\star)$ starts with the analysis of the atomic norm of ideal virtual measurements $\boldsymbol{Y}$, such that the parameters contained in $\boldsymbol{Y}$ can be continuously represented. According to the definition of $\|\boldsymbol{Y}\|_\mathcal{A}$ in (2.18), a natural objective to describe $\boldsymbol{Y}$ is performing the atomic decomposition of $\boldsymbol{Y}$ with the minimal number of atoms, i.e., minimizing $\|\boldsymbol{Y}\|_\mathcal{A}$. Alternatively, the relationships between $\boldsymbol{Y}$, $\boldsymbol{z}^\star$, and $\mathcal{T}(\boldsymbol{z}^\star)$ revealed in Corollaries 2.1 and 2.2 enable to reconstruct the interpolated virtual array covariance matrix $\mathcal{T}(\boldsymbol{z}^\star)$ by focusing on $\boldsymbol{z}^\star$, since the Hermitian Toeplitz covariance matrix depends only on its first column.

Based on the relationship established in (2.37), $\mathbf{z}^\star$ is equivalent to the virtual signals of the reference virtual array, and its corresponding atom set can thus be defined with respect to the steering vector of the reference virtual array as

$$\mathcal{A}_r = \left\{\mathbf{r}(\theta)|\theta \in [-90^\circ, 90^\circ]\right\}. \tag{2.40}$$

Note that the atom set $\mathcal{A}_r$ for $\mathbf{z}^\star$ relates to the atom set $\mathcal{A}$ for $\mathbf{Y}$ defined in (2.16) by setting the phase offset term in $\mathcal{A}$ as $\langle \mathbf{b}(\theta)\rangle_1 = 1$, which is consistent with the fact revealed in Corollary 2.2 that $\mathbf{z}^\star$ is actually the first column of matrix $\mathbf{Y}$. Accordingly, the atomic norm of the variable $\mathbf{z}$ can be represented as

$$\|\mathbf{z}\|_{\mathcal{A}_r} = \inf\left\{\sum_k p_k : \mathbf{z} = \sum_k p_k \mathbf{r}(\theta_k),\ p_k \geq 0\right\}. \tag{2.41}$$

By comparing the definitions of $\|\mathbf{Y}\|_{\mathcal{A}}$ in (2.18) and $\|\mathbf{z}\|_{\mathcal{A}_r}$ in (2.41), it is demonstrated that minimizing $\|\mathbf{Y}\|_{\mathcal{A}}$ can be equivalently transformed to minimizing $\|\mathbf{z}\|_{\mathcal{A}_r}$ by defining the atom set $\mathcal{A}_r$ based on $\mathcal{A}$. Therefore, we focus on minimizing the atomic norm of $\mathbf{z}$ for the reconstruction of the interpolated virtual array covariance matrix, which can be viewed as performing the atomic decomposition of $\mathbf{z}$ using the minimal number of atoms according to (2.41).

Taking matrix $\tilde{\mathbf{R}}_v$ in (2.39), which contains the partial correlation observations collected from the derived non-uniform virtual array $\mathbb{S}_V$, as the reference, an atomic norm minimization problem can be formulated for Hermitian Toeplitz covariance matrix reconstruction as

$$\begin{aligned} \min_{\mathbf{z}\in\mathbb{C}^L} \quad & \|\mathbf{z}\|_{\mathcal{A}_r} \\ \text{subject to} \quad & \|\mathcal{T}(\mathbf{z}) \circ \mathbf{G} - \tilde{\mathbf{R}}_v\|_F^2 \leq \varepsilon, \\ & \mathcal{T}(\mathbf{z}) \succeq \mathbf{0}, \end{aligned} \tag{2.42}$$

where $\mathbf{G} = \mathbf{g}\mathbf{g}^{\mathrm{T}} \in \mathbb{R}^{L\times L}$ is a binary matrix to distinguish the zero (interpolated) and nonzero (derived) statistics in $\tilde{\mathbf{R}}_v$ due to the initial virtual array interpolation, such that the nonzero elements in $\tilde{\mathbf{R}}_v$ are comparable with the corresponding elements in the reconstructed covariance matrix $\mathcal{T}(\mathbf{z})$. In (2.42), ε is a threshold to restrict the fitting error between the nonzero elements in $\tilde{\mathbf{R}}_v$ and those in the reconstructed covariance matrix $\mathcal{T}(\mathbf{z})$ projected onto $\mathbf{G}$. The PSD constraint on $\mathcal{T}(\mathbf{z})$ follows from its definition in (2.19). Alternatively, the optimization problem (2.42) can be reformulated as

$$\begin{aligned} \min_{\mathbf{z}\in\mathbb{C}^L} \quad & \frac{1}{2}\|\mathcal{T}(\mathbf{z}) \circ \mathbf{G} - \tilde{\mathbf{R}}_v\|_F^2 + \tau\|\mathbf{z}\|_{\mathcal{A}_r} \\ \text{subject to} \quad & \mathcal{T}(\mathbf{z}) \succeq \mathbf{0}, \end{aligned} \tag{2.43}$$

where τ is a regularization parameter to balance the fitting error and the atomic norm term.

Furthermore, with the PSD constraint $\mathcal{T}(\mathbf{z}) \succeq \mathbf{0}$, if $r = \mathrm{rank}(\mathcal{T}(\mathbf{z})) \leq L-1$, the Hermitian Toeplitz matrix $\mathcal{T}(\mathbf{z})$ can be uniquely decomposed via Vandermonde decomposition as [27]

$$\mathcal{T}(\mathbf{z}) = \sum_{k=1}^{r} p_k \mathbf{r}(\theta_k)\mathbf{r}^H(\theta_k), \tag{2.44}$$

which leads to

$$\mathrm{Tr}(\mathcal{T}(\mathbf{z})) = L\sum_{k=1}^{r} p_k. \tag{2.45}$$

Based on the definition of atom set $\mathcal{A}_r$ in (2.40), the relationship between the atomic norm term $\|z\|_{\mathcal{A}_r}$ and the trace of $\mathcal{T}(z)$ can be established as

$$\|z\|_{\mathcal{A}_r} = \mathrm{Tr}\left(\mathcal{T}(z)\right)/L. \tag{2.46}$$

Hence, the optimization problem (2.43) can be equivalently rewritten as

$$\begin{aligned} \min_{z \in \mathbb{C}^L} \quad & \frac{1}{2}\left\|\mathcal{T}(z) \circ G - \tilde{R}_v\right\|_F^2 + \mu \, \mathrm{Tr}\left(\mathcal{T}(z)\right) \\ \text{subject to} \quad & \mathcal{T}(z) \succeq \mathbf{0}, \end{aligned} \tag{2.47}$$

where $\mu = \tau/L$. The equivalent problem (2.47) is convex and can be solved using standard and highly efficient software tools based on the interior point methods. With the solution $\hat{z}$ of (2.47), the interpolated virtual array covariance matrix $\mathcal{T}(\hat{z})$ can be effectively reconstructed with the Hermitian Toeplitz PSD structure. The incorporation of the binary matrix G in (2.47) constrains the fitting error between the nonzero entries in $\tilde{R}_v$ and the optimized covariance matrix $\mathcal{T}(\hat{z})$ that are projected onto G, i.e., $\mathcal{T}(\hat{z}) \circ G$, such that all available correlation observations corresponding to the nonuniform virtual array $\mathbb{S}_V$ are utilized for denoising. Meanwhile, the optimization problem also simultaneously recovers the remaining unknown entries corresponding to the interpolated nominal sensors, which are initialized as zeros in $\tilde{R}_v$, i.e., $\hat{Z} = \mathcal{T}(\hat{z}) - \mathcal{T}(\hat{z}) \circ G$.

Since the reconstructed $\mathcal{T}(\hat{z})$ corresponds to the reference virtual array (i.e., a ULA), existing DOA estimation methods, including MUltiple SIgnal Classification (MUSIC)-based [8, 29–31], Estimation of Signal Parameters via Rotational Invariance Techniques (ESPRIT)-based [32, 33], and a series of sparsity-based techniques [34–36], can be exploited in the virtual domain for unambiguous DOA estimation. For instance, the MUSIC spatial spectrum is obtained as

$$f_{\mathrm{MUSIC}}(\theta) = \frac{1}{r^H(\theta) N_{\mathcal{T}(\hat{z})} N_{\mathcal{T}(\hat{z})}^H r(\theta)}, \tag{2.48}$$

where $N_{\mathcal{T}(\hat{z})}$ denotes the noise subspace of $\mathcal{T}(\hat{z})$. It is worth mentioning that the solution to problem (2.47), i.e., $\mathcal{T}(\hat{z})$, may not be low-rank in noisy and finite-snapshot cases. As such, the noise subspace $N_{\mathcal{T}(\hat{z})}$ is obtained by collecting the eigenvectors corresponding to the $L - K$ smallest eigenvalues of $\mathcal{T}(\hat{z})$ for the general case, where the number of sources K with $K < L$ is assumed known *a priori*. The DOA estimation can be obtained by searching the peaks of $f_{\mathrm{MUSIC}}(\theta)$.

The virtual array interpolation-based DOA estimation algorithm has the following key advantages. First, all the information contained in the nonuniform virtual array $\mathbb{S}_V$ is effectively utilized via array interpolation. Second, the reconstructed covariance matrix $\mathcal{T}(\hat{z})$ is strictly Hermitian Toeplitz, which follows the ideal covariance matrix structure of a ULA. Third, with the atomic norm minimization of the equivalent virtual signals, the formulated optimization problem is capable of reconstructing the interpolated virtual array covariance matrix in a gridless manner, where the basis mismatch problem is effectively avoided. Note that this algorithm is suitable for a general class of partially augmentable arrays, and the difference due to the diverse virtual array geometries is reflected in the representation of g and $\tilde{R}_v$.

2.2.4 Coarray Cramér–Rao Bound

When deriving the stochastic coarray Cramér–Rao bound (CRB) [37], although the reconstructed matrix $\mathcal{T}(\hat{z})$ behaves like the covariance matrix of a virtual array consisting of L sensors, the theoretical performance is still determined by the original coprime array. This is because no additional information is added during the initialization process described in (2.8). Thus, with the full utilization of the discontiguous virtual array $\mathbb{S}_V$, the Fisher information matrix for the proposed algorithm is a function of the coprime array covariance matrix $\mathbf{R}_x$, and the $(\tilde{i},\tilde{j})$th element can be represented as

$$\mathbf{FIM}_{\tilde{i},\tilde{j}} = T\text{Tr}\left[\mathbf{R}_x^{-1}\frac{\partial \mathbf{R}_x}{\partial \xi_{\tilde{i}}}\mathbf{R}_x^{-1}\frac{\partial \mathbf{R}_x}{\partial \xi_{\tilde{j}}}\right], \tag{2.49}$$

where $\xi_{\tilde{i}}$ denotes the $\tilde{i}$th element of the deterministic parameter vector ξ.

When the number of sources exceeds the number of physical sensors, however, the abovementioned Fisher information matrix is singular, resulting in the stochastic CRB being inapplicable. In view of this, we follow the vectorization process as in (2.5) and transform the Fisher information matrix into a virtual array-based form as [38]

$$\mathbf{FIM} = T\left[\text{vec}\left(\frac{\partial \mathbf{R}_x}{\partial \xi}\right)\right]^H\left(\mathbf{R}_x^{\mathrm{T}} \otimes \mathbf{R}_x\right)^{-1}\left[\text{vec}\left(\frac{\partial \mathbf{R}_x}{\partial \xi}\right)\right], \tag{2.50}$$

which remains non-singular within a much broader range of conditions. Hence, it overcomes the model mismatch issue of the stochastic CRB and presents a lower bound for the estimation error even when the number of sources exceeds the number of physical sensors.

In the formulated sparse array DOA estimation case, the deterministic parameter vector is defined as

$$\xi = \left[\boldsymbol{\theta}^{\mathrm{T}}, \mathbf{p}^{\mathrm{T}}, \sigma_n^2\right]^{\mathrm{T}}. \tag{2.51}$$

Accordingly, the Fisher information matrix can be specified as

$$\mathbf{FIM} = T\left[\frac{\partial \mathbf{y}_v}{\partial \xi}\right]^H\left(\mathbf{R}_x^{\mathrm{T}} \otimes \mathbf{R}_x\right)^{-1}\left[\frac{\partial \mathbf{y}_v}{\partial \xi}\right], \tag{2.52}$$

where

$$\frac{\partial \mathbf{y}_v}{\partial \xi} = \left[\frac{\partial \mathbf{y}_v}{\partial \theta_1}, \ldots, \frac{\partial \mathbf{y}_v}{\partial \theta_K}, \frac{\partial \mathbf{y}_v}{\partial p_1}, \ldots, \frac{\partial \mathbf{y}_v}{\partial p_K}, \frac{\partial \mathbf{y}_v}{\partial \sigma_n^2}\right] \tag{2.53}$$

with

$$\begin{aligned}
\frac{\partial \mathbf{y}_v}{\partial \theta_k} &= p_k\left[\frac{\partial \mathbf{a}^*(\theta_k)}{\partial \theta_k} \otimes \mathbf{a}(\theta_k) + \mathbf{a}^*(\theta_k) \otimes \frac{\partial \mathbf{a}(\theta_k)}{\partial \theta_k}\right],\\
\frac{\partial \mathbf{y}_v}{\partial p_k} &= \mathbf{a}^*(\theta_k) \otimes \mathbf{a}(\theta_k),\\
\frac{\partial \mathbf{y}_v}{\partial \sigma_n^2} &= \boldsymbol{i}.
\end{aligned} \tag{2.54}$$

Therefore, the CRB for the kth source can be obtained as

$$\text{CRB}(\theta_k) = \left[\mathbf{FIM}^{-1}\right]_{k,k}, 1 \le k \le K. \tag{2.55}$$

2.2.5 Simulation Results

The parameters of the simulation experiments are set as follows. We choose the pair of coprime integers $M = 3$ and $N = 5$ to deploy the coprime array, which yields a total number of $M + N - 1 = 7$ physical sensors located at $\{0, 3d, 5d, 6d, 9d, 10d, 12d\}$. The virtual array interpolation-based DOA estimation algorithm is compared to several popular DOA estimation algorithms using coprime arrays, namely, the Covariance Matrix Sparse Reconstruction (CMSR) algorithm [21], the Sparse Signal Reconstruction (SSR) algorithm [39], the Nuclear Norm Minimization (NNM) algorithm [14], the nuclear norm minimization with PSD constraint (NUC-PSD) algorithm [40], and the Maximum Entropy (ME) algorithm [40]. The sampling interval of the pre-defined sampling grids is selected to be 0.1° for the CMSR algorithm and the SSR algorithm. The regularization parameter μ for the CMSR algorithm, the SSR algorithm, and the proposed algorithm is empirically set to be 0.25. The optimization problems are solved using the CVX [41].

2.2.5.1 Comparison of Resolution

In this subsection, we compare the resolution by assuming two closely spaced uncorrelated sources impinging from the directions $\theta_1 = -0.5°$ and $\theta_2 = 0.5°$, whose signal-to-noise ratios (SNRs) are 30 dB. The normalized spatial spectra are compared in Fig. 2.4, where the number of snapshots is $T = 500$. Here, the vertical dashed lines denote thc actual directions of the incident sources.

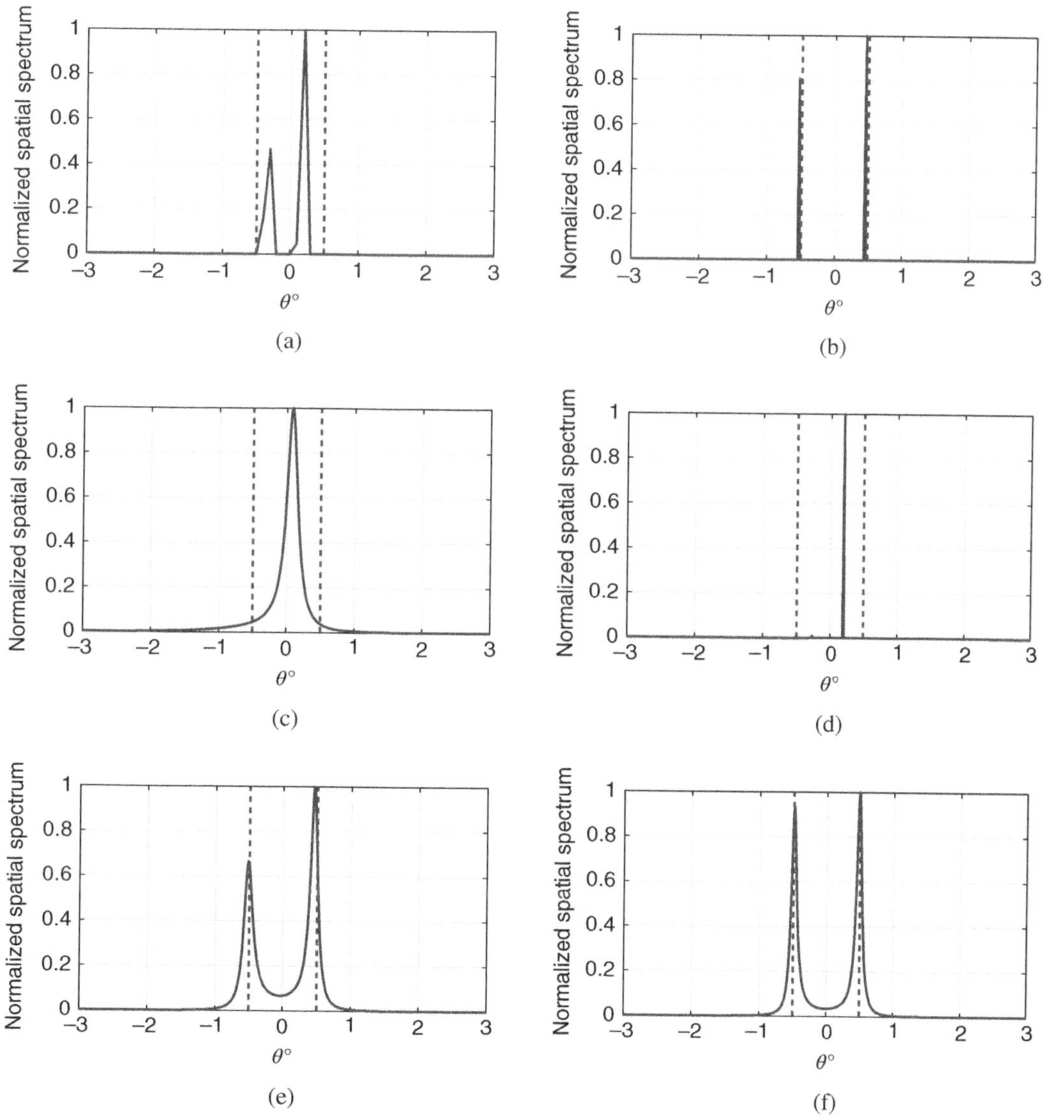

Figure 2.4 Resolution comparison in terms of the normalized spatial spectrum with the number of snapshots $T = 500$. (a) CMSR algorithm; (b) SSR algorithm; (c) ME algorithm; (d) NUC-PSD algorithm; (e) NNM algorithm; (f) Proposed algorithm.

It is observed in Fig. 2.4a that the estimation results of the CMSR method deviate from the actual source directions. It is because the CMSR algorithm only incorporates the contiguous part of the virtual array, i.e., $\mathbb{S}_C$ in Fig. 2.2c, and the reduced array aperture in the virtual domain affects the resolution. Since all the virtual signals vectorized from the sample covariance matrix are utilized in the SSR algorithm, its spatial spectrum presented in Fig. 2.4b has sharper peaks and more accurate estimates than the CMSR algorithm. The ME algorithm and the NUC-PSD algorithm fail to identify the two closely spaced sources. In contrast, both the NNM algorithm and the proposed algorithm are capable of resolving both peaks in the actual source directions.

When the number of snapshots is reduced to $T = 100$, as depicted in Fig. 2.5, all the tested algorithms suffer performance degradation to some extent due to the limited number of signal samples.

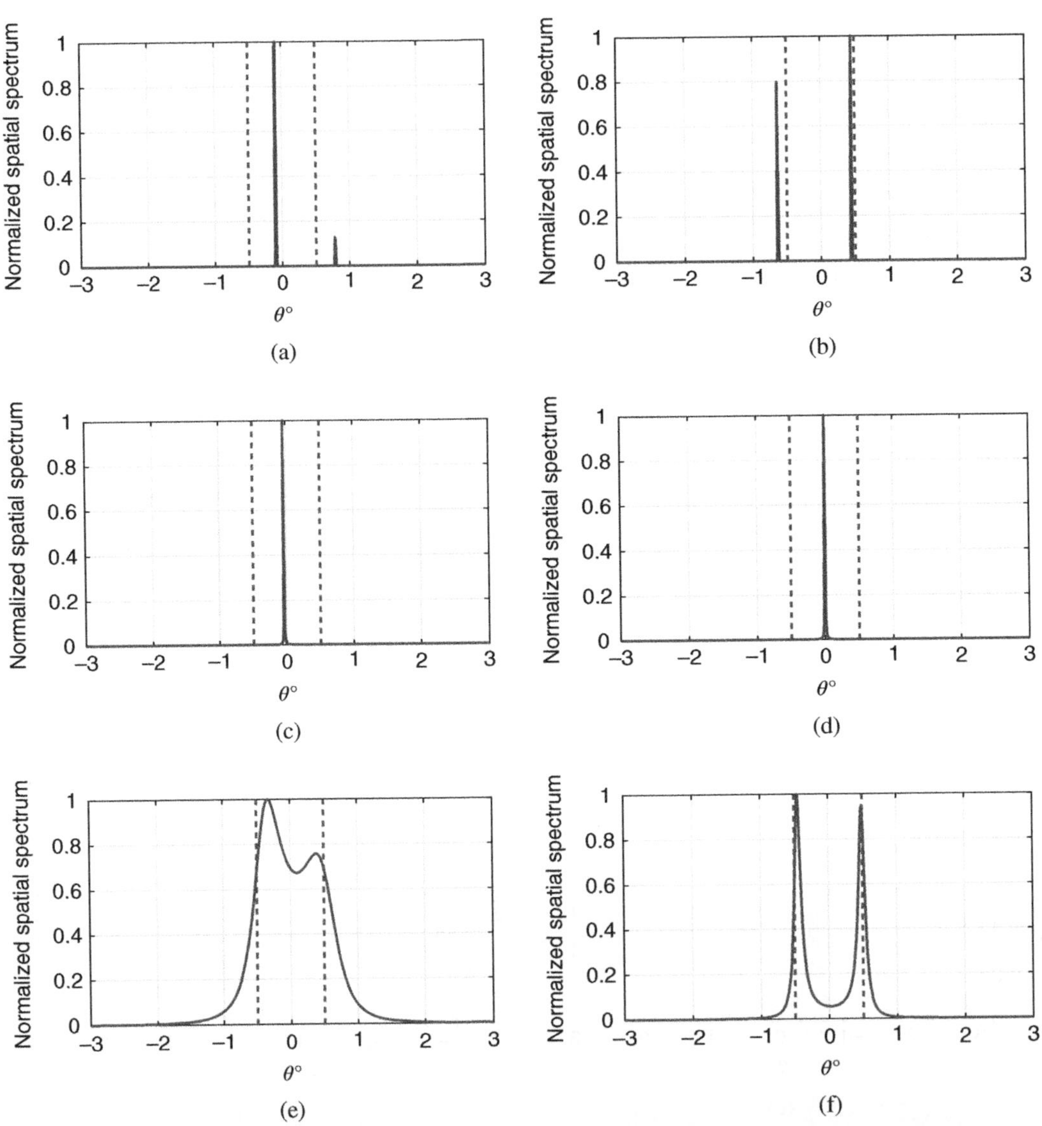

Figure 2.5 Resolution comparison in terms of the normalized spatial spectrum with the number of snapshots $T = 100$. (a) CMSR algorithm; (b) SSR algorithm; (c) ME algorithm; (d) NUC-PSD algorithm; (e) NNM algorithm; (f) Proposed algorithm.

It is observed in Fig. 2.5e that the peaks in the spatial spectrum of the NNM algorithm are no longer as sharp as those obtained from the $T = 500$ snapshots case as shown in Fig. 2.4e, and the DOA estimates deviate from the actual source directions. It is because the NNM algorithm recovers the covariance matrix of the interpolated virtual array based on the principle of matrix completion, which means that the partial correlation observations corresponding to the nonuniform virtual array $\mathcal{S}_V$ are retained in the optimized covariance matrix corresponding to the interpolated virtual array. Since the equivalent virtual signals are obtained from the sample covariance matrix, there exists an inherent bias due to the finite number of snapshots. Different from the NNM algorithm, the proposed algorithm retrieves the covariance matrix of the interpolated virtual array through covariance matrix reconstruction, where the partial correlation observations contained in $\tilde{\boldsymbol{R}}_v$ are taken as the reference. Therefore, the elements in the reconstructed covariance matrix $\mathcal{T}(\hat{\mathfrak{z}})$ may not be the same as those in $\tilde{\boldsymbol{R}}_v$. It is evident from Fig. 2.5f that the proposed algorithm achieves a better resolution than the others.

2.2.5.2 Comparison of DOFs

In this subsection, we compare the number of available DOFs of the tested algorithms. Assume that there are seven uncorrelated equal-power incident sources uniformly distributed in $\left[-50°, 50°\right]$ with SNR = 30 dB and $T = 500$. It can be seen in Fig. 2.6 that all the tested algorithms exhibit peaks around the actual directions by using only seven physical sensors, where the DOFs superiority using the coprime array is demonstrated.

When the number of incident sources is increased to nine, some sources are apparently missed by the CMSR algorithm as illustrated in Fig. 2.7a, because the CMSR algorithm picks only the contiguous part of the difference coarray $\mathcal{S}_C$ to use the spatial smoothing technique and the maximum achievable number of DOFs is seven. It is observed in Figs. 2.6b and 2.7b that there are several spurious peaks in the obtained sparse spatial spectrum of the SSR algorithm. In contrast, all the virtual array interpolation-based algorithms are capable of identifying the nine sources as exhibited in Fig. 2.7c through 2.7f, demonstrating that a high number of DOFs offered by the nonuniform virtual array is obtained via virtual array interpolation.

2.2.5.3 Comparison of Estimation Accuracy

In this subsection, we compare the DOA estimation performance of each algorithm using the root mean square error (RMSE) criterion. The RMSE is defined as

$$\text{RMSE} = \sqrt{\frac{1}{KQ}\sum_{k=1}^{K}\sum_{q=1}^{Q}\left(\hat{\theta}_k(q) - \theta_k\right)^2}, \tag{2.56}$$

where $\hat{\theta}_k(q)$ denotes the estimated DOA of the kth source θ_k in the qth Monte Carlo trial, and Q denotes the number of Monte Carlo trials.

We first consider the case of a single incident signal source. Assume that the direction of the incident source is randomly generated from the Gaussian distribution $\mathcal{N}(0°, (1°)^2)$ and remains fixed from snapshot to snapshot but changes from trial to trial. The number of snapshots is fixed at $T = 500$ when the input SNR varies, whereas the input SNR is fixed at 20 dB for each source when the number of snapshots varies. The DOAs are estimated by finding the spectrum peaks for the CMSR algorithm and the SSR algorithm, whereas the root-MUSIC [31] is performed on the optimized covariance matrix of the other gridless algorithms. The CRB obtained from (2.55) is also plotted. For each data point, the RMSE is calculated from $Q = 500$ Monte Carlo trials.

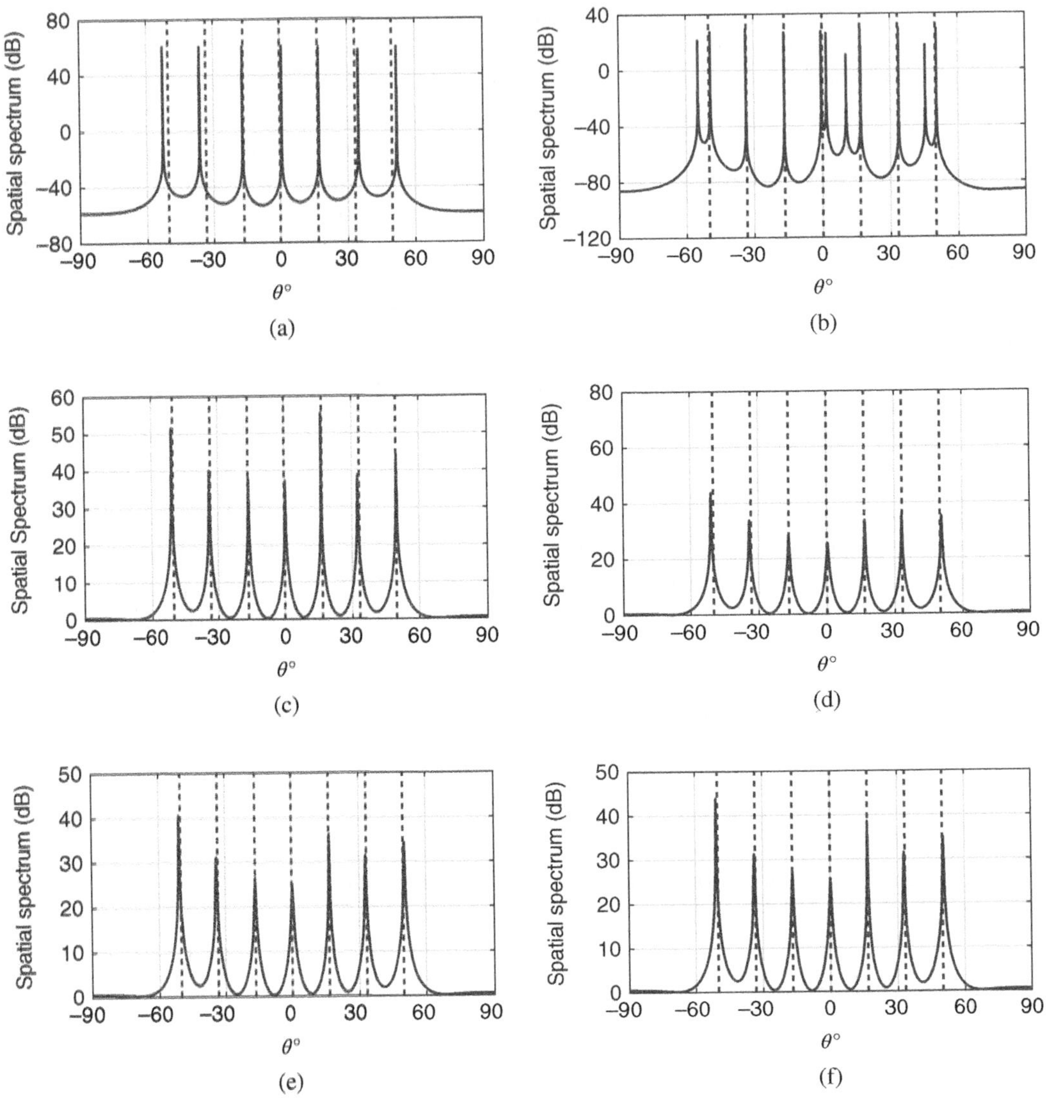

Figure 2.6 DOFs comparison in terms of the spatial spectrum for $K = 7$ sources. (a) CMSR algorithm; (b) SSR algorithm; (c) ME algorithm; (d) NUC-PSD algorithm; (e) NNM algorithm; (f) Proposed algorithm.

It is shown in Fig. 2.8a that the RMSE curves of both the CMSR algorithm and the SSR algorithm become relatively flat when the SNR is higher than 10 dB. The reason lies in that the fixed sampling interval for the pre-defined sampling grids leads to an inherent basis mismatch, which limits the estimation accuracy. In contrast, the gridless algorithms do not require pre-defined sampling grids, and thus, their performance is not limited by the fixed interval of sampling grids. Therefore, their RMSE performance is consistent with that manifested in the CRB when the input SNR increases. It is also observed in Fig. 2.8a that the gridless algorithms, especially the NNM algorithm, the NUC-PSD algorithm, and the proposed algorithm, achieve close RMSE performance. Similar performance comparison can also be found in Fig. 2.8b, where the number of snapshots varies. As

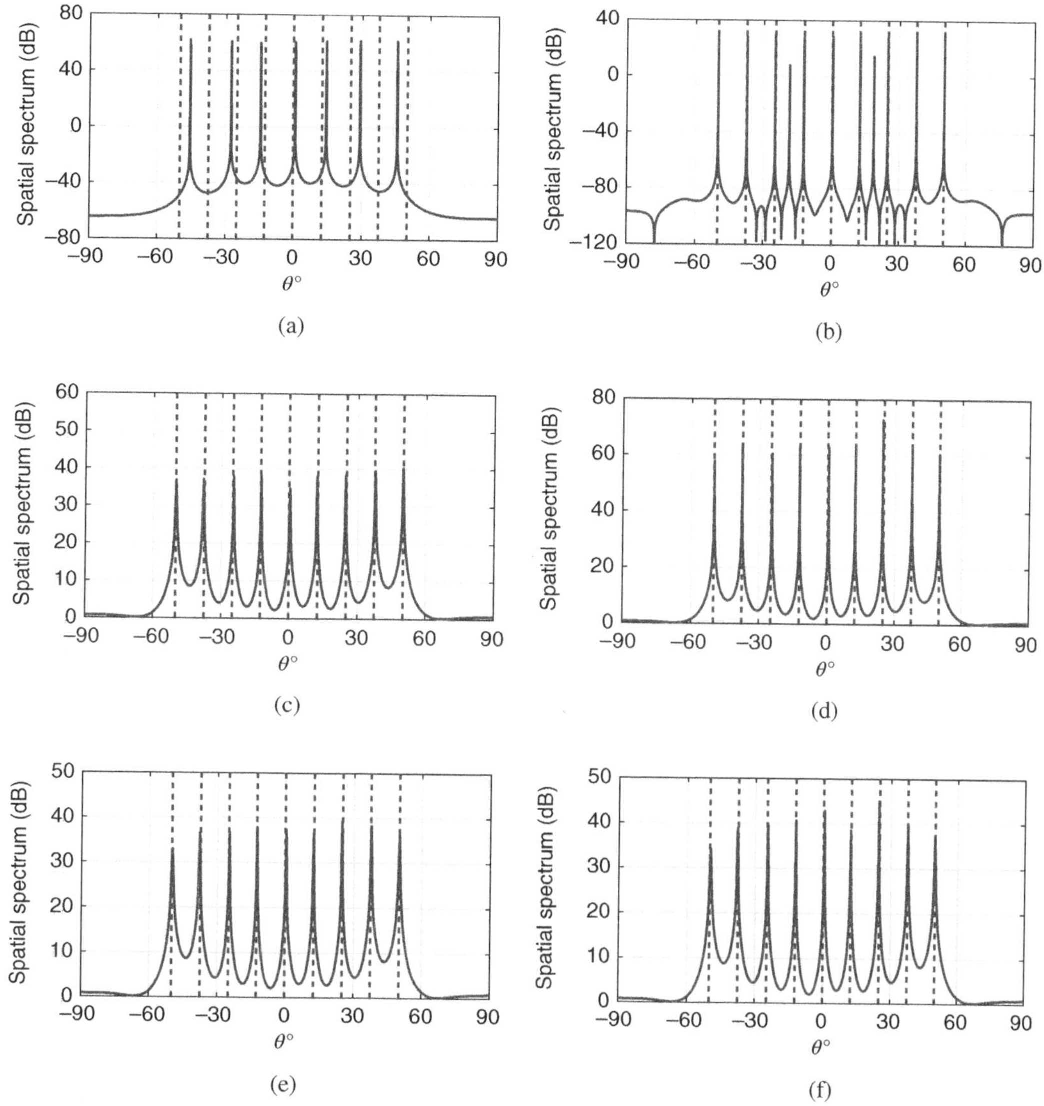

Figure 2.7 DOFs comparison in terms of the spatial spectrum for $K = 9$ sources. (a) CMSR algorithm; (b) SSR algorithm; (c) ME algorithm; (d) NUC-PSD algorithm; (e) NNM algorithm; (f) Proposed algorithm.

pointed out in subsection 2.2.5.1, the NNM algorithm and the proposed algorithm formulate the optimization problems based on different principles, namely, matrix completion and matrix reconstruction, respectively. It is shown in Fig. 2.8b that the RMSE of the proposed algorithm is slightly lower than that of the NNM algorithm.

Next, we evaluate the RMSE of the proposed algorithm in Fig. 2.9 in the case that the number of sources is equal to or greater than the number of physical sensors, where two cases are considered respectively for $K = 7$ and $K = 9$ sources uniformly distributed in $[-50°, 50°]$. The DOAs are estimated from the spatial spectra presented in Figs. 2.6 and 2.7 with the spectrum peak searching interval of $\Delta\theta = 0.1°$, and the RMSE is calculated from 500 Monte Carlo trials for each scenario.

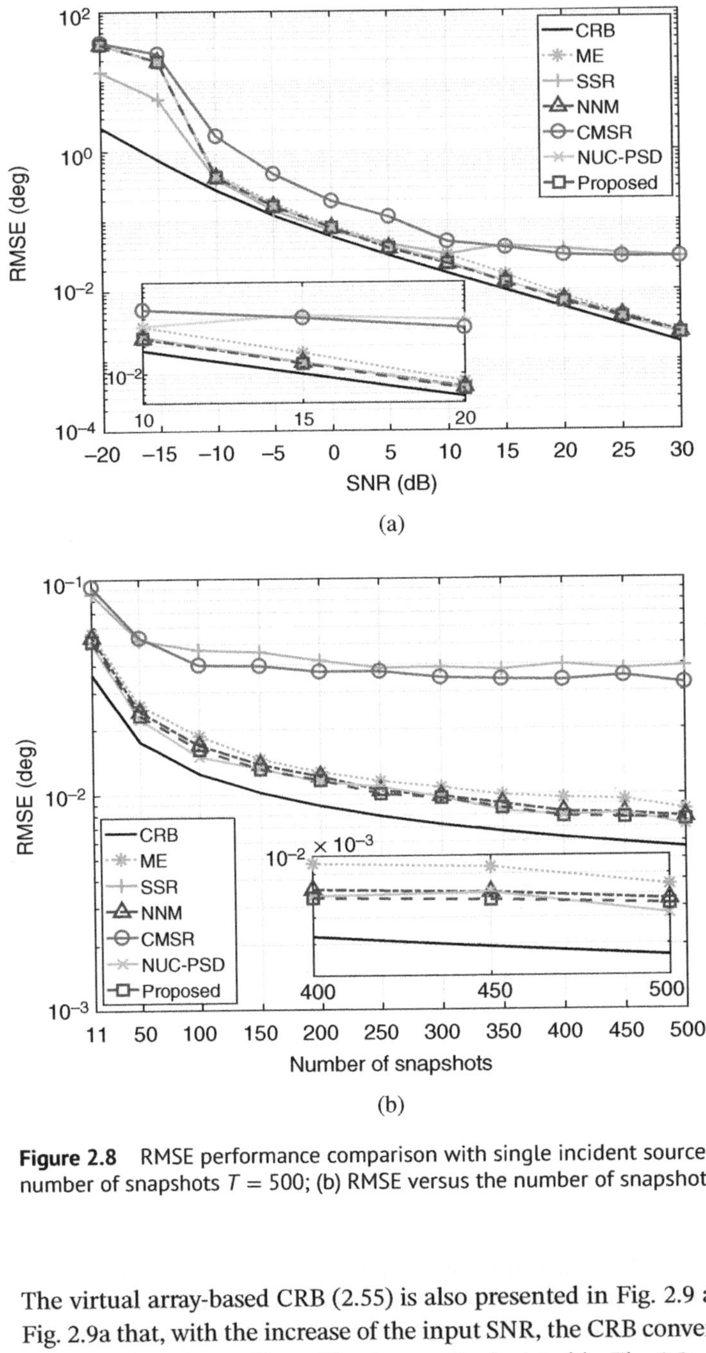

Figure 2.8 RMSE performance comparison with single incident source. (a) RMSE versus SNR with the number of snapshots $T = 500$; (b) RMSE versus the number of snapshots with SNR = 20 dB.

The virtual array-based CRB (2.55) is also presented in Fig. 2.9 as the reference. It is observed in Fig. 2.9a that, with the increase of the input SNR, the CRB converges to a constant when the input SNR is higher than 5 dB, unlike the results depicted in Fig. 2.8a where the CRB keeps decreasing with the input SNR. This is a typical *saturation* behavior [42]. Although the CRB for the seven sources case is lower than that for the nine sources case when the input SNR is relatively low, the CRB for the nine sources case converges to a lower value when input SNR is higher than 5 dB. It is demonstrated in Fig. 2.9 that the RMSE of the proposed algorithm has a similar trend with the CRB and exhibits the saturation behavior as the CRB in the asymptotic performance region.

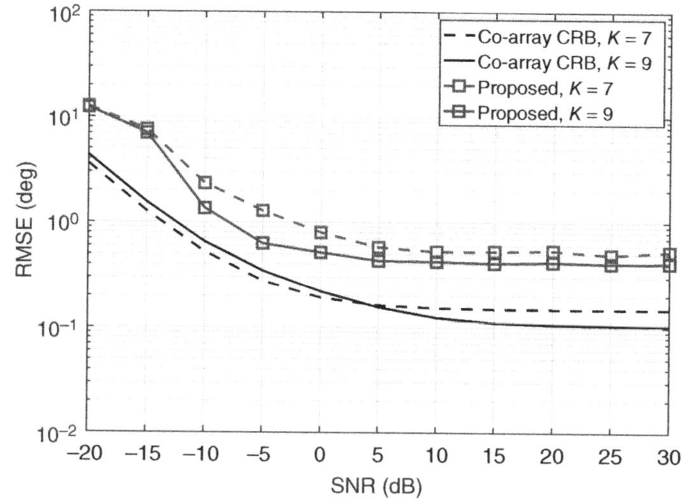

(a)

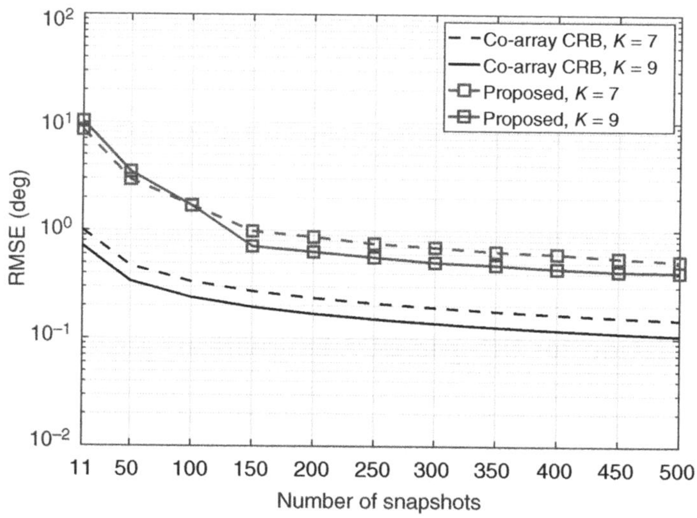

(b)

Figure 2.9 RMSE performance comparison with multiple incident sources. (a) RMSE versus SNR with the number of snapshots $T = 500$; (b) RMSE versus the number of snapshots with SNR $= 20$ dB.

2.2.5.4 Comparison of Computational Complexity

In this subsection, we compare the computational complexity measured by the computation time for 100 Monte Carlo trials on an Intel(R) Core(TM) i7-7600U CPU laptop with 16 GB RAM, where the sampling/searching interval is varied. According to Fig. 2.10, the computational complexity of the CMSR algorithm and the SSR algorithm increases exponentially when the sampling interval decreases. This is because the number of pre-defined dense sampling grids dramatically increases the computational cost when solving the corresponding optimization problems. In contrast, since the gridless algorithms formulate the optimization problem without pre-defined sampling grids, their computational costs are insensitive to the selection of the sampling interval. The ME

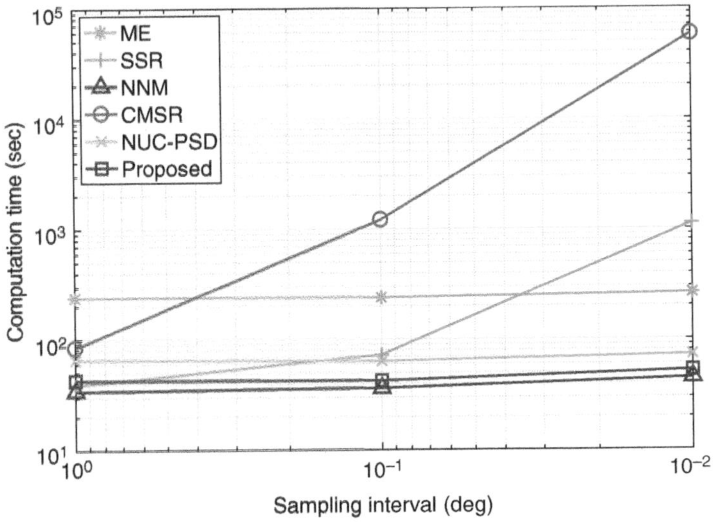

Figure 2.10 Computation time comparison with different sampling interval.

algorithm requires a higher computational complexity than the other gridless algorithms because it adopts a two-stage optimization process, whereas the NNM algorithm, the NUC-PSD algorithm, and the proposed algorithm have the same order of computational complexity. In particular, it is demonstrated in Fig. 2.10 that the proposed algorithm has a lower computational complexity than the NUC-PSD algorithm and requires a slightly higher computational complexity than the NNM algorithm.

2.3 Physical Array Interpolation for Off-grid DOA Estimation

While virtual array interpolation has become the mainstream solution for Nyquist-matched DOA estimation with sparse arrays, the contiguous coarray lags depend heavily on the original sparse array configuration, indicating that the virtual array geometry is quite sensitive to the deviation caused by physical sensors. In this regard, the idea of physical array interpolation [17] is proposed. It demonstrates the capability of increasing the number of DOFs which even exceeds the limit of the virtual array interpolation-based methods. In this section, we elaborate the coprime array interpolation algorithm for off-grid DOA estimation.

2.3.1 Physical Array Interpolation and Signal Model

We consider the physical coprime array $\mathbb{S}_P$ with the same configuration as in Fig. 2.1b of Section 2.2 and assume K uncorrelated narrowband signals impinging on the coprime array from directions $\theta = [\theta_1, \theta_2, \dots, \theta_K]^{\mathrm{T}}$. The corresponding received signal vector $\boldsymbol{x}(t)$ and the array manifold matrix $\boldsymbol{A}$ have been defined in Section 2.2. The objective of physical array interpolation is to interpolate the nonuniform coprime array $\mathbb{S}_P$ with additional interpolated sensors, as illustrated in Fig. 2.11. The resulting interpolated ULA

$$\mathbb{S}_V = \{i \mid 0 \leq i \leq \max(\mathbb{S}_P), i \in \mathbb{Z}\}, \tag{2.57}$$

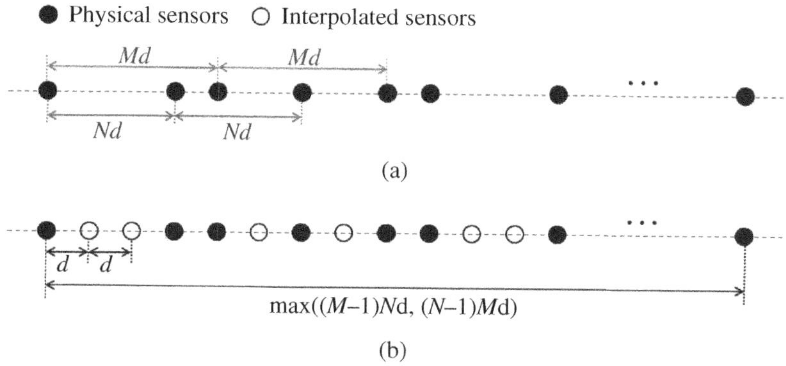

Figure 2.11 Illustration of array configurations. (a) The nonuniform coprime array $\mathcal{S}_\mathcal{P}$; (b) The interpolated ULA $\mathcal{S}_\mathcal{V}$.

has the same aperture as the coprime array $\mathcal{S}_\mathcal{P}$, where the indices $i \in \mathbb{Z}$ are consecutive integers, and $\mathcal{S}_\mathcal{P} \subset \mathcal{S}_\mathcal{V}$.

Similar to the virtual array interpolation depicted in (2.8), the interpolated sensors $\mathcal{S}_\mathcal{V} - \mathcal{S}_\mathcal{P}$, which are represented by the hollow circles in Fig. 2.11b, indicate that the corresponding received signals are practically unknown. Hence, the received signals of the interpolated ULA can be initialized as

$$\left[\boldsymbol{y}_{\mathcal{S}_\mathcal{V}}(t)\right]_i = \begin{cases} [\boldsymbol{x}(t)]_i, & i \in \mathcal{S}_\mathcal{P}, \\ 0, & i \in \mathcal{S}_\mathcal{V} - \mathcal{S}_\mathcal{P}. \end{cases} \tag{2.58}$$

Accordingly, we define a binary vector $\mathbf{g} \in \mathbb{R}^{|\mathcal{S}_\mathcal{V}|}$ to distinguish the sensors in the interpolated ULA $\mathcal{S}_\mathcal{V}$, whose elements are 1 for the physical sensors and 0 for the interpolated sensors. Then, the initialized received signals of the interpolated ULA $\boldsymbol{y}_{\mathcal{S}_\mathcal{V}}(t)$ can be related to its theoretical version $\boldsymbol{x}_{\mathcal{S}_\mathcal{V}}(t)$ as

$$\boldsymbol{y}_{\mathcal{S}_\mathcal{V}}(t) = \boldsymbol{x}_{\mathcal{S}_\mathcal{V}}(t) \circ \mathbf{g}. \tag{2.59}$$

Here, the theoretical received signals of the interpolated ULA can be modeled as

$$\boldsymbol{x}_{\mathcal{S}_\mathcal{V}}(t) = \sum_{k=1}^{K} \boldsymbol{a}_{\mathcal{S}_\mathcal{V}}(\theta_k) s_k(t) + \boldsymbol{n}_{\mathcal{S}_\mathcal{V}}(t) = \boldsymbol{A}_{\mathcal{S}_\mathcal{V}} \mathbf{s}(t) + \boldsymbol{n}_{\mathcal{S}_\mathcal{V}}(t), \tag{2.60}$$

where

$$\boldsymbol{a}_{\mathcal{S}_\mathcal{V}}(\theta_k) = \left[1, e^{-j\frac{2\pi}{\lambda}u_2 \sin(\theta_k)}, \dots, e^{-j\frac{2\pi}{\lambda}u_{|\mathcal{S}_\mathcal{V}|} \sin(\theta_k)}\right]^{\mathrm{T}} \tag{2.61}$$

is the steering vector of the interpolated ULA corresponding to the kth source θ_k, and $\boldsymbol{A}_{\mathcal{S}_\mathcal{V}} = \left[\boldsymbol{a}_{\mathcal{S}_\mathcal{V}}(\theta_1), \boldsymbol{a}_{\mathcal{S}_\mathcal{V}}(\theta_2), \dots, \boldsymbol{a}_{\mathcal{S}_\mathcal{V}}(\theta_K)\right] \in \mathbb{C}^{|\mathcal{S}_\mathcal{V}| \times K}$. Here, $u_\upsilon \in \mathcal{S}_\mathcal{V}, \upsilon = 1, 2, \dots, |\mathcal{S}_\mathcal{V}|$, denotes the position of the υth sensor in the interpolated ULA with $u_1 = 0$, and $\boldsymbol{n}_{\mathcal{S}_\mathcal{V}}(t) \in \mathbb{C}^{|\mathcal{S}_\mathcal{V}|} \sim \mathcal{CN}(\mathbf{0}, \sigma_n^2 \boldsymbol{I})$ is the additive white Gaussian noise vector corresponding to the $|\mathcal{S}_\mathcal{V}|$ sensors.

Based on the initialized received signal vector of the interpolated ULA $\boldsymbol{y}_{\mathcal{S}_\mathcal{V}}(t)$, its sample covariance matrix can be calculated as

$$\hat{\boldsymbol{R}}_{\boldsymbol{y}_{\mathcal{S}_\mathcal{V}}} = \frac{1}{T} \sum_{t=1}^{T} \boldsymbol{y}_{\mathcal{S}_\mathcal{V}}(t) \boldsymbol{y}_{\mathcal{S}_\mathcal{V}}^H(t). \tag{2.62}$$

The correlations in the rows and columns corresponding to the interpolated sensors are zeros, while the nonzero elements in $\hat{\boldsymbol{R}}_{\boldsymbol{y}_{\mathbb{S}_V}}$ are the correlations of the actual signals $\boldsymbol{x}_{\mathbb{S}_p}(t)$ received by the coprime array. In order to perform ULA-based DOA estimation, it is necessary to recover the unknown correlations in $\hat{\boldsymbol{R}}_{\boldsymbol{y}_{\mathbb{S}_V}}$ due to the initialized zero elements in $\boldsymbol{y}_{\mathbb{S}_V}(t)$, such that the recovered covariance matrix approaches its theoretical version

$$\boldsymbol{R}_{\boldsymbol{x}_{\mathbb{S}_V}} = \mathrm{E}\left[\boldsymbol{x}_{\mathbb{S}_V}(t)\boldsymbol{x}^H_{\mathbb{S}_V}(t)\right] = \boldsymbol{A}_{\mathbb{S}_V}\Sigma\boldsymbol{A}^H_{\mathbb{S}_V} + \sigma_n^2\boldsymbol{I}. \tag{2.63}$$

Here, $\Sigma = \mathrm{diag}\left(\sigma_1^2, \sigma_2^2, \dots, \sigma_K^2\right)$ with $\sigma_k^2 = \mathrm{E}\left[|s_k(t)|^2\right]$ denoting the power of the kth incident source on its diagonal.

2.3.2 Covariance Matrix Recovery for Off-grid DOA Estimation

The objective of covariance matrix recovery is to recover the desired covariance matrix $\boldsymbol{R}_{\boldsymbol{x}_{\mathbb{S}_V}}$ from $\hat{\boldsymbol{R}}_{\boldsymbol{y}_{\mathbb{S}_V}}$. However, it is challenging since a number of rows and columns are entirely missing in $\hat{\boldsymbol{R}}_{\boldsymbol{y}_{\mathbb{S}_V}}$ [43]. Encouragingly, the theoretical covariance matrix corresponding to uncorrelated signal received at the ULA has a Hermitian Toeplitz structure [44]. This property can be utilized to perform structured matrix recovery. On the other hand, the noise-free covariance matrix also exhibits a low-rank property, provided that the number of the incident sources is smaller than the number of sensors in the interpolated ULA. Based on these facts, the covariance matrix of the interpolated ULA can be recovered by solving the following optimization problem

$$\begin{aligned} \min_{\boldsymbol{z}} \quad & \mathrm{rank}\left(\mathcal{T}(\boldsymbol{z})\right) \\ \text{subject to} \quad & \left\|\left(\mathcal{T}(\boldsymbol{z}) - \hat{\boldsymbol{R}}_{\boldsymbol{y}_{\mathbb{S}_V}}\right) \circ \boldsymbol{G}\right\|_F^2 \leq \varepsilon, \\ & \mathcal{T}(\boldsymbol{z}) \succeq \boldsymbol{0}, \end{aligned} \tag{2.64}$$

where $\boldsymbol{G} = \boldsymbol{g}\boldsymbol{g}^{\mathrm{T}}$ is an $|\mathbb{S}_V| \times |\mathbb{S}_V|$ dimensional binary matrix distinguishing the known (nonzero) elements and the unknown (zero) elements in the initialized sample covariance matrix $\hat{\boldsymbol{R}}_{\boldsymbol{y}_{\mathbb{S}_V}}$, $\mathcal{T}(\boldsymbol{z}) \succeq \boldsymbol{0}$ is a Hermitian Toeplitz PSD matrix with $\boldsymbol{z} \in \mathbb{C}^{|\mathbb{S}_V|}$ as its first column, and ε is a user-defined parameter to constrain the fitting error.

In the above optimization problem, we try to recover a low-rank Hermitian Toeplitz PSD covariance matrix of the interpolated ULA, while minimizing the difference between the known correlations in the sample covariance matrix $\hat{\boldsymbol{R}}_{\boldsymbol{y}_{\mathbb{S}_V}}$ and the corresponding elements in the recovered covariance matrix $\mathcal{T}(\boldsymbol{z})$. By introducing the nuclear norm convex relaxation, the NP-hard rank minimization problem (2.64) can be reformulated as

$$\begin{aligned} \min_{\boldsymbol{z}} \quad & \left\|\left(\mathcal{T}(\boldsymbol{z}) - \hat{\boldsymbol{R}}_{\boldsymbol{y}_{\mathbb{S}_V}}\right) \circ \boldsymbol{G}\right\|_F^2 + \tau\|\mathcal{T}(\boldsymbol{z})\|_* \\ \text{subject to} \quad & \mathcal{T}(\boldsymbol{z}) \succeq \boldsymbol{0}, \end{aligned} \tag{2.65}$$

where τ is a regularization parameter to balance the fitting error and the nuclear norm. According to the fact that

$$\|\mathcal{T}(\boldsymbol{z})\|_* = \mathrm{Tr}\left(\sqrt{\mathcal{T}^H(\boldsymbol{z})\mathcal{T}(\boldsymbol{z})}\right), \tag{2.66}$$

the PSD constraint on $\mathcal{T}(\boldsymbol{z})$ enables to equivalently transform the nuclear norm minimization problem (2.65) to a trace minimization problem as

$$\begin{aligned}\min_{z} \quad & \left\| \left(\mathcal{T}(z) - \hat{R}_{y_{S_V}}\right) \circ G \right\|_F^2 + \tau \mathrm{Tr}\left(\mathcal{T}(z)\right)\\ \text{subject to} \quad & \mathcal{T}(z) \succeq \mathbf{0}.\end{aligned} \tag{2.67}$$

The optimization problem (2.67) is convex and can be efficiently solved via interior point methods.

The optimized covariance matrix $\mathcal{T}(\hat{z}) \in \mathbb{C}^{|S_V|\times|S_V|}$ is an estimate of $R_{x_{S_V}}$, i.e., the covariance matrix of the signals received by the interpolated ULA, where the incorporation of the ULA-based methods increases the number of achievable DOFs from $|S_P|$ to $|S_V|$. On the other hand, the matrix $\mathcal{T}(\hat{z})$ recovered in a gridless manner enables to effectively estimate off-grid sources. The candidate ULA-based methods for DOA estimation include the subspace-based methods, such as MUSIC [30], root-MUSIC [31], ESPRIT [32], and a series of sparsity-based methods [35, 36, 45–47]. Considering the trade-off between the estimation performance and the computational complexity, the root-MUSIC is a preferred choice to provide off-grid DOA estimation, whose estimation accuracy is not limited by the spectrum searching process.

It should be mentioned that, while the array interpolation-based DOA estimation algorithm is formulated on the systematically designed coprime array configurations, it can be readily extended to other sparse arrays, and the difference is reflected in the binary matrix G and the sample covariance matrix $\hat{R}_{y_{S_V}}$.

2.3.3 Push the Limit of Achievable Degrees-of-Freedom

Since the nonuniform coprime array is equivalently interpolated into an augmented ULA with the same array aperture through the physical array interpolation-based DOA estimation algorithm, the augmented ULA is expected to achieve additional DOFs compared to the original array.

In the simulations, the pair of coprime integers are selected to be $M = 3$ and $N = 5$ for the coprime array deployment. Accordingly, we have $|S_P| = 7$ and $|S_V| = 13$. The regularization parameter τ is set to 0.25. The DOFs performance of the proposed physical array interpolation-based DOA estimation algorithm is shown in Fig. 2.12 by depicting its spatial spectra based on $\mathcal{T}(\hat{z})$, where the incident sources are assumed to be uniformly distributed in $[-50°, 50°]$ with input SNR = 0 dB and $T = 500$.

For the case that the number of sources is $K = 9$, which exceeds the number of physical sensors as well as the number of maximum achievable DOFs for the coarray-based SS-MUSIC algorithm [8], it can be seen in Fig. 2.12a that the proposed physical array interpolation-based DOA estimation algorithm effectively resolves all the nine sources. When the number of sources is increased to $K = 12$, which is the maximum number of achievable DOFs for the subspace-based methods based on $\mathcal{T}(\hat{z})$, the spatial spectrum shown in Fig. 2.12b indicates that the proposed algorithm is still effective. Recalling the fact that only $|S_P| = 7$ physical sensors are utilized for receiving the signals, it is clear that additional DOFs are obtained by the proposed physical array interpolation-based DOA estimation algorithm.

It is worth mentioning that the coarray-based algorithms exploiting the contiguous part of the virtual array, such as the SS-MUSIC algorithm [8] and the CMSR algorithm [21], fail to perform effective DOA estimation in this scenario, since the contiguous coarray ranges only between −7 and 7, thus indicating that the maximum number of resolvable sources is 7. In contrast, the proposed physical array interpolation-based DOA estimation algorithm is capable of resolving all the off-grid sources with a higher number of DOFs.

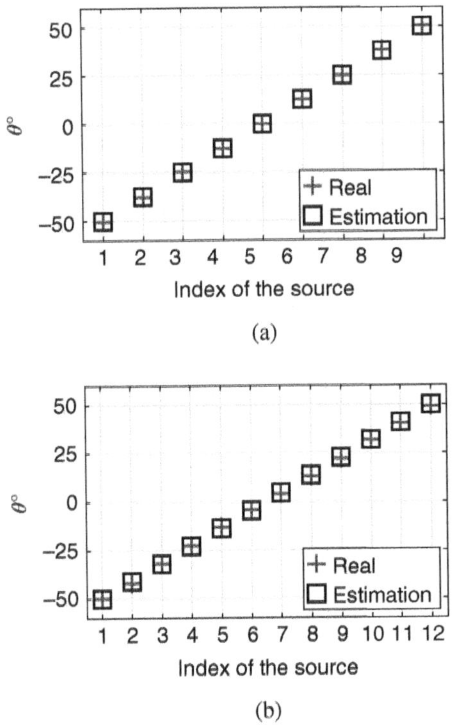

Figure 2.12 DOFs illustration of the proposed physical array interpolation-based coprime array DOA estimation algorithm. (a) $K = 9$; (b) $K = 12$.

2.3.4 Simulation Results

In this subsection, the performance of the proposed physical array interpolation-based DOA estimation algorithm is compared to two coarray-based DOA estimation algorithms, i.e., the SSR algorithm [39] and the gridless reconstruction-based virtual array interpolation algorithm [16, 48]. Consider nine equal-power sources impinging on the deployed coprime array from the directions uniformly distributed in $[-52.78°, 50.35°]$, where the number of off-grid sources exceeds the number of physical sensors. For a fair comparison, both the gridless reconstruction-based virtual array interpolation algorithm and the proposed physical array interpolation-based DOA estimation algorithm apply the root-MUSIC method to the optimized covariance matrix for DOA estimation. The pre-defined spatial sampling grid interval $\Delta\theta$ for the SSR algorithm is selected to be $0.1°$ and $0.01°$. The performance is compared in terms of the RMSE, and $Q = 500$ Monte Carlo trials are performed for each data point (either input SNR or the number of snapshots). The CRB is also plotted for reference.

The RMSE versus the input SNR is compared in Fig. 2.13a where the number of snapshots is $T = 500$. It is shown in Fig. 2.13a that the physical array interpolation-based algorithm outperforms the gridless reconstruction-based virtual array interpolation algorithm, and its performance trend is consistent with the CRB predictions. For the SSR algorithm, peaks of the irregular spurious spatial spectra introduced by the sparsity-based approaches account for the fluctuant trend when the input SNR is higher than 0 dB. Such performance trend does not improve when a denser grid with $\Delta\theta = 0.01°$ is used in the underlying multiple off-grid sources case. When the number of snapshots varies with input SNR = 10 dB, as shown in Fig. 2.13b, the physical array interpolation-based DOA estimation algorithm provides the best estimation accuracy in coping with off-grid sources, and its RMSE is close to the CRB when the number of snapshots is higher than 50.

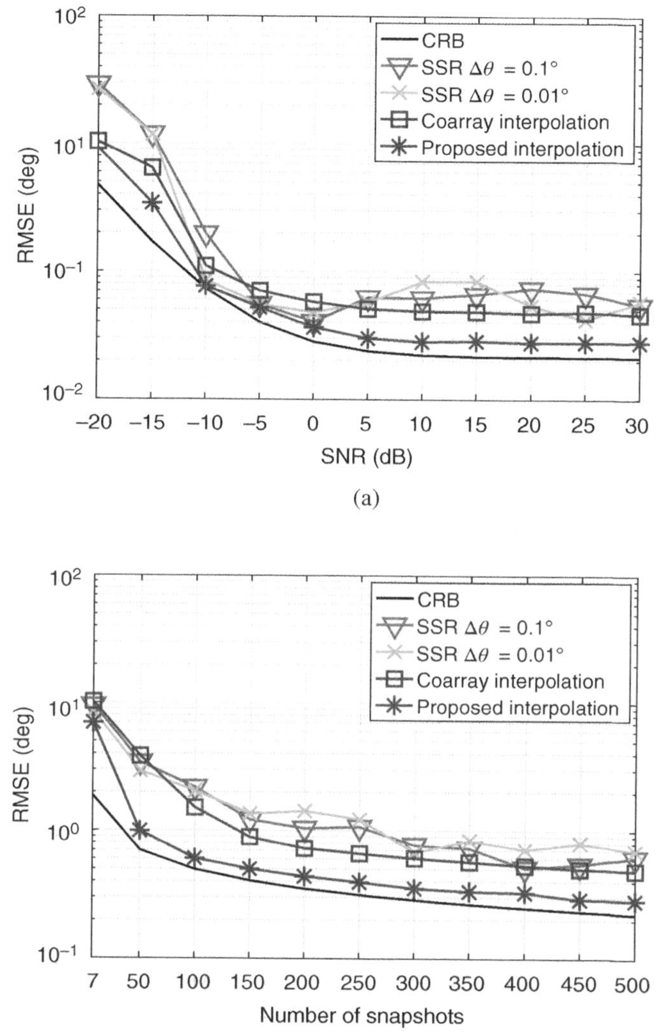

(a)

(b)

Figure 2.13 Comparison of DOA estimation performance. (a) RMSE versus SNR when $T = 500$; (b) RMSE versus number of snapshots when SNR = 10 dB.

2.4 Prospective Research Directions

The Sections 2.2 and 2.3 provide a thorough theory of sparse array interpolation, and the examples mainly focused on one-dimensional (1-D) prototype coprime arrays. In the following, several prospective research directions stemming from these results are outlined.

2.4.1 Interpolation-Aware Sparse Array Design

Utilizing the sparse array interpolation techniques, sensor arrays can be designed with holes either in their physical or virtual arrays. For example, the generalized nonredundant sparse array designs are optimized to avoid lag redundancies and, thereby, achieving the maximum number of unique lags a sparse array can offer [49]. Such designs consider the minimum hole array as a special case and offer flexibility in the array aperture and the minimum inter-sensor spacing while maintaining the desired redundancy-free feature. While the sensor positions are constrained on discretized grid positions, the array optimization methodology exploiting mixed integer linear programming technique is developed. Meanwhile, a fractal array design with a recursively expanded generator array is proposed to maximize the difference coarray [50]. On the other hand, sparse array motion has also emerged as an effective solution to fill the discontiguous difference coarray with increased number of DOFs [5, 51].

Figure 2.14 illustrates examples of different six-element sparse arrays. Compared to the coprime array and nested array, the generalized nonredundant sparse array design avoids any redundancies

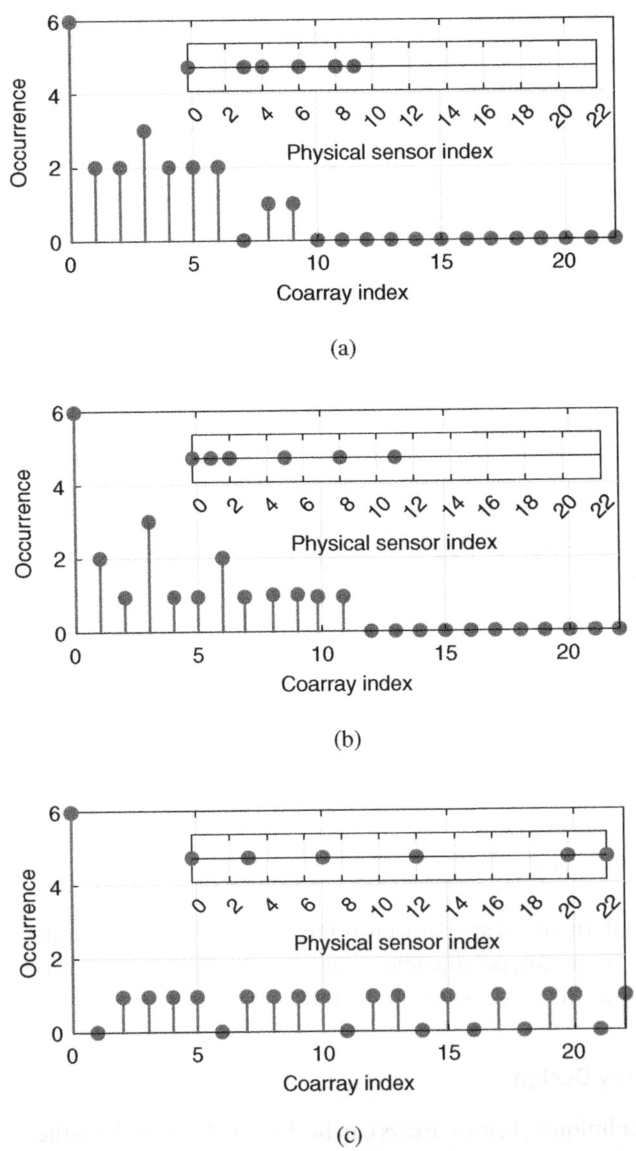

Figure 2.14 Difference lags obtained from six-element coprime array, nested array, and a nonredundant sparse array. The respective physical arrays are shown in the inset. (a) Coprime array; (b) Nested array; (c) Nonredundant array.

in nonzero lags and thus achieves the highest number of unique lags and a large virtual array aperture. This sparse array also removes lag-one entries so that mutual coupling effects are reduced. Note that the holes in positions 1, 6, 11, 14, 16, 18, and 21 can be effectively filled using the array interpolation techniques described in this chapter to render a contiguous difference coarray with non-negative lags between 0 and 22. In comparison, the maximum lags obtained from the coprime array and the nested array are respectively 9 and 11. As a result, the generalized nonredundant sparse array resolves much more sources than the coprime and nested arrays can offer.

2.4.2 Multi-dimensional Sparse Array Interpolation

With the increasing demand of perceiving high-dimensional signals in complex propagation environments, sparse arrays can be extended from 1-D configuration to two-dimensional (2-D) and even three-dimensional (3-D) configurations. Some examples are shown in Fig. 2.15 [52–56]. Accordingly, the coarray processing techniques are also extended to a high-dimensional space. By incorporating the coprime array structure and coprime frequency offsets, a 2-D space-frequency difference coarray corresponding to uniform virtual array and uniform virtual frequency offset is formed to increase the number of DOFs for joint DOA and range estimation [57].

2.4.3 Sparse Array Interpolation in Tensor Signal Processing

In the scenario of multi-dimensional array processing, the traditional matrix-based signal processing methods become less effective in handling the multi-dimensional undersampled signals, since they lack the ability to extract multiway features. As a result, their performance deviates from the theoretical performance bound.

As shown in Fig. 2.16, tensor, as an extension of matrix, can preserve the intrinsic structural information of multi-dimensional data. Therefore, tensor has been utilized for multi-dimensional array processing, and it is shown that tensor decompositions can estimate the underlying high-dimensional parameters in wireless communications and radar systems [58, 59]. In particular, the Vandermonde structure-constrained tensor decomposition method [60] and the Hermitian Toeplitz-constrained tensor reconstruction method [61] respectively achieve 2-D DOA estimation for transmit beamspace multiple-input multiple-output (MIMO) radar and coherent sources DOA estimation for uniform rectangular arrays.

More recently, sub-Nyquist tensor processing for DOA estimation is proposed to seek the performance breakthrough [62]. In particular, the sub-Nyquist *temporal* sampling is investigated in [63, 64], where temporally undersampled tensor signals are completed for DOA estimation.

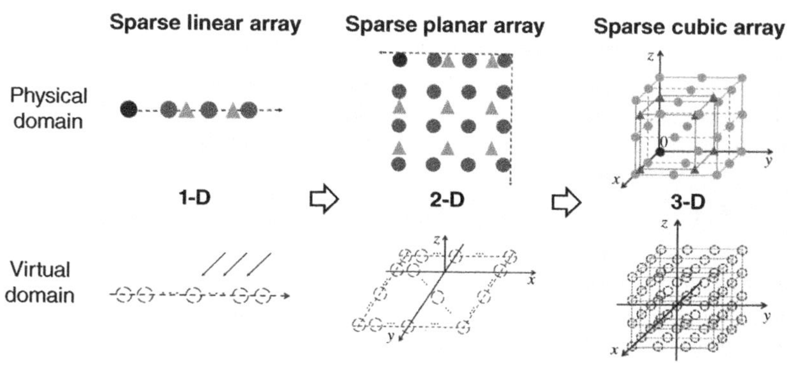

Figure 2.15 Illustration of multi-dimensional sparse arrays.

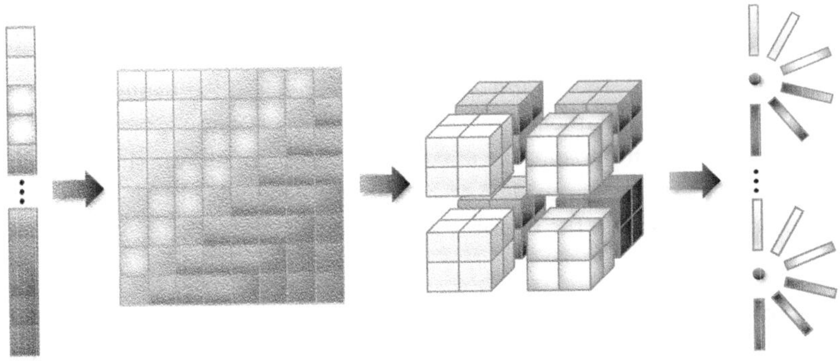

Figure 2.16 Development of tensor signal processing.

By incorporating tensor model with sub-Nyquist *spatial* sampling, a tensor-based sub-Nyquist radar equipped with a thinned array is proposed for DOA estimation [65], where the first-order sub-Nyquist tensor signals are recovered via tensorial compressed sensing. More recently, augmented multi-dimensional virtual arrays are derived for coarray tensor DOA estimation [66, 67], where the optimal source identifiability brought by coarray tensor processing was investigated. To interpolate a multi-dimensional discontiguous virtual array with slices of holes, a slice-missing coarray tensor completion algorithm is developed in [68, 69], which demonstrates further improvement to the DOA estimation performance compared to state-of-art matrix-based sparse array interpolation methods.

Acknowledgments

The work of C. Zhou and Z. Shi was supported in part by National Natural Science Foundation of China (No. 62271444, U21A20456, 61901413), Zhejiang Provincial Natural Science Foundation of China (No. LZ23F010007), Zhejiang University Education Foundation Qizhen Scholar Foundation, and the 5G Open Laboratory of Hangzhou Future Sci-Tech City.

References

1 Qin, S., Zhang, Y.D., and Amin, M.G. (2015). Generalized coprime array configurations for direction-of-arrival estimation. *IEEE Transactions on Signal Processing* 63 (6): 1377–1390.
2 Zhou, C., Gu, Y., Shi, Z., and Haardt, M. (2023). Structured Nyquist correlation reconstruction for DOA estimation with sparse arrays. *IEEE Transactions on Signal Processing* 71: 1849–1862.
3 Zheng, H., Zhou, C., Vorobyov, S.A. et al. (2023). Decomposed CNN for sub-Nyquist tensor-based 2-D DOA estimation. *IEEE Signal Processing Letters* 30: 708–712.
4 He, J., Li, L., and Shu, T. (2021). Sparse nested arrays with spatially spread square acoustic vector sensors for high-accuracy underdetermined direction finding. *IEEE Transactions on Aerospace and Electronic Systems* 57 (4): 2324–2336.
5 Qin, G., Amin, M.G., and Zhang, Y.D. (2019). DOA estimation exploiting sparse array motions. *IEEE Transactions on Signal Processing* 67 (11): 3013–3027.

6 Zhou, C., Gu, Y., Zhang, Y.D. et al. (2017). Compressive sensing-based coprime array direction-of-arrival estimation. *IET Communications* 11 (11): 1719–1724.

7 Abramovich, Y.I., Spencer, N.K., and Gorokhov, A.Y. (1999). Positive-definite Toeplitz completion in DOA estimation for nonuniform linear antenna arrays. II. Partially augmentable arrays. *IEEE Transactions on Signal Processing* 47 (6): 1502–1521.

8 Pal, P. and Vaidyanathan, P.P. (2011). Coprime sampling and the MUSIC algorithm. *Proceedings of the Digital Signal Processing and Signal Processing Education Meeting (DSP/SPE)*, Sedona, AZ, January 2011, 289–294.

9 Zhou, C., Shi, Z., Gu, Y., and Shen, X. (2013). DECOM: DOA estimation with combined MUSIC for coprime array. *Proceedings of the International Conference on Wireless Communications and Signal Processing (WCSP)*, Hangzhou, China, October 2013.

10 Liu, S., Mao, Z., Zhang, Y.D., and Huang, Y. (2021). Rank minimization-based Toeplitz reconstruction for DoA estimation using coprime array. *IEEE Communications Letters* 25 (7): 2265–2269.

11 Shen, Q., Liu, W., Cui, W. et al. (2019). Simplified and enhanced multiple level nested arrays exploiting high-order difference co-arrays. *IEEE Transactions on Signal Processing* 67 (13): 3502–3515.

12 Rajamäki, R. and Koivunen, V. (2021). Sparse symmetric linear arrays with low redundancy and a contiguous sum co-array. *IEEE Transactions on Signal Processing* 69: 1697–1712.

13 BouDaher, E., Jia, Y., Ahmad, F., and Amin, M.G. (2015). Multi-frequency co-prime arrays for high-resolution direction-of-arrival estimation. *IEEE Transactions on Signal Processing* 63 (14): 3797–3808.

14 Liu, C.-L., Vaidyanathan, P.P., and Pal, P. (2016). Coprime coarray interpolation for DOA estimation via nuclear norm minimization. *Proceedings of the IEEE International Symposium on Circuits and Systems* (ISCAS), Montréal, Canada, May 2016, 2639–2642.

15 Hosseini, S.M. and Sebt, M.A. (2017). Array interpolation using covariance matrix completion of minimum-size virtual array. *IEEE Signal Processing Letters* 24 (7): 1063–1067.

16 Zhou, C., Gu, Y., Fan, X. et al. (2018). Direction-of-arrival estimation for coprime array via virtual array interpolation. *IEEE Transactions on Signal Processing* 66 (22): 5956–5971.

17 Zhou, C., Gu, Y., Shi, Z., and Zhang, Y.D. (2018). Off-grid direction-of-arrival estimation using coprime array interpolation. *IEEE Signal Processing Letters* 25 (11): 1710–1714.

18 Zheng, Z., Huang, Y., Wang, W.-Q., and So, H.C. (2020). Direction-of-arrival estimation of coherent signals via coprime array interpolation. *IEEE Signal Processing Letters* 27: 585–589.

19 Goodman, N.R. (1963). Statistical analysis based on a certain multivariate complex Gaussian distribution (an introduction). *Annals of Mathematical Statistics* 34 (1): 152–177.

20 Vaidyanathan, P.P. and Pal, P. (2011). Sparse sensing with co-prime samplers and arrays. *IEEE Transactions on Signal Processing* 59 (2): 573–586.

21 Shi, Z., Zhou, C., Gu, Y. et al. (2017). Source estimation using coprime array: a sparse reconstruction perspective. *IEEE Sensors Journal* 17 (3): 755–765.

22 Tang, G., Bhaskar, B.N., Shah, P., and Recht, B. (2013). Compressed sensing off the grid. *IEEE Transactions on Information Theory* 59 (11): 7465–7490.

23 Tang, G., Bhaskar, B.N., and Recht, B. (2015). Near minimax line spectral estimation. *IEEE Transactions on Information Theory* 61 (1): 499–512.

24 Li, Y. and Chi, Y. (2016). Off-the-grid line spectrum denoising and estimation with multiple measurement vectors. *IEEE Transactions on Signal Processing* 64 (5): 1257–1269.

25 Rockafellar, R.T. (1970). *Convex Analysis*. Princeton, NJ: Princeton University Press.

26 Chandrasekaran, V., Recht, B., Parrilo, P.A., and Willsky, A.S. (2012). The convex geometry of linear inverse problems. *Foundations of Computational Mathematics* 12 (6): 805–849.

27 Stoica, P. and Moses, R. (2005). *Spectral Analysis of Signals*. Upper Saddle River, NJ: Prentice Hall.

28 Zhang, F. (2005). *The Schur Complement and Its Applications*. Springer.

29 Liu, C.-L. and Vaidyanathan, P.P. (2015). Remarks on the spatial smoothing step in coarray MUSIC. *IEEE Signal Processing Letters* 22 (9): 1438–1442.

30 Schmidt, R.O. (1986). Multiple emitter location and signal parameter estimation. *IEEE Transactions on Antennas and Propagation* 34 (3): 276–280.

31 Rao, B.D. and Hari, K.V.S. (1989). Performance analysis of root-MUSIC. *IEEE Transactions on Acoustics, Speech, and Signal Processing* 37 (12): 1939–1949.

32 Roy, R. and Kailath, T. (1989). ESPRIT–Estimation of signal parameters via rotational invariance techniques. *IEEE Transactions on Acoustics, Speech, and Signal Processing* 37 (7): 984–995.

33 Zhou, C. and Zhou, J. (2017). Direction-of-arrival estimation with coarray ESPRIT for coprime array. *Sensors* 17 (1779): 1–17.

34 Pal, P. (2018). Correlation awareness in low-rank models: sampling, algorithms, and fundamental limits. *IEEE Signal Processing Magazine* 35 (4): 56–71.

35 Wu, X., Zhu, W.-P., Yan, J., and Zhang, Z. (2018). Two sparse-based methods for off-grid direction-of-arrival estimation. *Signal Processing* 142: 87–95.

36 Wu, X., Zhu, W.-P., and Yan, J. (2016). Direction of arrival estimation for off-grid signals based on sparse Bayesian learning. *IEEE Sensors Journal* 16 (7): 2004–2016.

37 Stoica, P. and Nehorai, A. (1990). MUSIC, maximum likelihood, and Cramér–Rao bound: further results and comparisons. *IEEE Transactions on Acoustics, Speech, and Signal Processing* 38 (12): 2140–2150.

38 Liu, C.-L. and Vaidyanathan, P.P. (2017). Cramér–Rao bounds for coprime and other sparse arrays, which find more sources than sensors. *Digital Signal Processing* 61: 43–61.

39 Zhang, Y.D., Amin, M.G., and Himed, B. (2013). Sparsity-based DOA estimation using co-prime arrays. *Proceedings of the IEEE International Conference on Acoustics, Speech and Signal Processing (ICASSP)*, Vancouver, Canada, May 2013, 3967–3971.

40 Qiao, H. and Pal, P. (2017). Unified analysis of co-array interpolation for direction-of-arrival estimation. *Proceedings of the IEEE International Conference on Acoustics, Speech and Signal Processing (ICASSP)*, New Orleans, LA, March 2017, 3056–3060.

41 Grant, M. and Boyd, S. (2014). CVX: Matlab software for disciplined convex programming, version 2.1. http://cvxr.com/cvx (accessed 3 October 2023).

42 Wang, M. and Nehorai, A. (2017). Coarrays, MUSIC, and the Cramér–Rao bound. *IEEE Transactions on Signal Processing* 65 (4): 933–946.

43 Candès, E.J. and Plan, Y. (2010). Matrix completion with noise. *Proceedings of the IEEE* 98 (6): 925–936.

44 Wu, X., Zhu, W.-P., and Yan, J. (2017). A Toeplitz covariance matrix reconstruction approach for direction-of-arrival estimation. *IEEE Transactions on Vehicular Technology* 66 (9): 8223–8237.

45 He, Z.-Q., Shi, Z.-P., Huang, L., and So, H.C. (2015). Underdetermined DOA estimation for wideband signals using robust sparse covariance fitting. *IEEE Signal Processing Letters* 22 (4): 435–439.

46 Qin, G., Zhang, Y.D., and Amin, M.G. (2019). DOA estimation exploiting moving dilated nested arrays. *IEEE Signal Processing Letters* 26 (3): 490–494.

47 Wu, X., Zhu, W.-P., and Yan, J. (2018). A high-resolution DOA estimation method with a family of nonconvex penalties. *IEEE Transactions on Vehicular Technology* 67 (6): 4925–4938.

48 Fan, X., Zhou, C., Gu, Y., and Shi, Z. (2017). Toeplitz matrix reconstruction of interpolated coprime virtual array for DOA estimation. *Proceedings of the IEEE Vehicular Technology Conference (VTC)*, Sydney, Australia, June 2017.

49 Ahmed, A. and Zhang, Y.D. (2021). Generalized non-redundant sparse array designs. *IEEE Transactions on Signal Processing* 69: 4580–4594.

50 Cohen, R. and Eldar, Y.C. (2020). Sparse array design via fractal geometries. *IEEE Transactions on Signal Processing* 68: 4797–4812.

51 Qin, G. and Amin, M.G. (2021). Structured sparse array design exploiting two uniform subarrays for DOA estimation on moving platform. *Signal Processing* 180: 107872.

52 Nannuru, S., Gerstoft, P., Ping, G., and Fernandez-Grande, E. (2021). Sparse planar arrays for azimuth and elevation using experimental data. *The Journal of the Acoustical Society of America* 149 (1): 167–178.

53 Zheng, H., Shi, Z., Zhou, C. et al. (2021). Coupled coarray tensor CPD for DOA estimation with coprime L-shaped array. *IEEE Signal Processing Letters* 28: 1545–1549.

54 He, J., Shu, T., Li, L., and Truong, T.-K. (2022). Mixed near-field and far-field localization and array calibration with partly calibrated arrays. *IEEE Transactions on Signal Processing* 70: 2105–2118.

55 Zheng, H., Zhou, C., Shi, Z., and de Almeida, A.L.F. (2022). SubTTD: DOA estimation via sub-Nyquist tensor train decomposition. *IEEE Signal Processing Letters* 29: 1978–1982.

56 Wang, S., Li, S., Ren, B. et al. (2023). Convex optimization-based design of sparse arrays for 3-D near-field imaging. *IEEE Sensors Journal* 23 (9): 9640–9648.

57 Mao, Z., Liu, S., Zhang, Y.D. et al. (2022). Joint DoA-range estimation using space-frequency virtual difference coarray. *IEEE Transactions on Signal Processing* 70: 2576–2592.

58 Comon, P., Luciani, X., and de Almeida, A.L.F. (2009). Tensor decompositions, alternating least squares and other tales. *Journal of Chemometrics* 23 (7–8): 393–405.

59 Chen, H., Ahmad, F., Vorobyov, S.A., and Porikli, F. (2021). Tensor decompositions in wireless communications and MIMO radar. *IEEE Journal on Selected Topics in Signal Processing* 15 (3): 438–453.

60 Xu, F., Morency, M.W., and Vorobyov, S.A. (2022). DOA estimation for transmit beamspace MIMO radar via tensor decomposition with Vandermonde factor matrix. *IEEE Transactions on Signal Processing* 70: 2901–2917.

61 Zheng, H., Zhou, C., Shi, Z., and Gu, Y. (2022). Structured tensor reconstruction for coherent DOA estimation. *IEEE Signal Processing Letters* 29: 1634–1638.

62 Han, K. and Nehorai, A. (2014). Nested vector-sensor array processing via tensor modeling. *IEEE Transactions on Signal Processing* 62 (10): 2542–2553.

63 Huang, L.-T., de Almeida, A.L.F., and So, H.C. (2016). Target estimation in bistatic MIMO radar via tensor completion. *Signal Processing* 120: 654–659.

64 Liu, L., Gu, J.-F., and Wei, P. (2019). Joint DOA and frequency estimation with sub-Nyquist sampling. *Signal Processing* 154: 87–96.

65 Na, S., Mishra, K.V., Liu, Y. et al. (2019). TenDSuR: Tensor-based 4D sub-Nyquist radar. *IEEE Signal Processing Letters* 26 (2): 237–241.

66 Zheng, H., Zhou, C., Gu, Y., and Shi, Z. (2020). Two-dimensional DOA estimation for coprime planar array: A coarray tensor-based solution. *Proceedings of the IEEE International Conference on Acoustics, Speech and Signal Processing (ICASSP)*, Barcelona, Spain, May 2020, 4562–4566.

67 Zheng, H., Zhou, C., Shi, Z. et al. (2023). Coarray tensor direction-of-arrival estimation. *IEEE Transactions on Signal Processing* 71: 1128–1142.

68 Zheng, H., Zhou, C., de Almeida, A.L.F. et al. (2022). DOA estimation via coarray tensor completion with missing slices. *Proceedings of the IEEE International Conference on Acoustics, Speech and Signal Processing (ICASSP)*, Singapore, May 2022, 5053–5057.

69 Zheng, H., Shi, Z., Zhou, C. et al. (2023). Coarray tensor completion for DOA estimation. *IEEE Transactions on Aerospace and Electronic Systems* 59 (5): 5472–5486.

3

Wideband and Multi-frequency Sparse Array Processing

Fauzia Ahmad[1], *Peter Gerstoft*[2], *and Wei Liu*[3]

[1]*Department of Electrical and Computer Engineering, Temple University, Philadelphia, PA, USA*
[2]*NoiseLab, University of California San Diego, La Jolla, CA, USA*
[3]*School of Electronic Engineering and Computer Science, Queen Mary University of London, London, UK*

3.1 Introduction

This chapter introduces wideband and multi-frequency operation of sparse arrays for source direction-of-arrival (DOA) estimation and beamforming which have found a wide range of applications in radar, sonar, speech enhancement, and wireless communications. We focus on uncorrelated sources at different directions in the far-field of a sparse linear array. Unlike Chapters 1 and 2 where the discussion concentrated on monochromatic signals impinging on the array, we assume the sources to have a wide signal bandwidth or comprise multiple frequencies. The time-domain measurements at the sparse array are decomposed into multiple nonoverlapping narrowband components by using the discrete Fourier transform (DFT) to extract the received signals at each frequency bin of the source signal bandwidth.

We first concentrate on wideband signal processing within the group sparsity framework wherein we utilize information in all subbands comprising the full signal bandwidth. With J frequency bins covering the signal bandwidth, there is a J-fold increase in the number of unknowns due to the wideband model formulation. The increased complexity is remedied by the group-sparsity constraint which accounts for the fact that different frequency components from the same source correspond to the same direction. Wideband processing provides a significantly improved DOA estimation performance over monochromatic sparse reconstruction methods. Next, we exploit multi-frequency processing to use array observations at two or more frequencies in the signal bandwidth for introducing additional elements in the coarray. The resulting mutli-frequency coarray offers enhanced degrees-of-freedom (DOFs) beyond those available for single-frequency operation. We note that not all array elements are required to be operated at more than one frequency to avail the additional DOFs. This is in contrast to wideband operation where all elements are assumed to function at the full signal bandwidth. Finally, we introduce wideband beamforming with sparse arrays via the probabilistic model-based method of sparse Bayesian learning (SBL) to provide higher resolution and, in turn, improved source localization over conventional beamforming (CBF) [1].

We note that the algorithms discussed in this chapter are applicable to sparse array configurations with both uniform and nonuniform coarrays. The former include minimum redundancy [2–4] and nested arrays [5], while examples of the latter are coprime arrays [6] and minimum hole arrays [7, 8]. The signal models and the algorithms can also be readily extended to cater to planar array

Sparse Arrays for Radar, Sonar, and Communications, First Edition. Edited by Moeness G. Amin.

geometries [9] which are detailed in Chapter 5 of this book. Further, we focus on gridded methods as these are easy to implement and generally quite efficient. Nevertheless, there has been considerable interest in grid-less (where a grid is used initially, but solution is not confined to the grid) or grid-free (no grid is used) methods [10–12]. The chapter considers passive sensing and, as such, deals with difference coarrays. Active sensing and imaging deal with sum coarrays and is addressed in Chapters 9 and 11 of this book [13].

The chapter is organized as follows. In Section 3.2, we begin by describing the wideband signal model for sparse linear arrays, followed by introduction of low-complexity DOA estimation methods for a single-frequency operation and their wideband extensions under the group sparsity framework. Supporting simulation results are also provided therein. In Section 3.3, the multi-frequency signal model is presented and the multi-frequency DOA estimation is posed as a spare reconstruction problem under two separate cases, namely, sources with proportional and nonproportional spectra at the employed operational frequencies. Performance of these multi-frequency methods is evaluated through numerical simulations. The SBL algorithm for beamforming is discussed in Section 3.4, starting with single-frequency implementation and then extending it to wideband operation. Experimental results are also provided which compare the performance of the SBL with CBF using sparse array geometries. Suggestions for further reading are provided in Section 3.5. Section 3.6 contains the concluding remarks.

3.2 Wideband DOA Estimation

Although wideband DOA estimation can be performed in the time domain, such as the method developed under the convolutional sparse coding framework [14], a common approach entails applying DFT to the signals received at the array, thereby decomposing the wideband signals into different frequency bins or subbands [15–18]. A narrowband signal model can then be established at each frequency bin, provided that the number of DFT points is sufficiently large [19]. Under such a framework, all narrowband DOA estimation algorithms described in this book for sparse arrays can be applied directly to the measurements at each frequency comprising the wide bandwidth. However, one key issue arises in regard to combining the resulting DOA estimates at different frequencies in an effective manner, since simple averaging of the estimates over different frequencies may not work. One solution is to apply focusing to different frequencies [17, 18], such that different frequency components share the same frequency-independent steering vector. Unlike traditional focusing designed for physical arrays, no preliminary estimate of the DOAs is required [20, 21] due to the increased DOFs in coarray-based underdetermined DOA estimation. An alternate solution exploits group sparsity by recognizing the fact that different frequency components from the same source correspond to the same direction or in other words they share the same spatial support. In this section, we focus on the group-sparsity-based solution [22, 23].

3.2.1 Wideband Array Model

We consider an M-element linear sparse array with sensor locations $\{i_m d, m = 0, 1, \dots, M-1\}$, where i_m are integers with $i_0 = 0$ and d is the unit sensor spacing. We assume that K wideband signals, $s_k(t)$, impinge on the array from directions θ_k, $k = 1, 2, \dots, K$, respectively. The corresponding received signal at the mth sensor is expressed as

$$x_m(t) = \sum_{k=1}^{K} \delta(t - \tau_{m,k}) * s_k(t) + n_m(t), m \in \{0, \dots, M-1\}, \tag{3.1}$$

where $n_m(t)$ is the noise, $\tau_{m,k} = \frac{i_m d \sin\theta_k}{c}$ is the time delay of the kth signal with DOA θ_k at the mth sensor, c is the signal propagation speed, and "$*$" denotes the convolution operation.

With a sampling period T_s, the discrete-time received signal at the mth sensor can be expressed in a discrete convolution form as [24]

$$x_m[p] = \sum_{k=1}^{K} \left(\sum_{i=-\infty}^{+\infty} a_{i,m,k} s_k[p-i] \right) + n_m[p], \tag{3.2}$$

where $s_k[p]$ represents the pth sample of the kth source signal,

$$a_{i,m,k} = \mathrm{sinc}(i - \tau_{m,\theta_k}/T_s)\,, \tag{3.3}$$

and sinc() is the normalized sinc function, defined as

$$\mathrm{sinc}(v) = \sin(\pi v)/(\pi v). \tag{3.4}$$

Equation (3.2) can be transformed into matrix form as

$$\mathbf{x}[p] = \sum_{i=-\infty}^{+\infty} \mathbf{A}_i \mathbf{s}[p-i] + \mathbf{n}[p], \tag{3.5}$$

where

$$\begin{aligned}
\mathbf{s}[p] &= \left[s_0[p]\ s_1[p]\ \cdots\ s_{K-1}[p]\right]^{\mathrm{T}} \\
\mathbf{x}[p] &= \left[x_0[p]\ x_1[p]\ \cdots\ x_{M-1}[p]\right]^{\mathrm{T}} \\
\mathbf{n}[p] &= \left[n_0[p]\ n_1[p]\ \cdots n_{M-1}[p]\right]^{\mathrm{T}},
\end{aligned} \tag{3.6}$$

and the (m, k)th element of the $M \times K$ matrix $\mathbf{A}_i$ is $a_{i,m,k}$. We see that the mixing of impinging signals at the sensor array is convolutive for the wideband case [24, 25], while it is instantaneous in the narrowband case.

To obtain the equivalent signal model in the frequency domain, we assume availability of a total of P samples of the received signal vector $\mathbf{x}[p]$, which are processed as follows. First, the mth received signal $x_m[p], p = 0, 1, \ldots, P-1$, is divided into Q nonoverlapping groups of length $L = P/Q$, and an L-point DFT is applied to each group to obtain

$$X_m[l, q] = \sum_{i=0}^{L-1} x_m[L \cdot (q-1) + i] \cdot e^{-j\frac{2\pi}{L} il}\,, \tag{3.7}$$

with $q = 0, 1, \ldots, Q-1$, and $l = 0, 1, \ldots, L-1$. Then, the lth frequency bin/subband samples of the qth group are stacked to form a vector as

$$\mathbf{x}\left[l, q\right] = \left[X_1[l, q], X_2[l, q], \ldots, X_{M-1}[l, q]\right]^T. \tag{3.8}$$

Likewise, defining $S_k[l, q]$ and $N_m[l, q]$ as

$$S_k[l, q] = \sum_{i=0}^{L-1} s_k[L \cdot (q-1) + i] \cdot e^{-j\frac{2\pi}{L} il}\,, \tag{3.9}$$

$$N_m[l, q] = \sum_{i=0}^{L-1} n_m[L \cdot (q-1) + i] \cdot e^{-j\frac{2\pi}{L} il}\,, \tag{3.10}$$

we can form the source signal and noise vectors corresponding to the lth frequency bin of the qth group as $\mathbf{s}[l, q] = \left[S_1[l, q], \ldots, S_K[l, q]\right]^T$ and $\mathbf{n}[l, q] = \left[N_0[l, q], \ldots, N_{M-1}[l, q]\right]^T$. Then, the output signal model can be expressed in the frequency domain as

$$\mathbf{x}[l, q] = \mathbf{A}(l, \boldsymbol{\theta})\mathbf{s}[l, q] + \mathbf{n}[l, q], \tag{3.11}$$

where $\mathbf{A}(l,\boldsymbol{\theta}) = \left[\mathbf{a}(l,\theta_1), \dots, \mathbf{a}(l,\theta_K)\right]$ is the array manifold matrix at frequency ω_l corresponding to the lth frequency bin and $\mathbf{a}(l,\theta_k)$ is the $M \times 1$ steering vector at frequency ω_l and angle θ_k with its mth element given by $[\mathbf{a}(l,\theta_k)]_m = e^{-j\omega_l i_m d \sin(\theta_k)/c}$.

Assuming that the sources are uncorrelated and the noise is spatially and temporally white and uncorrelated from the sources, the correlation matrix of $\mathbf{x}[l,q]$ is given by

$$\begin{aligned}\mathbf{R}_{\mathbf{xx}}[l] &= \mathbb{E}\left\{\mathbf{x}[l,q]\mathbf{x}^H[l,q]\right\} \\ &= \sum_{k=1}^{K} \sigma_k^2[l]\mathbf{a}(l,\theta_k)\mathbf{a}^H(l,\theta_k) + \sigma_n^2[l]\mathbf{I}_M,\end{aligned} \tag{3.12}$$

where $\mathbb{E}\{\cdot\}$ denotes the expectation operator, $\mathbf{I}_M$ is the $M \times M$ identity matrix, $\sigma_k^2[l]$ is the power of the kth source signal at the lth frequency bin, and $\sigma_n^2[l]$ is the corresponding noise power, which is assumed to be the same for all sensors. In practice, with Q signal blocks for DFT, $\mathbf{R}_{\mathbf{xx}}[l]$ can be approximated by its finite sample estimate as

$$\mathbf{R}_{\mathbf{xx}}[l] \approx \hat{\mathbf{R}}_{\mathbf{xx}}[l] = \frac{1}{Q}\sum_{q=0}^{Q-1} \mathbf{x}[l,q]\mathbf{x}^H[l,q]. \tag{3.13}$$

Next, we vectorize $\mathbf{R}_{\mathbf{xx}}[l]$ as

$$\mathbf{z}[l] = \operatorname{vec}\left\{\mathbf{R}_{\mathbf{xx}}[l]\right\} = \widetilde{\mathbf{A}}[l]\widetilde{\mathbf{s}}[l] + \sigma_n^2[l]\widetilde{\mathbf{i}}_M, \tag{3.14}$$

where $\widetilde{\mathbf{A}}[l] = \left[\widetilde{\mathbf{a}}(l,\theta_1), \dots, \widetilde{\mathbf{a}}(l,\theta_K)\right]$ with $\widetilde{\mathbf{a}}(l,\theta_k) = \mathbf{a}^*(l,\theta_k) \otimes \mathbf{a}(l,\theta_k)$, "$\otimes$" is the Kronecker product, $\{\cdot\}^*$ denotes the complex conjugate operation, $\widetilde{\mathbf{s}}[l] = \left[\sigma_1^2[l], \dots, \sigma_K^2[l]\right]^T$, and $\widetilde{\mathbf{i}}_M$ is an $M^2 \times 1$ column vector obtained by vectorizing $\mathbf{I}_M$. The vector $\mathbf{z}[l]$ behaves as the received signal vector, corresponding to frequency ω_l, at a longer virtual array with sensor positions given by the coarray of the physical array, with $\widetilde{\mathbf{A}}[l]$ being its manifold matrix and $\widetilde{\mathbf{s}}[l]$ its equivalent source signal vector. Thus, (3.14) represents the frequency-domain coarray signal model. In this model, the source signals are replaced by their respective powers and, as such, act as mutually coherent sources, and the noise is deterministic.

3.2.2 Sparsity-Based DOA Estimation at a Single Frequency

At a single frequency ω_l, the model in (3.14) is the same as that of the coarray corresponding to a narrowband array. As such, the DOA estimation problem at each frequency can be solved by the coarray-MUSIC algorithm described in Chapter 1, provided that we focus only on the central uniform segment of the coarray [26]. To make full use of the available DOFs, we can employ the sparsity-based DOA estimation algorithm to estimate the DOAs at each frequency. To this end, we rewrite (3.14) as

$$\mathbf{z}[l] = \widetilde{\mathbf{A}}[l]\widetilde{\mathbf{s}}[l] + \sigma_n^2[l]\widetilde{\mathbf{i}}_M = \widetilde{\mathbf{A}}^{\circ}[l]\widetilde{\mathbf{s}}^{\circ}[l], \tag{3.15}$$

where $\widetilde{\mathbf{A}}^{\circ}[l] = \left[\widetilde{\mathbf{A}}[l], \widetilde{\mathbf{i}}_M\right]$ and $\widetilde{\mathbf{s}}^{\circ}[l] = \left[\widetilde{\mathbf{s}}^T[l], \sigma_n^2[l]\right]^T$. Next, we create a search grid of K_g potential incident angles $\theta_{g,1}, \dots, \theta_{g,K_g}$, with the corresponding manifold matrix given by $\widetilde{\mathbf{A}}_g[l] = \left[\widetilde{\mathbf{a}}(l,\theta_{g,1}), \dots, \widetilde{\mathbf{a}}(l,\theta_{g,K_g})\right]$, and construct a $K_g \times 1$ column vector $\widetilde{\mathbf{s}}_g[l]$ of potential source signal powers for the corresponding angles. Here, the subscript $\{\cdot\}_g$ is used to indicate matrices, vectors, or elements related to the generated search grid. In this and all subsequent sections in this chapter, we assume that the sources are located on the grid. Several methods can be used to modify the model in order to deal with off-grid sources [10, 12].

In line with (3.15), the following matrix and vector are created

$$\widetilde{\mathbf{A}}_g^{\circ}[l] = \left[\widetilde{\mathbf{A}}_g[l], \widetilde{\mathbf{i}}_M\right], \ \widetilde{\mathbf{s}}_g^{\circ}[l] = \left[\widetilde{\mathbf{s}}_g[l], \sigma_n^2[l]\right]^T. \tag{3.16}$$

Then, the sparsity-based DOA estimation problem for a single frequency can be formulated as [27]

$$\begin{aligned} \min \quad & \left\|\widetilde{\mathbf{s}}_g^{\circ}[l]\right\|_1 \\ \text{subject to} \quad & \left\|\mathbf{z}[l] - \widetilde{\mathbf{A}}_g^{\circ}[l]\widetilde{\mathbf{s}}_g^{\circ}[l]\right\|_2 \le \varepsilon, \\ & \tilde{s}_{g,k_g}^{\circ}[l] \ge 0, \quad 0 \le k_g \le K_g, \end{aligned} \tag{3.17}$$

where ε is the allowable error bound, $\|\cdot\|_1$ is the l_1 norm, $\|\cdot\|_2$ is the l_2 norm, and $\tilde{s}_{g,k_g}^{\circ}[l]$ represents the k_gth entry in the column vector $\widetilde{\mathbf{s}}_g^{\circ}[l]$.

Complexity reduction can be achieved by reducing the redundant entries in $\mathbf{R}_{\mathbf{xx}}[l]$ since there are many repeated correlation lags. After removing those redundancies, we have a new relationship which is similar to (3.15), but with significantly reduced number of observed virtual measurements [22, 28]

$$\check{\mathbf{z}}_{\mathbf{c}}[l] = \check{\mathbf{A}}_{\mathbf{c}}^{\circ}[l]\widetilde{\mathbf{s}}^{\circ}[l], \tag{3.18}$$

where $\check{\mathbf{z}}_{\mathbf{c}}[l]$ and $\check{\mathbf{A}}_{\mathbf{c}}^{\circ}[l]$ are the new coarray signal vector and manifold matrix, respectively. By creating the same search grid of K_g potential incident angles, the reduced-complexity DOA estimation problem can be formulated as

$$\begin{aligned} \min \quad & \left\|\widetilde{\mathbf{s}}_g^{\circ}[l]\right\|_1 \\ \text{subject to} \quad & \left\|\check{\mathbf{z}}_{\mathbf{c}}[l] - \check{\mathbf{A}}_{\mathbf{cg}}^{\circ}[l]\widetilde{\mathbf{s}}_g^{\circ}[l]\right\|_2 \le \varepsilon, \\ & \tilde{s}_{g,k_g}^{\circ}[l] \ge 0, \quad 0 \le k_g \le K_g, \end{aligned} \tag{3.19}$$

where $\check{\mathbf{A}}_{\mathbf{cg}}^{\circ}[l]$ is defined in a similar way as in (3.16).

For further complexity reduction, we note that only the zeroth lag in $\mathbf{R}_{\mathbf{xx}}[l]$ is related to the noise power $\sigma_n^2[l]$. Since noise is assumed to be spatially white, and as the estimation of noise power takes up one DOF, it can be removed for further complexity reduction. As a result, in (3.14), all the components of $\mathbf{z}[l]$ related to the zeroth lag in $\mathbf{R}_{\mathbf{xx}}[l]$ are removed. In essence, the noise component is removed and the resulting model is noise-free. We can further remove/merge the redundant entries in $\mathbf{R}_{\mathbf{xx}}[l]$ in the same manner as in (3.18), arriving at a new modified low-complexity DOA estimation formulation [22, 28]

$$\begin{aligned} \min \quad & \left\|\widetilde{\mathbf{s}}_g[l]\right\|_1 \\ \text{subject to} \quad & \left\|\mathbf{z}_{\mathbf{s}}[l] - \widetilde{\mathbf{A}}_{\mathbf{sg}}[l]\widetilde{\mathbf{s}}_g[l]\right\|_2 \le \varepsilon \\ & \tilde{s}_{g,k_g}[l] \ge 0, \quad 0 \le k_g \le K_g - 1, \end{aligned} \tag{3.20}$$

where $\mathbf{z}_{\mathbf{s}}[l]$ and $\widetilde{\mathbf{A}}_{\mathbf{sg}}[l]$ are defined accordingly.

The single-frequency problems in (3.17), (3.19), and (3.20) are referred to as "Narrowband 1," "Narrowband 2," and "Narrowband 3," respectively. These are convex in nature and can be solved by existing convex optimization toolboxes. The first K_g elements of the resultant column vector $\widetilde{\mathbf{s}}_g^{\circ}[l]$ yield the corresponding DOA estimates over K_g grid points.

3.2.3 Wideband DOA Estimation Based on Group Sparsity

The methods described in Section 3.2.2 are for a single frequency bin. For multiple frequency bins, we could apply any aforementioned method to the frequencies constituting the signal bandwidth one by one and then average the results to yield the final estimate. This is not effective and may cause serious performance degradation since some of the frequency bin estimates may not be as reliable as others due to varying signal-to-noise ratio (SNR) for different frequencies. A more effective approach is joint estimation of the DOAs across the entire frequency bandwidth based on the group sparsity concept, i.e., the DOAs corresponding to different frequencies share the same spatial support [22, 27–30].

Assume that the source signal bandwidth covers J frequency bins in the DFT domain. For each frequency bin $l_j \in \Phi_l$, $0 \leq j \leq J-1$, where Φ_l is the set of J frequency bin indices, the same search grid of K_g potential incident angles is created. First, we construct two matrices: a block diagonal matrix $\widetilde{\mathbf{B}}_g^{\circ}$ using $\widetilde{\mathbf{A}}_g^{\circ}[l_j]$, expressed as

$$\widetilde{\mathbf{B}}_g^{\circ} = \text{blkdiag}\left\{\widetilde{\mathbf{A}}_g^{\circ}[l_0], \widetilde{\mathbf{A}}_g^{\circ}[l_1], \dots, \widetilde{\mathbf{A}}_g^{\circ}[l_{J-1}]\right\}, \tag{3.21}$$

and a $(K_g+1) \times J$ matrix $\mathbf{R}_g^{\circ}$ using $\widetilde{\mathbf{s}}_g^{\circ}[l_j]$ with

$$\mathbf{R}_g^{\circ} = \left[\widetilde{\mathbf{s}}_g^{\circ}[l_0], \widetilde{\mathbf{s}}_g^{\circ}[l_1], \dots, \widetilde{\mathbf{s}}_g^{\circ}[l_{J-1}]\right]. \tag{3.22}$$

Then, we obtain the following coarray signal model

$$\widetilde{\mathbf{z}} = \widetilde{\mathbf{B}}_g^{\circ}\widetilde{\mathbf{r}}_g^{\circ}, \tag{3.23}$$

where $\widetilde{\mathbf{z}} = \left[\mathbf{z}^T[l_0], \dots, \mathbf{z}^T[l_{J-1}]\right]^T$ and $\widetilde{\mathbf{r}}_g^{\circ} = \text{vec}\left(\mathbf{R}_g^{\circ}\right)$ is a $(K_g+1)J \times 1$ vector obtained by vectorizing $\mathbf{R}_g^{\circ}$.

We use the row vector $\mathbf{r}_{g,k_g}^{\circ}$, $0 \leq k_g \leq K_g$, to denote the k_gth row of the matrix $\mathbf{R}_g^{\circ}$. Then, we form a new $(K_g+1) \times 1$ vector $\hat{\mathbf{r}}_g^{\circ}$ based on the l_2 norm of $\mathbf{r}_{g,k_g}^{\circ}$, $0 \leq k_g \leq K_g$, given by

$$\hat{\mathbf{r}}_g^{\circ} = \left[\left\|\mathbf{r}_{g,0}^{\circ}\right\|_2, \left\|\mathbf{r}_{g,1}^{\circ}\right\|_2, \dots, \left\|\mathbf{r}_{g,K_g}^{\circ}\right\|_2\right]^T. \tag{3.24}$$

Finally, the group-sparsity-based wideband DOA estimation extension of "Narrowband 1" is formulated as

$$\begin{aligned} \min_{\widetilde{\mathbf{r}}_g^{\circ}} \quad & \left\|\hat{\mathbf{r}}_g^{\circ}\right\|_1 \\ \text{subject to} \quad & \left\|\widetilde{\mathbf{z}} - \widetilde{\mathbf{B}}_g^{\circ}\widetilde{\mathbf{r}}_g^{\circ}\right\|_2 \leq \varepsilon, \\ & \widetilde{r}_{g,k_g}^{\circ} \geq 0, \quad 0 \leq k_g \leq (K_g+1)J-1, \end{aligned} \tag{3.25}$$

where $\widetilde{r}_{g,k_g}^{\circ}$ represents the k_gth element of the column vector $\widetilde{\mathbf{r}}_g^{\circ}$, and the nonzero entries in the first K_g elements of the column vector $\hat{\mathbf{r}}_g^{\circ}$ are the corresponding wideband DOA estimates over the K_g search grids. Along the same lines, the other two narrowband DOA estimation algorithms can be extended to the wideband case in a straightforward manner. We refer to these three wideband extensions as "Wideband 1," "Wideband 2," and "Wideband 3," respectively.

For existing DOA estimation methods employing both narrowband and wideband signals, an equivalent unit spacing satisfying $d \leq \lambda_{\min}/2$ with $\lambda_{\min}$ being the wavelength corresponding to the highest signal frequency is normally chosen to avoid spatial aliasing. By applying the group-sparsity-based methods, we can increase the spacing d to be larger than $\lambda_{\min}/2$, while still avoiding spatial aliasing [31, 32]. This is because the aliased locations for different frequencies are different and the group-sparsity-based methods will force a common sparse location across all frequencies, corresponding to the true location of an impinging signal. As the resolution of an array improves with increased aperture, this will lead to an improved performance for the wideband array. However, it is expected that when d is larger than some threshold value, the DOA estimation results will degrade, while when $d = \lambda_{\max}/2$, where the largest virtual array aperture can be achieved under the condition of no spatial aliasing only for the minimum frequency of interest, we can still perform effective DOA estimation.

3.2.4 Simulation Results

We consider a 10-element extended coprime array with its sensors at positions given by $[0, 3d, 5d, 6d, 9d, 10d, 12d, 15d, 20d, 25d]$. We assume that $K = 15$ uncorrelated wideband signals are impinging on the array from directions uniformly distributed between −60° and 60°. With the sampling frequency f_s twice the highest frequency of interest, the normalized frequencies of the impinging signals range from 0.5π to π rads, and the unit spacing $d = \lambda_{\min}/2$ with $\lambda_{\min} = 2c/f_s$. The total number of data samples is $P = 128{,}000$, and a 64-point DFT ($L = 64$) is applied. As a result, the number of data blocks is $Q = 2000$. The other parameters are $K_g = 3601$ and $\Phi_l = \{17, 18, \cdots, 31\}$ with $J = 15$.

For the first set of simulations, the input SNR is 0 dB and the allowable error bound ε is chosen to provide the best result for each method through trial and error. The DOA estimates for the single frequency ($l = 31$) obtained by the narrowband method in (3.17) and the corresponding wideband result based on the solution in (3.25) are shown in Fig. 3.1, where the dotted lines represent the true incident angles, while the solid lines represent the estimates. It is clear that all sources have been distinguished successfully. Similar results are obtained by the low-complexity solutions in (3.19) and (3.20), and their wideband extensions.

To compare the performance with respect to input SNR, the root mean square error (RMSE) results are shown in Fig. 3.2, obtained by averaging over 500 simulation runs. For each case, the three different solutions provide similar performance. However, the group-sparsity-based wideband extensions significantly outperform their narrowband counterparts.

Next, we provide an example where the single-frequency method fails while the group-sparsity-based algorithm still works. This would highlight the benefit of incorporating information from

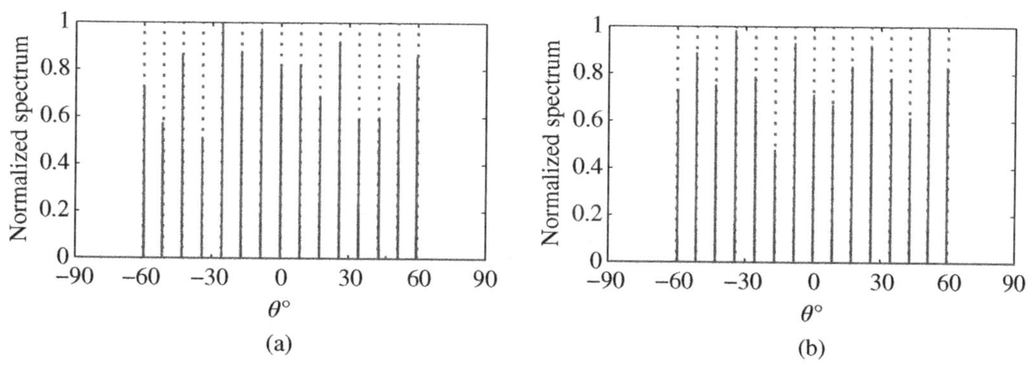

Figure 3.1 DOA estimation results obtained by the narrowband method and the group-sparsity based wideband method. (a) DOA estimation result for single frequency by the narrowband method in (3.17). (b) DOA estimation result obtained by the group-sparsity-based solution in (3.25). Source: Shen et al. [22]/IEEE.

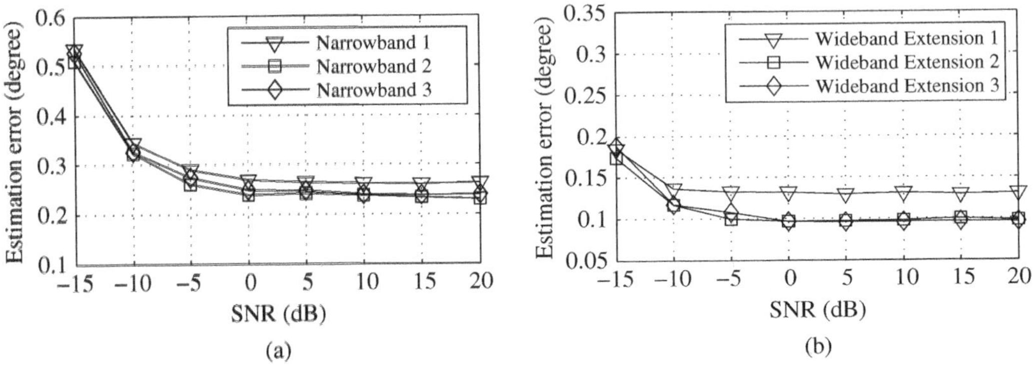

Figure 3.2 RMSEs versus input SNR obtained for a single frequency by "Narrowband 1," "Narrowband 2," and "Narrowband 3" and their wideband extensions "Wideband 1," "Wideband 2," and "Wideband 3." (a) RMSEs of different methods for single frequency. (b) RMSEs of different wideband extensions. Source: Shen et al. [22]/IEEE.

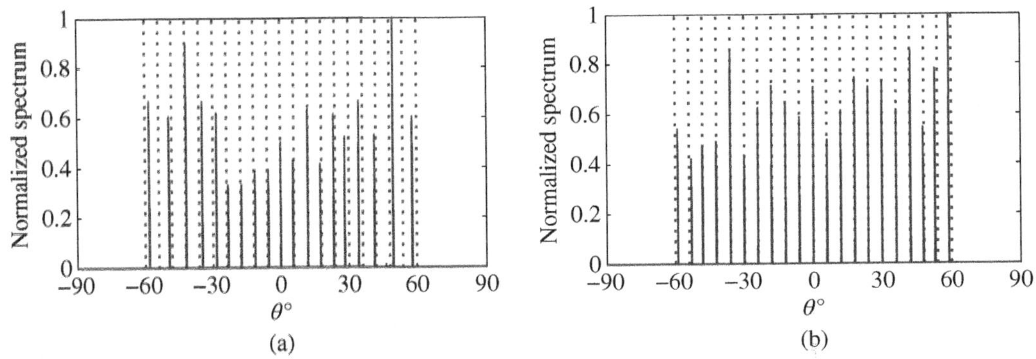

Figure 3.3 DOA estimation results obtained for a single frequency and multiple frequencies through group sparsity. (a) Single-frequency estimation result. (b) Group-sparsity estimation results. Source: Shen et al. [22]/IEEE.

multiple frequencies into the estimation process directly instead of through post-processing. The setting is the same as before except that now there are 21 sources uniformly distributed between −60° and 60°. Due to the increased signal number and reduced separation between DOAs of adjacent signals, the estimation task is tougher than the previous setting and, therefore, can illustrate performance differences more effectively. The results are shown in Fig. 3.3. Clearly, in this case, a simple average of the DOA estimates for different frequencies would not provide satisfactory performance.

As mentioned earlier, by increasing the unit spacing d beyond $\lambda_{\min}/2$, with $d = d_f \lambda_{\min}/2 = d_f c/f_s$ and d_f being the scaling factor, a better performance by the group-sparsity-based solution can be achieved without suffering from the spatial aliasing problem. To show this, the RMSE results versus d_f with input SNR fixed at 0 dB are depicted in Fig. 3.4. Given the frequency range from 0.5π to π, we have $\lambda_{\max} = 2\lambda_{\min}$, and $d = \lambda_{\max}/2$ corresponds to $d_f = 2$. As verified by Fig. 3.4, $d_f = 2$ still provides good performance since there is no aliasing at the lowest frequency of interest in this case. However, if we keep increasing d further, at some point, the gain due to the larger aperture size will be offset by the loss due to worsening aliasing problem. As shown in Fig. 3.4, $d_f = 2$, i.e., $d = \lambda_{\max}/2$, can be a good choice in practice.

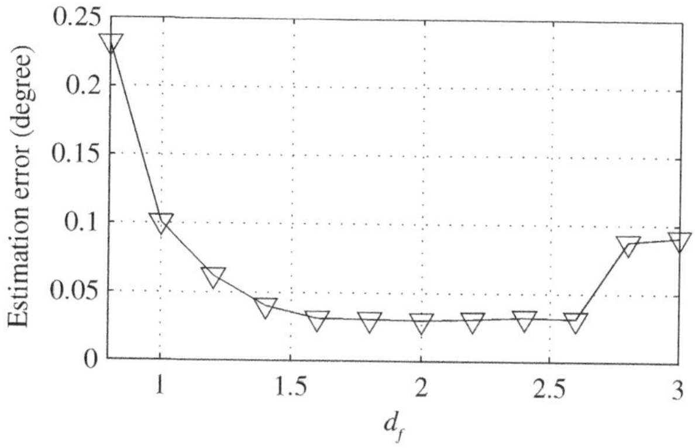

Figure 3.4 RMSEs versus d_f. Source: Shen et al. [22]/IEEE.

3.3 Multi-frequency DOA Estimation

Instead of availing the entire signal bandwidth, it is possible to achieve performance improvements over monochromatic operation by using the sparse array at two or more select frequencies in the signal bandwidth. Array operation at frequencies other than the design frequency (used to set the array unit spacing) generates additional virtual elements through coarray dilations [33]. The multi-frequency coarray is the union of these dilated coarrays corresponding to all operating frequencies, and thus, it is much richer (more densely populated) compared to the single-frequency coarray. As such, it offers an increase in the maximum number of resolvable sources over the single-frequency operation. Even dual-frequency operation can offer a reasonable increase in the DOFs for DOA estimation beyond those afforded by a single-frequency operation [34]. The chosen frequencies need not be uniformly spaced over the signal bandwidth. In fact, the frequencies should be carefully selected to reap the maximum benefits in terms of the richness of the combined coarray. In this section, we focus on sparsity-based solutions that exploit the multi-frequency coarray for DOA estimation. Both cases of proportional and nonproportional sources' spectra are considered. The former accommodates a single measurement vector model, which corresponds to a virtual array whose element positions are given by the multi-frequency coarray. Sparse reconstruction is then applied for estimating the directions of signal arrivals. Sources with nonproportional spectra admit the group-sparsity assumption as the sources maintain their DOAs even if their power varies with frequency [35], suggesting the application of a group sparse reconstruction.

3.3.1 Multi-frequency Signal Model

Let $\omega_{\text{ref}} \in \{\omega_{l_j}, l_j \in \Phi_l\}$ denote the reference frequency whose corresponding one-half wavelength is set as the unit spacing d of the physical sparse array. We consider operating the physical array at a set of $I(< J)$ different frequencies, $\Omega = \{\omega_{l_{j_i}} = \alpha_i \omega_{\text{ref}}, l_{j_i} \in \Phi_l, i = 0, 1, \dots, I-1\}$, where α_i is a constant. Note that it is not required for ω_{ref} to be one of the operational frequencies in Ω. If it is included in Ω, then the corresponding α_i assumes a unit value. Using (3.11), the received array signal vector at $\omega_{l_{j_i}}$ can be expressed as

$$\mathbf{x}[l_{j_i}, q] = \mathbf{A}(l_{j_i}, \theta)\mathbf{s}[l_{j_i}, q] + \mathbf{n}[l_{j_i}, q], \tag{3.26}$$

where the (m, k)th element of the $M \times K$ array manifold matrix $\mathbf{A}(l_{j_i}, \theta)$ is given by

$$[\mathbf{A}(l_{j_i}, \theta)]_{(m,k)} = \exp(-j\omega_{l_{j_i}} i_m d \sin(\theta_k)/c),$$
$$= \exp(-j\omega_{\text{ref}}\alpha_i i_m d \sin(\theta_k)/c). \tag{3.27}$$

We observe from (3.27) that the array manifold matrix at $\omega_{l_{j_i}}$ is equivalent to the array manifold matrix of a scaled version of the physical sparse array, with the position of the mth sensor in this scaled array given by $\alpha_i i_m d$. This, in turn, results in the coarray at $\omega_{l_{j_i}}$ to be a scaled version of the coarray at the reference frequency ω_{ref}, with α_i being the scaling factor [33]. If $\omega_{l_{j_i}}$ is higher than ω_{ref}, the coarray at $\omega_{l_{j_i}}$ is an expanded version of the one at ω_{ref}. On the other hand, for $\omega_{l_{j_i}}$ lower than ω_{ref}, the equivalent coarray at $\omega_{l_{j_i}}$ is a contracted version of that at the reference frequency. For illustration, we consider an extended coprime array with a total of 12 elements with sensor positions $[0d, 3d, 6d, 7d, 9d, 12d, 14d, 15d, 18d, 21d, 28d, 35d]$, where $d = \lambda_{\text{ref}}/2$. The corresponding coarray at ω_{ref} is shown in Fig. 3.5a. Operating the array at frequency $(8/7)\omega_{\text{ref}}$, which is larger than ω_{ref}, results in stretching the difference coarray of Fig. 3.5a, as shown in Fig. 3.5b. On the other hand, if the array is operated at a smaller frequency, say $6/7\omega_{\text{ref}}$, the difference coarray undergoes contraction as depicted in Fig. 3.5c.

For multi-frequency processing, we employ the normalized autocorrelation matrices at each of the I operational frequencies for DOA estimation [35, 36]. The (m, n)th element of the normalized autocorrelation matrix $\overline{\mathbf{R}}_{\mathbf{xx}}[l_{j_i}]$ is defined as

$$[\overline{\mathbf{R}}_{\mathbf{xx}}[l_{j_i}]]_{m,n} = \frac{[\mathbf{R}_{\mathbf{xx}}[l_{j_i}]]_{m,n}}{\frac{1}{M}\mathbb{E}\{\mathbf{x}[l_{j_i}]^H\mathbf{x}[l_{j_i}]\}},$$
$$\approx \frac{\widehat{\mathbf{R}}_{\mathbf{xx}}[l_{j_i}]}{\frac{1}{MQ}\sum_{q=0}^{Q-1}\mathbf{x}[l_{j_i}, q]^H\mathbf{x}[l_{j_i}, q]}. \tag{3.28}$$

It can be readily shown that in the normalized autocorrelation matrix $\overline{\mathbf{R}}_{\mathbf{xx}}[l_{j_i}]$, the source and noise powers are replaced by their normalized values, which can be expressed as

$$\overline{\sigma}_k^2[l_{j_i}] = \frac{\sigma_k^2[l_{j_i}]}{\sum_{k=1}^{K}\sigma_k^2[l_{j_i}] + \sigma_n^2[l_{j_i}]}, \quad \overline{\sigma}_n^2[l_{j_i}] = \frac{\sigma_n^2[l_{j_i}]}{\sum_{k=1}^{K}\sigma_k^2[l_{j_i}] + \sigma_n^2[l_{j_i}]}. \tag{3.29}$$

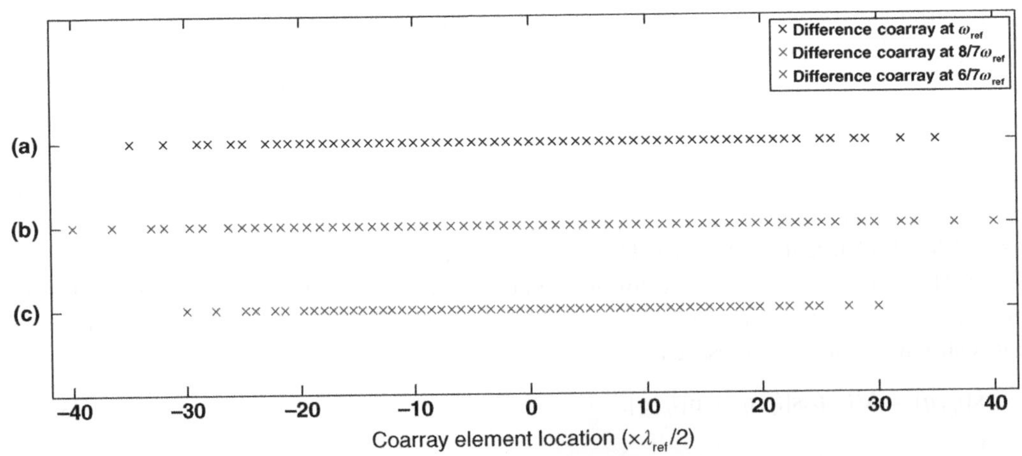

Figure 3.5 Difference coarray at various operational frequencies. Source: BouDaher et al. [35]/Springer Nature/CCBY 4.0/Public domain.

Similar to (3.14), we can express $\overline{\mathbf{R}}_{\mathbf{xx}}[l_{j_i}]$ in vectorized form as

$$\overline{\mathbf{z}}[l_{j_i}] = \text{vec}(\overline{\mathbf{R}}_{\mathbf{xx}}[l_{j_i}]) = \widetilde{\mathbf{A}}[l_{j_i}]\overline{\mathbf{s}}[l_{j_i}] + \overline{\sigma}_n^2[l_{j_i}]\widetilde{\mathbf{i}}_M, \tag{3.30}$$

where $\overline{\mathbf{s}}[l] = \left[\overline{\sigma}_1^2[l], \dots, \overline{\sigma}_K^2[l]\right]^T$. The vector $\overline{\mathbf{z}}[l_{j_i}]$ emulates observations at the coarray corresponding to $\omega_{l_{j_i}}$. The coarray measurement vectors, $\omega_{l_{j_i}}, i = 0, 1, \dots, I-1$, can be combined to establish an appropriate multi-frequency linear model that permits DOA estimation within the sparse reconstruction framework. We distinguish two cases of normalized source spectra. In the first case, we assume the normalized power of each source to be independent of frequency, i.e.,

$$\overline{\sigma}_k^2[l_{j_i}] = \overline{\sigma}_k^2, \ \forall\, k, i, \tag{3.31}$$

whereas the normalized source powers are allowed to vary with frequency in the second case.

3.3.2 DOA Estimation Under Proportional Spectra

Similar to Section 3.2.2, we create a search grid of K_g potential incident angles $\theta_{g,1}, \dots, \theta_{g,K_g}$. Then, (3.30) can be rewritten as

$$\overline{\mathbf{z}}[l_{j_i}] = \widetilde{\mathbf{A}}_g[l_{j_i}]\overline{\mathbf{s}}_g[l_{j_i}] + \overline{\sigma}_n^2[l_{j_i}]\widetilde{\mathbf{i}}_M, \tag{3.32}$$

where the columns of $\widetilde{\mathbf{A}}_g[l_{j_i}]$ are the coarray steering vectors at $\omega_{l_{j_i}}$ corresponding to the defined angles in the grid and $\overline{\mathbf{s}}_g[l_{j_i}]$ is a $K_g \times 1$ column vector of potential normalized source signal powers for the corresponding grid angles. The vector $\overline{\mathbf{s}}_g[l]$ is a K-sparse vector whose support corresponds to the source directions with the nonzero values equal to the normalized source powers. For a high SNR, a sufficient condition for (3.31) to hold is that the sources must have proportional spectra at the employed frequencies [35, 36]. That is,

$$\frac{\overline{\sigma}_{k_1}^2[l_{j_i}]}{\overline{\sigma}_{k_2}^2[l_{j_i}]} = \beta_{k_1,k_2}, \ k_1, k_2 \in \{1, 2, \dots, K\}, \ i \in \{0, 1, \dots, I-1\}. \tag{3.33}$$

This case is applicable, e.g., when the K sources have binary phase shift keying (BPSK) or chirp-like signals. Under proportional source spectra, the source vector $\overline{\mathbf{s}}[l_{j_i}]$ is no longer a function of $\omega_{l_{j_i}}$, i.e., $\overline{\mathbf{s}}[l_{j_i}] = \overline{\mathbf{s}} = [\overline{\sigma}_1^2, \overline{\sigma}_2^2, \dots, \overline{\sigma}_K^2]^{\mathrm{T}}$ for all i, which implies that $\overline{\mathbf{s}}_g[l_{j_i}] = \overline{\mathbf{s}}_g$ for all i. As such, the measurement vectors $\overline{\mathbf{z}}[l_{j_i}]$ at the I operating frequencies can be stacked to form a single $I(M^2) \times 1$ vector

$$\acute{\mathbf{z}} = \widetilde{\mathbf{C}}_g\overline{\mathbf{s}}_g + \acute{\mathbf{i}}, \tag{3.34}$$

where $\acute{\mathbf{z}} = [\overline{\mathbf{z}}[l_{j_0}]^{\mathrm{T}}, \dots, \overline{\mathbf{z}}[l_{j_{I-1}}]^{\mathrm{T}}]^{\mathrm{T}}$, $\acute{\mathbf{i}} = [\overline{\sigma}_n^2[l_{j_0}]\widetilde{\mathbf{i}}_M^{\mathrm{T}}, \dots, \overline{\sigma}_n^2[l_{j_{I-1}}]\widetilde{\mathbf{i}}_M^{\mathrm{T}}]^{\mathrm{T}}$, and the dictionary $\widetilde{\mathbf{C}}_g = [\widetilde{\mathbf{A}}_g[l_{j_0}^{\mathrm{T}}, \dots, \widetilde{\mathbf{A}}_g[l_{j_{I-1}}]^{\mathrm{T}}]^{\mathrm{T}}$. The measurement vector is equivalent to that of a virtual array, whose element positions are given by the combined coarrays at the I frequencies. It is noted that in the case of overlapping points in the I coarrays, either one measurement per coarray location or an averaged value of the multiple measurements that correspond to the same coarray location can be used. This results in a reduction in the dimensionality of $\acute{\mathbf{z}}$, along the same lines as discussed in Section 3.2.2. As such, the length of $\acute{\mathbf{z}}$ becomes equal to the total number of unique lags in the combined coarray. The dictionary matrix and the noise vector would be changed accordingly.

It should be noted that not all physical sensors must operate at all I frequencies. Situations may arise due to cost and hardware restrictions that only a few sensors can accommodate a diverse set of frequencies. The overall coarray is still the union of coarrays at the individual frequencies.

However, the coarray at each frequency may no longer be a scaled version of the coarray at the reference frequency.

Given the model in (3.34), DOA estimation can proceed in terms of sparse reconstruction by solving the following constrained minimization problem, similar to the single-frequency problem of (3.17),

$$\begin{aligned} \min \quad & \left\| \bar{\mathbf{s}}_g \right\|_1 \\ \text{subject to} \quad & \left\| \acute{\mathbf{z}} - \tilde{\mathbf{C}}_g \bar{\mathbf{s}}_g \right\|_2 \leq \varepsilon, \\ & \bar{s}_{g,k_g} \geq 0, \quad 0 \leq k_g \leq K_g - 1, \end{aligned} \tag{3.35}$$

where $\bar{s}_{g,k_g}$ represents the k_gth entry in the column vector $\bar{\mathbf{s}}_g$. The maximum number of resolvable sources using the method in (3.35) is equal to the number of unique positive lags in the combined coarray corresponding to the multiple frequencies.

3.3.3 DOA Estimation Under Nonproportional Spectra

When the source powers vary with frequency, the single measurement vector model of (3.35) is no longer applicable. However, the K sources have the same directions $[\theta_1, \theta_2, \dots, \theta_K]$ regardless of their power distribution with frequency. As such, the vectors $\bar{\mathbf{s}}_g[l_{j_i}], i = 0, \dots, I-1$, in (3.32) have a common support, which suggests the application of a group sparse reconstruction similar to the wideband case in Section 3.2.3. We, therefore, estimate the DOAs based on group sparsity for the nonproportional spectra case.

The received signal vectors $\bar{\mathbf{s}}_g[l_{j_i}], i = 0, \dots, I-1$, in (3.32) corresponding to the I frequencies are stacked to form a long vector,

$$\acute{\mathbf{z}} = \tilde{\mathbf{B}}_g \bar{\mathbf{r}}_g + \acute{\mathbf{i}}, \tag{3.36}$$

where the dictionary $\tilde{\mathbf{B}}_g = \text{blkdiag}\{\tilde{\mathbf{A}}_g[l_{j_0}], \dots, \tilde{\mathbf{A}}_g[l_{j_{I-1}}]\}$ and $\bar{\mathbf{r}}_g = \text{vec}(\mathbf{R}_g) = \text{vec}([\bar{\mathbf{s}}_g[l_{j_0}], \dots, \bar{\mathbf{s}}_g[l_{j_{I-1}}]])$. The vector $\bar{\mathbf{r}}_g$ is a group sparse vector where each group consists of the source powers corresponding to a specific direction at all operating frequencies. Similar to (3.25), the group sparse solution is obtained by solving the following optimization problem

$$\begin{aligned} \min_{\hat{\mathbf{r}}_g} \quad & \left\| \hat{\mathbf{r}}_g \right\|_1 \\ \text{subject to} \quad & \left\| \bar{\mathbf{z}} - \tilde{\mathbf{B}}_g \bar{\mathbf{r}}_g \right\|_2 \leq \varepsilon, \\ & \bar{r}_{g,k_g} \geq 0, \quad 0 \leq k_g \leq (K_g)I - 1, \end{aligned} \tag{3.37}$$

where $\hat{\mathbf{r}}_g = \left[\|\mathbf{r}_{g,0}\|_2, \dots, \|\mathbf{r}_{g,K_g}\|_2 \right]^{\mathrm{T}}$ and the row vector $\mathbf{r}_{g,k_g}, 0 \leq k_g \leq K_g$, denotes the k_gth row of the matrix $\mathbf{R}_g$. It is noted that the formulation in (3.37) results in a smaller number of achievable DOFs compared to the case where the sources have proportional spectra. The maximum number of resolvable sources is now limited by the number of observations or unique lags at each frequency [10].

3.3.4 Simulation Results

In this section, we present the sparsity-based DOA estimation results for multi-frequency coprime arrays and provide performance comparison with the single-frequency sparsity-based approach. We consider both cases of proportional and nonproportional source spectra. For all of the examples in this section, an extended coprime array configuration with six physical elements is considered. The six sensor positions are given by $[0, 2d, 3d, 4d, 6d, 9d]$ with $d = \lambda_{\text{ref}}/2$. The corresponding coarray consists of 17 virtual elements. The coarray aperture extends from $-9d$ to $9d$ with two holes at $\pm 8d$.

In the first example, sparse reconstruction is applied under dual-frequency operation. The physical array is operated at $I = 2$ frequencies, ω_{ref} and $(8/9)\omega_{\text{ref}}$. Sources with proportional spectra are assumed, and thus, the single measurement vector formulation of Section 3.3.2 can be used. The combined dual-frequency coarray, shown in Fig. 3.6a, has a total of 33 unique lags, which makes it capable of resolving up to 16 sources. However, this number is not achievable because of the high mutual coherence of the dictionary. Since some of the virtual sensors in the combined coarray are closely separated, leading to highly correlated observations, deterioration in performance is observed if the number of sources is increased beyond 11. We, therefore, consider 11 BPSK sources with proportional spectra, which are uniformly spaced between $-75°$ and $75°$. The SNR is set to 10 dB for the sources at the two frequencies, and the number of snapshots $Q = 1000$ at each frequency. The search space from $-90°$ and $90°$ is divided into $K_g = 181$ grid points. Figure 3.7a shows the normalized spectrum obtained using this method. It is evident that all sources are correctly resolved. The RMSE in this example is equal to 0.84°. A different choice of the two operational frequencies may reduce the mutual coherence, thereby permitting a larger number of sources to be estimated. For illustration, the second frequency is now set to $2\omega_{\text{ref}}$. By choosing a frequency which is an integer multiple of ω_{ref}, the combined coarray positions are guaranteed to be integer multiples of d. As a result, the minimum separation between two consecutive coarray elements is equal to d. The combined coarray, shown in Fig. 3.6b, has 13 unique positive lags, which means that the maximum number of resolvable sources is equal to 13. This is tested by considering 13 uniformly spaced sources between $-75°$ and $-75°$. The rest of the parameters are kept the same as in the previous example. Figure 3.7b shows the normalized spectrum using the dual-frequency sparse reconstruction method. It is evident that all the sources are correctly estimated. The RMSE is found to be 0.26° in this case. We note that single-frequency operation will fail in both considered scenarios, since the maximum number of resolvable sources using single-frequency sparse reconstruction is equal to the number of unique positive lags in the coarray which is eight in this case. This confirms the increase in the number of resolvable sources by using group sparse reconstruction compared to the single-frequency sparse reconstruction.

The next example examines the case when sources have nonproportional spectra. In this case, group sparse reconstruction is applied. The two operational frequencies are selected to be ω_{ref} and $2\omega_{\text{ref}}$. Eight sources with nonproportional spectra are considered. The SNR of all the sources at the first frequency is set to 10 dB. At the second frequency, the SNR of each source is a realization of a uniformly distributed random variable between 5 and 15 dB. This ensures that the sources have nonproportional spectra. The noise variance is set to unity at the two frequencies, and $Q = 1000$

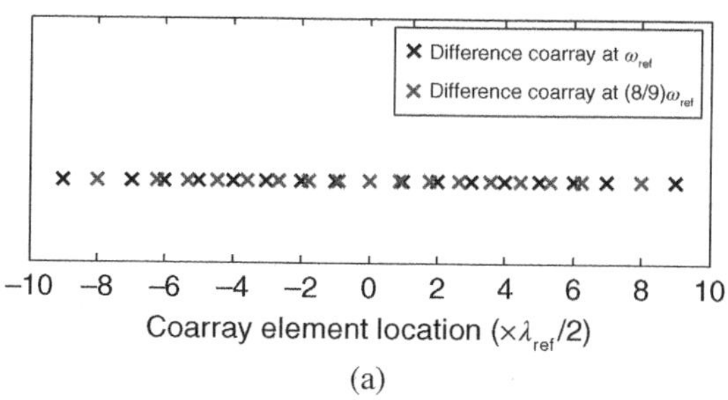

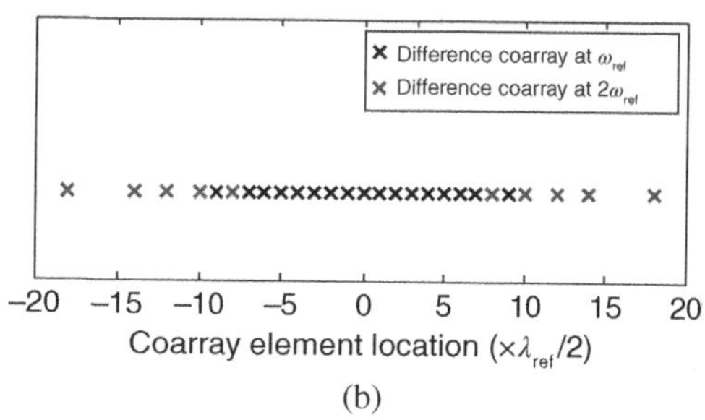

Figure 3.6 Dual-frequency coarray. (a) ω_{ref} and $8/9\omega_{\text{ref}}$. (b) ω_{ref} and $2\omega_{\text{ref}}$. Source: BouDaher et al. [35]/Springer Nature/CCBY 4.0/Public domain.

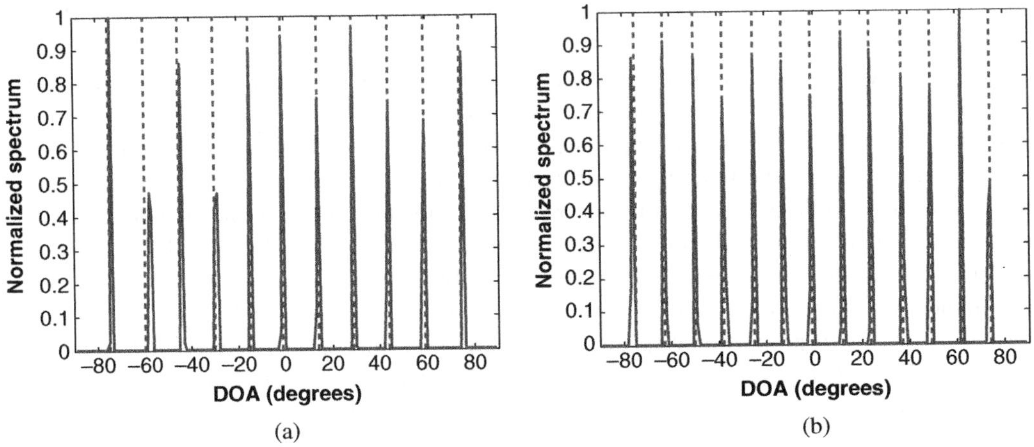

Figure 3.7 Dual-frequency sparse reconstructions results under proportional source spectra. (a) $K = 11$ sources. (b) $K = 13$ sources. Source: BouDaher et al. [35]/Springer Nature /CCBY 4.0/Public domain.

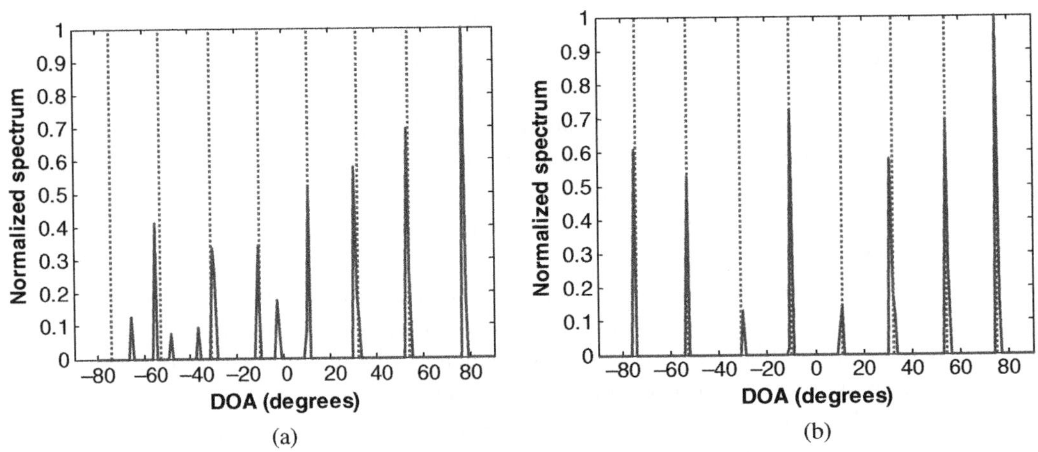

Figure 3.8 Dual-frequency sparse reconstructions results for nonproportional source spectra; $K = 8$ sources. (a) Sparse reconstruction. (b) Group sparse reconstruction. Source: BouDaher et al. [35]/Springer Nature/CCBY 4.0/Public domain.

snapshots per frequency are used. Figure 3.8 shows the normalized spectrum obtained using the formulation in Section 3.3.2 which mistakenly assumes proportional source spectra. Consequently, this method is expected to fail as evident in the spectrum of Fig. 3.8a. One of the sources is not resolved and several spurious peaks appear in the spectrum. The DOA estimation is next repeated using group sparse reconstruction which was discussed in Section 3.3.3. This method does not require the sources to have proportional spectra. The mean of the recovered spectra at the two employed frequencies is computed and shown in Fig. 3.8b. It can be seen that group sparse reconstruction is successful in localizing the DOAs of all the sources. The RMSE is found to be 0.6° in this case.

The final example examines the case where the source signals have overlapping spectra but do not share the same bandwidth. Group sparse reconstruction can still be used to perform DOA estimation. A total of 16 sources with nonproportional spectra are considered to be uniformly

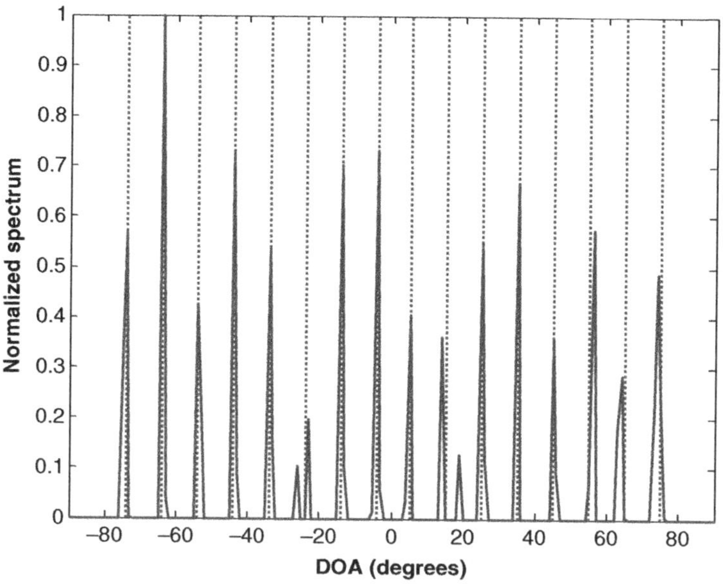

Figure 3.9 Multi-frequency group sparse reconstructions results with nonproportional source spectra. Source: BouDaher et al. [35]/Springer Nature/CCBY 4.0/Public domain.

spaced between −75° and 75°. Twenty uniformly spaced frequencies between ω_{ref} and $2\omega_{\text{ref}}$ are employed. The number of snapshots at each frequency is set to 1000. The SNR of each source at each frequency is chosen randomly between −5 and 5 dB, and then 30% of the source powers at the employed frequencies are randomly set to zero. Figure 3.9 shows the normalized spectrum using group sparse reconstruction. It is evident that all sources are correctly estimated. Some spurious peaks are present in the spectrum, and an increase in the estimates bias is obtained. The RMSE is found to be 0.61°.

3.4 Wideband SBL for Beamforming

Design of sparse arrays for wideband signals using convex optimization techniques is given in [37]. SBL [38, 39] is a sparse reconstruction technique that uses Bayesian framework to find sparse solution to an underdetermined linear problem. SBL offers many advantages over other array processing methods. These include: (i) Relatively smaller number of unknown parameters, since SBL estimates the covariance of the signal amplitudes rather than the amplitudes themselves; (ii) Computationally faster compared to convex optimization as iterative updates are employed for parameter estimation; and (iii) Superior performance with fewer snapshots over other array processing algorithms, since a sample correlation matrix computation is not performed. In this section, we focus on SBL to directly process observations from sparse linear arrays without constructing a coarray observation vector. We demonstrate the utility of sparse arrays in combination with wideband SBL by processing Noise Correlation 2009 (NC09) experiment data [40] with line arrays; simulations with planar arrays are available in [9].

We note that SBL as used here is a general algorithm and there is no special treatment for sparse arrays. The sparse array structure is inserted into the algorithm via the dictionary matrix **A**. Further, grid-less methods can be related to the SBL objective function [11], which for a single-frequency has been used for irregular arrays [41] and coprime arrays [42]. Furthermore, while we concentrate on SBL in this section, a variety of other sparse reconstruction methods, including basis pursuit [43] and matching pursuit [44], have been used in various physical applications for parameter estimation [10, 39, 45, 46] including DOA estimation [47–49].

3.4.1 SBL for Beamforming at a Single Frequency

We first consider operation at a single frequency, say the lth frequency, to describe the SBL algorithm [31, 38, 39, 50]. Similar to Section 3.2.2, we consider a search grid of K_g potential incident angles $\theta_{g_1}, \dots, \theta_{g_{K_g}}$. Using the Q snapshots at the lth frequency, we can express the multi-snapshot array observation matrix $\mathbf{X}_l$ as

$$\mathbf{X}_l = \mathbf{A}(l, \theta_g)\mathbf{S}_{g,l} + \mathbf{N}_l, \tag{3.38}$$

where $\mathbf{X}_l = [\mathbf{x}[l,0], \dots, \mathbf{x}[l, Q-1]], \mathbf{N}_l = [\mathbf{n}[l,0], \dots, \mathbf{n}[l, Q-1]]$ is the noise matrix, and $\mathbf{S}_{g,l} = [\mathbf{s}_g[l,0], \dots, \mathbf{s}_g[l, Q-1]]$ is the matrix of sparse source amplitudes with all columns sharing the same sparsity profile. The observations are assumed to be independent across the Q snapshots, which results in multi-snapshot likelihood function assuming the form

$$p(\mathbf{X}_l|\mathbf{S}_{g,l}) = \prod_{q=0}^{Q-1} p(\mathbf{x}[l,q]|\mathbf{s}_g[l,q]) = \prod_{q=0}^{Q-1} \mathcal{CN}(\mathbf{x}[l,q]; \mathbf{A}(l,\theta_g)\mathbf{s}_g[l,q], \sigma_n^2[l]\mathbf{I}_M), \tag{3.39}$$

where $\mathcal{CN}(\cdot)$ denotes a complex Gaussian distribution.

Prior: In SBL, the sparse source amplitudes are treated as zero-mean complex Gaussian random vectors with diagonal correlation matrix, $\Gamma_l = \text{diag}(\sigma_1^2[l], \dots \sigma_K^2[l]) = \text{diag}(\gamma_l)$. The prior model is given by

$$p(\mathbf{S}_{g,l}) = \prod_{q=0}^{Q-1} p(\mathbf{s}_g[l,q]) = \prod_{q=0}^{Q-1} \mathcal{CN}(\mathbf{s}_g[l,q]; \mathbf{0}, \Gamma_l). \tag{3.40}$$

Evidence: From the Gaussian prior in (3.40) and the likelihood in (3.39), the evidence $p(\mathbf{X}_l)$ is Gaussian and given by

$$p(\mathbf{X}_l) = \int p(\mathbf{S}_{g,l})p(\mathbf{X}_l|\mathbf{S}_{g,l})d\mathbf{S}_{g,l} = \prod_{q=0}^{Q-1} \mathcal{CN}(\mathbf{x}[l,q]; \mathbf{0}, \mathbf{R}_{\mathbf{xx}}[l]), \tag{3.41}$$

where $\mathbf{R}_{\mathbf{xx}}[l]$, defined in (3.12), can be expressed in matrix form for the considered grid as $\sigma_n^2[l]\mathbf{I}_M + \mathbf{A}(l,\theta_g)\Gamma_l\mathbf{A}^H(l,\theta_g)$. The SBL approach is to estimate the diagonal entries of Γ_l by maximizing the (log) evidence

$$\begin{aligned}(\hat{\sigma}_1^2[l], \dots, \hat{\sigma}_{K_g}^2[l]) &= \arg\max_{\gamma_l} \log p(\mathbf{X}_l) \\ &= \arg\max_{\gamma_l} \left\{ -\sum_{q=0}^{Q-1} (\mathbf{x}^H[l,q]\mathbf{R}_{\mathbf{xx}}{}^{-1}[l]\mathbf{x}[l,q] + \log|\mathbf{R}_{\mathbf{xx}}[l]|) \right\},\end{aligned} \tag{3.42}$$

where $|\cdot|$ denotes the determinant of its matrix argument. The objective function in (3.42) is non-convex with respect to the unknown γ_l. Differentiating the above objective function and equating

Table 3.1 SBL pseudocode: input consists of data $\mathbf{X}_l$, dictionary $\mathbf{A}(l, \boldsymbol{\theta}_g)$, and noise variance $\sigma_n^2[l]$.

SBL Algorithm

1. Input: $\mathbf{X}_l, \mathbf{A}(l, \boldsymbol{\theta}_g), \sigma_n^2[l]$ (optional)

2. Parameters: $\epsilon = 10^{-3}, N_t = 500$
3. Initialization: $\sigma_k^{2\,\text{old}} = \mathbf{a}^H(l, \theta_k)\mathbf{R}_{\mathbf{xx}}^{-1}[l]\mathbf{a}(l, \theta_k), \forall k$
4. **for** $i = 1$ to N_t
5. Compute: $\mathbf{R}_{\mathbf{xx}}[l] = \sigma_n^2[l]\mathbf{I}_M + \mathbf{A}(l, \boldsymbol{\theta}_g)\Gamma_l^{\text{old}}\mathbf{A}^H(l, \boldsymbol{\theta}_g)$
6. $\sigma_k^{2\,\text{new}}[l]$ update $\forall k$ using (3.43)
7. $\sigma_n^2[l]$ update using (3.44)
8. If $\frac{||\boldsymbol{\gamma}_l^{\text{new}} - \boldsymbol{\gamma}_l^{\text{old}}||_1}{||\boldsymbol{\gamma}_l^{\text{old}}||_1} < \epsilon$, **break**
9. $\boldsymbol{\gamma}_l^{\text{old}} = \boldsymbol{\gamma}_l^{\text{new}}, \Gamma_l^{\text{old}} = \text{diag}(\boldsymbol{\gamma}_l^{\text{new}})$
10. **end**

11. Output: $\boldsymbol{\gamma}_l, \sigma_n^2[l]$ (optional)

Convergence is controlled by the error threshold, ϵ, and maximum number of iterations N_t.

the derivatives to zero provides the fixed point update rule [38, 39, 50]

$$\begin{aligned}\sigma_k^{2\,\text{new}}[l] &= \sigma_k^{2\,\text{old}}[l]\frac{1}{Q}\frac{||\mathbf{X}^H[l]\mathbf{R}_{\mathbf{xx}}{}^{-1}[l]\mathbf{a}(l, \theta_k)||_2^2}{\mathbf{a}^H(l, \theta_k)\mathbf{R}_{\mathbf{xx}}{}^{-1}[l]\mathbf{a}(l, \theta_k)} \\ &= \sigma_k^{2\,\text{old}}[l]\frac{\text{Tr}[\widehat{\mathbf{R}}_{\mathbf{xx}}[l]\mathbf{R}_{\mathbf{xx}}{}^{-1}[l]\mathbf{a}(l, \theta_k)\mathbf{a}^H(l, \theta_k)\mathbf{R}_{\mathbf{xx}}{}^{-1}[l]]}{\mathbf{a}^H(l, \theta_k)\mathbf{R}_{\mathbf{xx}}{}^{-1}[l]\mathbf{a}(l, \theta_k)},\end{aligned} \tag{3.43}$$

where Tr[$\cdot$] denotes the trace of a matrix and $\widehat{\mathbf{R}}_{\mathbf{xx}}[l]$ is the sample correlation matrix defined in (3.13).

The noise variance, $\sigma_n^2[l]$, is also required to be estimated, if unknown. An estimate for $\sigma_n^2[l]$ can be obtained using the stochastic maximum likelihood method [48, 50] as

$$\hat{\sigma}_n^2[l] = \frac{\text{Tr}[(\mathbf{I}_M - \mathbf{A}_{\mathcal{M}}[l]\mathbf{A}_{\mathcal{M}}^{\dagger}[l])\widehat{\mathbf{R}}_{\mathbf{xx}}[l]]}{M - K}. \tag{3.44}$$

Here, $\mathcal{M}$ is the set of indices where $\boldsymbol{\gamma}_l$ is nonzero, i.e, the cardinality of $\mathcal{M} = K$, $\mathbf{A}_{\mathcal{M}}[l]$ is the matrix $\mathbf{A}(l, \boldsymbol{\theta}_g)$ restricted to columns corresponding to indices in $\mathcal{M}$, and $\mathbf{A}_{\mathcal{M}}^{\dagger}[l]$ denotes the Moore–Penrose pseudo-inverse matrix of $\mathbf{A}_{\mathcal{M}}[l]$. We note that this noise estimate is not valid for more sources than sensors ($K \geq M$). The estimate in (3.44) requires knowledge of the number of sources K, which can be obtained using model order selection methods [51], such as Akaiki Information Criterion and Bayesian Information Criterion.

The pseudocode of the SBL algorithm for beamforming is given in Table 3.1. The parameters ϵ and N_t are the convergence error threshold and number of iterations of the algorithm. The unknown vector $\boldsymbol{\gamma}_l$ is initialized to the un-normalized CBF output.

It has been shown [38] that the estimate of $\boldsymbol{\gamma}_l$ by SBL is sparse. Since the element $\sigma_k^2[l]$ of $\boldsymbol{\gamma}_l$ corresponds to the variance of the potential source at the kth grid point, $\sigma_k^2[l] = 0$ implies that $\mathbf{s}_{g,k}[l, q] = 0\ \forall q$. Hence, a sparse $\boldsymbol{\gamma}_l$ corresponds to a sparse source amplitude vector $\mathbf{s}_g[l, q]$. In other words, the sparsity profile of the source correlation diagonal γ_l corresponds to the source DOAs. Cramér–Rao bound analysis for SBL can be found in the literature [52, 53].

3.4.2 Wideband SBL for Beamforming

The SBL algorithm discussed in Section 3.4.1 was limited to single-frequency observations. We now extend the algorithm to process wideband measurements [31, 54, 55]. Let the observations be available at J frequency bins, $l_j \in \Phi_l, 0 \leq j \leq J-1$, covering the signal bandwidth. Then, we have

$$\mathbf{X}_{l_j} = \mathbf{A}(l_j, \theta_g)\mathbf{S}_{g,l_j} + \mathbf{N}_{l_j}, \tag{3.45}$$

where $\mathbf{X}_{l_j}$, $\mathbf{A}(l_j, \theta_g)$, $\mathbf{S}_{g,l_j}$, and $\mathbf{N}_{l_j}$ are the observations, array manifold, source amplitudes, and the noise matrices at the l_jth frequency, respectively. The observations $\mathbf{X}_{l_0:l_{J-1}} = [\mathbf{X}_{l_0}, \dots, \mathbf{X}_{l_{J-1}}]$ and the noise $\mathbf{N}_{l_0:l_{J-1}} = [\mathbf{N}_{l_0}, \dots, \mathbf{N}_{l_{J-1}}]$ are assumed to be independent across the J frequencies. Following (3.39), the likelihood for the l_jth frequency is given as

$$p(\mathbf{X}_{l_j}|\mathbf{S}_{g,l_j}) = \prod_{q=0}^{Q-1} \mathcal{CN}(\mathbf{x}[l_j, q]; \mathbf{A}(l_j, \theta_g)\mathbf{s}_g[l_j, q], \sigma_n^2[l_j]\mathbf{I}_M), \tag{3.46}$$

where $\sigma_n^2[l_j]$ is the noise variance at the l_jth frequency.

Prior: Assuming Γ_{l_j} to be the diagonal source correlation matrix at the l_jth frequency and source amplitudes to be independent across all frequencies, the priors are given as

$$p(\mathbf{S}_{g,l_j}) = \prod_{q=0}^{Q-1} \mathcal{CN}(\mathbf{s}_g[l_j, q]; \mathbf{0}, \Gamma_{l_j}), \quad j = 0, 1, \dots, J-1. \tag{3.47}$$

Without any further simplifications, this model has $J \times K_g$ unknown parameters, which need to be estimated. Also, since a separate correlation matrix Γ_{l_j} is assumed at each frequency, it fails to capture the shared sparsity that exists among $\mathbf{S}_{g,l_j}$ across the J frequencies. To remedy this situation, we consider the following two models:

A) Flat spectrum – This assumption implies that the source correlation matrix is the same across the J frequencies, i.e.,

$$\Gamma = \Gamma_{l_0} = \Gamma_{l_1} = \dots = \Gamma_{l_{J-1}}. \tag{3.48}$$

This reduces the number of unknown parameters to K_g while enforcing a common sparsity profile across all observations [31].

B) Colored spectrum – The flat spectrum model can be quite restrictive as not all sources are white in practice. To account for the colored source spectrum while enforcing common sparsity, we assume

$$\Gamma_{l_j} = c_j\Gamma, \quad j = 0, 1, \dots, J-1, \tag{3.49}$$

where c_j models the frequency-dependent spectrum while Γ models the shared sparsity profile. Here, we have $K_g + J$ unknown parameters. We note that this model assumes various sources to share the same spectral profile $c_j, j = 0, \dots, J-1$.

Evidence maximization: The multi-frequency evidence is given as

$$\begin{aligned} p(\mathbf{X}_{l_0:l_{J-1}}) &= \prod_{j=0}^{J-1} \int p(\mathbf{S}_{g,l_j})p(\mathbf{X}_{l_j}|\mathbf{S}_{g,l_j})d\mathbf{S}_{g,l_j} \\ &= \prod_{j=0}^{J-1}\prod_{q=0}^{Q-1} \mathcal{CN}(\mathbf{x}[l_j, q]; \mathbf{0}, \mathbf{R}_{\mathbf{xx}}[l_j]). \end{aligned} \tag{3.50}$$

Based on the prior model for the source correlation matrix, we have two possible solutions:

A) Flat spectrum – Estimate Γ only. Proceeding as for the single-frequency case, we maximize the (log) evidence

$$(\hat{\sigma}_1^2, \dots, \hat{\sigma}_{K_g}^2) = \arg\max_{\gamma} \log p(\mathbf{X}_{l_0:l_{J-1}})$$
$$= \arg\max_{\gamma} \left\{ -\sum_{j=0}^{J-1}\sum_{q=0}^{Q-1} \left(\mathbf{x}^H[l_j, q]\mathbf{R}_{\mathbf{xx}}{}^{-1}[l_j]\mathbf{x}[l_j, q] + \log |\mathbf{R}_{\mathbf{xx}}[l_j]|\right) \right\}. \tag{3.51}$$

We obtain the following update rule by differentiation

$$\sigma_k^{2\,\text{new}} = \sigma_k^{2\,\text{old}} \frac{\sum_{j=0}^{J-1} \text{Tr}[\widehat{\mathbf{R}}_{\mathbf{xx}}[l_j]\mathbf{R}_{\mathbf{xx}}{}^{-1}[l_j]\mathbf{a}(l_j, \theta_k)\mathbf{a}^H(l_j, \theta_k)\mathbf{R}_{\mathbf{xx}}{}^{-1}[l_j]]}{\sum_{j=0}^{J-1} \mathbf{a}^H(l_j, \theta_k)\mathbf{R}_{\mathbf{xx}}{}^{-1}[l_j]\mathbf{a}(l_j, \theta_k)}. \tag{3.52}$$

This is a multi-frequency generalization of (3.43) and the latter is easily obtained from (3.52) by setting $J = 1$. We refer to this wideband SBL as SBL MF-A.

B) Colored spectrum – Estimate c_j and Γ. Following the above procedure to estimate γ, we obtain the same update equation as (3.52) where the model correlation matrix $\mathbf{R}_{\mathbf{xx}}[l_j] = \sigma_n^2[l_j]\mathbf{I}_M + c_j\mathbf{A}(l_j, \theta_k)\Gamma\mathbf{A}^H(l_j, \theta_k)$ now includes c_j. The parameters c_j also have to be updated along with Γ. Differentiating the objective function with respect to c_j and equating the derivatives to zero results in the following update rule for c_j,

$$c_j^{\text{new}} = c_j^{\text{old}} \frac{\text{Tr}[\widehat{\mathbf{R}}_{\mathbf{xx}}[l_j]\mathbf{R}_{\mathbf{xx}}{}^{-1}[l_j]\mathbf{A}(l_j, \theta_k)\Gamma\mathbf{A}^H(l_j, \theta_k)\mathbf{R}_{\mathbf{xx}}{}^{-1}[l_j]]}{\text{Tr}[\widehat{\mathbf{R}}_{\mathbf{xx}}[l_j]\mathbf{A}(l_j, \theta_k)\Gamma\mathbf{A}^H(l_j, \theta_k)]}. \tag{3.53}$$

Since c_j and Γ appear in product form, we perform normalization such that $\sum_{j=0}^{J-1} c_j = 1$ in each iteration to avoid scaling ambiguity and numerical convergence issues. We refer to this wideband SBL as SBL MF-B.

Extensions of the pseudocode given in Table 3.1 for the two wideband SBL algorithms for beamforming are straightforward.

We note that an alternative is to consider the frequency-dependent prior model and solve for $\Gamma_{l_j}, j = 0, 1, \dots, J-1$. In this case, the obtained solution is the same as processing observations at each frequency independently. The effective γ can then be obtained as the average of γ_{l_j} across the J frequencies. However, when using sparse arrays, the location of the aliased peaks is frequency dependent. As such, jointly processing the observations at the J frequencies using wideband SBL can significantly suppress spatial aliasing [31], thus outperforming the averaged alternative.

3.4.3 Experimental Results

We use an acoustic data set, corresponding to the NC09 experiment [40], for evaluating the performance of the wideband SBL algorithms. The SBL code used in examples is available from GitHub [56]. The experiment was carried out on the Coronado Bank located south-west of Point Loma, San Diego, California, USA. As shown in Fig. 3.10, acoustic data over 80 – 1000 Hz bandwidth was recorded on equallyconfigured 16-element vertical hydrophone line arrays (VLAs) with VLA2 used for processing. The transect of the 52 m long oceanographic research vessel "New Horizon" is easterly of the VLA bearing line and passes each VLA at port side within ~100 m proximity. The vessel transits at a constant speed of 2.2 m/s. The first hydrophone of VLA2 is 7 m above the seafloor

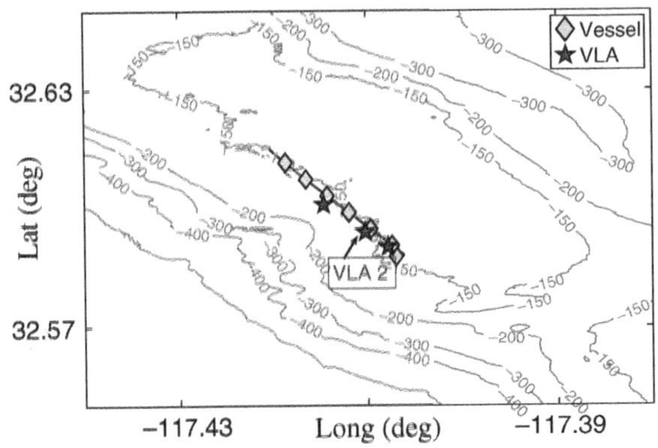

Figure 3.10 NC09 experiment map showing location of VLA2 (star) and trajectory (diamond) of the vessel over 30 minutes. The ship follows a path of approximately constant depth as indicated by the 150 m contour. Source: Nannuru et al. [57]/Acoustical Society of America.

and all 16 hydrophones are used for processing. The element spacing of $d = 1$ m corresponds to half-wavelength of a design frequency of 750 Hz at 1500 m/s sound speed.

The 30-minute data is first downsampled from 25 kHz to a sampling frequency of 5 kHz. A total of 4393 snapshots are computed continuously from the data with a Fast Fourier transform length of 4096 samples (0.8 seconds) and 50% overlap. Next, $Q = 6$ snapshots are designated to segments of ~2.9 seconds duration, resulting in 732 partially overlapping segments for the entire data set. The data in the segments are designated for further processing (e.g., $Q = 6$ snapshots are used to construct a correlation matrix). For CBF, a Kaiser window with $\beta = 4.7$ is applied to each snapshot. The average sound speed over the array is 1492 m/s obtained from a nearby conductivity, temperature, and depth (CTD) cast. The mixed surface layer (1504–1506 m/s) extends to a depth of 22 m, followed by a downward refracting profile.

We process the data using the full 16-element VLA as well as from its subarrays listed in Table 3.2. Specifically, we construct an eight-element extended coprime array, a seven-element nested array, a uniform array "ULA1" comprising the first 8 out of 16 elements of the VLA2, and a second uniform array "ULA2" consisting of every other sensor of VLA2. Both ULA1 and ULA2 have eight sensors each, but ULA2 has twice the aperture relative to ULA1. The aperture of coprime and nested arrays is comparable to that of ULA2. The specific choice of coprime and nested arrays is made to maximize their apertures using the 16-element VLA2. ULA1 was chosen to study the effect of a dense ULA but with limited aperture. The undersampled ULA2 was chosen to demonstrate the ability of wideband SBL (SBL MF1) to combat spatial aliasing. All the arrays have approximately the same number of sensors and hence the same array gain.

We use the wideband SBL discussed in Section 3.4.2 to process the experimental data. The noise estimate (3.44) is obtained with $K = 4$ because there are always at least four relatively strong sources

Table 3.2 The four sub-arrays used for processing NC09 experiment data.

Array	Sensor positions	Length
ULA1	$\{1, 2, 3, 4, 5, 6, 7, 8\}d$	$7d$
ULA2	$\{1, 3, 5, 7, 9, 11, 13, 15\}d$	$14d$
Co-prime	$\{1, 3, 5, 6, 7, 9, 11, 16\}d$	$15d$
Nested	$\{1, 2, 3, 4, 5, 10, 15\}d$	$14d$

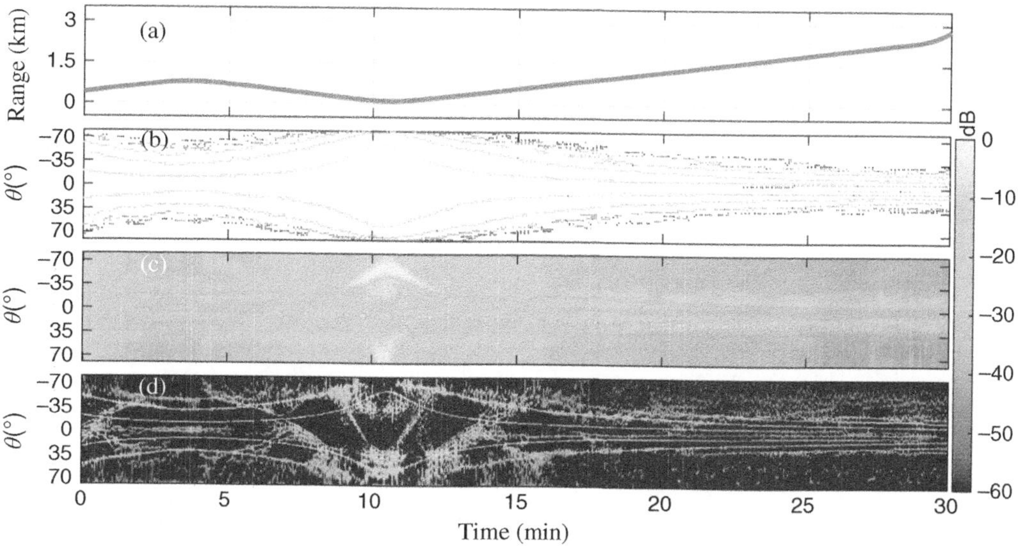

Figure 3.11 NC09: Evolution of DOAs with time as the ship moves. (a) Distance of ship from VLA (b) Simulated DOAs using Bellhop (c) CBF (d) SBL. For CBF, the spectra over 650–750 Hz are averaged, while for SBL 80–1000 Hz is used. Source: Nannuru et al. [57]/Acoustical Society of America.

due to multipath as seen from the Bellhop [58] model in Fig. 3.11b. Though the number of multipaths is changing over time, we use a fixed value of K for uniformity of processing across the whole 30-minute data.

The outputs of conventional beamformer and SBL for the full VLA2 data are shown in Fig. 3.11c and d. The spectrum of CBF is averaged over 650 – 750 Hz. For wideband SBL, 80 – 1000 Hz bandwidth is used. A reduced bandwidth is used for CBF because for lower frequencies, the resolution is low and higher frequencies suffer from aliasing. Increasing bandwidth for CBF will blur the image in Fig. 3.11c. SBL does not suffer from these issues as it simultaneously processes all frequencies giving high resolution and significantly suppresses aliasing [31]. The SBL results closely match the Bellhop model output in Fig. 3.11b.

The later half of the event (from 15 to 30 minutes) has more multipaths as the ship moves away from the array, as seen in Fig. 3.11a. We process data from the full array and the four sub-arrays (see Table 3.2) with the estimated angular source power versus time in Fig. 3.12. All sub-arrays will suffer from reduced array gain compared to the full array as the number of sensors is approximately halved. Only SBL results are shown as the resolution of CBF is poor due to incoherent spectrum averaging.

The coprime, nested, and ULA2 sub-arrays are able to resolve the sources as well as the full VLA2, see Fig. 3.12. This is highlighted around time 23 minutes, indicated by the vertical bar with the corresponding estimated γ versus angle depicted in Fig. 3.13. VLA2 can identify up to seven sources, whereas the coprime and ULA2 sub-arrays identify slightly fewer sources. The performance of ULA1 suffers significantly when there are more than four source directions. This is because ULA1 has the smallest aperture of all sub-arrays and thus lower resolution relative to others. In comparison, ULA2 has twice the aperture of ULA1. Though ULA2 will suffer from aliasing for some of the higher frequencies, these get suppressed due to wideband SBL processing and only the sparse support common across all frequencies is retained.

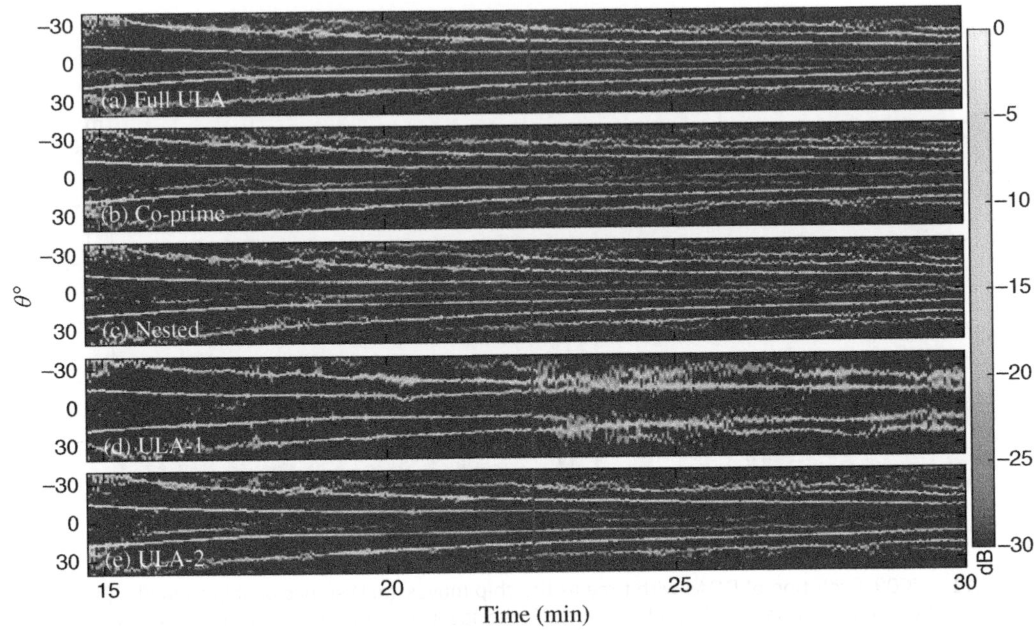

Figure 3.12 NC09: Evolution of DOAs with time (15–30 minutes portion from Fig. 3.11). The angular source powers are estimated using multi-frequency SBL (SBL MF1) for (a) full ULA, (b) coprime, (c) nested, (d) ULA1, and (e) ULA2. The beampatterns corresponding to the verticle line around 22 minutes are shown in Fig. 3.13. Source: Nannuru et al. [57]/Acoustical Society of America.

3.5 Suggested Further Reading

Sections 3.2–3.4 provided a synopsis of wideband and multi-frequency sparse array processing within the sparse reconstruction framework. In this section, we suggest references for further reading on certain aspects of wideband/multi-frequency processing that we did not elaborate upon.

Along the same lines as Section 3.3, operation at multiple frequencies can also be used to enable high-resolution DOA estimation methods, such as coarray MUSIC, to exploit all DOFs afforded by sparse arrays with nonuniform coarrays. In [34], measurements made at carefully chosen frequencies, in addition to the reference frequency, are used to fill in the missing lags, thereby extending the central uniform segment of the coarray. However, the increase in DOFs through operation at multiple frequencies restricts the sources to have proportional spectra at the considered frequencies. Under this constraint, multi-frequency measurements are used to construct a virtual correlation matrix corresponding to a filled uniform coarray at a single frequency. This virtual correlation matrix can then be employed for high-resolution DOA estimation.

Sparse arrays can achieve a higher number of DOFs compared with a ULA. To further reduce the number of physical sensors while maintaining a high number of DOFs, operation at multiple frequencies which satisfy a coprime relationship can be exploited. Ref. [59] employs a single sparse ULA to exploit signals with multiple coprime frequencies. This extends the coprime array concept to a joint spatio-spectral domain, thereby achieving a large coarray with a lower complexity. The signal DOAs are estimated using group-sparsity techniques. The Cramer-Rao bound is analyzed in [60] for ULA using two frequencies. It is interesting to note that the identifiability between two closely placed sources is improved when the phase response of the two frequency components differs.

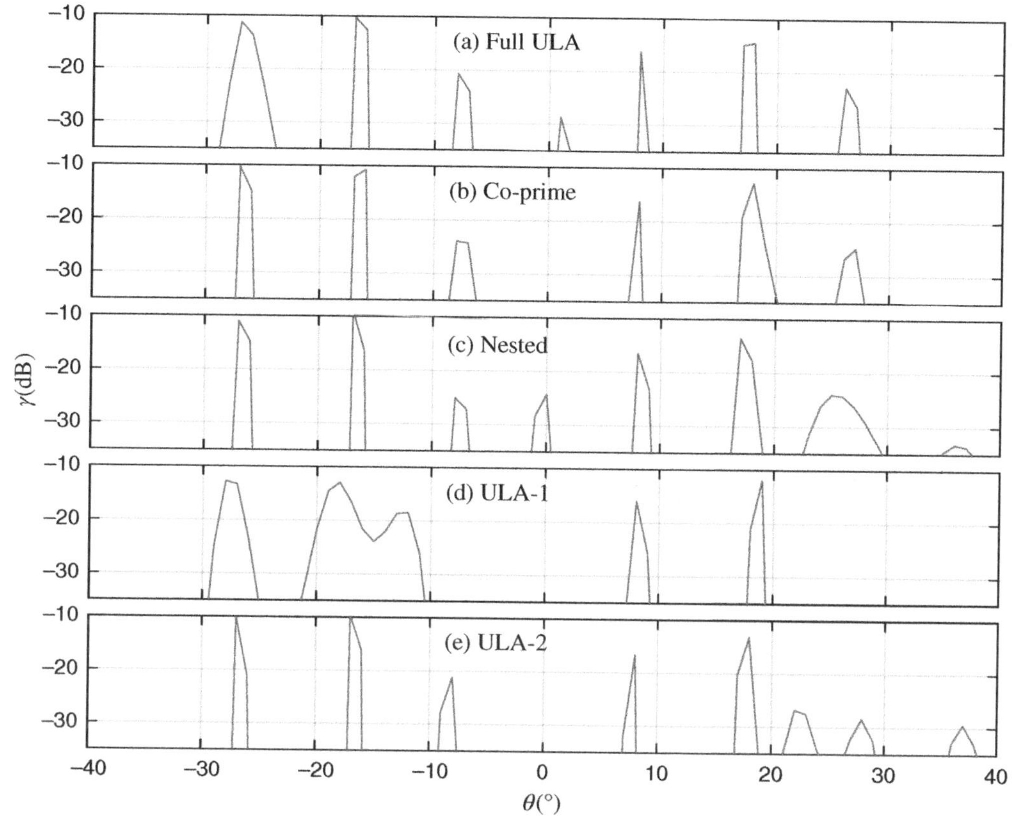

Figure 3.13 γ versus θ for the time step indicated by the verticle line around 22 minutes in Fig. 3.12. Source: Nannuru et al. [57]/Acoustical Society of America.

Design of multi-frequency sparse arrays is considered in [61] to offer zero lag redundancy in the rendered difference coarray, thus enabling high-resolution DOA estimation. Sensor interpolation is exploited to estimate the signal correlation matrix of an augmented ULA, thereby mitigating the effect of holes in the coarray and enhancing DOA estimation performance.

3.6 Conclusion

This chapter addressed DOA estimation and beamforming with sparse arrays through exploitation of wide signal bandwidths within the sparse reconstruction and Bayesian learning frameworks. First, a class of group-sparsity-based wideband algorithms were introduced for coarray-domain DOA estimation by simultaneously exploiting the information at different frequency bins comprising the signal bandwidth in the estimation process. Improved performance was achieved by wideband operation with sparse arrays whose unit spacing was increased beyond Nyquist sampling. Next, operation of sparse arrays at two or more frequencies was used to introduce additional virtual elements in the difference coarray. These added DOFs in the multi-frequency coarray were exploited via sparse reconstruction methods to resolve more sources than the narrowband coarray. Finally, SBL method was presented for wideband beamforming which was shown to have better resolution than CBF and, thus, it can resolve more sources. By making use of wideband observations with sparse arrays, SBL can significantly reduce spatial aliasing in beamforming. Illustrative simulation and experimental examples were provided validating the performance of the considered wideband and multi-frequency array processing methods.

References

1 Van Trees, H.L. (2002). *Optimum Array Processing: Part IV of Detection, Estimation, and Modulation Theory*. New York: Wiley.

2 Ishiguro, M. (1980). Minimum redundancy linear arrays for a large number of antennas. *Radio Science* 15 (06): 1163–1170.

3 Moffet, A. (1968). Minimum-redundancy linear arrays. *IEEE Transactions on Antennas and Propagation* 16 (2): 172–175.

4 Rossouw, M.J., Joubert, J., and McNamara, D.A. (1997). Thinned arrays using a modified minimum redundancy synthesis technique. *Electronics Letters* 33 (10): 826–827.

5 Pal, P. and Vaidyanathan, P.P. (2010). Nested arrays: a novel approach to array processing with enhanced degrees of freedom. *IEEE Transactions on Signal Processing* 58 (8): 4167–4181.

6 Vaidyanathan, P.P. and Pal, P. (2011). Sparse sensing with co-prime samplers and arrays. *IEEE Transactions on Signal Processing* 59 (2): 573–586.

7 Bloom, G.S. and Golomb, S.W. (1977). Applications of numbered undirected graphs. *Proceedings of the IEEE* 65 (4): 562–570.

8 Chambers, C., Tozer, T.C., Sharman, K.C., and Durrani, T.S. (1996). Temporal and spatial sampling influence on the estimates of superimposed narrowband signals: when less can mean more. *IEEE Transactions on Signal Processing* 44 (12): 3085–3098.

9 Nannuru, S., Gerstoft, P., Ping, G., and Fernandez-Grande, E. (2021). Sparse planar arrays for azimuth and elevation using experimental data. *The Journal of the Acoustical Society of America* 149 (1): 167–178.

10 Malioutov, D., Çetin, M., and Willsky, A.S. (2005). A sparse signal reconstruction perspective for source localization with sensor arrays. *IEEE Transactions on Signal Processing* 53 (8): 3010–3022.

11 Pote, R.R. and Rao, B.D. (2023). Maximum likelihood-based gridless DoA estimation using structured covariance matrix recovery and SBL with grid refinement. *IEEE Transactions on Signal Processing* 71: 802–815.

12 Tan, Z. and Nehorai, A. (2014). Sparse direction of arrival estimation using co-prime arrays with off-grid targets. *IEEE Signal Processing Letters* 21 (1): 26–29.

13 BouDaher, E., Ahmad, F., and Amin, M.G. (2015). Sparsity-based direction finding of coherent and uncorrelated targets using active nonuniform arrays. *IEEE Signal Processing Letters* 22 (10): 1628–1632. https://doi.org/10.1109/LSP.2015.2417807.

14 Wan, Z. and Liu, W. (2022). Time-domain wideband DOA estimation under the convolutional sparse coding framework. *IEEE Signal Processing Letters* 29: 274–278.

15 di Claudio, E.D. and Parisi, R. (2001). WAVES: weighted average of signal subspaces for robust wideband direction finding. *IEEE Transactions on Signal Processing* 49 (10): 2179–2191.

16 Su, G. and Morf, M. (1983). The signal subspace approach for multiple wide-band emitter location. *IEEE Transactions on Acoustics, Speech, and Signal Processing* 31 (6): 1502–1522.

17 Wang, H. and Kaveh, M. (1985). Coherent signal-subspace processing for the detection and estimation of angles of arrival of multiple wide-band sources. *IEEE Transactions on Acoustics, Speech, and Signal Processing* 33: 823–831.

18 Yoon, Y.S., Kaplan, L.M., and McClellan, J.H. (2006). TOPS: new DOA estimator for wideband signals. *IEEE Transactions on Signal Processing* 54 (6): 1977–1989.

19 Liang, Y.B., Cui, W., Shen, Q. et al. (2021). Cramér–Rao bound analysis of underdetermined wideband DOA estimation under the subband model via frequency decomposition. *IEEE Transactions on Signal Processing* 69: 4132–4148.

20 Cui, W., Shen, Q., Liu, W., and Wu, S. (2019). Low complexity DOA estimation for wideband off-grid sources based on re-focused compressive sensing with dynamic dictionary. *IEEE Journal of Selected Topics in Signal Processing* 13 (5): 918–930.

21 Shen, Q., Liu, W., Cui, W. et al. (2017). Focused compressive sensing for underdetermined wideband DOA estimation exploiting high-order difference coarrays. *IEEE Signal Processing Letters* 24 (1): 86–90.

22 Shen, Q., Liu, W., Cui, W. et al. (2015). Low-complexity direction-of-arrival estimation based on wideband co-prime arrays. *IEEE Transactions on Audio, Speech and Language Processing* 23: 1445–1456.

23 Shen, Q., Liu, W., Wang, L., and Liu, Y. (2020). Group sparsity based localization for far-field and near-field sources based on distributed sensor array networks. *IEEE Transactions on Signal Processing* 68: 6493–6508.

24 Liu, W. and Weiss, S. (2010). *Wideband Beamforming: Concepts and Techniques*. Chichester: Wiley.

25 Liu, W. (2012). Wideband beamforming for multi-path signals based on frequency invariant transformation. *International Journal of Automation and Computing* 9: 420–428.

26 Liu, C.L. and Vaidyanathan, P.P. (2015). Remarks on the spatial smoothing step in coarray MUSIC. *IEEE Signal Processing Letters* 22 (9): 1438–1442.

27 Zhang, Y.D., Amin, M.G., and Himed, B. (2013). Sparsity-based DOA estimation using co-prime arrays. *Proceedings of the IEEE International Conference on Acoustics, Speech, and Signal Processing*, 3967–3971, Vancouver, Canada, May 2013.

28 Shen, Q., Liu, W., Cui, W., and Wu, S.L. (2014). Low-complexity compressive sensing based DOA estimation for co-prime arrays. *Proceedings of the International Conference on Digital Signal Processing*, Hong Kong, August 2014.

29 Shen, Q., Liu, W., Cui, W. et al. (2014). Group sparsity based wideband DOA estimation for co-prime arrays. *Proceedings of the IEEE China Summit and International Conference on Signal and Information Processing*, Xi'an, China, July 2014.

30 Shen, Q., Liu, W., Cui, W., and Wu, S.L. (2016). Underdetermined DOA estimation under the compressive sensing framework: a review. *IEEE Access* 4: 8865–8878.

31 Nannuru, S., Gemba, K.L., Gerstoft, P. et al. (2019). Sparse Bayesian learning with multiple dictionaries. *Signal Processing* 159: 159–170.

32 Zhu, H., Leus, G., and Giannakis, G.B. (2011). Sparsity-cognizant total least-squares for perturbed compressive sampling. *IEEE Transactions on Signal Processing* 59 (5): 2002–2016.

33 Ahmad, F. and Kassam, S.A. (1998). Performance analysis and array design for wide-band beamformers. *SPIE Journal of Electronic Imaging* 7 (4): 825–838.

34 BouDaher, E., Jia, Y., Ahmad, F., and Amin, M.G. (2015). Multi-frequency co-prime arrays for high-resolution direction-of-arrival estimation. *IEEE Transactions on Signal Processing* 63 (14): 3797–3808.

35 BouDaher, E., Ahmad, F., and Amin, M.G. (2014). Sparse reconstruction for direction-of-arrival estimation using multi-frequency co-prime arrays. *EURASIP Journal on Advances in Signal Processing* 2014: 1–11. Art. ID: 168.

36 Moulton, J.L. and Kassam, S.A. (2009). Resolving more sources with multi-frequency coarrays in high-resolution direction-of-arrival estimation. *Proceedings of the 43rd Annual Conference on Information Sciences and Systems*, 772–777.

37 Hamza, S.A. and Amin, M.G. (2021). Sparse array beamforming design for wideband signal models. *IEEE Transactions on Aerospace and Electronic Systems* 57 (2): 1211–1226. https://doi.org/10.1109/TAES.2020.3037409.

38 Tipping, M.E. (2001). Sparse Bayesian learning and the relevance vector machine. *Journal of Machine Learning Research* 1: 211–244.

39 Wipf, D.P. and Rao, B.D. (2007). An empirical Bayesian strategy for solving the simultaneous sparse approximation problem. *IEEE Transactions on Signal Processing* 55 (7): 3704–3716.

40 Gemba, K.L., Sarkar, J., Cornuelle, B. et al. (2018). Estimating relative channel impulse responses from ships of opportunity in a shallow water environment. *The Journal of the Acoustical Society of America* 144 (3): 1231–1244.

41 Wagner, M., Park, Y., and Gerstoft, P. (2021). Gridless doa estimation and root-music for non-uniform linear arrays. *IEEE Transactions on Signal Processing* 69: 2144–2157.

42 Park, Y. and Gerstoft, P. (2022). Gridless sparse covariance-based beamforming via alternating projections including co-prime arrays. *The Journal of the Acoustical Society of America* 151 (6): 3828–3837.

43 Chen, S.S., Donoho, D.L., and Saunders, M.A. (2001). Atomic decomposition by basis pursuit. *SIAM Review* 43 (1): 129–159.

44 Mallat, S.G. and Zhang, Z. (1993). Matching pursuits with time-frequency dictionaries. *IEEE Transactions on Signal Processing* 41 (12): 3397–3415.

45 Gemba, K.L., Hodgkiss, W.S., and Gerstoft, P. (2017). Adaptive and compressive matched field processing. *The Journal of the Acoustical Society of America* 141 (1): 92–103.

46 Gerstoft, P., Mecklenbräuker, C.F., Seong, W., and Bianco, M. (2018). Introduction to compressive sensing in acoustics. *The Journal of the Acoustical Society of America* 143 (6): 3731–3736.

47 Gerstoft, P., Xenaki, A., and Mecklenbräuker, C.F. (2015). Multiple and single snapshot compressive beamforming. *The Journal of the Acoustical Society of America* 138 (4): 2003–2014.

48 Liu, Z.-M., Huang, Z.-T., and Zhou, Y.-Y. (2012). An efficient maximum likelihood method for direction-of-arrival estimation via sparse Bayesian learning. *IEEE Transactions on Wireless Communications* 11 (10): 1–11.

49 Xenaki, A., Gerstoft, P., and Mosegaard, K. (2014). Compressive beamforming. *The Journal of the Acoustical Society of America* 136 (1): 260–271.

50 Gerstoft, P., Mecklenbräuker, C.F., Xenaki, A., and Nannuru, S. (2016). Multi snapshot sparse Bayesian learning for DOA. *IEEE Signal Processing Letters* 23 (10): 1469–1473.

51 Austin, C.D., Moses, R.L., Ash, J.N., and Ertin, E. (2010). On the relation between sparse reconstruction and parameter estimation with model order selection. *IEEE Journal of Selected Topics in Signal Processing* 4 (3): 560–570.

52 Liu, C.-L. and Vaidyanathan, P.P. (2017). Cramér–Rao bounds for coprime and other sparse arrays, which find more sources than sensors. *Digital Signal Processing* 61: 43–61.

53 Prasad, R. and Murthy, C.R. (2013). Cramér–Rao-type bounds for sparse Bayesian learning. *IEEE Transactions on Signal Processing* 61 (3): 622–632.

54 Gemba, K.L., Nannuru, S., Gerstoft, P., and Hodgkiss, W.S. (2017). Multi-frequency sparse Bayesian learning for robust matched field processing. *The Journal of the Acoustical Society of America* 141 (5): 3411–3420.

55 Gemba, K.L., Nannuru, S., and Gerstoft, P. (2019). Robust ocean acoustic localization with sparse Bayesian learning. *IEEE Journal of Selected Topics in Signal Processing* 13 (1): 49–60.

56 Park, Y., Nannuru, S., Gemba, K., and Gerstoft, P. (2020). *SBL4 from NoiseLab*. GitHub. https://github.com/gerstoft/SBL (accessed 8 September 2023).

57 Nannuru, S., Koochakzadeh, A., Gemba, K.L. et al. (2018). Sparse Bayesian learning for beamforming using sparse linear arrays. *The Journal of the Acoustical Society of America* 144 (5): 2719–2729.

58 Porter, M.B. and Bucker, H.P. (1987). Gaussian beam tracing for computing ocean acoustic fields. *The Journal of the Acoustical Society of America* 82 (4): 1349–1359.

59 Qin, S., Zhang, Y.D., Amin, M.G., and Himed, B. (2017). DOA estimation exploiting a uniform linear array with multiple co-prime frequencies. *Signal Processing* 130: 37–46.

60 Guo, M., Zhang, Y.D., and Chen, T. (2018). Performance analysis for uniform linear arrays exploiting two coprime frequencies. *IEEE Signal Processing Letters* 25 (6): 838–842.

61 Zhang, S., Ahmed, A., Zhang, Y.D., and Sun, S. (2021). Enhanced DOA estimation exploiting multi-frequency sparse array. *IEEE Transactions on Signal Processing* 69: 5935–5946.

4

Sparse Arrays in Sample Starved Regimes: Algorithms and Performance Analysis

Piya Pal[1] *and Heng Qiao*[2]

[1] *Department of Electrical and Computer Engineering, University of California San Diego, La Jolla, CA, USA*
[2] *University of Michigan-Shanghai Jiao Tong University Joint Institute, Shanghai Jiao Tong University, Shanghai, China*

4.1 Introduction

Sparse sensor arrays such as nested [1] and coprime arrays [2] offer several benefits over the conventional and widely used Uniform Linear Array (ULA) [3–5] in terms of their ability to localize more sources than sensors, and higher resolution with the same number of physical sensing elements. One of the main features of these arrays is that their difference co-array [6] contains a ULA segment of length $\Theta(M^2)$. This difference co-array controls the correlation structure of the measurements, and can be utilized to resolve more sources than sensors. Besides applications in Direction-of-Arrival (DOA) estimation and super-resolution imaging [7, 8], the idea of the sparse array as a sampling operator has intimate connections to the important problem of compressive covariance sensing [3, 5, 9–11]. Statistical performance of sparse arrays in estimation of more source directions than sensors has been analyzed in terms of Cramér-Rao bounds [5, 12, 13], mean squared error (MSE) of estimators based on the Co-array MUSIC algorithm [5] and (nonlinear) Least Squares algorithms [4].

However, these results are asymptotic in nature (i.e., tight only when the number of snapshots $L \rightarrow \infty$) and are either loose or inapplicable when L is finite. Moreover, they also require the number of sources to be known apriori, or assume certain stringent conditions on correct identification of the signal subspace, which may be unrealistic in practice (or at best, very difficult to ensure with finite L). On the other hand, recent advances in analysis of super-resolution algorithms based on Total Variation (TV) norm and atomic norm minimization[1] cannot ensure absence of spurious spike estimates (since the dual polynomial exhibits multiple peaks in the so-called near and far regions with respect to the true source locations) and cannot be directly utilized to characterize the MSE of source localization [18, 19].

Sparse Arrays and MMV: As pointed out in [20, 21], the Multiple Measurement Vector (MMV) model used in compressed sensing, has its origins in sensor array and multichannel signal processing. Contrary to standard MMV algorithms [22, 23], these techniques aim to directly recover the sparse support (bypassing the recovery of the unknown vectors $\mathbf{x}[l]$) from the data correlation matrix $\frac{1}{L}\sum_{l=1}^{L}\mathbf{y}[l]\mathbf{y}^H[l]$ captured by a sparse array, under the assumption that the non-zero elements of $\mathbf{x}[l]$ are statistically uncorrelated. The idea of using such correlation priors on $\mathbf{x}[l]$ for source

1 Contrary to MMV models that assume sources to be located on a grid [8, 14–17], these algorithms assume a "gridless" model for source locations.

Sparse Arrays for Radar, Sonar, and Communications, First Edition. Edited by Moeness G. Amin.

localization has been of continued interest in MMV models for over a decade [14, 16, 17, 24–28], giving rise to powerful Bayesian algorithms for sparse signal recovery, such as the M-SBL framework [27, 29, 30].

In [31], the first non-asypmptotic guarantees were established for a generic class of *computationally efficient* algorithms for support recovery in the regime $s = \Theta(M^2)$. This analysis was made possible by observing how non-negativity (of the source powers) and sparse array geometry together enable unique support recovery in the regime $|\mathcal{S}| > M$ even *without enforcing any sparsity constraints* on the unknown signal. Under a "minimum separation condition" on the source locations [18, 32], *universal upper bounds on the estimation error of signal power were obtained, which would be satisfied by any algorithm that enforces the non-negativity constraint on the signal power*. Using this error bound on the estimated powers, a simple hard-thresholding technique was employed for recovering the support from the power estimates. By analyzing this simple yet effective method, the first non-asymptotic lower bounds on the number of snapshots were obtained, which guarantee the support recovery to be successful with high probability.

A major portion of the chapter is dedicated to an in-depth review of this result. Finally in Section 4.5, we consider the specific scenario of a single snapshot and investigate how the virtual array of a nested array can still be exploited to harness it superior resolution and avoid ambiguity in source localization. The ability to synthesize the virtual array with only a single snapshot can be especially lucrative for applications such as automotive radar and joint communication and radar sensing. In these applications, the sources or multi paths may be coherent and the environment is dynamic due to the high mobility of the sources. This can significantly limit the number of temporal snapshots available for source localization [33, 34]. It is therefore of great interest to understand to what extent, *deterministic* sparse array geometries can be successfully interpolated in order to leverage its virtual array. This question was recently addressed in [35] for a nested array geometry. We review this result in Section 4.5 and discuss its significance for understanding the performance of sparse arrays in extremely sample-starved regimes.

4.2 Background on Correlation-Aware Sparse Support Recovery with Sparse Arrays

The problem of source localization and direction-finding with sparse arrays has its natural counterpart in sparse signal processing or compressed sensing, where the goal is to recover high-dimensional sparse vectors from compressive linear measurements. Specifically, direction-of-arrival (DOA) estimation with sparse arrays can be cast as the problem of joint sparse support recovery from the so-called Multiple Measurement Vector (MMV) models, which is of significant interest in compressive sensing and sparse signal recovery [14, 15, 20–22, 36–39]. The goal here is to identify the common support (denoted by $\mathcal{S}$) of a set of L vectors $\mathbf{x}[l] \in \mathbb{C}^N, 1 \leqslant l \leqslant L$ using L (compressed) measurement vectors $\mathbf{y}[l] \in \mathbb{C}^M, 1 \leqslant l \leqslant L$ $(M < N)$ acquired using a common measurement matrix $\mathbf{A} \in \mathbb{C}^{M \times N}$ and contaminated with additive noise $\mathbf{w}[l]$

$$\mathbf{y}[l] = \mathbf{A}\mathbf{x}[l] + \mathbf{w}[l], \quad \text{Supp}(\mathbf{x}[l]) = \mathcal{S}, |\mathcal{S}| = s. \tag{4.1}$$

The MMV model is originally inspired by measurement models in sensor array processing [14, 21] where $\mathbf{y}[l]$ denotes the lth temporal sample (or snapshot) of the measurements collected at an array of M sensors. The angular range of interest is discretized into a grid of size N, so that the common support $\mathcal{S}$ of the sparse vectors $\mathbf{x}[l]$ indicates the source directions (assumed to be "on-grid"), and the non-zero elements of $\mathbf{x}[l]$ represent the (time-varying) source amplitudes. The following assumptions on the correlation of $\mathbf{X}_L$ are standard in Bayesian approaches (such as Sparse Bayesian Learning) to sparse support recovery from MMV models [27, 37, 39, 40]:

- **[(A1)] Uncorrelated Source Signals.** The source signals $\mathbf{x}[l], 1 \leqslant l \leqslant L$ are assumed to be zero-mean, independent and identically distributed (i.i.d.) random vectors. Furthermore, the non-zero elements of $\mathbf{x}[l]$ are assumed to be statistically uncorrelated
$$\mathbb{E}(\mathbf{x}[l]) = \mathbf{0}, \forall l, \quad \mathbb{E}\left([\mathbf{x}[l]]_i [\mathbf{x}[m]]_k^*\right) = p_i^\star \delta[i-k]\delta[l-m], \quad i,k \in \mathcal{S}, \tag{4.2}$$
where $p_i^\star$ is the power of ith component, $\delta[\cdot]$ is the Kronecker delta function and * denotes conjugate.
- **[(A2)] Noise.** The additive noise vectors $\{\mathbf{w}[l]\}_{l=1}^L$ are statistically uncorrelated with $\mathbf{X}_L$ and distributed as
$$\mathbf{w}[l] \overset{i.i.d}{\sim} \mathcal{CN}(\mathbf{0}, \sigma^2\mathbf{I}).$$
The noise power σ^2 is assumed to be *unknown*.
- **[(A3)] Measurement Matrix.** The measurement matrix $\mathbf{A}$ is assumed to be a fixed deterministic matrix, representing the array manifold of a sparse array (as described later). The theoretical guarantees will be developed for such a deterministic $\mathbf{A}$.[2]

4.2.1 Fundamental Limits: Is $|\mathcal{S}|^2 > M$ Achievable?

The goal in MMV problems is to detect the common support $\mathcal{S}$ from $\mathbf{Y}_L$. Most MMV algorithms achieve this by first recovering $\mathbf{X}_L$ and then detecting its common support [14, 23, 36]. However, the need to exactly recover $\mathbf{X}_L$ first automatically restricts the size of recoverable support to $|\mathcal{S}| < M$ [21, 22, 36, 42]. In particular, it is well-known that $\mathbf{X}_L$ can be exactly recovered if [21, 22]
$$s < \frac{\text{Krank}(\mathbf{A}) + \text{rank}(\mathbf{Y}_L)}{2}. \tag{4.3}$$
Since Krank$(\mathbf{A}) \leqslant M$ and rank$(\mathbf{Y}_L) \leqslant M$, (4.3) implies that $s < M$. It has been shown in [20, 37, 39] that by exploiting correlation priors on $\mathbf{X}_L$ (given by Assumption **(A1)**), it is possible to recover $\mathcal{S}$ from the correlation matrix $\mathbf{R}_{\mathbf{yy}} = \mathbb{E}(\frac{1}{L}\mathbf{Y}_L\mathbf{Y}_L^H)$ of the measurements *without having to first estimate* $\mathbf{X}_L$. The idea behind such "correlation-aware" techniques for support recovery is also closely related to an important body of research on compressive covariance sampling [3, 10, 11, 39, 43] where the goal is to recover the covariance matrix of $\mathbf{X}_L$ from its compressed sketch $\mathbf{R}_{\mathbf{yy}}$. The main idea behind these methods is to design the measurement matrix appropriately so that it becomes possible to recover supports of size as large as $s = \Theta(M^2)$. To see this, notice that Assumption **(A1)** implies
$$\mathbf{R}_{\mathbf{yy}} = \mathbf{A}\text{diag}(\mathbf{p}^\star)\mathbf{A}^H + \sigma^2\mathbf{I}, \tag{4.4}$$
where $\mathbf{p}^\star \in \mathbb{R}^N$ is a *sparse vector* satisfying
$$[\mathbf{p}^\star]_i = \begin{cases} p_i^\star, & i \in \mathcal{S} \\ 0 & \text{otherwise.} \end{cases} \tag{4.5}$$
In other words, the support of $\mathbf{p}^\star$ is exactly $\mathcal{S}$. Hence, one can recover $\mathcal{S}$ by first finding a sparse $\mathbf{p}^\star$ that satisfies (4.4). Assuming that[3] the noise power σ^2 is known, this amounts to solving the following l_0 quasi-norm minimization problem
$$\min_{\mathbf{z}\succeq\mathbf{0}} \|\mathbf{z}\|_0 \quad \text{s.t.} \quad (\mathbf{A}^* \odot \mathbf{A})\mathbf{z} = \text{vec}(\mathbf{R}_{\mathbf{yy}} - \sigma^2\mathbf{I}) \qquad (P0)_{Co}. \tag{4.6}$$

---

2 The results herein deviate from usual guarantees in compressed sensing which rely on the measurement matrix $\mathbf{A}$ being random with appropriate distributions [41].
3 This assumption will be relaxed later.

where we used the fact that $\text{vec}(\mathbf{R}_{\mathbf{yy}}) = (\mathbf{A}^* \odot \mathbf{A})\mathbf{p}^\star + \sigma^2 \text{vec}(\mathbf{I})$. Using standard results from compressed sensing [41], it follows that the solution to $(P0)_{Co}$ will be unique, and identical to $\mathbf{p}^\star$ if

$$s \leqslant \frac{1}{2}\ \text{Krank}(\mathbf{A}^* \odot \mathbf{A}). \tag{4.7}$$

Hence, the Kruskal-rank of $\mathbf{A}^* \odot \mathbf{A}$ (and not of $\mathbf{A}$) controls the maximum size of support that can be recovered by solving $(P0)_{Co}$. Since the matrix $\mathbf{A}^* \odot \mathbf{A}$ is of size $M^2 \times N$, (i.e., it has the same number of columns as $\mathbf{A}$ but its number of rows is M times larger), its Kruskal-rank can be as large as $\Theta(M^2)$. This is precisely where suitable sparse arrays such as a nested or coprime array [1, 2] offer a constructive way to realize such measurement matrices, thanks to their large and contiguous difference sets, which are examined in detail in Section 4.2.2.

4.2.2 Role of Difference Sets

In DOA estimation and super-resolution microscopy,[4] the location of each point source of interest (which can represent a far-field target or a molecule) can be mapped to a spatial frequency $\omega \in [0, 2\pi)$. By discretizing the range of spatial frequencies into N bins as $\omega_n = 2\pi n/N, 0 \leqslant n \leqslant N-1$, one obtains the measurement matrix $\mathbf{A}$ as [7, 8, 16, 32, 38, 44, 45]

$$[\mathbf{A}]_{m,n} = e^{-jd_m\frac{2\pi n}{N}}, \quad 1 \leqslant m \leqslant M, 0 \leqslant n \leqslant N-1, \tag{4.8}$$

where d_m is an integer denoting the location of the mth sensor (or pixel), normalized with respect to the wavelength of the impinging narrowband waveform. This discretized model for source localization has been widely used, especially in conjunction with MMV models [8, 16, 38, 39, 44, 46, 47].

Under assumption (**A1**), we know from (4.7) that the Kruskal-rank of $\mathbf{A}_{\text{KR}} = \mathbf{A}^* \odot \mathbf{A}$ dictates the size of recoverable support. Upon close inspection, it can be verified that $\mathbf{A}_{\text{KR}}$ is given by

$$[\mathbf{A}_{\text{KR}}]_{(m-1)M+l,n} = e^{j(d_m-d_l)\frac{2\pi n}{N}}, \quad 1 \leqslant m, l \leqslant M. \tag{4.9}$$

Comparing (4.8) and (4.9), each column of $\mathbf{A}_{\text{KR}}$ can be thought of as the steering vector (corresponding to a candidate spatial frequency) of a *virtual* array (also called the difference co-array) whose elements are located at positions given by the following difference set [1, 3, 6, 44, 48, 49]

$$\mathbb{D}_{\text{diff}} \triangleq \{d_m - d_l, 1 \leqslant m, l \leqslant M\}. \tag{4.10}$$

The structure of $\mathbb{D}_{\text{diff}}$ and the number of its distinct elements directly control the Kruskal-rank of $\mathbf{A}_{\text{KR}}$, and therefore the maximum size of recoverable support [39]. If the set of sensor locations $\mathbb{S} = \{d_m\}_{m=1}^{M}$ is carefully designed, then it is possible to ensure that its difference set $\mathbb{D}_{\text{diff}}$ has $\Theta(M^2)$ distinct *consecutive integers*. In other words, the virtual array (whose element locations are given by $\mathbb{D}_{\text{diff}}$) will have a ULA segment with $\Theta(M^2)$ elements. This is precisely the main idea behind the design of sparse arrays. Sparse arrays such as minimum redundancy arrays [6], nested arrays [1, 50], coprime arrays [2], and their generalizations and extensions [44] are carefully designed so that their difference sets satisfy this property. In the following, nested and coprime arrays are revisited and compared against the popular ULA, with respect to the Kruskal-rank of the Khatri–Rao product of their array manifolds.

1. **ULA**: For a uniform linear array,

   $$d_i = i - 1, \quad 1 \leqslant i \leqslant M.$$

   It can be easily seen that in this case

   $$\mathbb{D}_{\text{diff}} = \{m, -M+1 \leqslant m \leqslant M-1\},$$

4 For ease of exposition, we consider a one-dimensional source localization problem, although the results can be directly extended to two dimensions.

and hence $|\mathbb{D}_{\text{diff}}| = 2M - 1$. In fact, it can be verified that in this case $\text{Krank}(\mathbf{A}_{\text{KR}}) = 2M - 1$. Hence, (4.7) implies the following upper bound on the size of the recoverable support $s < M$. This is consistent with the popular belief that the number of resolvable directions is ultimately upper-bounded by the number of sensors or antennas.

2. **Nested Array**: For a nested array [1], the set of sensor locations is given by

$$\{d_i\}_{i=1}^M = \{m - 1, 1 \leqslant m \leqslant M/2\} \cup \{(M/2 + 1)n - 1, 1 \leqslant n \leqslant M/2\}. \tag{4.11}$$

The difference set is

$$\mathbb{D}_{\text{diff}} = \{m, -M^2/4 - M/2 + 1 \leqslant m \leqslant M^2/4 + M/2 - 1\}. \tag{4.12}$$

Since $\mathbb{D}_{\text{diff}}$ contains *all consecutive integers* within the above range, $\mathbf{A}_{\text{KR}}$ is a Vandermonde matrix with $\frac{M^2}{2} + M - 1$ distinct rows. Hence,

$$\text{Krank}(\mathbf{A}_{\text{KR}}) = \frac{M^2}{2} + M - 1. \tag{4.13}$$

Hence the Kruskal-rank of $\mathbf{A}_{\text{KR}}$ for nested arrays is as large as $\Theta(M^2)$, whereas that for ULA is $2M - 1$.

3. **Coprime Array**: A coprime array is an array of $M = 2M_1 + M_2 - 1$ sensors, where M_1 and M_2 are coprime integers. The locations of the sensors are given by

$$\{d_i\}_{i=1}^M = \{M_1 m, 0 \leqslant m \leqslant M_2 - 1\} \cup \{M_2 n, 1 \leqslant n \leqslant 2M_1 - 1\}. \tag{4.14}$$

The difference set in this case contains all integers in the range

$$\{n, -M_1M_2 \leqslant n \leqslant M_1M_2\}.$$

Thus, the rows of $\mathbf{A}_{\text{KR}}$ contain a Vandermonde submatrix with $2M_1M_2 + 1$ distinct rows, implying

$$\text{Krank}(\mathbf{A}_{\text{KR}}) \geq 2M_1M_2 + 1. \tag{4.15}$$

For $M_2 = M_1 + 1$, we have $M = 3M_1$ and $\text{Krank}(\mathbf{A}_{\text{KR}}) = \Theta(M^2)$.

From the above discussions, it is clear that for nested and coprime arrays, we can find an integer M_{ca} such that (i) $\mathbb{D}_{\text{diff}}$ contains all *consecutive integers* in the range $-M_{ca}$ and M_{ca}, *and* (ii) M_{ca} is as large as $\Theta(M^2)$. The same is true for generalizations of nested and coprime arrays, such as super nested [51], and generalized coprime arrays [44]. More generally, for all these arrays, $\mathbb{D}_{\text{diff}}$ contains a ULA segment given by the set

$$\mathbb{U} = \{n, -M_{ca} \leqslant n \leqslant M_{ca}\}, \quad M_{ca} = \Theta(M^2) \tag{4.16}$$

This prompts us to unify the treatment of nested arrays, coprime arrays (and their extensions) by defining a family of "Order-Optimal Sparse Arrays (OOSA)":

Definition 4.1 ***(OOSA)*** An array is an OOSA if its difference set $\mathbb{D}_{\text{diff}}$ contains a subset $\mathbb{U}$ of the form (4.16) with $M_{ca} = \Theta(M^2)$.

The criterion for any array to be an OOSA is that its difference set should contain a ULA segment of size $\Theta(M^2)$. Since the maximum size of $\mathbb{D}_{\text{diff}}$ can only be $\Theta(M^2)$, these arrays are called "Order-Optimal" since the ULA segment $\mathbb{U}$ is of maximum possible size order-wise with respect to M.

Remark 4.1 ***Virtual Array Manifold corresponding to*** $\mathbb{U}$ In Section 4.3 (also the subsections therein), non-asymptotic support recovery guarantees will be derived for the general case when **A** corresponds to an OOSA (which includes the nested, the coprime and their extensions). From (4.8) and the structure of the difference set (4.10) of any OOSA, it can be seen that $\mathbf{A}_{\text{KR}} \in \mathbb{C}^{M^2 \times N}$ contains a Vandermonde sub-matrix $\mathbf{A}_{\text{KR}}^{\mathbb{U}} \in \mathbb{C}^{2M_{ca}+1 \times N}$ with $2M_{ca} + 1$ rows, whose elements are given by

$$[\mathbf{A}_{\text{KR}}^{\mathbb{U}}]_{m+M_{ca},n} = e^{-j2\pi mn/N}$$
$$-M_{ca} \leqslant m \leqslant M_{ca}, 0 \leqslant n \leqslant N-1. \tag{4.17}$$

In other words, $\mathbf{A}_{\text{KR}}^{\mathbb{U}}$ represents the *virtual array manifold* corresponding to the largest central ULA segment $\mathbb{U}$ in the difference set of an OOSA.

4.3 Universal Recovery Guarantees for OOSA: The Role of Non-negativity

The non-asymptotic guarantees presented in this chapter depart from the usual analysis of sparse support recovery that relies on mutual coherence or Restricted Isometry Property (RIP). Instead, the analysis is enabled by the fact that besides being sparse, $\mathbf{p}^\star$ is also *non-negative.* By utilizing (i) the non-negativity of $\mathbf{p}^\star$ (or positivity of its non-zero elements) and (ii) the geometry of sparse arrays as captured by the matrix $\mathbf{A}_{\text{KR}} = \mathbf{A}^* \odot \mathbf{A}$, it is possible to obtain universal guarantees for stable reconstruction of $\mathbf{p}^\star$ (as a function of L) that will be obeyed by a large class of correlation-aware support recovery algorithms (to be introduced later) in the regime of interest $s = \Theta(M^2)$.

4.3.1 Why Positivity Alone Suffices

In order to understand why the positivity of $\mathbf{p}^\star$ can alone ensure its exact reconstruction from $M = \Omega(\sqrt{s})$ measurements (without even explicitly using its sparsity), we first revisit the case when we have perfect knowledge of the covariance matrix $\mathbf{R}_{\mathbf{yy}} = \mathbb{E}\left(\frac{1}{L}\mathbf{Y}_L\mathbf{Y}_L^H\right)$ of the measurements. As discussed earlier, the non-negative l_0 quasi-norm minimization problem $(P0)_{Co}$ from [39] can uniquely recover $\mathbf{p}^\star$ with support as large as $s = \Theta(M^2)$, when **A** represents the manifold of a suitable sparse array. We now strengthen this result from [39] by showing that in this case, there is actually no need to minimize the l_0 quasi-norm, (or, explicitly enforce sparsity) to recover $\mathbf{p}^\star$. As long as $s \leqslant M_{ca}$, non-negativity of $\mathbf{p}^\star$ alone is sufficient to uniquely recover it from $\mathbf{R}_{\mathbf{yy}}$ *even when the noise power* σ^2 *is unknown*, as given by the following result.

Theorem 4.1 *[31]* Suppose we are given the exact covariance matrix $\mathbf{R}_{\mathbf{yy}} = \mathbb{E}(\frac{1}{L}\mathbf{Y}_L\mathbf{Y}_L^H)$. Even when the noise power σ^2 is unknown, it is possible to exactly recover $\mathbf{p}^\star$ (using a variant of the Co-array MUSIC algorithm [52]) as long as $s \leqslant M_{ca}$.

Proof: The proof utilizes a technique similar to Co-array MUSIC [1, 52] in order to establish uniqueness. We provide a complete proof in order to illustrate why non-negativity alone suffices, which is one of the main messages of this chapter. Since **A** represents an OOSA, $\mathbf{A}_{\text{KR}} = \mathbf{A}^* \odot \mathbf{A}$ contains a Vandermonde sub-matrix $\mathbf{A}_{\text{KR}}^{\mathbb{U}}$ given by (4.17). We further select the last $M_{ca} + 1$ rows of $\mathbf{A}_{\text{KR}}^{\mathbb{U}}$, and collect them in a matrix $\mathbf{V} \in \mathbb{C}^{M_{ca}+1 \times N}$ given by

$$\mathbf{V}_{m,n} = e^{-j2\pi mn/N}, 0 \leqslant m \leqslant M_{ca}, 0 \leqslant n \leqslant N-1.$$

Let $\mathbf{v_y} \triangleq \text{vec}(\mathbf{R_{yy}})$. From (4.24), we can extract an appropriate sub-vector of $\mathbf{v_y}$, viz. $\mathbf{v_y^{\mathbb{U}+}} \in \mathbb{C}^{M_{ca}+1}$ such that

$$\mathbf{v_y^{\mathbb{U}+}} = \mathbf{r_y^{\mathbb{U}+}} + \sigma^2\mathbf{e}_1 = \mathbf{V}_S\mathbf{p}_S^\star + \sigma^2\mathbf{e}_1. \tag{4.18}$$

where $\mathbf{e}_1 = [1, 0, \cdots, 0]^T \in \mathbb{R}^{M_{ca}+1}$, the matrix $\mathbf{V}_S \in \mathbb{C}^{M_{ca}+1\times s}$ consists of the columns of $\mathbf{V}$ indexed by S, and $\mathbf{p}_S^\star$ is the vector of non-zero entries of $\mathbf{p}^\star$. Given $\mathbf{R_{yy}}$, the Co-array MUSIC algorithm [52] produces an estimate $\hat{S}$ of the support S as follows:

Step 1: Extract the vector $\mathbf{v_y^{\mathbb{U}+}}$ from $\mathbf{R_{yy}}$ and construct a Hermitian Toeplitz matrix $\mathbf{T} \in \mathbb{C}^{(M_{ca}+1)\times(M_{ca}+1)}$ with $\mathbf{v_y^{\mathbb{U}+}}$ as the first column.

Step 2: Let $\hat{\sigma}^2$ be the smallest singular value of $\mathbf{T}$, and $\mathbf{U} \in \mathbb{C}^{M_{ca}+1\times K}$ be an orthonormal basis for $\text{Null}(\mathbf{T} - \hat{\sigma}^2\mathbf{I})$ where $K = \dim(\text{Null}(\mathbf{T} - \hat{\sigma}^2\mathbf{I}))$. Construct the set $\hat{S} \subset [0, N-1]$ such that

$$\hat{S} = \{i \in [0, N-1] \quad \text{s.t.} \quad \|\mathbf{U}^H\mathbf{v}_i\|_2 = 0\}. \tag{4.19}$$

Step 3: Return $\hat{S}$ and $\hat{\sigma}$ as estimates for S and σ.

We show that as long as $s \leqslant M_{ca}$, we will have $\hat{S} = S, \quad \hat{\sigma} = \sigma$. Firstly, it can be verified that

$$\mathbf{T} = \mathbf{V}_S \text{ diag}(\mathbf{p}_S^\star)\mathbf{V}_S^H + \sigma^2\mathbf{I}. \tag{4.20}$$

Since $\mathbf{p}_S^\star > \mathbf{0}$ and $s \leqslant M_{ca}$, the smallest singular value of $\mathbf{T}$ is σ^2, implying that $\hat{\sigma} = \sigma$ and $\text{rank}(\mathbf{T} - \hat{\sigma}^2\mathbf{I}) = s$. Therefore, from (4.20), we have

$$\mathbf{V}_S^H\mathbf{u} = \mathbf{0}, \quad \forall\mathbf{u} \in \text{Null}(\mathbf{T} - \hat{\sigma}^2\mathbf{I}), \tag{4.21}$$

which implies that $S \subseteq \hat{S}$. Suppose there exists $k_0 \in \hat{S}, k_0 \notin S$. Then, we must have

$$[\mathbf{V}_S, \mathbf{v}_{k_0}]^H\mathbf{u} = \mathbf{0}, \forall\mathbf{u} \in \text{Null}(\mathbf{T} - \hat{\sigma}^2\mathbf{I}). \tag{4.22}$$

Now $[\mathbf{V}_S, \mathbf{v}_{k_0}] \in \mathbb{C}^{(M_{ca}+1)\times(s+1)}$ is a row-Vandermonde matrix and hence its $s+1$ columns must be linearly independent since $s \leqslant M_{ca}$. However, (4.22) implies that these $s+1$ columns should also be orthogonal to $\text{Null}(\mathbf{T} - \hat{\sigma}^2\mathbf{I})$. Since $\text{rank}(\mathbf{T} - \hat{\sigma}^2\mathbf{I}) = s$, the orthogonal complement of $\text{Null}(\mathbf{T} - \hat{\sigma}^2\mathbf{I})$ has dimension s and cannot contain $s+1$ independent vectors. Hence such a k_0 cannot exist, implying $\hat{S} = S$. Once the support and σ are correctly recovered, the vector $\mathbf{p}_S^\star$ can be recovered as the unique solution to the system of equations (4.18), since $\mathbf{V}_S$ is full-column rank. This concludes the proof. □

Remark 4.2 When $N > M^2$ (i.e., when $\mathbf{A}_{\text{KR}}$ is a fat matrix), $\text{vec}(\mathbf{R_{yy}}) = \mathbf{A}_{\text{KR}}\mathbf{z} + \sigma^2 \text{ vec}(\mathbf{I})$ represents an underdetermined system of equations in $\mathbf{z}$ with possibly infinite solutions. However, Theorem 4.1 shows that by utilizing the fact that $\mathbf{p}^\star$ is non-negative, it is possible to exactly recover it even when the noise power is unknown. Therefore, there is no need to minimize the l_0 quasi-norm in $(P0)_{Co}$, or explicitly utilize the sparsity of $\mathbf{p}^\star$. Since for an OOSA, $M_{ca} = \Theta(M^2)$, we can uniquely recover supports of size as large as $\Theta(M^2)$.

Remark 4.3 The fact that non-negative constraints (or more generally, conic constraints) can lead to unique signal recovery in noiseless underdetermined problems has been previously utilized in compressed sensing [53–55], super-resolution imaging [32] and phase retrieval [56]. However, these results do not exploit the structure of the correlation matrix $\mathbf{R_{yy}}$ of the measurements (especially when acquired using a sparse array), and therefore cannot guarantee the recovery of supports of size $s > M$. In contrast, the preceding result shows that using sparse arrays and the non-negativity of $\mathbf{p}^\star$, it is possible to uniquely recover the support from $\mathbf{R_{yy}}$ (without explicit minimization) even when $s = \Theta(M^2)$.

4.3.2 Stable Recovery in the Regime $|S| > M$ with Correlation Estimates: Preliminaries

Theorem 4.1 shows that it is possible to exactly recover $\mathbf{p}^\star$ even when $s \gg M$ assuming we have *exact knowledge of the correlation matrix* $\mathbf{R_{yy}}$. In practice however, we can only *estimate* $\mathbf{R_{yy}}$ from a finite number of (L) of multiple measurement vectors (MMV) $\mathbf{y}[1], \dots \mathbf{y}[L]$. An important question therefore is

"Can the positivity of the non-zero elements of $\mathbf{p}^\star$ still be exploited in order to derive non-asymptotic performance guarantees for support recovery algorithms in the regime $s = \Theta(M^2)$?"

A somewhat surprising fact is that this is indeed possible, *even without explicitly utilizing the sparsity of* $\mathbf{p}^\star$. Let $\hat{\mathbf{r}}_\mathbf{y} = \text{vec}(\hat{\mathbf{R}}_\mathbf{yy}) \in \mathbb{C}^{M^2}$ and $\mathbf{r_y} \triangleq \text{vec}\left(\mathbf{R_{yy}} - \sigma^2\mathbf{I}\right) = \mathbf{A}_{\text{KR}}\mathbf{p}^\star$. Define

$$\Delta_\mathbf{r} \triangleq \hat{\mathbf{r}}_\mathbf{y} - \mathbf{r_y} = \underbrace{\text{vec}(\hat{\mathbf{R}}_\mathbf{yy} - \mathbf{R_{yy}})}_{\Delta_f} + \sigma^2 \text{ vec}(\mathbf{I}), \tag{4.23}$$

where Δ_f represents the finite sample correlation estimation error and the second term accounts for the *unknown* noise power σ^2.

Extracting the ULA segment from Difference Co-array: From Remark 4.1 and (4.17), it is clear that one can construct a sub-vector $\mathbf{r}_\mathbf{y}^{\mathbb{U}} \in \mathbb{C}^{2M_{ca}+1}$ such that $\mathbf{r}_\mathbf{y}^{\mathbb{U}} = \mathbf{A}_{\text{KR}}^{\mathbb{U}}\mathbf{p}^\star$. The same $2M_{ca} + 1$ indices can be extracted from $\hat{\mathbf{r}}_\mathbf{y}$ to obtain the vector $\hat{\mathbf{r}}_\mathbf{y}^{\mathbb{U}}$ such that

$$\hat{\mathbf{r}}_\mathbf{y}^{\mathbb{U}} = \mathbf{r}_\mathbf{y}^{\mathbb{U}} + \Delta_\mathbf{r}^{\mathbb{U}} = \mathbf{A}_{\text{KR}}^{\mathbb{U}}\mathbf{p}^\star + \Delta_\mathbf{r}^{\mathbb{U}}, \tag{4.24}$$

where $\Delta_\mathbf{r}^{\mathbb{U}}$ is composed of the corresponding indices of $\Delta_\mathbf{r}$.

4.3.3 Universal Upper Bounds on Error with Non-negative Constraint When $|S| > M$

Given any $\mathbf{p}^\# \in \mathbb{C}^N$ such that $\mathbf{p}^\# \succeq \mathbf{0}$, we will now derive a *universal upper bound* on the estimation error $\|\mathbf{p}^\star - \mathbf{p}^\#\|_2$ in terms of $\|\mathbf{A}_{\text{KR}}^{\mathbb{U}}(\mathbf{p}^\star - \mathbf{p}^\#)\|_2$. To begin, define a set of *non-negative sparse signals* whose supports obey a certain "separation condition." Such separation condition (without non-negative constraint) has been utilized in recent times for analyzing the performance of super-resolution imaging algorithms [10, 18, 19, 32].[5]

Definition 4.2 ***(Set of Non-negative Signals Obeying Separation Condition)*** Given N and M_{ca}, define the set $\mathcal{P}_{\text{sep}}^+$

$$\mathcal{P}_{\text{sep}}^+ \triangleq \{\mathbf{p} \in \mathbb{C}^N \mid \mathbf{p} \succeq \mathbf{0}, \phi\left(\frac{k}{N}, \frac{l}{N}\right) \geqslant \frac{2}{M_{ca}}, \forall k \neq l \in \text{Supp}(\mathbf{p})\} \tag{4.25}$$

where $\phi(\cdot,\cdot) : \mathbb{R}^2 \to \mathbb{R}_+$ is a wrap-around distance function [18] such that for $\forall \mu_1, \mu_2 \in [0, 1]$

$$\phi(\mu_1, \mu_2) \triangleq \min\left(|\mu_1 - \mu_2|, |\mu_1 + 1 - \mu_2|, |\mu_2 + 1 - \mu_1|\right). \tag{4.26}$$

The set $\mathcal{P}_{\text{sep}}^+$ consists of non-negative sparse signals such that the indices belonging to their support are separated by at least $\frac{2N}{M_{ca}}$.[6] The main result is given by the following theorem,

---

5 It is to be noted that in the absence of noise, classical methods such as MUSIC or root-MUSIC do not require any separation condition. However, in presence of noise and with finite snapshots, the separation between source angles controls the estimation error.

6 The separation condition can be relaxed to the more general Rayleigh Regularity condition [32]. However, for ease of exposition, in this paper we assume that $\mathbf{p}^\star$ obeys the separation condition.

which holds for any $\mathbf{p}^{\star} \in \mathcal{P}_{\text{sep}}^{+}$. As shown below, the proof utilizes a key Lemma from [32]. The separation condition used in [32] automatically restricts the sparsity to be smaller than M (i.e. $s < M$). Theorem 4.2 significantly generalizes this result by showing that it can be used to derive performance guarantees for a larger class of algorithms, even in the regime $s > M$ by exploiting the difference co-array of an OOSA.

Theorem 4.2 *[31]* Suppose $\mathbf{p}^{\star} \in \mathcal{P}_{\text{sep}}^{+}$, $M_{ca} \geq 128$[7] and $N \geq 3.03(2M_{ca}+1)$. Given any $\mathbf{p}^{\#} \geq \mathbf{0}$, we have

$$\|\mathbf{p}^{\#} - \mathbf{p}^{\star}\|_1 \leqslant \frac{1-\rho}{\rho}\|\mathbf{A}_{\text{KR}}^{\mathbb{U}}(\mathbf{p}^{\#} - \mathbf{p}^{\star})\|_2, \tag{4.27}$$

where $\rho = 0.0295\left(\frac{M_{ca}}{N}\right)^2$.

Proof: Consider the matrix $\tilde{\mathbf{A}}_{\text{KR}}^{\mathbb{U}} \in \mathbb{C}^{N\times N}$ obtained by zero-padding $\mathbf{A}_{\text{KR}}^{\mathbb{U}}$ as follows:

$$\tilde{\mathbf{A}}_{\text{KR}}^{\mathbb{U}} = [\mathbf{0}^T_{(\lceil N/2\rceil - M_{ca}-1)\times N}, (\mathbf{A}_{\text{KR}}^{\mathbb{U}})^T, \mathbf{0}^T_{(\lfloor N/2\rfloor - M_{ca})\times N}]^T. \tag{4.28}$$

It can be easily verified that $\tilde{\mathbf{A}}_{\text{KR}}^{\mathbb{U}} = \mathbf{D}_N\mathbf{F}_N$ where the elements of $\mathbf{F}_N \in \mathbb{C}^{N\times N}$ are given by $[\mathbf{F}_N]_{k,l} = e^{-j2\pi kl/N}$, $-\lceil N/2\rceil + 1 \leqslant k \leqslant \lfloor N/2\rfloor, 0 \leqslant l \leqslant N-1$ and $\mathbf{D}_N = \text{diag}(\hat{d}_{-\lceil N/2\rceil+1}, \cdots \hat{d}_{\lfloor N/2\rfloor})$ is a diagonal matrix with

$$\hat{d}_m = \begin{cases} 1, & -M_{ca} \leqslant m \leqslant M_{ca} \\ 0, & \text{otherwise.} \end{cases}$$

We utilize the following lemma (Lemma 1 in [32]) adapted to our measurement model. □

Lemma 4.1 *[32]* If $\mathbf{p}^{\star} \in \mathcal{P}_{sep}^{+}$, $M_{ca} \geq 128$ and $N \geq 3.03(2M_{ca}+1)$, then for any non-negative vector $\hat{\mathbf{p}} \in \mathbb{R}^N$ and $\mathbf{h} = \hat{\mathbf{p}} - \mathbf{p}^{\star}$, there exists $\mathbf{q} = [q_0, \cdots, q_{N-1}]^T \in \mathbb{R}^N$ such that

$$\tilde{\mathbf{A}}_{\text{KR}}^{\mathbb{U}}\mathbf{q} = \mathbf{F}_N\mathbf{q}, \quad \|\mathbf{q}\|_\infty \leqslant 1$$

where $\rho = 0.0295\left(\frac{M_{ca}}{N}\right)^2$ and

$$q_l = 0 \text{ if } h_l < 0; \;\; q_l > 2\rho \text{ if } h_l \geq 0$$

Given the existence of the vector $\mathbf{q}$ from Lemma 4.1, define $\tilde{\mathbf{q}} \triangleq \mathbf{q} - \rho\mathbf{1}$. Let $\mathbf{h} = \mathbf{p}^{\#} - \mathbf{p}^{\star}$. Then, we have

$$|\langle\tilde{\mathbf{q}}, \mathbf{h}\rangle| = |\sum_{l=0}^{N-1}\tilde{q}_l h_l| = \sum_{l=0}^{N-1}|\tilde{q}_l||h_l| \geq \rho\|\mathbf{h}\|_1. \tag{4.29}$$

On the other hand, using properties of $\mathbf{q}$ from Lemma 4.1, it follows that

$$\tilde{\mathbf{A}}_{\text{KR}}^{\mathbb{U}}(\mathbf{q} - \rho\mathbf{1}) = \mathbf{F}_N(\mathbf{q} - \rho\mathbf{1}).$$

Then, we have

$$\begin{aligned}
|\langle\tilde{\mathbf{q}}, \mathbf{h}\rangle| &= |\langle\frac{1}{N}\mathbf{F}_N^H\tilde{\mathbf{A}}_{KR}^{\mathbb{U}}\tilde{\mathbf{q}}, \mathbf{h}\rangle| = |\langle\frac{1}{N}\mathbf{F}_N^H\mathbf{D}_N\mathbf{F}_N\tilde{\mathbf{q}}, \mathbf{h}\rangle| \\
&= |\langle\tilde{\mathbf{q}}, \frac{1}{N}\mathbf{F}_N^H\tilde{\mathbf{A}}_{\text{KR}}^{\mathbb{U}}\mathbf{h}\rangle| \leqslant \|\tilde{\mathbf{q}}\|_2\|\frac{1}{N}\mathbf{F}_N^H\tilde{\mathbf{A}}_{\text{KR}}^{\mathbb{U}}\mathbf{h}\|_2 \\
&= \|\tilde{\mathbf{q}}\|_2\frac{1}{\sqrt{N}}\|\tilde{\mathbf{A}}_{\text{KR}}^{\mathbb{U}}\mathbf{h}\|_2 \leqslant (1-\rho)\|\mathbf{A}_{\text{KR}}^{\mathbb{U}}\mathbf{h}\|_2,
\end{aligned} \tag{4.30}$$

7 For nested array, this holds if $M \geq 22$.

where we used the facts that $1-\rho > \rho$, $\mathbf{F}_N^H\mathbf{F}_N = N$, $\|\tilde{\mathbf{q}}\|_\infty \leqslant 1-\rho$ and $\|\mathbf{A}_{\mathrm{KR}}^{\mathbb{U}}\mathbf{h}\|_2 = \|\tilde{\mathbf{A}}_{\mathrm{KR}}^{\mathbb{U}}\mathbf{h}\|_2$. Combining (4.29) and (4.30), we have

$$\|\mathbf{h}\|_1 \leqslant \frac{1-\rho}{\rho}\|\mathbf{A}_{\mathrm{KR}}^{\mathbb{U}}\mathbf{h}\|_2,$$

which completes the proof.

Notice that Theorem 4.2 only requires $\mathbf{p}^\star$ to obey the separation condition; there is no constraint on the other vector $\mathbf{p}^\#$ except requiring it to be non-negative. The Section 4.3.4 demonstrates that the upper bound (4.27) can be used to bound the estimation error of *the following class of optimization problems*, where $\mathbf{p}^\#$ is interpreted as the optimal solution to the respective problem.

4.3.4 Stability Guarantees for Generic Correlation-Matching Techniques

As a direct application of Theorem 4.2, consider the following generic class of optimization problems for estimating $\mathbf{p}^\star$.

$$\min_{\mathbf{z}} f(\mathbf{z}) \qquad (P_{\text{Co-den}})$$
$$\|\hat{\mathbf{r}}_{\mathbf{y}}^{\mathbb{U}} - \mathbf{A}_{\mathrm{KR}}^{\mathbb{U}}\mathbf{z}\|_2 \leqslant \epsilon, \quad \mathbf{z} \succeq \mathbf{0}.$$

The non-negative constraint is natural since the optimization variable $\mathbf{z}$ corresponds to the unknown signal powers. The parameter $\epsilon > 0$ captures the effect of (finite sample) covariance estimation error.[8] The following corollary to Theorem 4.2 shows how (4.27) can be used to obtain stability guarantees for $(P_{\text{Co-den}})$.

Corollary 4.1 Suppose the objective $f(\mathbf{z})$ of the optimization problem $(P_{\text{Co-den}})$ is a continuous function and $\epsilon \geq \|\Delta_{\mathbf{r}}^{\mathbb{U}}\|_2$. Assume the conditions of Theorem 4.2 hold. If $\mathbf{p}^\#$ is a solution to $(P_{\text{Co-den}})$, then

$$\|\mathbf{p}^\# - \mathbf{p}^\star\|_1 \leqslant \frac{1-\rho}{\rho}(\|\Delta_{\mathbf{r}}^{\mathbb{U}}\|_2 + \epsilon). \tag{4.31}$$

Proof: Since $\epsilon \geq \|\Delta_{\mathbf{r}}^{\mathbb{U}}\|_2$, the feasible set of $(P_{\text{Co-den}})$ contains $\mathbf{p}^\star$ and is non-empty. Moreover, since the objective function $f(\mathbf{z})$ of $(P_{\text{Co-den}})$ is continuous and the constraint set $\{\mathbf{z} \succeq \mathbf{0}, \|\hat{\mathbf{r}}_{\mathbf{y}}^{\mathbb{U}} - \mathbf{A}_{\mathrm{KR}}^{\mathbb{U}}\mathbf{z}\|_2 \leqslant \epsilon\}$ is compact, the minimizer $\mathbf{p}^\#$ exists [58]. The result then directly follows from Theorem 4.2 via the triangle inequality

$$\|\mathbf{A}_{\mathrm{KR}}^{\mathbb{U}}(\mathbf{p}^\star - \mathbf{p}^\#)\|_2 \leqslant \|\hat{\mathbf{r}}_{\mathbf{y}}^{\mathbb{U}} - \mathbf{A}_{\mathrm{KR}}^{\mathbb{U}}\mathbf{p}^\star\|_2 + \|\hat{\mathbf{r}}_{\mathbf{y}}^{\mathbb{U}} - \mathbf{A}_{\mathrm{KR}}^{\mathbb{U}}\mathbf{p}^\#\|_2. \qquad \square$$

Remark 4.4 Notice that $(P_{\text{Co-den}})$ actually represents a *class of optimization problems* which can be either convex or non-convex depending on the choice of $f(\mathbf{z})$. The stability guarantee from Corollary 4.1 applies to the estimate produced by each optimization problem from this class. In fact, the bound (4.31) is independent of the specific choice of $f(z)$ and actually applies to any point $\mathbf{p}^\#$ belonging to the feasible set.

---

8 We call this optimization framework $(P_{\text{Co-den}})$ (where "Co-den" is abbreviation for "correlation-aware denoising") since it resembles the popular basis pursuit denoising [57], except that $\hat{\mathbf{r}}_{\mathbf{y}}^{\mathbb{U}}$ is the estimated correlation of the measurements.

Remark 4.5 ***Exact Reconstruction and Size of Support*** In absence of noise (i.e., $\sigma = 0$) and correlation estimation error (i.e., $\Delta_f = 0$), Corollary 4.1 ensures exact recovery of $\mathbf{p}^\star$. In this case, the only assumption is that $\mathbf{p}^\star$ should obey the separation condition, which in turn determines the size of the recoverable support as follows. The separation condition dictates that any two indices in S are separated by at least $2N/M_{ca}$. This implies that the size of the support must satisfy $|S| \leqslant M_{ca}/2$. Since $M_{ca} = \Theta(M^2)$ for an OOSA, Corollary 4.1 indicates that it is possible to recover well-separated supports of size $\Theta(M^2)$.

Remark 4.6 ***On Error Amplification and Stability*** The upper bound (4.31) on the estimation error $\|\mathbf{p}^\# - \mathbf{p}^\star\|_1$ is an amplified version of the covariance estimation error $\|\Delta_\mathbf{r}^{\mathbb{U}}\|_2$, where the amplification factor is $(1-\rho)/\rho \approx (1/\rho) = c(N/M_{ca})^2$, c being a universal constant. Such amplification of noise during signal reconstruction has been recently observed/analyzed in the context of super-resolution imaging [32], where the goal is to recover an N-dimensional sparse signal (with both high and low frequency components) from only K *low-frequency* DFT coefficients. The noise amplification factor (NAF) in such cases is known to scale as $(N/K)^2$ [18, 32, 59]. From the structure of $\mathbf{A}_{\text{KR}}^{\mathbb{U}}$ as given by (4.17), we can easily identify the vector $\mathbf{A}_{\text{KR}}^{\mathbb{U}}\mathbf{p}^\star$ as the $K = 2M_{ca}+1$ low-frequency DFT coefficients (symmetrically located around 0) of $\mathbf{p}^\star$. Hence the error bound is consistent with known results from super-resolution image reconstruction.

4.4 Support Recovery with High Probability: How Many Snapshots Suffice?

Theorem 4.2 shows that it is possible to perform stable reconstruction of $\mathbf{p}^\star$ from the estimated correlation matrix, by exploiting its non-negativity. However, in many applications (such as direction-of-arrival in radar, source localization using EEG measurements), it is the support S of $\mathbf{p}^\star$ that carries the information of interest. Hence, it is crucial to understand how the error bound in (4.27) can be utilized to overcome this weakness and ensure support recovery in the regime $|S| = \Theta(M^2)$. Notice that (4.31) is derived in terms of the covariance estimation error $\|\Delta_\mathbf{r}^{\mathbb{U}}\|_2$. Since we estimate the covariance matrix using a finite number L of snapshots, $\Delta_\mathbf{r}$ is a random variable whose distribution depends on L. Hence, the guarantees of Theorem 4.2 are actually *conditioned on* $\Delta_\mathbf{r}$. In this section, probabilistic guarantees on recovering S in the regime $|S| > M$ are derived by utilizing the distribution of $\Delta_\mathbf{r}$, which will explicitly reveal the role of the number of measurement vectors L. To make the analysis tractable, the following additional assumptions are made on the distribution of $\mathbf{y}[l]$ similar to [5, 27, 40]:

- **[(A4)]** The measurements $\{\mathbf{y}[l]\}_{l=1}^{L}$ are zero mean i.i.d complex Gaussian random vectors distributed as $\mathbf{y}[l] \sim \mathcal{CN}(\mathbf{0}, \mathbf{R}_{\mathbf{yy}})$.

4.4.1 Characterizing the Snapshot Requirement for Support Recovery with High Probability

In order to recover S from the estimate $\mathbf{p}^\#$, the following simple element-wise Hard-Thresholding operator is employed $\mathcal{H}_T : \mathbb{R}^N \to \mathbb{R}^N$ (with threshold $T \in \mathbb{R}_+$)

$$[\mathcal{H}_T(\mathbf{x})]_i = \begin{cases} x_i & \text{If } x_i > T \\ 0 & \text{If } x_i \leqslant T. \end{cases}$$

Define

$$p_{\min} \triangleq \min_{i \in S} p_i^\star. \tag{4.32}$$

The following theorem provides explicit conditions on $p_{\min}$, L, as well as the choices of parameters ϵ in ($P_{\text{Co-den}}$) and T, which ensure recovery of S (with high probability in L) by applying the Hard-Thresholding operator $\mathcal{H}_T$ on the solution of ($P_{\text{Co-den}}$). The guarantees continue to hold in the regime $M < |S| \leqslant M_{ca}$, where $M_{ca} = \Theta(M^2)$ for any OOSA.

Theorem 4.3 *[31]* Consider the MMV model (4.1) satisfying the assumptions **(A1–A4)**. Further assume that $\mathbf{p}^\star \in \mathcal{P}_{\text{sep}}^+$, $M_{ca} \geq 128$, $N \geq 3.03(2M_{ca} + 1)$ and

$$p_{\min} > 4\frac{1-\rho}{\rho}\sigma^2 \tag{4.33}$$

$$L > \frac{32M^2(\|\mathbf{p}^\star\|_1 + \sigma^2)^2}{(p_{\min} - 4\frac{1-\rho}{\rho}\sigma^2)^2}\left(\frac{1-\rho}{\rho}\right)^2. \tag{4.34}$$

If $\mathbf{p}^\#$ is a solution to ($P_{\text{Co-den}}$) (where $f(\mathbf{z})$ can be any continuous function), then, with probability at least $1 - 2e^{-2c\sqrt{L}}$, we have

$$\text{Supp}\left(\mathcal{H}_T(\mathbf{p}^\#)\right) = S, \tag{4.35}$$

provided that we choose the parameter ϵ of ($P_{\text{Co-den}}$), and the threshold T from the following ranges:

$$\sigma^2 + \frac{\sqrt{2}M(\|\mathbf{p}^\star\|_1 + \sigma^2)}{\sqrt{L}} \leqslant \epsilon < \left(\frac{\rho}{1-\rho}\right)\frac{p_{\min}}{4}, \tag{4.36}$$

and

$$\left(\frac{1-\rho}{\rho}\right)(\epsilon + \frac{\sqrt{2}M(\|\mathbf{p}^\star\|_1 + \sigma^2)}{\sqrt{L}} + \sigma^2) \leqslant T$$
$$< p_{\min} - \left(\frac{1-\rho}{\rho}\right)(\epsilon + \frac{\sqrt{2}M(\|\mathbf{p}^\star\|_1 + \sigma^2)}{\sqrt{L}} + \sigma^2). \tag{4.37}$$

Proof: The proof combines the universal error bound (4.27) with certain concentration inequalities on the finite-sample estimation error $\Delta_\mathbf{r}$. Please refer to [31] for details. □

Remark 4.7 ***Guaranteed Support Recovery with finite*** L ***when*** $|S| > M$: Theorem 4.3 shows that as long as L is larger than the threshold (4.34) and the minimum source power $p_{\min}$ is large enough compared to the noise power σ^2 (as given by (4.33)), it is possible to *exactly recover* supports that obey the *separation condition.* In particular, such supports can be recovered by first solving ($P_{\text{Co-den}}$) (with any continuous objective $f(\mathbf{z})$) with the parameter ϵ tuned to be in the range (4.36), followed by hard-thresholding with the threshold T chosen from the range (4.37). The size of the recoverable support is implicitly controlled by the *separation condition* and as discussed in Section 4.3.3, the size can be as large as $s \leqslant M_{ca} = \Theta(M^2)$. Hence, for any OOSA (such as nested and coprime arrays), it is indeed possible to recover supports of size $\Theta(M^2)$ with overwhelming probability (that increases to 1 exponentially in $\sqrt{L}$).

4.4.2 Tightness of the Upper Bound

The preceding guarantees on support recovery build on the universal upper bound (4.27) on the distance between $\mathbf{p}^\star \in \mathcal{P}_{sep}^+$ (obeying the separation condition) and any *non-negative* vector $\mathbf{p}^\#$ that serves as an estimate of $\mathbf{p}^\star$. As discussed in Remark 4.6 and Theorem 4.2, for large grid sizes (in particular, when $N/(2M_{ca}+1) \geq 3.03$) the upper bound essentially scales as

$$\|\mathbf{p}^\star - \mathbf{p}^\#\|_1 \leqslant A(N,M)\|\mathbf{A}_{\mathrm{KR}}^{\mathbb{U}}(\mathbf{p}^\star - \mathbf{p}^\#)\|_2, \tag{4.38}$$

where

$$A(N,M) = 34\frac{N^2}{M_{ca}^2}$$

represents the amplification factor, i.e., the factor by which $\|\mathbf{A}_{\mathrm{KR}}^{\mathbb{U}}(\mathbf{p}^\star - \mathbf{p}^\#)\|_2$ gets amplified in the reconstruction/estimation error $\|\mathbf{p}^\star - \mathbf{p}^\#\|_1$. Notice that (4.27) is valid for any non-negative pair of signals $(\mathbf{p}^\star, \mathbf{p}^\#)$ as long as $\mathbf{p}^\star$ also obeys the separation condition. Hence, in order to show tightness of the upper bound, one needs to construct a specific pair of vectors $(\mathbf{p}_1^\star, \mathbf{p}_1^\#)$ with $\mathbf{p}_1^\star, \mathbf{p}_1^\# \geq \mathbf{0}$ and $\mathbf{p}_1^\star \in \mathcal{P}_{\mathrm{sep}}^+$, such that

$$\|\mathbf{p}_1^\star - \mathbf{p}_1^\#\|_2 \geq CN^2\|\mathbf{A}_{\mathrm{KR}}^{\mathbb{U}}(\mathbf{p}_1^\star - \mathbf{p}_1^\#)\|_2, \tag{4.39}$$

where C is a constant (possibly depending only on M which is assumed to be constant). Establishing such a *lower bound* that also scales quadratically in N, will establish that the exponent of N in the upper bound (4.38) is indeed tight. The following theorem shows that such $\mathbf{p}_1^\star, \mathbf{p}_1^\#$ satisfying (4.39) indeed exist.

Theorem 4.4 *[31]* There exist $\mathbf{p}_1^\star \in \mathcal{P}_{\mathrm{sep}}^+$ and $\mathbf{p}_1^\# \geq \mathbf{0}$ such that whenever $M_{ca} \geq 128$ and $N \geq 3.03(2M_{ca}+1)$, we have

$$\|\mathbf{p}_1^\star - \mathbf{p}_1^\#\|_1 \leqslant C_1(M)N^2\|\mathbf{A}_{\mathrm{KR}}^{\mathbb{U}}(\mathbf{p}_1^\star - \mathbf{p}_1^\#)\|_2$$

and

$$\|\mathbf{p}_1^\star - \mathbf{p}_1^\#\|_1 \geq C_2(M)N^2\|\mathbf{A}_{\mathrm{KR}}^{\mathbb{U}}(\mathbf{p}_1^\star - \mathbf{p}_1^\#)\|_2, \tag{4.40}$$

where $C_1(M)$ and $C_2(M)$ are constants (which are functions of M) given by

$$C_1(M) = \frac{34}{M_{ca}^2}, \quad C_2(M) = \frac{1}{KM_{ca}^{2.5}}$$

with $K = \pi^2\sqrt{\frac{7}{15}} + 0.1973\pi^3$.

Proof: The upper bound is due to Theorem 4.2. For the lower bound, refer to [31]. □

The explicit construction of $\mathbf{p}_1^\star$ and $\mathbf{p}_1^\#$ in [31] shows that the exponent of N in $A(N,M)$ is indeed tight and for fixed M

$$\frac{\|\mathbf{p}_1^\star - \mathbf{p}_1^\#\|_1}{\|\mathbf{A}_{\mathrm{KR}}^{\mathbb{U}}(\mathbf{p}_1^\star - \mathbf{p}_1^\#)\|_2} = \Theta(N^2). \tag{4.41}$$

The construction of $\mathbf{p}_1^\star$ and $\mathbf{p}_1^\#$ also sheds further light into what kind of source power and its estimate may result in the worst-case estimation error. In the proof of Theorem 4.4, $\mathbf{p}_1^\star$ is constructed as a single non-negative spike at index l and hence it automatically satisfies the separation condition. The vector $\mathbf{p}_1^\#$, on the other hand, consists of two non-negative spikes symmetrically placed on either side of the true spike at indices $l-1$ and $l+1$. This represents a classical scenario where two very closely spaced spikes can be misdetected as a single spike, or when a single spike splits into two closely spaced neighboring spikes [60]. Under these scenarios, Theorem 4.4 shows that the estimation error indeed grows as $\Theta(N^2)$.

4.4.3 Numerical Experiments

We now illustrate the performance of sparse arrays for support recovery in the regime $s > M$ through a series of numerical experiments. The correlation-aware support recovery framework is implemented with $\mathbf{A}$ chosen as the array manifold of a nested array (an OOSA) given by (4.11). The joint support S is generated uniformly at random obeying the separation condition in Def. 4.2. The non-zero elements of $\mathbf{p}^\star$ are assumed to be all equal to 10 (i.e., $p_{\min} = 10$) and the noise power is assumed to be $\sigma = 1$. We consider two metrics for evaluating performance: (i) Power Estimation Error and (ii) Success Rate of Support Recovery. Similar to [21, 36, 40, 42], we claim that support recovery is successful if the support of the s largest non-zero entries of $\mathbf{p}^\#$ is identical to the true support S. Notice that $(P_{\text{Co-den}})$ actually represents a class of optimization problems, which can be convex or non-convex depending on the choice of $f(\mathbf{z})$. In this paper, we consider the following three specific instances of this family

- Case 1: $f(\mathbf{z}) = \|\mathbf{z}\|_1$. We call this problem $(P1_{\text{Co-den}})$ and implement it using standard CVX package.
- Case 2: $f(\mathbf{z}) = \|\mathbf{z}\|_0$. We call this problem $(P0_{\text{Co-den}})$. Since it is a non-convex problem, we approximate it using a reweighted l_1 minimization heuristic [61].
- Case 3: $f(\mathbf{z}) = \|\mathbf{z}\|_{1/2}$. We call this $(P_{1/2,\text{Co-den}})$. This is a non-convex quasi-norm minimization problem with non-negative constraints, which we solve using a majorization–minimization-based heuristic [62].

4.4.3.1 Power Estimation Error and the Universal Upper Bound

We first compare the normalized power estimation error $\|\mathbf{p}^\# - \mathbf{p}^\star\|_1/\|\mathbf{p}^\star\|_1$(denoted by NE) for each of $(P1_{\text{Co-den}})$, $(P0_{\text{Co-den}})$, and $(P_{1/2,\text{Co-den}})$ against the theoretical upper bound from (4.31). For this experiment, we let $M = 12, N = 200$, and $s = 20$. We vary the number of snapshots L and compute the empirical mean covariance perturbation error $\|\boldsymbol{\Delta}_\mathbf{r}\|_2$ for each L by averaging over 100 runs. Fig. 4.1 shows the mean normalized power estimation error (NE) as a function of the (empirical) mean covariance perturbation $\|\boldsymbol{\Delta}_\mathbf{r}\|_2$, establishing the validity of the upper bound (4.31) for both convex and non-convex problems.

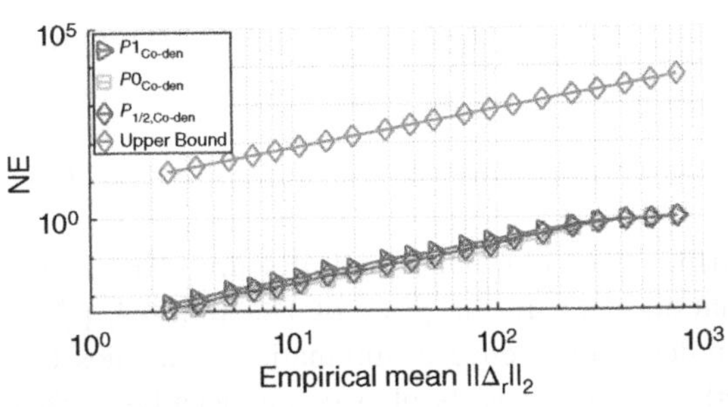

Figure 4.1 Mean Normalized Power Estimation Error (NE) and the Upper Bound (4.31) as a function of $\|\boldsymbol{\Delta}_\mathbf{r}\|_2$. Here $M = 12$, $N = 200$, $s = 20$, and $\mathbf{A}$ is generated using a nested array with $M_{ca} = 41$. The results are averaged over 100 runs.

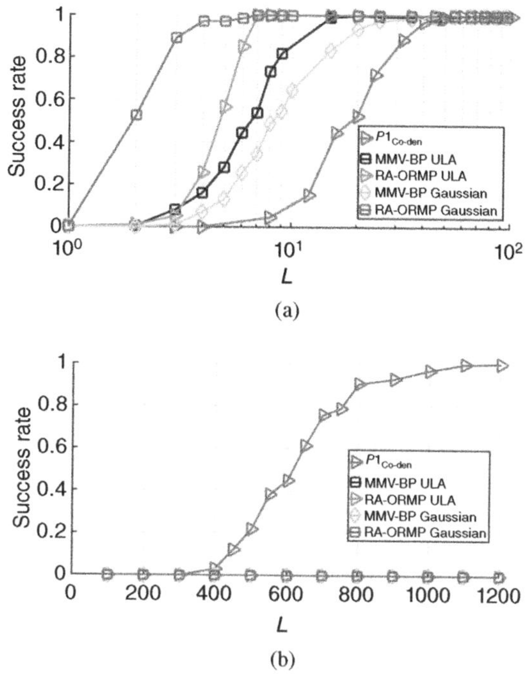

Figure 4.2 Probability of successful support recovery as a function of L. Here, $M = 24, N = 400$, and the results are averaged over 300 runs. (a) $s = 8 < M = 24$. (b) $s = 72 > M = 24$.

4.4.3.2 Comparison of Support Recovery as a function of L and s

We compare the success rate of $(P1_{\text{Co-den}})$ (using nested arrays), against two conventional MMV algorithms: MMV-BP [23] and RA-ORMP [21].[9] Similar to the experimental settings in [21],[10] we implement these algorithms either using a random Gaussian $\mathbf{A}$ with i.i.d zero-mean unit-variance elements, or the manifold of a ULA with elements given by $\mathbf{A}_{m,n} = e^{-j2\pi mn/N}, 1 \leqslant m \leqslant M, 0 \leqslant n \leqslant N-1$. Let $M = 24, N = 400$. Fig. 4.2a shows the success rate of the algorithms as a function of L in two distinct regimes (a) $s = 8 < M$ and (b) $s = 72 > M$. When $s < M$, Fig. 4.2a shows that RA-ORMP with random Gaussian measurement matrix $\mathbf{A}$ has the highest success rate for a given L. On the other hand, when $s > M$, MMV-BP, and RA-ORMP completely fail to localize the sources since they do not use the correlation structure of the data. However, $(P1_{\text{Co-den}})$ successfully recovers the sources in this regime with probability > 0.9 when $L > 800$. Notice that the number of snapshots (L) needed to ensure success rate of 1 is higher in this case than when $s < M$. Fig. 4.3 shows the success rate of all the algorithms as a function of s for a fixed $L = 100$. As expected, MMV-BP and RA-ORMP are unable to recover supports of size $s > M = 24$ while $(P1_{\text{Co-den}})$ is able to recover supports of size significantly larger than 25.

4.4.3.3 Comparison with Vector Approximate Message Passing

Very recently, the authors of [63] proposed a new sparse signal reconstruction algorithm called Vector Approximate Message Passing (VAMP). Compared to traditional Approximate Message Passing [64], VAMP is applicable to a larger class of measurement matrices $\mathbf{A}$. In this section, we conduct numerical experiments to compare $(P1_{\text{Co-den}})$ with VAMP, especially in the regime $s > M$. We implemented VAMP using the code from the website of the authors of [63] and used similar

9 We implemented MMV-BP using the SPGL1 solver for MATLAB available at http://www.cs.ubc.ca/mpf/spgl1/.
10 Since MMV-BP and RA-ORMP do not utilize correlation of the measurements, they are unable to exploit the difference co-array of sparse arrays and hence are unsuitable to be used with such measurement matrices.

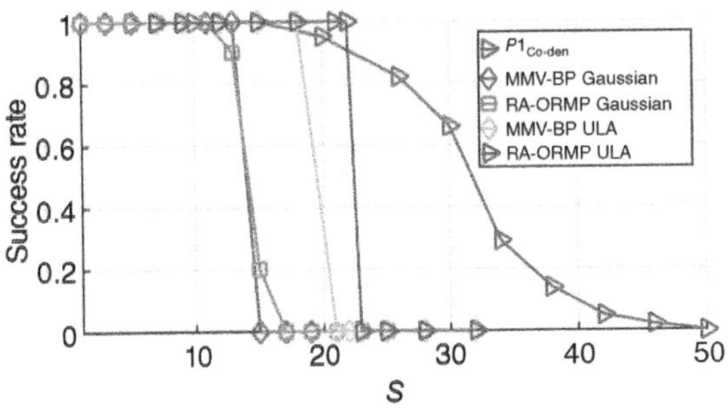

Figure 4.3 Probability of successful support recovery as a function of sparsity s. Here, $M = 24, N = 300, L = 100$. For nested array, we have $M_{ca} = 155$. Results are averaged over 100 runs.

simulation settings as suggested by the demo code. The non-zero entries of signal and noise are generated as zero-mean i.i.d Gaussian random variables. The nonzero signal powers are assumed to be equal to p, while the noise power is denoted by σ^2. We implemented VAMP with two types of measurement matrices: (i) **A** representing a nested array and constructed according to (4.8), and (ii) **A** with i.i.d Gaussian entries as defined in the demo code of [63]. VAMP returns an estimate for $\mathbf{X}_L$ and we estimate the support by retaining the indices of s rows of this estimate with the largest l_2 norm. We consider two simulation settings:

- **Case 1:** We use i.i.d Gaussian **A** for VAMP and choose $N = 200, p = 10,000, \sigma^2 = 1, L = 1000$.
- **Case 2:** We use **A** corresponding to a nested array for VAMP and choose $N = 200, p = 1000, \sigma^2 = 1, L = 100$.

For both cases, we implement ($P1_{\text{Co-den}}$) using nested arrays. The probability of successful support recovery as a function of sparsity s is plotted in Fig. 4.4. From the simulations, it appears that when **A** corresponds to a nested array, VAMP fails to detect support in the regime $s > M = 150$. On the other hand, when **A** is an i.i.d. Gaussian matrix, both algorithms are able to recover supports of size larger than M. For example, both ($P1_{\text{Co-den}}$) and VAMP can recover supports of size larger than 100 with $M = 36$ and $M = 80$ measurements respectively.

4.4.3.4 Phase Transition

We study phase transition of ($P1_{\text{Co-den}}$), MMV-BP, and RA-ORMP by plotting their success rates as functions of both M and s. As before, we implement ($P1_{\text{Co-den}}$) using nested arrays, while we use a ULA measurement matrix for MMV-BP and RA-ORMP. We fix $N = 600, L = 2000$. For each (M, s) pair, we compute the success rate over 100 Monte Carlo runs. Figure 4.5 shows the average success rate with respect to the number of sensors M and number of sources s, where black pixels

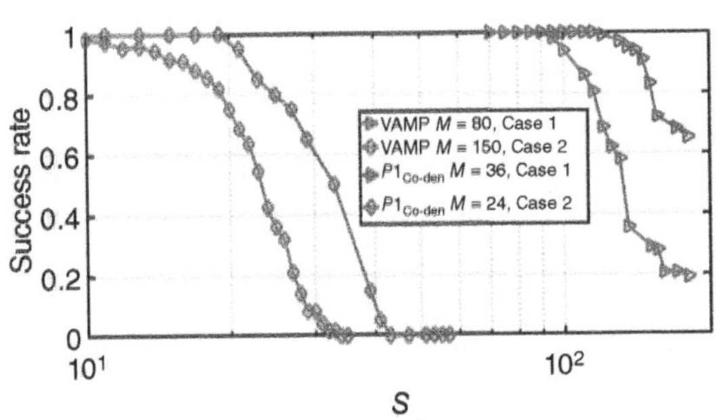

Figure 4.4 Comparative probability of successful support recovery for ($P1_{\text{Co-den}}$) and VAMP as a function of sparsity s. Results are averaged over 100 runs.

Figure 4.5 Phase transition of success rate as function of sparsity s and number of measurements M: (a) $(P1_{\text{Co-den}})$, (b) RA-ORMP, and (c) MMV-BP. White pixels indicate perfect recovery and black pixels denote total failure. Here $L = 2000, N = 600$ and the results are averaged over 50 runs. The overlaid red curve represents $s = 0.18M^2$ in (a) and $s = M$ in (b) and (c).

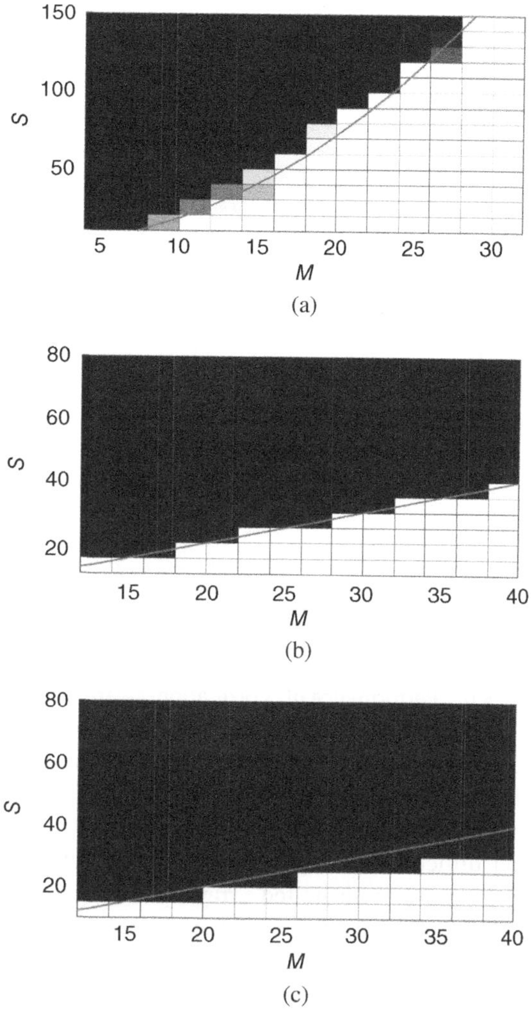

represent zero success rate and white pixels denote 100% success rate. For comparison, we also overlay the curve $s = 0.18M^2$ on the phase transition plot of $(P1_{\text{Co-den}})$ which roughly matches the phase transition boundary and validates our claim that it is possible to localize $s = \Theta(M^2)$ sources with nested arrays. The phase transition regions of MMV-BP and RA-ORMP, on the other hand, show that the number of resolvable sources scales linearly with M, matching their theoretical performance limits [21, 23].

4.4.3.5 Achievability of Upper Bound

In Theorem 4.4, we showed that for fixed M, there exist $\mathbf{p}_1^\star$ and $\mathbf{p}_1^\#$ such that (4.41) holds. In this section, we demonstrate this scaling with respect to N by plotting $\|\mathbf{A}_{\text{KR}}^{\mathbb{U}}(\mathbf{p}_1^\star - \mathbf{p}_1^\#)\|_2$ against N for fixed $\|\mathbf{p}_1^\star - \mathbf{p}_1^\#\|_1$. We generate $\mathbf{p}_1^\star$ and $\mathbf{p}_1^\#$ according to the proof of Theorem 4.4, such that $\mathbf{p}_1^\star$ contains a single spike and $\mathbf{p}_1^\#$ consists of two neighboring equal power spikes given by Appendix D in [31]. Figure 4.6 clearly shows that $\|\mathbf{A}_{\text{KR}}^{\mathbb{U}}(\mathbf{p}_1^\star - \mathbf{p}_1^\#)\|_2$ indeed scales as $\Theta(N^2)$.

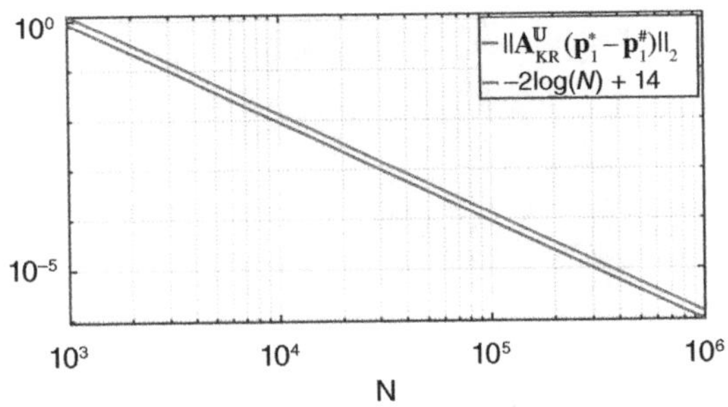

Figure 4.6 Log–log plot of $\|\mathbf{A}^{\mathrm{U}}_{\mathrm{KR}}(\mathbf{p}^{\star}_1 - \mathbf{p}^{\#}_1)\|_2$ as a function of N, for fixed $\|\mathbf{p}^{\star}_1 - \mathbf{p}^{\#}_1\|_1 = 2$. Here $\mathbf{p}^{\star}_1$ and $\mathbf{p}^{\#}_1$ are constructed according to the proof of Theorem 4.4. The plot is overlaid with the function $-2\log(N) + 14$. The two plots share the same slope of -2, demonstrating the order-wise tightness of (4.27) with respect to N.

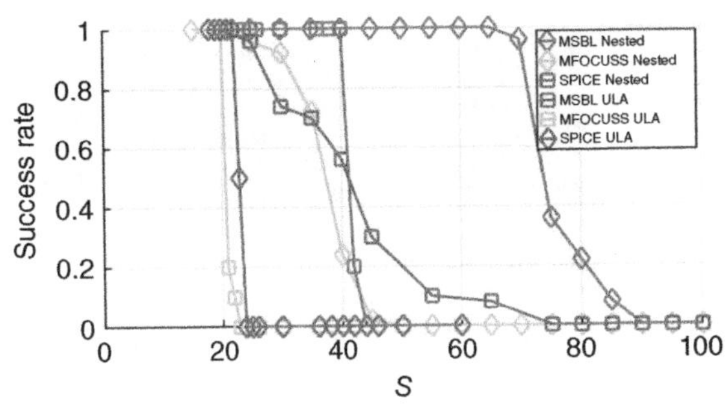

Figure 4.7 Success rate of M-SBL, M-FOCUSS and SPICE as a function of sparsity s. Here $M = 24, N = 300, L = 100$. For nested array, $M_{ca} = 155$. The results are averaged over 200 runs.

4.4.3.6 Performance of "Correlation-Aware" Algorithms for MMV Models

In our final set of experiments, we demonstrate the superior performance of a family of related "correlation-aware" algorithms for MMV models, namely M-SBL [27], M-FOCUSS [14] and SPICE [16] that also utilize the correlation of the measurements, and are potentially capable of resolving more sources than sensors when used with OOSA. However, these algorithms are non-convex and difficult to analyze, especially with finite L. Nonetheless, they show excellent empirical performance with nested arrays, as demonstrated in Fig. 4.7. We fix $M = 24, N = 300, L = 100$, and plot the success rate of these three algorithms as a function of L, for both ULA and nested array (with same number of sensors). It can be seen that MSBL, MFOCUSS, and SPICE have significantly higher success rate with nested arrays, and cannot resolve more sources than sensors with ULA. Hence, Fig. 4.7 empirically demonstrates that exploiting the difference co-array of OOSA can help these algorithms identify more sources than sensors with finite measurements. A rigorous *non-asymptotic* analysis of this behavior is a very interesting problem for future research. We conclude this section by noting that successful support recovery for the on-grid measurement model also implies exact identification of the source directions, whenever they coincide with the grid points. If the source directions do not lie on the grid, then the grid size may introduce bias, and limit the achievable resolution for "gridless" direction estimation. Indeed, extending the current non-asymptotic analysis to the gridless setting is a major direction of ongoing and future research.

4.5 Single-Snapshot Virtual Array Interpolation: Deterministic Guarantees

We finally study the problem of completing the virtual array of a nested array with a single snapshot. Existing bodies of work aim at "completing a virtual ULA" with the same aperture as the sparse array, by estimating/interpolating the missing measurements with a single snapshot [65].

The virtual measurements can then be used for diverse tasks such as beamforming and source localization, aided by the enhanced resolution of the filled aperture of the virtual ULA [34, 65–68]. A popular approach is to synthesize the virtual ULA measurements by using low-rank Toeplitz or Hankel matrix completion [66, 67, 69]. Indeed, the virtual measurements can be arranged in the form of a *low-rank* Hankel/Toeplitz matrix, and the measurements acquired by the sparse array only reveal certain entries of this matrix. In practice, for computational tractability, the rank constraint is often relaxed to a suitable convex surrogate, such as the nuclear norm or atomic norm [34, 66, 67, 69].

Although the aforementioned algorithms can also be applied for nested virtual array completion with only one snapshot, there exists a disconnect between theory and practice. In order to develop guarantees for *deterministic* sparse array completion using nuclear norm minimization, one often needs to invoke certain coherence conditions on the virtual Toeplitz/Hankel matrix, and utilize specific graph-based array designs [34, 67]. On the other hand, theoretical guarantees for atomic norm minimization typically assume *randomized* sparse arrays, and require the source locations to satisfy a certain minimum separation even in the absence of noise [18, 65, 70]. These results therefore do not apply to deterministic spatial samplers such as nested arrays.

Tight necessary and sufficient conditions for single-snapshot virtual array interpolation of a *deterministic* sparse array remained an open question, until it was recently resolved in [35] for a nested array. In [35], the first necessary and sufficient conditions were provided for such virtual array interpolation to be provably successful in the absence of noise. The results exploit the geometry of nested arrays, and do not depend on the source configuration or on the separation between sources. In the following, we will review and discuss this result.

4.5.1 Matrix Completion with Nested Array

Consider K far-field narrowband sources (with wavelength λ) impinging from directions $\{\theta_i\}_{i=1}^{K}$ on a one-dimensional nested array with $P = 2M$ sensors whose locations are given by $\mathbb{S}_{\text{nest}}$, where $\mathbb{S}_{\text{nest}} := \mathbb{S}_1 \cup \mathbb{S}_2$ is the union of integer sets $\mathbb{S}_1 = \{m-1\}_{m=1}^{M+1}$ and $\mathbb{S}_2 = \{m(M+1)-1\}_{m=2}^{M}$. The signal received at the nested array is given by:

$$\mathbf{y}_{\text{nest}} = \mathbf{A}_{\mathbb{S}_{\text{nest}}}(\boldsymbol{\omega})\mathbf{x} + \mathbf{n}, \tag{4.42}$$

where $\mathbf{x} \in \mathbb{R}^K$ denotes real-valued[11] (deterministic) source signals and $\mathbf{n}$ is an additive noise term. The normalized spatial frequencies are denoted by $\boldsymbol{\omega} = \{\omega_k\}_{k=1}^{K}$, with $\omega_i = \pi \sin(\theta_i)$.

The difference set $\mathbb{D}_{\mathbb{S}_{\text{nst}}}$ of $\mathbb{S}_{\text{nst}}$ is defined as $\mathbb{D}_{\mathbb{S}_{\text{nst}}} := \{m-n, m, n \in \mathbb{S}_{\text{nst}}\}$. It is well known that the set of non-negative elements in $\mathbb{D}_{\mathbb{S}_{\text{nst}}}$ are given by $\mathbb{U} := \{0, 1, \cdots, N-1\}$ where $N = M(M+1)$[1]. In the absence of noise, we can rewrite (4.42) as

$$\mathbf{y}_{\text{nest}} = \mathbf{S}_{\text{nst}}\mathbf{y}_{\text{full}}, \quad \mathbf{y}_{\text{full}} := \mathbf{A}_{\mathbb{U}}(\boldsymbol{\omega})\mathbf{x}, \tag{4.43}$$

where $\mathbf{S}_{\text{nest}}$ is a row-selection matrix given by:

$$\left[\mathbf{S}_{\text{nst}}\right]_{i,j} = \begin{cases} 1, & \text{if } d_i + 1 = j, d_i \in \mathbb{S}_{\text{nest}} \\ 0, & \text{otherwise.} \end{cases}$$

The vector $\mathbf{y}_{\text{full}}$ is a "*virtual measurement*," received by the virtual array $\mathbb{U}$, *with identical source configurations* (same DOAs $\boldsymbol{\omega}$ and source signal $\mathbf{x}$).

---

11 A similar setting with real source signals has been considered in [71]. In future, we will extend our theoretical results for the complex case.

Key Question: We are interested in the problem of "sparse array interpolation" with only a *single snapshot*, where the goal is to estimate $\mathbf{y}_{\text{full}}$ from $\mathbf{y}_{\text{nest}}$. As discussed earlier, theoretical guarantees for matrix-completion or atomic norm minimization based virtual array synthesis do not readily extend to nested arrays. This raises the open question: *What are the necessary and sufficient conditions under which rank-minimization with nested arrays leads to exact virtual array completion?*

4.5.2 Guaranteed Single Snapshot Interpolation with Nested Matrix Completion

Consider the noiseless measurement model (4.42) with $\mathbf{n} = \mathbf{0}$. From (4.43), it can be seen that when $\mathbf{x}$ is real, the matrix $\mathcal{T}_N(\mathbf{y}_{\text{full}})$ admits the following Vandermonde decomposition:

$$\mathcal{T}_N(\mathbf{y}_{\text{full}}) = \mathbf{A}_{\mathbb{U}}(\boldsymbol{\omega})\text{diag}(\mathbf{x})\mathbf{A}_{\mathbb{U}}^H(\boldsymbol{\omega}). \tag{4.44}$$

Consider the rank-minimization problem

$$\min_{\mathbf{u}\in\mathbb{C}^N} \text{rank}[\mathcal{T}_N(\mathbf{u})] \quad \text{subject to } \mathbf{S}_{\text{nest}}\mathbf{u} = \mathbf{y}_{\text{nest}}. \tag{P1}$$

The following theorem provides necessary and sufficient conditions under which perfect interpolation is possible (in absence of noise) by solving (P1).

Theorem 4.5 *[35]* Consider the measurement model (4.42) with $\mathbf{n} = \mathbf{0}$. If $K \leqslant M$, then problem (P1) has a unique solution $\mathbf{u}^\star$ satisfying $\mathbf{u}^\star = \mathbf{y}_{\text{full}} = \mathbf{A}_{\mathbb{U}}(\boldsymbol{\omega})\mathbf{x}$, for every $\boldsymbol{\omega}$ and $\mathbf{x}$. Conversely if $K > M$, there exist source configurations with K source angles $\boldsymbol{\omega}_0 \in [-\pi, \pi)^K$, and amplitudes $\mathbf{x}_0 \in \mathbb{R}^K$, such that one can find a vector $\hat{\mathbf{y}}$, with $\hat{\mathbf{y}} \neq \mathbf{y}_{\text{full}}$ (where $\mathbf{y}_{\text{full}} = \mathbf{A}_{\mathbb{U}}(\boldsymbol{\omega}_0)\mathbf{x}_0$), satisfying

$$\mathbf{S}_{\text{nest}}\hat{\mathbf{y}} = \mathbf{S}_{\text{nest}}\mathbf{y}_{\text{full}}, \quad \text{rank}\left(\mathcal{T}_N(\hat{\mathbf{y}})\right) \leqslant K. \tag{4.44}$$

Proof: For a complete proof, please refer to [35]. The converse result constructs a source configuration for which exact interpolation for nested array will provably fail. □

Theorem 4.1 guarantees that when $K \leqslant M$, it is possible to perfectly interpolate the missing sensors in a nested array, by solving the rank-minimization problem (P1) *regardless of the separation* between the sources. In fact, it can be shown that the sufficient condition broadly applies to any sparse array with a ULA segment of length at least K. It also shows that, even with a single snapshot, a nested array can identify $O(M)$ sources (by applying any subspace based technique on the output of (P1)). As will be shown in simulations, the enhanced resolution of the large filled aperture of the nested array leads to superior performance in the presence of noise. In the future, we will analyze how the outer ULA $\mathbb{S}_2$ controls this noisy interpolation error.

Theoretical conditions under which nested arrays will provably fail to identify sources with one snapshot were also unavailable.[12] An important contribution of Theorem 4.5 is to settle this question by showing that the sufficient condition is also necessary. This is done by constructing ambiguous source configurations that exploit the specific geometry of nested arrays. While Theorem 4.5 applies to noiseless measurements, a future direction is to obtain corresponding guarantees for noisy matrix completion by relating the maximum possible interpolation error to the noise power, under suitable assumptions on the minimum separation between source directions.

12 Most existing works focus on the multi-snapshot setting [1, 72, 73].

4.5.3 Numerical Examples

We solve a relaxed version of (P1) by relaxing the rank with nuclear norm[13] and by replacing the equality constraint by a norm constraint $\|\mathbf{S}_{\text{nest}}\mathbf{u} - \mathbf{y}_{\text{nest}}\|_2 \leqslant \epsilon$, assuming that the noise is bounded as $\|\mathbf{n}\|_2 \leqslant \epsilon$. We call this approach Toeplitz Completion (TC). We first show the benefits of interpolation in beamforming with nested arrays using a single snapshot. We consider noiseless measurements acquired by a nested array with $P = 12$ sensors, comprised of three sources with amplitudes $\mathbf{x} = [1, -1, 1]$. In Fig. 4.8, we plot the beam pattern obtained by interpolating the nested array measurements using TC, and then beamforming with the interpolated measurements for two different DOA configurations. For comparison, we plot the beam pattern obtained from beamforming with the physical nested array (without interpolation). We also perform interpolation with Atomic Norm Minimization [65] and plot the resulting beam pattern in Fig. 4.8 (last row). Due to large side lobes of the nested array, the source locations are not distinguishable when using the physical measurements. On the other hand, using the interpolated signal produced by TC, we can identify three closely spaced sources. Beamforming with the interpolated measurements using ANM fails to resolve sources with small separation (Fig. 4.8a), and succeeds only when the separation is large enough (Fig. 4.8b).

We next study the DOA estimation error and the interpolation error in presence of noise. We consider a nested array with $P = 12$ sensors and $K = 5$ sources with spatial frequencies $\omega = \{\pi/20 + 0.1k\}_{k=0}^{4}$, and fixed amplitudes $\mathbf{x} = [1, -1, 1, 1, -1]^{\mathrm{T}}$. The additive noise is generated as $\mathbf{w} \sim \mathcal{U}(-\sigma/2, \sigma/2)$ and the SNR $= 10\log(1/\sigma^2)$ is controlled by varying σ. In Fig. 4.9a we plot the RMSE of DOA estimates (computed over 200 trials) as a function of SNR, by performing Root-MUSIC on the output of TC. We also compare against Successive cancellation beamforming (SC-Beam) [74], atomic norm minimization (ANM) [75], and Hankel Completion (HC) [67], all of which permit single-snapshot DOA estimation with (arbitrary) sparse arrays, although their performances vary.

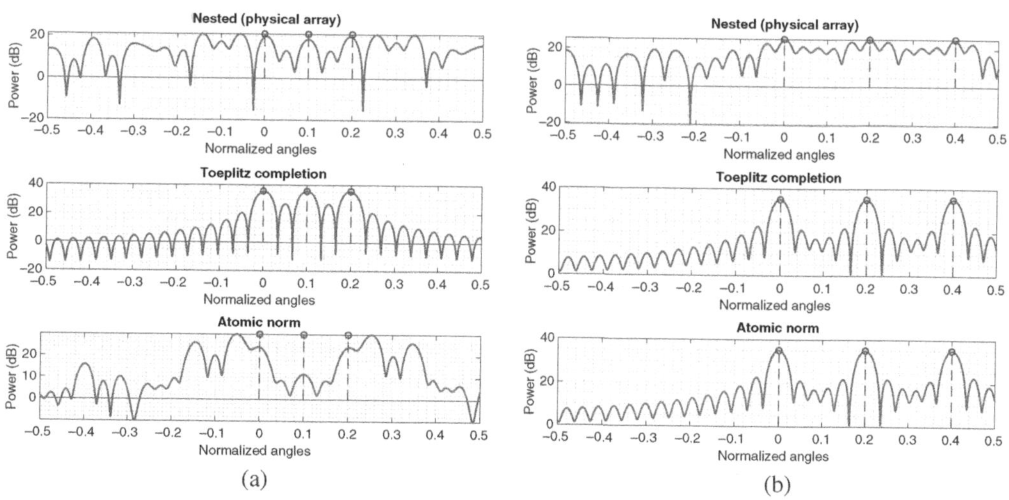

Figure 4.8 Comparison of beamforming on the nested array and interpolated virtual array with $K = 3$ sources located at (a) $\boldsymbol{\omega} = \{0.0, 0.1, 0.2\}$ and (b) $\boldsymbol{\omega} = \{0.0, 0.2, 0.4\}$. The total number of sensors is $P = 15$ and the interpolation was performed up to $N = 64$ sensors corresponding to the aperture of the nested array.

13 Nuclear norm is just one among many approaches to replace "rank" by a suitable surrogate in order to attain computational tractability.

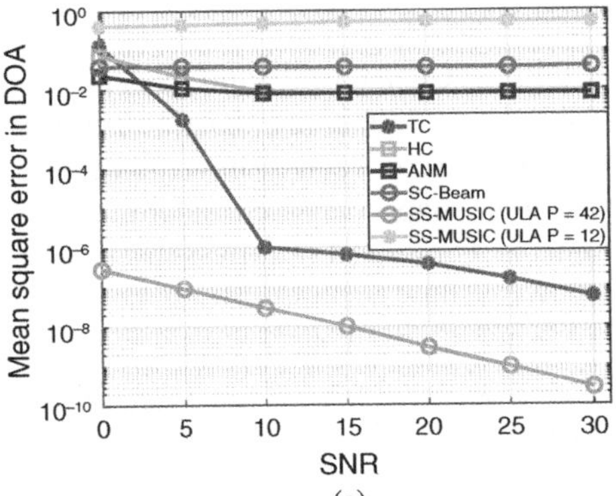

(a)

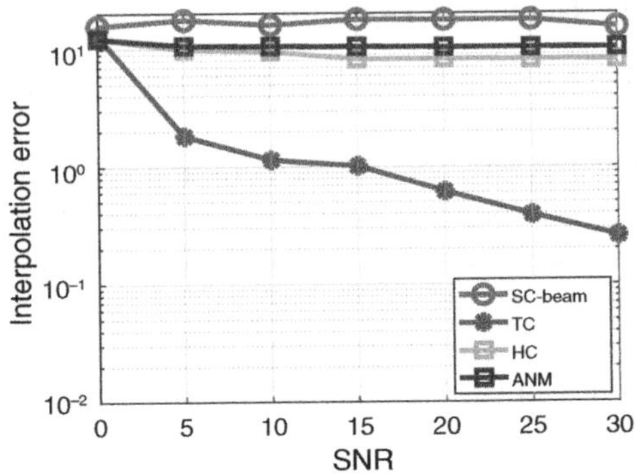

(b)

Figure 4.9 (a) MSE in DOA vs. SNR for $K = 5$ sources with a nested array with $P = 12$ sensors and ULA (with $P = 12$ and $P = 42$). (b) Normalized interpolation error versus SNR for interpolating up to $N = 42$ sensors corresponding to the aperture of the $P = 12$ sensor nested array. The proposed method is labeled as TC.

In Fig. 4.9b we also plot the interpolation error $\|\hat{\mathbf{y}}_{\text{full}} - \mathbf{y}_{\text{full}}\|_2/N$ versus SNR where $\hat{\mathbf{y}}_{\text{full}}$ is the estimated virtual measurement. For algorithms such as SBL and SC-Beam that do not perform explicit interpolation, we generate the interpolated signal using the DOA and source amplitude estimates as $\hat{\mathbf{y}}_{\text{full}} = \mathbf{A}_{\mathbb{U}}(\hat{\boldsymbol{\omega}})\hat{\mathbf{x}}$. In both plots, we observe that the RMSE of TC decays sharply with SNR while the other algorithms exhibit saturation. It is to be noted that the theoretical guarantees for these algorithms (if available) do not necessarily apply to deterministic sampling patterns such as nested arrays. Therefore, these techniques may fail to correctly identify all K sources (especially with small separation) with nested arrays, leading to saturation. The steady decay in the RMSE of TC with increasing SNR is consistent with Theorem 4.1, which guarantees that exact interpolation is possible with nested arrays with $K \leqslant M$ sources by seeking low-rank solutions.

4.6 Concluding Remarks and Future Directions

Given the relevance of sparse arrays to emerging applications such as low-complexity hybrid receiver design for millimeter-wave communication [76], autonomous sensing, and joint radar-communication system design [34, 67], characterization of spatiotemporal trade-offs in various operational regimes will continue to be of significant theoretical and practical interest. The analysis framework presented in this chapter can be extended (albeit with nontrivial modifications and new analytical techniques) to gridless source configurations, and to active sensing modalities [77], where one can also utilize waveform diversity as an additional design parameter. This also paves the way to develop new *adaptive and sequential* sensing strategies (inspired by sparse arrays) [78] that can strike a judicious balance between spatial and temporal degrees of freedom at low SNR.

References

1 Pal, P. and Vaidyanathan, P.P. (2010). Nested arrays: a novel approach to array processing with enhanced degrees of freedom. *IEEE Transactions on Signal Processing* 58 (8): 4167–4181. https://doi.org/10.1109/TSP.2010.2049264.

2 Vaidyanathan, P.P. and Pal, P. (2011). Sparse sensing with co-prime samplers and arrays. *IEEE Transactions on Signal Processing* 59 (2): 573–586. https://doi.org/10.1109/TSP.2010.2089682.

3 Romero, D., Ariananda, D.D., Tian, Z., and Leus, G. (2016). Compressive covariance sensing: structure-based compressive sensing beyond sparsity. *IEEE Signal Processing Magazine* 33 (1): 78–93. https://doi.org/10.1109/MSP.2015.2486805.

4 Sedighi, S., Mysore, R.B.S., Maleki, S., and Ottersten, B. (2018). Consistent least squares estimator for co-array-based DoA estimation. *2018 IEEE 10th Sensor Array and Multichannel Signal Processing Workshop (SAM)*, 524–528. https://doi.org/10.1109/SAM.2018.8448477.

5 Wang, M. and Nehorai, A. (2017). Coarrays, music, and the Cramér–Rao bound. *IEEE Transactions on Signal Processing* 65 (4): 933–946. https://doi.org/10.1109/TSP.2016.2626255.

6 Moffet, A. (1968). Minimum-redundancy linear arrays. *IEEE Transactions on Antennas and Propagation* 16 (2): 172–175. https://doi.org/10.1109/TAP.1968.1139138.

7 Qiao, H. and Pal, P. (2017). Performance limits of covariance-driven super resolution imaging. *2017 51st Asilomar Conference on Signals, Systems, and Computers*, 681–685, 2017. https://doi.org/10.1109/ACSSC.2017.8335429.

8 Solomon, O., Eldar, Y.C., Mutzafi, M., and Segev, M. (2019). SPARCOM: Sparsity based super-resolution correlation microscopy. *SIAM Journal on Imaging Sciences* 12 (1): 392–419.

9 Pal, P. (2018). Correlation awareness in low-rank models: sampling, algorithms, and fundamental limits. *IEEE Signal Processing Magazine* 35 (4): 56–71. https://doi.org/10.1109/MSP.2018.2827108.

10 Qiao, H. and Pal, P. (2017). Gridless line spectrum estimation and low-rank Toeplitz matrix compression using structured samplers: a regularization-free approach. *IEEE Transactions on Signal Processing* 65 (9): 2221–2236. https://doi.org/10.1109/TSP.2017.2659644.

11 Romero, D., López-Valcarce, R., and Leus, G. (2015). Compression limits for random vectors with linearly parameterized second-order statistics. *IEEE Transactions on Information Theory* 61 (3): 1410–1425. https://doi.org/10.1109/TIT.2015.2394784.

12 Koochakzadeh, A. and Pal, P. (2016). Cramér–Rao bounds for underdetermined source localization. *IEEE Signal Processing Letters* 23 (7): 919–923. https://doi.org/10.1109/LSP.2016.2569504.

13 Liu, C.-L. and Vaidyanathan, P.P. (2017). Cramér–Rao bounds for coprime and other sparse arrays, which find more sources than sensors. *Digital Signal Processing* 61: 43–61.

14 Cotter, S.F., Rao, B.D., Engan, K., and Kreutz-Delgado, K. (2005). Sparse solutions to linear inverse problems with multiple measurement vectors. *IEEE Transactions on Signal Processing* 53 (7): 2477–2488. https://doi.org/10.1109/TSP.2005.849172.

15 Jin, Y. and Rao, B.D. (2013). Support recovery of sparse signals in the presence of multiple measurement vectors. *IEEE Transactions on Information Theory* 59 (5): 3139–3157. https://doi.org/10.1109/TIT.2013.2238605.

16 Stoica, P., Babu, P., and Li, J. (2011). SPICE: A sparse covariance-based estimation method for array processing. *IEEE Transactions on Signal Processing* 59 (2): 629–638. https://doi.org/10.1109/TSP.2010.2090525.

17 Zheng, J. and Kaveh, M. (2013). Sparse spatial spectral estimation: a covariance fitting algorithm, performance and regularization. *IEEE Transactions on Signal Processing* 61 (11): 2767–2777. https://doi.org/10.1109/TSP.2013.2256903.

18 Candès, E.J. and Fernandez-Granda, C. (2014). Towards a mathematical theory of super-resolution. *Communications on Pure and Applied Mathematics* 67 (6): 906–956.

19 Tang, G., Bhaskar, B.N., and Recht, B. (2015). Near minimax line spectral estimation. *IEEE Transactions on Information Theory* 61 (1): 499–512. https://doi.org/10.1109/TIT.2014.2368122.

20 Balkan, O., Kreutz-Delgado, K., and Makeig, S. (2014). Localization of more sources than sensors via jointly-sparse Bayesian learning. *IEEE Signal Processing Letters* 21 (2): 131–134. https://doi.org/10.1109/LSP.2013.2294862.

21 Davies, M.E. and Eldar, Y.C. (2012). Rank awareness in joint sparse recovery. *IEEE Transactions on Information Theory* 58 (2): 1135–1146. https://doi.org/10.1109/TIT.2011.2173722.

22 Chen, J. and Huo, X. (2006). Theoretical results on sparse representations of multiple-measurement vectors. *IEEE Transactions on Signal Processing* 54 (12): 4634–4643. https://doi.org/10.1109/TSP.2006.881263.

23 Tropp, J.A. (2006). Algorithms for simultaneous sparse approximation. Part II: Convex relaxation. *Signal Processing* 86 (3): 589–602.

24 Guo, M., Zhang, Y.D., and Chen, T. (2018). DOA estimation using compressed sparse array. *IEEE Transactions on Signal Processing* 66 (15): 4133–4146. https://doi.org/10.1109/TSP.2018.2847645.

25 Kim, J.M., Lee, O.K., and Ye, J.C. (2012). Compressive music: revisiting the link between compressive sensing and array signal processing. *IEEE Transactions on Information Theory* 58 (1): 278–301. https://doi.org/10.1109/TIT.2011.2171529.

26 Liu, K. and Zhang, Y.D. (2018). Coprime array-based DOA estimation in unknown nonuniform noise environment. *Digital Signal Processing* 79: 66–74.

27 Wipf, D.P. and Rao, B.D. (2007). An empirical Bayesian strategy for solving the simultaneous sparse approximation problem. *IEEE Transactions on Signal Processing* 55 (7): 3704–3716. https://doi.org/10.1109/TSP.2007.894265.

28 Zhang, Y., Zhou, G., Jin, J. et al. (2016). Sparse Bayesian classification of EEG for brain–computer interface. *IEEE Transactions on Neural Networks and Learning Systems* 27 (11): 2256–2267. https://doi.org/10.1109/TNNLS.2015.2476656.

29 Hu, N., Sun, B., Zhang, Y. et al. (2017). Underdetermined DOA estimation method for wideband signals using joint nonnegative sparse Bayesian learning. *IEEE Signal Processing Letters* 24 (5): 535–539. https://doi.org/10.1109/LSP.2017.2673850.

30 Wipf, D.P., Owen, J.P., Attias, H.T. et al. (2010). Robust Bayesian estimation of the location, orientation, and time course of multiple correlated neural sources using meg. *NeuroImage* 49 (1): 641–655.

31 Qiao, H. and Pal, P. (2019). Guaranteed localization of more sources than sensors with finite snapshots in multiple measurement vector models using difference co-arrays. *IEEE Transactions on Signal Processing* 67 (22): 5715–5729.

32 Morgenshtern, V.I. and Candes, E.J. (2016). Super-resolution of positive sources: the discrete setup. *SIAM Journal on Imaging Sciences* 9 (1): 412–444.

33 Fortunati, S., Grasso, R., Gini, F. et al. (2014). Single-snapshot DOA estimation by using compressed sensing. *EURASIP Journal on Advances in Signal Processing* 2014 (1): 1–17.

34 Sun, S. and Zhang, Y.D. (2021). 4D automotive radar sensing for autonomous vehicles: a sparsity-oriented approach. *IEEE Journal of Selected Topics in Signal Processing* 15 (4): 879–891.

35 Sarangi, P., Hücümenoğlu, M.C., and Pal, P. (2022). Single-snapshot nested virtual array completion: necessary and sufficient conditions. *IEEE Signal Processing Letters* 29: 2113–2117. https://doi.org/10.1109/LSP.2022.3213140.

36 Eldar, Y.C. and Mishali, M. (2009). Robust recovery of signals from a structured union of subspaces. *IEEE Transactions on Information Theory* 55 (11): 5302–5316. https://doi.org/10.1109/TIT.2009.2030471.

37 Koochakzadeh, A., Qiao, H., and Pal, P. (2018). On fundamental limits of joint sparse support recovery using certain correlation priors. *IEEE Transactions on Signal Processing* 66 (17): 4612–4625. https://doi.org/10.1109/TSP.2018.2858211.

38 Malioutov, D., Cetin, M., and Willsky, A.S. (2005). A sparse signal reconstruction perspective for source localization with sensor arrays. *IEEE Transactions on Signal Processing* 53 (8): 3010–3022. https://doi.org/10.1109/TSP.2005.850882.

39 Pal, P. and Vaidyanathan, P.P. (2015). Pushing the limits of sparse support recovery using correlation information. *IEEE Transactions on Signal Processing* 63 (3): 711–726. https://doi.org/10.1109/TSP.2014.2385033.

40 Zhang, Z. and Rao, B.D. (2011). Sparse signal recovery with temporally correlated source vectors using sparse Bayesian learning. *IEEE Journal of Selected Topics in Signal Processing* 5 (5): 912–926. https://doi.org/10.1109/JSTSP.2011.2159773.

41 Foucart, S. and Rauhut, H. (2013). *A Mathematical Introduction to Compressive Sensing*. Springer.

42 Lee, K., Bresler, Y., and Junge, M. (2012). Subspace methods for joint sparse recovery. *IEEE Transactions on Information Theory* 58 (6): 3613–3641. https://doi.org/10.1109/TIT.2012.2189196.

43 Chen, Y., Chi, Y., and Goldsmith, A.J. (2015). Exact and stable covariance estimation from quadratic sampling via convex programming. *IEEE Transactions on Information Theory* 61 (7): 4034–4059. https://doi.org/10.1109/TIT.2015.2429594.

44 Qin, S., Zhang, Y.D., and Amin, M.G. (2015). Generalized coprime array configurations for direction-of-arrival estimation. *IEEE Transactions on Signal Processing* 63 (6): 1377–1390. https://doi.org/10.1109/TSP.2015.2393838.

45 Solomon, O., Mutzafi, M., Segev, M., and Eldar, Y.C. (2018). Sparsity-based super-resolution microscopy from correlation information. *Optics Express* 26 (14): 18238–18269.

46 Hyberg, P., Jansson, M., and Ottersten, B. (2004). Array interpolation and bias reduction. *IEEE Transactions on Signal Processing* 52 (10): 2711–2720. https://doi.org/10.1109/TSP.2004.834402.

47 Zhang, Y.D., Amin, M.G., and Himed, B. (2013). Sparsity-based DOA estimation using co-prime arrays. *2013 IEEE International Conference on Acoustics, Speech and Signal Processing*, pages 3967–3971. https://doi.org/10.1109/ICASSP.2013.6638403.

48 Hoctor, R.T. and Kassam, S.A. (1990). The unifying role of the coarray in aperture synthesis for coherent and incoherent imaging. *Proceedings of the IEEE* 78 (4): 735–752. https://doi.org/10.1109/5.54811.

49 Leech, J. (1956). On the representation of 1, 2,…, n by differences. *Journal of the London Mathematical Society* 1 (2): 160–169.

50 Pal, P. and Vaidyanathan, P.P. (2012). Nested arrays in two dimensions, Part I: Geometrical considerations. *IEEE Transactions on Signal Processing* 60 (9): 4694–4705. https://doi.org/10.1109/TSP.2012.2203814.

51 Liu, C.-L. and Vaidyanathan, P.P. (2016). Super nested arrays: linear sparse arrays with reduced mutual coupling – Part I: Fundamentals. *IEEE Transactions on Signal Processing* 64 (15): 3997–4012. https://doi.org/10.1109/TSP.2016.2558159.

52 Pal, P. and Vaidyanathan, P.P. (2011). Coprime sampling and the music algorithm. *2011 Digital Signal Processing and Signal Processing Education Meeting (DSP/SPE)*, 289–294. https://doi.org/10.1109/DSP-SPE.2011.5739227.

53 Donoho, D.L. and Tanner, J. (2005). Sparse nonnegative solution of underdetermined linear equations by linear programming. *Proceedings of the National Academy of Sciences of the United States of America* 102 (27): 9446–9451.

54 Fuchs, J.-J. (2005). Sparsity and uniqueness for some specific under-determined linear systems. *Proceedings. (ICASSP '05). IEEE International Conference on Acoustics, Speech, and Signal Processing, 2005*, volume 5, v/729–v/732. https://doi.org/10.1109/ICASSP.2005.1416407.

55 Wang, M., Xu, W., and Tang, A. (2011). A unique "nonnegative" solution to an underdetermined system: from vectors to matrices. *IEEE Transactions on Signal Processing* 59 (3): 1007–1016. https://doi.org/10.1109/TSP.2010.2089624.

56 Demanet, L. and Hand, P. (2014). Stable optimizationless recovery from phaseless linear measurements. *Journal of Fourier Analysis and Applications* 20: 199–221.

57 Chen, S.S., Donoho, D.L., and Saunders, M.A. (2001). Atomic decomposition by basis pursuit. *SIAM Review* 43 (1): 129–159.

58 Rudin, W. (1976). *Principles of Mathematical Analysis*, vol. 3. New York: McGraw-Hill.

59 Donoho, D.L. (1992). Superresolution via sparsity constraints. *SIAM Journal on Mathematical Analysis* 23 (5): 1309–1331.

60 Qiao, H. and Pal, P. (2017). Understanding the role of positive constraints in sparse bilinear problems. *2017 IEEE 7th International Workshop on Computational Advances in Multi-Sensor Adaptive Processing (CAMSAP)*, 1–5. https://doi.org/10.1109/CAMSAP.2017.8313205.

61 Candes, E.J., Wakin, M.B., and Boyd, S.P. (2008). Enhancing sparsity by reweighted ℓ_1 minimization. *Journal of Fourier Analysis and Applications* 14: 877–905.

62 Qiao, H. and Pal, P. (2019). A non-convex approach to non-negative super-resolution: theory and algorithm. *ICASSP 2019 - 2019 IEEE International Conference on Acoustics, Speech and Signal Processing (ICASSP)*, 4220–4224. https://doi.org/10.1109/ICASSP.2019.8683415.

63 Rangan, S., Schniter, P., and Fletcher, A.K. (2019). Vector approximate message passing. *IEEE Transactions on Information Theory* 65 (10): 6664–6684. https://doi.org/10.1109/TIT.2019.2916359.

64 Donoho, D.L., Maleki, A., and Montanari, A. (2009). Message-passing algorithms for compressed sensing. *Proceedings of the National Academy of Sciences of the United States of America* 106 (45): 18914–18919.

65 Tang, G., Bhaskar, B.N., Shah, P., and Recht, B. (2013). Compressed sensing off the grid. *IEEE Transactions on Information Theory* 59 (11): 7465–7490.

66 Liu, S., Mao, Z., Zhang, Y.D., and Huang, Y. (2021). Rank minimization-based Toeplitz reconstruction for DoA estimation using coprime array. *IEEE Communications Letters* 25 (7): 2265–2269.

67 Sun, S. and Petropulu, A.P. (2020). A sparse linear array approach in automotive radars using matrix completion. *ICASSP 2020-2020 IEEE International Conference on Acoustics, Speech and Signal Processing (ICASSP)*, 8614–8618. IEEE.

68 Zhou, C., Gu, Y., Fan, X. et al. (2018). Direction-of-arrival estimation for coprime array via virtual array interpolation. *IEEE Transactions on Signal Processing* 66 (22): 5956–5971.

69 Chen, Y. and Chi, Y. (2013). Spectral compressed sensing via structured matrix completion. *International Conference on Machine Learning*, 414–422. PMLR.

70 Da Costa, M.F. and Dai, W. (2018). A tight converse to the spectral resolution limit via convex programming. *2018 IEEE International Symposium on Information Theory (ISIT)*, 901–905. IEEE.

71 Ma, Y., Cao, X., and Wang, X. (2019). Off-grid DOA estimation with arbitrary-spaced linear array using single snapshot. *2019 IEEE Radar Conference (RadarConf)*, 1–6. IEEE.

72 Sarangi, P., Hücümenoğlu, M.C., and Pal, P. (2021). Beyond coarray MUSIC: harnessing the difference sets of nested arrays with limited snapshots. *IEEE Signal Processing Letters* 28: 2172–2176.

73 Wang, M., Zhang, Z., and Nehorai, A. (2017). Performance analysis of coarray-based MUSIC and the Cramér–Rao bound. *2017 IEEE International Conference on Acoustics, Speech and Signal Processing (ICASSP)*, 3061–3065. IEEE.

74 Ma, Y., Cao, X., Wang, X. et al. (2021). Multi-source off-grid DOA estimation with single snapshot using non-uniform linear arrays. *Signal Processing* 189: 108238.

75 Bhaskar, B.N., Tang, G., and Recht, B. (2013). Atomic norm denoising with applications to line spectral estimation. *IEEE Transactions on Signal Processing* 61 (23): 5987–5999.

76 Haghighatshoar, S. and Caire, G. (2017). Massive MIMO channel subspace estimation from low-dimensional projections. *IEEE Transactions on Signal Processing* 65 (2): 303–318. https://doi.org/10.1109/TSP.2016.2616336.

77 Rajamäki, R. and Koivunen, V. (2021). Sparse symmetric linear arrays with low redundancy and a contiguous sum co-array. *IEEE Transactions on Signal Processing* 69: 1697–1712.

78 Shahsavari, S., Sarangi, P., Hücümenoğlu, M.C., and Pal, P. (2022). Ada-JSR: Sample efficient adaptive joint support recovery from extremely compressed measurement vectors. In *ICASSP 2022 - 2022 IEEE International Conference on Acoustics, Speech and Signal Processing (ICASSP)*, 9077–9081. https://doi.org/10.1109/ICASSP43922.2022.9747053.

5

Sparse Sensor Arrays for Two-dimensional Direction-of-arrival Estimation

Ali H. Muqaibel[1,2] *and Saleh A. Alawsh*[2]

[1] *Electrical Engineering Department, King Fahd University of Petroleum and Minerals, Dhahran 31261, Saudi Arabia*
[2] *Center for Communication Systems and Sensing, King Fahd University of Petroleum and Minerals, Dhahran 31261, Saudi Arabia*

5.1 Introduction

Direction-of-arrival (DOA) estimation is a valuable application of array signal processing that has received increasing interest in the past periods. There are many applications for DOA estimation in sonar, radar [1], satellite navigation [2], and multiple input multiple output (MIMO) systems [3]. DOA estimation is broadly used in radio frequency and acoustics domains with similar array model and processing perceptions. Factors including but not limited to coherence of sources, noise distribution, signal-to-noise ratio (SNR), and the array geometry affect the estimation performance.

Several sensor arrays have been investigated for DOA estimation like one-dimensional (1D) linear arrays (LAs), and two-dimensional (2D) planar arrays. An important property of antenna arrays is its resolution capacity, i.e., the maximum number of resolvable waves impinging on the array [4, 5]. For many years, the linear equispaced array (LES) [6], more commonly known as the uniform linear array (ULA) was dominant for developing DOA estimation algorithms. Only $N_t - 1$ sources can be resolved with a ULA that uses N_t sensors [5]. The ULA can have larger estimation capability if higher order statistics like the fourth-order cumulants (FOC) are used [7, 8]. The resolution capacity for linear arrays can be increased by reducing the redundancy of the inter-sensor[1] spacing, where sparse arrays came to the scene. Sparse antenna arrays have nonuniform inter-element spacings that are integer multiple of a fundamental distance, $d = \lambda/2$, where λ is the signal wavelength.

There is an increased interest in 2D and three-dimensional (3D) arrays that generally result in a much finer location resolution. The authors in [9] presented a comparison between the uniform rectangular array (URA) and the L-shaped array. Comprehensive reviews were published for 1D and 2D-DOA estimation [10] and 2D sparse antenna arrays [11]. The review in [10] focused on search-free methods, and algorithms that work for arbitrary arrays. The authors in [12] presented a tutorial about direction finding but did not designate specific antenna array geometries. There are also other comparative studies [13–16] and reviews in DOA estimation [17, 18]. Most of the researchers considered DOA estimation of sources in the far-field. That is, sources far away from the Fresnel region of the antenna array [2].

Lately, there has been a growing interest in proposing specific 2D array geometries and appropriate 2D-DOA estimation algorithms. There is no clear design roadmap, as more configurations

1 In this chapter, the word sensor refers to the element of the array; thus, sensors and array elements are used interchangeably.

Sparse Arrays for Radar, Sonar, and Communications, First Edition. Edited by Moeness G. Amin.

keep appearing. This chapter aims to capture and put in perspective existing work in 2D sparse array design. It delivers an overview of 2D sparse array geometries that go in tandem with special DOA estimation algorithms, which can therefore surpass the performance of arbitrary sparse sensor array geometries with generic DOA estimation algorithms. We help identify the design gaps and avoid unnecessary minor modifications. The rest of this chapter is organized as follows: the essentials of 2D-DOA estimation are presented in Section 5.2. Then, Section 5.3 describes notable 2D sparse geometries [19–98] for DOA estimation. Comparative evaluation is presented in Section 5.4, and finally the chapter is summarized in Section 5.5. It is important to note that Chapter 1 of this book provides the background in 1-D sparse arrays which this chapter builds on.

The following scientific representations are used: for a matrix $\mathbf{A}$, $\mathbf{A}^*$ is its complex conjugate, $\mathbf{A}^{\mathrm{T}}$ is its transpose, and $\mathbf{A}^{\mathrm{H}}$ is the Hermitian. The operator $\mathbb{E}$ is the expectation operator, $\otimes$ is the Kronecker product, $\odot$ is the Khatri–Rao product, $\lfloor x \rfloor$ is the floor of x, bold-face lower-case letters (e.g., $\mathbf{a}$) stands for vectors, bold-face upper-case letters (e.g., $\mathbf{A}$) stands for matrices, and the overhead hat (as in $\hat{\theta}$) denotes the estimated version of the intended variable.

5.2 Two-Dimensional DOA Estimation Essentials

The subsequent broad questions are important in illustrating any 2D-DOA estimation algorithm, some of them are extended from the 1D case:

- Can the algorithm detect more signals than the number of antenna elements? [99].
- Are the existing physical elements fully exploited? [100, 101]. For instance, algorithms based on propagator method (PM) usually only detect sources less than half the number of elements [19].
- What are the ranges of angles the 2D algorithm successfully covers? For example, some algorithms cannot detect high elevation angles [102].
- Can the antenna array do well in real environments? For instance, would the array work fine in the presence of mutual coupling [67], model mismatch [103, 104], and sensor failure [105, 106]?

The required aperture of an antenna array is generally related to the maximum number of resolvable sources, and the inter-element spacing is related to aliasing of some angle estimates [15]. If the linear sparse array has a continuous hole-free coarray, then it can detect the same number of sources as a ULA with similar aperture [15]. This can be extended to 2D-DOA estimation with uniform rectangular coarray. The following subsections present the system model for 2D-DOA estimation, then define the 2D coarray concept.

5.2.1 2D System Model

The 2D system model describes the signals emerging from the sources to be localized, the antenna array, and the utilized signal processing. Signals can be in the far-field or the near-field, polarized or nonpolarized, coherent or incoherent, correlated or uncorrelated. The antenna array could be 1D, 2D, or 3D, uniform or sparse or neither (see Section 5.3). The sensors can be omnidirectional or directional, polarized or nonpolarized. The properties to be estimated could be the 1D-DOA or 2D-DOA, polarization or range (distance to the source). The aforementioned cases are not comprehensive and further attention is required for practical executions. For instance, mutual coupling, sensor failure, array calibration, and sensor characteristics mismatch are some physical hindrances for successful deployment of a functioning system. A 20-element URA is shown in Fig. 5.1 to demonstrate the important variables.

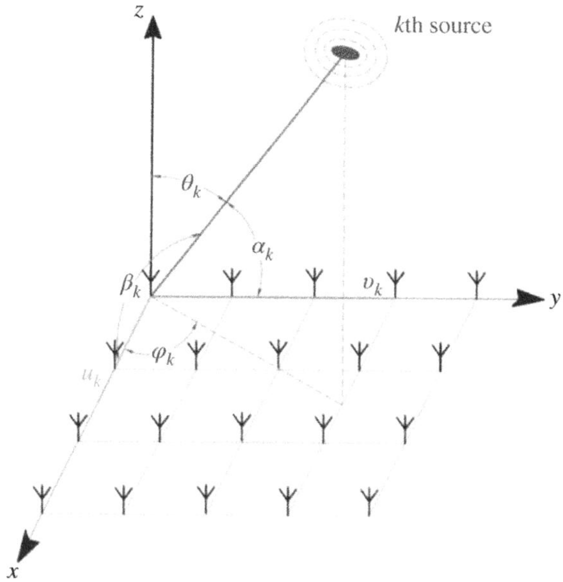

Figure 5.1 2D-DOA signal model with a twenty-element URA.

The following symbols are assumed and used in this chapter unless otherwise stated.

- T is the number of samples,
- K is the number of narrow-band impinging sources and k is used for the kth source (i.e., $k = 1, 2, \ldots, K$),
- N_i and M_i are used for the number of sensors in the linear subarray i (the precise meaning is explained for different individual array geometries),
- N_t is the total number of sensors in the array,
- d_x, d_y, and d_z are the unit spacings along the x-axis, y-axis, and z-axis, respectively, often $d_x = d_y = d_z$,
- $\phi_k \in [0, 2\pi]$ is the azimuth angle of the kth source (measured from the positive x-axis), and $\theta_k \in [0, \pi]$ is the elevation angle (measured from the positive z-axis); these angles are demonstrated in Fig. 5.1,
- $\alpha_k \in [0, \pi]$ is the angle between the line connecting the kth source with the origin and the positive y-axis, and $\beta_k \in [0, \pi]$ is the angle between the line joining the kth source with the origin and the positive x-axis.

Furthermore, the next direction cosines are used in the literature to simplify some expressions.

$$u_k := \cos(\beta_k) = \sin(\theta_k)\cos(\phi_k) \tag{5.1}$$

$$v_k := \cos(\alpha_k) = \sin(\theta_k)\sin(\phi_k) \tag{5.2}$$

where α_k and β_k are named *electric angles*, and they simplify many analytical expressions. The next geometrical variables are also used [67]

$$\overline{\theta}_k = \frac{d}{\lambda}\sin(\theta_k)\cos(\phi_k) \tag{5.3}$$

$$\overline{\phi}_k = \frac{d}{\lambda}\sin(\theta_k)\sin(\phi_k) \tag{5.4}$$

where λ is the wavelength. Consider a 2D-DOA estimation system for far-field nonpolarized signals. The received signal at the output of a 2D antenna array at a time t in the presence of a single source is expressed as [107, 108]

$$\mathbf{x}(t) = \mathbf{a}(\theta_k, \phi_k)s(t) + \mathbf{n}(t) \tag{5.5}$$

where $\mathbf{a}(\theta_k, \phi_k) \in \mathbb{C}^{N_t \times 1}$ is the array steering vector with entries $e^{j2\pi(\overline{\theta}_k p_x + \overline{\phi}_k p_y)}$ for all $(p_x, p_y) \in \mathbb{S}$ and $\mathbb{S}$ is the set of sensor locations, $s(t)$ is the transmitted signal, $\mathbf{x}(t) \in \mathbb{C}^{N_t \times 1}$ for N_t sensors, and $\mathbf{n}(t) \in \mathbb{C}^{N_t \times 1}$ is the additive noise vector, that is assumed to be spatially and temporally white Gaussian noise and uncorrelated at each array element, that is $\mathbb{E}[\mathbf{n}(t)\mathbf{n}^H(t)] = \sigma^2\mathbf{I}$, with σ^2 being the noise power and $\mathbf{I}$ is the identity matrix.

In the presence of K sources, the model can be expressed as

$$\begin{bmatrix} x_1(t) \\ x_2(t) \\ x_3(t) \\ \vdots \\ x_{N_t}(t) \end{bmatrix} = [\mathbf{a}(\theta_1, \phi_1)\ \mathbf{a}(\theta_2, \phi_2)\ \ldots\ \mathbf{a}(\theta_K, \phi_K)] \times \begin{bmatrix} s_1(t) \\ s_2(t) \\ \vdots \\ s_K(t) \end{bmatrix} + \begin{bmatrix} n_1(t) \\ n_2(t) \\ n_3(t) \\ \vdots \\ n_{N_t}(t) \end{bmatrix}. \tag{5.6}$$

Then taking T time samples (snapshots), the model can be written compactly as

$$\mathbf{X} = \mathbf{AS} + \mathbf{N} \tag{5.7}$$

where $\mathbf{X} \in \mathbb{C}^{N_t \times T}$, $\mathbf{A} \in \mathbb{C}^{N_t \times K}$, $\mathbf{S} \in \mathbb{C}^{K \times T}$, and $\mathbf{N} \in \mathbb{C}^{N_t \times T}$.

The source is considered in the far-field if it lies beyond the Fraunhofer distance, $2D^2/\lambda$, where D is the aperture of the array, and λ is the wavelength of the source [63, 109].

5.2.2 Terminology of 2D Arrays

Since this chapter considers 2D antenna arrays which can be represented in a rectangular grid, it is appropriate to describe the sensor locations as integers, where the normalized unit spacing is $d = \lambda/2$. The unit spacing is the minimum distance between any pair of antenna elements. In this chapter, the set of antenna elements is represented by $\mathbb{S}$.

Mutual coupling is usually modeled by a mutual coupling matrix (MCM) $\mathbf{C} \in \mathbb{C}^{N_t \times N_t}$ that is multiplied by the array manifold matrix $\mathbf{A} \in \mathbb{C}^{N_t \times K}$. Theoretically, the MCM should be specific to each set of sensors, taking into consideration the physical dimensions of the sensors/antennas and feeding points [103]. The inter-element spacing is generally the dominant factor. In fact, the behavior of the MCM can be empirically approximated as a function of sensor separations only [110]. The MCM in 2D can be assumed to be a B-banded symmetric Toeplitz matrix [111]

$$\langle \mathbf{C} \rangle_{\mathbf{p}_1, \mathbf{p}_2} = \begin{cases} c(\|\mathbf{p}_1 - \mathbf{p}_2\|_2), & \|\mathbf{p}_1 - \mathbf{p}_2\|_2 < B \\ 0, & \text{otherwise} \end{cases} \tag{5.8}$$

where $\langle \mathbf{C} \rangle_{\mathbf{p}_1, \mathbf{p}_2}$ represents the value of $\mathbf{C}$ at a sensor pair $\mathbf{p}_1, \mathbf{p}_2 \in \mathbb{S}$, $\|.\|_2$ is the l_2-norm, B is the maximum separation between coupled sensors, and $c(.)$ represents the mutual coupling coefficient. Furthermore, it is assumed that

$$c(0) = 1, \ \ \& \ \ \left|\frac{c(k)}{c(l)}\right| = \frac{l}{k}, \ \ \text{for } l, k > 0 \tag{5.9}$$

5.2.3 Coarrays in 2D

An essential step utilized in most DOA estimation algorithms with sparse arrays is finding the coarray of the physical array. For the case of uncorrelated sources, the difference coarray is of interest, and is defined as follows [99].

Definition 5.1 ***(Difference Coarray)*** If a set $\mathbb{S}$ represents the set of ordered pairs of points representing the coordinates of physical sensors in an antenna array, then the difference coarray set is given by

$$\mathbb{D} = \{\mathbf{m}|\mathbf{m} = \mathbf{p}_1 - \mathbf{p}_2, \forall \mathbf{p}_1, \mathbf{p}_2 \in \mathbb{S}\} \tag{5.10}$$

$\mathbf{p}_1$ and $\mathbf{p}_2$ can be viewed as vectors in $\mathbb{R}^{2\times 1}$ denoting sensor locations. The maximum number of virtual sensor locations in the coarray is $N_t(N_t - 1) + 1$, irrespective of the array geometry [112]. The difference coarray in most cases is symmetric with respect to the origin, and it is important since it appears naturally in the cross correlation between two sensors in the array, i.e., when second-order statistics (SOS) are used [82].

Definition 5.2 ***(Hole-Free Coarray)*** If a set $\mathbb{D}$ denotes a coarray on a rectangular integer grid, and the minimum and maximum elements are represented as min $\mathbb{D}$ and max $\mathbb{D}$. If all possible combinations of pairs in [min $\mathbb{D}$, max $\mathbb{D}$] exist in the coarray, then the coarray is hole-free. Instead, if $\mathbb{U}$ denotes a URA which comprises the origin, and $\mathbb{D} = \mathbb{U}$, then $\mathbb{D}$ is a hole-free.

This definition infers that any element in $\mathbb{U}$ but not in $\mathbb{D}$ is a hole. It is often preferred to get a continuous (hole-free) coarray. Reasons include: better identifiability of DOAs using estimation algorithms like multiple signal classification (MUSIC) with spatial smoothing [112, 113]; and $\mathcal{O}\left(N_t^2\right)$ uncorrelated DOAs can be identified with a coarray of $\mathcal{O}(N_t)$ sensors. This almost holds for all 2D antenna arrays [67, 114]. It is possible to extend the hole-free definition to the 3D arrays [67]. The weight function is another important definition for the coarray, which gives an insight on the consequence of mutual coupling on DOA estimation [67, 110].

Definition 5.3 ***(Sum Coarray)*** If a set $\mathbb{S}$ denotes the set of ordered pairs of points representing the coordinates of physical sensors in a sensor array, then the sum coarray set is given by

$$\mathbb{D}_S = \{\mathbf{m}|\mathbf{m} = \pm(\mathbf{p}_1 + \mathbf{p}_2), \forall \mathbf{p}_1, \mathbf{p}_2 \in \mathbb{S}\} \tag{5.11}$$

The difference-sum coarray is the combination of both, the difference and the sum coarrays, that is $\mathbb{D}_{\text{Diff-Sum}} = \mathbb{D} \cup \mathbb{D}_S$ [77]. There is also the notion of the difference coarray of the sum co-array which is extendable to 2D configurations [115].

Definition 5.4 ***(Weight Function)*** If a set $\mathbb{S}$ represents the set of physical sensors' positions in an antenna array, and $\mathbb{D}$ represents the difference coarray, then

$$w(\mathbf{m}) = w(m_x, m_y) = |\{\mathbf{p}_1, \mathbf{p}_2 \in \mathbb{S}^2 | \mathbf{p}_1 - \mathbf{p}_2 = \mathbf{m}\}| \tag{5.12}$$

represents the number of sensor pairs with separation $\mathbf{m} \in \mathbb{D}$. Note that $m = (m_x, m_y)$ is a two-component vector, and $|.|$ denotes the cardinality.

For instance, the most important weights are $w(0, 1)$, $w(1, 0)$, and then $w(1, 1)$, $w(1, -1)$, since the first two count the elements with unity inter-element spacing, and the other two for $\sqrt{2}$ inter-element spacing. The first two weights reflect the number of elements having inter-element spacing of unity, which is the smallest possible inter-element spacing. The difference is in the axis over which this weight is calculated. An array geometry with the lower values for $w(0, 1)$, $w(1, 0)$, $w(1, 1)$, and $w(1, -1)$ is usually less vulnerable to mutual coupling degradation.

5.3 Sparse Array Geometries for 2D-DOA Estimation

This section presents a comprehensive overview of the sparse array geometries used for 2D-DOA estimation. The geometries are categorized, as shown in Fig. 5.2, into five broad categories: (i) arrays constructed using parallel linear arrays, (ii) arrays constructed using nonparallel linear arrays, (iii) arrays constructed using interleaved rectangular arrays, (iv) arrays constructed using conformal arrays, and (v) other geometries. Furthermore, a timeline of the usage of arrays from each category is shown in Fig. 5.3. Note that some references may deal with more than

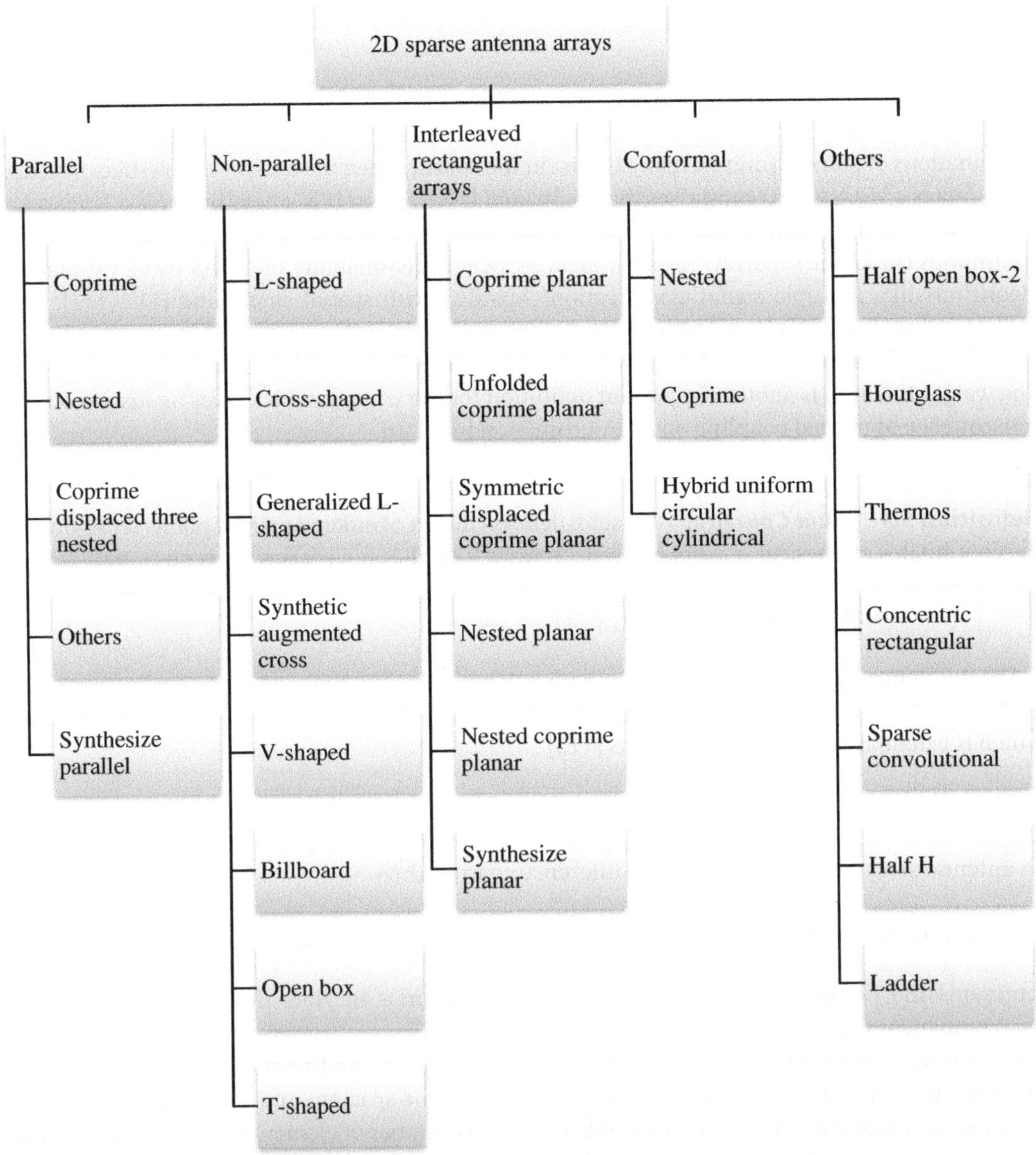

Figure 5.2 Classification of 2D sparse antenna arrays.

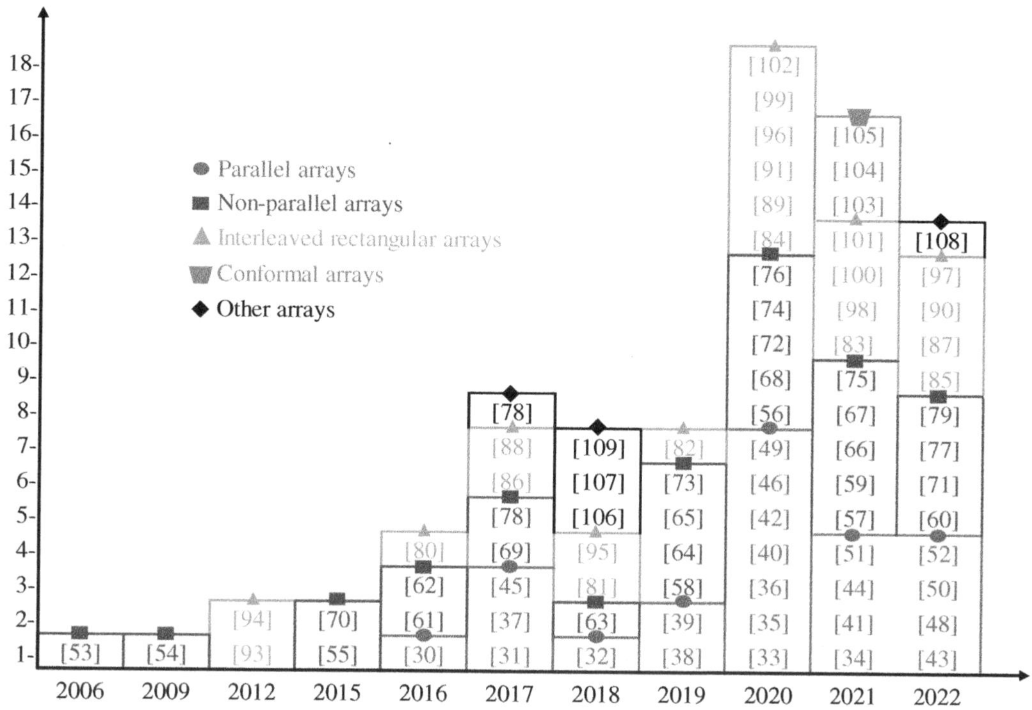

Figure 5.3 Timeline for previous work on 2D sparse array geometries.

one geometry such as the work of Liu and Vaidyanathan [67]. Some of these geometries are named slightly different by some authors. For example, the parallel coprime array (PCA) is called coprime parallel array in [20, 21], yet the rest of the authors use PCA [19, 22, 25, 26]. This name also prevents confusing the PCA with the coprime planar array (CPA). The knowledge of sparse linear arrays, especially the nested array (NA) and the coprime array (CA) is important since many 2D antenna arrays are constructed via a combination of these and probably the ULA.

This section illustrates the antenna arrays demonstrated in Fig. 5.2 from left to right. The adopted classification categories based on the building blocks of the 2D array are:

(1) **Parallel linear arrays**, which include PCA, three PCA (TPCA), parallel nested array (PNA), coprime-displaced three parallel nested array (CDTPNA), other parallel structures, and parallel arrays with motion. These are discussed in Section 5.3.1.
(2) **Nonparallel linear arrays**, which contain various sparse L-shaped arrays, various sparse cross-shaped array, synthetic augmented cross array (SACA), sparse V-shaped arrays, billboard array, open box array (OBA), and T-shaped array (TSA). These are discussed in Section 5.3.2.
(3) **Interleaved rectangular arrays**, which comprise CPA, generalized coprime planar array (GCPA), unfolded coprime planar array (UCPA), symmetric displaced coprime planar array (SDCPA), nested planar array (NPA), nested coprime planar array (NCPA), and other planar arrays with motion. These are discussed in Section 5.3.3.
(4) **Conformal arrays**, which cover nested conformal array (NCA), coprime conformal array (CCA), and hybrid uniform circular cylindrical array (UCyA). These are discussed in Section 5.3.4.

(5) **Other arrays,** which include half OBA with two layers (HOBA-2), hourglass array, thermos array, concentric rectangular array (CcRA), extended sparse convolutional array (ESCA), half H array (HHA), and ladder array (LAA). These are discussed in Section 5.3.5.

Since this chapter focuses on sparse array geometries, it is useful to mention that the term sparse can have different appearances in the context of DOA estimation:

(1) **Sparse sensor array:** where sensors are placed on some point of a uniform grid (thinned array). Some sensors may have a unit inter-sensor spacing and the rest greater than unity (assuming a normalized fundamental inter-element spacing d). This meaning is used in this chapter.
(2) **Uniform sparse sensor array:** where the spacing of all elements are identical and greater than unity. This is a special case of the earlier one, and essentially the resultant array is a uniform array. In case of linear arrays, that resultant array is referred to as sparse ULA, some examples are available in [116].
(3) **Sparse recovery:** a method from compressive sensing (CS) framework used in some DOA estimation algorithms.

Section 5.3.1 starts with parallel arrays as building blocks for 2D arrays. Then, nonparallel, interleaved, conformal, and other arrays follow.

5.3.1 Parallel Arrays

This subsection describes 2D sparse arrays made of two, or more, parallel arrays. Particularly, the PCA, the TPCA, the PNA, CDTPNA, and parallel arrays with motion are considered. Furthermore, other reported parallel arrays that do not fall under any of these classes are also mentioned.

5.3.1.1 Parallel Coprime Array (PCA)

This subsection presents an overview of PCA geometry [19–22]. The structure of PCA is shown in Fig. 5.4a, where two ULAs are in parallel (spaced by d_x). One ULA has M_1 elements with $M_2 d_y$ inter-element spacing (shown in boxes), and the other has M_2 elements with $M_1 d_y$ inter-element spacing (shown in circles). In this geometry, M_1 and M_2 are coprime integers with $M_1 < M_2$, and the example of Fig. 5.4a assumes $M_1 = 4$ and $M_2 = 5$. The set of array elements is [21]

$$\mathbb{S} = \{(0, mM_2 d_y)|0 \leq m \leq M_1 - 1\} \cup \{(d_x, nM_1 d_y)|0 \leq n \leq M_2 - 1\} \tag{5.13}$$

where $d_x = d_y = d = \lambda/2$, with λ being the source wavelength. This array has $N_t = M_1 + M_2$ sensors. Two extra elements at $(\lambda, 0)$ and $(-\lambda/2, 0)$ were added in [23] along the vertical axis to extend the aperture of the PCA. This raises the total number of elements to $N_t = M_1 + M_2 + 2$. An example with $M_1 = 4$ and $M_2 = 5$ is shown in Fig. 5.4b. A modified PCA was considered in [24] where the number of elements is doubled to $2M_1$, while the spacing remains the same, $M_2 d_y$, and the other ULA is intact. In this case, the total number of sensors is $N_t = 2M_1 + M_2$. Figure 5.4c illustrates this configuration where $M_1 = 4$ and $M_2 = 5$. The symmetric coprime parallel array [20] is constructed using symmetrical subarrays, i.e., concatenating two arrays of [24]. The first subarray has a total of $4M_1 - 1$ sensors spaced by M_2 units and the second has $2M_2 - 1$ senors spaced by M_1 units. This leads to a total of $N_t = 4M_1 + 2M_2 - 2$ sensors. Figure 5.4d shows an example with $M_1 = 4$ and $M_2 = 5$.

5.3.1.2 Three Parallel Coprime Array (TPCA)

This subsection introduces an extension to the PCA using a third parallel linear array (LA) [25–27]. A modified version of the PCA is the three parallel coprime structure presented in [25], where

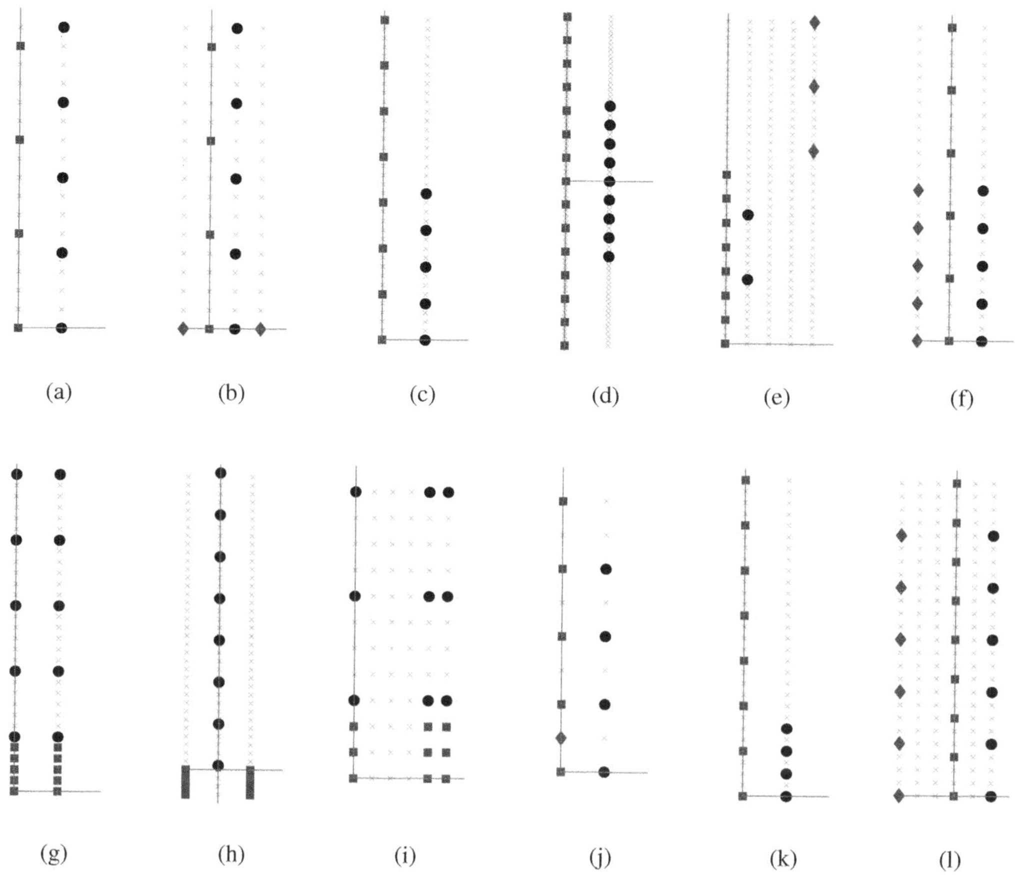

Figure 5.4 Examples of 2D sparse arrays constructing using parallel linear arrays. (a) PCA, (b) PCA with two extra sensors, (c) Modified PCA, (d) Symmetric PCA, (e) PCA with displaced ULA, (f) PCA with parallel ULA, (g) PNA, (h) ThPNA, (i) CDTPNA, (j) 2-ULAs with extra sensor, (k) 2-ULAs with different spacing, and (l) 3-ULAs.

the third LA is located after the end of the other two LAs as shown in Fig. 5.4e. The first ULA has M_2 elements with inter-element spacing $M_1 d_y$ (shown in boxes) and the other two ULAs have inter-element spacing $M_2 d_y$. The second ULA is spaced by d_x from the y-axis and has $M_1 - 1$ elements, whereas the third ULA is spaced by Ld_x, $L > d_x$ from the second ULA and has M_1 elements. The example shown in Fig. 5.4e assumes $M_1 = 3$, $M_2 = 8$, and $L = 3$. The set of elements is [25]

$$\mathbb{S} = \{(0, mM_1 d_y) | 0 \le m \le M_2 - 1\} \cup \{(d_x, nM_2 d_y) | 1 \le n \le M_1 - 1\} \\ \cup \{(Ld_x + d_x, pM_2 d_y + M_1 M_2 d_y) | 0 \le p \le M_1 - 1\} \tag{5.14}$$

This configuration has a total of $N_t = 2M_1 + M_2 - 1$ sensors. The optimal selection in this case is to have $2M_1$ and M_2 as close as possible [25].

A three parallel coprime array can be constructed by adding an extra LA to the PCA in (5.13). The middle subarray has $2M_1$ elements with inter-element spacing of $M_2 d_y$, and the outer coprime LAs are spaced by $d_x \le \lambda/2$ from the y-axis and has M_2 elements with inter-element spacing of $M_1 d_y$, where $d_y \le \lambda/2$. A total of $N_t = 2(M_1 + M_2)$ sensors are defined by the following set

$$\mathbb{S} = \{(0, mM_2 d_y) | 0 \le m \le 2M_1 - 1\} \cup \{(d_x, nM_1 d_y) | 0 \le n \le M_2 - 1\} \\ \cup \{(-d_x, nM_1 d_y) | 0 \le n \le M_2 - 1\} \tag{5.15}$$

An example is given in Fig. 5.4f where $M_1 = 4$ and $M_2 = 5$.

5.3.1.3 Parallel Nested Array (PNA)

Parallel nested arrays were reported in [28–30]. All three articles utilized two identical parallel nested linear arrays (NLAs) [112] (shown in Fig. 5.4g with $N_1 = N_2 = 4$). Assuming each LA has N elements, the total number of elements is $2N$. The set of antenna elements is

$$\mathbb{S} = \{(0, md_y)|m \in \mathfrak{g}\} \cup \{(d_x, nd_y)|n \in \mathfrak{g}\}, \mathfrak{g} = \{0, 1, \dots, N_1, 2(N_1+1)-1, \dots, N_2(N_1+1)-1\} \quad (5.16)$$

The three parallel nested arrays (ThPNA) on the other hand uses three parallel ULAs. Two dense subarrays are located along the negative side of the y-axis and shifted by $d = \pm\lambda/2$ from the origin along the x-axis [31]. The third is sparse subarray located along the positive side of the y-axis and shifted by $0.5d$ from the origin along the y-axis. Each of the dense subarrays has N_1 elements and the sparse subarray has $N_2 = 2N_1$ elements spaced by $(N_1+1)d$. The set of antenna elements is

$$\mathbb{S} = \{(d_x, -md_y)|0 \le m \le N_1 - 1\} \cup \{(-d_x, -nd_y)|0 \le n \le N_1 - 1\} \cup \{(0, pd_y)|0 \le p \le N_2\} \quad (5.17)$$

Figure 5.4h shows an example of the ThPNA with $N_1 = 4$ and $N_2 = 8$.

5.3.1.4 Coprime-displaced Three Parallel Nested Arrays (CDTPNA)

Consider three parallel nested arrays [32, 33], the first one is located along the y-axis starting from the origin, the second one is shifted along the x-axis by L_2, and the third one is shifted along the x-axis by L_3. The dense subarray in each one has N_1 elements spaced by d, while the sparse subarray of each one has $N_2 = N - N_1$ elements spaced by $(N_1+1)d$. The two shifts are taken to be pairwise coprime integers which can enhance the array aperture, that is $L_1 = M_1 d$ and $L_3 = M_2 d$ with M_1 and M_2 being coprime integers. The CDTPNA has elements located at

$$\mathbb{S} = \mathbb{S}_1 \cup \mathbb{S}_2 \cup \mathbb{S}_3 \quad (5.18)$$

where $\mathbb{S}_i = \{(L_i, \mathbb{Z})\}, i = 1, 2, 3$ with $L_1 = 0$, and $\mathbb{Z} = \{z_i, i = 1, 2, \dots, N\} = \{0, 1, \dots, N_1, \dots, N_2(N_1+1)-1\}$. The array has a total of $N_t = 3N = 3(N_1+N_2)$ sensors. Figure 5.4i shows an example of the CDTPNA with $N_1 = N_2 = 3$, $M_1 = 4$, and $M_2 = 5$.

5.3.1.5 Other Parallel Arrays

Other than the PCA or the PNA, two parallel ULAs [34, 35] and three parallel ULAs [37] were proposed. Zheng et al. [34] suggest two parallel ULAs with an extra element. The two ULAs are spaced by $\lambda/2$, and each of the two ULAs has inter-element spacing of λ, and the extra sensor is placed at $\lambda/2$, as shown in Fig. 5.4j, where the first ULA has $N+1$ elements (shown in squares) and the second ULA has N elements (shown in circles). An example is given in Fig. 5.4j, where $N = 4$ and $N_t = 2(N+1) = 10$ sensors.

Two parallel ULAs with different spacing was considered in [35]. One is located along the z-axis and other is shifted by $d_x = \lambda/2$. The first and the second subarrays have M and N elements, respectively. The elements within the first and the second ULAs are spaced respectively by $d_z = \lambda/2$ and qd_z, with $N = M/q$ and $q \ge 2$ [35]. Figure 5.4k shows an example of this configuration with $M = 8$, $q = 2$, and $N = 4$.

Three parallel ULAs of two different spacings were also considered [37]. One subarray is located along the z-axis, while the other two are spaced by pd_x and $-(p+1)d_x$ along the x-axis, $p \ge 2$. Each of the subarrays on the sides has M elements spaced by $(q+1)d_z$, $q \ge 2$. The other subarray has $N \ge M$ elements spaced by qd_z. Therefore, the array has a total of $N_t = 2M + N$ elements with $d_x = d_z = \lambda/2$. Figure 5.4l shows an example of this configuration with $M = 6$, $N = 9$, $p = 2$, and $q = 3$.

5.3.1.6 Parallel Arrays with Motion

A 2D nested array can be formed by moving the 1D nested array [38]. The resultant configuration after movement is similar to that in [28–30]. The array has the original nested subarray and the one after movement located at a distance $d = v\varepsilon = \lambda/2$, where v is the velocity of movement and ε is the duration. The resultant configuration achieves a hole-free difference coarray.

The aperture can be extended by moving the two parallel arrays along their axes at a constant speed v [39]. The 2D moving sampling scheme (MSS) is defined as the moving displacements along y-axis. It depends on the total number of platform motions synthesized, and the designed MSS unit spacing. The resultant array has elements located at the original two parallel array and its shifted versions. Due to the large aperture, this structure achieves higher degrees of freedom (DOF) compared with existing structures with the same number of physical sensors.

A 2D configuration can be constructed by combining two parallel dilated arrays (DAs) [40]. An q-DA is equivalent to multiplying each sensor's position of a known array by a positive integer, a dilation factor, q. This operation shall reduce mutual coupling effect at the expense of increased aperture size. By exploiting array motion, the number of detectable sources cannot exceed three [40]. To overcome this, multiple shifted arrays with displacements of half wavelength are utilized instead of only one [41]. Therefore, a high DOF with $q \geq 4$ can be achieved where the number of shifted arrays is a function of q [41]. The cost for this improvement is increasing the required aperture size.

Table 5.1 shows the important weight functions for parallel coprime arrays. The arrays are compared with similar or comparable number of elements. This can be achieved with different structures and selections. The total number of elements is 12, 13, or 14. It is evident that the weight values are small because of the large spacing. The highest weight value is 3, where two extra sensors added to the coprime array [23]. Configurations with three parallel subarrays have larger weights compared with others.

Table 5.2 shows the important weight functions for parallel nested and other arrays. All nested structures have relatively large weight values because of the dense subarrays. Configurations generated by motion [38–41] are not listed due to synthetization, which may lead to unfair comparison. These types eventually have lower weight values but there is a delay in getting the measured signals.

A 2D-DOA estimation using CS and least square technique [19] is presented in Fig. 5.5 for the for PCA, PCA with doubling the number of elements in one subarray, PCA with doubling the

Table 5.1 Weight functions of parallel coprime arrays.

References	N_t Formula	M_1, M_2	N_t	$w(1,0)$	$w(0,1)$	$w(1,1)$	$w(1,-1)$
[19]	$2M_1 + M_2$	4, 5	13	1	0	1	1
[20]	$4M_1 + 2M_2 - 2$	2, 3	12	1	0	2	2
[21]	$M_1 + M_2$	6, 7	13	1	0	1	1
[22]	$M_1 + M_2$	6, 7	13	1	0	1	1
[23]	$M_1 + M_2 + 2$	5, 6	13	3	0	1	1
[24]	$2M_1 + M_2$	4, 5	13	1	0	1	1
[25]	$2M_1 + M_2 - 1$	3, 8	13	0	0	1	1
[26]	$2(M_1 + M_2)$	3, 4	14	2	0	2	2
[27]	$2(M_1 + M_2)$	3, 4	14	2	0	2	2

Table 5.2 Weight functions of parallel nested and other arrays.

Arrays	References	N_t formula	Var. values	N_t	$w(1, 0)$	$w(0, 1)$	$w(1, 1)$	$w(1, -1)$
Nested	[28]	$2(N_1 + N_2)$	3, 3	12	6	6	3	3
	[29]	$2(N_1 + N_2)$	3, 3	12	6	6	3	3
	[30]	$2(N_1 + N_2)$	3, 3	12	6	6	3	3
ThPNA	[31]	$4N_1$	3, 6	12	4	0	0	0
CDTPNA	[32]	$3(N_1 + N_2)$	2, 2	12	4	6	2	2
	[33]	$3(N_1 + N_2)$	2, 2	12	4	6	2	2
Others	[34]	$2M + 1$	6	13	6	2	1	1
	[35]	$N + M$	8, 4	12	2	0	0	0
	[37]	$2M + N$	3, 6	12	0	0	0	0

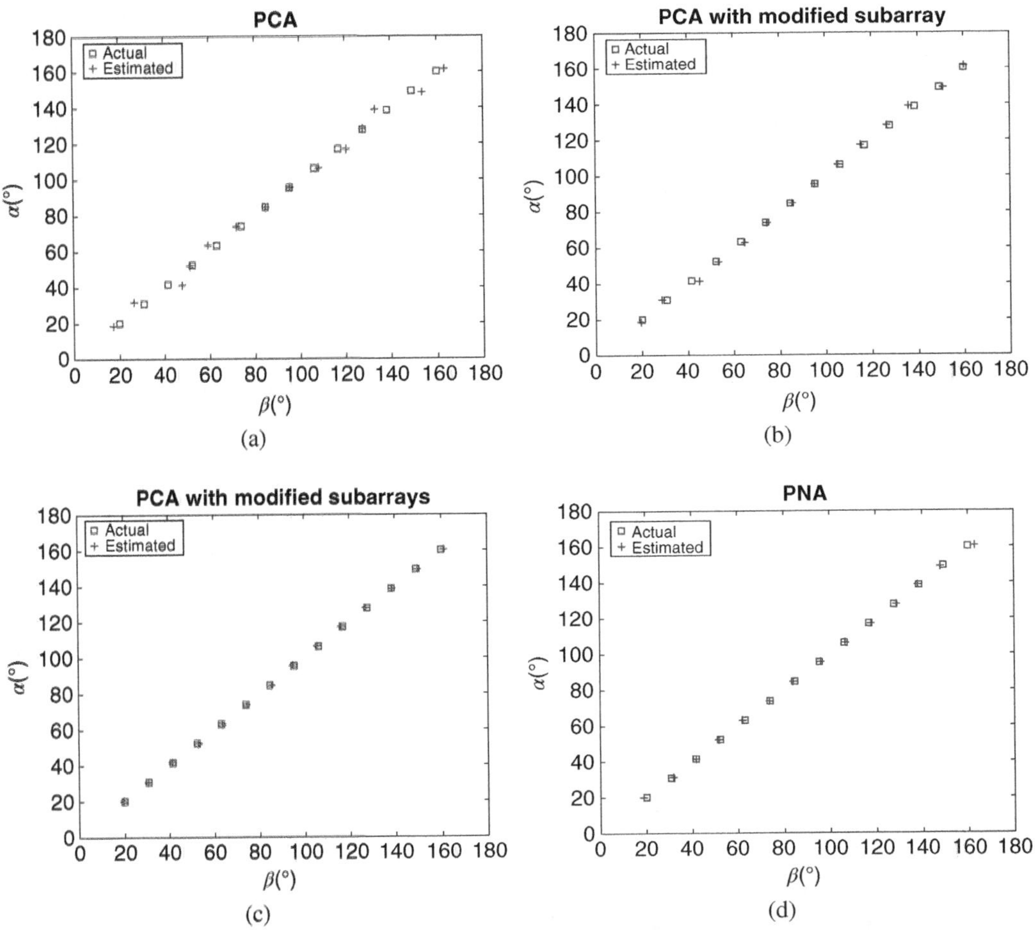

Figure 5.5 2D-DOA estimation for parallel arrays (a) PCA, (b) PCA with modified subarray, (c) PCA with modified subarrays , and (d) PNA, based on Yall1 algorithm with SNR = 0 dB, $T = 1000$ samples, $K = 14$ sources, and $H = 100$ Monte Carlo trials.

number of elements in both subarrays, and PNA. The following parameters are selected. The selected coprime pairs (M_1, M_2) are (5, 6), (3, 5), (3, 4). Note that the subarrays share more than one element apart from the first one when doubling the number elements in both subarrays, i.e., $N_t < 2M_1 + 2M_2 - 1$. In case of PNA, the parameters are $N_1 = N_2 = 5$. Therefore, the total number of elements is $N_t = 20$, expect for the latter PCA $N_t = 24$. A total of $K = 14$ far field uniformly distributed sources between $(\alpha, \beta) = (20°, 20°)$ and $(160°, 160°)$ is assumed when SNR $= 0\,$dB. The spacing between the two parallel subarrays is $d_x = d_y = d = \lambda/2$. A total of $T = 1000$ samples is assumed to estimate the covariance matrix and the number of Monte Carlo trials is 100. The virtual lag locations contribute to the estimation process when relying on the difference coarray. The PNA realizes a hole-free difference coarray. The largest number of unique lags is achieved by PNA followed by PCA with modified subarray, PCA with modified subarrays, and PCA.

It is evident that the arrays detect most of the sources. The best performance is achieved with PNA. Due to the large number of elements, the PCA with modified subarrays has better estimation compared with that when modifying one subarray. Note how errors start occurring at the range edge because of the ambiguity. The arrays are compared in Section 5.4.1 using the achieved uniform DOF (uDOF).

5.3.2 Nonparallel Linear Arrays

This subsection describes 2D arrays made of nonparallel linear arrays, namely: the L-shaped, cross-shaped, generalized L-shaped array with odd–even locations (GLA-OEL), SACA, V-shaped arrays. Furthermore, the billboard, OBA, and TSA are also mentioned for completeness.

5.3.2.1 L-Shaped Array

The L-shaped array is arguably the most straightforward extension from 1D to 2D arrays, since it involves two orthogonal ULAs. This chapter only considers orthogonal sparse linear arrays like [42–61]. Note that many works on L-shaped and cross-shaped arrays use the x–z plane for the placing the sensor array, as in the example in Fig. 5.6a.

A sparse L-shaped array [50, 51] can be constructed using two orthogonal two-level nested arrays [112] along the x- and z-axes. As shown in Fig. 5.6a, the first (dense) ULA of each nested array has $N_1 = 3$ elements spaced by $d_1 = d_x = d_y$ and the other (sparse) ULA has $N_2 = 3$ elements with wider $d_2 = (N_1 + 1)d_1$ inter-element spacing.

The L-shaped array was also considered in monostatic MIMO radar. Researchers assumed two NAs on each axis for both of the transmitter and the receiver, relying on the difference and sum coarrays [55]. This array attains $\mathcal{O}(M_1M_2 + N_1N_2)$ DOF using $\mathcal{O}(M_1 + M_2 + N_1 + N_2)$ elements, where M_1 (N_1) is the number of elements of the dense subarray at the transmitter (receiver) and M_2 (N_2) is the number of elements of the sparse subarray at the transmitter (receiver).

An L-shaped array can be constructed using four nested arrays, two nested arrays on the y-axis and another two along the z-axis [56]. On each axis, two nested arrays are concatenated, separated, and rotated, see Fig. 5.6b. The array is constructed usually with large number of elements due to four nested arrays.

Another approach considers an M-ULA along the x-axis with $d_x = \lambda/2$ spacing, $(M+1)$-ULA along the z-axis with $d_z = \lambda$ spacing, and an extra sensor at $\lambda/2$ [58]. An example of this array with $M = 5$ is shown in Fig. 5.6c, this leads to a total of $N_t = 2M + 1$ sensors.

Coprime arrays were also utilized in the legs of the L-shaped array [44–48]. The prototype coprime array [46] and coprime array doubling the number of elements in one subarray [44, 45, 48] were proposed. The elements were even doubled in both subarrays [47]. In addition, unfolded coprime array was also considered in [49]. Figure 5.6d shows a 2D L-shaped coprime

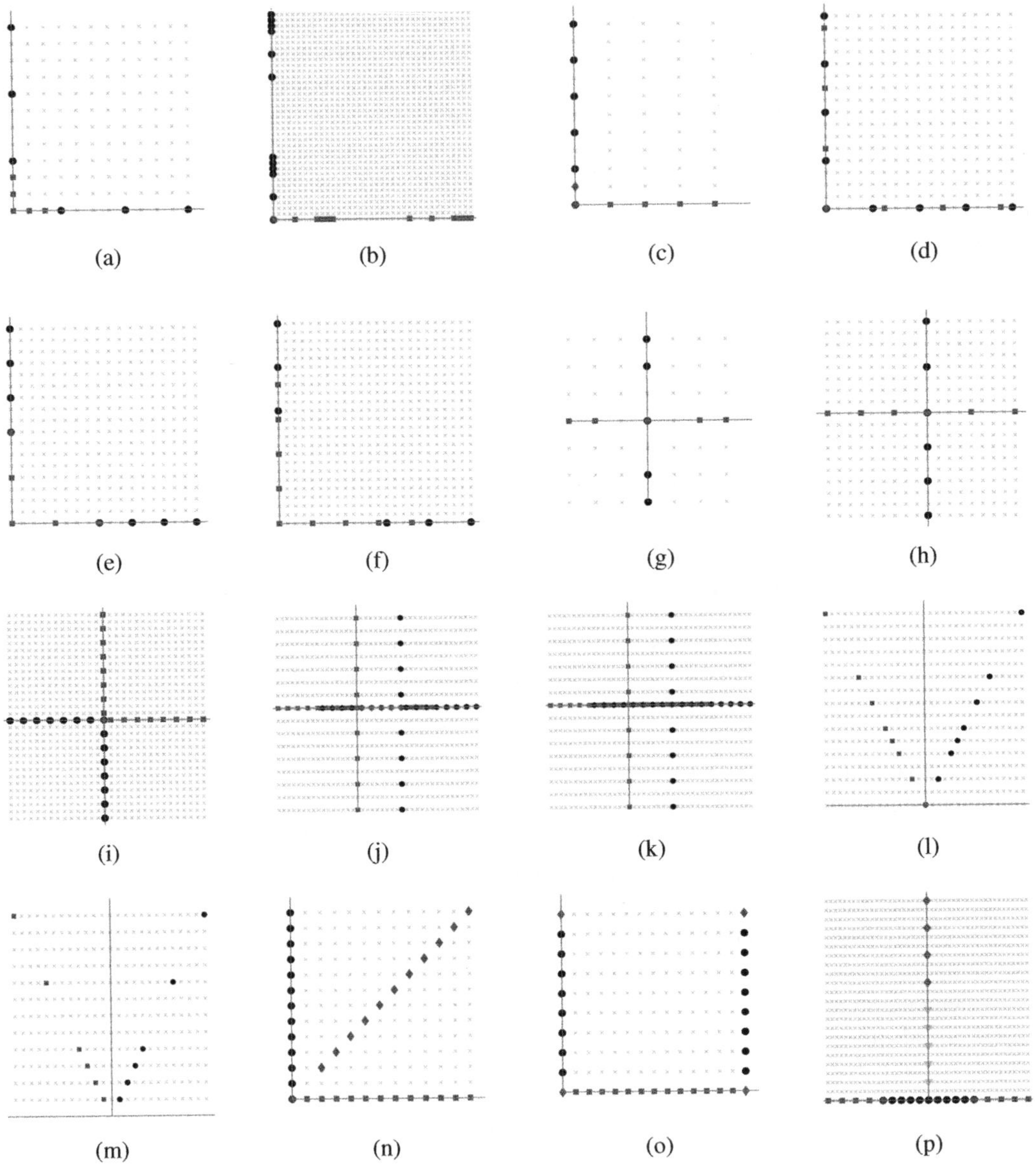

Figure 5.6 Examples of nonparallel sparse arrays. (a) Two NAs, (b) Four NAs, (c) Two ULAs with extra sensor, (d) Two orthogonal CAs and (e) UCLsA, (f) OIAC, (g) Two sparse ULAs and (h) UCLsA, (i) GLA-OEL, (j) SACA, (k) ISACA, (l) VCA, (m) VNA, (n) Billboard, (o) OBA, and (p) TSA.

array using two identical prototype coprime subarrays with $M_1 = 4$ and $M_2 = 5$. The unfolded coprime L-shaped array (UCLsA) consists of two unfolded coprime subarrays in each axis. Within each axis, the two coprime subarrays (ULAs) are spread out in opposite directions. An example is shown in Fig. 5.6e with $M_1 = 3$ and $M_2 = 4$. Each configuration presented in Fig. 5.6e has a total of $N_t = 2(M_1 + M_2 - 1) - 1$ sensors.

The optimized interleaved array configuration (OIAC) is constructed using two sparse LAs along each axis. Each sparse LA is made up of two interleaved subarrays (IAs) [57]. The first subarray has

N elements spaced by $R_1 = (M+1)d$ and the second subarray has M elements spaced by $R_2 = R_1 + d$ with a shift of $L = (N-P)R_1 - Md$, where $P = \lfloor M/2 \rfloor$ and $N \geq M \geq 2$. The OIAC has a total of $N_t = 2(M+N) - 1$ sensors. An OIAC with $N_t = 15$ elements is shown in Fig. 5.6f with $M = 3$ and $N = 5$.

5.3.2.2 Cross-shaped Array

Wu and Zhu [63] considered a uniform cross array, and a sparse symmetric cross array. For the uniform cross array, two orthogonal ULAs are used with $N_x = 2M_x + 1$ and $N_y = 2M_y + 1$ elements on the x- and y-axes, respectively. The set of array elements can be written as

$$\mathbb{S} = \{(md, 0) | -M_x \leq m \leq M_x\} \cup \{(0, nd) | -M_y \leq n \leq M_y\} \tag{5.19}$$

where d is the fundamental spacing. To construct the sparse symmetric cross array, each ULA is replaced by a symmetric sparse linear array. In this case, the set of elements becomes

$$\mathbb{S} = \{(md, 0) | m \in \mathbb{G}_x\} \cup \{(0, nd) | n \in \mathbb{G}_y\} \tag{5.20}$$

where $\mathbb{G}_x$ is a symmetric set (about the origin) and is a subset of $\{-M_x, \ldots, 0, \ldots, M_x\}$ and $\mathbb{G}_y$ is a subset of $\{-M_y, \ldots, 0, \ldots, M_y\}$. An example of this structure when $M_x = M_y = 3$ and $\mathbb{G}_x = \mathbb{G}_y = \{-3, -2, 0, 2, 3\}$ is shown in Fig. 5.6g where boxes represent $\mathbb{G}_x$ and circles represents $\mathbb{G}_y$. The two linear arrays share a common element at the origin.

A cross-shaped array can be also generated by folding one subarray in each axis to the negative side of any L-shaped-based coprime structure. Although its final shape is cross, the resultant configuration was named as the UCLsA [64]. The UCLsA is constructed using two prototype coprime subarrays. Each subarray has $M_1 + M_2 - 1$ elements. The UCLsA has a total of $2(M_1 + M_2) - 3$ located at

$$\mathbb{S} = \{(p_x d_x, 0) | p_x \in \mathbb{S}_c\} \cup \{(0, p_y d_y) | p_y \in \mathbb{S}_c\} \tag{5.21}$$

where $d_x = d_y = d = \lambda/2$, $\mathbb{S}_c = \{m_1 M_2 \cup m_2 M_1\}$, $0 \leq m_1 \leq M_1 - 1$, and $-(M_2 - 1) \leq m_2 \leq 0$. An example of this configuration is shown in Fig. 5.6h with $M_1 = 3$ and $M_2 = 4$. The array can be also viewed as moving the reference element in [49] of both original coprime arrays to the origin.

5.3.2.3 Generalized L-shaped Array with Odd–Even Locations (GLA-OEL)

Given an L-shaped array with the largest virtual URA, then relocating some elements from two legs to the negative sides of the axes can generate a cross-shaped array with the same virtual URA. A proper selection for the location of the moved elements is a key. This shall reduce mutual coupling effect but increase the required aperture. Given two positive integers N_x and N_y, the GLA-OEL has elements located as in [65]. The illustrated structure implies that all adjacent elements are spaced by two units except at the origin, as shown in Fig. 5.6i with $N_x = N_y = 15$. The sum and difference coarrays are both considered [65].

5.3.2.4 Synthetic Augmented Cross Array (SACA)

Two cross-shaped arrays share one branch can form sparse planar array on a moving platform named as SACA [66]. The DOF may suffer from severe degradation because the longest ULA cannot be generated. Meanwhile, DOF monotonically increase for the even moving distance [66].

The largest DOF can be achieved when the sensors are moved to complement every hole [66]. Noted that such an array cannot steadily achieve the largest DOF. Therefore, we need to further reduce the redundancy and improve the configuration. To overcome this, the improved ACA (IACA) is used to generate the improved SACA (ISACA). Figure 5.6j and k demonstrates a SACA and ISACA with $N_x = 16$, $N_y = 8$, and $\mathbf{m} = [9\ 0]$. The boxes indicate the original sensor locations, while the circles indicate the new locations after synthesizing the array.

5.3.2.5 V-shaped Array

A generalization to the L-shaped array is the V-shaped array, where the 90° angle between the two LAs of the L-shaped array becomes $\Omega < 90°$. Elbir [62] proposed V-shaped coprime array (VCA) and V-shaped nested array (VNA). An example of VCA with $M_1 = 2, M_2 = 5$ and VNA $N_1 = 3, N_2 = 3$ are shown in Fig. 5.6l and m, respectively, with $\Omega = 53.28°$. The author reported that the V-shaped array can resolve the same number of sources with less number of sensors when compared to CPA or GCPA (discussed in Section 5.3.3.1).

5.3.2.6 Billboard Array

The billboard array is constructed using an L-shaped array with an additional LA at 45° from both legs of the L-shaped array. An example of this array is shown in Fig. 5.6n. This array enjoys a hole-free difference coarray [67].

5.3.2.7 Open Box Array (OBA)

Three orthogonal LAs can be used to construct the OBA. The OBA can also be considered as two L-shaped arrays sharing a LA. An example OBA when $N_x = 15$ and $N_y = 10$ is shown in Fig. 5.6o where boxes represent $\mathbb{G}_1$, circles represent $\mathbb{H}_1$ and $\mathbb{H}_2$, and diamonds represent $\mathbb{F}$. For two integers N_x and N_y, the set of sensor locations can be expressed as [67]

$$\mathbb{S} = \mathbb{G}_1 \cup \mathbb{H}_1 \cup \mathbb{H}_2 \cup \mathbb{F} \tag{5.22}$$

where

$$\mathbb{G}_1 = \{(l_x, 0) | l_x \in \mathfrak{g}_1\}, \mathfrak{g}_1 = \{1, 2, \ldots, N_x - 2\} \tag{5.23}$$

$$\mathbb{H}_1 = \{(0, l_y) | l_y \in \mathfrak{h}_1\}, \mathfrak{h}_1 = \{1, 2, \ldots, N_y - 2\} \tag{5.24}$$

$$\mathbb{H}_2 = \{(N_x - 1, l_y) | l_y \in \mathfrak{h}_2\}, \mathfrak{h}_2 = \{1, 2, \ldots, N_y - 2\} \tag{5.25}$$

$$\mathbb{F} = \{(0, 0), (N_x - 1, 0), (0, N_y - 1), (N_x - 1, N_y - 1)\} \tag{5.26}$$

The difference coarray is given by

$$\mathbb{D} = \{(m_x, m_y) \in \mathbb{Z}^2 | -N_x + 1 \le m_x \le N_x + 1, -N_y + 1 \le m_y \le N_y + 1\} \tag{5.27}$$

which is a URA. Generalizations of OBA are discussed in Section 5.3.5.1. In particular, the generalizations will ensure the same continuous difference coarray in (5.27) while increasing inter-element spacing, thus reducing mutual coupling effects.

The two arrays (billboard and OBA) are constructed using ULAs, and the performance of both is generally exceeded by three newly developed arrays which are: HOBA-2 [67], the hourglass array [67], and the thermos array [95], discussed in Sections 5.3.5.1, 5.3.5.2, and 5.3.5.3, respectively.

5.3.2.8 T-shaped Array (TSA)

A TSA can be constructed with a vertical subarray located along the z-axis and a horizontal subarray along both sides of x-axis [68]. The TSA consists of a generalized unfolded coprime array and a symmetrically unfolded coprime linear array. It also can be seen as the extended structure of conventional L-shaped array by adding one leg. The element locations of the TSA are given by the following set

$$\mathbb{S} = \mathbb{S}_v \cup \mathbb{S}_h \tag{5.28}$$

where $\mathbb{S}_v$ represents the vertical subarray and $\mathbb{S}_h$ denotes the horizontal subarray. The vertical and the horizontal subarrays have $N_y = N_{y1} + N_{y2}$ and $N_x = 2N_{x1} + 2N_{x2} + 1$ elements respectively, which are located at

$$\begin{aligned}\mathbb{S}_v &= \{(0, 0, \tilde{d} + M_1 n_{y1} d_z) | 0 \leq n_{y1} \leq N_{y1} - 1\} \\ &\cup \{(0, 0, \tilde{d} + ((N_{y1} - 1)M_1 + n_{y2} N_1) d_z) | 1 \leq n_{y2} \leq N_{y2}\}\end{aligned} \tag{5.29}$$

$$\begin{aligned}\mathbb{S}_h &= \{(-((N_{x2} + 1)M_2 + N_2 n_{x1}) d_x, 0, 0) | 0 \leq n_{x1} \leq N_{x1} - 1\} \\ &\cup \{(M_2 n_{x2} d_x, 0, 0) | - N_{x2} - 1 \leq n_{x2} \leq N_{x2} + 1\} \\ &\cup \{(((N_{x2} + 1)M_2 + N_2 n_{x1}) d_x, 0, 0) | 0 \leq n_{x1} \leq N_{x1} - 1\}\end{aligned} \tag{5.30}$$

where $\tilde{d} > \lambda/2$, $d_z = d_x = \lambda/2$, (M_1, N_1) and (M_2, N_2) are two pairs of coprime integers, and the TSA has a total of $N_t = 2N_x + N_y + 1$ sensors. When $N_{y2} = M_1$ and $N_{y1} = N_1$, the horizontal subarray becomes an unfolded coprime array. In addition, the vertical subarray can be viewed as an extended version of unfolded coprime array. The major difference is that the minimum inter-element spacing in not limited by half-wavelength [68]. A TSA is shown in Fig. 5.6p, where $N_{x1} = N_{y1} = 5$, $N_{x2} = N_{y2} = 4$, $M_1 = M_2 = 2$, $N_1 = N_2 = 3$, and $\tilde{d} = 2$.

The discussed nonparallel arrays are compared in Section 5.4.1 using their achieved uDOF. Other billboard and T-shaped configurations based on coprime, nested, and super-nested arrays are recently proposed in [117].

5.3.3 Interleaved Rectangular Arrays

This section describes 2D arrays made by interleaving two rectangular arrays like the CPA, GCPA, and the UCPA. In addition, other planar structures like NPA, NCPA, and other planar arrays with motion are examined.

5.3.3.1 Coprime Planar Array (CPA)

Coprime planar array results from interleaving two uniform square subarrays [69–74]. The GCPA [75–79] is a generalization of CPA where the number of elements along the x- and y-axes need not be equal. Some details of 2D-DOA estimation algorithms using these arrays are outlined.

A CPA is constructed using two interleaved uniform square subarrays. There are $M_1 \times M_1$ and $M_2 \times M_2$ elements in subarrays 1 and 2, respectively. Since the two subarrays share a common element at the origin, the total number of elements is $M_1^2 + M_2^2 - 1$. The set of sensor elements is [69–73]

$$\mathbb{S} = \{(md_1, nd_1) | 0 \leq m, n \leq M_1 - 1\} \cup \{(pd_2, qd_2) | 0 \leq p, q \leq M_2 - 1\} \tag{5.31}$$

where $d_1 = M_2 d$, $d_2 = M_1 d$, $d = \lambda/2$, and m, n, p, q are integers. Figure 5.7a shows a CPA with $M_1 = 3$ and $M_2 = 4$, where the elements within the first and second planar subarrays are shown with square and circle markers, respectively.

The GCPA structure [75, 76] has more DOF and produces better DOA estimates for the same number of sensors as CPA. GCPA relaxes the condition of using uniform square subarrays and allows uniform rectangular subarrays. An example of GCPA is shown in Fig. 5.7b. GCPA is made up using two interleaved URAs having the dimensions of $M_1 \times N_1$ and $M_2 \times N_2$, where $M_1 = 3$, $N_1 = 4$, $M_2 = 4$, and $N_2 = 5$. Again, M_1 and M_2 are coprime integers, the same is applicable for N_1 and N_2. Here, M_1 (M_2) is the number of elements of the first (second) URA along the x-axis, N_1 (N_2) is the number of elements of the first (second) URA along the y-axis, and $N_1 = M_1$, $N_2 = M_2$ in the

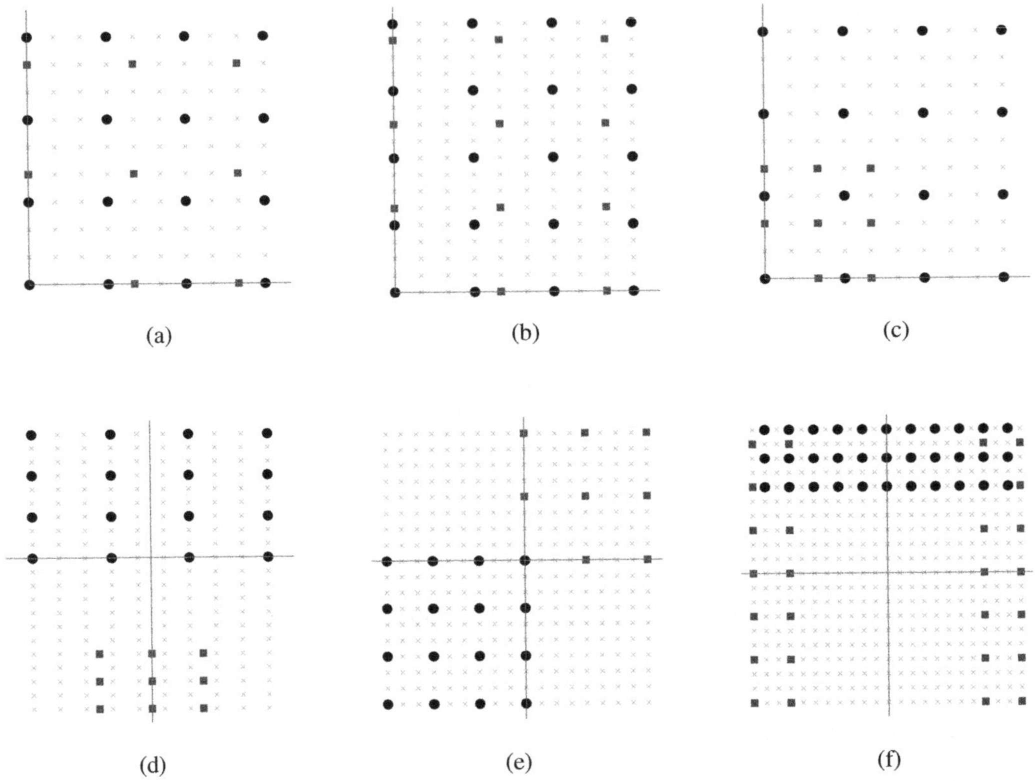

Figure 5.7 Examples of 2D sparse arrays constructed using planar subarrays of coprime-based structure. (a) CPA, (b) GCPA, (c) CAACS, (d) CATSS, (e) UCPA, and (f) SDCPA.

case of restricting URAs to become uniform square subarrays. In the case of GCPA, the set of array elements becomes

$$\mathbb{S} = \{(mM_2d_x, nN_2d_y)|0 \leq m \leq M_1 - 1, 0 \leq n \leq N_1 - 1\} \cup \{(pM_1d_x, qN_1d_y)|0 \leq p \leq M_2 - 1, 0 \leq q \leq N_2 - 1\} \tag{5.32}$$

where d_x (d_y) is the unit spacing along the x- (y-) axis and is often chosen to be $\lambda/2$. The same structure was studied in [77–79], but the number of elements in one URA was doubled.

The coprime array with a compressed subarray (CAACS) [80] can be constructed by applying compression for one subarray within the CPA in [69–73]. If M_1 can be factorized into two integers, $M_1 = q\widetilde{M}_1, 1 < q < M_1$. Then the spacing of subarray is reduced by a factor q, that is $d_1 = \widetilde{M}_1 d$. This implies that $\widetilde{M}_1$ is also a coprime with M_2. Let's assume that $M_1 = 4$, $M_2 = 3$, and $q = 2$. The resultant CAACS is shown in Fig. 5.7c. Note that the aperture size is the same as its counterpart CPA. This operation increases the uDOF compared with CPA at the expense of increased mutual coupling. Motivated by the redundancy that appears in the difference and sum coarrays, another 2D structure was proposed. The coprime array with two separated subarrays (CATSS) [80] can be constructed by applying a suitable separation and shift between the two subarrays of CAACS, see Fig. 5.7d. The compression factor and the amount of shift have a direct impact on the DOF at the expense of increased mutual coupling and aperture size. This is achieved using the difference and sum coarrays.

A virtual CPA (VCPA) can be generated by moving the prototype coprime LA at certain velocity, v [74]. Consider a coprime LA of two subarrays located along the x-axis with M_1 and M_2 elements spaced respectively by M_2d and M_1d. It is assumed that this array is moved along y-axis at a fixed velocity, v, and snapped at $t+m_1M_2\tau$ and $t+m_2M_1\tau$, where m_1 and m_2 are two positive integers with $v\tau = d$. This configuration realizes $M_1^2 + M_1^2 - 1$ DOF using $N_t = M_1^2 + M_2^2 - 1$ elements [74]. In [81], one CPA was used as a transmitter and multiple CPAs with certain spacing from the transmitter were used as receivers in monostatic MIMO radar.

5.3.3.2 Unfolded Coprime Planar Array (UCPA)

Instead of interleaving two uniform square arrays (subarrays) to construct the CPA in Section 5.3.3.1, one of the uniform square arrays is unfolded, that is, flipped across the x-axis or the y-axis. Therefore, the two subarrays lie in different quadrants. Fixing one subarray at the first quadrant, Zheng et al. [84] investigated the performance of 2D-DOA estimation when the other subarray is in the three other quadrants. When one subarray is flipped across the line $y = -x$, the performance is better than other flipped structures and the traditional CPA [84]. Figure 5.7e shows this case where the first (second) subarray has $M_1 \times M_1$ ($M_2 \times M_2$) sensor elements, and the set of array elements is

$$\mathbb{S} = \{(md_1, nd_1)|0 \leq m, n \leq M_1 - 1\} \cup \{(pd_2, qd_2)| - (M_2 - 1) \leq p, q \leq 0\} \tag{5.33}$$

where $d_1 = M_2d$, $d_2 = M_1d$, and $d = \lambda/2$.

5.3.3.3 Symmetric Displaced Coprime Planar Array (SDCPA)

The SDCPA is constructed using three sparse uniform rectangular arrays with a certain horizontal and vertical displacement [86]. It can be viewed as two bilaterally symmetric displaced CPAs with one shared subarray. Consider a CPA structure consisting of two subarrays, whose sensor positions are given as

$$\mathbb{S} = \mathbb{S}_1 \cup \mathbb{S}_2 \tag{5.34}$$

where $\mathbb{S}_1$ and $\mathbb{S}_2$ are two sets of the sensor locations of the two subarrays expressed as

$$\mathbb{S}_1 = \{(p_{x1}, p_{y1})|p_{x1} \in \mathfrak{x}_1, p_{x1} \in \mathfrak{h}_1\} \tag{5.35}$$

$$\mathbb{S}_2 = \{(p_{x2}, p_{y2})|p_{x1} \in \mathfrak{x}_2, p_{x2} \in \mathfrak{h}_2\} \tag{5.36}$$

where $\mathfrak{x}_1$ and $\mathfrak{h}_1$ denote respectively the x- and y-coordinate sets of the sensor locations in $\mathbb{S}_1$, and $\mathfrak{x}_2$ and $\mathfrak{h}_2$ denote respectively the x- and y-coordinate sets of the sensor locations in $\mathbb{S}_2$. For symmetric subarrays in the vertical directions, $\mathfrak{x}_1$ and $\mathfrak{x}_2$ are symmetric about element 0, while the sensors in $\mathbb{S}_1$ and $\mathbb{S}_2$ are arranged in the horizontal direction. The x-coordinates of sensor locations can be expressed by a set of integers as

$$\mathfrak{x} = \mathfrak{x}_1 \cup \mathfrak{x}_2 \tag{5.37}$$

where

$$\mathfrak{x}_1 = \mathfrak{x}_{11} \cup \mathfrak{x}_{12} = \{L_1 + M_1n|n \in [0, N_1 - 1]\} \cup \{-L_1 - M_1n|n \in [0, N_1 - 1]\} \tag{5.38}$$

$$\mathfrak{x}_2 = \{N_1m - N_1(2M_1 - 1)|m \in [0, 4M_1 - 2]\} \tag{5.39}$$

with M_1 and N_1 being coprime integers, and $L_1 = k_1N_1$ ($k_1 \in \mathbb{Z}^+$) is the displacement factor in the horizontal direction. Whereas the y-coordinates of sensor locations can be expressed by a set of integers as

$$\mathfrak{h} = \mathfrak{h}_1 \cup \mathfrak{h}_2 \tag{5.40}$$

where

$$\mathfrak{h}_1 = \{N_2 m - N_2(2M_2 - 1) | m \in [0, 4M_2 - 2]\} \tag{5.41}$$

$$\mathfrak{h}_2 = \{L_2 + M_2 n | n \in [0, N_2 - 1]\} \tag{5.42}$$

where M_2 and N_2 being coprime integers, and $L_2 = k_2 N_2 (k_2 \in \mathbb{Z}^+)$. The maximum uDOF is achieved when $L_1 = \left(2M_1 - 1 - \left\lfloor \frac{M_1}{2} \right\rfloor\right) N_1$ and $L_2 = \left(2M_2 - 1 - \left\lfloor \frac{M_2}{2} \right\rfloor\right) N_2$. Figure 5.7f presents an example of this configuration with $M_1 = 3$, $M_2 = 2$, $N_1 = 2$, $N_2 = 3$, $L_1 = 8$, and $L_2 = 6$. The array has closed form expressions for sensor locations, the achieved DOF, and suffers less from mutual coupling compared with other CPAs [86]. The weight functions are $w(0, 1) = 4$, $w(1, 0) = 2$, $w(1, 1) = 2$, and $w(1, -1) = 2$.

The central URA is extracted from the difference coarray together with the corresponding measurements and used for 2D-DOA estimation. Spatial smoothing followed by the 2D unitary estimation of signal parameters via rotational invariance techniques (ESPRIT) algorithm [118] are used. A total of $K = 16$ far field uniformly distributed sources [67] is assumed when SNR = 0 dB. The CPA is constructed with $M_1 = 5$ and $M_2 = 6$. The GCPA has $M_1 = 3$, $N_1 = 7$, $M_2 = 5$, and $N_2 = 8$. CAACS uses $M_1 = 6$, $M_2 = 5$, and $q = 3$. Finally, the SDCPA is constructed using $M_1 = 2$, $M_2 = 3$, $N_1 = 2$, $N_2 = 3$, $L_1 = 8$, and $L_2 = 6$. Thus, the SDCPA has $N_t = 59$ sensors and the remaining configurations have $N_t = 60$ sensors. The unit spacing is set to $d = \lambda/2$. A total of $T = 1000$ samples is assumed to estimate the covariance matrix and the number of Monte Carlo trials is $H = 100$. The mutual coupling parameters are $c_1 = 0.25$, $B = 5$, and $c_l = \frac{c_1}{l} e^{j\pi(l-1)/4}$ [67]. Figure 5.8 shows the actual angles; indicated by circles; and the estimated angles; indicated by dots. Unlike the lower subplots, no mutual coupling is considered in the upper subplots. All arrays detect all sources when there is no mutual coupling. On the other hand, only SDCPA detects all sources in the

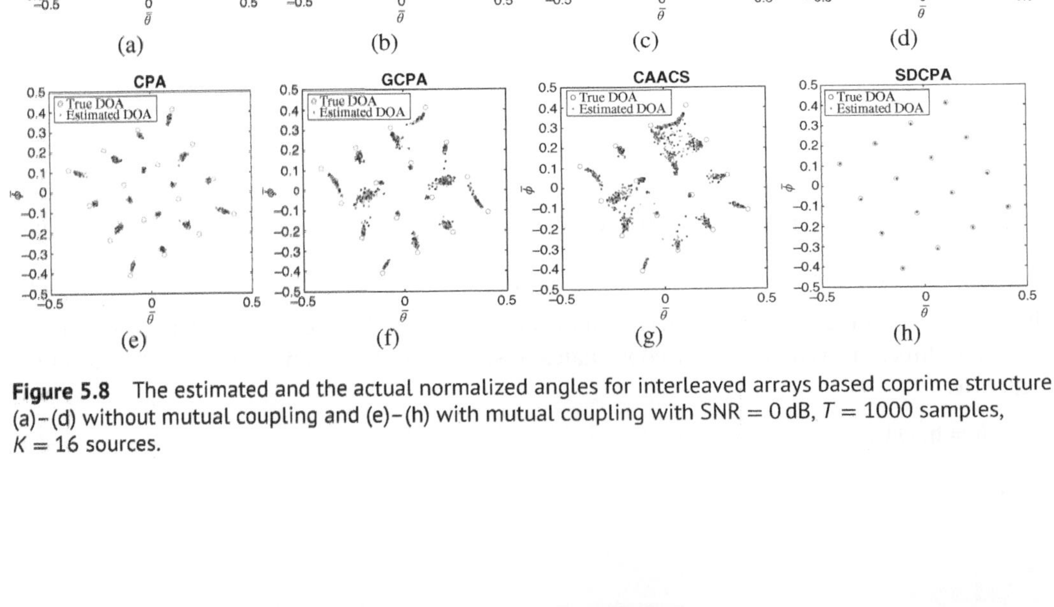

Figure 5.8 The estimated and the actual normalized angles for interleaved arrays based coprime structure (a)–(d) without mutual coupling and (e)–(h) with mutual coupling with SNR = 0 dB, $T = 1000$ samples, $K = 16$ sources.

presence of mutual coupling. The performance of the estimation process is related to the coarrays, because the weight functions for these configurations are the same, that is $w(1, 0) = w(0, 1) = w(1, 1) = w(1, -1) = 2$. The SDCPA defers in only one weight, that is $w(0, 1) = 4$.

5.3.3.4 Nested Planar Array (NPA)

The NPA was proposed by Pal and Vaidyanathan [82, 83] as an extension to the 1D nested array, and it has $\mathcal{O}(N_1 N_2)$ difference coarray with only $\mathcal{O}(N_1 + N_2)$ sensors. Here N_1 denotes the number of sensors in a dense grid, and N_2 is the physical sensors in a sparse grid. For example, using two linear nested arrays with $N_1 = N_2 = 3$, the nested planar array looks like Fig. 5.9a. The authors also showed that with effective selection of N_1 and N_2, $\mathcal{O}\left(N_t^2\right)$ sources can be identified using the difference coarray.

5.3.3.5 Nested Coprime Planar Array (NCPA)

The structure of NCPA is quite involved to describe, despite being simple in design. Among the motivations reported by Si et al. [85] is that their NCPA can outperform all other CPAs, namely GCPA [75], CPA [69–71], and UCPA [84] in terms DOF. The NCPA also produces two VCPAs that are used to obtain two sets of DOA estimation that are linked together to get the unique angles.

The NCPA is constructed from two sparse planar arrays, unlike the other types: GCPA, CPA, and UCPA which are constructed from uniform planar arrays. Each sparse planar array is constructed using NPA. Each NPA is composed of a dense subarray and a sparse subarray. An example of this structure with $M_1 = 2$ and $M_2 = L = 3$ is shown in Fig. 5.9b. The NPA is constructed from the elements of two orthogonal nested linear arrays (NLAs) and the elements at the intersections of lines, in the first quadrant, emerging from each sensor element and orthogonal to the nested linear array. Mathematically, the set of array elements is

$$\mathbb{S} = \{(m, n) | m, n \in \mathbb{S}_{\text{NPA1}}\} \cup \{(m, n) | m, n \in \mathbb{S}_{\text{NPA2}}\} \tag{5.43}$$

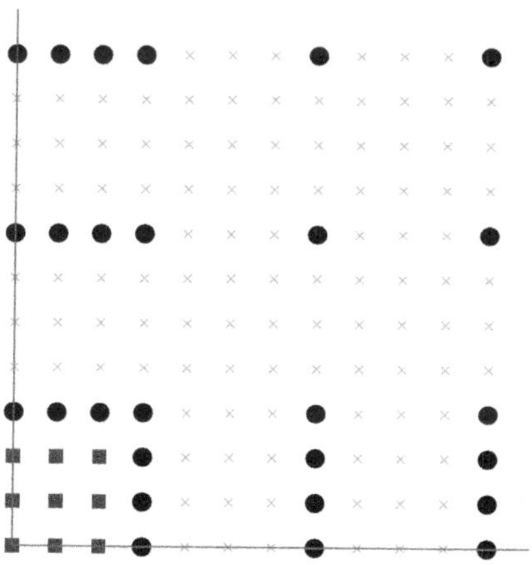

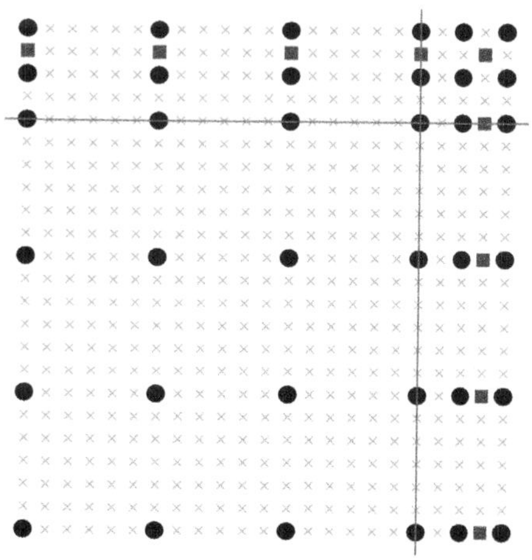

(a) (b)

Figure 5.9 Examples of 2D sparse arrays constructed using planar subarrays of nested based-structure. (a) NPA and (b) NCPA.

This formula shows that the NCPA are made up of NPAs. The NPAs are made up of two orthogonal NLAs and other sensors given as

$$\mathbb{S}_{\text{NPA1}} = \{-L+1, -L+2, \dots, 0\} \times M_1M_2d \cup \{0, 1, \dots, M_1-1\} \times M_2d, \quad (5.44)$$

$$\mathbb{S}_{\text{NPA2}} = \{-L+1, -L+2, \dots, 0\} \times M_1M_2d \cup \{0, 1, \dots, M_2-1\} \times M_1d, \quad (5.45)$$

where M_1, M_2 are coprime positive integers, and L is an integer.

5.3.3.6 Planar Arrays with Motion

Moving any arbitrary LA multiple times can form a new planar array called vertical synthetic planar array (VSPA) [87]. Consider an arbitrary sparse LA with N_s sensors located along the x-axis moving at a constant velocity v vertically along the y-axis. The sensors of the original sparse array are located at $\mathbb{S}_i = \{(x, 0)|x \in \mathbf{p}_0\}$, with $\mathbf{p}_0 = \{p_1, p_2, \dots, p_{N_s}\}$ being the sensor positions. The array moves a distance d after a duration of ε, where $d = v\varepsilon = \lambda/2$. The sensor locations of the LA in the ith measurement are expressed as [87]

$$\mathbb{S} = \mathbb{S}_1 \cup \mathbb{S}_2 \cup \dots \cup \mathbb{S}_{M_s} \quad (5.46)$$

where $\mathbb{S}_i = \{(x, l_i)|x \in \mathbf{p}_0, l_i \in \mathbf{p}_m\}$ for $i = 1, 2, \dots, M_s$, with M_s being the total measurements taken when the array moves vertically, and $\mathbf{p}_m = \{l_1, l_2, \dots, l_{M_s}\}$ is called the moving trail. The array is named as "VSPA-origin-trail." The VSPA has a total of $N_t = N_sM_s$ elements. The term "origin" represents the structure of the original linear array, and the term "trail" represents the structure of the moving trail. By controlling the set of the moving trail as per the most candidate sparse linear arrays, CA, NA, VSPA-CA-CA, and VSPA-NA-NA can be generated, respectively. In the former case, each row and column in the planar array is a CA. On the other hand, each row and column of VSPA-NA-NA is an NA. A mixed of both may be also generated, that is VSPA-CA-NA or VSPA-NA-CA, and VSPA-MRA-MRA was studied as well in [87].

Figure 5.10 shows different VSPAs generated by moving CA, NA, super nested array (SNA), or MRA along the y-axis. The circles denote the original array geometry and the squares represent the moving trail. Coprime array with $M_1 = 3$ and $M_2 = 4$ is used to construct the arrays in Fig. 5.10a and c. Nested and super nested arrays with $N_1 = N_2 = 3$ is used to construct the arrays in Fig. 5.10b, d, and e. A minimum redundancy array with [0, 1, 6, 9, 11, 13] is used to construct the array in Fig. 5.10f. The total number of sensors for the underlaying arrays is $N_s = 6$ and there are $M_s = 5$ trails. Therefore, each configuration has a total of $N_t = 36$ sensors. Due to nested and super nested structures, VSPA-NA-NA and VSPA-SNA-SNA achieve hole-free difference coarrays. The difference coarray of VSPA-CA-NA is equivalent to that of the VSPA-NA-CA but rotated by 90°. The VSPA-CA-CA realizes the smallest central URA.

VSPA like structures were proposed in [88–90] but without introducing any motion to synthesized the arrays. Specifically, two arrays; named as the generalized planar sparse array (GPSA) [88]; similar to VSPA-NA-NA and VSPA-MRA-MRA were proposed in [87]. The rectangular generalized minimum redundancy array (RGMRA) [89] is similar to VSPA-MRA-MRA [87]. The rectangular hybrid-nested array (RHNA) and rectangular super-nested array (RSNA) [90] are constructed using MRA-SNA and SNA-SNA, respectively. The last MRA (row) was reversed to reduce the mutual coupling effect between adjacent sparse linear arrays in [89, 90]. Structures based on NA, MRA, and SNA realize hole-free difference coarrays.

The performance of different VSPAs is studied next without mutual coupling. Arrays with holes need pre-processing to capture the measurements corresponding to the central URA. Then spatial smoothing followed by 2D unitary ESPRIT algorithm can be used [118]. The VSPAs presented

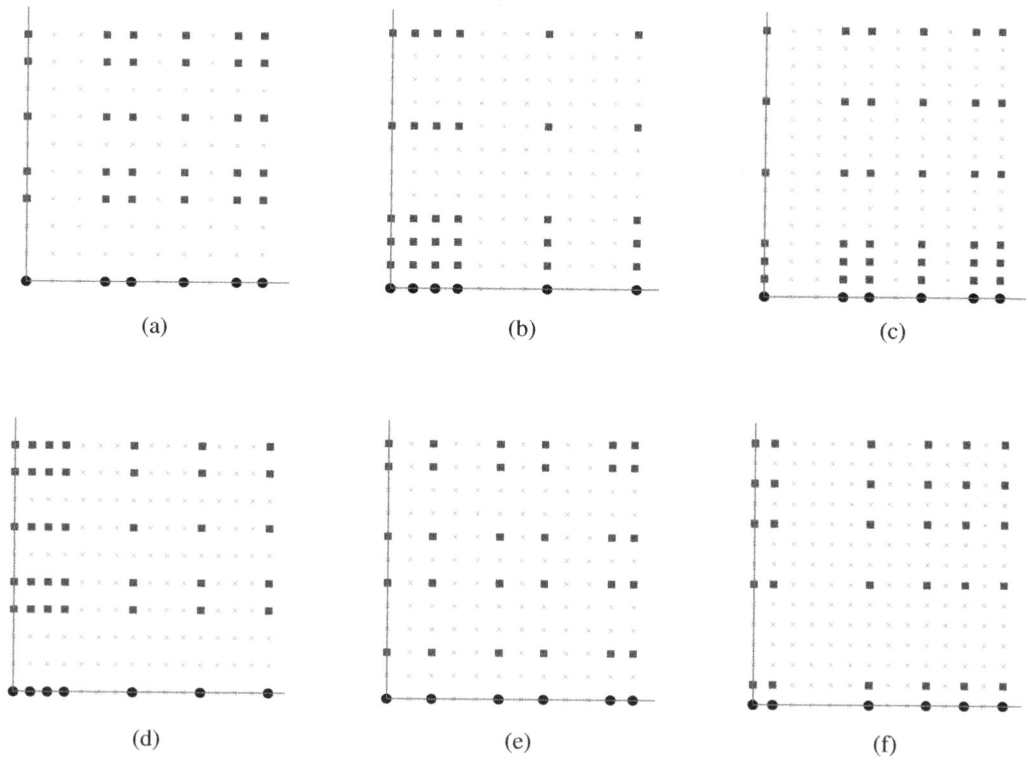

Figure 5.10 Examples of 2D sparse arrays constructed using VSPA. (a) VSPA-CA-CA, (b) VSPA-NA-NA, (c) VSPA-CA-NA, (d) VSPA-NA-CA, (e) VSPA-SNA-SNA, and (f) VSPA-MRA-MRA.

in Fig. 5.10 are considered, where $N_t = 36$ sensors in the presence of $K = 36$ far field uniformly distributed sources [67]. The SNR $= 0\,\text{dB}$ and unit spacing $d = \lambda/2$. A total of $T = 1000$ samples is assumed and the number of Monte Carlo trials is $H = 100$. Figure 5.11 shows the actual angles versus the estimated angles. It is evident that all arrays detect all sources except VSP-CA-CA since it has the smallest central URA. Few sources are missing in different iterations in the case of VSPA-CA-NA and VSPA-NA-CA. CPAs and NPAs have different way of construction where there is closed form expressions for sensor locations. On the other hand, sensor locations within the VSPAs depend on the selected original LA and moving trail. They require a smaller number of sensors for the same aperture compared with CPAs and NPAs. The achieved uDOF of these arrays are compared in Section 5.4.1.

The authors in [91] proposed to move the original 2D NPA [82, 83] several times to construct a 2D multi-level dilated nested array (MLDiNA). The original sensor locations of the MLDiNA are similar to that of NPA with the inter-element spacing between adjacent sensors of each row being scaled by q, where q is an integer. The array can have a hole-free difference coarray and has at least two times more DOF than the NPA by properly selecting the number of movements, that is $q \leq 2L + 1$ where L is the array level [91]. This approach is good for stationary environments.

5.3.4 Conformal Arrays

Conformal arrays follow or conform to a pronounced shape, for example circular, rectangular, etc. This subsection presents few conformal shaped arrays including nested conformal arrays, CCAs,

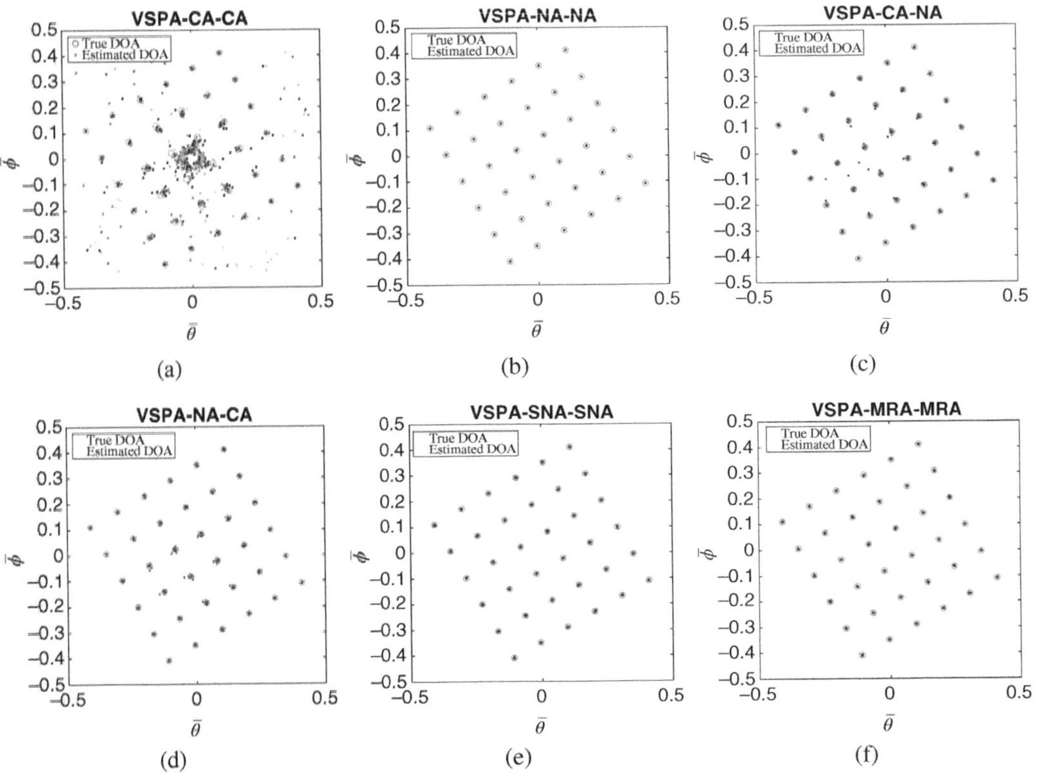

Figure 5.11 The estimated and the actual normalized angles for (a) VSPA-CA-CA, (b) VSPA-NA-NA, (c) VSPA-CA-NA, (d) VSPA-NA-CA, (e) VSPA-SNA-SNA, and (f) VSPA-MRA-MRA with SNR = 0 dB, $T = 1000$ samples, $K = 36$ sources.

and hybrid uniform circular cylindrical array. There are also uniform conformal arrays, but they are not sparse.

Two N-elements nested parallel subarrays along a cylinder with radius r covering an angular sector of 90° are combined to form nested conformal arrays (NCAs) [92]. The 2D estimation problem is converted into two 1D problems in the coarray domain, where the elevation angles are estimated through sparse reconstruction, and then the azimuth angles are estimated by the least squares [92].

A coprime circular conformal array (C3MA) is constructed by staking L number of coprime circular microphone (sensor) arrays (CCMA). In [93], two sparse microphone subarrays consisting of M_1 and M_2 microphones, respectively, where M_1 and M_2 are coprime integers. Assuming the reference microphone of CCMA of radius r, the angular distance Ω between two microphones is given by

$$\Omega_n = \begin{cases} \frac{\Omega_n}{M_1}, & n \bmod 2 = 0 \\ \frac{\Omega(n-1)}{M_2}, & n \bmod 2 = 1 \end{cases} \tag{5.47}$$

where $n = 1, \ldots, N$ and N is the number of microphones in CCMA such that $M_1 + M_2 - 1 = N$. The C3MA has a total of $N_t = 2N = 2(M_1 + M_2 - 1)$ sensors.

The hybrid uniform circular cylindrical array (UCyA) is constructed using L vertically positioned UCAs [94]. There are M elements in each UCA, therefore the UCyA has a total of $N_t = LM$ sensors. With r being the radius of each UCA, the vertical elements are spaced by $d = \lambda/2$, and $\varphi_m = 2\pi(m-1)/M$ denotes the difference of the central angles between the mth element and the first element of each UCA.

5.3.5 Other 2D Arrays

This subsection describes additional 2D sparse arrays that do not readily fall under any of the preceding categories. In particular, the half OBA with two layers (HOBA-2), the hourglass array, the thermos array, and the CcRA are considered. Additionally, ESCA, HHA, and LAA are presented.

5.3.5.1 Half Open Box Array-2 (HOBA-2)

The HOBA-2 is an extension of the HOBA which is a special case of the partial open box array (POBA). The POBA is a systematic redistribution of elements in the OBA (Section 5.3.2.7) to reduce closely spaced elements which decreases mutual coupling. An example of this array with $N_x = 16$ and $N_y = 12$ is shown in Fig. 5.12a, where circles denote $\mathbb{H}_1$, $\mathbb{H}_2$, squares represent $\mathbb{G}_1$, $\mathbb{G}_2$, and diamonds stand for $\mathbb{F}$. If some elements of $\mathbb{G}_1$ of the OBA (5.22) are redistributed on the opposite (empty) side of the OBA, the elements of POBA can be written as [67]

$$\mathbb{S} = \mathbb{G}_1 \cup \mathbb{G}_2 \cup \mathbb{H}_1 \cup \mathbb{H}_2 \cup \mathbb{F} \tag{5.48}$$

where

$$\mathbb{G}_1 = \{(l_x, 0) | l_x \in \mathfrak{g}_1\}, \mathfrak{g}_1 = \{1, 2, \ldots, N_x - 2\} \tag{5.49}$$

$$\mathbb{G}_2 = \{(l_x, N_y - 1) | l_x \in \mathfrak{g}_2\}, \mathfrak{g}_2 = \{1, 2, \ldots, N_x - 2\} \tag{5.50}$$

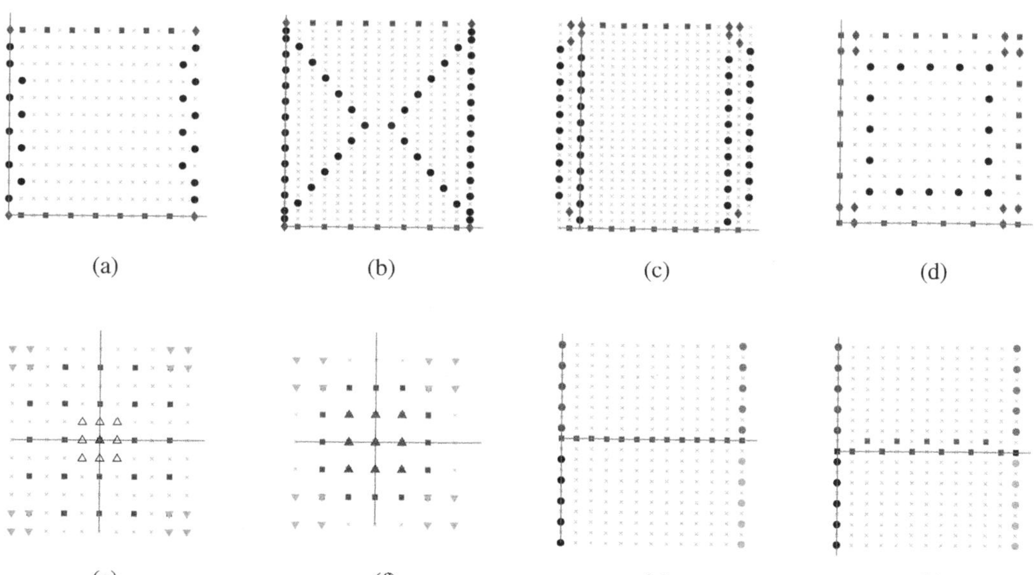

Figure 5.12 Additional examples of 2D sparse arrays. (a) HOBA-2, (b) Hourglass array, (c) Thermos array, (d) CcRA, (e) ESCA with $N_1 = 2, N_2 = 3$, (f) ESCA with $N_1 = N_2 = 2$, (g) HHA, and (h) LAA.

where $\mathfrak{g}_1$ and $\mathfrak{g}_2$ are subsets of $\{1, 2, \ldots, N_x - 2\}$ with $|\mathfrak{g}_1| + |\mathfrak{g}_2| = N_x - 2$. Note that POBA is still defined using N_x and N_y as the OBA, and $\mathbb{H}_1$, $\mathbb{H}_2$, and $\mathbb{F}$ are as defined in (5.24), (5.25), and (5.26), respectively. If $\mathfrak{g}_1$ and $\mathfrak{g}_2$ are defined as

$$\mathfrak{g}_1 = \{1 + 2l | 0 \leq l \leq \lfloor (N_x - 3)/2 \rfloor\}, \tag{5.51}$$

$$\mathfrak{g}_2 = \{N_x - 1 - 2l | 1 \leq l \leq \lfloor (N_x - 2)/2 \rfloor\}. \tag{5.52}$$

The resultant array is called HOBA. Similar argument can be applied to the other ULAs (on the right and left) of the OBA, which results in the partial open box array with L levels (POBA-L)

$$\mathbb{S} = \mathbb{G}_1 \cup \mathbb{G}_2 \cup \left(\bigcup_{l=1}^{L} \mathbb{H}_{1,l} \cup \mathbb{H}_{2,l} \right) \cup \mathbb{F} \tag{5.53}$$

where each of $\mathfrak{g}_1$ and $\mathfrak{g}_2$ is a partition of $\{1, 2, \ldots, N_x - 2\}$, $\{\mathfrak{h}_{1,l}\}_{l=1}^{L}$ is a partition of $\{1, 2, \ldots, N_y - 2\}$, and $\mathfrak{h}_{2,l} = N_y - 1 - \mathfrak{h}_{1,l}$ for $l = 1, 2, \ldots, L$. A third positive integer $L \leq N_x/2$ is required to design POBA-L. If L is chosen to be 2, $\mathfrak{g}_1$ and $\mathfrak{g}_2$ as in (5.51), (5.52) and

$$\mathfrak{h}_{1,1} = \{1 + 2l | 0 \leq l \leq \lfloor (N_y - 3)/2 \rfloor\} \cup \{N_y - 2\}, \tag{5.54}$$

$$\mathfrak{h}_{1,2} = \{2l | 1 \leq l \leq \lfloor (N_y - 3)/2 \rfloor\}, \tag{5.55}$$

the array is called HOBA-2. The next subsection introduces the hourglass array, which is a POBA-L array.

5.3.5.2 Hourglass Array

Hourglass array is a special case of POBA [67]. The merits of hourglass arrays include (i) having the same coarray as the OBA (using the same aperture), (ii) having a hole-free difference coarray (based on the previous virtue), and (iii) reduced mutual coupling compared with the OBA. Figure 5.12b shows hourglass array with $N_x = 15$ and $N_y = 27$.

To validate the robustness of HOBA-2 and hourglass arrays in presence of mutual coupling, the authors compared the arrays with billboard, OBA, and URA using 2D unitary ESPRIT without modifications to account for mutual coupling. With small number of samples, it was shown that hourglass achieves the best performance compared to others [67].

5.3.5.3 Thermos Array

Sun et al. [95] proposed the thermos array as an improvement to the hourglass array in terms of reducing mutual coupling. Since the elements with vertical inter-sensor spacing of d are about five times larger than the horizontal ones, they proposed to reduce the vertical spacings. Two parameters are enough to design a thermos array: $N_x \in \mathbb{Z}^+$ and $N_y \in \mathbb{Z}^+$ which yield the total number of sensors as $N_x + 2N_y - 2$. An example of a thermos array is shown in Fig. 5.12c with $N_x = 15$ and $N_y = 27$, where the circles represent $\mathbb{S}_1$, $\mathbb{S}_2$, $\mathbb{R}_1$, $\mathbb{R}_2$, boxes denote $\mathbb{B}$, $\mathbb{T}$, and diamonds denote $\mathbb{F}$. The thermos array has a rectangular shape and is designed using six ULAs with inter-element spacing of $2d$ plus four or ten more sensors at the corners depending on the number of elements in the y direction, N_y. If N_y is even, ten elements are required, and four if N_y is odd. The set of array elements $\mathbb{S}$ can be found by taking the union of a top, bottom, two left, and two right ULAs plus the extra sensors at the corner denoted by $\mathbb{F}$ [95].

5.3.5.4 Concentric Rectangular Array (CcRA)

The authors in [96] proposed the CcRA which can be thought of as a modification of the boundary array in [119]. A CcRA is shown in Fig. 5.12d with $N_x = N_y = 12$, where squares stand for $\mathbb{G}_o$, and circles stand for $\mathbb{G}_i$. The total number of sensors is $2(N_x + N_y)$. Although the paper does not directly implement this array for 2D-DOA estimation, it is cited as it proposes a sparse 2D array. The hourglass array is sparser than the CcRA, and both have a hole-free difference coarray. However, the CcRA has a contiguous sum coarray, unlike the hourglass array [96]. In addition, all the elements in the CcRA are essential [100], that is removing any element affects the coarray.

5.3.5.5 Extended Sparse Convolutional Array (ESCA)

A 2D ESCA can be constructed using three subarrays [97]. The sensor locations of the ESCA with N_t sensors are given as

$$\mathbb{S} = \mathbb{S}_1 \cup \mathbb{S}_2 \cup \mathbb{S}_3 \tag{5.56}$$

where

$$\mathbb{S}_1 = \{(x, y), x, y \in \{-(N_1 - 1), \dots, 0, \dots, N_1 - 1\}\} \tag{5.57}$$

$$\mathbb{S}_2 = \{(x, y), x, y \in \{-(N_2 - 1)N_1, \dots, -(N_2 - 1), 0, (N_2 - 1), \dots, (N_2 - 1)N_1\}\} \tag{5.58}$$

$$\mathbb{S}_3 = \{(x, y), x, y \in \{-(N - 1), \dots, -(N - N_1), (N - N_1), \dots, N - 1\}\} \tag{5.59}$$

where $N_1, N_2 \in \mathbb{N}^+$, $N = N_1N_2$, and N is a nonprime integer. Subarray 1 and 2 have $(2N_1 - 1)^2$ and $(2N_1 + 1)^2$ sensors, respectively. Both have a common sensor at (0, 0), and eight extra common sensors whenever $N_1 = N_2$. Subarray 3 has $4N_1^2$ sensors. Subarray 2 and subarray 3 have four common sensors at $\{(x, y) \mid x, y \in \{-(N_2 - 1)N_1, (N_2 - 1)N_1\}\}$, that are $(-(N_2 - 1)N_1, -(N_2 - 1)N_1)$, $(-(N_2 - 1)N_1, (N_2 - 1)N_1)$, $((N_2 - 1)N_1, -(N_2 - 1)N_1)$, and $((N_2 - 1)N_1, (N_2 - 1)N_1)$. Therefore, the ESCA has a total of $N_t = 3\left(4N_1^2 - 1\right)$ sensors when $N_1 \neq N_2$ and $N_t = 3\left(4N_1^2 - 1\right) - 8$ sensors when $N_1 = N_2$. Figure 5.12f and g illustrate the ESCA for the two cases, $N_1 = 2$, $N_2 = 3$, and $N_1 = N_2 = 3$, where tringles, squares, reversed triangles represent the sensor locations of subarray 1, 2, and 3, respectively. This configuration realizes a hole-free difference coarray of size $(4N - 2N_1 - 1)^2$ [97]. Therefore, the array achieves $\mathcal{O}(16N_1^4)$. DOF when $N_1 = N_2$.

5.3.5.6 Half H Array (HHA) and Ladder Array (LAA)

Two different configurations were proposed by redistributing elements of the dense subarray of the OBA to the outer diagonal straight line [98]. This was based on the central URA with the maximum 2D aperture in the difference coarray. Those arrays have similar performance as the OBAs. Given two positive integers $N_x \geq 3$ and $N_y \geq 2$, a HHA is defined as a planar sparse array with physical sensors located on the grids [98]

$$\mathbb{S} = \mathbb{H} \cup \mathbb{V}_{11} \cup \mathbb{V}_{12} \cup \mathbb{V}_{21} \cup \mathbb{V}_{22} \tag{5.60}$$

where

$$\begin{cases} \mathbb{H} = \{(n_x, 0) | n_x \in S_x\}, S_x = \{0, 1, \dots, N_x - 1\} \\ \mathbb{V}_{11} = \{(0, n_y) | n_y \in v_{11}\}, v_{11} = \left\{1 + 2l | 0 \leq l \leq \left\lfloor \frac{N_y - 2}{2} \right\rfloor\right\} \\ \mathbb{V}_{12} = \{(N_x - 1, -n_y) | n_y \in v_{12}\}, v_{12} = \left\{2l | 1 \leq l \leq \left\lfloor \frac{N_y - 1}{2} \right\rfloor\right\} \\ \mathbb{V}_{21} = \{(N_x - 1, n_y) | n_y \in v_{21}\}, v_{21} = v_{11} \\ \mathbb{V}_{22} = \{(0, -n_y) | n_y \in v_{22}\}, v_{22} = v_{12} \end{cases}$$

It's implied that $v_{11} \cup v_{12} = v_{21} \cup v_{22} = S_y = \{1, 2, \dots, N_y - 1\}$.

Table 5.3 Weight functions of different sparse arrays.

Array	References	Central URA	Uniform square array	$w(1, 0)$	$w(0, 1)$	$w(1, 1)$	$w(1, -1)$
HOBA-2	[67]	59×31	31×31	2	6	13	13
Hourglass	[67]	29×53	29×29	2	8	5	5
Thermos	[95]	61×29	29×29	3	3	3	2
CcRA	[96]	31×31	31×31	11	11	7	9
HHA	[98]	59×31	31×31	29	2	1	1
LAA	[98]	59×31	31×31	3	2	14	14

The LAA was proposed to improve the high value of $w(1, 0)$ obtained by the HHA [98]. Figure 5.12g and h show HHA and LAA with $N_x = 30$ and $N_y = 16$.

Table 5.3 shows the important weight functions for different sparse arrays. The two parameters N_x and N_y are selected such that all configurations have $N_t = 60$ sensors. Specifically, $N_x = N_y = 15$ for CcRA, and $N_x = 30$ and $N_y = 16$ for the remaining configurations. It is evident that the thermos

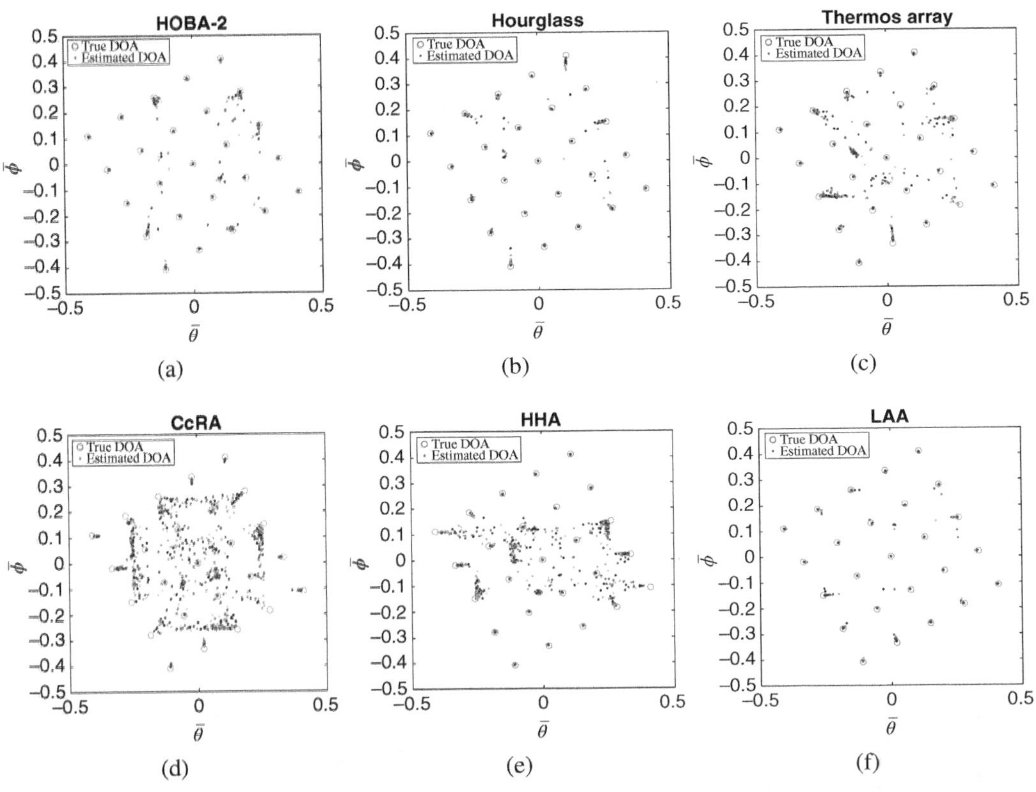

Figure 5.13 2D-DOA estimation for (a) HOBA-2, (b) Hourglass, (c) Thermos, (d) CcRA, (e) HHA, and (f) LAA, based on 2D unitary ESPIRT algorithm in the presence of mutual coupling with SNR = 0 dB, $T = 1000$ samples, $K = 25$ sources, and $H = 100$ Monte Carlo trials.

array and CcRA have smallest and largest values, respectively, compared with others. The LAA has lower weight functions compared with HHA, as it was proposed mainly to improve the HHA. Using coarray concept, the arrays can estimate large number of sources greater than the total number of sensors. The considered configurations have hole-free difference coarrays, except the thermos array. The smallest URA is realized by the CcRA, while other realize comparable coarray sizes as shown in Table 5.3.

To assess the estimation capability of the arrays presented in Table 5.3, the following parameters are selected. A 2D-DOA estimation using spatial smoothing and 2D unitary ESPRIT algorithm [118] is used. A total of $K = 25$ far field uniformly distributed sources [67] is assumed when SNR $= 0$ dB, and $N_t = 60$ sensors. The unit spacing is set to $d = \lambda/2$. A total of $T = 1000$ samples is assumed to estimate the covariance matrix and the number of Monte Carlo trials is $H = 100$. The mutual coupling parameters are $c_1 = 0.25$, $B = 5$, and $c_l = \frac{c_1}{l} e^{j\pi(l-1)/4}$ [67]. Figure 5.13 shows the estimated angles (indicated by circles) and the actual angles (indicated by dots). The best performance is achieved with hourglass array and LAA as shown in Fig. 5.13b and f. Due to mutual coupling, CcRA and HHA miss some sources at certain number of trails. Although they have different features, HOBA-2, hourglass arrays, HHA, and LAA have the same uDOF, see Table 5.4. This is larger than that achieved by thermos array by 4. ESCA has the smallest uDOF among the configurations in Section 5.3.5.

5.4 Comparative Evaluation

This section presents a general comparison of the mentioned arrays and highlights some important differences. This should be useful in predicting the performance and complexity of a sensor geometry; thus, aiding the decision of selecting a sensor geometry for 2D-DOA estimation. It can also serve as a starting point for researchers interested in exploring or designing new sparse planar arrays in more detail.

5.4.1 DOF and Number of Sensors

Table 5.4 lists the number of elements in each array geometry and the achievable DOF. For instance, the three references [28–30] that used parallel nested arrays are comparable, and the one using FOC [30] has higher DOF as expected. Furthermore, some geometries have different DOF when the parameters specifying the array are even or odd.

Arrays listed in Table 5.4 are based on the classification presented in Section 5.3. It is generally expected that, within each category, new geometries realize higher DOF than older ones. As an example, Zheng et al. [84] stated that their UCPA can exceed the DOF of the earlier works by Wu et al. [69] and Zheng et al. [75]. Similarly, Si et al. [85] showed that their NCPA can exceed the DOF of the previous CPA variants: CPA [69], GCPA [75], and UCPA [84]. Another example is the thermos array [95] which can surpass the DOF of the hourglass array when $N_y > N_x + 1$. Additionally, another interesting note might be the equal DOF of the arrays: OBA, HOBA-2, hourglass [67], HHA, and LAA [98]. In this case, this is an advantage for the newer arrays (HOBA-2 and hourglass), since these arrays retain the hole-free difference coarray of the OBA, yet they reduce mutual coupling.

Table 5.4 Comparison between physical sensors and achievable degrees of freedom for sparse array geometries.

	Array	References	Number of sensors, N_t	Max DOF
		[19]	$2M_1 + M_2$	$(4M_1^2 + 4M_1M_2 + M_2^2 - 1)/8$
		[20]	$4M_1 + 2M_2 - 2$	$4M_1M_2 - 1$
		[21]	$M_1 + M_2$	$\mathcal{O}\left((M_1 + M_2)^2\right)$
		[22]	$M_1 + M_2$	—
	Coprime	[23]	$M_1 + M_2 + 2$	$3M_1M_2 + M_1 + M_2 - 6 = n_{ul}$
				$2M_1M_2 + 2(M_1 + M_2) - 1 = n_{cl}, M_1 \geq 3$
				$2M_1M_2 + 2M_2 - 1 = n_{cl}, M_2 > M_1 = 2$
		[24]	$2M_1 + M_2$	$M_1(M_2 + 1)$
		[25]	$2M_1 + M_2 - 1$	$\lfloor(4M_1^2 + 4M_1M_2 + M_2^2 - 1)/8\rfloor$
		[26]	$2(M_1 + M_2)$	$2M_1(M_2 + 1) - 1$
		[27]	$2(M_1 + M_2)$	$2M_1(M_2 + 1) - 1$
		[28]		$N_2(N_1 + 1) - 1$
Parallel	Nested	[29]	$2N, N = N_1 + N_2$	$2N_2(N_1 + 1) - 2$
		[30]		$6N_2(N_1 + 1) - 4$
	ThPNA	[31]	$4N_1$	$4N_1^2$
	CDTPNA	[32]	$3N = 3(N_1 + N_2)$	—
		[33]		
		[34]	$2M + 1$	—
	Other	[35]	$N + M$	—
		[37]	$2M + N$	—
	1D-NA-motion	[38]	$2(N_1 + N_2)$	$2(N_2 + 1)N_1 - 1$
	TPA-MSS	[39]	$2(M + 1)N$	$N_{PA}N_{\text{MSS}} - 1$
	3-DA	[40]	$2N$	3DOF_o
	q-DN	[41]	$2N$	$q\text{DOF}_o$

Nonparallel	L MRA	[42]	$M+N-1$	—
	L 2 ||	[43]	$4N+1$	N
		[44]	$2(2M_1+M_2)-3$	M_1M_2
		[45]	$2(2M_1+M_2)-3$	M_1M_2
	L coprime	[46]	$2(M_1+M_2)-3$	—
		[47]	$2(2M_1+2M_2)-3$	—
		[48]	$2(2M_1+M_2)-3$	$\lfloor (4(M_1M_2+M_1)-3)/2 \rfloor$
		[49]	$2(M_1+M_2)-3$	—
		[50]	$2N-1, N=N_1+N_2$	$\begin{cases} \frac{3}{2}N^2-2, & N \text{ even} \\ \frac{3}{2}N^2-N-\frac{1}{2}, & N \text{ odd} \end{cases}$
	L nested	[51]		$\frac{N^2}{4}+\frac{N}{2}-1$
		[52]	$N_{x1}+N_{x2}+N_{y1}+N_{y2}$	$(2N_{x1}N_{x2}+2N_{x2}-1)(2N_{y1}N_{y2}+2N_{y1}+2N_{y2}+1)$
		[53]	$4N_1-2$	$4N_1^2-2$
		[54]	$2N-1, N=N_1+N_2$	$0.25(N^2-1)+0.5(N-1)$
		[55]	$2(M_1+M_2+N_1+N_2-1)$	$\mathcal{O}(M_1M_2N_1N_2)$
		[56]	$4(N_1+N_2)$	$\frac{1}{2}(N_1+N_2)^2+N_1+N_2+2L-1, 0<L<N_1$
	OIAC	[57]	$2(M+N)-1$	—
	L sULA+	[58]	$2M+1$	—
	L Other	[59]	M_x+M_z+1	—
	L MRA	[60]	$2N_{MRA}$	—
	L random	[61]	$M'+N'$	—
	VCA	[62]	$2(2M_1+M_2-1)-1$	M_1M_2
	VNA	[62]	$2N, N=N_1+N_2$	$0.25N^2+N+2-1$
	Cross	[63]	N_x+N_y-1	$\min\{N_x, N_y\}$
		[64]	$2(M_1+M_2-1)$	—

(Continued)

Table 5.4 (Continued)

	Array	References	Number of sensors, N_t	Max DOF
Interleaved Rectangular	GLA-OEL	[65]	$N_x + N_y + 1$	$(2N_x + 1)(2N_y + 1)$
	ISACA	[66]	$N_x + N_y + 1$	$\left(4\left\lfloor \frac{N_x+1}{2} \right\rfloor + 3\right) \times (2N_y + 1)$
	Billboard	[67]	$3(N_x - 1)$	—
	OBA	[67]	$N_x + 2N_y - 2$	$(2N_x - 1) \times (2N_y - 1)$
	TSA	[68]	$N_{y1} + N_{y2} + 2N_{x1}N_{x2} + 1$	—
	CPA	[69]	$M_1^2 + M_2^2 - 1$	$\min\{M_1^2, M_2^2\} - 1$
	CPA	[70]	$M_1^2 + M_2^2 - 1$	$\min\{M_1^2 + M_2^2 - M_1 - 1, M_1^2 + M_2^2 - M_2 - 1\} - 1$
	CPA	[71]	$M_1^2 + M_2^2 - 1$	$\min(M_{s2}^2, M_{s1}^2 - 1) + M_1^2 - 2$, $M_{s1} = \left\lceil \frac{M_1+1}{2} \right\rceil, M_{s2} = M_1 + 1 - M_{s1}$
	CPA	[72]	$M_1^2 + M_2^2 - 1$	$(2M_1 + 2M_2 - 1)^2$
	CPA	[73]	$M_1^2 + M_2^2 - 1$	$2M_1(M_2 - 1) + 1$
	VCPA	[74]	$M_1^2 + M_2^2 - 1$	$M_1^2 + M_2^2 - 1$
		[75]	$M_1N_1 + M_2N_2 - 1$	$\min\{M_1N_1, M_2N_2\} - 1$
		[76]		—
	GCPA	[77]	$4M_1M_2 + N_1N_1 - 1$	$2(M_1N_1 + M_1 + N_1 - 1)(M_2N_2 + M_2 + N_2 - 1) - (2N_1 - 1)(2N_2 - 1)$
		[78]	$4M_1M_2 + N_1N_2 - 1$	$(M_1N_1 + M_1 + N_1 - 1)(M_2N_2 + M_2 + N_2 - 1)$
		[79]	$4M_1N_1 + M_2N_2 - 1$	$J_x \times J_y, J_x = 3M_1M_2 - M_1 - M_2 + 1, J_y = 3N_1N_2 - N_1 - N_2 + 1$
	CAACS	[80]	$M_1^2 + M_2^2 - 1$	$(M_1M_2 - (\check{M}_1 - 1)(M_2 - 1))^2$
	CATSS	[80]	$M_1^2 + M_2^2$	$(M_1M_2 - (\check{M}_1 - 1)(M_2 - 1))(2M_1M_2 + 2l - 1)$, $l = M_1M_2 - (M_2 - 1)\check{M}_1 - 1$
	Nested	[82]	$(N_1 + N_2)^2$	Check [82] and [83]
		[83]		
	UCPA	[84]	$M_1^2 + M_2^2 - 1$	$M_1^2 + M_2^2 - 1$

NCPA	[85]	$(L+M_1-1)^2+(L+M_2-1)^2-L^2$	$\min\{L^2M_1^2, L^2M_2^2\}$
SDCPA	[86]	$2N_1(4M_2-1)+N_2(4M_1-1)-2$	$(2U_x+1)(2U_y+1)$, $U_x=\left(4M_1-1-\left\lfloor\frac{M_1}{2}\right\rfloor\right)N_1-1, U_y=\left(4M_2-1-\left\lfloor\frac{M_2}{2}\right\rfloor\right)N_2-1$
VSPA-CA-CA	[87]	$(M_1+M_2-1)M_s$	$(2M_1+2M_2-1)^2$
VSPA-NA-NA	[87]	$(M_1+M_2)M_s$	$(2N_2(N_1+1)-1)^2$
VSPA-CA-NA	[87]	$(M_1+M_2-1)M_s=(M_1+M_2)M_s$	$(2M_1+2M_2-1)(2N_2(N_1+1)-1)$
VSPA-NA-CA	[87]	$(M_1+M_2-1)M_s=(M_1+M_2)M_s$	$(2N_2(N_1+1)-1)(2M_1+2M_2-1)$
VSPA-MRA-MRA	[87]	N_sM_s	—
GPSA NA-NA	[88]	M_sN	$\left(\frac{1}{2}(N^2-2)+N\right)^2$, N: even, $\left(\frac{1}{2}(N^2-1)+N\right)^2$, N: odd
GPSA MRA-MRA	[88]	M_sN	DOF^2_{MRA}
RGMRA	[89]	M_xM_y	—
RHNA	[90]	M_xM_y	—
RSNA	[90]	M_xM_y	—
MLDiNA	[91]	$(L+1)\left(p_1p_2+\left(2N_1^{(s)}+1\right)N_2^{(s)}\right)$, $p_1p_2=\det(\mathbf{P})$	$\left(2\left[q\left(p_1N_1^{(s)}+\frac{p_1-1}{2}\right)+L\right]+1\right)\times\left(\left(p_2N_2^{(s)}-1\right)+1\right)$

(Continued)

Table 5.4 (Continued)

	Array	References	Number of sensors, N_t	Max DOF
Conformal	NCAs	[92]	$2(N_1 + N_2)$	—
	C3MA	[93]	$2(M_1 + M_2 - 1)$	—
	UCyA	[94]	$N_{vd}N_{hd} + N_{vs}N_{hs} - 1$	$\mathcal{O}((2N_{vd}N_{vs} - 1)N_{hdc})$
Other	HOBA-2	[67]	$N_x + 2N_y - 2$	$(2N_x - 1) \times (2N_y - 1)$
	Hourglass	[67]	$N_x + 2N_y - 2$	$(2N_x - 1) \times (2N_y - 1)$
	Thermos	[95]	$N_x + 2N_y - 2$	$(2N_x + 1) \times (2N_y - 3)$
	CcRA	[96]	$2(N_x + N_y)$	—
	ESCA	[97]	$3\left(4N_1^2 - 1\right)$	$(4N - 2N_1 - 1)^2$
	HHA	[98]	$N_x + 2N_y - 2$	$(2N_x - 1) \times (2N_y - 1)$
	LAA	[98]	$N_x + 2N_y - 2$	$(2N_x - 1) \times (2N_y - 1)$

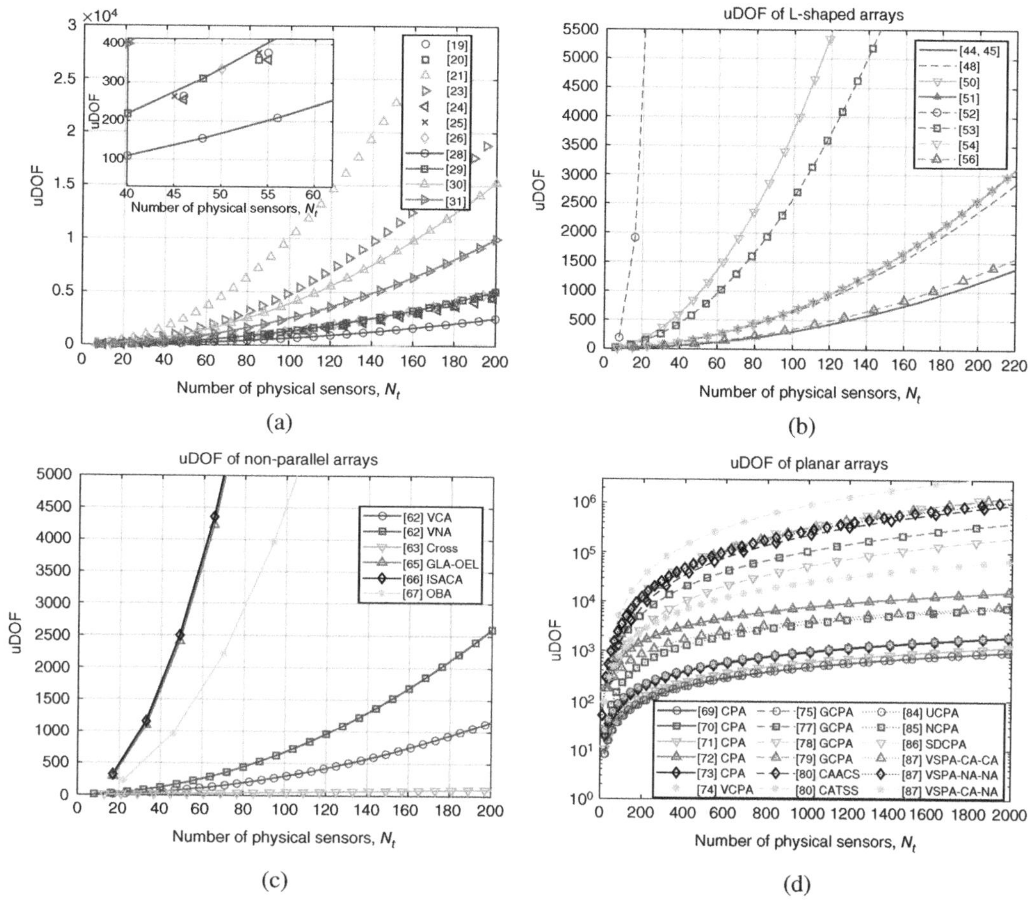

Figure 5.14 uDOF as a function of number of sensors for (a) parallel arrays, (b) L-shaped arrays, (c) nonparallel arrays, and (d) interleaved rectangular arrays.

To visualize the comparison and have clear perception, the achieved DOF could be plotted versus the total number of elements for each configuration. The focus here is only on the uDOF. Some configurations may have excellent uDOF but they suffer from mutual coupling and/or they require very large aperture size. The analytical expressions presented in Table 5.4 for selected configurations are plotted in Fig. 5.14. Similar structures are grouped for fair comparison.

Figure 5.14a shows the uDOF as a function of the total number of elements for parallel arrays. Solid lines with markers indicate nested structures and markers alone indicate coprime structures. The largest uDOF is achieved with coprime array in [21], due to exploiting both temporal and spatial information. Actually, the work in [21] didn't give an exact expression for the uDOF as Table 5.4 illustrates. Coprime array with two extra sensors [23], and nested array with FOC [30] have large uDOF as well. The aperture size of the former increases and the latter uses the FOC which is not the case for remaining configurations. The smallest uDOF is achieved in [28]. Arrays in [19, 20, 24–26, 29] have comparable performance.

The uDOF of the L-shaped coprime and nested arrays is plotted in Fig. 5.14b. Lines with markers represent those related to nested arrays. Because of the exploited FOC, the array in [52] has the largest uDOF. The arrays in [50, 53] have also large uDOF. The cross-correlation matrix of different subarrays is the main reason in [53]. While in [50], the reason is related to the conjugate symmetry property of the signal autocorrelation function with different delays. The nested arrays in [51] and [54] have almost the same uDOF. The worst performance is realized with coprime array in [44, 45], which is close to nested array in [56].

The other nonparallel sparse arrays are compared in Fig. 5.14c. ISACA [66] realizes the largest uDOF due to the large aperture size, where the two overlapped cross subarrays span all quadrants. GLA-OEL [65] has very close performance because of exploiting both the difference and the sum coarrays. VNA has better performance compared with its counterpart VCA [62]. The holes that appear in the difference coarray of VCA degrade the performance. The cross array [63] has the smallest uDOF.

The uDOF for interleaved rectangular arrays are presented in Fig. 5.14d. CATSS realizes the largest performance among all considered configurations, since it makes use of both the difference and sum coarrays [80]. The SDCPA [86], GCPA [79], CAACS [80], VSPA-NA-NA [87] have large and comparable performance. Similar performance is achieved with NCPA [85] and VSPA-CA-CA [87]. Moreover, CPA [70, 73], VCPA [74], UCPA [84] have equivalent performance, as all are based on coprime structure. The same is applicable for [69] CPA and [75] GCPA.

5.4.2 Aperture Size and Mutual Coupling

Table 5.5 shows a comparison of the same arrays in Table 5.4, but now shows the aperture, the estimated mutual coupling sensitivity based on the number of sensors with close inter-element spacing (MC-1), and whether the reference uses peak search or FOC which give an indication of the computational complexity. Only four references considered computationally complex FOC in the algorithm [30, 42, 48, 52]. FOC requires that the sources to be Gaussian and theoretically eliminates the Gaussian noise, and usually leads to much larger DOF compared to methods based on SOS. The authors in [71] exceed the performance of some FOC-based methods. In general, array geometries exploiting nested arrays have higher number of sensors closely-spaced, and more vulnerable to mutual coupling degradation. Furthermore, some of the recent arrays like hourglass or thermos arrays have low mutual coupling sensitivity. Other notes on Table 5.5 include:

- It is worth mentioning that comparing the DOF offered by a method using FOC to another method using SOS may not be fair. Not all arrays are expected to work well with FOC.
- The last two columns show the length and width of the array aperture. While in general, arrays based on coprime linear arrays have the same expressions for the aperture; some differences exist if authors use different coprime variants of the coprime linear array, or if the LA on the two directions are different. Note that it is not easy to write an expression for the aperture if the LA is an MRA. In many cases, the aperture is square, and the same expression appears in both columns. However, for the work in [55–57], the expression is made to span both columns to save space.

Table 5.5 Comparison between array geometries in terms of computational complexity, mutual coupling, and aperture.

	Array	References	PS[1]	FOC	MC-1[2]	Aperture length	Aperture width
Parallel	Coprime	[19]	x	x	Low	1	$M_2(2M_1-1)$
		[20]	x	x	Low	1	$2M_2(2M_1-1)$
		[21]	√	x	Low	1	$M_2(M_1-1)$
		[22]	1D	x	Low	1	$M_2(M_1-1)$
		[23]	1D	x	Low	3	$(M_2-1)M_1$
		[24]	1D	x	Low	1	$M_2(2M_1-1)$
		[25]	x	x	Low	$1+L$	$M_2(2M_1-1)$
		[26]	√	x	Low	1	$M_2(2M_1-1)$
		[27]	1D	x	Low	2	$M_2(2M_1-1)$
	Nested	[28]	x	x	High	1	$N_2(N_1+1)-1$
		[29]	x	x	High	1	$N_2(N_1+1)-1$
		[30]	x	√	High	1	$N_2(N_1+1)-1$
	ThPNA	[31]	1D	x	High	1	$(M_2-1)M_1+(M_1-1)+0.5$
	CDTPNA	[32] [33]	1D	x	High	M_1+M_2	$N_2(N_1+1)-1$
	Other	[34]	x	x	High	1	$M+1$
		[35]	x	x	High	qN	1
		[37]	x	x	Low	$2p+1, p\geq 2$	M
	1D-NA- motion	[38]	1D	x	High	1	$(M_1+1)M_2-1$
	TPA-MSS	[39]	x	x	High	1	$h_N d_{PA}+Q_M d_{MSS}$
	3-DA	[40]	x	x	Low	1	$3p_N$
	q-DA	[41]	x	x	High	1	$p_N q+R$

(Continued)

Table 5.5 (Continued)

	Array	References	PS[1]	FOC	MC-1[2]	Aperture length	Aperture width
Nonparallel	L MRA	[42]	x	√	Low	—	—
	L 2 ∥	[43]	x	x	High	$N-1$	$N-1$
		[44]	x	x	Low	$M_2(2M_1-1)$	$M_1(M_2-1)$
		[45]	1D	x	Low	$M_2(2M_1-1)$	$M_2(2M_1-1)$
	L coprime	[46]	1D	x	Low	$M_1(M_2-1)$	$M_1(M_2-1)$
		[47]	x	x	Low	$M_1(2M_2-1)$	$M_1(2M_2-1)$
		[48]	x	√	Low	$M_2(2M_1-1)$	$M_2(2M_1-1)$
		[49]	1D	x	Low	$(M_2-1)M_1+(M_1-1)M_2$	$(M_2-1)M_1+(M_1-1)M_2$
		[50]	x	x	High	$N_2(N_1+1)-1$	$N_2(N_1+1)-1$
		[51]	x	x	High	$N_2(N_1+1)-1$	$N_2(N_1+1)-1$
		[52]	x	√	High	$2N_{x2}(N_{x1}+1)-1$	$2N_{y2}(N_{y1}+1)-2$
	L nested	[53]	x	x	High	$2N_1-1$	$2N_1-1$
		[54]	2D	x	High	$N_2(N_1+1)-1$	$N_2(N_1+1)-1$
		[55]	1D	x	High	$D_{TX}=(2M_2(M_1+1)-1)(M_2(M_1+1)-1), D_{RX}=N_2(N_1+1)-1$	
		[56]	x	x	High	$(N_2(N_1+1)-1)+A, A\gg\lambda/2$	
	OIAC	[57]	1D	x	Medium	$\left(M+N-\left\lfloor\frac{M}{2}\right\rfloor-2\right)(M+1)M$	
	L sULA+	[58]	x	x	Medium	$M-1$	M
	L Other	[59]	x	x	Medium	M_x	M_y
	L MRA	[60]	1D	x	Low	—	—
	L random	[61]	1D	x	Medium	$p_{M'}$	$p_{N'}$
	VCA	[62]	1D	x	Low	$2M_1(M_2-1)\sin\left(\frac{\Omega}{2}\right)$	$M_1(M_2-1)\cos\left(\frac{\Omega}{2}\right)$
	VNA	[62]	1D	x	High	$2(N_2(N_1+1)-1)\sin\left(\frac{\Omega}{2}\right)$	$(N_2(N_1+1)-1)\cos\left(\frac{\Omega}{2}\right)$
	Cross	[63]	x	x	Medium	$2(M_x-1)$	$2(M_y-1)$
		[64]	1D	x	Low	$(M_2-1)M_1+(M_1-1)M_2$	$(M_2-1)M_1+(M_1-1)M_2$

Interleaved rectangular	GLA-OEL	[65]	x	x	Low	$1+2\left\lfloor \frac{N_x-1}{2} \right\rfloor + 2\left\lfloor \frac{N_x}{2} \right\rfloor$	$1+2\left\lfloor \frac{N_y-1}{2} \right\rfloor + 2\left\lfloor \frac{N_y}{2} \right\rfloor$
	ISACA	[66]	x	x	Low	$2\left\lfloor \frac{N_x+1}{2} \right\rfloor + 2\left\lfloor \frac{N_x}{2} \right\rfloor$	$1+2\left\lfloor \frac{N_y-1}{2} \right\rfloor + 2\left\lfloor \frac{N_y}{2} \right\rfloor$
	Billboard	[67]	x	x	High	$N_x - 1$	$N_y - 1$
	OBA	[67]	x	x	High	$N_x - 1$	$N_y - 1$
	TSA	[68]	x	x	Low	$2(M_2N_{x2} + N_2N_{x1})$	$N_{y2}N_1 + N_{y1}M_1$
	CPA	[69]	2D[3]	x	Low	$M_1(M_2-1)$	$M_1(M_2-1)$
	CPA	[70]	x	x	Low	$M_1(M_2-1)$	$M_1(M_2-1)$
	CPA	[71]	x	≳4	Low	$M_1(M_2-1)$	$M_1(M_2-1)$
	CPA	[72]	x	x	Low	$(M_2-1)M_1$	$(M_2-1)M_1$
	CPA	[73]	x	x	Low	$M_2(M_1-1)$	$M_2(M_1-1)$
	VCPA	[74]	2D	x	Low	$(M_2-1)M_1$	$(M_2-1)M_1$
		[75]	2D	x	Low	$M_1(M_2-1)$	$N_1(M_2-1)$
		[76]	2D	x	Low	$\max((M_1-1)M_2, (M_2-1)M_1)$	$\max((N_1-1)N_2, (N_2-1)N_1)$
	GCPA	[77]	2D	x	Low	$(2M_1-1)N_1$	$(2M_2-1)N_2$
		[78]	2D	x	Low	$2M_1N_1$	$2M_2N_2$
		[79]	2D	x	Low	$\max((2M_1-1)M_2, (M_2-1)M_1)$	$\max((2N_1-1)N_2, (N_2-1)N_1)$
	CAACS	[80]	x	x	Low	$(M_1-1)M_2$	$(M_1-1)M_2$
	CATSS	[80]	x	x	Low	$(M_1-1)M_2$	$(M_1-1)M_2 + l + (M_2-1)\widetilde{M}_1$
	Nested	[82] [83]	2D	x	High	$N_2(N_1+1)-1$	$N_2(N_1+1)-1$
	UCPA	[84]	1D	x	Low	$M_1(M_2-1)+M_2(M_1-1)$	$M_1(M_2-1)+M_2(M_1-1)$
	NCPA	[85]	2D	x	Medium	$M_1(M_2-1)+M_1M_2(L-1)$	$M_1(M_2-1)+M_1M_2(L-1)$
	SDCPA	[86]	x	x	Low	$2(L_1+M_1)\ L_1 = k_1N_1$	$2(L_2+M_2)+N_2\ L_2 = k_2N_2$
	VSPA- CA-CA	[87]	x	x	Low	$(M_2-1)M_1$	$(M_2-1)M_1$

(Continued)

Table 5.5 (Continued)

	Array	References	PS[1]	FOC	MC-1[2]	Aperture length	Aperture width
	VSPA- NA-NA	[87]	x	x	High	$N_2(N_1+1)-1$	$N_2(N_1+1)-1$
	VSPA- CA-NA	[87]	x	x	High	$(M_2-1)M_1$	$N_2(N_1+1)-1$
	VSPA- NA-CA	[87]	x	x	High	$N_2(N_1+1)-1$	$(M_2-1)M_1$
	VSPA- MRA-MRA	[87]	x	x	Low	—	—
	GPSA NA-NA	[88]	x	x	High	$N_2(N_1+1)-1$	$N_2(N_1+1)-1$
	GPSA MRA-MRA	[88]	x	x	Low	—	—
	RGMRA	[89]	x	x	Low	—	—
	RHNA	[90]	x	x	Low	$N_2(N_1+1)-1$	—
	RSNA	[90]	x	x	Low	$N_2(N_1+1)-1$	$N_2(N_1+1)-1$
	MLDiNA	[91]	2D	x	High	$(1+qL)D_{x-2DNA}$	$(1+qL)D_{y-2DNA}$
Conformal	NCAs	[92]	1D	x	High	$N_2(N_1+1)-1$	$2r$
	C3MA	[93]	x	x	Low	r	r
	UCyA	[94]	x	x	High	$N_2(N_1+1)-1$	$2r$
Other	HOBA-2	[67]	x	x	Low	N_x-1	N_y-1
	Hourglass	[67]	x	x	Low	N_x-1	N_y-1
	Thermos	[95]	2D	x	Low	N_x+3	N_y-2
	CcRA	[96]	—	—	Low	N_x	N_y
	ESCA	[97]	2D	x	Medium	N_1N_2-1	N_1N_2-1
	HHA	[98]	x	x	High	N_x-1	$1+2\left\lfloor\frac{N_y-2}{2}\right\rfloor+2\left\lfloor\frac{N_y-1}{2}\right\rfloor$
	LAA	[98]	x	x	Low	N_x-1	$1+2\left\lfloor\frac{N_y-2}{2}\right\rfloor+2\left\lfloor\frac{N_y-1}{2}\right\rfloor$

Notes: [1]PS: peak search. [2]MC-1: mutual coupling approximate classification based on number of elements with unity inter-sensor spacing. 2D[3]: partial spectral search (PSS). [4]Means the paper does not propose a method based on FOC but compares results with a method based on FOC.

5.5 Summary

This chapter presented a comprehensive overview of sparse arrays for 2D-DOA estimation. General designs like cross-shaped, L-shaped, T-shaped, V-shaped, half H, ladder, hourglass, thermos, concentric rectangular, convolutional, nested planar, and coprime planar were classified into parallel arrays, nonparallel arrays, interleaved rectangular arrays, conformal arrays, and other planar arrays. The considered configurations were compared in terms of the required number of sensors, DOF, algorithm used, aperture size, weight functions, and 2D-DOA estimation. A fair comparison should not overlook the aperture size associated with the improved estimation performance. Arrays with rectangular aperture tend to have large DOF, like the CATSS, hourglass, thermos, half H, or ladder arrays. Also, GCPA shows large DOF compared to CPA. Arrays with relatively more complicated closed-form expressions seem to enjoy more DOF, like the hourglass or thermos or half H or ladder arrays.

For the considered configurations, the following can be concluded about uDOF. Compressing the inter-element spacing can improve the uDOF but at the expense of increased mutual coupling. Some nonessential elements may be relocated to fill specific holes in the coarray. Expanding or displacing certain subarrays can improve the resolution and the DOF but at the expense of increased aperture size. The required aperture is directly proportional to the size of the resultant coarray. The largest uDOF for parallel arrays is achieved with coprime array in [21], where both temporal and spatial information were used. Because of the FOC, the L-shaped array in [52] has the best uDOF among the nonparallel arrays. Additionally, ISACA [66] has large uDOF due to the large aperture size. For interleaved rectangular arrays, CATSS enjoys the largest performance because of utilizing both the difference and sum coarrays [80] in addition to the large aperture.

2D antenna arrays based on coprime or super nested structures have low weight functions. Consequently, they are not much affected by the presence of mutual coupling. On the other hand, those based on nested arrays suffer from large mutual coupling effect because of the dense subarray. The presented structured review should help in predicting the DOA estimation performance and required complexity; thus, facilitating design and selection of sensor array geometry for 2D-DOA. We note that this chapter has focused on structured sparse arrays, where the configurations are independent of the environment. Two-dimensional sparse array design for optimum beamforming and direction finding is discussed in Chapter 6 of this book and [120].

References

1 Krim, H. and Viberg, M. (1996). Two decades of array signal processing research. *IEEE Signal Processing Magazine* 13 (4): 67–94.

2 Amin, M.G., Wang, X., Zhang, Y.D. et al. (2016). Sparse Arrays and Sampling for Interference Mitigation and DOA Estimation in GNSS. *Proc. IEEE* 104 (6): 1302–1317. https://doi.org/10.1109/JPROC.2016.2531582.

3 Björnson, E., Sanguinetti, L., Wymeersch, H. et al. (2019). Massive MIMO is a reality—what is next?: five promising research directions for antenna arrays. *Digital Signal Processing: A Review Journal* 94: 3–20. https://doi.org/10.1016/j.dsp.2019.06.007.

4 De Graaf, S.R. and Johnson, D.H. (1984). Optimal linear arrays for narrow—band beanforming. In: *International Conference on Acoustics, Speech, and Signal Processing*, 4081–4084.

5 Bresler, Y. and Macovski, A. (1986). On the number of signals resolvable by a uniform array. *IEEE Transactions on Acoustics* ASSP-34 (6): 1361–1375. https://doi.org/10.1109/OCEANS.2010.5664086.

6 Wylie, M.P., Roy, S., and Messer, H. (1994). Joint DOA estimation and phase calibration of linear equispaced (LES) arrays. *IEEE Transactions on Signal Processing* 42 (12): 3449–3459. https://doi.org/10.1109/78.340779.

7 Parot, B. and Friedlander, B. (1991). Direction finding algorithms based on high-order statistics. *IEEE Transactions on Signal Processing* 39 (9): 2016–2024.

8 Swami, A. and Mendel, J.M. (1991). Cumulant-based approach to the harmonic retrieval and related problems. *IEEE Transactions on Signal Processing* 39 (5): 1099–1109.

9 Mazlout, S., Ben Salah, M.B., and Samet, A. (2017). Comparative study of two array configurations for 2D-DOA estimation in LS-MIMO systems. In: *6th Internationl Conference on Communications and Networking (COMNET)*, Hammamet, Tunis, 1–6. https://doi.org/10.1109/COMNET.2017.8285597.

10 Gershman, A.B., Rübsamen, M., and Pesavento, M. (2010). One- and two-dimensional direction-of-arrival estimation: an overview of search-free techniques. *Signal Processing* 90 (5): 1338–1349. https://doi.org/10.1016/j.sigpro.2009.12.008.

11 Aboumahmoud, I., Muqaibel, A., Alhassoun, M., and Alawsh, S. (2021). A review of sparse sensor arrays for two- dimensional direction-of-arrival estimation. *IEEE Access* 9: 92999–93017. https://doi.org/10.1109/ACCESS.2021.3092529.

12 Zoubir, A.M., Koivunen, V., Chakhchoukh, Y., and Muma, M. (2012). Robust estimation in signal processing: a tutorial-style treatment of fundamental concepts. *IEEE Signal Processing Magazine* 29 (4): 61–80. https://doi.org/10.1109/MSP.2012.2183773.

13 Khedekar, S. and Mukhopadhyay, M. (2018). Analysis of estimation of direction of arrival by comparative study. *Materials Today: Proceedings* 5 (1): 1696–1703. https://doi.org/10.1016/j.matpr.2017.11.266.

14 Ahmed, A., Hamza, S.A., and Tufail, M. (2015). Comparative analysis of fourth order cumulant based ESPRIT algorithms. In: *12th International Conference on Frontiers of Information Technology*, Islamabad, Pakistan, 133–138. https://doi.org/10.1109/FIT.2014.33.

15 Adhikari, K. and Drozdenko, B. (2019). Comparison of MUSIC variants for sparse arrays. In: *IEEE National Conference on Aerospace and Electronics*, Dayton, OH, USA, 398–405. https://doi.org/10.1109/NAECON46414.2019.9058169.

16 Alawsh, S.A., Muqaibel, A.H., and Sharawi, M.S. (2019). Two-dimensional DOA estimation using parallel multi-level prime arrays. In: *2019 2nd IEEE Middle East North Africa Communications Conference MENACOMM 2019*, Manama, Bahrain, 1–6. https://doi.org/10.1109/MENACOMM46666.2019.8988552.

17 Shen, Q., Liu, W., Cui, W., and Wu, S. (2016). Underdetermined DOA estimation under the compressive sensing framework: a review. *IEEE Access* 4: 8865–8878. https://doi.org/10.1109/ACCESS.2016.2628869.

18 Liang, Y., Liu, W., Shen, Q. et al. (2020). A review of closed-form cramér-rao bounds for DOA estimation in the presence of gaussian noise under a unified framework. *IEEE Access* 8: 175101–175124. https://doi.org/10.1109/ACCESS.2020.3026203.

19 Qin, S., Zhang, Y.D., and Amin, M.G. (2016). Two-dimensional DOA estimation using parallel coprime subarrays. In: *IEEE Sensor Array and Multichannel Signal Processing Workhop (SAM)*, Rio Janerio, Brazil, 1–4.

20 Shi, J., Hu, G., Zhang, X., and Gong, P. (2017). Sum and difference coarrays based 2-D DOA estimation with co-prime parallel arrays. In: *9th International Conference on Wireless Communication Signal Processing (WCSP)*, Nanjing, China, 1–4. https://doi.org/10.1109/WCSP.2017.8171037.

21 Sun, F., Lan, P., Gao, B.I.N., and Zhang, G. (2018). An efficient dictionary learning-based 2-D DOA estimation without pair matching for co-prime parallel arrays. *IEEE Access* 6: 8510–8518.

22 Sun, F., Ouyang, S., Lan, P., and Li, F. (2020). Reduced dimensional 2-D DOA estimation via least partial search with automatic pairing for parallel co-prime arrays. In: *IEEE 11th Sensor Array and Multichannel Signal Processing Workshop (SAM)*, Hangzhou, China, 1–5.

23 Li, J., Zhao, J., Ding, Y. et al. (2021). An improved co-prime parallel array with conjugate augmentation for 2-D DOA estimation. *IEEE Sensors Journal* 21 (20): 23400–23411. https://doi.org/10.1109/JSEN.2021.3106382.

24 Hou, C., Fang, C., Deng, Z. et al. (2020). Two-dimensional direction-of-arrival and polarization parameter estimation using parallel co-prime polarization sensitive array. *IEEE Access* 8: 6566–6574. https://doi.org/10.1109/ACCESS.2019.2957546.

25 Qin, S., Zhang, Y.D., and Amin, M.G. (2020). Improved two-dimensional DOA estimation using parallel coprime arrays. *Signal Processing* 172: 1–9. https://doi.org/10.1016/j.sigpro.2019.107428.

26 Gong, P. and Zhang, X. (2017). Three-parallel co-prime array configuration for two-dimensional DOA estimation. In: *9th International Conference on Wireless Communications and Signal Processing (WCSP)*, Nanjing, China, 1–5.

27 Si, W., Wang, Y., and Zhang, C. (2019). Three-parallel co-prime polarization sensitive array for 2-D DOA and polarization estimation via sparse representation. *IEEE Access* 7: 15404–15413. https://doi.org/10.1109/ACCESS.2019.2894624.

28 Li, X., Zhang, W., Shu, T., and He, J. (2019). Sensor signal processing two-dimensional direction finding with parallel nested arrays using DOA matrix method. *IEEE Sensors Letters* 3 (7): 1–4.

29 Zheng, Z. and Mu, S. (2020). Two-dimensional DOA estimation using two parallel nested arrays. *IEEE Communications Letters* 24 (3): 568–571. https://doi.org/10.1109/LCOMM.2019.2958903.

30 He, J., Li, L., and Shu, T. (2021). 2-D direction finding using parallel nested arrays with full co-array aperture extension. *Signal Processing* 178: 1–6. https://doi.org/10.1016/j.sigpro.2020.107795.

31 Li, J. and Zhang, X. (2020). Two-dimensional grid-less angle estimation based on three parallel nested arrays. *Signal Processing* 173: 1–8.

32 Sun, F., Wu, J., and Lan, P. (2022). Improved two-dimensional doa estimation for three-parallel nested arrays. *Electronics Letters* 58 (9): 369–371. https://doi.org/10.1049/ell2.12463.

33 Sun, F. and Lan, P. (2021). Efficient two-dimensional DOA estimation for coprime-displaced three parallel nested arrays. In: *6th International Conference on Communications and Image Signal Processing (CCISP)*, Chengdu, China, 393–397.

34 Zheng, Z., Yang, Y., Wang, W.Q., and Zhang, S. (2018). Two-dimensional direction estimation of multiple signals using two parallel sparse linear arrays. *Signal Processing* 143: 112–121. https://doi.org/10.1016/j.sigpro.2017.08.013.

35 Liu, S. and Zhao, J. (2020). 2D DOA estimation algorithm with increased degrees of freedom for two parallel linear arrays. *China Communications* 17 (6): 101–108.

36 LIN, S., Zhou, Y., Zou, W. et al. (2022). An algorithm for single snapshot 2D-DOA estimation based on a three-parallel linear array model. *IEICE Transactions on Fundamentals of Electronics, Communications and Computer Sciences* E105.A (4): 673–681. https://doi.org/10.1587/transfun.2021eap1074.

37 Liu, S., Zhao, J., Wu, D. et al. (2022). A high-precision 2D DOA estimation algorithm by a large-space three-parallel array for reducing mutual coupling. *Digital Signal Processing* 122: 1–9.

38 Li, J., Ma, P., Zhang, X., and Zhao, G. (2020). Improved DFT algorithm For 2D DOA estimation based on 1D nested array motion. *IEEE Communications Letters* 1–4. https://doi.org/10.1109/LCOMM.2020.2997030.

39 Zhu, C. and Wang, Y. (May 2022). Two-dimensional DOA estimation based on two-parallel arrays exploiting nonuniform array motions. In: *IEEE 5th International Conference on Electronics Technology (ICET)*, Chengdu, China, vol. 1, 723–727.

40 Li, S. and Zhang, X.-P. (2021). Dilated arrays : a family of sparse arrays with increased uniform degrees of freedom and reduced mutual coupling on a moving platform. *IEEE Transactions on Signal Processing* 69: 3367–3382.

41 Li, S., Zhang, X.P., Ruan, D.L., and Zeng, H. (2022). Increasing the uniform degrees of freedom for moving Q-dilated arrays. *IEEE Internet of Things Journal* 10 (1): 194–209. https://doi.org/10.1109/JIOT.2022.3200955.

42 Jian, C., Wang, S., and Lin, L. (2006). 2-D DOA estimation by minimum-redundancy linear array. In: *8th International Conference on Signal Processing*, Guilin,China, vol. 1, 2–5. https://doi.org/10.1109/ICOSP.2006.344526.

43 He, J. and Liu, Z. (2009). Extended aperture 2-D direction finding with a two-parallel-shape-array using propagator method. *IEEE Antennas and Wireless Propagation Letters* 8 (1): 323–327. https://doi.org/10.1109/LAWP.2009.2015224.

44 Liu, Q., Yi, X., Jin, L., and Chen, W. (2015). Two dimensional direction of arrival estimation for co-prime L-shaped array using sparse reconstruction. In: *8th International Congress on Image Signal Processing (CISP)*, Shenyang, China, 1499–1503.

45 Elbir, A.M. (2020). L-shaped coprime array structures for DOA estimation. *Multidimensional Systems and Signal Processing* 31 (1): 205–219. https://doi.org/10.1007/s11045-019-00657-4.

46 Wang, K., Shi, L., and Chen, T. (2021). Two-dimensional separable gridless direction-of-arrival estimation based on finite rate of innovation. *IEEE Access* 9: 17275–17283. https://doi.org/10.1109/ACCESS.2021.3054660.

47 Lian, X.C., Li, S.B., Chen, H.H. et al. (2021). DOA estimation based on CNN in L-type mutual matrix. In: *6th International Conference on Communication on Image Signal Processing (CCISP)*, Chengdu, China, vol. 2019, 1–6. https://doi.org/10.1109/CCISP52774.2021.9639357.

48 Zheng, H., Zhou, C., Wang, Y. et al. (2021). Sample fourth-order cumulant tensor denoising for DOA estimation with coprime L-shaped array. In: *55th Asilomar Conference on Signals, Systems and Computers*, Pacific Grove, CA, USA, 1668–1672. https://doi.org/10.1109/IEEECONF53345.2021.9723340.

49 Zhang, Q., Li, J., Tang, Y. et al. (2022). An efficient 2-D DOA estimation method for unfolded coprime L-shaped array. *IEEE Systems Journal* 1–4.

50 Dong, Y.-Y., Dong, C.-X., Zhu, Y.-T. et al. (2016). Two-dimensional DOA estimation for L-shaped array with nested subarrays without pair matching. *IET Signal Processing* 10 (9): 1112–1117. https://doi.org/10.1049/iet-spr.2015.0252.

51 Niu, C., Zhang, Y., and Guo, J. (2016). Interlaced double-precision 2-D angle estimation algorithm using L-shaped nested arrays. *IEEE Signal Processing Letters* 23 (4): 522–526. https://doi.org/10.1109/LSP.2016.2537826.

52 Li, X., Ren, S., Liu, J., and Wang, W. (2018). Augmented L-shaped nested array based on the fourth-order difference co-array Concept. In: *IEEE 10th Sensor Array Multichannel Signal Processing Workshop (SAM)*, Sheffield, UK, vol. 4, 31–35. https://doi.org/10.1109/SAM.2018.8448823.

53 Gao, X., Hao, X., Li, P., and Li, G. (2019). An improved two-dimensional direction-of-arrival estimation algorithm for L-shaped nested arrays with small sample sizes. *Sensors* 19 (9): https://doi.org/10.3390/s19092176.

54 Yang, Y., Mao, X., Hou, Y., and Jiang, G. (2020). 2-D DOA estimation via correlation matrix reconstruction for nested L-shaped array. *Digital Signal Processing* 98: 102623. https://doi.org/10.1016/j.dsp.2019.102623.

55 Li, Z. and Zhang, X. (2021). Monostatic MIMO radar with nested L-shaped array: configuration design, DOF and DOA estimation. *Digital Signal Processing* 108: 1–11. https://doi.org/10.1016/j.dsp.2020.102883.

56 Liao, Y., Lin, Y., and Zhao, R. (2021). 2D-DOA estimation using L-shaped distributed augmented nested array. In: *13th International Conference on Wireless Communications and Signal Processing*. WCSP 2021. https://doi.org/10.1109/WCSP52459.2021.9613225.

57 Wu, F., Cao, F., Ni, X. et al. (2020). L-shaped sparse array structure for 2-D DOA estimation. *IEEE Access* 8: 140030–140037. https://doi.org/10.1109/access.2020.3012685.

58 Yang, Y., Zheng, Z., Yang, H. et al. (2017). A novel 2-D DOA estimation method via sparse L-shaped array. In: *2nd IEEE International Conference on Computer Communication (ICCC)*, Chengdu, China, 1865–1869. https://doi.org/10.1109/CompComm.2016.7925025.

59 Gu, J., Zhu, W., and Swamy, M.N.S. (2015). Joint 2-D DOA estimation via sparse L-shaped array. *IEEE Transactions on Signal Processing* 63 (5): 1171–1182. https://doi.org/10.1109/TSP.2015.2389762.

60 Chen, H., Chen, W., Wang, Y. et al. (2022). A low computational cost lightning mapping algorithm with a nonuniform L-shaped array: principle and verification. *IEEE Transactions on Geoscience and Remote Sensing* 60: 1–10. https://doi.org/10.1109/TGRS.2021.3139388.

61 Mirza, H.A.L.I., Asif, M., Raja, Z. et al. (2020). A robust multi sample compressive sensing technique for DOA estimation using sparse antenna array. *IEEE Access* 8: 140848–140861. https://doi.org/10.1109/ACCESS.2020.3011597.

62 Elbir, A.M. (2019). V-shaped sparse arrays for 2-D DOA estimation. *Circuits, Systems, and Signal Processing* 38 (6): 2792–2809. https://doi.org/10.1007/s00034-018-0991-5.

63 Wu, X. and Zhu, W.P. (2020). Single far-field or near-field source localization with sparse or uniform cross array. *IEEE Transactions on Vehicular Technology* 69 (8): 9135–9139. https://doi.org/10.1109/TVT.2020.2998128.

64 Zhang, Z., Guo, Y., Huang, Y., and Zhang, P. (2021). A 2-D DOA estimation method with reduced complexity in unfolded coprime L-shaped array. *IEEE Systems Journal* 15 (1): 407–410. https://doi.org/10.1016/s0965-206x(98)80009-0.

65 Li, X., Wang, X., Wang, W., and Ren, S. (2020). Generalized L-shaped array based on the difference and sum coarray concept. *IEEE Access* 8: 140456–140466. https://doi.org/10.1109/access.2020.3012527.

66 Li, X., Wang, W., Wang, X., and Ren, S. (2022). Synthetic sparse planar array design for two-dimensional DOA estimation. *Digital Signal Processing* 120: 1–11.

67 Liu, C.L. and Vaidyanathan, P.P. (2017). Hourglass arrays and other novel 2-D sparse arrays with reduced mutual coupling. *IEEE Transactions on Signal Processing* 65 (13): 3369–3383. https://doi.org/10.1109/TSP.2017.2690390.

68 Liu, S., Zhao, J., Wu, D. et al. (2022). 2D DOA estimation by a large-space T-shaped array. *Digital Signal Processing* 130: 1–11. https://doi.org/10.1016/j.dsp.2022.103699.

69 Wu, Q., Sun, F., Lan, P. et al. (2016). Two-dimensional direction-of-arrival estimation for co-prime planar arrays : a partial spectral search approach. *IEEE Sensors Journal* 16 (14): 5660–5670.

70 Zhang, D., Zhang, Y., Zheng, G. et al. (2018). Two-dimensional direction of arrival estimation for coprime planar arrays via polynomial root finding technique. *IEEE Access* 6: 19540–19549. https://doi.org/10.1109/ACCESS.2018.2821919.

71 Xu, H., Wang, D., Ba, B. et al. (2019). Direction-of-arrival estimation for both uncorrelated and coherent signals in coprime array. *IEEE Access* 7: 18590–18600. https://doi.org/10.1109/ACCESS.2019.2896979.

72 Yang, X., Wang, Y., and Charge, P. (2021). Hole locations and a filling method for coprime planar arrays for DOA estimation. *IEEE Communications Letters* 25 (1): 157–160.

73 Lu, A., Guo, Y., Li, N., and Yang, A. (2020). Efficient gridless 2-D direction-of-arrival estimation for coprime array based on decoupled atomic norm minimization. *IEEE Access* 8: 57786–57795.

74 Chu, J., Zhang, Z., Huang, Y., and Guo, Y. (2022). A novel approach of 2-D DOA estimation by employing coprime linear array motion. *IEEE Communications Letters* 26 (1): 69–73. https://doi.org/10.1109/LCOMM.2021.3123976.

75 Zheng, W., Zhang, X., and Zhai, H. (2017). Generalized coprime planar array geometry for 2-D DOA estimation. *IEEE Communications Letters* 21 (5): 1075–1078. https://doi.org/10.1109/LCOMM.2017.2664809.

76 Zheng, H., Zhou, C., Shi, Z., and Yan, C. (2022). Joint coprime weights optimization for sub-nyquist tensor beamforming. In: *IEEE Radar Conference (RadarConf22)*, New York City, NY, USA, 1–6. https://doi.org/10.1109/radarconf2248738.2022.9764278.

77 Shi, J., Hu, G., Zhang, X. et al. (2017). Sparsity-based two-dimensional DOA estimation for coprime array: from sum-difference coarray viewpoint. *IEEE Transactions on Signal Processing* 65 (21): 5591–5604. https://doi.org/10.1109/TSP.2017.2739105.

78 Zheng, H., Zhou, C., Gu, Y., and Shi, Z. (2020). Two-dimensional DOA estimation for coprime planar array : a coarray tensor-based solution. In: *IEEE International Conference on Acoustics Speech Signal Processing (ICASSP)*, Barcelona, Spain, 4562–4566.

79 Zheng, H., Zhou, C., De Almeida, L.F. et al. (2022). DOA estimation via coarray tensor completion with missing slices. In: *IEEE International Conference on Acoustics Speech Signal Processing (ICASSP)*, Singapore, Singapore, 5053–5057.

80 Wang, G., Fei, Z., Ren, S., and Li, X. (2020). Improved 2D coprime array structure with the difference and sum coarray concept. *Electronics* 9 (2): 1–25. https://doi.org/10.3390/electronics9020273.

81 Qiu, S., Ma, X., Zhang, R. et al. (2023). A dual-resolution unitary ESPRIT method for DOA estimation based on sparse co-prime MIMO radar. *Signal Processing* 202: 1–15.

82 Pal, P. and Vaidyanathan, P.P. (2012). Nested arrays in two dimensions, part I: geometrical considerations. *IEEE Transactions on Signal Processing* 60 (9): 4694–4705. https://doi.org/10.1109/TSP.2012.2203815.

83 Pal, P. and Vaidyanathan, P.P. (2012). Nested arrays in two dimensions, part Ii: application in two dimensional array processing. *IEEE Transactions on Signal Processing* 60 (9): 4706–4718. https://doi.org/10.1109/TSP.2012.2203815.

84 Zheng, W., Zhang, X., Xu, L., and Zhou, J. (2018). Unfolded coprime planar array for 2D direction of arrival estimation: an aperture-augmented perspective. *IEEE Access* 6: 22744–22753. https://doi.org/10.1109/ACCESS.2018.2828837.

85 Si, W., Zeng, F., Qu, Z., and Peng, Z. (2020). Two-dimensional DOA estimation via a novel sparse array consisting of coprime and nested subarrays. *IEEE Communications Letters* 24 (6): 1266–1270.

86 Tong, W., Zheng, Z., Wang, W., and So, H.C. (2022). Symmetric displaced coprime planar array for two-dimensional direction-of-arrival estimation. *IEEE Sensors Journal* 22 (23): 23221–23231. https://doi.org/10.1109/JSEN.2022.3215728.

87 Wang, G., Fei, Z., and Ren, S. (2021). 2D DOA estimation exploiting vertical synthetic planar arrays. *IEEE Access* 9: 3497–3507. https://doi.org/10.1109/ACCESS.2020.3047686.

88 Peng, Z., Si, W., and Zeng, F. (2020). 2-D DOA estimation for L-shaped sparse array via joint use of spatial and temporal information. *IEEE Communications Letters* 24 (9): 1934–1938.

89 Geng, W., Changxiao, C., Yi, H. et al. (2021). 2-D DOA estimation based on rectangular generalized minimum redundancy array via partial grid covariance vector sparse reconstruction. In: *2021 IEEE 6th International Conference on Intelligent Computing Signal Processing*. ICSP 2021, 859–864. https://doi.org/10.1109/ICSP51882.2021.9408984.

90 Geng, W., Changxiao, C., Jun, H. et al. (2021). Two-dimensional direction-of-arrival estimation based on augmented lagrange multiplier for multiband integrated sparse array. In: *2021 IEEE International Conference on Information Communication and Software Engineering*. ICICSE 2021, vol. 1, 179–184. https://doi.org/10.1109/ICICSE52190.2021.9404108.

91 Zhang, W., Liu, G., and Zhou, Y. (2020). DOA estimation for 2D multi-level dilated nested array. In: *IEEE 5th International Conference on Signal Image Processing*. Nanjing, China, vol. 1, 598–602.

92 Fu, M., Zheng, Z., and Wang, W.Q. (2021). 2-D DOA estimation for nested conformal arrays via sparse reconstruction. *IEEE Communications Letters* 25 (3): 980–984. https://doi.org/10.1109/LCOMM.2020.3039104.

93 Gohil, R.P., Routray, G., and Hegde, R.M. (2021). Learning based method for robust DOA estimation using co-prime circular conformal microphone array. In: *2021 National Conference on Communications NCC 2021*. https://doi.org/10.1109/NCC52529.2021.9530130.

94 Lin, Z., Lv, T., Ni, W. et al. (2021). Nested hybrid cylindrical array design and DoA estimation for massive IoT networks. *IEEE Journal on Selected Areas in Communications* 39 (4): 919–933. https://doi.org/10.1109/JSAC.2020.3018833.

95 Sun, L., Yang, M., and Chen, B. (2018). Thermos array: two-dimensional sparse array with reduced mutual coupling. *International Journal of Antennas and Propagation* 2018: 1–8. https://doi.org/10.1155/2018/3624514.

96 Rajamaki, R. and Koivunen, V. (2018). Sparse active rectangular array with few closely spaced elements. *IEEE Signal Processing Letters* 25 (12): 1820–1824.

97 Wang, Z., Liu, Y., Song, R. et al. (2022). Sparse convolutional array for DOA estimation. *EURASIP Journal on Advances in Signal Processing* 2022 (1): 1–15. https://doi.org/10.1186/s13634-022-00904-0.

98 Ren, S., Li, X., Luo, X., and Wang, W. (2018). Extensions of open box array with reduced mutual coupling. *IEEE Sensors Journal* 18 (13): 5475–5484.

99 Hoctor, R.T. and Kassam, S.A. (1990). The unifying role of the coarray in aperture synthesis for coherent and incoherent imaging. *Proceedings of the IEEE* 78 (4): 735–752. https://doi.org/10.1109/5.54811.

100 Liu, C.L. and Vaidyanathan, P.P. (2018). Maximally economic sparse arrays and cantor arrays. In: *IEEE 7th International Workshop on Computational Advances in Multi-Sensor Adaptive Processing (CAMSAP)*, Curacao, 1–5. https://doi.org/10.1109/CAMSAP.2017.8313139.

101 Zhu, C., Wang, W.Q., Chen, H., and So, H.C. (2015). Impaired sensor diagnosis, beamforming, and DOA estimation with difference co-array processing. *IEEE Sensors Journal* 15 (7): 3773–3780. https://doi.org/10.1109/JSEN.2015.2399510.

102 Li, J., Zhang, X., and Chen, H. (2012). Improved two-dimensional DOA estimation algorithm for two-parallel uniform linear arrays using propagator method. *Signal Processing* 92 (12): 3032–3038. https://doi.org/10.1016/j.sigpro.2012.06.010.

103 Friedlander, B. (2018). Antenna array manifolds for high-resolution direction finding. *IEEE Transactions on Signal Processing* 66 (4): 923–932. https://doi.org/10.1109/TSP.2017.2778683.

104 Huang, H., Fau, M., and Zoubir, A.M. (2019). Block sparsity-based DOA estimation with sensor gain and phase uncertainties. In: *27th European Signal Processing Conference (EUSIPCO)*, A Coruna, Spain, 1–5. https://doi.org/10.23919/EUSIPCO.2019.8903140.

105 Liu, C.L. and Vaidyanathan, P.P. (2019). Robustness of difference coarrays of sparse arrays to sensor failures - part I: a theory motivated by coarray music. *IEEE Transactions on Signal Processing* 67 (12): 3213–3226. https://doi.org/10.1109/TSP.2019.2912877.

106 Liu, C.L. and Vaidyanathan, P.P. (2019). Robustness of difference coarrays of sparse arrays to sensor failures - Part II: array geometries. *IEEE Transactions on Signal Processing* 67 (12): 3227–3242. https://doi.org/10.1109/TSP.2019.2912877.

107 Schmidt, R. (1986). Multiple emitter location and signal parameter estimation. *IEEE Transactions on Antennas and Propagation* 34 (3): 276–280.

108 Abramovich, Y.I., Spencer, N.K., Gorokhov, A.Y. et al. (1997). Positive-definite toeplitz completion in DOA estimation for nonuniform linear antenna arrays-Part II: partially augmentable arrays. *IEEE Transactions on Signal Processing* 45 (3): 820.

109 Orfanidis, S.J. (2008). Waves and antennas electromagnetic. *Media* 2: 687–688. http://www.ece.rutgers.edu/~orfanidi/ewa/.

110 Liu, C.L. and Vaidyanathan, P.P. (2016). Super nested arrays: linear sparse arrays with reduced mutual coupling-Part I: fundamentals. *IEEE Transactions on Signal Processing* 64 (16): 4203–4217. https://doi.org/10.1109/TSP.2016.2558167.

111 Friedlander, B. and Weiss, A.J. (1991). Direction finding in the presence of mutual coupling. *IEEE Transactions on Antennas and Propagation* 39 (3): 273–284. https://doi.org/10.1109/8.76322.

112 Pal, P. and Vaidyanathan, P.P. (2010). Nested arrays: a novel approach to array processing with enhanced degrees of freedom. *IEEE Transactions on Signal Processing* 58 (8): 4167–4181. https://doi.org/10.1109/TSP.2010.2049264.

113 Qin, S., Zhang, Y.D., and Amin, M.G. (2015). Generalized coprime array configurations for direction-of-arrival estimation. *IEEE Transactions on Signal Processing* 63 (6): 1377–1390.

114 Jiang, T., Sidiropoulos, N.D., and Ten Berge, J.M.F. (2001). Almost-sure identifiability of multidimensional harmonic retrieval. *IEEE Transactions on Signal Processing* 49 (9): 1849–1859. https://doi.org/10.1109/78.942615.

115 E. Boudaher, F. Ahmad, and M. G. Amin, "Sparsity-based direction finding of coherent and uncorrelated targets using active nonuniform arrays," *IEEE Signal Process. Lett.* vol. 22, no. 101628–1632, 2015https://doi.org/10.1109/LSP.2015.2417807.

116 Vaidyanathan, P.P. and Pal, P. (2012). Direct-MUSIC on sparse arrays. In: *International Conference on Signal Processing and Communication (SPCOM)*, Bangalore, India, vol. 3, 1–5. https://doi.org/10.1109/SPCOM.2012.6289992.

117 Alawsh, S.A., Mohamed, M.H., Mahmoud, I.B.A. et al. (2023). Sparse billboard and T-shaped arrays for two- dimensional direction of arrival estimation. *IEEE Open Journal of Signal Processing* 4: 322–335. https://doi.org/10.1109/OJSP.2023.3278593.

118 Zoltowski, M.D., Haardt, M., and Mathews, C.P. (1996). Closed-form 2-D angle estimation with rectangular arrays in element space or beamspace via unitary ESPRIT. *IEEE Transactions on Signal Processing* 44 (2): 316–328.

119 Coviello, C.M., Kozick, R.J., Hurrell, A. et al. (2012). Thin-film sparse boundary array design for passive acoustic mapping during ultrasound therapy. *IEEE Transactions on Ultrasonics, Ferroelectrics, and Frequency Control* 59 (10): 2322–2330. https://doi.org/10.1109/TUFFC.2012.2457.

120 Hamza, S.A. and Amin, M.G. (2019). *Two Dimensional Sparse Array Design for Wideband Signals, Big Data Learning, Analytics and Applications*, vol. 10989, 135–142. The International Society of Optics and Photonics.

6

Sparse Array Design for Direction Finding Using Deep Learning

Kumar Vijay Mishra[1], Ahmet M. Elbir[2], and Koichi Ichige[3]

[1] *Computational and Information Sciences Directorate (CISD), United States DEVCOM Army Research Laboratory, Adelphi, MD, USA*
[2] *Interdisciplinary Centre for Security, Reliability and Trust (SnT), University of Luxembourg, Esch-sur-Alzette, Luxembourg*
[3] *Department of Electrical and Computer Engineering, Yokohama National University, Yokohama, Japan*

6.1 Introduction

Phased sensor arrays have become widely prevalent in various applications such as radar, sonar, communications, acoustics, sounding, and ultrasound [1–4]. They allow electronic beam steering without requiring any mechanical motion, providing high spatial selectivity and the ability to reject interference. However, according to the Nyquist–Shannon theorem, the array must allow for at least two signal samples in a single spatial period, which is the same as the array's operating wavelength [5]. Otherwise, *spatial aliasing* occurs resulting in multiple main-lobes appearing in the beampattern, which reduces the directivity and affects the accuracy of estimating bearings or directions-of-arrival (DoAs) of unknown sources or targets [6]. To avoid this, conventional phased sensor arrays employ a uniform linear array (ULA), wherein sensors must be separated *at most* by half-wavelength spacing. However, as the number of sensors increases, the associated complexity, size, and cost of arrays become unacceptable. As a result, there is growing interest in thinned or sparse sensor arrays, which offer significantly reduced hardware [7–9].

Uniformly removing sensors from a ULA produces *grating lobes* or copies of the main lobe. This effect can be suppressed by removal of sensors at random locations giving rise to *thinned* or *sparse* arrays [10, 11]. However, randomly thinned arrays are accompanied by increased sidelobe levels, which hampers direction finding. Further, the number of elements in an array determines its degrees-of-freedom (DoFs) or the number of sources whose bearings can be determined by the sensor array. A full or filled sensor array with N elements has more DoFs than a sparse array with only $\mathcal{O}(\sqrt{N})$ sensors [1, 8]. But, with suitable parameter recovery algorithms [9, 12–14], sparse arrays have been shown to exhibit negligible performance degradation and reduced mutual coupling in direction finding [15–17] and spatial filtering [18]. Most of these algorithms employ sparse recovery techniques, including compressed sensing [19].

6.1.1 Prior Art and Historical Notes

Sparse array literature could be traced back to the seminal works of Y. T. Lo [10, 11], who studied the probabilistic properties (e.g., the probability of the array pattern crossing a specified sidelobe level) of random arrays. Other studies published around the same time [20–22] focused on approximating

Sparse Arrays for Radar, Sonar, and Communications, First Edition. Edited by Moeness G. Amin.

the analytic expressions of desired antenna patterns with few nonuniformly spaced elements but these were optimal only for smaller arrays. This was followed by sparse array designs using vastly different procedures such as dynamic programming [23], iterative least squares [24], nonlinear optimization [25], genetic algorithms [8], and simulated annealing (SA) [26]. These techniques are usually limited by slow speed of optimization, high sidelobes, nonoptimality for large arrays, and high dependence of final element placement based on the choices made in previous iterations.

Other geometries of nonuniformly spaced sparse arrays that have low sidelobes were also proposed; see, e.g., [27] for an overview, and references therein. For example, [28] describes a fractal-based quasi-random array that combines properties of both random and periodic arrays such that its peak sidelobe level (PSL) is lower than a purely random array. A space-tapered array [29, 30] places antennas with increasing inter-element spacing as one moves away from the aperture's center thereby achieving low PSL but higher integrated sidelobe levels (ISLs) than random arrays.

A sparse array is hole-free or *restricted* if its difference coarray is a ULA, otherwise it is a *general* array. A sparse array with N sensors and the aperture defined by the integer grid $[-M, M]$ may also be characterized by its *redundancy ratio* $R = N(N-1)/(2M) \geq 1$, which is the number of pairs of antennas divided by the aperture [31]. Using this metric, zero redundancy arrays (ZRAs) [32], minimum redundancy arrays (MRAs) [31], and low redundancy arrays (LRAs) [33] were proposed. These minimize spatial aliasing while maintaining a reasonably constant ISL. The ZRAs have $R = 1$ but they exist for only $N \leq 4$ [34]. the MRAs, which achieve the lowest R, select the missing elements such that the resulting sparse array has all possible inter-element spacings of the full array. Leech's bounds [35] for MRAs stipulate $1.217 \leq R \leq 1.674$ for $N \to \infty$. Optimal solutions for $N \leq 11$ were provided in [35] and, using exhaustive computer-aided search, expanded later to $N \leq 26$ [36]. For larger N, the procedure becomes increasingly complex [37] and there have been efforts to meet the Leech's lower bound through LRAs [33]. These problems have led to the popularity of more structured sparse designs such as co-prime arrays [38–40]. While these configurations provide a closed form of sensor positions and offer enhanced DoFs for parameter estimation, they are not applicable to arbitrary number of antenna elements. Detailed discussion of different sparse array structures with fixed geometries is provided in Chapter 1 of this book.

6.1.2 Learning-Based Approaches

In general, searching for an optimal sparse sensor array is a combinatorial problem [31], whose computational complexity increases with the number of sensors. Since a closed-form solution is difficult to come by, several sub-optimal (although mathematically tractable) solutions have been proposed [13, 16, 41–44]. Recently, there has been a growing interest in utilizing learning-based approaches in sparse sensor communications [18, 45] and signal processing [46, 47].

Deep learning (DL) has been shown to offer better computational efficiency than a combinatorial search [18, 48]. DL has been extremely effective in addressing challenging problems such as speech recognition, visual object recognition, and language processing [49, 50]. It offers advantages such as low computational complexity while tackling optimization-based or combinatorial search problems, and the ability to generate new features from a limited set of features available in a training set [49, 51].

In a typical DL-based sparse array design (Fig. 6.1), the antenna array data is fed into the DL network, which yields the indices of the selected antennas at the output. To train the DL model, the input datasets vary and may comprise of the raw array output [18], covariance matrix [52], or the autocorrelation function [53] of the array data. The output layer of the model is usually a classification layer, wherein each class corresponds to a possible set of subarray indices. In some

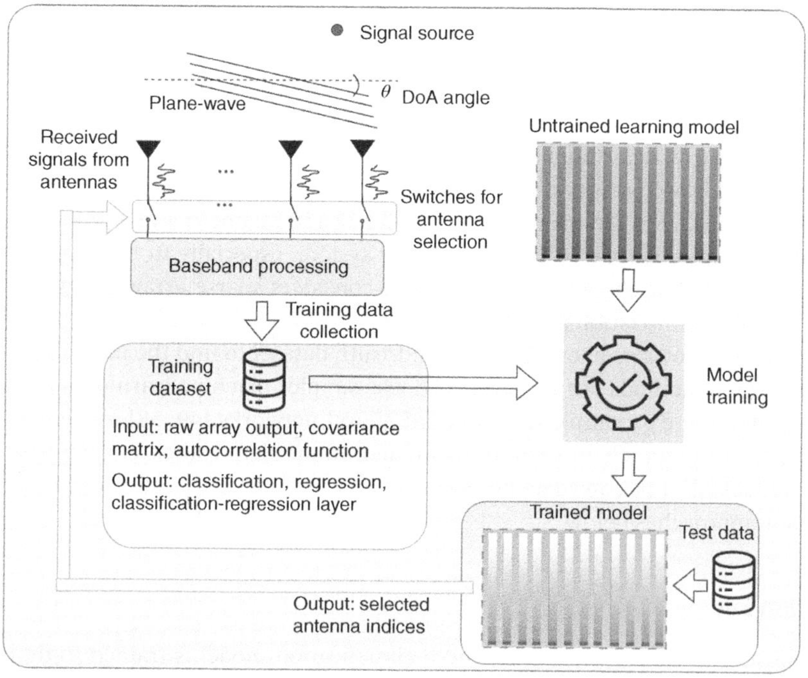

Figure 6.1 Illustration of DL-based antenna selection.

case, the output is a regression layer, whose values are either 0 or 1 indicating the (un)selected antennas [46].

The rest of the chapter is organized as follows: In the next section, we review general DL design procedures for sparse array design and DoA estimation. Then, Section 6.3 describes a major use case of this technique in cognitive radar, where a dynamic transmit antenna selection is required for the changing target environment. A DL-based approach is able to yield sparse (sub-)arrays in real-time for this application. When insufficient training data are available for a new antenna geometry, then Section 6.4 presents a transfer learning (TL) approach to utilize a trained network from a different antenna configuration. For planer or two-dimensional (2-D) large sparse arrays, usually a combination of SA and DL may be more appropriate, as detailed in Section 6.5. We then consider DL for sparse array applications in wireless communications by combining antenna selection with massive multiple-input multiple-output (MIMO) beamforming (Section 6.6) and integrated sensing and communications (ISAC) [54] (Section 6.7). Finally, we summarize our key observations and unaddressed challenges in Section 6.8.

Notations: Throughout the chapter, we reserve boldface lowercase and uppercase letters for vectors and matrices, respectively. The ith element of a vector $\mathbf{y}$ is $y(i)$ while the (i,j)th entry of the matrix $\mathbf{Y}$ is $[\mathbf{Y}]_{i,j}$. We denote the transpose and Hermitian by $(\cdot)^T$ and $(\cdot)^H$, respectively. The functions $\angle\{\cdot\}$, $\mathbb{R}\text{e}\{\cdot\}$ and $\mathbb{I}\text{m}\{\cdot\}$ designate the phase, real and imaginary parts of a complex argument, respectively. The combination of selecting K terms out of M is denoted by $\binom{M}{K} = \frac{M!}{K!(M-K)!}$. The Hadamard (point-wise) product is written as $\odot$. The functions $\mathrm{E}\{\cdot\}$ and max give the statistical expectation and maximum value of the argument, respectively. The notation $x \sim \mathrm{u}\{[u_l, u_u]\}$ means a random variable drawn from the uniform distribution over $[u_l, u_u]$ and $x \sim \mathcal{CN}(\mu_x, \sigma_x^2)$ represents the complex normal distribution with mean μ_x and variance σ_x^2.

6.2 General Design Procedures

DL-based design for sparse arrays comprise diverse procedures that differ in performance criteria, performance metrics and input/output data types for the learning model (Fig. 6.2). For instance, in wireless communications, sparse arrays manifest as antenna selection problem to reduce the complexity of common communications tasks such as channel estimation, beamforming, resource allocation, and user localization [42, 55]. On the other hand, other sparse geometries (e.g., coprime and nested arrays) are used to improve the DoF of the overall system, especially in radar applications [16, 38, 56, 57]. We note that chapter 15 of this book considers sparse array design for communication platforms from optimization perspective.

The data-driven nature of DL allows us to employ a ground-truth dataset to find the best sparse array configuration. While some methods use DL, others choose simpler machine learning methods with shallow learning models, e.g., simple feed-forward neural networks [49, 58]. Although defining the optimum sparse array search as a binary segmentation problem is theoretically possible, current DL-based models do not consider the problem as a multilabel classification task. We summarize some major approaches below.

6.2.1 Antenna Selection Setups

The potential array setups are translated into labels and a classification model is trained to find the optimum array structure. If the array setups are required to be in a specific configuration, e.g., minimum hole array, MRA, or nested arrays, either a constraint is added to the loss function to guarantee a desired architecture, or the set of possible sparse array designs is reduced to a set of sparse arrays following further constraints by a searching algorithm beforehand. The learned sparse array can be tested with any of the DoA estimation methods. In [46], a convolutional neural network (CNN) is used on a dataset composed of an $N \times N \times 3$ matrix representing real, imaginary, and angle components of a complex covariance matrix as input and each possible subarray for a given

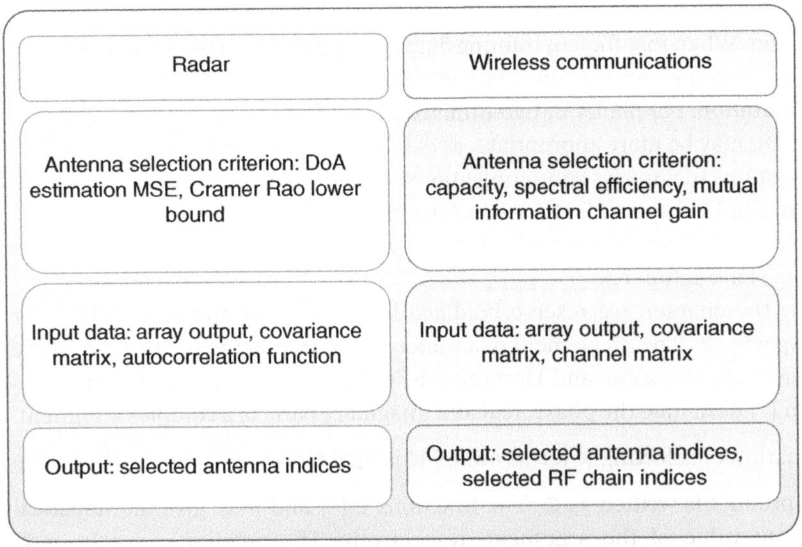

Figure 6.2 Design procedures for DL-based antenna selection for the applications in radar and wireless communications.

ULA as a class label as output. The problem is defined as the minimization of the Cramér–Rao bound (CRB), which limits the minimum variance of the correlation matrix in the DoA estimation problem. The learning model returns an appropriate subarray for individual received data, and thus works as a cognitive antenna array. The model is intended to be used as a cognitive selector, meaning that the model is used in real-time to determine the best sparse array to use in analysis by reading the incoming signal and changing the selected sparse array at a given period.

The model in [46] returns suboptimal 1-D arrays. The final arrays include elements at the end of the ULA grid without including surrounding elements. This issue results in coarray having many holes, causing high peak sidelobes and reducing the performance. To tackle this problem, [59] restricted the available antenna subarrays to the ones with minimal hole setup. Here, k-essential property is used to eliminate sensors from a ULA, which shows whether the coarray changes from the ULA or not when the elements are removed. The M-N sensors that are not essential can thus be discarded. The results show that the estimated subarrays provide a better beampattern with suppressed peak side lobes and a narrower main lobe. Estimated sparse arrays of this model also show much lower root mean-squared error (RMSE) values compared to [46].

Among model-based learning approaches, [60] performed antenna selection from a CRB minimization problem defined as a weight vector selection that gives the minimum CRB for a given number of nonzero weights at the output of the learning model. To make the original problem tractable, Dinkelbach approximation was used and solved by applying *adv*anced *grad*ient descent (AdaGrad) algorithm on a loss function.

6.2.2 DoA Estimation Setups

Some DL-based DoA estimation models include sparsity of the array design in their architecture in order to exploit the advantages of the sparse arrays. Model-based methods require the covariance matrix to be full, which is impossible in sparse array structures. DL-based methods specifically designed to work with sparse arrays are proven to be powerful enough to overcome this issue and learn an accurate mapping between sparse covariance matrices and arrival angles [46, 61]. These models can also utilize the coarray principle and detect more signal sources than the number of sensors in the sparse array.

In [62], a two-layer CNN was trained to estimate DoA angles. The array structure is a combination of different subarrays with the same number of elements. The position between different subarrays is assumed to be unknown while only the relative position of elements in each subarray is known. The model input is the fusion of the estimated covariance matrix of each subarray created with an element-wise sum and vectorization while the output yields the DoA angles. It is shown that the learning model can learn to detect more signal sources than the number of elements thanks to employing sparse arrays.

A deep CNN model called RFDOA-Net was designed in [63] to estimate the DoA angles for unmanned aerial vehicles with a complex architecture using multi-scale feature processing. The learning model was trained with a synthetic dataset with different signal-to-noise-ratio (SNR) values over a five-element sparse coarray and used raw antenna inputs converted into an input matrix of antenna elements with respect to snapshots. The output data was a vector of the probability distribution of the DoA angles.

In [53], CNN-based DoA estimation was studied with nested arrays. The input of the CNN was a proxy spectrum obtained from the autocorrelation of the nested array output. The output layer of the CNN included the angular spectra, of which the highest peaks corresponds to the source directions [52].

Some models are designed to capture both azimuth and elevation angles with a 3-D sparse array structure. In [64], CNN was applied on the 2-D readings of a coprime circular conformal (stacked array arrangement) microphone array (C3MA) to estimate the DoA angles of audio signals. The model took the cross-correlation value of individual microphones as input while the output yields the 2-D DoA angles.

6.3 Cognitive Sparse Array Design for DoA Estimation

We now examine a DL-based antenna selection procedure in detail for a cognitive radar application. The DL techniques directly fit this setting because the antenna selection problem can be considered as a classification problem, wherein each subarray designates a class. We leverage DL to consider a relatively large scale of the selection problem wherein the feature maps can be extracted to train the network for different array geometries. Unlike random array thinning where a fixed subarray is used for all scans, we select a new subarray based on the received data. In contrast to [44, 65], *we also assume that the target DoA angle is unknown while choosing the array elements.*

We employ a CNN, whose input data are the covariance samples of the received array signal. The CNN architecture models the selection of K best antennas out of M as a classification problem wherein each class denotes an antenna subarray. In order to create the training data, we choose those subarrays which estimate DoA with the lowest minimal bound on the mean-squared-error (MSE) so that the selected subarray can yield the best performance for the given scenario. We consider minimization of CRB as the performance benchmark in generating training sets for 1-D ULA and 2-D geometries such as uniform circular arrays (UCA) and randomly deployed arrays (RDA). For ULAs, the network is also trained with data obtained by minimizing Bayesian bounds such as the Bobrovsky–Zakai bound (BZB) and Weiss–Weinstein bound (WWB) on DoA [66] because these bounds provide better estimates of MSE at low SNRs. In particular, BZB-based selection has been shown to have the ability to control the sidelobes and avoid ambiguity in DoA estimation [67, 68].

6.3.1 Signal Model

Consider a phased array antenna with M elements where each element transmits a pulsed waveform $s(t_i)$ towards a Swerling Case 1 point target for $i = 1, \dots, T$, where T is the number of data snapshots. Since we are interested only in DoA recovery, the range and Doppler measurements are not considered and target's complex reflectivity is set to unity. Assumption of a Swerling I model implies that the target parameters remain constant for the duration of the scan. We characterize the target through its DoA $\Theta = (\theta, \phi)$ where θ and ϕ denote, respectively, the elevation and the azimuth angles with respect to the radar. The radar's pulse repetition interval and operating wavelength are, respectively, T_s and $\lambda = c_0/f_0$, where $c_0 = 3 \times 10^8$ m/s is the speed of light and $f_0 = \omega_0/2\pi$ is the carrier frequency.

To further simplify the geometries, we suppose that the targets are far enough from the radar so that the received signal wavefronts are effectively planar over the array. The array receives a narrowband signal reflected from a target located in the far-field of the array at Θ. Then, the output of the sensor array is [69]

$$\mathbf{y}(t_i) = \mathbf{a}(\Theta)s(t_i) + \mathbf{n}(t_i), \quad 1 \leq i \leq T, \tag{6.1}$$

where T is the number of snapshots, $\mathbf{y}(t_i) = [y_1(t_i), \dots, y_M(t_i)]^T$ and $y_m(t_i)$ denotes the output of the mth sensor for the ith snapshot, $\mathbf{n}(t_i) = [n_1(t_i), \dots, n_M(t_i)]^T$ is the noise vector and

$n_m(t_i)$ is zero-mean spatially and temporarily white Gaussian noise with variance σ_n^2, $\mathbf{a}(\Theta) = [a_1(\Theta), \ldots, a_M(\Theta)]^T$ is the $M \times 1$ steering vector. The mth element of $\mathbf{a}(\Theta)$ is

$$a_m(\Theta) = \exp\left\{-j\frac{2\pi}{\lambda}\mathbf{p}_m^T\mathbf{r}(\Theta)\right\}, \tag{6.2}$$

where $\mathbf{r}(\Theta)$ depends on the source direction as

$$\mathbf{r}(\Theta) = [\cos(\phi)\sin(\theta), \sin(\phi)\sin(\theta), \cos(\theta)]^T, \tag{6.3}$$

and $\mathbf{p}_m = [x_m, y_m, z_m]^T$ is the position of the mth sensor in the Cartesian coordinate system. Expanding the inner product $\mathbf{p}_m^T\mathbf{r}(\Theta)$ in the array steering vector gives $a_m(\Theta) = \exp\left\{-j\frac{2\pi}{\lambda}(p_{x_m}\mu + p_{y_m}\nu + p_{z_m}\xi)\right\}$, where $\mu = \cos(\phi)\sin(\theta)$, $\nu = \sin(\phi)\sin(\theta)$, and $\xi = \cos(\theta)$. Evidently, $a_m(\Theta)$ is a multi-dimensional harmonic. Once the frequencies μ, ν, and ξ in different directions are estimated, the DoA angles are obtained using the relations

$$\theta = \tan^{-1}\left(\frac{\nu}{\mu}\right), \quad \phi = \cos^{-1}\left(\frac{\xi}{\sqrt{\mu^2+\nu^2+\xi^2}}\right), \tag{6.4}$$

with the usual ambiguity in $[0, 2\pi]$. In case of a linear array, there is only one parameter in the steering vector whereas two parameters are involved in planar and three-dimensional arrays.

In the context of sparse array selection, our goal is to choose the "best" K sensors in an M-element array in the sense that the lowest statistical MSE, i.e., the CRB is achieved [69, 70]. Overall, $C = \binom{M}{K} = \frac{M!}{K!(M-K)!}$ possible subarray choices are available. Therefore, we can treat sensor selection as a classification problem with C classes. It seems impractical to visit all possible subarray configurations to arrive at the best subarray candidate. However, it has been shown [18, 46, 61] that many subarray candidates yield the same CRB level because of the nonunique placement of sensors within the array. Hence, the distinct number of subarrays is very small, which makes this technique very practical. Note that the literature suggests other statistical bounds [66] for DoA estimation but a closed-form solution of only CRB is available for higher dimensional arrays.

We observe Θ as the inner product $\mathbf{p}_m^T\mathbf{r}(\Theta)$. The exponential form of $a_m(\Theta)$ suggests that this is a multi-dimensional spatial harmonic whose frequencies (and hence, DoAs) can be extracted through conventional as well as sparse reconstruction algorithms [9, 46]. The uniqueness of spatial harmonic retrieval [71] is directly related to the number of sensors in the array. For a URA of size $M_1 \times M_2$, at least $M_1M_2 - \min(M_1, M_2)$ sensors are required for a perfect DoA retrieval in a noiseless setting. Hence, in any sparse sensor array selection, K must satisfy these guarantees.

In general, the target's position changes little during consecutive scans while a phased array can switch very fast from one antenna configuration to the other. Here, we consider the following scan strategy for the radar: at the beginning (the very first scan), all M antennas are active and the received signal from this scan is fed to the network. Our goal is to find an optimal antenna array for the next scan in which only K antennas will be used. The radar continues to use this subarray for a few subsequent scans. After surveying the target scene with this optimal subarray for a predetermined number of scans, the radar switches back to the full array for a single scan. The received signal from this full array scan is then used to find a new, optimal subarray for the subsequent few scans. The frequency of choosing a new subarray can be decided off-line based on the nature of the target and analysis of previous observations. This switching of elements between different scans is a cognitive operation because a new array is determined in every few scans based on received echoes from the target scene.

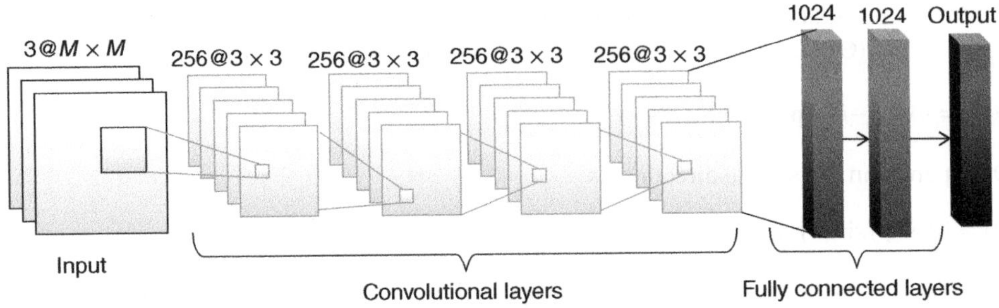

Figure 6.3 Structure of the CNN for sensor selection. Source: [72]/IEEE.

6.3.2 Antenna Selection via Deep Learning

A DL network (see Fig. 6.3) is defined as a nonlinear mapping which categorizes and clusters the input data. Let $\mathcal{D} = \{\mathcal{D}^{(1)}, \dots, \mathcal{D}^{(D)}\}$ and $\mathcal{Y} = \{\mathcal{Y}^{(1)}, \dots, \mathcal{Y}^{(Y)}\}$ denote the input and output labels for a dataset. Then, the deep classification network is represented as $\Sigma(\mathcal{D}) = \mathcal{Y}$ mapping the input data to the output labels which represent the best subarray indices.

Assume that an antenna subarray composed of K antennas is to be selected from an M-element antenna array. There are $C = \binom{M}{K}$ possible choices. This can be viewed as a classification problem with C classes each of which represents a different subarray. Let $\mathcal{P}_k^c = \{p_{x_k}^c, p_{y_k}^c, p_{z_k}^c\}$, $k = 1, \dots, K$, be the set of the kth antenna coordinates in the cth subarray. Then, the cth class consisting of the positions of all elements in the cth subarray is

$$\mathcal{S}_c = \{\mathcal{P}_1^c, \dots, \mathcal{P}_K^c\}, \tag{6.5}$$

and all classes are given by the set

$$\mathcal{S} = \{\mathcal{S}_1, \mathcal{S}_2, \dots, \mathcal{S}_C\}. \tag{6.6}$$

Next, we design a CNN to solve this classification problem by selecting the positions of the antenna subarray that provides the best DoA estimation performance. Note that, for an operational radar, generation of an artificial or simulated dataset to train the DL network is not necessary. Instead, the network can train itself with the data acquired by the radar during previous scans. During the test phase, the DoA angles are unknown to the network. The CNN accepts the features from the estimated covariance matrix and outputs a new array. This stage is, therefore, cognitive because the radar is adapting the antenna array in response to the received signal. In the following, we present the details of input and output design of the deep network.

6.3.2.1 Input Data

The input to the DL network are the covariance matrices of the received signal. In particular, we use the real, imaginary and the phase information of the covariance matrix. Let $\mathbf{X}$ be an $M \times M \times 3$ real-valued matrix with 3 "channel." Hence, we have $\mathcal{D}^{(i)} = \mathbf{X}$ for ith input instant. Specifically, we define the (i,j)th entry of the first and the second "channel" of the input data as $[\mathbf{X}_{(:,:,1)}]_{i,j} = \mathrm{Re}\{[\mathbf{R}]_{i,j}\}$ and $[\mathbf{X}_{(:,:,2)}]_{i,j} = \mathrm{Im}\{[\mathbf{R}]_{i,j}\}$, respectively, where $\mathbf{R}$ is the array covariance matrix as

$$\mathbf{R} = \frac{1}{T}\sum_{i=1}^{T} \mathbf{y}(t_i)\mathbf{y}^H(t_i). \tag{6.7}$$

Similarly, the third "channel" is given by $[\mathbf{X}_{(:,:,3)}]_{i,j} = \angle\{[\mathbf{R}]_{i,j}\}$. Although real and imaginary inputs are sufficient to describe the complex covariance matrix, feeding a third quantity such as phase (or magnitude) lets the network know that the first two inputs are related to each other.

6.3.2.2 Labeling

We treat the sensor selection problem as a classification problem with C classes. The class label comprises the positions of the sensor subarray corresponding to that class. Let $\mathcal{P}_c^{(k)} = \{x_k{}^{(c)}, y_k{}^{(c)}, z_k{}^{(c)}\}$ be the set of sensor coordinates in the cth subarray for $k = 1, \dots, K$. Then the positions of the sensors for the cth class form the set $\mathcal{Y}_c = \{\mathcal{P}_c^{(1)}, \dots, \mathcal{P}_c^{(K)}\}$. Therefore, the set of all classes is $\mathcal{Y} = \{\mathcal{Y}_1, \mathcal{Y}_2, \dots, \mathcal{Y}_C\}$.

In order to select the best subarrays in $\mathcal{Y}$, we compute the CRB for each element of $\mathcal{Y}$ as $c = 1, \dots, C$. Consider the $K \times 1$ subarray output

$$\mathbf{y}_c(t_i) = \mathbf{a}_c(\Theta)s(t_i) + \mathbf{n}_c(t_i), \tag{6.8}$$

where $\mathbf{a}_c(\Theta) \in \mathbb{C}^K$ denotes the array steering vector corresponding to the subarray with position set $\mathcal{Y}_c$. Let $\mathbf{R}_c = \frac{1}{T}\sum_{i=1}^{T} \mathbf{y}_c(t_i)\mathbf{y}_c^H(t_i)$ be the $K \times K$ subarray sample covariance matrix for the $K \times 1$ subarray output $\mathbf{y}_c(t_i)$. We denote the partial derivatives of $\mathbf{a}_c(\Theta)$ with respect to θ and ϕ by $\dot{\mathbf{a}}_c(\theta) = \frac{\partial \mathbf{a}_c(\Theta)}{\partial_\theta}$ and $\dot{\mathbf{a}}_c(\phi) = \frac{\partial \mathbf{a}_c(\Theta)}{\partial_\phi}$, respectively. The signal and noise variances are σ_s^2 and σ_n^2, respectively.

The CRBs for θ and ϕ in a single source scenario are [69]

$$\kappa(\theta, \mathcal{Y}_c) = \frac{\sigma_n^2}{2T\,\mathbb{R}\mathrm{e}\left\{\Pi_\theta \odot (\sigma_s^4 \mathbf{a}_c^H(\Theta)\mathbf{R}_c^{-1}\mathbf{a}_c(\Theta))\right\}}, \tag{6.9}$$

$$\kappa(\phi, \mathcal{Y}_c) = \frac{\sigma_n^2}{2T\,\mathbb{R}\mathrm{e}\left\{\Pi_\phi \odot (\sigma_s^4 \mathbf{a}_c^H(\Theta)\mathbf{R}_c^{-1}\mathbf{a}_c(\Theta))\right\}}, \tag{6.10}$$

where

$$\Pi_\theta = \dot{\mathbf{a}}_c^H(\theta)\left[\mathbf{I}_K - \frac{\mathbf{a}_c(\Theta)\mathbf{a}_c^H(\Theta)}{K}\right]\dot{\mathbf{a}}_c(\phi), \tag{6.11}$$

$$\Pi_\phi = \dot{\mathbf{a}}_c^H(\phi)\left[\mathbf{I}_K - \frac{\mathbf{a}_c(\Theta)\mathbf{a}_c^H(\Theta)}{K}\right]\dot{\mathbf{a}}_c(\theta). \tag{6.12}$$

We define the absolute CRB [73] for the directions Θ and $\mathcal{Y}_c$ as the RMS value

$$\kappa(\Theta, \mathcal{Y}_c) = \frac{1}{\sqrt{2}}[\kappa(\theta, \mathcal{Y}_c)^2 + \kappa(\phi, \mathcal{Y}_c)^2]^{1/2}. \tag{6.13}$$

For simplicity, we select $\sigma_s^2 = 1$ and define the signal-to-noise ratio in the training data as $\mathrm{SNR}_{\mathrm{TRAIN}} = 10\log_{10}(\sigma_s^2/\sigma_n^2)$.

Once $\kappa(\Theta, \mathcal{Y}_c)$ is computed for $c = 1, \dots, C$, the best subarray label $\mathcal{B}_{\bar{c}}$ is

$$\mathcal{B}_{\bar{c}} = \arg\min_{c=1,\dots,C} \kappa(\Theta, \mathcal{Y}_c). \tag{6.14}$$

Here, the subscript $(\cdot)_{\bar{c}}$ denotes the index of best subarrays, $\bar{c} = 1, \dots, \bar{C}$, where $\bar{C}$ is the number of best subarrays. As K increases, C becomes very large. This makes the classification operation very difficult. However, experiments reveal that most of the sensor subarrays yield the same $\kappa(\Theta, \mathcal{Y}_c)$ because the nonunique sensor positions are common in many subarray combinations. Hence, $\bar{C} \ll C$ implying that only a handful of classes yield the lowest estimation errors [18, 46]. In Table 6.1, we present the comparison of C and $\bar{C}$ for a UCA with $M = 16$ antennas. We note that $\bar{C}$ is very small, which leads to an effective classification performance. After computing all best subarray indices, we finally construct the best subarray set as $\mathcal{B} = \{\mathcal{B}_1, \dots, \mathcal{B}_{\bar{C}}\}$, where $\mathcal{B} \subset \mathcal{Y}$.

Algorithm 6.1 Training data generation

Input: Sensor positions $\{\mathbf{p}_m\}_{m=1}^{M}$, K, T, number of data realizations L, number of directions P and $\text{SNR}_{\text{TRAIN}}$.

Output: Training data $\mathcal{T}$ with dimensions $\{M \times M \times 3 \times LP, LP\}$.

Generate P DoA angles $\Theta_p = (\theta_p, \phi_p)$ for $p = 1, \ldots, P$.

for $1 \leq p \leq P$ **do**

  for $1 \leq l \leq L$ **do**

  Generate the array output $\{\mathbf{y}^{(l,p)}(t_i)\}_{i=1}^{T}$ as

$$\mathbf{y}^{(l,p)}(t_i) = \mathbf{a}(\Theta_p)s^{(l,p)}(t_i) + \mathbf{n}^{(l,p)}(t_i),$$

  for $s^{(l,p)}(t_i) \sim \mathcal{CN}(0, \sigma_s^2)$, $\mathbf{n}^{(l,p)}(t_i) \sim \mathcal{CN}(0, \sigma_n^2\mathbf{I})$.

  Construct all $K \times 1$ subarray output configurations $\mathbf{y}_c^{(l,p)}(t_i)$ as in (6.8) from $\mathbf{y}^{(l,p)}(t_i)$ for $c = 1, \ldots, C$.

  Compute $\kappa(\Theta_p, \mathcal{Y}_c)$ for $c = 1, \ldots, C$ by using the covariance matrices $\mathbf{R}_c^{(l,p)}$.

  Using $\kappa(\Theta_p, \mathcal{Y}_c)$, find the best subarray index as $\mathcal{B}_{\bar{c}}^{(l,p)}$ from (6.14).

  Compute the full array covariance matrix $\mathbf{R}^{(l,p)}$ from $\mathbf{y}^{(l,p)}(t_i)$, $i = 1, \ldots, T$.

  Construct the input data $\mathbf{X}^{(l,p)}$ as

$$\begin{aligned}
[\mathbf{X}_{(:,:,1)}^{(l,p)}]_{i,j} &= \text{Re}\{[\mathbf{R}^{(l,p)}]_{i,j}\},\\
[\mathbf{X}_{(:,:,2)}^{(l,p)}]_{i,j} &= \text{Im}\{[\mathbf{R}^{(l,p)}]_{i,j}\},\\
[\mathbf{X}_{(:,:,3)}^{(l,p)}]_{i,j} &= \angle\{[\mathbf{R}^{(l,p)}]_{i,j}\}.
\end{aligned}$$

  Design the output label as $z^{(l,p)} = \mathcal{B}_{\bar{c}}^{(l,p)}$.

  end for l

end for p

Construct training data by concatenating the input-output pairs: $\mathcal{T} = \{(\mathbf{X}^{(1,1)}, z^{(1,1)}), (\mathbf{X}^{(1,2)}, z^{(2,1)}), \ldots, (\mathbf{X}^{(1,L)}, z^{(L,1)}), (\mathbf{X}^{(2,1)}, z^{(1,2)}) \ldots, (\mathbf{X}^{(P,L)}, z^{(L,P)})\}$.

Table 6.1 Number of classes C and the reduced number of classes $\bar{C}$ for a UCA with $M = 16$.

	$K = 3$	$K = 4$	$K = 5$	$K = 6$	$K = 7$	$K = 8$
C	560	1820	4368	8008	11,440	12,870
$\bar{C}$	16	10	16	11	16	16

Algorithm 1 lists the steps to generate the training data by incorporating the input and labels, as discussed above. The training data is then fed to the deep network represented by $\Sigma(\cdot) : \mathbb{R}^{M \times M \times 3} \rightarrow \mathcal{Y}$ that maps the input data $\mathbf{X}$ to the corresponding class in $\mathcal{Y}$.

6.3.2.3 Network Architecture

Figure 6.3 illustrates the deep network architecture for sensor selection. For a multi-layer network, the nonlinear function $\Sigma(\cdot)$ is represented by the inner layers as

$$\Sigma(D) = f^{(15)}\left(f^{(14)}(\ldots f^{(2)}(f^{(1)}(D)))\right) = \mathcal{Y}, \tag{6.15}$$

where the first layer $f^{(1)}$ is the input layer and $f^{(i)}_{i\in\{2,4,6,8\}}$ denote the convolutional layers, each of which has 256 filters of size 3×3. The arithmetic operation of a single filter of a convolutional layer is defined for an arbitrary input $\overline{\mathbf{X}} \in \mathbb{R}^{d_x \times d_x \times V_x}$ and output $\overline{\mathbf{Y}} \in \mathbb{R}^{d_y \times d_y \times V_y}$ as

$$\overline{\mathbf{Y}}_{p_y,v_y} = \sum_{p_k,p_x} \langle \overline{\mathbf{W}}_{v_y,p_k}, \overline{\mathbf{X}}_{p_x} \rangle, \tag{6.16}$$

where $d_x \times d_y$ is the size of the convolutional kernel, $V_x \times V_y$ is the size of the response of a convolutional layer, $\overline{\mathbf{W}}_{v_y,v_k} \in \mathbb{R}^{V_x}$ denotes the weights of the v_yth convolutional kernel, and $\overline{\mathbf{X}}_{p_x} \in \mathbb{R}^{V_x}$ is the input feature map at spatial position p_x. Hence, we define p_x and p_k as the 2-D spatial positions in the feature maps and convolutional kernels, respectively [74].

The 10th and 12th layer are fully connected with 1024 units whose 50% is randomly selected during training to avoid overfitting. A fully connected layer maps an arbitrary input $\overline{\mathbf{x}} \in \mathbb{R}^{U_x}$ to the output $\overline{\mathbf{y}} \in \mathbb{R}^{U_y}$ by using the weights $\overline{\mathbf{W}} \in \mathbb{R}^{U_x \times U_y}$. Then, the u_y-th element of the output of the layer is the inner product

$$\overline{\mathbf{y}}_{u_y} = \langle \overline{\mathbf{W}}_{u_y}, \overline{\mathbf{x}} \rangle = \sum_i [\overline{\mathbf{W}}]^{\mathrm{T}}_{u_y,i} \overline{\mathbf{x}}_i, \tag{6.17}$$

for $u_y = 1, \ldots, U_y$ and $\overline{\mathbf{W}}_{u_y}$ is the u_y-th column vector of $\overline{\mathbf{W}}$, and $U_x = U_y = 1024$ is selected for $f^{(14)}$.

After each convolutional and fully connected layers (i.e., $f^{(i)}_{i\in\{3,5,7,9,11,13\}}$), there is a rectified linear unit (ReLU) layer where $\mathrm{ReLU}(x) = \max(0, x)$. The ReLU layers are powerful in constructing the nonlinearity of the deep network as well as providing non-negative output at the output layers, which is very useful for classification networks. The 14th layer has a classification layer with $\overline{C}$ units, where a softmax function is used to obtain the probability distribution of the classes. The softmax layer is defined for an arbitrary input $\overline{\mathbf{x}} \in \mathbb{R}^D$ as $\mathrm{softmax}(\overline{x}_i) = \frac{\exp\{\overline{x}_i\}}{\sum_{i=1}^{D} \exp\{\overline{x}_i\}}$. The last layer $f^{(15)}$ is the classification layer.

6.3.3 Numerical Experiments

We now investigate antenna selection performance of various array geometries such as UCAs and RDAs as shown in Fig. 6.4. The classification accuracies of training and validation data for UCA and RDA are listed in Table 6.2. We can see that the learning model achieves high accuracy when M is large and $\mathrm{SNR}_{\mathrm{TEST}} \geq 10\,\mathrm{dB}$.

Figure 6.5 shows the classification performance of the CNN for $J_{\mathrm{TEST}} = 100$ Monte Carlo trials. Figure 6.5 also shows the performance of the noisy test data when the network is trained

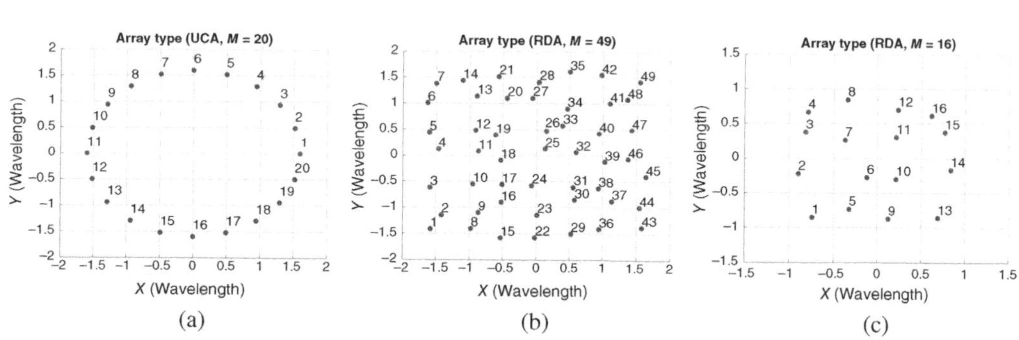

Figure 6.4 Placement of antennas for (a) UCA with $M = 20$ elements, (b) RDA with $M = 49$ and (c) RDA with $M = 16$. Source: [46]/John Wiley & Sons.

Table 6.2 The accuracy percentages for training and validation datasets in 1-D and 2-D scenario

	1-D, UCA with $M = 20, K = 6$		**1-D, UCA with $M = 45, K = 5$**	
SNR_{TRAIN}	Training	Validation	Training	Validation
10 dB	65.2%	68.7%	98.7%	97.8%
15 dB	98.1%	98.5%	99.0%	98.7%
20 dB	99.2%	99.5%	100%	99.7%
25 dB	99.4%	99.8%	100%	100%
30 dB	100%	100%	100%	100%
inf dB	100%	100%	100%	100%
	1-D, RDA with $M = 49, K = 5$		**2-D, RDA with $M = 16, K = 6$**	
SNR_{TRAIN}	Training	Validation	Training	Validation
10 dB	97.8%	95.7%	8.1%	10.7%
15 dB	99.9%	98.6%	60.1%	63.2%
20 dB	97.5%	98.1%	80.8%	80.6%
25 dB	100%	100%	88.9%	89.2%
30 dB	100%	100%	82.6%	83.2%
inf dB	100%	100%	85.0%	83.9%

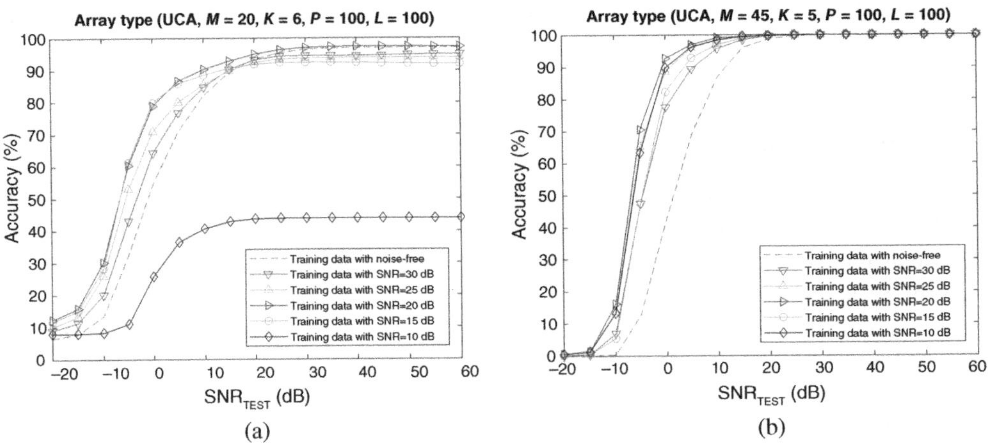

Figure 6.5 Performance of test dataset using CNN with respect to SNR_{TEST} for a UCA with (a) $M = 20$, $K = 6$, and (b) $M = 45$, $K = 5$, respectively. Source: [46]/John Wiley & Sons.

with noise-free dataset; its performance degrades especially at low SNR levels. These observations imply that noisy training datasets should be used for robust classification performance with the test data. On the other hand, when the training data are corrupted with strong noise content (e.g., $SNR_{TRAIN} \leq 10$ dB), then despite using the noisy training data, the CNN model does not recover from poor performance at low SNR_{TEST} regimes. The performance at low SNRs can be improved when the size of the array increases and, as a result, the input data is huge and the SNR is enhanced due to large M. As an example, Fig. 6.5b illustrates the performance of the network for UCA with

$M = 45$ and $K = 5$, where the network provides high accuracy for a wide range of SNR_{TEST} compared to the scenario in Fig. 6.5a.

DoA estimation performance of the DL-based scheme is also presented. The CNN approach is compared with support vector machine (SVM) and random antenna selection (RAS) algorithms. The selected antenna subarrays from CNN and SVM are inserted to the beamforming technique [75] for DoA estimation. As a traditional technique, we consider the RAS algorithm where, instead of all subarray candidates, a number of subarray geometries are realized randomly (i.e., 1000 realizations) and their beamforming spectra is obtained by a search algorithm [76]. We also added the full array performance where $M = K$ for comparison. In Fig. 6.6a, the results are given for a UCA with $M = 20$ and $K = 6$ antennas to be selected. Here, "best subarray" denotes the beamforming performance of the subarray that gives the lowest CRB. It can be seen that CNN provides better performance as compared to SVM (32% more accurate) and RAS (72%) and it approaches the performance of the "best subarray" as expected from the accuracy results given in Fig. 6.5a. SVM performs poorer due to its lower antenna selection accuracy. We present 2-D DoA estimation results in Fig. 6.6b for RDA with $M = 16$ and $K = 6$. Similar observations are obtained for the 2-D case as compared to the 1-D scenario.

We further compare the DoA estimation performance of the selected subarrays with full array ($M = K$) performance in both Fig. 6.6a and b. While there is a gap between subarray and the full array performances, antenna selection provides less computation and cost.

We also analyze the effect of the performance metrics on the antenna selection and DoA estimation accuracy by employing the simplest and most common geometry of a ULA. For creating the training data, we employed three bounds: CRB, BZB, and WWB [66]. The network was trained for $M = 10$, $K = 4$, $L_{\text{TRAIN}} = 100$ snapshots, $T_{\text{TRAIN}} = 100$ signal and noise realizations, and $P_{\text{TRAIN}} = 100$ DoA angles. The number of uniformly spaced azimuthal grid points are set to $P_{\phi} = 100$. For test mode, we fed the network with data corresponding to $P_{\text{TEST}} = 100$ DoA angles different than the ones used in the training phase but keeping the values of M, K, L, and T same as in the training. Fig. 6.7a shows the percentage of times a particular antenna index appears as part of the optimal array in the output over $J_{\text{TEST}} = T_{\text{TEST}}P_{\text{TEST}}$ trials with different performance metrics used during training. As seen here, when the CNN is trained with data created from the CRB, the classifier output arrays usually consists of the elements at the extremities. However, the

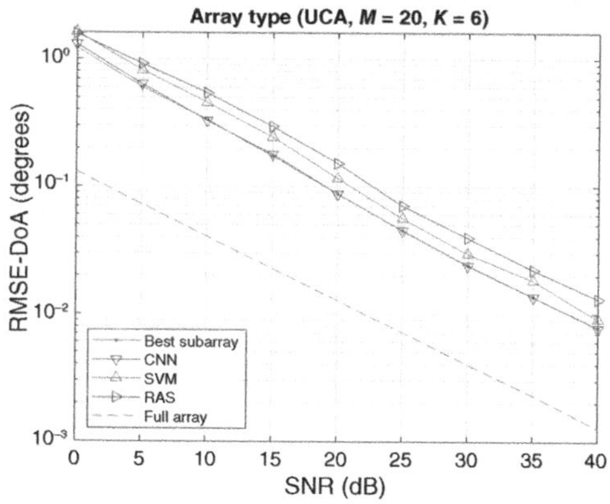

(a)

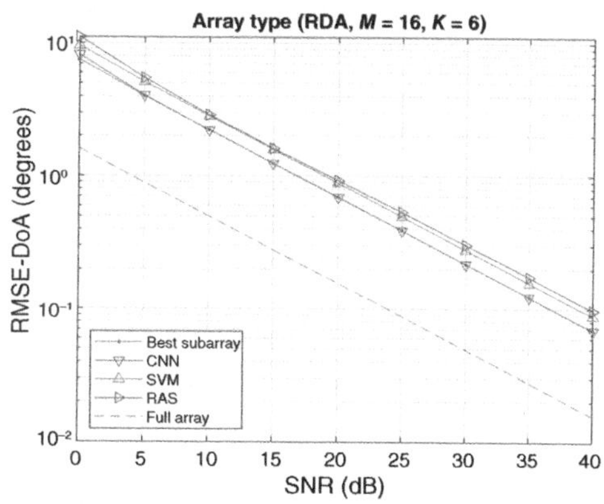

(b)

Figure 6.6 DoA estimation performance with respect to SNR. $\text{SNR}_{\text{TRAIN}} = 20$ dB. The antenna geometry is (a) UCA with $M = 20$ and $K = 6$, (b) RDA with $M = 16$ and $K = 6$. Source: [46]/John Wiley & Sons.

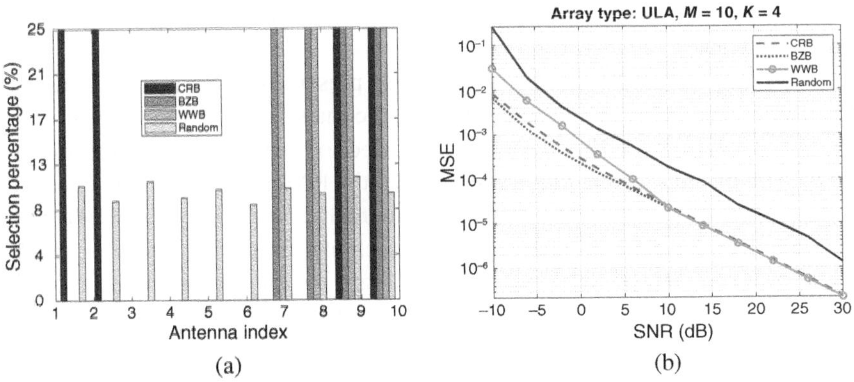

Figure 6.7 (a) Antenna selection percentage over $J_{TEST} = 10,000$ trials. (b) MSE of DoA for selected subarrays. The array geometry is a ULA with $M = 10$ and $K = 4$. Source: [46]/John Wiley & Sons.

network trained on BZB and WWB usually selects arrays with elements close to each other leading to low sidelobe levels. Also shown here is the random selection wherein each element is chosen with approximately 10% selection rate. We provide the DoA estimation performance of the antenna subarrays selected by the network for different values of test data SNRs in Fig. 6.7b. We observe that, compared to the DL-based approach, the random thinning results in inferior DoA estimation due to small array aperture. Among various bounds, the MSE is somewhat similar at high SNR regimes with the BZB faring better than CRB at low SNRs.

6.4 TL for Sparse Arrays

The CNN architecture presented in Section 6.3 is designed for a specific array geometry and is, therefore, inapplicable to different array configurations without significant re-training with new data. This arises from the assumption that the data used for training and testing are drawn from the same or similar distribution, which is difficult to guarantee in real world. In DL, this problem is called *domain mismatch* [77]. On the other hand, labeling sufficient training data for all possible application domains is prohibitive. It has been shown [78] that it is possible to establish a reasonable model by exploiting the labeled data drawn from another sufficiently labeled *source domain* which is closer to describing similar contents of the *target domain*. This *domain adaptation* (DA) enables knowledge transfer across domains. Lately, TL has emerged as an effective domain adaptation technique, wherein the DL network learns domain-invariant models across source and target domains [79], and has been applied to processing of image, bio-medical, radar, and speech signals.

Apart from a domain mismatch problem, networks such as a CNN also suffer from the need of a large training database. When only limited labeled data is available, the CNN fails to optimally select the sensor subarrays. Since CNN objective functions are highly nonconvex and convergence of optimization algorithms to a global optimum is not guaranteed, training with only large data could increase the probability of convergence. Alternatively, training forms such as convolutional autoencoders (CAE) [80] and TL are employed in data-limited applications. Some studies [81] suggest that TL outperforms CAE especially when the sample sizes are very small.

In TL-based antenna selection, we address the domain mismatch between various array geometries and lack of massive training data by developing a more efficient, deep TL-based sensor subarray selection approach. Indeed, sufficient datasets required to support the level of training

CNNs need are unavailable, expensive or impossible to extract in many real-world sensor array applications. We apply TL to enable the network in selecting sensor subarrays accurately even when limited labeled data are available. In particular, we transfer the features in the training data from one array geometry to a different array configuration. For example, we use a CNN trained with a URA to select sensors in a UCA. This domain transfer is advantageous when a large off-line database is required for system identification or calibration [82].

Conventionally, DoA estimation across various geometries is performed with array transformation and array interpolation techniques [83]. However, for multiple targets and complex geometries, these techniques are difficult to come by [84, 85]. TL-based approach is helpful in overcoming such limitations.

When compared with the domain transfer in shallow TL techniques [86], such as classification based on SVM, a deep TL approach combines DA with the power of a deep network to learn the explanatory factors of variations in data and reduce the mismatch between the marginal distributions across array geometries. In Fig. 6.8, we define the source (target) data and labels as $\mathcal{D}_S$ ($\mathcal{D}_T$) and $\mathcal{Y}_S$ ($\mathcal{Y}_T$), respectively. We train the source network CNN_S, which learns the nonlinear relationship between $\mathcal{D}_S$ and $\mathcal{Y}_S$ as

$$\mathcal{Y}_S = \Sigma_S(\mathcal{D}_S), \tag{6.18}$$

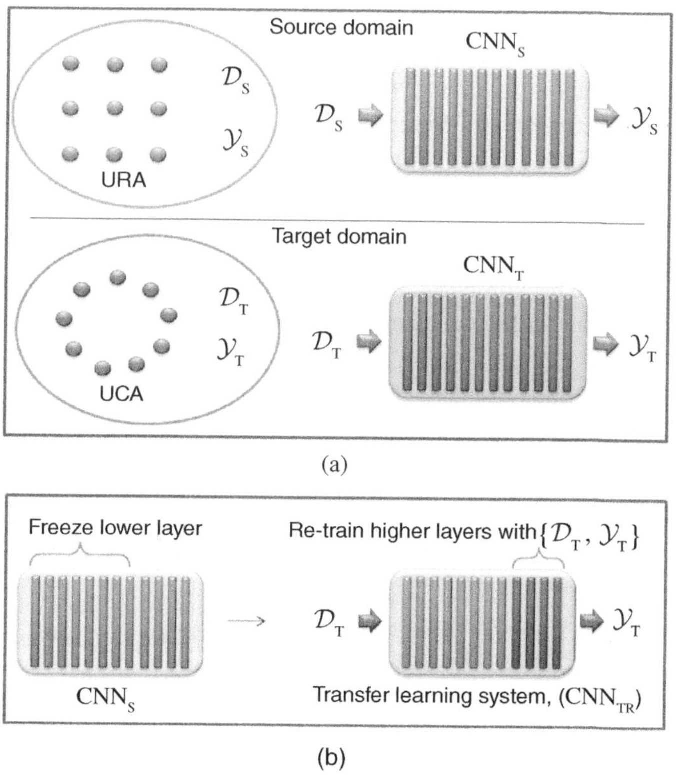

Figure 6.8 The representation of the source and target domains for knowledge transfer from, for example, URA to UCA configuration. (a) Source (top) and target (bottom) domain data with corresponding learning networks CNN_S and CNN_T, respectively. (b) In deep TL, lower layers of CNN_S are frozen and only higher layers are re-trained with the target domain (UCA) data $\{\mathcal{D}_T, \mathcal{Y}_T\}$ to transfer sensor selection knowledge from source domain (URA). Source: [72]/IEEE.

where $\Sigma_S(\cdot)$ is the nonlinear function that constructs the mapping between the data and labels in the source domain. In (6.18), the label data are the positions of the best subarray sensors as $\mathcal{Y}_S = \{\mathcal{Y}_S^{(1)}, \mathcal{Y}_S^{(2)}, \dots, \mathcal{Y}_S^{(S)}\}$ where $\mathsf{S} = |\mathcal{Y}_S| = |\mathcal{D}_S|$ is the number of elements in the source domain. Furthermore, $\mathcal{D}_S$ is the collection of covariance matrices of the array outputs of the source array geometry, i.e.,

$$\mathcal{D}_S = \{\mathcal{D}_S^{(1)}, \mathcal{D}_S^{(2)}, \dots, \mathcal{D}_S^{(S)}\}, \tag{6.19}$$

where $\mathcal{D}_S^{(i)} = \mathbf{X}_S$ which is constructed from the source domain covariance matrix

$$\mathbf{R}_S = \frac{1}{T}\sum_{i=1}^{T} \mathbf{y}_S(t_i)\mathbf{y}_S^H(t_i), \tag{6.20}$$

where $\mathbf{y}_S(t_i)$ denotes the array output of the source data.

Similarly, the target domain data and labels are $\mathcal{D}_T = \{\mathcal{D}_T^{(1)}, \mathcal{D}_T^{(2)}, \dots, \mathcal{D}_T^{(T)}\}$ and $\mathcal{Y}_T = \{\mathcal{Y}_T^{(1)}, \mathcal{Y}_T^{(2)}, \dots, \mathcal{Y}_T^{(T)}\}$, respectively, where $\mathsf{T} = |\mathcal{D}_T|$ and $\mathbf{R}_T = \frac{1}{T}\sum_{i=1}^{T} \mathbf{y}_T(t_i)\mathbf{y}_T^H(t_i)$. For the target network CNN_T, we have

$$\mathcal{Y}_T = \Sigma_T(\mathcal{D}_T). \tag{6.21}$$

The TL framework assumes that the source domain has much larger dataset than the target domain, i.e., $\mathsf{S} \gg \mathsf{T}$. This implies that CNN_S will turn out to be a well-trained deep network whereas CNN_T has poor mapping performance and does not reflect the same mapping profile as CNN_S. To improve the performance of CNN_T, the key idea is to use the sensor selection ability of the pre-trained network CNN_S even if it is trained with different array data [87]. This is achieved by re-training CNN_S with the target domain data $\{\mathcal{D}_T, \mathcal{Y}_T\}$ while freezing the lower layers (i.e., convolutional layers) of CNN_S.[1] The new deep transfer network is CNN_{TR} (Fig. 6.8b). The lower layers are kept intact or *frozen* because they are generally domain invariant[2] and hence, harbor the bulk of sensor selection knowledge. The higher layers, however, are largely domain variant such that when new labels are added to the problem (i.e., $\mathcal{Y}_S$ is replaced with $\mathcal{Y}_T$), they require re-training. This approach accelerates the computation of the gradient in the backpropagation stage. Furthermore, it allows us to enlarge the feature space of the deep network without causing large error on the already-learned features [87].

6.4.1 Knowledge Transfer Across Different Array Geometries

Once CNN_S (i.e., $\Sigma_S(\cdot)$) is trained with the source domain data, we freeze the weights in the $\{2, 4, 6, 8\}$th layers (i.e., the convolutional layers) to preserve the sensor selection ability of the deep network before transferring it to the target domain. We construct the TL network such that

$$\Sigma_{TR}(\mathcal{D}_T) = f^{(15)}\left(f^{(14)}(\dots \tilde{f}^{(2)}(f^{(1)}(\mathcal{D}_T)))\right) = \mathcal{Y}_T, \tag{6.22}$$

where the frozen layers are $\tilde{f}^{(i)}_{i\in\{2,4,6,8\}}$. Algorithm 2 lists these steps of the TL approach.

1 We do not freeze the layers $\{3, 5, 7\}$, because they are ReLU layers with no weight to freeze.
2 "Domain invariance" implies that when new labels are added to the network, the lower layers remain unaffected even though the problem has changed.

Algorithm 6.2 Transfer learning for sensor selection

Input: $\{\mathcal{D}_{\mathrm{S}}, \mathcal{Y}_{\mathrm{S}}\}, \{\mathcal{D}_{\mathrm{T}}, \mathcal{Y}_{\mathrm{T}}\}$

Output: $\mathrm{CNN}_{\mathrm{TR}}$.

1: Train $\mathrm{CNN}_{\mathrm{S}}$ with $\{\mathcal{D}_{\mathrm{S}}, \mathcal{Y}_{\mathrm{S}}\}$.

2: Construct TL network $\mathrm{CNN}_{\mathrm{TR}}$ whose convolutional layers are designated the same as of $\mathrm{CNN}_{\mathrm{S}}$, i.e., $f_{\mathrm{TR}}{}^{(i)}_{i\in\{2,4,6,8\}} = f_{\mathrm{S}}{}^{(i)}_{i\in\{2,4,6,8\}}$.

3: Train the remaining layers of the TL network with $\{\mathcal{D}_{\mathrm{T}}, \mathcal{Y}_{\mathrm{T}}\}$. Then, use $\mathrm{CNN}_{\mathrm{TR}}$ for sensor selection for target domain data.

6.4.2 Deep Network Realization and Training

For training, we used stochastic gradient descent algorithm with momentum 0.9 and updated the network parameters at learning rate 0.01 and mini-batch size of 512. The loss function was the cross-entropy cost

$$\mathrm{C_E} = -\frac{1}{\overline{\mathsf{T}}} \sum_{t=1}^{\overline{\mathsf{T}}} \sum_{c=1}^{\overline{C}} \left[\chi_c^{(t)} \ln \eta_c^{(t)} + (1 - \chi_c^{(t)}) \ln(1 - \eta_c^{(t)}) \right], \tag{6.23}$$

where $\overline{\mathsf{T}}$ is the length of the dataset and $\{\eta_c^{(t)}, \chi_c^{(t)}\}_{t=1,c=1}^{\overline{\mathsf{T}},\overline{C}}$ is the input–output pair for the classification layer. It is worth noting that the cost function in (6.23) can be defined in terms of RMSE of DoA estimation procedure. However, this makes the training process problem-dependent. During training, the training data is shuffled for each epoch until training is terminated. Further, 80% and 20% of all generated data are chosen for training and validation datasets, respectively. The training rate is reduced by a factor of 0.9 after each 10 epochs. The training stops when the validation accuracy does not improve for three consecutive epochs.

To train $\mathrm{CNN}_{\mathrm{S}}$, we collected array data for $P_{\mathrm{S}} = 100$ equally spaced direction in the sector $\tilde{\Theta} = [0°, 359°]$ azimuth plane and $L_{\mathrm{S}} = 100$ noisy data realizations with $T = 100$ data snapshots. During training, we set $\sigma_s^2 = 1$ and use different SNR levels, namely, $\mathrm{SNR_{TRAIN}} \in \{15, 20, 25\}$ dB. Hence, the total training data length is $3L_{\mathrm{S}}P_{\mathrm{S}} = 30,000$. Once $\mathrm{CNN}_{\mathrm{S}}$ is trained, the $\mathrm{CNN}_{\mathrm{TR}}$ is constructed by following the steps in Algorithm 2. For the above-mentioned settings with $M = 16$ and $K = 6$, the training time for $\mathrm{CNN}_{\mathrm{S}}$, $\mathrm{CNN}_{\mathrm{T}}$ are approximately 40 and 5 minutes respectively, whereas the TL network $\mathrm{CNN}_{\mathrm{TR}}$ needs only 5 seconds to be trained.

6.4.3 Performance in Source Domain

We first present the performance of the CNN approach for the source domain case where different array geometries such as URA and UCA are considered with different array settings. In particular, we consider sensor arrays with half wavelength sensor spacing for both UCA and URA. When $\mathrm{CNN}_{\mathrm{S}}$s are trained for different arrays, we obtained above 90% validation accuracy for the training data in all cases. In the prediction stage, the DoA angles are generated uniformly at random in the interval $\tilde{\Theta}$ so that the DoA angles in the training and prediction are selected from the same distribution. After feeding $\mathrm{CNN}_{\mathrm{S}}$ with these input data, the selected subarrays are obtained from the output for each scenario. Then, the sensor outputs of corresponding subarrays are employed for DoA estimation using MUSIC (MUltiple SIgnal Classification) algorithm [88]. During the simulations in

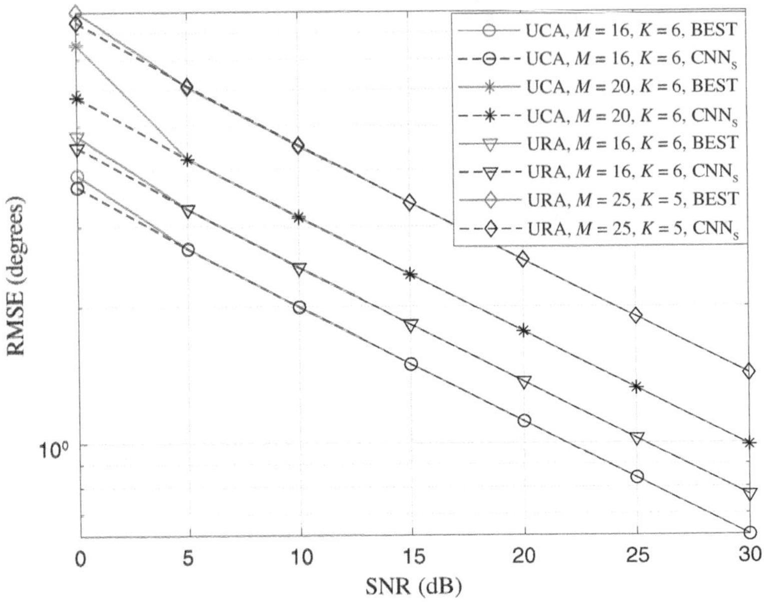

Figure 6.9 DoA estimation performance for CNN_S for different array geometries. Source: [72]/IEEE.

the prediction state, the network is tested for different SNR levels for $J_T = 100$ Monte Carlo trials. Figure 6.9 shows the RMSE in DoA estimation, i.e.,

$$\text{RMSE} = \left(\frac{1}{J_T}\sum_{j=1}^{J_T}(\hat{\phi}^{(j)} - \phi)^2\right)^{\frac{1}{2}}, \tag{6.24}$$

where $\hat{\phi}^{(j)}$ and ϕ denote the estimated and true DoA angles, respectively. We compare the DoA estimation performance of CNN_S with the best subarray that provides the lowest CRB. Figure 6.9 demonstrates that CNN_S asymptotically follows the best subarray performance.

6.4.4 Performance for TL

In order to evaluate the TL performance, we trained CNN_S with different sizes of datasets and then constructed CNN_{TR} from CNN_S for sensor selection. We considered URA and UCA geometries with $M = 16$, $K = 6$ for source and target domains, respectively. Figure 6.10 shows the sensor selection accuracy

$$\text{Accuracy}(\%) = \frac{\mathsf{U}}{\mathsf{V}} \times 100, \tag{6.25}$$

where V is the total number of input datasets in which the model identified the best subarrays correctly U times. In Fig. 6.10a, the target domain $\mathcal{D}_T$ are generated for $P_T = 10$ grid points in $\tilde{\Theta}$ and $L_T = 10$ and we varied P_S from 5 to 150 for $\mathcal{D}_S$ with $L_S = 100$. For all three networks, the performance of CNN_T is fixed because $\mathcal{D}_T$ does not change during the simulations. When P_S is very small (i.e., < 10), CNN_{TR} performs even worse than CNN_T. However, as P_S increases, CNN_{TR} and CNN_S exhibit higher selection accuracy. For large source datasets, e.g., $P_S \in [80, 120]$, CNN_{TR} outperforms CNN_T by a large margin because of the learned and transferred features from CNN_S. The increase in P_S does not necessarily improve the sensor selection performance because when the training data

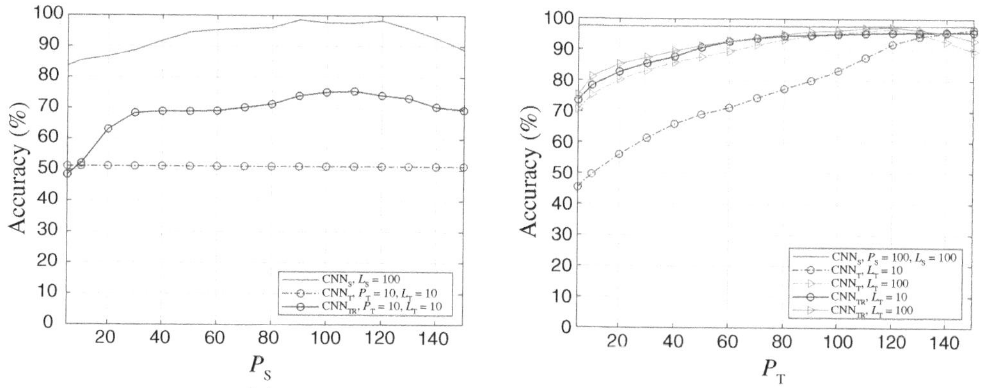

(a) (b)

Figure 6.10 Performance of CNN$_S$, CNN$_T$ and CNN$_{TR}$ versus the number of DoA angles. Sensor selection accuracy is given with respect to (a) P_S when $L_S = 100$, $L_T = 10$, and (b) P_T when $P_S = 100$, $L_S = 100$. SNR$_{TRAIN}$ = 15 dB. Source: [72]/IEEE.

are densely sampled (i.e., P_S is high) the deep network cannot distinguish the input data of different directions and produce inaccurate classification output. These results suggest that CNN$_S$ needs to be trained with at least $S = P_S L_S = 40 \cdot 100 = 4000$ to provide satisfactory accuracy (e.g., above 90%). As a result, $P_S = 100$ is a reasonable choice for TL, wherein the target dataset 1000 times smaller, i.e., $\frac{S}{T} = \frac{L_S P_S}{L_T P_T} = 1000$. In Fig. 6.10b, we repeat the same analysis for CNN$_T$ where we assume that CNN$_S$ is well-trained with $P_S = 100$ and $L_S = 100$. Then, we sweep P_T similarly for both $L_T = 10$ and $L_T = 100$. We can see that when $L_T = 100$, CNN$_T$ quickly reaches maximum similar to CNN$_S$ as illustrated in Fig. 6.10a. In this case, the improvement gained by TL is incremental because CNN$_T$ is already well-trained. However, if small dataset is used, i.e., $L_T = 10$, then it requires larger P_T to reach high accuracy. Expectedly, this analysis shows that TL provides reasonable improvement if the target dataset is relatively small, i.e., $T = P_T L_T \leq 1000$ ($S = 10,000$). In other words, when T is high there is no need to use TL. Therefore, in the following experiments, we select $P_T = L_T = 10$ and employ TL to improve the performance.

Table 6.3 lists the validation accuracy of CNN$_T$ and CNN$_{TR}$ for different TL scenarios. We consider TL between UCA and URA as well as the perturbed array geometries denoted by $\overline{\text{UCA}}$ and $\overline{\text{URA}}$. In a perturbed array geometry, the mth sensor position is selected uniformly at random as $\{\tilde{x}_m, \tilde{y}_m, \tilde{z}_m\} \sim \mathcal{N}(\{x_m, y_m, z_m\}, (\lambda/4)^2)$ for each instance of the training data. It is evident that the sensor selection accuracy of CNN$_{TR}$ is approximately 20% higher than CNN$_T$.

Table 6.3 Training validation accuracy (%) for different TL scenarios.

	Validation accuracy (%)	
TL scenario (Source → Target)	**CNN$_T$**	**CNN$_{TR}$**
UCA → URA, $M = 16, K = 6$	54.9	70.1
URA → UCA, $M = 16, K = 6$	42.3	79.8
UCA → $\overline{\text{UCA}}$, $M = 20, K = 6$	63.1	98.8
URA → $\overline{\text{URA}}$, $M = 25, K = 5$	55.2	77.4

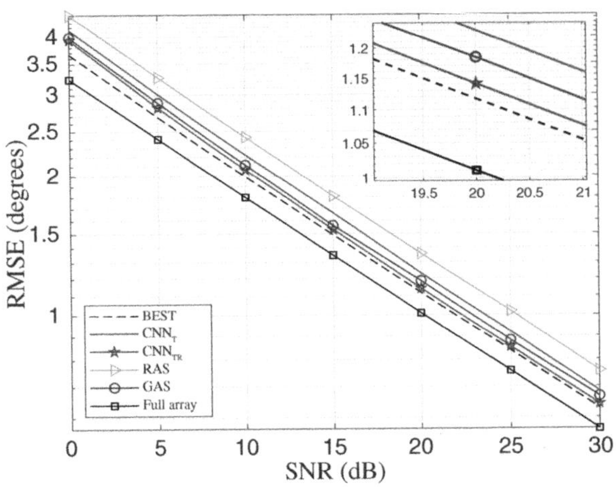

Figure 6.11 DoA estimation performance when source domain has URA; target domain has UCA geometry when $M = 16$, $K = 6$. Source: [72]/IEEE.

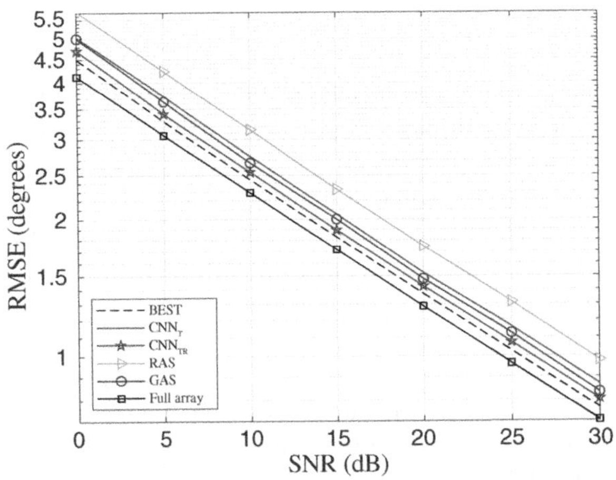

Figure 6.12 DoA estimation performance when source domain has UCA; target domain has URA geometry when $M = 16$, $K = 6$. [72]/IEEE.

We further assessed the DoA estimation performance of the selected subarrays for target domain data. For $M = 16$, $K = 6$, Figs. 6.11 and 6.12 depict the performance for URA→UCA and UCA→URA scenarios, respectively. We compared the sensor selection performance of CNN$_T$ and CNN$_{TR}$ with greedy-based antenna selection (GAS) [44], random selection (RAS) as well as the fully array performance. As expected, we see that the full array has the lowest SNR due to large array aperture. We observe that CNN$_{TR}$ closely follows the performance of the best subarray. The CNN$_{TR}$ exhibits approximately 4%, 8.5% and 23% lower RMSE as compared to GAS, CNN$_T$ and RAS, respectively. It is worth noting that RAS has no rule on selecting the antennas while GAS is a greedy-based suboptimum method seeking the best subarray based on the CRB information [44]. These results establish the effectiveness of TL for DoA estimation with sensor selection. The superior performance of CNN$_{TR}$ is because of the learned and transferred features from source domain data via CNN$_S$.

6.5 Large Planar Sparse Array Design with SA-Assisted DL

The SA-assisted DL-based approach was originally presented in [89, 90] to optimize sensor positions while minimizing the mutual coupling between the sensors. It was extended to 2-D sparse array design in [91]. Here, given an initial array as $\mathcal{Z}_{\text{init}}$, the optimized array $\mathcal{Z}_{\text{SA}}$ is obtained by minimizing

$$\mathcal{Z}_{\text{SA}} = \arg\min_{\mathcal{Z}_{\text{init}}} k_o = \sum_{i=1}^{M}\sum_{j=i+1}^{M} \frac{1}{||m_i - m_j||_2}, \tag{6.26}$$

$$\text{such that} \quad ||m_i - m_j||_2 \leqslant B,$$

where $m_i, m_j \in \mathcal{Z}_{\text{init}}$, M is the number of sensors, $||\cdot||_2$ is the l_2-norm of a vector and B is the mutual coupling coefficient upper-bound. The SA algorithm is used as an initialization step for the DL model, mainly to generate sparse subarrays with large physical apertures and well-distributed sensors instead of an enumeration approach.

Figure 6.13 summarizes the step-by-step procedure of the flow diagram. Note that this SA-based approach can be extended to any planar array configuration, and the same applies to the whole SA-based initialization method [91–98]. The best subarrays are selected from $\mathcal{H}_{\text{sa}}$ with the lowest CRB values as labels.

The whole algorithm is summarized in the Algorithm 3. The inputs are as follows: the total number of given antennas M, the number of antennas to be selected K, the number of snapshots T, the number of different DOA angles D, the number of signals and noise realizations P and the SNR. Moreover, the elements of $\mathcal{V}$ are chosen from $\mathcal{H}_{\text{sa}}$ which is calculated using the SA-based optimization method as shown in Fig. 6.13 rather than the enumeration of the entire combinations like in $\mathcal{H}$. The SNR used for calculation of the covariance matrices is denoted as $\text{SNR}_{\text{TRAIN}}$.

The performance of the SA-assisted DL-based antenna selection approach is evaluated through simulation. The problem of selecting $K = 16$ sensors out of a $N = 42 = 6 \times 7$ sensor URA is considered. The details of training and testing specifications can be found in [91].

The estimation accuracy of the SA-assisted DL-based method is evaluated in comparison to the conventional approach using the CNN models trained. Figure 6.14 shows the array configuration of predicted 2-D sparse arrays. Particularly, Fig. 6.14a illustrates the parent 42-element URA whereas Fig. 6.14b–d show the realized 16-element DL-based 2D sparse array using the conventional and DL-based method respectively.

Figure 6.13 Improved SA algorithm for random 2-D sparse subarray generation and optimization.

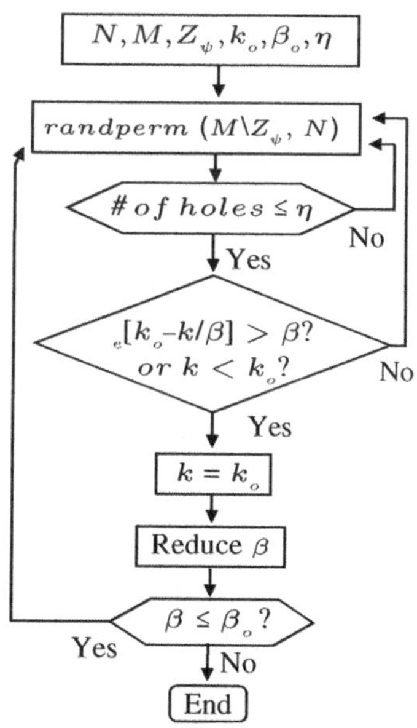

Algorithm 6.3 Training dataset generation method

Input: N, M, T, K, P, and $\mathrm{SNR}_{\mathrm{TRAIN}}$

Output: Training data $\mathcal{D}_{\mathrm{TRAIN}}$

Generate $\mathcal{H}_{\mathrm{sa}}$ as shown in Fig. 6.13.

Sample D DOA angles $\{\Theta_d\}_{d=1}^{D}$.

Compute P different realizations of subarray output, $\{X_d^i\}_{i=1}^{P}$ for $d = 1, \ldots, d$

$X_d^i = [\mathbf{x}_d^i(1), \mathbf{x}_d^i(2), \ldots, \mathbf{x}_d^i(T)]$,

where $\mathbf{x}_d^i(t) = a(d)s^{(i)}(t) + n^{(i)}(t)$,

$s^{(i)}(t) \sim \mathcal{CN}(0, \sigma_s^2 I)$ and $n^{(i)}(t) \sim \mathcal{CN}(0, \sigma_n^2)$

Calculate sample covariance matrix $\hat{R}$ and $N \times N$ covariance matrices $R_{h_s}^{(i,d)}$ for $h_s = 1, 2, \ldots, \bar{\mathcal{H}}_{\mathrm{sa}}$.

Compute $C\left(\Theta, \mathcal{Z}_{h_s}\right)$ for all $h_s \in \mathcal{H}_{\mathrm{sa}}$ and select subarrays.

Create input-output data pairs as $(\hat{R}^{(i,d)}, u_d^{(i)})$ for $d = 1, 2, \ldots, D$ and for $i = 1, 2, \ldots, P$.

Connect the input-output pairs to form the training dataset as

$\mathcal{D}_{\mathrm{TRAIN}} = \Big[(\hat{R}^{(1,1)}, u_1^{(1)}), (\hat{R}^{(2,1)}, u_1^{(2)}), \ldots$

$, (\hat{R}^{(P,1)}, u_1^{(P)}), (\hat{R}^{(1,2)}, u_2^{(1)}), \ldots, (R^{(P,D)}, u_D^{P})\Big]$

where the size of the training dataset is $\mathcal{R} = PD$.

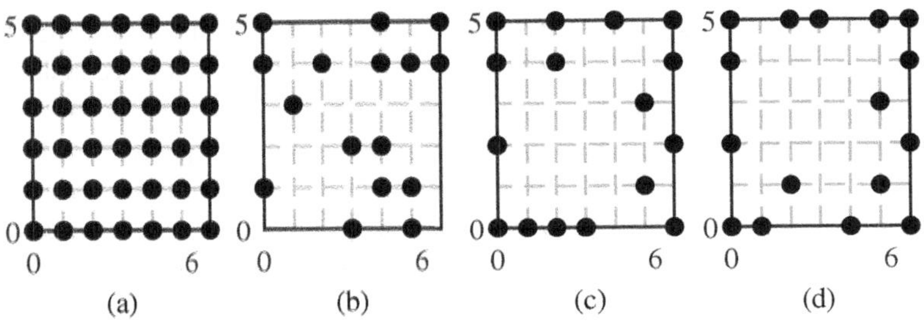

Figure 6.14 The array configurations of (a) parent 42-sensor URA, (b) the conventional DL-based array (16-sensors), (c) the conventional SA-based array (16sensors) and (d) the DL-based array (16sensors). Note that the dots denote physical sensors.

Here, the performance of the realized 2-D sparse arrays is evaluated. In particular, we explore the behavior of the RMSE as a function of SNR and the number of snapshots. The RMSE is defined as

$$\mathrm{RMSE} = \sqrt{\frac{1}{\mathcal{T}D}\sum_{i=1}^{\mathcal{T}}\sum_{d=1}^{D}\left[(\tilde{\phi}_d^i - \phi_d) + (\tilde{\theta}_d^i - \theta_d)\right]^2}, \tag{6.27}$$

where $\tilde{\phi}_d^i, \tilde{\theta}_d^i$ and ϕ_d, θ_d denote the estimated and true d-th DOAs in the ith trial, respectively.

Table 6.4 lists the parameters used to compute RMSE with respect to SNR (Example #1) and the number of snapshots (Example #2), respectively. Moreover, the 2-D-ESPRIT algorithm is used to estimate the sources. However, if the DCA of the realized 2-D arrays has holes in the coarray, the resulting virtual 2-D array becomes irregular. Therefore, the spatial-smoothing DOA estimation method such as 2-D-ESPRIT cannot be applied. This is the case as the 2-D-ESPRIT algorithm requires a URA array structure for spatial smoothing pre-processes. As a result, a nuclear norm minimization (NNM) approach [92] is applied to fill the holes to restore a standard 2-D configuration.

Table 6.4 2-D DOA estimation simulation parameters.

Parameters	Simulation examples	
	Example #1	**Example #2**
# of sources D	9	
# of trials, $\mathcal{T}$	1000	
# of snapshots	500	10 to 1000
SNR [dB]	−20 to 10	0
DOAs (θ, ϕ) [deg.]	(10, 255), (45, 300), (20, 345), (50, 210), (70, 0), (0, 30), (60, 165), (5, 120), (30, 79)	
Estimator	2-D-ESPRIT	

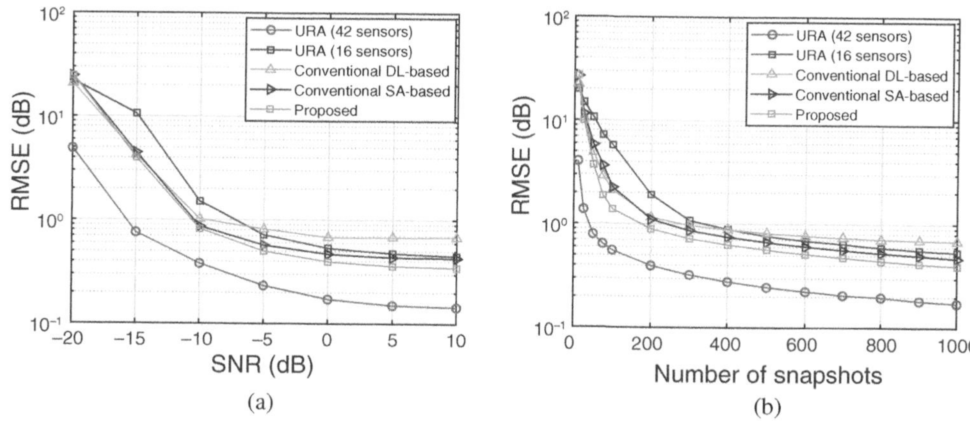

Figure 6.15 RMSE performance versus (a) SNR and (b) number of snapshots for the realized DL-based sparse array compared to URA, conventional DL-based array, and SA-based array.

Figure 6.15 shows the DOA estimation performance of the realized DL-based 2-D sparse array compared to the parent URA and other 2-D sparse arrays. It can be observed in Fig. 6.15a that the URA with 42 sensors has better performance overall due to the large physical aperture. In contrast, the performance of the DL-based 2-D sparse array realized using the DL-based method has better than a URA with 16 sensors and slightly lower than that of the parent URA. Besides, the 2-D sparse array performs better than the conventional SA-based sparse array. Moreover, the conventional DL-based performed poorly as compared to both the DL-based array and 16-sensor URA.

In Fig. 6.15b, we can observe a similar trend where the performance of the DL-based array is bounded by the parent URA and the 16-sensor URA as the number of snapshots increases. The parent URA has the lowest RMSE values, whereas the 16-sensor URA has higher RMSE values than the DL-based array. Also, the conventional DL-based array performs better than the 16-sensor URA in lower numbers of snapshots. However, the RMSE values degrade as the number of snapshots increases. These examples demonstrate that the DL method can thin 2-D arrays cognitively to a manageable size without considerable loss of DOA estimation resolution.

6.6 DL-Based Sparse Array Design for Hybrid Beamforming

DL may also be used to simultaneously select antenna elements and design hybrid beamformers for massive MIMO wireless communications [18]. Here, we employ a CNN for each task. As before, the element selection problem is cast as classification [93]. The hybrid beamformer design is included in this DL framework by exploiting the structure of analog beamformers which are obtained by minimizing the cost between hybrid and unconstrained beamformers. The optimization problem is cast jointly with the antenna selection problem and it is solved by MATLAB-based Manopt algorithm [94] via manifold optimization (MO) [95].

Among the prevalent architectures for sparse array beamformers, a popular scheme is to employ a predetermined subarray with N_{RS} antennas selected from a full array of N_R elements (Fig. 6.16a). Each subarray feeds into a fullyconnected phase shifter network of size N_R^{RF} with a single RF chain. This has the complexity of phase shifters but the antenna selection process is not optimized. Another common receiver architecture feeds the antennas directly to the RF chains thereby eliminating the phase shifters completely. Here, each RF chain is connected to the N_R antennas of which N_{RS} elements are selected using switches (Fig. 6.16b). In this case, the entries of the combiner matrix are either 1 or 0 to indicate the selected or unselected antennas, respectively. This is the simplest structure with no phase shifters. However, the antenna selection is not optimized and the elements are determined by simply choosing the largest absolute values in each column of the channel matrix. Finally, Fig. 6.16c shows a receiver that employs a switching network with phase shifters. In this system, a subarray with N_{RS} antennas is selected from a full array comprising N_R antennas. The subarray is connected to a phase shifter network of size N_R^{RF} which may apply an optimization procedure for antenna selection to achieve greater efficiency.

In this formulation, a CNN accepts channel matrix as input and provides the subarray that maximizes the spectral efficiency (SE). Once the antenna selection (CNN_{AS}) is finalized (see Fig. 6.17), the corresponding partial channel matrix is fed to a second CNN (CNN_{RF}) which then chooses the best RF beamformer and constructs the corresponding baseband beamformer. To train both CNN models, different realizations of the channel matrix are used and the input data are labeled by the selected subarray/RF chains with the highest (SE).

. Even though DL network structures require channel matrix as an input, precise knowledge of this matrix is only required in the training stage to obtain the labels of the network. In the prediction

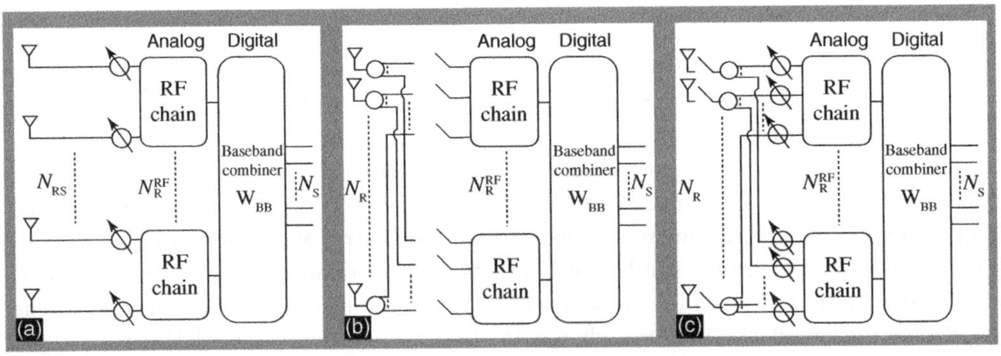

Figure 6.16 Receiver architectures with antenna selection for single user mmWave MIMO systems. (a) Scheme 1: Fixed subarray with fully-connected phase shifters. (b) Scheme 2: Switching network without optimized antenna selection and no phase shifters. (c) Scheme 3: Switching network with optimized antenna selection and phase shifters. Source: [18]/IEEE.

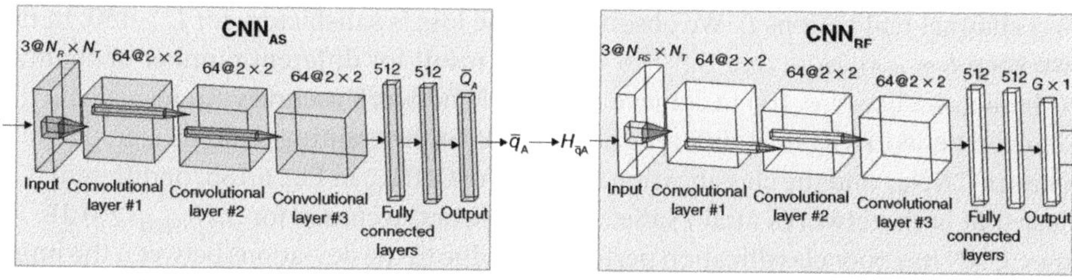

Figure 6.17 The CNN architecture for antenna selection and RF beamformer design. Source: [18]/IEEE.

stage, where the RF beamformers are estimated, precise channel knowledge is not necessary. Both CNNs are trained with channel matrices generated for different user location, channel gains and number of user clusters. Furthermore, each realization of channel matrix in the training data is corrupted by synthetic noise so that the performance of the learning network does not deteriorate with noisy test inputs.

Figure 6.18 summarizes the assessment of the performance of unquantized CNN. Here, we set $N = 100$, $N_\mathrm{T} = N_\mathrm{R} = 64$, and $N_\mathrm{S} = 4$. Figure 6.18a shows the validation loss of $\mathrm{CNN_{RF}}$ against the

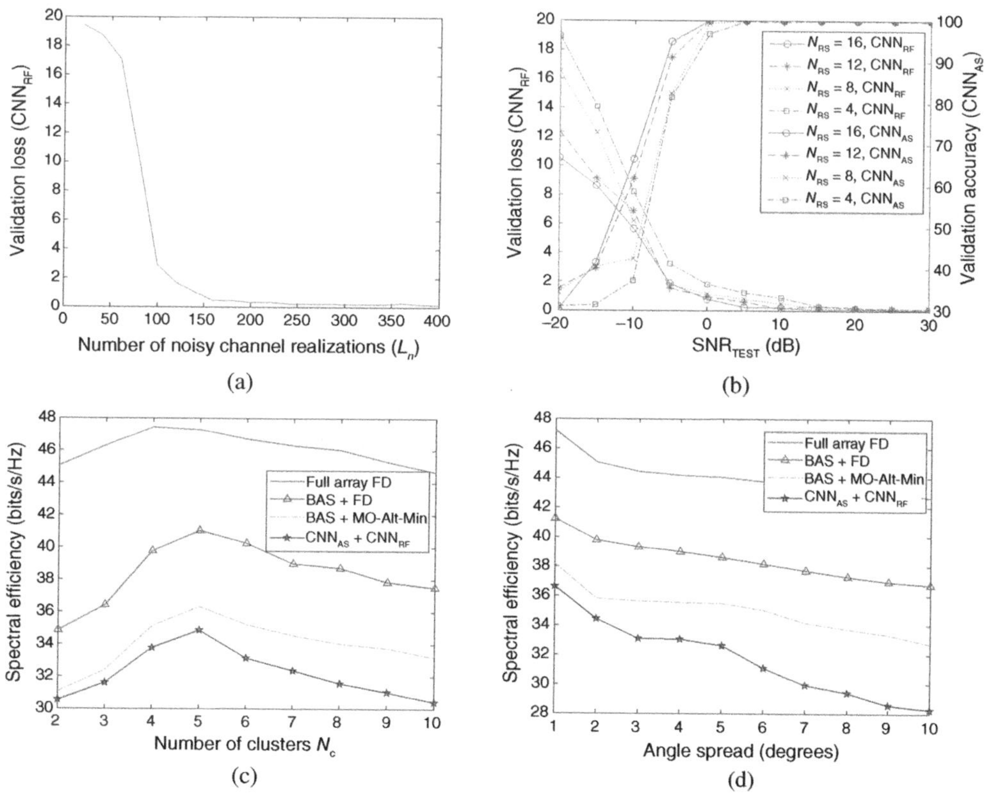

Figure 6.18 Performance of CNN. (a) Validation loss versus number of channel realizations L_n. (b) Validation loss for $\mathrm{CNN_{RF}}$ and validation accuracy for $\mathrm{CNN_{AS}}$. (c) Spectral efficiency versus number of clusters in when the angle spread is $\sigma_\Theta^2 = 5°$. (d) Spectral efficiency versus angle spread when the number of paths $N_c = 4$. We set the number of antennas as $N_\mathrm{T} = N_\mathrm{R} = 64$ with $N_\mathrm{RS} = 16$ selected antennas. Source: [18]/IEEE.

number of noisy channel realizations L. We observe that the loss is satisfactory for $L \geq 150$; in the simulations, we keep $L = 200$. Note that this is a common result for different number of channel realizations N. In fact, we use $N = 100$ to achieve reasonable network accuracy. To investigate the performance of deep networks against different noise levels in the training data, we demonstrate the validation loss of CNN_{RF} and the classification accuracy of CNN_{AS} in Fig. 6.18b for different N_{RS} values. It is clear that both networks attain satisfactory network accuracy for $\text{SNR}_{\text{TEST}} \geq 0$ dB. At low SNR regimes, CNN has poor classification performance due to the deviations between the input and the channel matrices used in the training data. In order to make CNN more robust to noisy inputs, we draw the training data for multiple $\text{SNR}_{\text{TRAIN}}$ levels. Nevertheless, noise in the training data expectedly limits the performance since the network cannot distinguish the input data if it is corrupted too much. This issue is also reported in [46, 96] for the multiple $\text{SNR}_{\text{TRAIN}}$ case.

The channel statistics are important parameters that change in very short time in mm-Wave channels. Hence, in the training stage, we feed the network with channel realizations of different N_c values. To further investigate the performance with respect to different channel statistics, we compare the algorithms in Fig. 6.18c and d for different N_c and σ_{Θ}^2, respectively. The CNN framework provides robust performance against different channel statistics. The effectiveness of the DL techniques can be attributed to training the network with several channel statistics and adding synthetic noise for multiple $\text{SNR}_{\text{TRAIN}}$ values. As a result, the DL network provides robust performance against the changes in channel statistics without a need to be re-trained. However, when there is a change in the *full array* system parameters such as N_T, N_R, N_{RS}, and N_S, the network does need to be re-trained because these parameters directly dictates the dimensions of the network input and output layers.

6.7 Deep Sparse Arrays for ISAC

To jointly access and manage the limited electromagnetic spectrum, ISAC has been envisaged as a technology that will use the same transmit/receive hardware for both radar and communications [54, 97]. In ISAC (Fig. 6.19), the aim is to select the subarray that provides both the best radar sensing and communications performance. In particular, the hybrid beamformers corresponding to the selected subarray is desgned while maximizing the beamforming performance toward both radar targets and communications users. Due to the combinatorial nature of the antenna selection

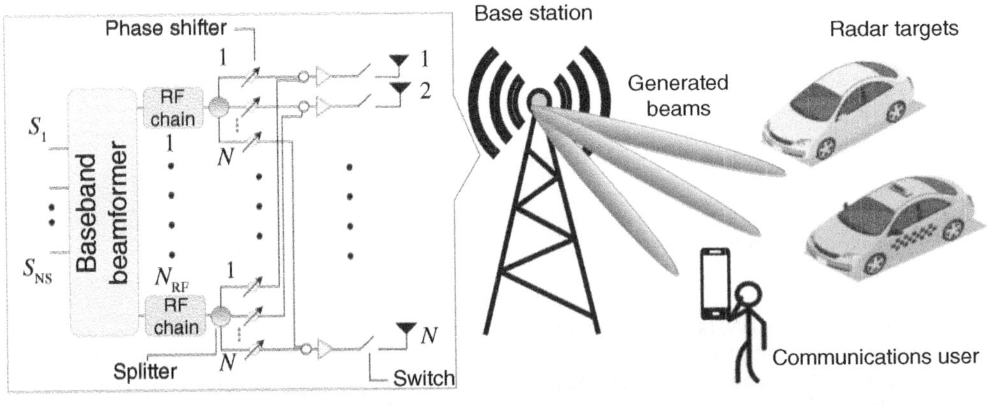

Figure 6.19 ISAC hybrid beamforming scenario with antenna selection.

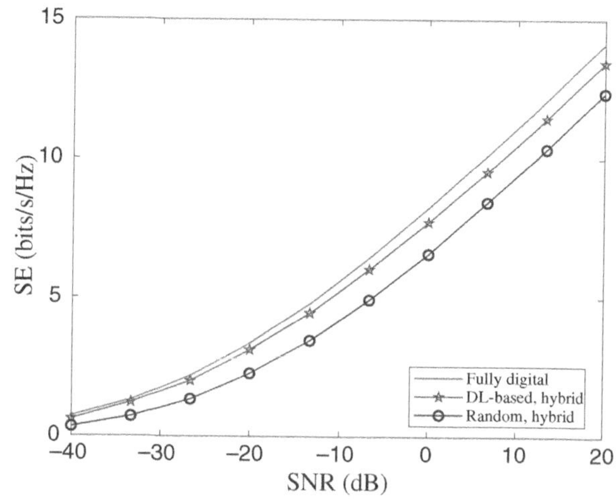

Figure 6.20 ISAC hybrid beamforming performance with antenna selection.

problem, the computational complexity of the joint (antenna selection and beamforming) problem is high. To combat this challenging problem, a low-complexity approach is devised, wherein the consecutive antennas in the whole array are partitioned into small groups so that the number of possible subarray configurations can be reduced. An optimization problem is formulated for jointly selecting the *best* subarray and design the hybrid analog and digital beamformers for the ISAC scenario. A DL-based approach can be applied to further lower the computational burden of the design problem. In this case, the input of the CNN is the combination of both communications (channel matrix) and sensing (radar-only beamformer) related data. The output of the learning model is the combination of the digital beamformer and the sparse analog beamformer.

Figure 6.20 shows the SE of the joint antenna selection and beamformer design in the ISAC scenario. DL-based CNN achieves a performance very close to the fully digital ISAC beamformer [97], which is computed in accordance with the selected sparse array by the CNN. In comparison, the random selection has a poorer SE.

6.8 Summary

In this chapter, we reviewed DL-based sparse array design principles, procedures, and applications with a focus on direction finding. In general, this problem may be viewed as either a classification or regression in the DL framework. Our DL-based design offers the advantages of feature engineering and prediction-stage computational efficiency. This is particularly useful for dynamic sparse array design scenarios such as in cognitive radar, where a new sparse array must be selected to adapt to the changes in the target environment. Here, we employed a CNN to find the antenna subarray that yields the minimum CRB of the DoA estimation error. Further, antenna selection knowledge already learned at a pre-trained model may be used for a different antenna array geometry using TL methods. This has applications in metacognitive radars [98]. For 2-D or planar sparse arrays, we illustrated the use of SA-assisted DL algorithms. There are a large number of deep sparse array applications in wireless communications, wherein antenna selection is generally employed to reduce the complexity of common functions such as beamforming, channel estimation, and ISAC.

There are many interesting open challenges that need to be tackled for practical and futuristic implementations of DL-based sparse array techniques. Training a learning model still requires solving complex optimization problems for each training data sample offline prior to training.

For instance, the learning models are pivoted by a reasonable trade-off between an efficient implementation and satisfactory beampattern (radar sensing) and SE (communications) performance. Practical implementations may employ compressed or quantized model parameters to reduce model complexity. Further, data/model parallelization is useful for accelerating training times. It is noted that Chapter 7 of this book discusses the formulation and DL-based implementation of optimum sparse array beamforming maximizing the output signal-to-interference and noise ratio [*,**].

Acknowledgments

K. V. M. acknowledges support from the National Academies of Sciences, Engineering, and Medicine via Army Research Laboratory Harry Diamond Distinguished Fellowship.

References

1 Elbir, A.M., Mishra, K.V., Vorobyov, S.A., and Heath, R.W. Jr. (2023). Twenty-five years of advances in beamforming: from convex and nonconvex optimization to learning techniques. *IEEE Signal Processing Magazine* 40 (4): 118–131.

2 Shenoy, R.P. (1993). Phased array antennas. In: *Advanced Radar Techniques and Systems* (ed. G. Galati), 598–821. Peter Peregrinus.

3 Frank, J. and Richards, J.D. (2008). Phased array radar antennas. In: *Radar Handbook*, 3e (ed. M.I. Skolnik), 13.1–13.74. McGraw-Hill Education.

4 Herd, J.S. and Conway, M.D. (2015). The evolution to modern phased array architectures. *Proceedings of the IEEE* 104 (3): 519–529.

5 Haupt, R.L. (2015). *Timed Arrays: Wideband and Time Varying Antenna Arrays*. Wiley.

6 Johnson, D.H. (1982). The application of spectral estimation methods to bearing estimation problems. *Proceedings of the IEEE* 70 (9): 1018–1028.

7 Linebarger, D.A., Sudborough, I.H., and Tollis, I.G. (1993). Difference bases and sparse sensor arrays. *IEEE Transactions on Information Theory* 39 (2): 716–721.

8 Haupt, R.L. (1994). Thinned arrays using genetic algorithms. *IEEE Transactions on Antennas and Propagation* 42 (7): 993–999.

9 Mishra, K.V., Kahane, I., Kaufmann, A., and Eldar, Y.C. (2017). High spatial resolution radar using thinned arrays. *IEEE Radar Conference*, 1119–1124.

10 Lo, Y.T. (1964). A mathematical theory of antenna arrays with randomly spaced elements. *IEEE Transactions on Antennas and Propagation* 12 (3): 257–268.

11 Agrawal, V.D. and Lo, Y.T. (1972). Mutual coupling in phased arrays of randomly spaced antennas. *IEEE Transactions on Antennas and Propagation* 20 (3): 288–295.

12 Wang, X., Chen, Z., Ren, S., and Cao, S. (2017). DOA estimation based on the difference and sum coarray for coprime arrays. *Digital Signal Processing* 69: 22–31.

13 Sedighi, S., Bhavani Shankar, M.R., Mishra, K.V., and Ottersten, B. (2019). Optimum design for sparse FDA-MIMO automotive radar. *Asilomar Conference on Signals, Systems, and Computers*, 913–918.

14 Lv, W., Mishra, K.V., and Chen, S. (2023). Co-pulsing FDA radar. *IEEE Transactions on Aerospace and Electronic Systems* 59 (2): 1107–1126.

15 BouDaher, E., Ahmad, F., Amin, M.G., and Hoorfar, A. (2017). Mutual coupling effect and compensation in non-uniform arrays for direction-of-arrival estimation. *Digital Signal Processing* 61: 3–14.

16 Liu, C.L. and Vaidyanathan, P.P. (2016). Super nested arrays: linear sparse arrays with reduced mutual coupling; Part I: Fundamentals. *IEEE Transactions on Signal Processing* 64 (15): 3997–4012.

17 Liu, C.L. and Vaidyanathan, P.P. (2017). Hourglass arrays and other novel 2-D sparse arrays with reduced mutual coupling. *IEEE Transactions on Signal Processing* 65 (13): 3369–3383.

18 Elbir, A.M. and Mishra, K.V. (2020). Joint antenna selection and hybrid beamformer design using unquantized and quantized deep learning networks. *IEEE Transactions on Wireless Communications* 19 (3): 1677–1688.

19 Foucart, S. and Rauhut, H. (2013). *A Mathematical Introduction to Compressive Sensing*. Birkhäuser.

20 Unz, H. (1962). Nonuniform arrays with spacings larger than one wavelength. *IRE Transactions on Antennas and Propagation* 10 (5): 647–648.

21 Ishimaru, A. and Chen, Y.S. (1965). Thinning and broadbanding antenna arrays by unequal spacings. *IEEE Transactions on Antennas and Propagation* AP-13 (1): 34–42.

22 Sandler, S.S. (1960). Some equivalences between equally and unequally spaced arrays. *IRE Transactions on Antennas and Propagation* 8 (5): 496–500.

23 Skolnik, M.I., Nemhauser, G., and Sherman, J.W. (1964). Dynamic programming applied to unequally spaced arrays. *IEEE Transactions on Antennas and Propagation* 12 (1): 35–43.

24 Redlich, R.W. (1973). Iterative least-squares synthesis of nonuniformly spaced linear arrays. *IEEE Transactions on Antennas and Propagation* 21 (1): 106–108.

25 Schjær-Jacobsen, H. and Madsen, K. (1976). Synthesis of nonuniformly spaced arrays using a general nonlinear minimax optimisation method. *IEEE Transactions on Antennas and Propagation* 24 (4): 501–506.

26 Murino, V., Trucco, A., and Regazzoni, C.S. (1996). Synthesis of unequally spaced arrays by simulated annealing. *IEEE Transactions on Signal Processing* 44 (1): 119–123.

27 Haupt, R.L. (2010). *Antenna Arrays: A Computational Approach*. Wiley-IEEE Press.

28 Kim, Y. and Jaggard, D.L. (1986). The fractal random array. *Proceedings of the IEEE* 74 (9): 1278–1280.

29 Lo, Y.T. and Lee, S.W. (1966). A study of space-tapered arrays. *IEEE Transactions on Antennas and Propagation* 14 (1): 22–30.

30 Willey, R. (1962). Space tapering of linear and planner arrays. *IRE Transactions on Antennas and Propagation* 10 (4): 369–377.

31 Moffet, A. (1968). Minimum-redundancy linear arrays. *IEEE Transactions on Antennas and Propagation* 16 (2): 172–175.

32 Arsac, J. and Danjon, A. (1955). Nouveau róseau pour l'observation radio astronomique de labrillance sur le soleil à 9359 MC. *Comptes Rendus Hebdomadaires Des Seances De L Academie Des Sciences* 240 (9): 942–945. (In French).

33 Camps, A., Cardama, A., and Infantes, D. (2001). Synthesis of large low-redundancy linear arrays. *IEEE Transactions on Antennas and Propagation* 49 (12): 1881–1883.

34 Bracewell, R.N. (1966). Optimum spacings for radio telescopes with unfilled apertures. *URSI Assembly*, 243.

35 Leech, J. (1956). On the representation of 1, 2, …, *n* by differences. *Journal of the London Mathematical Society* 1 (2): 160–169.

36 Schwartau, F., Schröder, Y., Wolf, L., and Schoebel, J. (2021). Large minimum redundancy linear arrays: systematic search of perfect and optimal rulers exploiting parallel processing. *IEEE Open Journal of Antennas and Propagation* 2: 79–85.

37 Ishiguro, M. (1980). Minimum redundancy linear arrays for a large number of antennas. *Radio Science* 15 (6): 1163–1170.

38 Vaidyanathan, P.P. and Pal, P. (2010). Sparse sensing with co-prime samplers and arrays. *IEEE Transactions on Signal Processing* 59 (2): 573–586.

39 Qin, S., Zhang, Y.D., and Amin, M.G. (2015). Generalized coprime array configurations for direction-of-arrival estimation. *IEEE Transactions on Signal Processing* 63 (6): 1377–1390.

40 Zheng, W., Zhang, X., Li, J., and Shi, J. (2021). Extensions of co-prime array for improved DOA estimation with hole filling strategy. *IEEE Sensors Journal* 21 (5): 6724–6732.

41 Kozick, R.J. and Kassam, S.A. (1991). Linear imaging with sensor arrays on convex polygonal boundaries. *IEEE Transactions on Systems, Man, and Cybernetics* 21 (5): 1155–1166.

42 Duman, T.M. and Ghrayeb, A. (2007). Antenna selection for MIMO systems. In: *Coding for MIMO Communication Systems*, 287–315. Wiley. https://onlinelibrary.wiley.com/doi/10.1002/9780470724347.ch11.

43 Joshi, S. and Boyd, S. (2009). Sensor selection via convex optimization. *IEEE Transactions on Signal Processing* 57 (2): 451–462.

44 Godrich, H., Petropulu, A.P., and Poor, H.V. (2012). Sensor selection in distributed multiple-radar architectures for localization: a knapsack problem formulation. *IEEE Transactions on Signal Processing* 60 (1): 247–260.

45 Elbir, A.M. and Mishra, K.V. (2019). Robust hybrid beamforming with quantized deep neural networks. *IEEE International Workshop on Machine Learning for Signal Processing*, 1–6.

46 Elbir, A.M., Mishra, K.V., and Eldar, Y.C. (2019). Cognitive radar antenna selection via deep learning. *IET Radar, Sonar & Navigation* 13: 871–880.

47 Yu, D. and Deng, L. (2011). Deep learning and its applications to signal and information processing [Exploratory DSP]. *IEEE Signal Processing Magazine* 28 (1): 145–154.

48 Elbir, A.M. and Mishra, K.V. (2019). Deep learning design for joint antenna selection and hybrid beamforming in massive MIMO. *IEEE International Symposium on Antennas and Propagation and USNC-URSI Radio Science Meeting*, 1585–1586.

49 Lecun, Y., Bengio, Y., and Hinton, G. (2015). Deep learning. *Nature* 521 (7553): 436–444.

50 Bengio, Y., Courville, A., and Vincent, P. (2013). Representation learning: a review and new perspectives. *IEEE Transactions on Pattern Analysis and Machine Intelligence* 35 (8): 1798–1828.

51 Gao, Y., Hu, D., Chen, Y., and Ma, Y. (2017). Gridless 1-b DOA estimation exploiting SVM approach. *IEEE Communications Letters* 21 (10): 2210–2213.

52 Elbir, A.M. (2020). DeepMUSIC: multiple signal classification via deep learning. *IEEE Sensors Letters* 4 (4): 1–4.

53 Kulkarni, P. and Vaidyanathan, P.P. (2021). Feature engineering for DOA estimation using a convolutional neural network, for sparse arrays. *Asilomar Conference on Signals, Systems, and Computers*, 246–250.

54 Mishra, K.V., Bhavani Shankar, M.R., Koivunen, V. et al. (2019). Toward millimeter wave joint radar-communications: a signal processing perspective. *IEEE Signal Processing Magazine* 36 (5): 100–114.

55 Demir, Ö.T. and Tuncer, T.E. (2016). Antenna selection and hybrid beamforming for simultaneous wireless information and power transfer in multi-group multicasting systems. *IEEE Transactions on Wireless Communications* 15 (10): 6948–6962.

56 BouDaher, E., Ahmad, F., and Amin, M.G. (2015). Sparsity-based direction finding of coherent and uncorrelated targets using active nonuniform arrays. *IEEE Signal Processing Letters* 22 (10): 1628–1632. https://doi.org/10.1109/LSP.2015.2417807.

57 Pal, P. and Vaidyanathan, P.P. (2010). Nested arrays: a novel approach to array processing with enhanced degrees of freedom. *IEEE Transactions on Signal Processing* 58 (8): 4167–4181.

58 O'Shea, T. and Hoydis, J. (2017). An introduction to deep learning for the physical layer. *IEEE Transactions on Cognitive Communications and Networking* 3 (4): 563–575.

59 Wandale, S. and Ichige, K. (2021). Design of sparse arrays via deep learning for enhanced DOA estimation. *EURASIP Journal on Advances in Signal Processing* 2021 (1): 17–13.

60 Aboutanios, E., Nosrati, H., and Wang, X. (2021). Online antenna selection for enhanced DOA estimation. *IEEE International Conference on Acoustics, Speech and Signal Processing*, 8468–8472.

61 Elbir, A.M., Mulleti, S., Cohen, R. et al. (2019). Deep-sparse array cognitive radar. *IEEE International Conference on Sampling Theory and Applications*, 1–5.

62 Pavel, S.R., Chowdhury, M.W.T.S., Zhang, Y.D. et al. (2021). Machine learning-based direction-of-arrival estimation exploiting distributed sparse arrays. *Asilomar Conference on Signals, Systems, and Computers*, 241–245.

63 Akter, R., Doan, V.S., Huynh-The, T., and Kim, D.S. (2021). RFDOA-Net: an efficient ConvNet for RF-based DOA estimation in UAV surveillance systems. *IEEE Transactions on Vehicular Technology* 70 (11): 12209–12214.

64 Gohil, R.P., Routray, G., and Hegde, R.M. (2021). Learning based method for robust DOA estimation using co-prime circular conformal microphone array. *National Conference on Communications*, 1–6.

65 Roy, V., Chepuri, S.P., and Leus, G. (2013). Sparsity-enforcing sensor selection for DOA estimation. *IEEE International Workshop on Computational Advances in Multi-Sensor Adaptive Processing*, 340–343.

66 Renaux, A., Forster, P., Larzabal, P. et al. (2008). A fresh look at the Bayesian bounds of the Weiss-Weinstein family. *IEEE Transactions on Signal Processing* 56 (11): 5334–5352.

67 Tabrikian, J., Isaacs, O., and Bilik, I. (2016). Cognitive antenna selection for DOA estimation in automotive radar. *IEEE Radar Conference*, 1–5.

68 Isaacs, O., Tabrikian, J., and Bilik, I. (2015). Cognitive antenna selection for optimal source localization. *IEEE International Workshop on Computational Advances in Multi-Sensor Adaptive Processing*, 341–344.

69 Stoica, P. and Nehorai, A. (1990). MUSIC, maximum likelihood, and Cramér-Rao bound: further results and comparisons. *IEEE Transactions on Acoustics, Speech, and Signal Processing* 38 (12): 2140–2150.

70 Friedlander, B. and Weiss, A.J. (1991). Direction finding in the presence of mutual coupling. *IEEE Transactions on Antennas and Propagation* 39 (3): 273–284.

71 Nion, D. and Sidiropoulos, N.D. (2010). Tensor algebra and multidimensional harmonic retrieval in signal processing for MIMO radar. *IEEE Transactions on Signal Processing* 58 (11): 5693–5705.

72 Elbir, A.M. and Mishra, K.V. (2020). Sparse array selection across arbitrary sensor geometries with deep transfer learning. *IEEE Transactions on Cognitive Communications and Networking* 7 (1): 255–264.

73 Ye, Z. and Liu, C. (2008). 2-D DOA estimation in the presence of mutual coupling. *IEEE Transactions on Antennas and Propagation* 56 (10): 3150–3158.

74 Cheng, J., Wu, J., Leng, C. et al. (2018). Quantized CNN: a unified approach to accelerate and compress convolutional networks. *IEEE Transactions on Neural Networks and Learning Systems* 29 (10): 4730–4743.

75 Dudgeon, D.E. (1977). Fundamentals of digital array processing. *Proceedings of the IEEE* 65 (6): 898–904.

76 Athley, F. (2001). Optimization of element positions for direction finding with sparse arrays. *IEEE Signal Processing Workshop on Statistical Signal Processing*, 516–519.

77 Duan, L., Tsang, I.W., and Xu, D. (2012). Domain transfer multiple kernel learning. *IEEE Transactions on Pattern Analysis and Machine Intelligence* 34 (3): 465–479.

78 Kulis, B., Saenko, K., and Darrell, T. (2011). What you saw is not what you get: domain adaptation using asymmetric kernel transforms. *IEEE Conference on Computer Vision and Pattern Recognition*, 1785–1792.

79 Pan, S.J. and Yang, Q. (2010). A survey on transfer learning. *IEEE Transactions on Knowledge and Data Engineering* 22 (10): 1345–1359.

80 Vincent, P., Larochelle, H., Lajoie, I. et al. (2010). Stacked denoising autoencoders: learning useful representations in a deep network with a local denoising criterion. *Journal of Machine Learning Research* 11: 3371–3408.

81 Seyfioğlu, M.S. and Gürbüz, S.Z. (2017). Deep neural network initialization methods for micro-Doppler classification with low training sample support. *IEEE Geoscience and Remote Sensing Letters* 14 (12): 2462–2466.

82 Chen, B., Zhong, Z., Ai, B., and Chen, X. (2011). Comparison of antenna arrays for MIMO system in high speed mobile scenarios. *IEEE Vehicular Technology Conference - Spring*, 1–5.

83 Hyberg, P., Jansson, M., and Ottersten, B. (2004). Array interpolation and bias reduction. *IEEE Transactions on Signal Processing* 52 (10): 2711–2720.

84 Rubsamen, M. and Gershman, A.B. (2008). Direction-of-arrival estimation for nonuniform sensor arrays: from manifold separation to Fourier domain MUSIC methods. *IEEE Transactions on Signal Processing* 57 (2): 588–599.

85 Liu, Y., Chen, H., Peng, Z., and Fang, J. (2019). DOA estimation for mixed circular and noncircular signals by using the conversion relationship between URAs and a virtual ULA. *IEEE Sensors Letters* 3 (11): 1–4.

86 Pan, S.J., Tsang, I.W., Kwok, J.T., and Yang, Q. (2010). Domain adaptation via transfer component analysis. *IEEE Transactions on Neural Networks* 22 (2): 199–210.

87 Yosinski, J., Clune, J., Bengio, Y., and Lipson, H. (2014). How transferable are features in deep neural networks? *Advances in Neural Information Processing Systems 27 (NIPS 2014)*, 3320–3328.

88 Schmidt, R.O. (1986). Multiple emitter location and signal parameter estimation. *IEEE Transactions on Antennas and Propagation* 34 (3): 276–280.

89 Nakamura, S., Iwazaki, S., and Ichige, K. (2021). Optimization and hole interpolation of 2-D sparse arrays for accurate direction-of-arrival estimation. *IEICE Transactions on Communications* 104 (4): 401–409.

90 Wandale, S. and Ichige, K. (2021). 2D sparse array selection via deep learning. *International Symposium on Antennas and Propagation*, 445–446.

91 Wandale, S. and Ichige, K. (2021). Simulated annealing assisted sparse array selection utilizing deep learning. *IEEE Access* 9: 156907–156914.

92 Nakamura, S., Iwazaki, S., and Ichige, K. (2019). Optimum 2D sparse array and its interpolation via nuclear norm minimization. *IEEE International Symposium on Circuits and Systems*, 1–5.

93 Elbir, A.M. and Mishra, K.V. (2019). Deep learning design for joint antenna selection and hybrid beamforming in massive MIMO. *IEEE International Symposium on Antennas and Propagation USNC/URSI National Radio Science Meeting*, 1–2.

94 Boumal, N., Mishra, B., Absil, P.A., and Sepulchre, R. (2014). Manopt, a Matlab toolbox for optimization on manifolds. *Journal of Machine Learning Research* 15: 1455–1459.

95 Yu, X., Shen, J., Zhang, J., and Letaief, K.B. (2016). Alternating minimization algorithms for hybrid precoding in millimeter wave MIMO systems. *IEEE Journal of Selected Topics in Signal Processing* 10 (3): 485–500.

96 Elbir, A.M. (2019). CNN-based precoder and combiner design in mmWave MIMO systems. *IEEE Communications Letters* 23 (7): 1240–1243.

97 Elbir, A.M., Mishra, K.V., and Chatzinotas, S. (2021). Terahertz-band joint ultra-massive MIMO radar-communications: model-based and model-free hybrid beamforming. *IEEE Journal of Selected Topics in Signal Processing* 15 (6): 1468–1483.

98 Mishra, K.V., Shankar, M.B., and Ottersten, B. (2020). Toward metacognitive radars: concept and applications. *IEEE International Radar Conference*, 77–82.

7

Sparse Array Design for Optimum Beamforming Using Deep Learning

Syed A. Hamza[1], *Kyle Juretus*[2], *and Moeness G. Amin*[3]

[1] *Electrical Engineering Department, School of Engineering, Widener University, Chester, PA, USA*
[2] *ECE Department, College of Engineering, Villanova University, Villanova, PA, USA*
[3] *Center for Advanced Communications (CAC), College of Engineering, Villanova University, Villanova, PA, USA*

7.1 Motivation

A typical sensor array processing system has main costs associated with the sensors, transceivers, and processing hardware. The bulk of the cost is attributed to transceivers consisting of expensive (RF) chains utilizing a cascade of electronic components to implement low noise amplification, mixing, analogue to digital conversion, filtering, and detection tasks. Sparse array design reduces system transceiver costs by limiting the transceiver RF chains through strategic sensor selection, thereby reducing the hardware and processing complexity [1–5]. In part due to the benefit of reducing transceiver cost, sparse array design has been used in a multitude of sensor signal processing tasks, such as Multiple Input Multiple Output (MIMO) communications ([6–10]), radar ([11–15]), sonar [16], satellite navigation, radio telescopes, speech enhancement, and medical imaging applications ([12, 17, 18]). For a given number of transceiver chains, the performance gains in using sparse arrays stem from their inherent ability of providing the additional degrees of freedom, such as sensor locations, to achieve pre-defined metrics. A variety of methods exist to assess the performance of a sparse array design, but the methods can generally be categorized into environment-independent and environment-dependent design [11]. Environment-dependent designs require array reconfigurability since the receiver performance largely depends on the operating environment, which may change according to the source and interference signals and locations [19, 20]. This is in contrast to environment-independent sparse arrays whose configurations follow certain formulas and seek to attain structured sparse configurations with extended aperture co-arrays [21–23]. As such, environment-independent design has the driving objective to enable direction of arrival (DOA) estimation of more sources than the available physical sensors, including the common nested and coprime array structures [21, 22, 24, 25].

Unlike a fixed environmental-independent design, environment-dependent sparse array design can be enhanced by optimizing the signal-to-interference and noise ratio (SINR), which has a direct bearing on improving target detection and localization for radar signal processing, increasing throughput or channel capacity for MIMO wireless communication systems, and enhancing resolution capability in medical imaging [26–28]. Maximizing signal-to-noise ratio (MaxSNR) and MaxSINR criteria have been shown to be important performance criteria for beamforming

Sparse Arrays for Radar, Sonar, and Communications, First Edition. Edited by Moeness G. Amin.

performance and interference mitigation. For sparse array design, the MaxSINR beamforming performance depends on the selected positions of the sensors as well as the locations and the powers of the desired source and interferers in the field of view (FOV) [29–31]. It is noted that with sparse arrays, the optimum Capon beamforming, in achieving MaxSINR, must not only find the optimum weights, as commonly used in uniform arrays, but also the optimum array configuration [27]. This is clearly an entwined optimization problem and requires attaining maximum SINR considering all possible sparse array configurations.

The process of optimizing for SINR is part of the perception–action cycle (PAC), where the environment is first analyzed and then the sparse array configuration is updated. This can provide additional capability to cognitive radar ([32, 33]) operation, which continuously interacts with the environment and updates the radar parameters through the acquired knowledge. For instance, in MIMO radar, the "perception–action" cognition approach typically entails adaptive optimizations of the transmit waveforms and power allocation [34–36]. In addition to these parameters, array reconfiguration (hardware) can significantly improve system performance beyond that achieved with fixed antenna positions [35, 37]. This chapter focuses on dynamic RF environments where the sources in the FOV of the array can rapidly change their locations over time, thus requiring fast array reconfigurability under two constraints: (i) antenna switching should not lead to an increase in the hardware complexity [6–8] and (ii) optimal array configuration and beamforming weights should be determined in a sufficiently short time consistent with the employed PAC of cognitive sensing [38].

While rapid array reconfigurability techniques exist to switch the attached sensors of the array to the transceivers, the system still has to determine what sensor positions are optimal to maximize SINR. Environment-dependent sparse array design algorithms generally follow two different approaches. The approaches based on prior knowledge of interference parameters, essentially require that the real time interference parameters, such as DOAs and respective SINRs, are either known or estimated apriori in real time. The other approach is more practical as it does not require the information of the interfering environment, which is the case in Capon beamforming. In both cases, iterative algorithms have been proposed to optimize the sparse array beamformer design. Although convex based optimization algorithms, such as semidefinite relaxation (SDR) and successive convex approximation (SCA) have been developed to yield sparse configurations with desirable beamforming performances [2, 39], real time implementations of these algorithms remain limited due to the relatively high computation cost. The problem becomes more pronounced in rapidly changing environments which result from temporal and spatial nonstationary behaviors of the sources in the FOV.

In this chapter, machine learning is utilized for sparse network design in order to reduce the perception action time of the system by replacing the costly optimization algorithms with an interference pass of a neural network. The rest of the chapter is organized as follows: In Section 7.2, an overview of the related work and contributions of this chapter are provided. The problem formulation for maximizing the output SINR is described in Section 7.3. Section 7.4 analyzes the different optimization algorithms available for sparse array design, and examines their efficiency and applicability to training data generation. Section 7.5 describes machine-learning-based methods for sparse array design and the challenges of utilizing machine learning for sparse array design. Section 7.6 provides simulation results for three different machine-learning network architectures. Section 7.7 describes the future directions for the field to ensure machine-learning methods can be utilized in real-world radar systems and Section 7.8 provides concluding remarks.

7.2 Contributions

In this chapter, an overview of sparse array beamformer design using supervised machine learning is provided. Machine learning has been previously proposed in multiple different steps of the PAC. Deep learning techniques have recently been studied for DOA estimation and efficient beamforming for communication systems in [40–45]. The overarching premise in the related work is to avoid solving a difficult optimization problem and shifting the problem complexity to neural network training, which may be carried out offline. Specifically, thinning the array for direction finding was performed in [41] using a convolutional neural network (CNN), where the problem is cast as multi-class classifications where the labels are thinned arrays, i.e., subarrays. Labeling is done based on the Cramer–Rao Lower Bound with the network input representing the estimated covariance matrix. The work in [45] relaxes the constraints of requiring a single target and the full array to be active at certain radar scans that were assumed in [41], applying the work to DOA estimation utilizing deep learning.

Main Contributions: Unlike, the aforementioned work which implements sparse array design mainly for DOA estimation, this chapter deals with sparse array optimization for MaxSINR-based beamforming. The main contributions of the chapter are,

(1) An overview of optimization-based methods for sparse array design and their run-time complexity, including SDR, SCA, and the sparse beamformer spectral analysis (SBSA) algorithm. The effects of run-time complexity and the effect that has on the PAC and training data generation is discussed.

(2) The Dual Neural Network (DNN)-based approach is developed to configure a Capon-based data driven sparse beamformer by learning via enumeration based optimization strategies as well as SBSA design. The design is achieved through a direct mapping of the sensor data correlations to the optimum sparse array configuration for a given "look direction." This methodology utilizes the merits of the data dependent designs, through online implementation, and exploits the benefits of assuming prior information of the interfering environment by efficient training with low computational complexity. This design can also be used for the case of limited data snapshots and is extendable to robust adaptive beamforming due to uncertainty regarding the look direction DOA as well as array calibration errors.

(3) Fast-switching that avoids hardware complexity is paramount to a realizable PAC. In this chapter, sparse array design is completed under the constraint of binary switching per RF chain as opposed to a fully switched network, which assumes that all antenna locations can be activated through all RF chains. It is noted that binary switching greatly simplifies antenna switching, but limits the number of possible configurable sparse arrays. Specifically, two binary switching strategies are considered and their achievable sparse array classification rates and SINR using a CNN and a MLP structure are examined. These switching strategies are assessed in terms of their respective classification rates and the highest attainable SINR.

7.3 Problem Formulation

The block diagram of Fig. 7.1 depicts the essence of sparse array beamforming. Dark circles indicate selected antennas that are connected to the front-end receivers through a multiplexing process. Consider a desired source and L independent interfering sources whose signals impinge on a linear

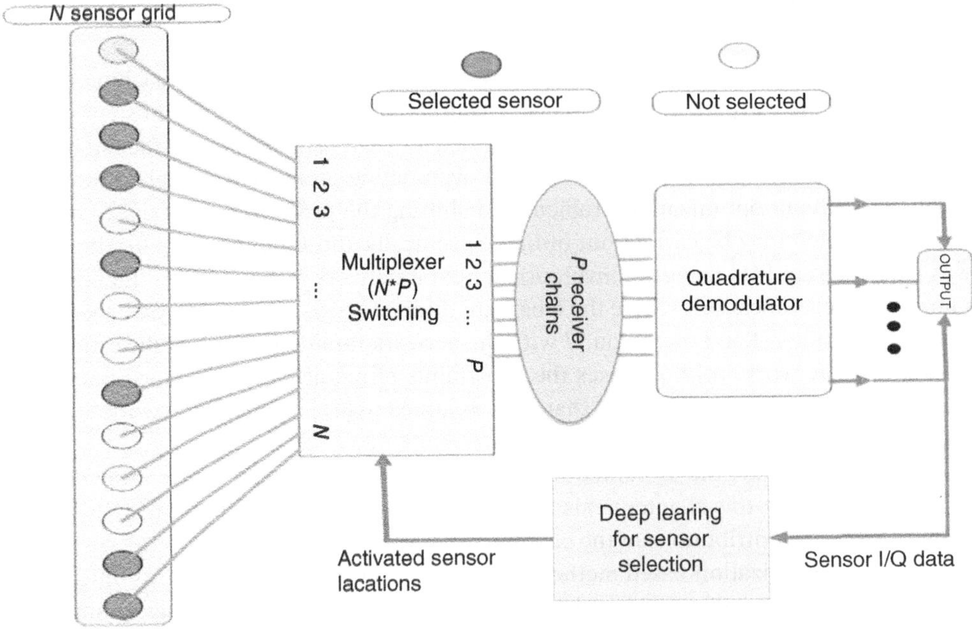

Figure 7.1 Block diagram of adaptive switched sensor beamformer.

array with N uniformly placed sensors. The baseband data received at the array at sampling instant n is then given by,

$$\mathbf{x}[n] = (\alpha[n])\mathbf{s}(\theta) + \sum_{l=1}^{L}(\beta_l[n])\mathbf{v}(\theta_l) + \mathbf{n}[n], \tag{7.1}$$

where, $\mathbf{s}(\theta_k)$ and $\mathbf{v}(\theta_l) \in \mathbb{C}^N$ are the steering vectors corresponding to the direction of arrival, θ or θ_l of the desired source and lth interference, respectively, and are defined as follows,

$$\mathbf{s}(\theta) = [1 \;\; e^{j(2\pi/\lambda)d\cos(\theta)} \ldots e^{j(2\pi/\lambda)d(N-1)\cos(\theta)}]^T, \tag{7.2}$$

where d is the inter-element spacing and $(\alpha[n], \beta_l[n]) \in \mathbb{C}$ are the complex amplitudes of the incoming baseband signals [46]. The additive Gaussian noise $\mathbf{n}[n] \in \mathbb{C}^N$ has variance σ_t^2. The elements of the received data vector $\mathbf{x}[n]$ are combined linearly by the N-sensor beamformer that strives to maximize the output SINR. The output signal $y[n]$ of the optimum beamformer for maximum SINR is given by [47],

$$y[n] = \mathbf{w}_o^H \mathbf{x}[n], \tag{7.3}$$

where $\mathbf{w}_o$ is the optimum weight vector resulting in the maximum output SINR. The SINR is given by the following expression,

$$\text{SINR} = \frac{\mathbf{w}^H \mathbf{R}_s \mathbf{w}}{\mathbf{w}^H \mathbf{R}_{i+n} \mathbf{w}}. \tag{7.4}$$

For statistically independent signals, the desired source correlation matrix is $\mathbf{R}_s = \sigma^2 \mathbf{s}(\theta)\mathbf{s}^H(\theta)$, where $\sigma^2 = E\{\alpha[n]\alpha^H[n]\}$. Likewise, the interference and noise correlation matrix, $\mathbf{R}_{i+n} = \sum_{l=1}^{L}(\sigma_l^2 \mathbf{v}(\theta_l)\mathbf{v}^H(\theta_l)) + \sigma_t^2 \mathbf{I}_{N\times N}$, with $\sigma_l^2 = E\{\beta_l[n]\beta_l^H[n]\}$ being the power of the lth interfering source.

In order to maximize the SINR expression in (7.4), the numerator is constrained and the denominator term is minimized as follows,

$$\begin{aligned} \underset{\mathbf{w}\in\mathbb{C}^N}{\text{minimize}} \quad & \mathbf{w}^H \mathbf{R}_{i+n} \mathbf{w}, \\ \text{s.t.} \quad & \mathbf{w}^H \mathbf{R}_s \mathbf{w} = 1. \end{aligned} \tag{7.5}$$

The problem in (7.5) can be written equivalently by replacing $\mathbf{R}_{i+n}$ with the received data covariance matrix, $\mathbf{R}_{\mathbf{xx}} = \mathbf{R}_s + \mathbf{R}_{i+n}$ as follows [47],

$$\begin{aligned} \underset{\mathbf{w}\in\mathbb{C}^N}{\text{minimize}} \quad & \mathbf{w}^H \mathbf{R}_{\mathbf{xx}} \mathbf{w}, \\ \text{s.t.} \quad & \mathbf{w}^H \mathbf{R}_s \mathbf{w} \geq 1. \end{aligned} \tag{7.6}$$

It is noted that the equality constraint in (7.5) is relaxed in (7.6) due to the inclusion of the constraint as part of the objective function, and as such, (7.6) converges to the equality constraint. Additionally, the optimal solution in (7.6) is invariant up to uncertainty in the absolute power of the desired source. In practice, the actual source parameters can deviate from the perceived ones. This discrepancy is typically mitigated, to an extent, by pre-processing the received data correlation matrix through diagonal loading or tapering the correlation matrix [27].

There exists a closed form solution to maximize the SINR expression in (7.4) and is given by $\mathbf{w}_o = \mathscr{P}\{\mathbf{R}_{i+n}^{-1}\mathbf{R}_s\} = \mathscr{P}\{\mathbf{R}_{\mathbf{xx}}{}^{-1}\mathbf{R}_s\}$ [47]. The operator $\mathscr{P}\{.\}$ computes the principal eigenvector of the input matrix, and $\mathbf{R}_{\mathbf{xx}} = \mathbf{R}_s + \mathbf{R}_{i+n}$. Substituting $\mathbf{w}_o$ into (7.4) yields the corresponding optimum output SINR_o,

$$\text{SINR}_o = \Lambda_{\max}\{\mathbf{R}_{i+n}^{-1}\mathbf{R}_s\}. \tag{7.7}$$

Accordingly, the optimum output SINR_o is given by the maximum eigenvalue ($\Lambda_{\max}$) associated with the product of the inverse of interference plus noise correlation matrix and the desired source correlation matrix. This is the general expression for point and distributed sources and gives the similar expression in the point source case. Therefore, the performance of the optimum beamformer for maximizing the output SINR is directly related to the correlation matrix of the desired source and that of interference-plus-noise.

7.4 Efficient Generation of Training Data for Optimum Beamforming

There are two different schemes for training machine-learning models to perform sparse array design. In the first scheme, as described in Fig. 7.2, an enumeration technique is used to generate the training labels for any given sensor correlation function. In this case, the MaxSINR array configuration is obtained by sifting through all possible sparse configurations and choosing the best performing array topology. Although the training data is generated offline, it becomes infeasible to obtain an optimum configuration even for a moderately sized array due to the enormous number of sensor permutations.

Three sparse array optimization algorithms, including SDR, SCA, and SBSA-based design, are discussed in this section. The third scheme, SBSA, is designed to reduce the computation complexity of the design problem, allowing for more feasible training data generation times. The proposed SBSA scheme considers the corresponding array spatial spectrum and incorporates the sensor correlation lag redundancy for determining the desired array structure, which translates to a considerably faster time as compared to SDR and SCA algorithms which rely on convex relaxation techniques. While SBSA substantially decreases the time to solve for the array configuration, the run time still has a large effect on the perception acton cycle.

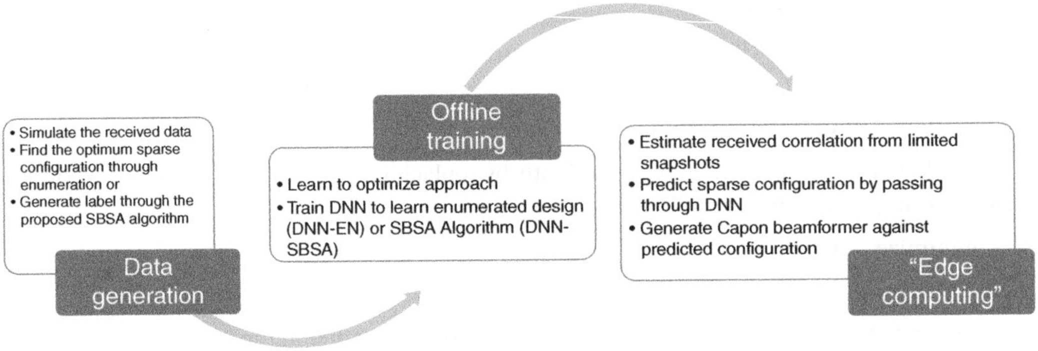

Figure 7.2 Overview of the proposed sparse array design approach using machine learning.

7.4.1 Sparse Array Design Through the SDR Algorithm

The problem of locating the maximum principal eigenvalue among all the correlation matrices associated with P sensors to select is a combinatorial optimization problem. The constraint optimization (7.6) can be re-formulated for optimum sparse array design by incorporating an additional constraint on the cardinality of the weight vector;

$$\begin{aligned} \underset{\mathbf{w}\in\mathbb{C}^N}{\text{minimize}} \quad & \mathbf{w}^H\mathbf{R}_{\mathbf{xx}}\mathbf{w}, \\ \text{s.t.} \quad & \mathbf{w}^H\mathbf{R}_s\mathbf{w} \geq 1, \\ & \|\mathbf{w}\|_0 = P. \end{aligned} \tag{7.8}$$

Here, $\|.\|_0$ is a vector norm which determines the cardinality of the weight vector $\mathbf{w}$. It is assumed that an estimate of all the filled co-array correlation lags corresponding to the correlation matrix of the full aperture array is available. The problem expressed in (7.8) can be relaxed to induce sparsity in the beamforming weight vector $\mathbf{w}$ without placing a hard constraint on the specific cardinality of $\mathbf{w}$, as follows [48];

$$\begin{aligned} \underset{\mathbf{w}\in\mathbb{C}^N}{\text{minimize}} \quad & \mathbf{w}^H\mathbf{R}_{\mathbf{xx}}\mathbf{w} + \mu(\|\mathbf{w}\|_1), \\ \text{s.t.} \quad & \mathbf{w}^H\mathbf{R}_s\mathbf{w} \geq 1. \end{aligned} \tag{7.9}$$

Here, $\|.\|_1$ is the sparsity inducing l_1-norm which calculates the sum of the magnitudes of the weight vector and μ is a parameter to control the desired sparsity in the solution. Even though the relaxed problem expressed in (7.9) is not exactly similar to that of (7.8), it is well known that l_1-norm regularization has been an effective tool for recovering sparse solutions in many diverse formulations [49, 50]. The problem in (7.9) can be penalized instead by the weighted l_1-norm function which is a well known sparsity promoting formulation [51],

$$\begin{aligned} \underset{\mathbf{w}\in\mathbb{C}^N}{\text{minimize}} \quad & \mathbf{w}^H\mathbf{R}_{\mathbf{xx}}\mathbf{w} + \mu(\|(\mathbf{b}^i \circ |\mathbf{w}|)\|_1), \\ \text{s.t.} \quad & \mathbf{w}^H\mathbf{R}_{\mathbf{s}}\mathbf{w} \geq 1. \end{aligned} \tag{7.10}$$

where, "$\circ$" denotes the element wise product, "$|.|$" is the modulus operator and $\mathbf{b}^i \in \mathbb{R}^N$ is the regularization re-weighting vector at the ith iteration. Therefore, (7.10) is the sequential optimization methodology, where the regularization re-weighting vector $\mathbf{b}^i$ is typically chosen as an inverse function of the beamforming weight vector obtained at the previous iteration. This, in turn, suppresses the sensors corresponding to smaller beamforming weights, thereby encouraging sparsity in an iterative fashion. The weighted l_1-norm function in (7.10) is replaced by the l_1-norm squared function which does not alter the regularization property of the weighted l_1-norm function [9],

$$\begin{aligned} \underset{\mathbf{w}\in\mathbb{C}^N}{\text{minimize}} \quad & \mathbf{w}^H\mathbf{R}_{\mathbf{xx}}\mathbf{w} + \mu(\|(\mathbf{b}^i\circ|\mathbf{w}|)\|_1^2), \\ \text{s.t.} \quad & \mathbf{w}^H\mathbf{R}_s\mathbf{w} \geq 1. \end{aligned} \tag{7.11}$$

The semidefinite formulation (SDP) of the above problem can then be realized by re-expressing the quadratic form, $\mathbf{w}^H\mathbf{R}_{\mathbf{xx}}\mathbf{w} = \text{Tr}(\mathbf{w}^H\mathbf{R}_{\mathbf{xx}}\mathbf{w})$ which is equal to $\text{Tr}(\mathbf{R}_{\mathbf{xx}}\mathbf{W})$, where Tr(.) is the trace of the matrix. Similarly, the regularization term $\|(\mathbf{b}^i\circ|\mathbf{w}|)\|_1^2 = (|\mathbf{w}|^T\mathbf{b}^i)((\mathbf{b}^i)^T|\mathbf{w}|) = |\mathbf{w}|^T\mathbf{B}^i|\mathbf{w}| = \text{Tr}(\mathbf{B}^i|\mathbf{W}|)$. Here, $\mathbf{W} = \mathbf{w}\mathbf{w}^H$ and $\mathbf{B}^i = \mathbf{b}^i(\mathbf{b}^i)^T$ is the regularization re-weighting matrix at the ith iteration. Utilizing these quadratic expressions in (7.11) yields the following problem [9, 52, 53],

$$\begin{aligned} \underset{\mathbf{W}\in\mathbb{C}^{N\times N},\tilde{\mathbf{W}}\in\mathbb{R}^{N\times N}}{\text{minimize}} \quad & \text{Tr}(\mathbf{R}_{\mathbf{xx}}\mathbf{W}) + \mu\text{Tr}(\mathbf{B}^i\tilde{\mathbf{W}}), \\ \text{s.t.} \quad & \text{Tr}(\mathbf{R}_s\mathbf{W}) \geq 1, \\ & \tilde{\mathbf{W}} \geq |\mathbf{W}|, \\ & \mathbf{W} \succeq 0, \ \text{Rank}(\mathbf{W}) = 1. \end{aligned} \tag{7.12}$$

The function "|.|" returns the absolute values of the entries of the matrix, "$\geq$" is the element wise comparison and "$\succeq$" denotes the generalized matrix inequality. The auxiliary matrix $\tilde{\mathbf{W}} \in \mathbb{R}^{N\times N}$ implements the weighted l_1-norm squared regularization along with the re-weighting matrix $\mathbf{B}^i$. The rank constraint in (7.12) is nonconvex and therefore needs to be removed. The rank relaxed approximation works well for the underlying problem. In case the solution matrix is not rank 1, randomization can be utilized to harness rank 1 approximate solutions [54]. Alternatively, one could minimize the nuclear norm of $\mathbf{W}$, as a surrogate for the l_1-norm in the case of matrices, to induce sparsity in the eigenvalues of $\mathbf{W}$ and promote rank one solutions [55, 56]. The resulting rank relaxed semidefinite program (SDR) is given by;

$$\begin{aligned} \underset{\mathbf{W}\in\mathbb{C}^{N\times N},\tilde{\mathbf{W}}\in\mathbb{R}^{N\times N}}{\text{minimize}} \quad & \text{Tr}(\mathbf{R}_{\mathbf{xx}}\mathbf{W}) + \mu\text{Tr}(\mathbf{B}^i\tilde{\mathbf{W}}), \\ \text{s.t.} \quad & \text{Tr}(\mathbf{R}_s\mathbf{W}) \geq 1, \\ & \tilde{\mathbf{W}} \geq |\mathbf{W}|, \\ & \mathbf{W} \succeq 0. \end{aligned} \tag{7.13}$$

In general, QCQP is NP hard and cannot be solved in polynomial time. The formulation in (7.13) is clearly convex, in terms of unknown matrices, as all the other correlation matrices involved are guaranteed to be positive semidefinite. The sparsity parameter μ largely determines the cardinality of the solution beamforming weight vector. To ensure a selection of P sensors, an appropriate value of μ is typically found by carrying a binary search over the probable range of μ. After achieving the desired cardinality, the reduced size thinned correlation matrix $\mathbf{R}_{\mathbf{xx}}$ is formed corresponding to the nonzero values of $\tilde{\mathbf{W}}$. The reduced dimension SDR is now solved with setting $\mu = 0$, yielding the optimum beamformer $\mathbf{w}_o = \mathscr{P}\{\mathbf{W}\}$. The above SDR can be solved to an arbitrary small accuracy ζ, by employing interior point methods involving the worst case complexity of $\mathcal{O}\{\max(K, N)^4 N^{(1/2)} \log(1/\zeta)\}$ [54].

7.4.1.1 Modified Re-weighting for Fully Augmentable Hybrid Array

Formulation (7.13) penalizes all the sensor weights rather judiciously in an effort to optimize the objective function. This is referred to as the free-design approach. On the other hand, the hybrid

Algorithm 7.1: Proposed algorithm to achieve desired cardinality of optimal weight vector

Input: Data correlation matrix $\mathbf{R}_{\mathbf{xx}}$, N, P, look direction DOA's θ_k, hybrid selection vector $\mathbf{z}$.
Output: P sensor beamforming weight vector $\mathbf{w}_o$,
Initialize ϵ.
Initialize μ_{lower}, μ_{upper} (Initializing lower and upper limits of sparsity parameter range for binary search for desired cardinality P)
FSDR: Initialize $\mathbf{B} = \mathbf{z}\mathbf{z}^T$.
NFSDR: For optimum array design without the augmentability constraint, initialize $\mathbf{z}$ to be all ones vector, $\mathbf{B} = \mathbf{z}\mathbf{z}^T$ (all ones matrix).
Perturbed-NFSDR: Locate the sensor i such that, if not selected, results in the minimum compromise of the objective function. Perturb $\mathbf{z}$ at position i, $\mathbf{z}(i) = \mathbf{z}(i) + \gamma$, afterwards calculating $\mathbf{B} = \mathbf{z}\mathbf{z}^T$.
while (Cardinality of $\mathbf{w}_o \neq P$) **do**
  Update μ through binary search.
  for (Typically requires five to six iterations) **do**
    Run the SDR of (7.13)
    Update the regularization weighting matrix $\mathbf{B}$ according to (7.15).
  end for
end while
After achieving the desired cardinality, run SDR for reduced size correlation matrix corresponding to nonzero values of $\tilde{\mathbf{W}}$ and $\mu = 0$, yielding, $\mathbf{w}_o = \mathscr{P}\{\mathbf{W}\}$.
return $\mathbf{w}_o$

sparse array design, penalizes only some sensor weights, leaving the remaining sensors to assume prefixed positions. These positions are chosen to guarantee full augmentatbility of the sparse array, i.e., provide the ability to estimate the autocorrelation at all spatial lags across the array aperture. This provides the means for thinning the array and carrying out sparse optimization at all times. In order to discriminate the prefixed sensors from those which are available for design, the weighted formulation is adopted as follows. For the case without the full augmentability constraint the regularization re-weighting matrix $\mathbf{B}$ is initialized unweighted, i.e., by the all ones matrix and the m, nth element of $\mathbf{B}$ is iteratively updated as follows [51],

$$\mathbf{B}^{i+1}_{m,n} = \frac{1}{|\mathbf{W}^{i}_{m,n}| + \epsilon}. \tag{7.14}$$

The parameter ϵ avoids the unwanted case of division by zero, though its choice is fairly independent to the performance of the iterative algorithm but at times very small values of ϵ can result in the algorithm getting trapped in a local minima. For the hybrid array design, the re-weighting matrix is instead initialized as an outer product of hybrid selection vector $\mathbf{z}$. The hybrid selection vector $\mathbf{z}$ is an N dimensional vector containing binary entries of zero and one, where, zeros correspond to the pre-selected sensors and ones correspond to the remaining sensors to be selected. Hence, the cardinality of $\mathbf{z}$ is equal to the difference of the total number of available sensors and the number of pre-selected sensors. This modified re-weighting approach ensures that the sensors corresponding to the pre-selected configuration is not penalized as part of the regularization, hence, $\mathbf{B} = \mathbf{z}\mathbf{z}^T$, finds solutions that incorporate the pre-selected array topology. The modified penalizing weight update for the hybrid array design can be expressed as;

$$\mathbf{B}^{i+1} = (\mathbf{z}\mathbf{z}^T) \oslash (|\mathbf{W}^{i}| + \epsilon). \tag{7.15}$$

The symbol "⊘" denotes element wise division. For the hybrid design, 7.15 is proposed with appropriate selection of **z**, as explained above. The array designed without the augmentability consideration is the special case of 7.15 with **z** being an all ones vector. The pseudo-code for controlling the sparsity of the optimal weight vector $\mathbf{w}_o$ is summarized in Algorithm 1.

7.4.2 Sparse Array Design Through SCA Algorithm

The objective function and quadratic constraint in (7.8) are interchanged, transforming into equivalent formulation as follows,

$$\begin{aligned} \underset{\mathbf{w}\in\mathbb{C}^N}{\text{maximize}} \quad & \mathbf{w}^H\mathbf{R}_s\mathbf{w}, \\ \text{s.t.} \quad & \mathbf{w}^H\mathbf{R}_{xx}\mathbf{w} \leq 1 \\ & ||\mathbf{w}||_0 = P. \end{aligned} \tag{7.16}$$

In general, the beamforming weight vector is complex valued, however the quadratic functions are real. The real and imaginary entries of the optimal weight vector are typically decoupled, permitting the involvements of only real unknowns. This is achieved through concatenating the beamforming weight vector and defining the respective correlation matrices [57],

$$\tilde{\mathbf{R}}_s = \begin{bmatrix} \text{real}(\mathbf{R}_s) & -\text{imag}(\mathbf{R}_s) \\ \\ \text{imag}(\mathbf{R}_s) & \text{real}(\mathbf{R}_s) \end{bmatrix}, \qquad \tilde{\mathbf{w}} = \begin{bmatrix} \text{real}(\mathbf{w}) \\ \\ \text{imag}(\mathbf{w}) \end{bmatrix} \tag{7.17}$$

$$\tilde{\mathbf{R}}_{\mathbf{XX}} = \begin{bmatrix} \text{real}(\tilde{\mathbf{R}}_{\mathbf{XX}}) & -\text{imag}(\tilde{\mathbf{R}}_{\mathbf{XX}}) \\ \\ \text{imag}(\tilde{\mathbf{R}}_{\mathbf{XX}}) & \text{real}(\tilde{\mathbf{R}}_{\mathbf{XX}}) \end{bmatrix}. \tag{7.18}$$

Replacing $\mathbf{R}_s$ and $\mathbf{R}_x$ by $\tilde{\mathbf{R}}_s$ and $\tilde{\mathbf{R}}_{\mathbf{XX}}$ respectively, (7.16) can be expressed in terms of real variables,

$$\begin{aligned} \underset{\tilde{\mathbf{w}}\in\mathbb{R}^{2N}}{\text{maximize}} \quad & \tilde{\mathbf{w}}'\tilde{\mathbf{R}}_s\tilde{\mathbf{w}} \\ \text{s.t.} \quad & \tilde{\mathbf{w}}'\tilde{\mathbf{R}}_{\mathbf{XX}}\tilde{\mathbf{w}} \leq 1 \\ & ||\mathbf{w}||_0 = P. \end{aligned} \tag{7.19}$$

The quadratic constraint clearly has the convex feasibility region, however, there is still a non-convex constraint involving the l_0 norm. In order to realize the convex feasible region, the l_0 norm is typically relaxed to the l_1 norm, which has been effectively used in many sparse recovery applications. The maximization problem is first transformed to a minimization in order to move the l_1 norm constraint to the objective function and realize a sparse solution. This is achieved by reversing the sign of the entries of the desired source correlation matrix $\overline{\mathbf{R}}_s = -\tilde{\mathbf{R}}_s$,

$$\begin{aligned} \underset{\tilde{\mathbf{w}}\in\mathbb{R}^{2N}}{\text{minimize}} \quad & \tilde{\mathbf{w}}'\overline{\mathbf{R}}_s\tilde{\mathbf{w}} \\ \text{s.t.} \quad & \tilde{\mathbf{w}}'\tilde{\mathbf{R}}_{\mathbf{XX}}\tilde{\mathbf{w}} \leq 1 \\ & ||\mathbf{w}||_0 = P. \end{aligned} \tag{7.20}$$

To convexify the objective function, the concave objective is iteratively approximated through successive linear approximation. The approximation coefficients $\mathbf{m}^i$ and b^i, are updated iteratively $\mathbf{m}^{i+1} = 2\overline{\mathbf{R}}_s\tilde{\mathbf{w}}^i$, $b^{i+1} = -\tilde{\mathbf{w}}^{i\prime}\overline{\mathbf{R}}_s\tilde{\mathbf{w}}^i$ by first order approximation, resulting in the following form,

$$\begin{aligned} \underset{\tilde{\mathbf{w}}\in\mathbb{R}^{2N}}{\text{minimize}} \quad & \mathbf{m}^{i\prime}\tilde{\mathbf{w}} + b^i \\ \text{s.t.} \quad & \tilde{\mathbf{w}}'\tilde{\mathbf{R}}_{\mathbf{XX}}\tilde{\mathbf{w}} \leq 1 \\ & ||\mathbf{w}||_0 = P. \end{aligned} \tag{7.21}$$

Algorithm 7.2: SCA for sparse array beamforming

Input: Received data sparse correlation matrix $\mathbf{R}_{\mathbf{xx}}$, look direction DOA θ.
Output: P sensor beamforming weight vector
Initialization:
Initialize the beamforming vectors randomly to find $\mathbf{m}$ and b. Initialize ϵ, $\mu = 0$.
while (Solution does not converge corresponding to $\mu = 0$) **do**
  Run Eq. (7.24).
end while
Initialize $\mathbf{h}^i$ = all ones vector, Binary vector for hybrid design.
Select μ (Binary search)
while (Beamforming weight vector is not P sparse) **do**
  Run Eq. (7.24). (for initial iteration use $\mathbf{m}^i$ and b^i from previous while loop)
  Update the regularization weighting parameter, $\mathbf{h}^{i+1}(k) = \dfrac{1}{||\tilde{\mathbf{w}}_k^i||_2 + \epsilon}$, Update $\mathbf{m}^i$ and b^i
end while
After achieving the desired cardinality, analytically solve for $\tilde{\mathbf{w}}$ corresponding to the selected sensor locations, yielding, optimal weight vector.
return Optimal weight vector $\mathbf{w}_o$

Finally, the nonconvex l_0 norm is relaxed through minimizing the mixed $l_{1-\infty}$ norm to recover sparse solutions,

$$\begin{aligned} \underset{\tilde{\mathbf{w}} \in \mathbb{R}^{2N}}{\text{minimize}} \quad & \mathbf{m}^{i\prime}\tilde{\mathbf{w}} + b^i + \mu\left(\sum_{k=1}^{N} ||\tilde{\mathbf{w}}_k||_\infty\right) \\ \text{s.t.} \quad & \tilde{\mathbf{w}}'\tilde{\mathbf{R}}_{\mathbf{xx}}\tilde{\mathbf{w}} \le 1. \end{aligned} \tag{7.22}$$

The summation implements the l_1 norm that is minimized as a convex surrogate of the l_0 norm. The vector $\tilde{\mathbf{w}}_k \in \mathbb{R}^2$ has two entries containing the real and imaginary parts of the beamforming weight corresponding to the kth sensor. The $||.||_\infty$ selects the maximum entry of $\tilde{\mathbf{w}}_k$ and discourages the real and imaginary entries concurrently. This is because not selecting a sensor implies the simultaneous removal of both the real and corresponding imaginary entries in the final solution vector. The sparsity parameter μ is set to zero for the first few initial iterations to allow the solution to converge to an optimal solution for the full array elements. The sparsity parameter μ by itself does not guarantee the final solution to be P sparse. To guarantee a P sparse solution, the optimization problem is solved successively against different values of μ. The values of μ are typically given by a binary search over the possible upper and lower limit of μ until the algorithm converges to P sensors [9].

Similar to the approach in Section 7.4.1.1, a few sensors locations might be prefixed to ensure the availability of the full data correlation matrix as required to carry out the proposed optimization algorithm, allowing the use of the weighted formulation, in turn modifying (7.22) as follows,

$$\begin{aligned} \underset{\tilde{\mathbf{w}} \in \mathbb{R}^{2N}}{\text{minimize}} \quad & \mathbf{m}^{i\prime}\tilde{\mathbf{w}} + b^i + \mu\left(\sum_{k=1}^{N} \mathbf{h}(k)||\tilde{\mathbf{w}}_k||_\infty\right) \\ \text{s.t.} \quad & \tilde{\mathbf{w}}'\tilde{\mathbf{R}}_{\mathbf{x}}\tilde{\mathbf{w}} \le 1. \end{aligned} \tag{7.23}$$

The weighting vector $\mathbf{h}$ is a binary vector with $1's$ and $0's$ entries. The entries corresponding to the prefixed sensor locations are set to 0 while the remaining entries are initialized to 1. In this way, the partial penalization is implemented in (7.23) that ensures the sparsity is not enforced to the prefixed locations. The weighted penalization can easily be extended to the reweighting formulation, which can further promote sparsity and facilitates the P sparse solution [51]. This is achieved by iteratively updating the weighting vector $\mathbf{h}$ [58, 59],

$$\begin{aligned} \underset{\tilde{\mathbf{w}}\in\mathbb{R}^{2N}}{\text{minimize}} \quad & \mathbf{m}^{\mathbf{i}\prime}\tilde{\mathbf{w}} + b^i + \mu\left(\sum_{k=1}^{N}\mathbf{h}^i(k)\|\tilde{\mathbf{w}}_k\|_\infty\right) \\ \text{s.t.} \quad & \tilde{\mathbf{w}}'\tilde{\mathbf{R}}_{\mathbf{x}}\tilde{\mathbf{w}} \leq 1. \end{aligned} \tag{7.24}$$

The re-weighting vector $\mathbf{h}^i$, at the ith iteration, is updated as an inverse function of the beamforming weights at the present iteration,

$$\mathbf{h}^{i+1}(k) = \frac{1}{\|\tilde{\mathbf{w}}_k^i\|_2 + \epsilon}. \tag{7.25}$$

This relatively suppresses the low magnitude weights in the next iteration to accelerate sparsity. The parameter ϵ avoids the case of division by zero. The reweighting is applied to both the freely designed array and the hybrid design. For the former, the vector $\mathbf{h}$ is initialized to the all $1'$s vector and updated iteratively. However, to preserve the prefixed sensor locations for the hybrid design, the entries of $\mathbf{h}^i$ corresponding to the prefixed locations must remain zero for all iterations, while the remaining entries are initialized to 1 and updated as explained above. The procedure is summarized in Algorithm 2.

7.4.3 SBSA Design

Several different approaches have been developed to mitigate the computational expense of the combinatorial search by either exploiting prior given information on the interference parameters, such as respective DOAs, or employing data-dependent algorithms realized through the SDR and SCA algorithms as explained above [5, 30, 60].

These algorithms, however, have high computational costs, impeding real time implementations, especially in applications involving fast dynamic environments. In the context of designing sparse arrays using machine learning, the optimum sparse arrays for different operating environments constitute labels for the network training data. These labels can be generated off-line through enumeration, accounting for all possible source and interference scenarios. For a large array, however, this approach becomes computationally challenging. In order to mitigate this problem, an efficient technique to generate neural network training samples using SBSA is proposed. This design has desired performance and low computational complexity.

7.4.3.1 The Role of Spare Configuration in MaxSINR

As implied by (7.7), the optimum sparse array for MaxSINR design depends on both the beamforming weights and the sparse array configuration. Therefore, the problem of interference mitigation is not a cascade design of the optimum configuration followed by the optimum weights, or vice versa. Rather, it is an entwined task, in essence, calling for joint optimization of the beamforming weights and sparse array configuration. Albeit not optimum, the cascade design approach is not entirely without merits. It can offer a unique insight into the problem and address the suitability of a given sparse array configuration in canceling the interfering signals.

In order to shed more light on the role of array configuration in optimum sparse beamforming, it is recognized that the problem formulation developed in Section 7.3 is valid irrespective of the array configuration and holds true for the compact Uniform Linear Array (ULA) or any generic sparse configuration. The beamformer output signal, $y(t) = \mathbf{w}^H\mathbf{x}$, for a sparse beamformer can also be written as $\mathring{y}(t) = \mathring{\mathbf{w}}^H \mathbf{x}$, where the typeset "∘" indicates that the corresponding vector is sparse with few zero entries. The sparse beamformer $\mathring{y}(t)$ can be rewritten, equivalently, as $\mathring{y}(t) = \mathring{\mathbf{w}}^H \{\mathring{\mathbf{z}} \odot \mathbf{x}\}$. The entries of $\mathring{\mathbf{z}} \in \mathbb{R}^N$ are either 1's or 0's depending on whether the corresponding sensor location is active or inactive respectively. The point-wise multiplication ($\odot$) of the received vector $\mathbf{x}$ with a sparse selection vector $\mathring{\mathbf{z}}$ sets the corresponding entries of the received signal to zero, which amounts to the zero beamforming weights of $\mathring{\mathbf{w}}$.

Therefore, sparse beamforming can be viewed as pre-processing the received signal by point-wise multiplication, prior to applying the optimum weights. The impact of such multiplication on the performance of the subsequent, or cascade, design of the beamformer weights is analyzed, $\mathring{\mathbf{w}}$. Let $\mathbf{X} = \mathcal{F}(\mathbf{x})$ and $\mathbf{Z} = \mathcal{F}(\mathring{\mathbf{z}})$ denote the Discrete Fourier Transform (DFT) of the input data and the selection vector, respectively. The DFT of the point-wise multiplication of these two vectors is given by the circular convolution of the corresponding DFTs, i.e., $\mathcal{F}(\mathring{\mathbf{z}} \odot \mathbf{x}) = \mathbf{X} \circledast \mathbf{Z}$, where $\circledast$ denotes circular convolution. From (7.1) and the linearity property of DFT, $\mathcal{F}(\mathring{\mathbf{z}} \odot \mathbf{x}) = \alpha\mathbf{S} \circledast \mathbf{Z} + \sum_{l=1}^{L} \beta_l \mathbf{V}_l \circledast \mathbf{Z} + \mathbf{N} \circledast \mathbf{Z}$ is obtained, where $\mathbf{V}_l$ is the DFT of the lth interference vector. The N components of $\mathbf{X}$ represent the contributions to orthogonal beamformers, each pointing towards a certain spatial direction. The convolution with $\mathbf{Z}$ alters these components as well as changes the contribution of the lth interference to the signal received from the desired source location. The proposed scheme gauges the overlap between the spatial spectra of the desired signal and the interfering signals as a function of the underlying sparse configuration. A desirable array configuration would minimize such overlapping, enabling the beamformer weights to effectively remove the interference, while maintaining the desired signal.

Towards this end, we propose a design metric $\Omega(\mathbf{Z})$ based on the weighted sum of the spatial spectrum (denoted by $|.|^2$) of the individual interfering sources scaled by the desired signal spatial spectrum as follows,

$$\Omega(\mathbf{Z}) = \sum_{l=1}^{L} \{\alpha^2 |\mathbf{S} \circledast \mathbf{Z}|^2\} \odot \{\beta_l^2 |\mathbf{V}_l \circledast \mathbf{Z}|^2\}. \tag{7.26}$$

Equation (7.26) performs element-wise scaling of the interfering powers in the DFT domain. Therefore, if the interfering signal power, after convolution is concentrated in the DFT bins different from those occupied primarily by the desired signal, then the point-wise product would assume low values. Conversely, if there is a significant overlap between the results of the two convolutions, then the objective function $\Omega(\mathbf{Z})$ is significantly higher.

The spatial spectrum can be estimated by computing the DFT of the autocorrelation function of the corresponding signal. It is worth noting that, for a given sparse configuration, the autocorrelation function of the selection vector assumes a specific redundancy of the autocorrelation lags. Therefore, unlike the structured sparse array design that seeks to maximize the contiguous correlation lags, MaxSINR sparse design is guided by the DFT of the autocorrelation sequence of the lag redundancy.

The following example illustrates the proposed approach. Consider an eight-element sparse array on the 24-point equally spaced grid locations that are potentially available for sensor selections. The minimum spacing among the sensors is $d = \lambda/2$. Consider a source signal located at 60° and four unwanted interferences located at 154°, 55°, 117°, and 50° with the interference-to-noise-ratio (INR) ranging from 10 − 20 dB. The sparse array configuration achieving the best SINR performance is found through enumeration and shown in Fig. 7.3. The associated correlation lag redundancy of this configuration and the corresponding spatial spectrum are depicted in Figs. 7.4 and 7.5, respectively. The spatial spectrum of the desired source at 60° prior to convolution is

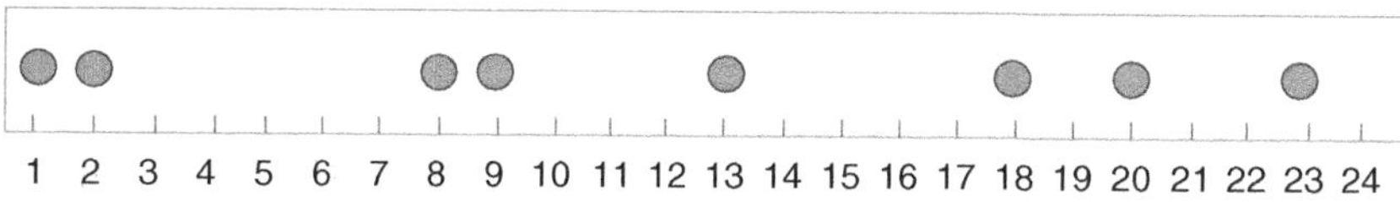

Figure 7.3 Eight element sparse array configuration.

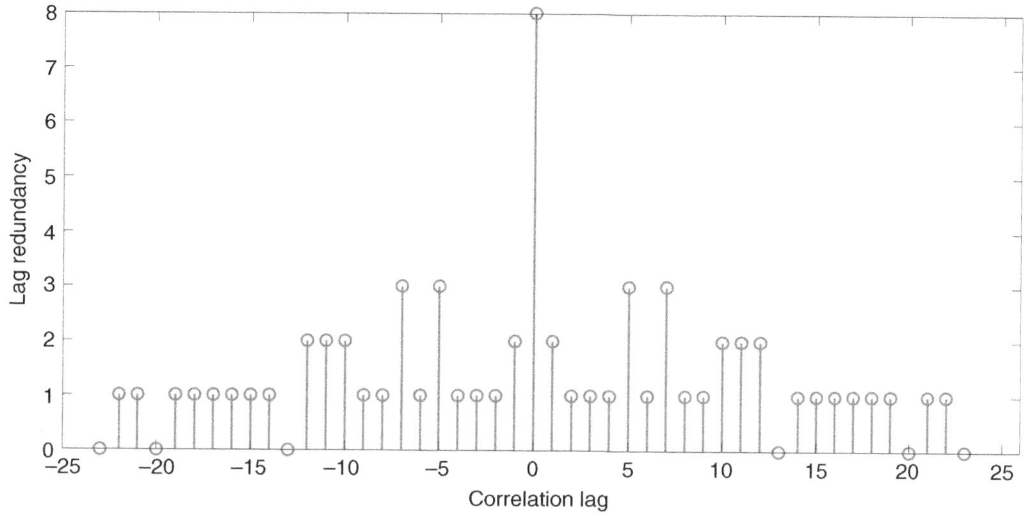

Figure 7.4 Lag redundancy of the sparse array shown in Fig. 7.3.

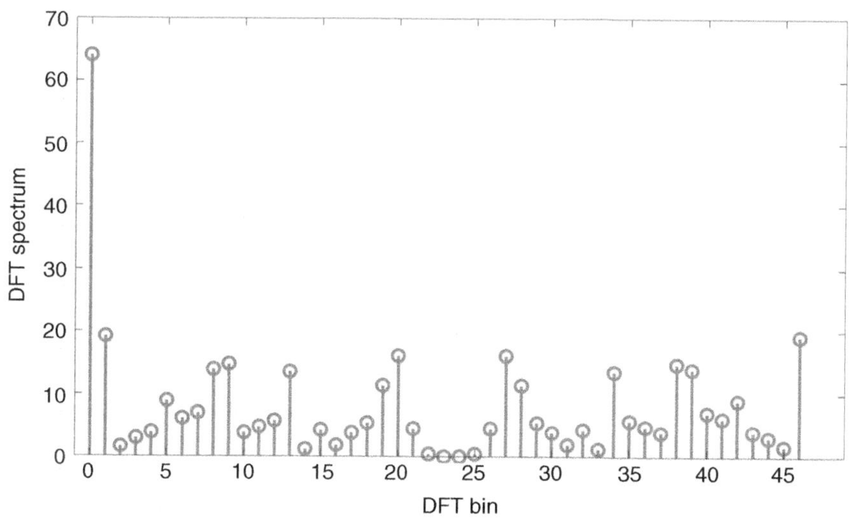

Figure 7.5 DFT of the lag redundancy.

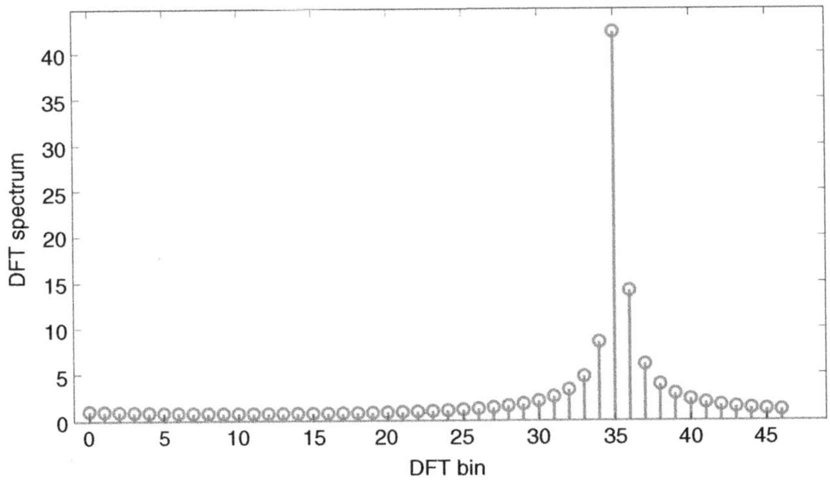

Figure 7.6 Power spectrum of the desired signal.

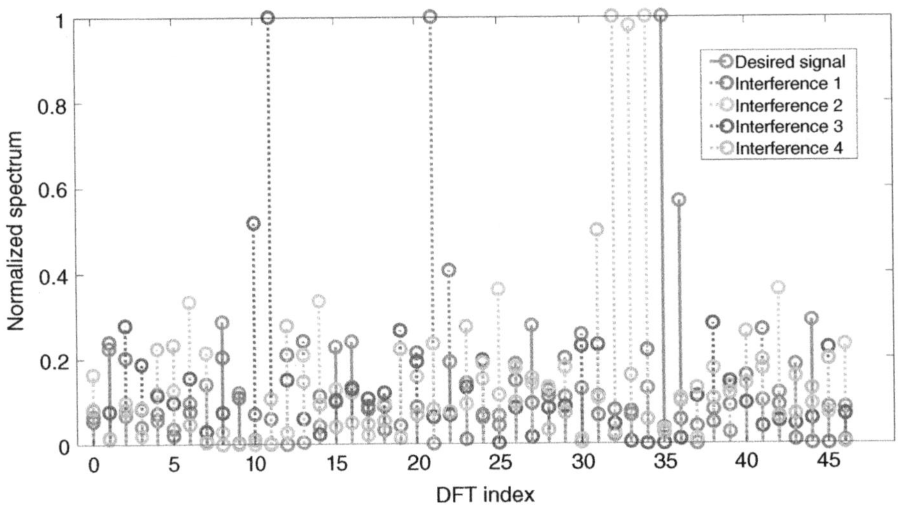

Figure 7.7 Explanation of the proposed objective criterion for the optimum array configuration shown in Fig. 7.3.

depicted in the Fig. 7.6. The normalized result of the two convolved spectra in Figs. 7.5, 7.6 is shown as a solid line in Fig. 7.7.

The normalized spectrum for each interfering signal is also shown as a dotted line in the same figure. Figure 7.8 plots the normalized spectrum for the sparse array worst case scenario for comparison purposes. Note that the maximum of the convolved desired signal spectrum occurs at the 35th DFT bin. For the best case scenario, all convolved interfering signals assume minimum power at the aforementioned DFT position. This is in contrast to the worst case scenario where there is considerable interference power at the peak of the desired source. Apart form the maximum location, it is noted that for the best case design, there is minimum overlapping between the desired signal and interfering signals at the DFT bin locations, which is clearly in contrast to the worst case design.

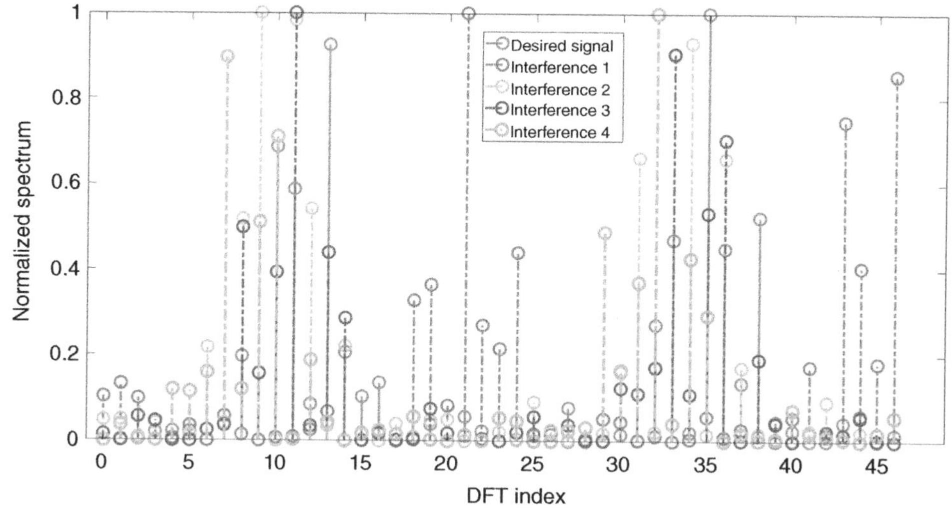

Figure 7.8 Explanation of the proposed objective criterion for the worst possible array configuration.

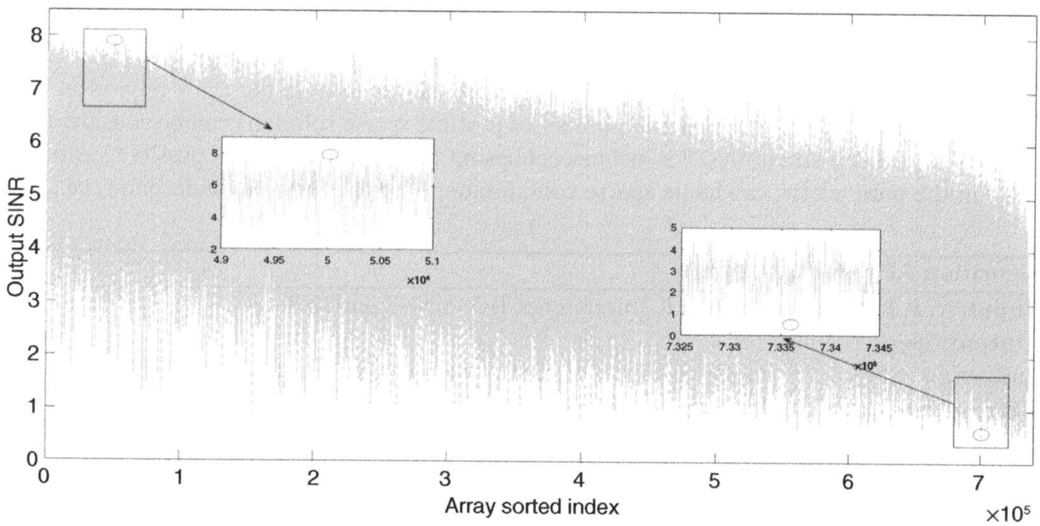

Figure 7.9 Plot of the proposed objective criterion in ascending order.

In order to further understand the offerings of the proposed approach, Fig. 7.9 plots the SINR performance of all possible sparse configurations after sorting the array configurations in ascending order of the output of the proposed objective function of (7.26). It is clear that the average SINR in the plot is higher and more desirable towards the left side where the objective function is minimum. It is important to note that the best enumerated result of MaxSINR performance does not correspond to the array configuration with the smallest objective function. This is because, the optimum sparse configuration also depends on the beamformer weights to minimize the interfering signals. This is also clear from the high variance of the curve. Similarly, the worst performing array is very close to the right side of the curve where the proposed objective function is high due to the strong overlap of the desired and undesired signal spectra. At each iteration, the proposed algorithm involves performing 1 FFT operation and $L + 1$ convolutional operations, each having a computational complexity of $\mathcal{O}(N\,logN)$.

In lieu of enumerations, and for efficient generations of DNN training data, the objective function (7.26) can be minimized through an iterative algorithm that implements successive sensor selections, hence deciding on one sensor location at a time. For the initial iteration, a sensor location is chosen randomly on an N grid points. For the ith subsequent iteration, the proposed objective is evaluated at the remaining $N - i$ locations and then selecting the sensor location that yields the minimum objective function. The procedure is iterated P times for selecting the P locations from N possible locations. Due to high variance of the curve in Fig. 7.11, it is best to initialize the algorithm with different sensor locations and find the corresponding configurations for each initialization, and eventually consider the one with the best SINR performance. The steps of the above algorithm are detailed in Algorithm 3. It is noted that aside from the efficient generations of machine-learning training data in the underlying problem, this approach can itself be used in general as a stand alone method to determine appropriate array configurations if prior information of the interference parameters is provided.

7.4.4 Summary of Data Generation Approaches

This section discussed the different sparse array design approaches, including SDR, SCA, hybrid sparse array design, and SBSA design.

Although the training phase is entirely offline, it is infeasible to train the DNN relying on the enumerated results. This is because the number of possible sparse solutions can be considerably large even for modest size arrays. For instance, choosing 14 elements out of 24 possible locations results in the order of 10^6 candidate sparse configurations for each training data point, i.e., for

Algorithm 7.3: SBSA algorithm

Input: N, P, Look direction DOA θ, Interference DOAs, SNR and INRs
Output: Sparse beamformer $\mathbf{w}_o$
Initialize $\mathbf{z} = [0\ 1\ 0 \ldots 0]$ where all entries of $\mathbf{z}$ are zero except an arbitrarily selected entry.
Compute the spatial spectrum of desired source and interfering signals.
for $\mathrm{j} = 1$ to P-1
for $\mathrm{i} = 1$ to P-1-j
Select the ith sensor from P-1-j remaining locations.
Compute the lag redundancy of this sparse array consisting of $\mathrm{j} + 1$ sensor.
Compute the spatial spectrum of the $\mathrm{j} + 1$ sensor sparse array.
Convolve the spatial spectrum of the $\mathrm{j} + 1$ sensor sparse array with the spectrum of the desired source and the interfering sources.
Compute the overlapping power in the spatial spectra by computing the proposed metric in (7.26).
end for
Select the ith sensor from the inner for loop which results in the minimum overlapping power computed by (7.26).
Update $\mathbf{z}$ by setting the $\mathrm{j}th$ location in $\mathbf{z}$ to 1.
end for
After finding the sparse configuration find $\mathbf{w}_o$ by running (7.6) for reduced size correlation matrix while ignoring the sensor locations corresponding to $\mathbf{z}$.

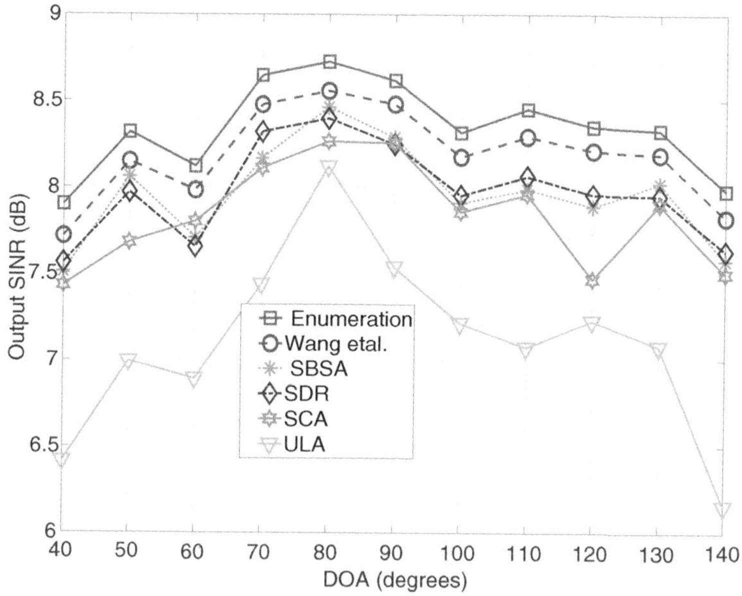

Figure 7.10 Performance comparison of proposed SBSA algorithm with other state-of-the-art algorithms.

each environment scenario. Instead of enumeration, the SBSA algorithm can be used to reduce the number of candidate configurations that need to be examined. The SBSA design algorithm is compared with the other state-of-the-art sparse array optimization algorithms in terms of average SINR performance in Fig. 7.10. It shows the Monte Carlo simulations where each point on the horizontal axis represents scenarios where the desired source DOA is fixed with an SNR of 0 dB, and three interference DOAs are generated through a uniform random distribution with the three interference DOAs equally likely on a grid ranging from 20° to 160°. The INRs of these sources are uniformly drawn from 0 to 20 dB and eight antenna locations out of possible $N = 16$ equally spaced locations with inter-element spacing of $\lambda/2$ are chosen. On average, the proposed SBSA algorithm consistently provided superior SINR performance which is around 1.2 dB sub-optimal to the average SINR computed through enumeration. However, the computational complexity of enumeration is very high as selecting 8 out of 16 antennas requires complex matrix operations such as eigenvalue decomposition for over 12,836 possible sparse array configurations. The proposed SBSA algorithm underperforms the algorithm proposed in [31]. However, the algorithm in [31] is based on a convex relaxation approach and has significantly higher computational cost as compared to the SBSA algorithm. On average, the proposed algorithm shows better SINR performance as compared to the SDR [2] and SCA algorithms [39], which is depicted in Fig. 7.10. The SCA algorithm relies on successive linear approximation of the objective function as opposed to the quadratic implementation of the SDR, thereby further deteriorating the performance. It is important to note that the proposed SBSA algorithm and algorithm in [31] both require the apriori knowledge of the interference steering vectors. Contrary to this, the SDR and SCA algorithms provide a tradeoff between lack of apriori knowledge and SINR performance by directly operating on the data correlation matrix. The SINR performance for the compact ULA is also shown in the Fig. 7.10, further highlighting the utility of sparse array design. On a 1.4 GHz Quad-Core Intel Core i5 computer running MATLAB 2021, the average run time of the proposed algorithm is 25 ms as opposed to 0.8 seconds for algorithm in [31] and 1.5 and 1.1 seconds for SDR and SCA algorithms, respectively. The computational complexity of the proposed algorithm scales as $O(N \log N)$ as compared to considerably higher order complexity for convex relaxation algorithms, which makes the proposed algorithm more viable for problems with higher number (N) of antenna elements. The enumeration approach is NP hard and cannot be solved in polynomial time. The SDR and SCA algorithms have to successively solve the optimization problem for different values of μ, leading to a significant runtime cost.

The high computational costs of SDR, SCA, and the hybrid sparse array algorithms not only impede real time implementations for environment-dependent fast PAC, the computational cost is also detrimental to training data generation times for machine-learning-based sparse array design approaches. In order to mitigate this problem, the SBSA algorithm can be utilized to decrease training data generation time via the iterative-based algorithm with a computational complexity of $\mathcal{O}(N \log N)$. On a 1.4 GHz Quad-Core Intel Core i5 computer running MATLAB 2021, the average run time for selecting 10 antenna locations out 20 possible locations, for the SBSA algorithm is 250 ms as opposed to 4.75 seconds for SDR and 2.1 seconds for SCA algorithm. The reduction in computation cost greatly assists training times, but is still a significant cost for near real time PAC implementations.

7.5 Machine-Learning Methods for Sparse Array Design

This section presents the design considerations and challenges associated with utilizing machine learning for sparse array design. Additionally, results of utilizing machine learning for sparse array design are presented for multiple scenarios.

There are a variety of challenges when shifting from an analytical design approach, such as those discussed in Section 7.4, to a machine-learning-based approach. Specially, this section will discuss the issues of generalization, a noisy input–output space, input data format, obtaining full correlation data, and network architectures.

7.5.1 Generalization

The analytical approaches of Section 7.4 are applicable for any input correlation matrix, producing the optimum sparse array design for any given auto correlation matrix input. However, a machine-learning approach will only have a finite number of training samples to learn from and as the size of the sparse array grows, the percent of the input–output space represented by the training data will decrease. This means that a large percentage of the input–output space will not be seen by a neural network during training. Therefore, for effective learning, it is important that the machine-learning model can generalize to a broader class than that represented by the finite number of drawn training examples.

From the Capon beamforming perspective presented in this chapter, a given arrangement of a desired source direction, interference DOAs, and respective SNR/INRs constitute just one particular example. A class, in this case, is defined by any arbitrary permutation of the interference DOAs and respective powers while keeping the desired source DOA fixed. The learning task of the network is, therefore, to learn from a dataset, characterized by a repository of different training examples and corresponding optimum sparse array predictions.

An important question arises to whether a stronger notion of generalization could be achieved by generalizing over all possible desired source DOAs, instead of training for a fixed desired source. To answer this query, note that for a given desired source and interference setting, the received array data, and thereby the data correlation, remains the same even if the desired signal and one of the interferers is switched. However, the Capon beamformer should yield a different configuration due the directional constraint being changed. Therefore, instead of relying entirely on the information of the received data correlation, it is imperative to incorporate the knowledge of the desired source or look direction, which is always assumed in the Capon beamforming formulation. For the machine-learning network, this information can either be incorporated by exclusively training the DNN for each desired source DOA or the desired source DOA can be incorporated as an additional input feature to the network. The networks in this chapter will adopt the approach of focusing on a single source DOA.

7.5.2 Noisy Input–Output Space

Another challenge when trying to create a sparse array design algorithm when utilizing machine learning is that the training data input–output space can be very noisy, meaning that two very similar auto correlation matrices can result in very different optimum array designs, in terms of Hamming distance. The similarity in input but drastic change in output makes it challenging for the network to learn, as it is getting conflicting signals for how to adjust the network weights.

There are different ways to help mediate this issue. One is to smooth the output space, meaning that the sparse array design is limited in how many sensors can change for small perturbations in input parameters, such as interferer angles or power. While this would help the machine-learning algorithm, it may also reduce the accuracy of the machine-learning approach when compared with an analytical approach, such as SDR or SCA enumeration-based approaches.

Another approach is to change the output encoding of the network. Instead of directly outputting the array structure as a binary output from the machine-learning model, where the maximum N out of M outputs are considered “on” or 1s, the possible arrays structures can be encoded into classes. Through array symmetry and practical switching considerations [61–63], such as those discussed in Section 7.5.5.2, the number of possible output array structures is greatly reduced. This helps smooth out the noise in the training data and thus helps the network learn the underlying sparse array design problem. This section will show architectures that utilize symmetry and binary switching to reduce the number of classes.

7.5.3 Input Data Format

The format and dimensionality of the data going into the machine-learning network is also an important consideration. The auto correlation data has both real and imaginary components. For a CNN, the real and imaginary parts can be separated out into different dimensions. For a MLP network, the real and imaginary values have to be either interleaved, or be placed one after another.

Prior networks for different tasks [41] also include the phase as an input to the network as well. For the CNN based networks, this can again be added as an additional dimension. The MLP based network could also have the phase data interleaved or concatenated to the real and imaginary data.

The modality of the input data can also be varied. Instead of utilizing the auto-correlation matrix as the input to the network, the FFT can be taken and utilized as the input or ancillary input to the network.

7.5.4 Obtaining the Full Correlation Matrix

The sparse array design can only have a few active sensors at a time, in essence, making it difficult to furnish the correlation values corresponding to the inactive sensor locations. However, for the scope of this chapter, it is assumed that an estimate of all the correlation lags corresponding to the full aperture array are available to input for prediction. This can typically be achieved by estimating missing correlation lags through sequential antenna switching or employing a low rank matrix completion strategy that permits the interpolation of the missing correlation lags [39]. Therefore, generalization in this context means that the learned machine-learning model works on different interference settings which can change according to changing environment conditions.

7.5.5 Network Architectures

Three different network architectures are explored in this section. The input modalities and output encoding utilized vary across the networks. This section will present the different network architectures and compare and contrast the implementations. Section 7.6 will present the results obtained from using each network architecture.

7.5.5.1 Dual Network Architecture

The first network architecture, shown in Fig. 7.11, utilizes two input modalities including the correlation matrix and the fast Fourier transform (FFT) input. Since the network utilizes two different input modalities, it will be referred to as the DNN throughout this work. The correlation array is of size $N \times N \times 2$ and is obtained by computing the real and imaginary parts of the elements of an $N \times N$ correlation matrix. The FFT is obtained by computing the M point FFT applied to the first row of the received correlation sequence ($\mathbf{r}_{\mathbf{x}}(n-1) = \mathbf{R}_{\mathbf{xx}}(1, n)$). The M point FFT is computed by appropriately zero padding the N values of the correlation sequence. The real and imaginary values of the FFT are then concatenated into a vector of size $2M$.

The correlation matrix is passed through a five-layer CNN consisting of 16 filters in each layer, employing 3×3 convolution. The FFT vector is passed to a MLP network with five-hidden layers as shown in Fig. 7.11. The outputs of the two networks corresponding to both the data inputs are concatenated and then inputted to a dense layer, followed by an output layer of size T, where T represents the number of possible *symmetric* array configurations. By utilizing only the symmetric array representations, the number of possible outputs for the network is reduced to 30 for $N = 12$

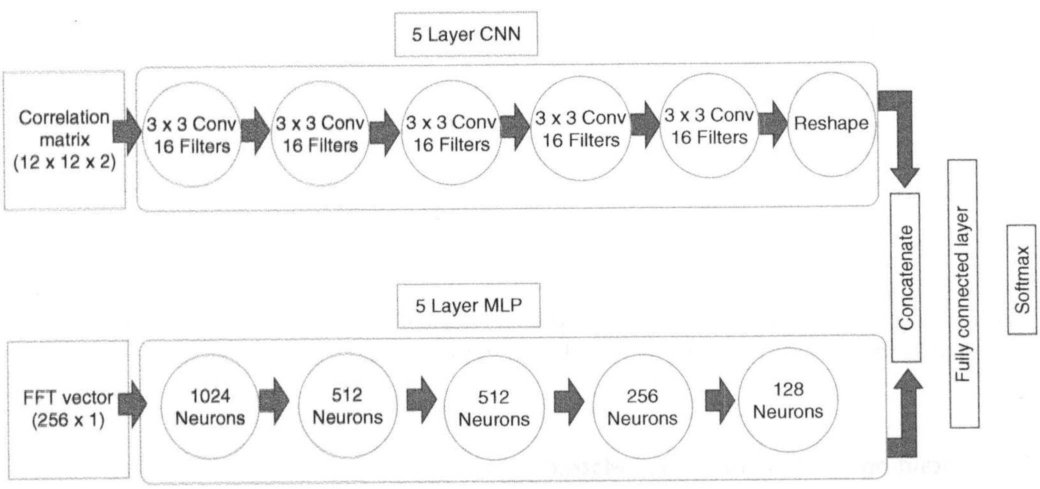

Figure 7.11 Architecture of the dual neural network (DNN). Both the data correlation matrix and FFT are utilized as inputs to the network.

and $M = 6$. Reducing the number of output possibilities allows for a one hot encoded vector to represent the array structure, where the location of a single "1" in the output vector indicates a specific sparse array topology. The use of a one hot encoded output, rather than directly outputting the sparse array design, is chosen in part in an attempt to reduce the IO noise that was observed in the data generation, as discussed in Section 7.5.2. All the internal layers of the network utilize a ReLU activation function, as opposed to the output layer, which utilizes the softmax activation function.

The two other network architectures utilize constraints on the network determined by the switches utilized in the sparse array implementation. More specifically, a binary switching strategy is utilized where each switch connects to only two sensors in the array. The specific switching strategies are important to consider for the network architecture as the strategies alter the number of possible output classes.

7.5.5.2 Binary Switching Strategies

There are a variety of possible ways to connect N sensors to the $N/2$ switches. This section focuses on two binary switching strategies, which are shown in Fig. 7.12, and consist of an adjacent and separate strategies, for selection of N out of $M = 2N$ antennas. Strategy-1 pairs adjacent antennas via a binary switch, whereas Strategy-2 divides the M-element array into two N-element subarrays and pairs two subarray antennas, separated by N grid points [7].

The total number of possible sparse array configurations for both strategies is 2^N. However, under a far-field assumption, linear sparse arrays, which are related by spatial shifts along the baseline in either direction, lead to the same optimum SINR. Additionally, sparse arrays that are generated from each other by a simple reflection around their center points also provide the same SINR. For illustration, Fig. 7.12c depicts three sparse arrays with $M = 12$ and $N = 6$ for Strategy-2. The top and middle configurations are related through a right-shift of two units, while the top and the bottom configuration are related via a reflection around the center point. The redundant cases are removed, meaning that only the arrays that produce a unique SINR are considered. Removing the redundant cases results in 36 classes when $M = 12$ and $N = 6$ for the adjacent Strategy-1 and 20 classes for separate Strategy-2. As mentioned previously, this reduction in classes helps to smooth out the input–output space to help overcome the challenge of sparse array design having lots of noise, as discussed in Section 7.5.2.

7.5.5.3 Binary Switching Network Architectures

In addition to switching to a binary switching strategy, the two binary switching network architectures forego the FFT input modality and focus solely on the covariance matrix. The main rationale for removing the FFT input is the consideration of the PAC complexity. Computing the FFT not only adds an additional processing step to the input, but also requires a larger network, influencing the inference time of the network itself.

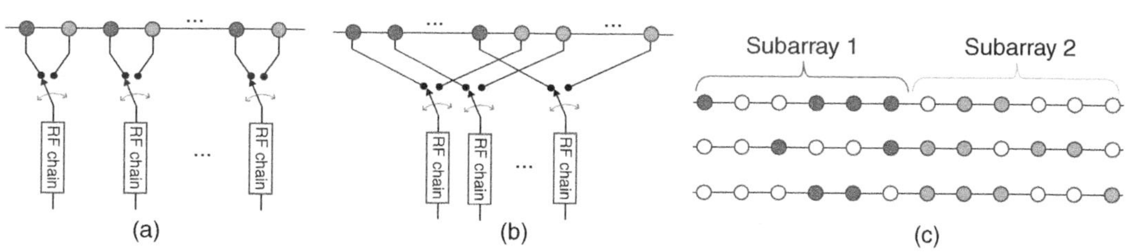

Figure 7.12 (a) Binary switching Strategy-1: Adjacent antennas; (b) Binary switching Strategy-2: Separate subarrays; (c) Examples of sparse arrays that yield the same SINR under far-field conditions for Strategy-2.

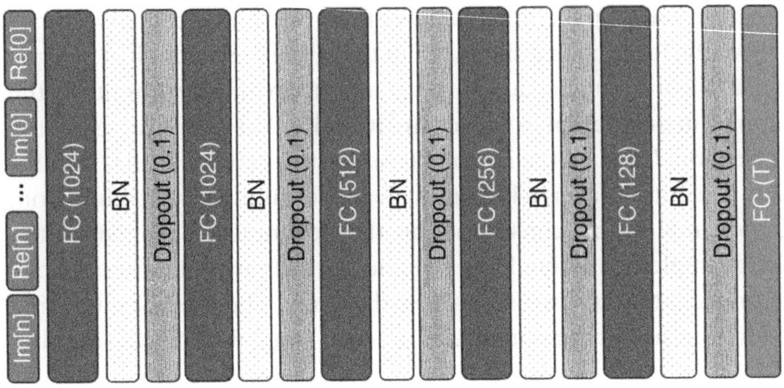

Figure 7.13 MLP architecture.

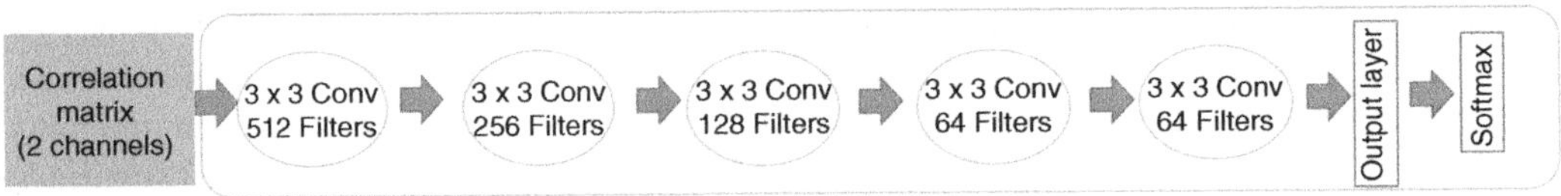

Figure 7.14 CNN architecture.

Two different network architectures are considered for each binary switching strategy. One network is a MLP and the other is a CNN. The MLP network is shown in Fig. 7.13, where five fully connected (FC) layers of respective sizes 1024, 1024, 512, 256, and 128 are used. The output layer has T outputs where T corresponds to the number of sparse array classes for the binary switching strategies. The inputs to the MLP network are the real and imaginary parts of the autocorrelation sequence of the received data, leading to a size of $2M$. The first four layers use the rectified linear unit (ReLU) activation function. Batch normalization and 10% dropout layers are used after each of the first four FC layers. The output layer uses a softmax activation function, with the outputs of the network being one-hot encoded such that the location of a "1" indicates a specific sparse array topology.

The employed CNN is shown in Fig. 7.14. The input is of size $2M^2$ and comprises the real and imaginary parts of the $M \times M$ correlation matrix. The output layer is of size T and represents the possible switched array configurations. The network input is passed through a five-layer CNN implementing 512, 256, 128, 64, and 64 filters of size 3×3. Batch normalization and 10% dropout layers are used after each convolution layer. The network output is also a one-hot encoded vector. The ReLU activation function is used for all layers except the output layer, which uses a softmax function.

7.6 Simulation Results

The performance of the DNN, MLP, and CNN architectures described in Section 7.5.5 are analyzed in this section. Additionally, the description of the data generation procedures is provided.

7.6.1 DNN Simulation Performance

7.6.1.1 Data Generation

The received data is generated in the following manner. For a given desired source location, the ith training realization is simulated by randomly selecting L interfering signals from a DOA grid as long as the minimum spacing among the interferes is maintained. The grid spans the FOV from 10° to 170°, with a grid spacing of 1°. The interferers are allocated random powers uniformly distributed with INR from 0 to 20 dB. To simulate an off-grid DOA, uniform random noise with a range of 0° to 0.5° is added to the interference angles. For this given scenario, the data correlation matrix corresponding to an N element array is computed. This matrix includes the contribution of the desired source signal. The corresponding optimum configuration is found through enumeration and also through the proposed SBSA algorithm.

It is noted that, in this case, the optimum configuration is computed by assuming the interference angles lie on the grid. This is because the performance of the sparse configuration is largely unaffected by a small change of 0.5°, given that the beamforming weights utilize the exact DOAs as expressed in the received correlation values. Thc process is repeated 50,000 times against a given desired source DOA to generate the training data set.

For the two interference scenario, the training examples are simulated by selecting two interfering signals from a DOA grid spanning from 10° to 170°, with a grid spacing of 1°. The minimum spacing among the interferes is maintained to be 4°. This results in 3440 possible DOA combinations. In total, 50,000 examples are generated by varying the INR from 0 to 20 dB and perturbing the interference DOA with a uniform noise with a range of 0° to 0.5°. Since there are 50,000 training examples and 3440 possible DOA combinations, this exposes the DNN to 15(50,000/3,440) examples of a given DOA combination.

Likewise the testing data is generated off grid by adding uniform noise of 0.5° to possible interference DOAs. The SNR of the desired signal is kept at 0 dB for all these scenarios.

7.6.1.2 Results

The results obtained for the DNN network utilize a crossentropy loss function and the ADAM algorithm [64] for parameter optimization. An exponential learning rate is used with the initial learning rate set to $1E-4$. Weight initialization is performed via Xavier initialization.

The results presented for the DNN architecture select $P = 6$ antennas from $N = 12$ possible equally spaced locations with an inter-element spacing of $\lambda/2$. Three different simulation scenarios are considered, including a single interference, two interference sources, and a case with a mix of one and two interference sources. The mixed interference scenario covers the case where the number of interference signals are not precisely known. The 3000 sample testing data includes added uniform noise to the grid locations to ensure the dataset varies from the training dataset.

Figure 7.15 shows the output SINR performance comparisons for different array configurations in the single interference scenario. The horizontal axis is the DOA of the desired point source, and the performance is computed at six different source DOAs between 45° and 135°. The results are presented in Fig. 7.15 and are obtained by using an unlimited number of data snapshots, i.e., exact spatial correlation values. Figure 7.15 compares eight different cases referred to as EN, EN-SYM, EN-SYM-DNN, SBSA, SBSA-DNN, ULA, Center Sparse, and Random. EN represents the average SINR performance achieved through enumeration of all $P = 6$ and $N = 12$ sparse array configurations, representing the upper bound of performance. EN-SYM represents the MaxSINR achieved through enumeration over all possible symmetric arrays. EN-SYM-DNN represents the performance achieved by the DNN. The ULA and "Center Sparse" results represent the average SINR performance of array configurations shown in Fig. 7.20a and b, respectively. Random implies the average SINR performance of a randomly chosen array from all possible configurations.

It is evident from Fig. 7.15 that for the single interference scenario, the EN, EN-SYM, and EN-SYM-DNN results have nearly identical performance. This demonstrates that, for the underlying problem, utilizing the symmetric array configurations provides a nearly identical performance opportunity as utilizing all possible configurations. For this single interference scenario, the

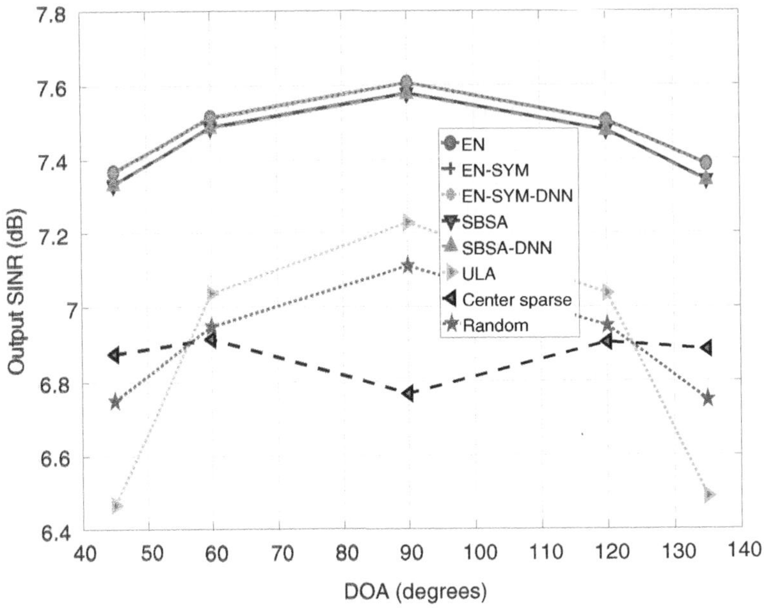

Figure 7.15 Performance comparison for Scenario 1 considering one interference DOA.

EN-SYM-DNN has a classification accuracy of 99.4%, meaning the model almost exclusively chooses the correct symmetric array class.

7.6.2 DNN-Based SBSA Design

As discussed in Section 7.4, the complexity of traditional optimization algorithms to determine the optimal sparse array can be prohibitively large when the size of the array increases. One solution was to utilize the SBSA aglorithm. The SBSA algorithm is constrained to generate symmetric arrays by concurrently considering two symmetric sensor locations, at each iteration, rather than minimizing the cost function at each sensor location. Figure 7.15 shows that the SBSA design performs close (0.001 dB trade off) to the performance of the EN and EN-SYM designs found through enumeration. The DNN trained on the SBSA data (SBSA-DNN) achieves a classification accuracy of 98.7% and effectively matches the SINR performance of SBSA. The SBSA-DNN-based sparse array designs are superior to the ULA, Center Sparse, and Random array configurations by more than 0.5 dB on average.

The SBSA and SBSA-DNN are evaluated for Scenario 2, which involves two interference signals, and Scenario 3, which represents a mix of one and two interference sources, as shown in Figs. 7.16 and 7.17, respectively. In both scenarios, the DNN achieves an average classification accuracy of approximately 94%. However, this corresponds to an average performance degradation of less than 0.028 dB as compared to the enumerated design (EN). It is evident from Figs. 7.15, 7.16, and 7.17 that the proposed SBSA-DNN design yields significant gains over a compact ULA, center sparse array, and randomly selected array topology.

The confusion matrices for Scenario 3 are shown in Figs. 7.18 and 7.19, respectively, where Fig. 7.18 has a 45° DOA and Fig. 7.19 has a 90° DOA. The results demonstrate that a few symmetric configurations are employed more often than the other configurations. For example, the array structure shown in Fig. 7.20b, referred to as the "Center Sparse" array, is the most frequent

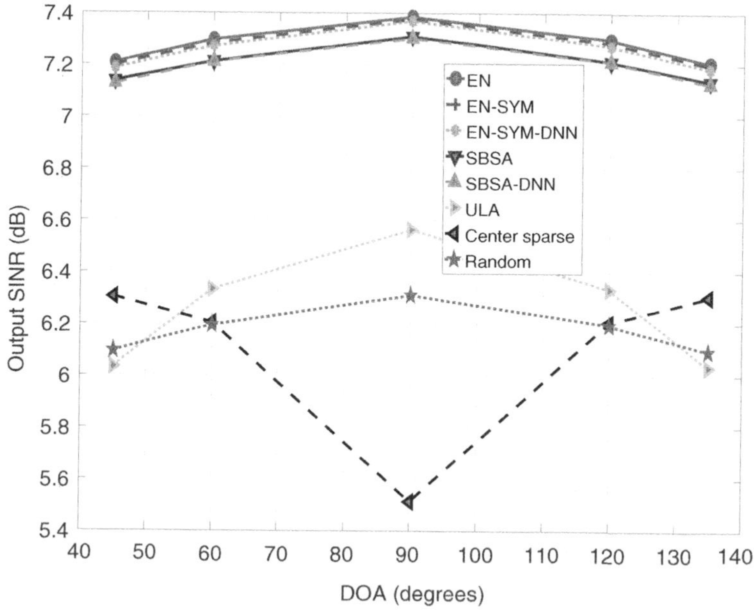

Figure 7.16 Performance comparison for Scenario 2 considering two interference DOAs.

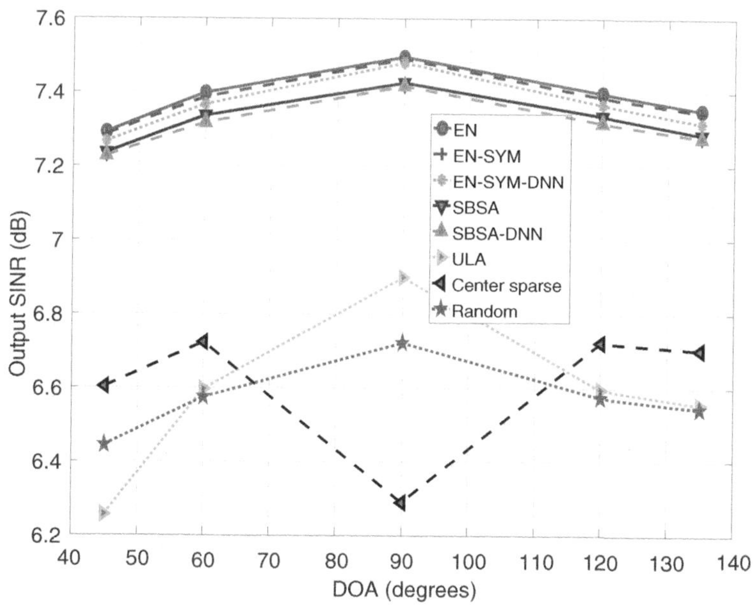

Figure 7.17 Performance comparison for Scenario 3 considering a mix of one or two interference DOAs.

configuration, followed by the array depicted in Fig. 7.20c. It is noted that the configuration in Fig. 7.20d is used only two times when the desired source is at 45° but is configured more often when the source look direction is at 90°. The same holds true for the configurations shown in Fig. 7.20e and f, which sparingly emerge when the source look direction is 90°, but increase in frequency when the desired source is at 45°.

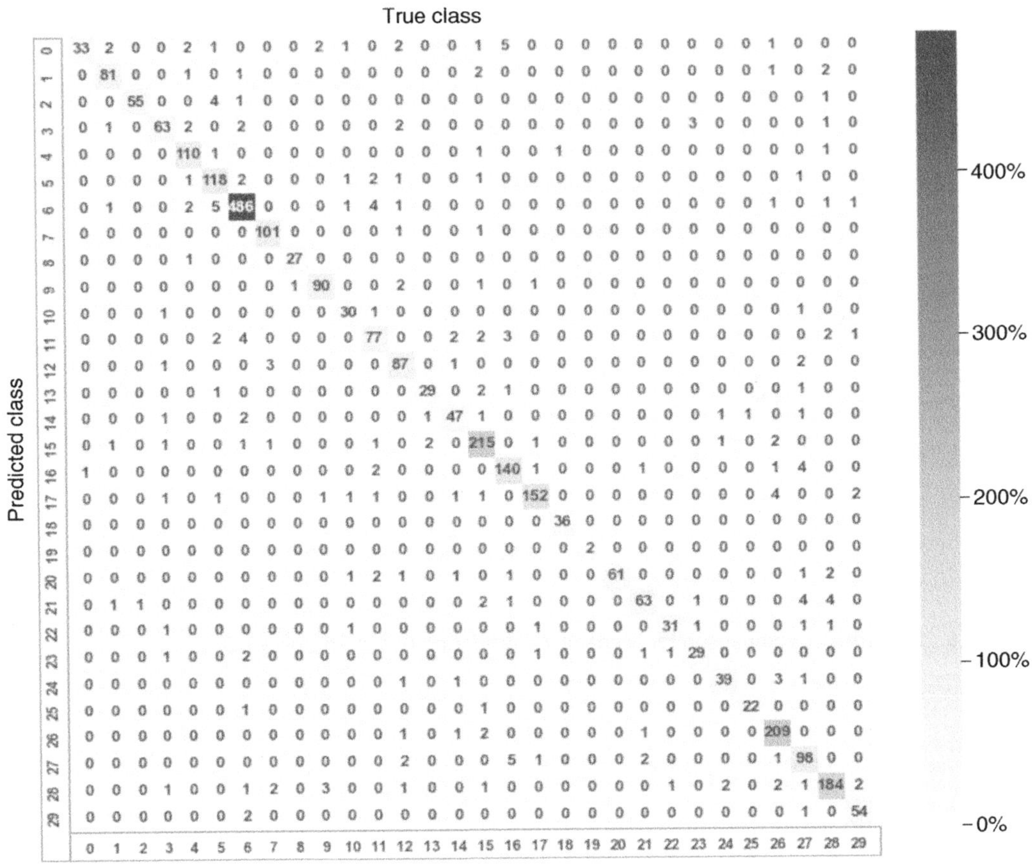

Figure 7.18 Confusion matrix for Scenario 3 considering a mix of one or two interference DOAs and a desired source at 45°.

7.6.3 MLP and CNN Simulation Performance

7.6.3.1 Dataset Generation

MATLAB simulation was used to generate the datasets for experimentation. Both switching strategies from Fig. 7.12 were simulated. Enumeration of all possible sparse array configurations was used to determine the best possible performing array in terms of SINR within the possibilities of binary switching as well as to determine the MaxSINR for all possible array configurations. The source was fixed at 90° with a signal-to-noise ratio (SNR) of 0 dB. Datasets for single and two interferers were created, with interference DOAs ranging from 10° to 170° and INR randomly selected between −5 and 20 dB. A 1° grid was utilized which yielded 161 distinct DOAs for the single interferer case, and 3280 unique pairs of DOAs with a minimum separation of 4° for the two interferers case. Each set of unique DOAs was then repeated until the desired number of samples for the corresponding dataset was reached, which is 10,000 samples in case of a single interferer and 50,000 samples for the case of two interferers. Each sample was then paired with a random INR in the −5 to 20 dB range.

The generation of the test datasets utilized the same procedure, with the exception that each interferer DOA was perturbed by δ_l, with δ_l uniformly distributed over $[0°, 1°]$. Therefore, the test

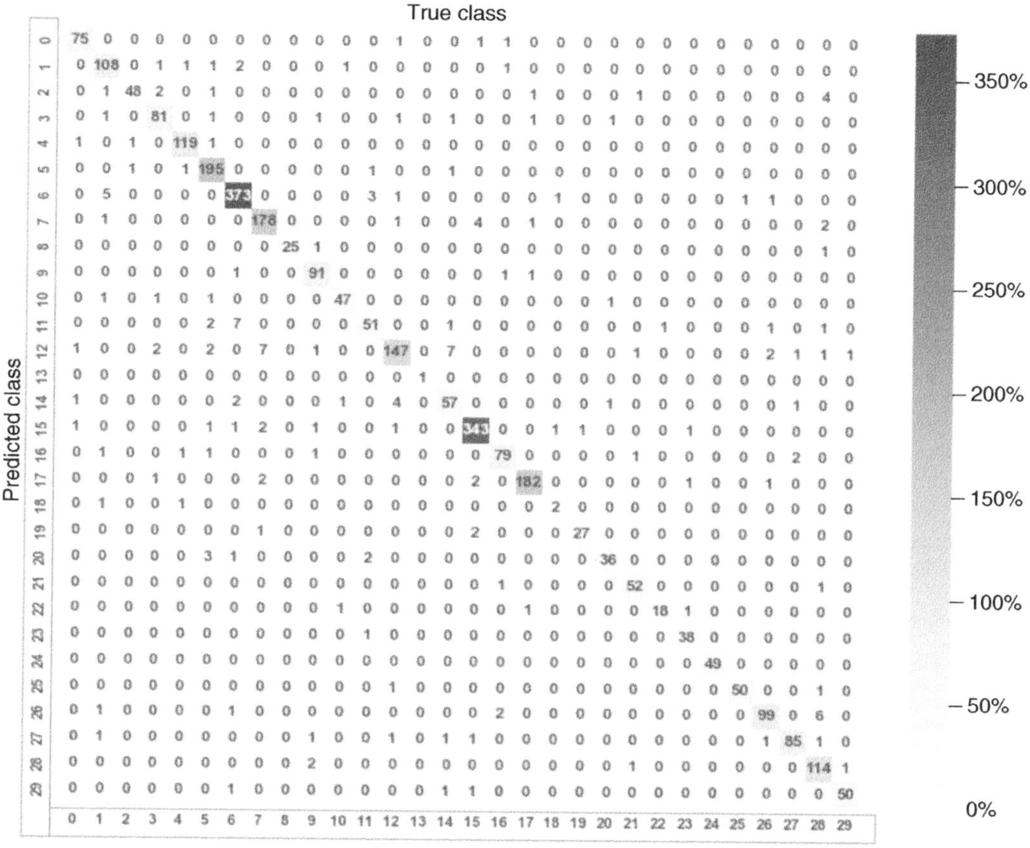

Figure 7.19 Confusion matrix for Scenario 3 considering a mix of one or two interference DOAs and a desired source at 90°.

dataset is simulated off-grid with the minimum spacing between two interferers reduced to 3° versus on-grid training dataset having minimum spacing of 4°. The size of the testing datasets was 5000 for the single interferer case and 10,000 samples for scenario with two interferers.

7.6.3.2 Results

The network training utilized a batch size of 512 for the MLP and 64 for the CNN with a binary cross entropy loss function. The ADAM optimizer [64] was used with cyclic learning rate [65] schedule of $1/(2^{(x-1)})$, where x is the epoch number during training. The initial learning rate of the cyclic schedule was set to 10^{-4} and the maximum learning rate was set to 10^{-2}. Up to 100 epochs were used for training, with an early stopping parameter set on the validation accuracy of the network and a patience of 40 epochs.

The results of the evaluation are presented in Tables 7.1 and 7.2. The term "MaxSINR" in the table refers to the maximum possible output SINR, determined through enumeration, over all possible sparse array configurations of N antenna locations chosen from M available sensor locations. The "Max Switched SINR" denotes the maximum possible output SINR, found through enumeration, over all possible binary switched array configurations. The term "SINR" is associated with the SINR offerings of the proposed DL framework, whereas "ULA SINR" is the output SINR achieved by a compact ULA, shown in Fig. 7.20a. All reported SINR values are averaged over the entire test dataset.

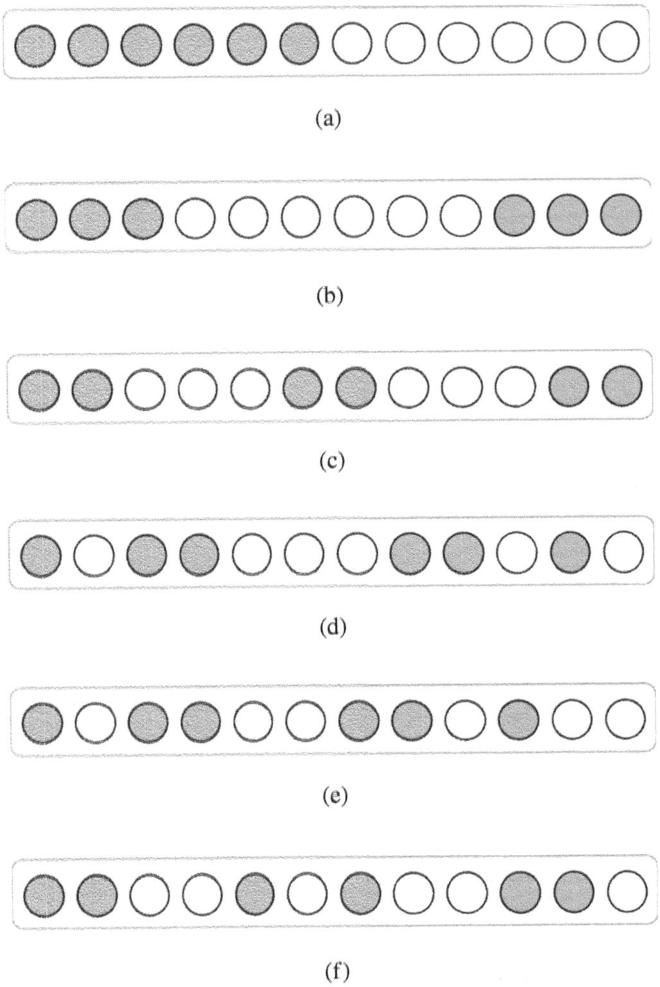

Figure 7.20 Symmetric sparse arrays (Dark circles shows antenna selected, Light circles show antenna element not active). (a) Uniform linear array (ULA). (b) Center sparse array-sparse configuration configured most frequently. (c) Second most frequently configured sparse array after center sparse configuration (d) Least configured sparse array when desired source is at 45°. (e) and (f) Sparse arrays configured less frequently when desired source is at 90°.

From Table 7.1, it can be seen that binary switching Strategy-1 resulted in a classification accuracy of 91.14% when evaluated with a single interferer on the MLP network, whereas the CNN achieves 91.5% accuracy. Both networks achieved 7.586 dB SINR across the 5000 testing samples versus the Max Switched SINR of 7.587 dB obtained via enumeration.

Additionally, a compact ULA configuration yielded an SINR of 7.217 dB, demonstrating the benefit of sparse array reconfigurability on SINR performance. If all possible sparse array configurations are considered, and not just those under Strategy-1, then the MaxSINR for the testing samples is 7.628 dB, indicating that the array performance decreased for some testing samples under Strategy-1.

For the two interferers model, shown in Table 7.2 under switching Strategy-1, the accuracy of the MLP and CNN models decreased to 82.12% and 84.65%, respectively, with respective SINR values of

Table 7.1 MLP and CNN accuracy with considered binary switching strategies for a single interfer.

Network	Accuracy	SINR (dB)	Max Switched SINR (dB)	ULA SINR (dB)	MaxSINR (dB)
MLP - Strategy-1	91.14%	7.586	7.587	7.217	7.628
MLP - Strategy-2	95.98%	7.614	7.615		
CNN - Strategy-1	91.5%	7.586	7.587		
CNN - Strategy-2	96.2%	7.614	7.615		

Table 7.2 MLP and CNN accuracy with considered binary switching strategies with two interfers.

Network	Accuracy	SINR (dB)	Max Switched SINR (dB)	ULA SINR (dB)	MaxSINR (dB)
MLP – Strategy-1	82.12%	7.302	7.323	6.601	7.423
MLP – Strategy-2	84.65%	7.344	7.362		
CNN – Strategy-1	83.91%	7.314	7.323		
CNN – Strategy-2	87.32%	7.355	7.362		

7.302 and 7.314 dB. The Max Switched SINR, in this case, is 7.323 dB on the 10,000 sample test set. If the binary switching constraint is removed, then the MaxSINR is 7.42 dB. Similar to the single interferer case, even when the network chooses an alternate "incorrect" sparse array configuration, the SINR performance closely matches the Max Switched SINR.

For the single interferer case, Strategy-2 achieved 95.98% classification accuracy for the MLP and 96.2% using the CNN, both providing an SINR of 7.614 dB on the 5000 sample test set. The Max Switched SINR for the test set under Strategy-2 was 7.615 dB, demonstrating the nearly optimum selection of array configurations for the single interferer case. Comparing the two switching strategies, Strategy-2 is able to achieve a higher Max Switched SINR of 7.615 dB versus 7.587 dB of Strategy-1, demonstrating a benefit in the sparse array configurations possible under Strategy-2.

Finally, from Table 7.2, observe that the MLP and CNN networks, evaluated on the two interferers dataset under Strategy-2, were able to achieve an accuracy of 84.65% and 87.32% respectively, with respective SINR of 7.344 dB and 7.355 dB. The Max Switched SINR over the test dataset in this case is 7.362 dB. The MaxSINR with no switching constraints is 7.42 dB, showing the network was able to again nearly match the Max Switched SINR for Strategy-2 which is very close to the MaxSINR considering all sparse array configurations. The benefit of Strategy-2 is demonstrated when comparing to the ULA SINR of 6.601 dB. Again, Strategy-2 provides a higher Max Switched SINR of 7.362 dB versus 7.323 dB of Strategy-1.

7.6.4 Comparison of the Network Architectures

The DNN, MLP, and CNN architectures presented in Section 7.5.5 are very different from one another. The differences in training data generation procedures utilized between the networks makes it difficult to directly compare SINR performance. However, all three models perform significantly better than the ULA models and closely match the performance of the enumerated sparse array design results. While the performance of each of the different architectures is promising, the drastically different architectures will influence the PAC.

Each network architecture was evaluated for inference time by applying 10,000 data point samples for the $M = 12$ and $N = 6$ case on a 2.4 GHz Quad-Core Intel Core i5 processor. The DNN network had an average inference time of 12.59 ms. Additionally, the DNN architecture requires the FFT of the first row of the auto correlation matrix to be computed, which adds an additional 0.186 ms to the processing time, for a total of 12.75 ms. The MLP architecture is much less complex than the DNN architecture, which is evident in the inference time as well. The inference time of the MLP network was 7.12 ms on average. Finally, the CNN network architecture had an average inference time of 10.37 ms over the 10,000 runs. For comparison, the SDR algorithm required 4.75 seconds to solve a 10 out of 20 sparse array configuration. The machine-learning approaches thus represent an approximately 395× reduction in run time, representing a large improvement in PAC time.

The selection of an architecture then comes down to performance, in regards to SINR, and inference time. As such, the small gains in the CNN network architecture over the MLP network may not be worth the 31.3% increase in PAC time, providing an important network architecture design space exploration consideration.

7.7 Future Directions

While machine-learning methods show promise to reduce the PAC of sparse array radar systems, as discussed in Section 7.6, there are multiple considerations for future exploration in order to utilize machine learning for sparse array design in real world production systems.

7.7.1 Multiple Direction of Arrivals

The first area of expansion is to switch the networks from training on a single DOA angle, to a single network being applicable to all desired source angles. Since it is not feasible to store a network for every DOA angle, a production system will require a network that can take in the autocorrelation matrix and information regarding the DOA to produce the sparse array design. How the DOA is fed into the network and the scaling between the autocorrelation matrix and DOA information are all important future considerations.

7.7.2 Utilizing Limited Snapsots

The presented results for the machine-learning networks also all have assumed access to unlimited snapshots. To mimic a real-world scenario, the number of snapshots to produce the autocorrelation data must be limited. The effects of the limited snapshots must be studied on the network to analyze the robustness of the network to limited snapshots.

7.7.3 Missing Correlation Data

This work assumed access to the full correlation matrix, but the given sparse array structure may lead to missing correlation data. Methods to estimate the correlation of the full array from sparse array data already exist, including those in References [66] and [67]. Such methods can be consider to generate the necessary correlation data to the machine-learning models. One such method is

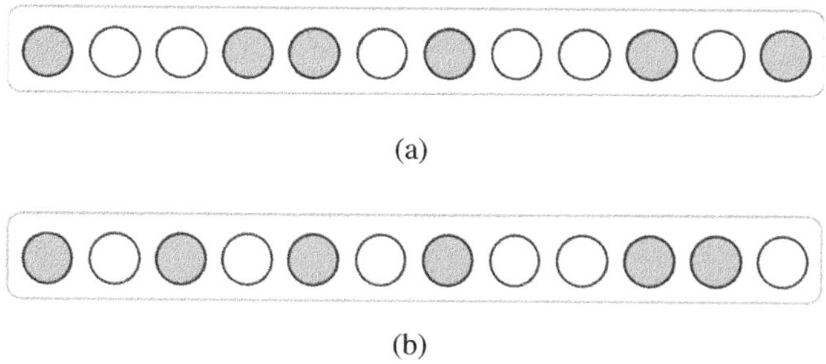

Figure 7.21 Two sparse array combinations with nonoverlapping single missing lags, including (a) with lag 11 missing and (b) with lag 10 missing.

matrix completion [66], which interpolates the data of all receiving antennas. In a dynamic environment, utilizing this method would add to processing overhead and effect the PAC time.

Another approach, specific to the underlying problem, is to sequentially estimate the missing data correlation values over different subarrays which are configured through binary switching. In this case, the missing correlation lags are estimated by switching to two or more array configurations. In essence, for two configurations, the number of snapshots acquired to estimate each correlation lag value is reduced to half. For example, Fig. 7.21 shows two possible sparse arrays configured through adjacent antenna switching. Both configurations are missing only one nonoverlapping correlation lag, namely lag 10 and 11. This paves the way for jointly estimating the full correlation matrix over two sets of snapshots collected sequentially from configurations shown in Fig. 7.21.

Lastly, there are some binary switching strategies that are able to produce configurations where all of the correlation lags are available. Discovering these configurations would mean that no switching would have to occur to achieve the full correlation data. Limiting the training to these array configurations could possibly improve the number of snapshots processed or the PAC time. However, the SINR effects of only utilizing those configurations requires further analysis and validation.

The other option is to have the machine-learning networks learn on data with missing correlation information, meaning the sparse array design could be completed without requiring the sparse array to switch to an alternate configuration. If accuracy can be maintained, such a strategy would allow for more binary switching strategies to be employed, limiting any effect on SINR performance. Additionally, if the network accuracy can be maintained with missing correlation data, it will help to validate the robustness of the network in real-world scenarios where there will be be noise to the correlation data.

7.7.4 Rapid Dynamic Environments

Another future direction is dynamic RF environments where the sources in the array's FOV can rapidly change their locations over time, thus requiring fast array reconfigurability. To this end, sparse array design can be posed as maximizing the average SINR over an angular sector containing the desired source directions. It is noted that each source direction likely demands a different array configuration as well as different set of beamformer weights. This entails switching the antennas rapidly for a small change in the source direction; a process that could prove difficult and infeasible to implement. In order to avoid a prolonged PAC and very frequent sensor switching within each cycle, sensor locations can be optimized to obtain a common sparse configuration for all source directions within predefined angular sectors with the flexibility to have angular, rather sector, dependent beamforming weights.

Sparse array design through fast switching that avoids hardware complexity, insertion loss, cross-talk distortion and array calibration errors is paramount to a realizable PAC. Since, adaptive beamforming algorithms can be extremely sensitive to slight errors in array characteristics. In addition to fully analyze the aforementioned real-world variations, robust sparse array design needs to be explored.

7.8 Conclusions

This chapter considered machine learning for sparse array design with the objective of maximizing the beamformer output SINR for a desired source in an interference active scenario. The main benefit of utilizing machine learning for sparse array design is to improve the PAC time of the system. Traditional optimizing algorithms like SDR and SCA are complex and difficult to scale to large array designs. While SBSA helps improve the complexity, and is vital to efficient training data generation, the time to calculate an optimal sparse array design is still significantly larger than the inference time of the machine-learning algorithms.

Three different network architectures were demonstrated in this chapter, including a DNN that utilized autocorrelation and FFT data as inputs, as well as MLP and CNN networks that solely utilized autocorrelation data and constrained the problem to two different binary switching strategies. All three networks showed promising results for a fixed DOA and significantly outperformed a ULA in terms of SINR. The promising results in terms of accuracy and inference time demonstrate that machine learning can be an effective sparse array design technique and with the future work considerations presented in the chapter, will be able to facilitate sparse array design in real-world systems.

References

1 Asano, Y., Ohshima, S., Harada, T. et al. (2001). Proposal of millimeter-wave holographic radar with antenna switching. *2001 IEEE MTT-S International Microwave Sympsoium Digest*, Volume 2, 1111–1114, May 2001. https://doi.org/10.1109/MWSYM.2001.967086.

2 Hamza, S.A. and Amin, M.G. (2019). Hybrid sparse array beamforming design for general rank signal models. *IEEE Transactions on Signal Processing* 67 (24): 6215–6226. https://doi.org/10.1109/TSP.2019.2952052.

3 Yang, L., Liwan, L., Weifeng, P. et al. (2001). Signal processing method for switch antenna array of the FMCW radar. *Proceedings of the 2001 IEEE Radar Conference (Cat. No.01CH37200)*, 289–293, May 2001. https://doi.org/10.1109/NRC.2001.922993.

4 Lee, M., Katkovnik, V., and Kim, Y. (2004). System modeling and signal processing for a switch antenna array radar. *IEEE Transactions on Signal Processing* 52 (6): 1513–1523. https://doi.org/10.1109/TSP.2004.827204.

5 Sheinvald, J. and Wax, M. (1999). Direction finding with fewer receivers via time-varying preprocessing. *IEEE Transactions on Signal Processing* 47 (1): 2–9. https://doi.org/10.1109/78.738234.

6 Alkhateeb, A., Nam, Y.-H., Zhang, J., and Heath, R.W. (2016). Massive MIMO combining with switches. *IEEE Wireless Communications Letters* 5 (3): 232–235. https://doi.org/10.1109/LWC.2016.2522963.

7 Gao, X., Edfors, O., Tufvesson, F., and Larsson, E.G. (2015). Multi-switch for antenna selection in massive MIMO. *2015 IEEE Global Communications Conference (GLOBECOM)*, 1–6. https://doi.org/10.1109/GLOCOM.2015.7417765.

8 Gao, Y., Vinck, H., and Kaiser, T. (2018). Massive MIMO antenna selection: switching architectures, capacity bounds, and optimal antenna selection algorithms. *IEEE Transactions on Signal Processing* 66 (5): 1346–1360. https://doi.org/10.1109/TSP.2017.2786220.

9 Mehanna, O., Sidiropoulos, N.D., and Giannakis, G.B. (2013). Joint multicast beamforming and antenna selection. *IEEE Transactions on Signal Processing* 61 (10): 2660–2674.

10 Nosrati, H., Aboutanios, E., Wang, X., and Smith, D. (2022). Switch-based hybrid beamforming for massive MIMO communications in mmwave bands. *Signal Processing* 200: 108659.

11 Amin, M.G., Vaidyanathan, P.P., Zhang, Y.D., and Pal, P. (2017). Editorial for coprime special issue. *Digital Signal Processing* 61 (Suppl. C): 1–2. https://doi.org/10.1016/j.dsp.2016.12.006. Special Issue on Coprime Sampling and Arrays.

12 Van Cappellen, W.A., Wijnholds, S.J., and Bregman, J.D. (2006). Sparse antenna array configurations in large aperture synthesis radio telescopes. *2006 European Radar Conference*, 76–79. https://doi.org/10.1109/EURAD.2006.280277.

13 Godrich, H., Petropulu, A.P., and Poor, H.V. (2012). Sensor selection in distributed multiple-radar architectures for localization: a knapsack problem formulation. *IEEE Transactions on Signal Processing* 60 (1): 247–260. https://doi.org/10.1109/TSP.2011.2170170.

14 Joshi, S. and Boyd, S. (2009). Sensor selection via convex optimization. *IEEE Transactions on Signal Processing* 57 (2): 451–462. https://doi.org/10.1109/TSP.2008.2007095.

15 Rajamäki, R. and Koivunen, V. (2021). Sparse symmetric linear arrays with low redundancy and a contiguous sum co-array. *IEEE Transactions on Signal Processing* 69: 1697–1712. https://doi.org/10.1109/TSP.2021.3057982.

16 Chavali, V., Wage, K.E., and Buck, J.R. (2014). Coprime processing for the Elba Island sonar data set. *2014 48th Asilomar Conference on Signals, Systems and Computers*, 1864–1868. https://doi.org/10.1109/ACSSC.2014.7094791.

17 Amin, M.G., Wang, X., Zhang, Y.D. et al. (2016). Sparse arrays and sampling for interference mitigation and DOA estimation in GNSS. *Proceedings of the IEEE* 104 (6): 1302–1317. https://doi.org/10.1109/JPROC.2016.2531582.

18 Lockwood, G.R., Talman, J.R., and Brunke, S.S. (1998). Real-time 3-D ultrasound imaging using sparse synthetic aperture beamforming. *IEEE Transactions on Ultrasonics, Ferroelectrics, and Frequency Control* 45 (4): 980–988. https://doi.org/10.1109/58.710573.

19 Wang, X., Aboutanios, E., Trinkle, M., and Amin, M.G. (2014). Reconfigurable adaptive array beamforming by antenna selection. *IEEE Transactions on Signal Processing* 62 (9): 2385–2396.

20 Wang, X., Amin, M., and Cao, X. (2017). Analysis and design of optimum sparse array configurations for adaptive beamforming. *IEEE Transactions on Signal Processing* 66 (2): 340–351. https://doi.org/10.1109/TSP.2017.2760279.

21 Pal, P. and Vaidyanathan, P.P. (2010). Nested arrays: a novel approach to array processing with enhanced degrees of freedom. *IEEE Transactions on Signal Processing* 58 (8): 4167–4181.

22 Qin, S., Zhang, Y.D., and Amin, M.G. (2015). Generalized coprime array configurations for direction-of-arrival estimation. *IEEE Transactions on Signal Processing* 63 (6): 1377–1390.

23 Qin, G. and Amin, M.G. (2021). Structured sparse array design exploiting two uniform subarrays for DOA estimation on moving platform. *Signal Processing* 180: 107872. https://doi.org/10.1016/j.sigpro.2020.107872.

24 BouDaher, E., Ahmad, F., and Amin, M.G. (2015). Sparsity-based direction finding of coherent and uncorrelated targets using active nonuniform arrays. *IEEE Signal Processing Letters* 22 (10): 1628–1632. https://doi.org/10.1109/LSP.2015.2417807.

25 Moffet, A. (1968). Minimum-redundancy linear arrays. *IEEE Transactions on Antennas and Propagation* 16 (2): 172–175. https://doi.org/10.1109/TAP.1968.1139138.

26 Goldsmith, A. (2005). *Wireless Communications*. New York: Cambridge University Press.

27 Li, J., Stoica, P., and Wang, Z. (2003). On robust Capon beamforming and diagonal loading. *IEEE Transactions on Signal Processing* 51 (7): 1702–1715. https://doi.org/10.1109/TSP.2003.812831.

28 Ṽan Trees, H.L. (1992). *Detection, Estimation, and Modulation Theory: Radar-Sonar Signal Processing and Gaussian Signals in Noise*. Melbourne, FL: Krieger Publishing Co., Inc.

29 Roy, V., Chepuri, S.P., and Leus, G. (2013). Sparsity-enforcing sensor selection for DOA estimation. *2013 5th IEEE International Workshop on Computational Advances in Multi-Sensor Adaptive Processing (CAMSAP)*, 340–343, December 2013.

30 Sidiropoulos, N.D., Davidson, T.Ñ., and Luo, Z.-Q. (2006). Transmit beamforming for physical-layer multicasting. *IEEE Transactions on Signal Processing* 54 (6): 2239–2251.

31 Wang, X., Amin, M.G., Wang, X., and Cao, X. (2017). Sparse array quiescent beamformer design combining adaptive and deterministic constraints. *IEEE Transactions on Antennas and Propagation* 65 (11): 5808–5818.

32 Greco, M.S., Gini, F., Stinco, P., and Bell, K.L. (2018). Cognitive radars: on the road to reality: progress thus far and possibilities for the future. *IEEE Signal Processing Magazine* 35: 112–125.

33 Haykin, S. (2006). Cognitive dynamic systems. *Proceedings of the IEEE* 94: 1910–1911. https://doi.org/10.1109/JPROC.2006.886014.

34 Bergin, J.S., Guerci, J.R., Guerci, R.M., and Rangaswamy, M. (2015). MIMO clutter discrete probing for cognitive radar. *2015 IEEE Radar Conference (RadarCon)*, 1666–1670. https://doi.org/10.1109/RADAR.2015.7131266.

35 Karamalis, P.D., Skentos, N.D., and Kanatas, A.G. (2004). Selecting array configurations for MIMO systems: an evolutionary computation approach. *IEEE Transactions on Wireless Communications* 3 (6): 1994–1998. https://doi.org/10.1109/TWC.2004.837447.

36 Sharaga, N., Tabrikian, J., and Messer, H. (2015). Optimal cognitive beamforming for target tracking in MIMO radar/sonar. *IEEE Journal of Selected Topics in Signal Processing* 9: 1440–1450. https://doi.org/10.1109/JSTSP.2015.2467354.

37 Huleihel, W., Tabrikian, J., and Shavit, R. (2013). Optimal adaptive waveform design for cognitive MIMO radar. *IEEE Transactions on Signal Processing* 61: 5075–5089. https://doi.org/10.1109/TSP.2013.2269045.

38 Martone, A. and Amin, M. (2021). A view on radar and communication systems coexistence and dual functionality in the era of spectrum sensing. *Digital Signal Processing* 119: 103135. https://doi.org/10.1016/j.dsp.2021.103135.

39 Hamza, S.A. and Amin, M.G. (2020). Sparse array design for maximizing the signal-to-interference-plus-noise-ratio by matrix completion. *Digital Signal Processing* 105: 102678. https://doi.org/10.1016/j.dsp.2020.102678.

40 Elbir, A. and Mishra, K. (2020). Joint antenna selection and hybrid beamformer design using unquantized and quantized deep learning networks. *IEEE Transactions on Wireless Communications* 19: 1677–1688.

41 Elbir, A., Mishra, K.V., and Eldar, Y. (2019). Cognitive radar antenna selection via deep learning. *IET Radar, Sonar & Navigation* 13 (6): 871–880.

42 Elbir, A.M. (2020). DeepMUSIC: multiple signal classification via deep learning. *IEEE Sensors Letters* 4 (4): 1–4. https://doi.org/10.1109/LSENS.2020.2980384.

43 Elbir, A.M. and Mishra, K.V. (2021). Sparse array selection across arbitrary sensor geometries with deep transfer learning. *IEEE Transactions on Cognitive Communications and Networking* 7 (1): 255–264. https://doi.org/10.1109/TCCN.2020.2999811.

44 Elbir, A.M., Mishra, K.V., Shankar, M.R.B., and Ottersten, B. (2021). A family of deep learning architectures for channel estimation and hybrid beamforming in multi-carrier mm-Wave massive MIMO. *IEEE Transactions on Cognitive Communications and Networking* 8 (2): 642–656. https://doi.org/10.1109/TCCN.2021.3132609.

45 Wandale, S. and Ichige, K. (2021). Design of sparse arrays via deep learning for enhanced DOA estimation. *EURASIP Journal on Applied Signal Processing* 2021 (1): 17. https://doi.org/10.1186/s13634-021-00727-5.

46 Stoica, P. and Moses, R.L. (1997). *Introduction to Spectral Analysis*. Upper Saddle River, NJ: Prentice Hall. ISBN: 0132584190.

47 Shahbazpanahi, S., Gershman, A.B., Luo, Z.-Q., and Wong, K.M. (2003). Robust adaptive beamforming for general-rank signal models. *IEEE Transactions on Signal Processing* 51 (9): 2257–2269.

48 Donoho, D.L. (2006). For most large underdetermined systems of linear equations the minimal l_1-norm solution is also the sparsest solution. *Communications on Pure and Applied Mathematics* 59 (6): 797–829.

49 Bruckstein, A.M., Donoho, D.L., and Elad, M. (2009). From sparse solutions of systems of equations to sparse modeling of signals and images. *SIAM Review* 51 (1): 34–81. https://doi.org/10.1137/060657704.

50 Yang, A.Y., Sastry, S.S., Ganesh, A., and Ma, Y. (2010). Fast l_1-minimization algorithms and an application in robust face recognition: a review. *2010 IEEE International Conference on Image Processing*, 1849–1852, September 2010. https://doi.org/10.1109/ICIP.2010.5651522.

51 Candès, E.J., Wakin, M.B., and Boyd, S.P. (2008). Enhancing sparsity by reweighted $l_1$minimization. *Journal of Fourier Analysis and Applications* 14 (5): 877–905.

52 Bengtsson, M. and Ottersten, B. (1999). Optimal downlink beamforming using semidefinite optimization. *Proceedings of the 37th Annual Allerton Conference on Communication, Control, and Computing*, 987–996.

53 Hamza, S.A., Amin, M.G., and Fabrizio, G. (2018). Optimum sparse array beamforming for general rank signal models. *2018 IEEE Radar Conference (RadarConf18)*, 1343–1347, April 2018. https://doi.org/10.1109/RADAR.2018.8378759.

54 Luo, Z.Q., Ma, W.K., So, A.M.C. et al. (2010). Semidefinite relaxation of quadratic optimization problems. *IEEE Signal Processing Magazine* 27 (3): 20–34. https://doi.org/10.1109/MSP.2010.936019.

55 Mohan, K. and Fazel, M. (2012). Iterative reweighted algorithms for matrix rank minimization. *Journal of Machine Learning Research* 13 (1): 3441–3473.

56 Recht, B., Fazel, M., and Parrilo, P.A. (2010). Guaranteed minimum-rank solutions of linear matrix equations via nuclear norm minimization. *SIAM Review* 52 (3): 471–501. https://doi.org/10.1137/070697835.

57 Ibrahim, M.S., Konar, A., Hong, M., and Sidiropoulos, N.D. (2018). Mirror-Prox SCA algorithm for multicast beamforming and antenna selection. *2018 IEEE 19th International Workshop on Signal Processing Advances in Wireless Communications (SPAWC)*, 1–5, June 2018. https://doi.org/10.1109/SPAWC.2018.8445845.

58 Nai, S.E., Ser, W., Yu, Z.L., and Chen, H. (2011). Beampattern synthesis for linear and planar arrays with antenna selection by convex optimization. *IEEE Transactions on Antennas and Propagation* 58: 3923–3930.

59 Fuchs, B. (2014). Application of convex relaxation to array synthesis problems. *IEEE Transactions on Antennas and Propagation* 62 (2): 634–640. https://doi.org/10.1109/TAP.2013.2290797.

60 Demir, Ö.T. and Tuncer, T.E. (2014). Alternating maximization algorithm for the broadcast beamforming. *2014 22nd European Signal Processing Conference (EUSIPCO)*, 1915–1919.

61 Hamza, S.A. and Amin, M.G. (2023). Sparse array design for optimum beamfoming using deep learning. *IEEE Transactions on Aerospace and Electronic Systems* 1–12. https://doi.org/10.1109/TAES.2023.3285201.

62 Hamza, S.A., Juretus, K., Amin, M.G., and Ahmad, F. (2023). Deep learning sparse array design using binary switching configurations. *ICASSP 2023 - 2023 IEEE International Conference on Acoustics, Speech and Signal Processing (ICASSP)*, 1–5. https://doi.org/10.1109/ICASSP49357.2023.10094662.

63 Hamza, S.A., Juretus, K., Amin, M.G., and Ahmad, F. (2023). Deep learning sparse array design considering binary switching and missing coarray lags. *2023 International Symposium on Signals, Circuits and Systems (ISSCS)*, 1–4. https://doi.org/10.1109/ISSCS58449.2023.10190882.

64 Kingma, D.P. and Ba, J. (2014). Adam: a method for stochastic optimization. *arXiv preprint arXiv:1412.6980.*

65 Smith, L.N. (2017). Cyclical learning rates for training neural networks. *2017 IEEE Winter Conference on Applications of Computer Vision (WACV).*

66 Huang, H., Miao, Y., Gong, Y., and Liao, B. (2019). Toeplitz matrix completion for direction finding using a modified nested linear array. *ICASSP 2019 – 2019 IEEE International Conference on Acoustics, Speech and Signal Processing (ICASSP)*, 4474–4478. https://doi.org/10.1109/ICASSP.2019.8682693.

67 Liu, C., Vaidyanathan, P.P., and Pal, P. (2016). Coprime coarray interpolation for DOA estimation via nuclear norm minimization. *2016 IEEE International Symposium on Circuits and Systems (ISCAS)*, 2639–2642, May 2016. https://doi.org/10.1109/ISCAS.2016.7539135.

8

Sensor Placement for Distributed Sensing

Geert Leus[1], Mario Coutino[2], and Sundeep Prabhakar Chepuri[3]

[1] *Faculty of Electrical Engineering, Mathematics and Computer Science (EEMCS), Delft University of Technology, Delft, The Netherlands*
[2] *Netherlands Organisation for Applied Scientific Research (TNO), The Hague, The Netherlands*
[3] *Department of Electrical Communication Engineering (ECE), Indian Institute of Science, Bengaluru, India*

Sensing the environment is of crucial importance in many different applications, such as autonomous driving, weather prediction, and hazard prevention to name a few. To improve the sensing outcome, sensor networks are often considered because they are capable of observing the environment from different viewpoints. Sensors are generally not cheap though; think for instance, about radar stations or satellites. As a result, we want to make optimal use of a limited amount of sensors, and thus sensor placement becomes a critical problem.

Sensor placement can be considered as an instance of sparse sensing, the more general field concerned with allocating a limited amount of resources to tackle a specific statistical inference task [1]. Besides sensor placement, it encompasses applications such as antenna selection and (non-uniform) sub-Nyquist sampling [2–5]. The basic idea of sparse sensing is that there is a large number of potential candidate measurements from which only a limited amount is selected due to cost considerations. Sparse sensing has been optimized for several different inference tasks, such as estimation [6–10] and detection [11–15], which are the two tasks we will focus on this chapter. Different performance measures for these tasks have been considered ranging from the Cramér–Rao bound or minimum mean square estimation error, over frame potential, to the log likelihood ratio. Both linear as well as nonlinear observation models have been considered. Moreover, different solution approaches have been proposed. The two most popular ones are convex relaxation and greedy selection methods.

What is not often treated in sensor placement is the fact that the measurements can be conditionally dependent [10, 14]. This can for instance occur when the candidate measurements are corrupted by interference from undesired sources which can be modeled as correlated noise. This complicates matters significantly and makes the sensor placement problem for estimation and detection tasks tough to handle. We show in this chapter how to tackle this scenario for both the convex relaxation and greedy selection methods. For simplicity of presentation, we only focus on the linear model in this chapter, but it can be easily extended to nonlinear models as well. Furthermore, we assume additive Gaussian noise, although that can also be relaxed in some cases.

In Section 8.1, we introduce the sensor placement problem for a linear data model with additive Gaussian noise. As mentioned earlier, we will not only assume uncorrelated noise, but also handle the correlated case. Furthermore, the convex relaxation as well as the greedy selection methods will be presented. Finally, a field estimation and detection example will be introduced that will be used

Sparse Arrays for Radar, Sonar, and Communications, First Edition. Edited by Moeness G. Amin.

throughout the chapter for illustrative purposes. Sections 8.2 and 8.3 will then respectively discuss in detail the estimation and detection problem. We will introduce the related performance metrics and illustrate how the convex relaxation and greedy selection methods can be employed for these measures. The field estimation and detection example allows us to demonstrate the behavior of the solutions as well as the related trade-offs.

8.1 Data Model

In this section, we will detail the setup that will be considered throughout this chapter. We assume that K sensors can be distributed over M candidate positions, and the goal is to place these sensors in some optimal fashion. The candidate positions can represent the only available sensor positions, e.g., rooftops of buildings for a distributed radar application, or they are obtained by gridding some continuous area. The number of sensors we have available is generally much smaller than M, i.e., $K \ll M$

The potential measurement $x_m \in \mathbb{R}$ that is taken at position $m = 1, 2, \dots, M$ is assumed to be linearly dependent on some parameter vector $\boldsymbol{\theta} \in \mathbb{R}^N$ corrupted by addition noise $n_m \in \mathbb{R}$, i.e.,

$$x_m = \mathbf{h}_m^\top \theta + n_m, \tag{8.1}$$

where $\mathbf{h}_m$ collects the linear coefficients related to the mth measurement. Stacking the different candidate measurements into $\mathbf{x} = [x_1, \dots, x_M]^\top$, we can write

$$\mathbf{x} = \mathbf{H}\boldsymbol{\theta} + \mathbf{n}, \tag{8.2}$$

where $\mathbf{H} = [\mathbf{h}_1, \dots, \mathbf{h}_M]^\top$ and $\mathbf{n} = [n_1, \dots, n_M]^\top$. Throughout, we will assume that the noise is Gaussian distributed with zero mean, i.e., $\mathbb{E}\{\mathbf{n}\} = \mathbf{0}$, and covariance matrix Σ, i.e., $\mathbb{E}\{\mathbf{n}\mathbf{n}^\top\} = \Sigma$. So in summary, the data model can be expressed as $\mathbf{x} \sim \mathcal{N}(\mathbf{H}\boldsymbol{\theta}, \Sigma)$.

Note that although such a linear Gaussian model is restrictive, some of the results in this chapter can be extended to nonlinear non-Gaussian models as well. In the nonlinear case, we can view (8.2) as a local linear model around a specific $\boldsymbol{\theta}$ where $\mathbf{H}$ represents the Jacobian matrix of the nonlinear model in $\boldsymbol{\theta}$. This will generally make $\mathbf{H}$ dependent on $\boldsymbol{\theta}$ which will require some adjustments to the proposed methods. The Gaussianity assumption, on the other hand, is required for the correlated noise case (non-diagonal Σ), but can easily be dropped for the uncorrelated noise case (diagonal Σ).

We cannot use all M measurements in $\mathbf{x}$ since we only have $K \ll M$ sensors available. To select the "best" K measurements, we will adopt the selection vector $\mathbf{w} \in \{0, 1\}^M$ whose mth entry is one if the position m is selected or zero otherwise, and whose total number of non-zero entries is K, i.e., $\|\mathbf{w}\|_0 = K$. The selected measurements can then be represented by

$$\mathbf{y} = \Phi(\mathbf{w})\mathbf{x} = \Phi(\mathbf{w})\mathbf{H}\boldsymbol{\theta} + \Phi(\mathbf{w})\mathbf{n}, \tag{8.3}$$

where $\Phi(\mathbf{w})$ is the submatrix of the $M \times M$ identity matrix $\mathbf{I}_M$ that is obtained by removing the zero rows from $\mathrm{diag}(\mathbf{w})\mathbf{I}_M$. Note thereby that

$$\Phi(\mathbf{w})\Phi^\top(\mathbf{w}) = \mathbf{I}_K, \quad \Phi^\top(\mathbf{w})\Phi(\mathbf{w}) = \mathrm{diag}(\mathbf{w}). \tag{8.4}$$

The problem we are tackling in this chapter is how to design the "best" $\mathbf{w}$ under the constraints $\mathbf{w} \in \{0, 1\}^M$ and $\|\mathbf{w}\|_0 = K$. What we mean by "best" is determined by some inference performance metric $f : \{0, 1\}^M \to \mathbb{R}$. Depending on the inference task, this metric can be the estimation error or miss detection probability for instance. Hence, the problem we need to solve is

$$\underset{\mathbf{w} \in \{0,1\}^M}{\arg\min}\, f(\mathbf{w}) \text{ s.t. } \|\mathbf{w}\|_0 = K. \tag{8.5}$$

This is known as the cardinality constrained (CC) problem.

In some cases, though, we might want to keep K as small as possible as long as we can guarantee a certain inference performance. In that case, we can formulate a performance constrained (PC) problem as

$$\arg\min_{\mathbf{w}\in\{0,1\}^M} \|\mathbf{w}\|_0 \text{ s.t.} f(\mathbf{w}) \leq \lambda, \tag{8.6}$$

where λ denotes the considered performance bound. Both problems are equivalent in that for every K in (8.5), there exists a related λ in (8.6).

Solving these problems would require a complex combinatorial search which quickly becomes intractable. As a result, in this chapter, we will focus on convex relaxation and greedy selection approaches. The specific inference tasks we will aim at are

- Estimation of $\boldsymbol{\theta}$.
- Detection of $\boldsymbol{\theta}$, where $\boldsymbol{\theta}$ can be known or not.

In both cases, we assume the knowledge of $\mathbf{H}$ and the noise covariance matrix Σ. Before we delve into the details of the estimation and detection inference tasks, we briefly overview the general idea of the convex relaxation and greedy selection methods.

8.1.1 Solution Approaches

A common approach to tackle the CC and PC problems in (8.5) and (8.6) is adopting convex relaxation. More specifically, the l0-norm will be replaced by its convex hull, which is given by the l1-norm, whereas the Boolean constraint is replaced by a box constraint. Assuming further that $f(\cdot)$ is a convex function, potentially obtained after convexifying the desired performance measure, the CC and PC problems can respectively be relaxed to the following convex problems

$$\arg\min_{\mathbf{w}\in[0,1]^M} f(\mathbf{w}) \text{ s.t. } \|\mathbf{w}\|_1 = K. \tag{8.7}$$

and

$$\arg\min_{\mathbf{w}\in[0,1]^M} \|\mathbf{w}\|_1 \text{ s.t.} f(\mathbf{w}) \leq \lambda. \tag{8.8}$$

Note that the constraint $\|\mathbf{w}\|_1 = K$ in (8.7) can be considered convex, since due to $\mathbf{w} \in [0,1]^M$ it can be written as $\mathbf{1}^\mathsf{T}\mathbf{w} = 1$. For the same reason, we can also replace $\|\mathbf{w}\|_1$ in (8.8) by $\mathbf{1}^\mathsf{T}\mathbf{w}$.

Convex problems have been well-studied, although not all forms are necessarily easy to solve. Furthermore, it is not always possible to quantify the suboptimality of the convex relaxation. Specific convex performance measures $f(\cdot)$ for the estimation and detection problem will be discussed in Sections 8.2 and 8.3. Note that an approximate Boolean solution for $\mathbf{w}$ can be obtained from the solution of the above problems by means of thresholding or randomization [1].

Another method to handle the CC and PC problems is greedy selection. The idea behind this approach is rather intuitive and basically selects one sensor at a time (myopic method), either to add it to or to remove it from the previously selected sensors. Suppose that $\mathcal{X}$ represents the set of currently active or inactive sensors. Then a greedy approach would iteratively add to $\mathcal{X}$ the sensor that is obtained as

$$s^* = \arg\max_{s\notin\mathcal{X}} f(\mathcal{X} \cup s),$$

where $f(\mathcal{X})$ is the set function that correctly translates the performance measure $f(\mathbf{w})$. Note that this translation will depend on the fact whether $\mathcal{X}$ represents the active or inactive set of sensors.

The CC problem is now handled by iterating till we have K active sensors, whereas for the PC problem the iterations will stop when the desired performance has been reached. Interestingly, if $f(\mathcal{X})$ is submodular, monotone nondecreasing, and normalized (i.e., $f(\emptyset) = 0$), then the greedy solution is near optimal and approaches the optimal cost with a factor $1 - 1/e$, where e is the Euler number [16].

8.1.2 Running Example

The running example that we will consider throughout this chapter[1] is that of spatial field estimation/detection. To simplify the presentation, we will consider a one-dimensional spatial field with candidate sensor positions that are uniformly spaced over the considered area. We further consider a spatially correlated field that is non-stationary (i.e., the correlation depends on the sensor location). More specifically, we assume the spatial correlation matrix of the field is given by

$$\mathbf{R} = \begin{bmatrix} 1 & \delta_1 & \delta_1^2 & \dots & \delta_1^{M-1} \\ \delta_1 & 1 & \delta_2 & \dots & \delta_2^{M-2} \\ \delta_1^2 & \delta_2 & 1 & \dots & \delta_3^{M-3} \\ \vdots & \vdots & \vdots & \ddots & \vdots \\ \delta_1^{M-1} & \delta_2^{M-2} & \delta_3^{M-3} & \dots & 1 \end{bmatrix}.$$

To further simplify the setup and reduce the number of variables, we take $0 < \delta_1 = \delta < 1$ and $\delta_i = \delta_{i-1} - \kappa$, with $0 < \kappa < \delta/M$. This structure will allow us to analyze how the developed selection methods behave as a function of different correlation structures. Note however that we do not assume a stochastic signal in (8.3), but a deterministic one. That is why we further assume the observed deterministic field is somehow "smooth" within the class of fields characterized with correlation matrix $\mathbf{R}$. In other words, we define $\mathbf{H}$ as the matrix that stacks the $N \ll M$ unit-norm eigenvectors of $\mathbf{R}$ with the largest eigenvalues, i.e., $\mathbf{R} = \mathbf{H}\Lambda\mathbf{H}^\top + \mathbf{H}_n\Lambda_n\mathbf{H}_n^\top$, with $\Lambda \in \mathbb{R}^{N\times N}$ the diagonal matrix of the N strongest eigenvalues of $\mathbf{R}$ and $\Lambda_n \in \mathbb{R}^{(M-N)\times(M-N)}$ the diagonal matrix collecting the remaining eigenvalues. For the specific values of $M = 30$, $N = 3$, $\delta = 0.9$ and $\kappa = 0.01$, the spatial correlation matrix $\mathbf{R}$ and related measurement matrix $\mathbf{H}$ are depicted in Fig. 8.1.

As far as the additive Gaussian noise is concerned, two scenarios will be studied in this chapter: uncorrelated and correlated noise. In case of uncorrelated noise, we simply assume $\Sigma = \sigma^2\mathbf{I}$. The correlated noise scenario is more complex. There we only focus on spatially stationary noise but we will assume a decaying correlation function as well as a constant correlation function. In the first case, the noise covariance matrix is given by

$$\Sigma = \sigma^2 \begin{bmatrix} 1 & \gamma & \gamma^2 & \dots & \gamma^{M-1} \\ \gamma & 1 & \gamma & \dots & \gamma^{M-2} \\ \gamma^2 & \gamma & 1 & \dots & \gamma^{M-3} \\ \vdots & \vdots & \vdots & \ddots & \vdots \\ \gamma^{M-1} & \gamma^{M-2} & \gamma^{M-3} & \dots & 1 \end{bmatrix}, \tag{8.9}$$

whereas for the second case, we use

$$\Sigma = \sigma^2[(1-\gamma)\mathbf{I} + \gamma\mathbf{1}\mathbf{1}^T]. \tag{8.10}$$

In both cases, we take $\gamma < 1$. Clearly, when $\gamma = 0$ we obtain the uncorrelated noise scenario. Note that in all scenarios the noise variance is given by σ^2.

1 Software to reproduce results from the chapter are available at https://github.com/spchepuri/sensorplacement.git.

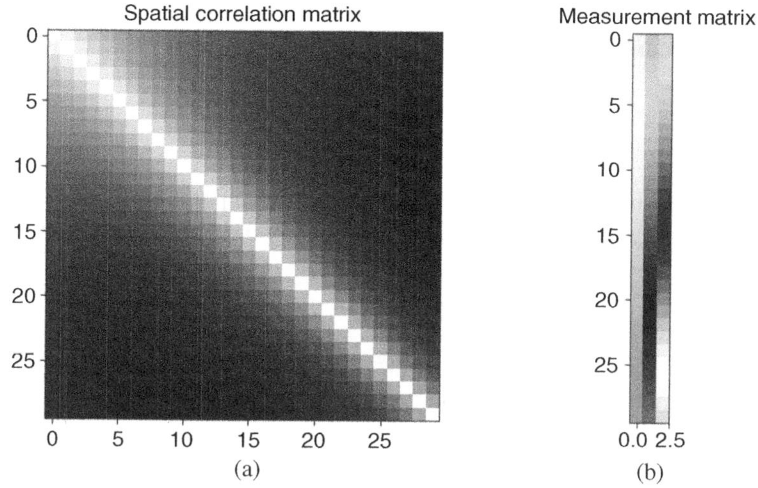

Figure 8.1 Illustration of the running example that will be used throughout this chapter. (a) is a representation of the spatial correlation matrix **R** whereas the related measurement matrix **H** is depicted in (b). The chosen parameters are $M = 30$, $N = 3$, $\delta = 0.9$, and $\kappa = 0.01$.

8.2 Distributed Estimation

Consider the setting where the fusion center, which gathers a subset of sensor data, performs an estimate of the unknown parameter vector $\boldsymbol{\theta}$. Let us denote the estimator as $\hat{\theta}(\mathbf{w})$, where we recall that $\mathbf{w}$ is the selection vector. The performance of such parameter estimation problems is characterized by the error covariance matrix denoted by $\mathbf{E}(\mathbf{w}) = \mathbb{E}\{[\theta - \hat{\theta}(\mathbf{w})][\theta - \hat{\theta}(\mathbf{w})]^\top\}$. Since $\hat{\theta}(\mathbf{w})$ (and hence $\mathbf{E}(\mathbf{w})$) depends on the subset of measurements acquired via $\mathbf{w}$, various scalar measures of $\mathbf{E}(\mathbf{w})$ are optimized with respect to (w.r.t.) the selection variable $\mathbf{w}$. Next, we discuss some of the popular choices of performance measures.

8.2.1 Estimation Optimality Criteria

To measure the quality of estimation, the following scalar measures can be used.

- *A-optimality criterion or mean squared error*: The mean squared error in estimating θ is

$$f(\mathbf{w}) = \mathbb{E}\{\|\theta - \hat{\theta}\|^2\} = \text{trace}(\mathbf{E}(\mathbf{w})), \tag{8.11}$$

  which is also the sum of the eigenvalues of $\mathbf{E}(\mathbf{w})$.
- *D-optimality criterion or volume of the confidence ellipsoid*: The η-confidence ellipsoid is the minimum volume ellipsoid that contains the estimation error $\theta - \hat{\theta}$ with probability η. It can be shown to be proportional to the determinant of $\mathbf{E}$:

$$f(\mathbf{w}) = \log\det(\mathbf{E}(\mathbf{w})). \tag{8.12}$$

- *E-optimality criterion or worst-case error variance*: Another criterion related to the volume of the confidence ellipsoid is the worst-case variance of the estimation error:

$$f(\mathbf{w}) = \max_{\|\mathbf{u}\|=1} \mathbf{u}^\top \mathbf{E}(\mathbf{w})\mathbf{u} = \lambda_{\max}(\mathbf{E}(\mathbf{w})), \tag{8.13}$$

  where $\lambda_{\max}(\cdot)$ denotes the maximum eigenvalue of its argument.

- *Worst-case coordinate error variance*: The worst-case coordinate error variance is the largest diagonal entry of the error covariance matrix $\mathbf{E}(\mathbf{w})$, i.e.,

$$f(\mathbf{w}) = \max_{1 \leq i \leq M} [\mathbf{E}(\mathbf{w})]_{ii}, \tag{8.14}$$

where $[\mathbf{E}(\mathbf{w})]_{ii}$ denotes the ith diagonal entry of $\mathbf{E}(\mathbf{w})$.

Although there is no general answer to how one performance measure compares with another, all the above scalar measures are equally reasonable because they quantify the estimation quality.

8.2.2 Uncorrelated Observations

Suppose the noise covariance matrix is diagonal, i.e., $\Sigma = \text{diag}(\sigma_1^2, \dots, \sigma_M^2)$ with $\sigma_1^2 = \cdots = \sigma_M^2 = \sigma^2$ being a special case. Since the parameter $\boldsymbol{\theta}$ is assumed deterministic, we can then focus on the maximum likelihood (ML) estimate, which is given by

$$\hat{\boldsymbol{\theta}} = \left(\sum_{m=1}^{M} w_m \sigma_m^{-2} \mathbf{h}_m \mathbf{h}_m^{\mathsf{T}} \right)^{-1} \sum_{m=1}^{M} w_m x_m \mathbf{h}_m.$$

The estimation error $\boldsymbol{\theta} - \hat{\boldsymbol{\theta}}$ then has zero mean, i.e., the estimate is unbiased, and covariance matrix

$$\mathbf{E}(\mathbf{w}) = \left(\sum_{m=1}^{M} w_m \sigma_m^{-2} \mathbf{h}_m \mathbf{h}_m^{\mathsf{T}} \right)^{-1}. \tag{8.15}$$

We can observe the explicit role of $\mathbf{w}$ in (8.15) for selecting the measurements that improve the conditioning of $\mathbf{E}(\mathbf{w})$.

Now we describe the sparse sampler design problem for the setting with uncorrelated observations using the above described scalar measures of $\mathbf{E}(\mathbf{w})$ and the CC problem formulation (8.7).

For the D-optimality criterion, (8.7) specializes to

$$\begin{aligned} &\arg\max_{\mathbf{w} \in [0,1]^M} \ \log\det \sum_{m=1}^{M} w_m \sigma_m^{-2} \mathbf{h}_m \mathbf{h}_m^{\mathsf{T}} \\ &\text{subject to } \mathbf{1}^{\mathsf{T}} \mathbf{w} = K, \end{aligned} \tag{8.16}$$

which is a convex problem as the objective is a concave function in $\mathbf{w}$, the sum constraint is linear, and the w_m variables are restricted to the interval $[0, 1]$. This problem can be solved using any one of the standard convex solvers. Similarly, we obtain convex problems for the A-optimality and E-optimality criteria as

$$\begin{aligned} &\arg\min_{\mathbf{w} \in [0,1]^M} \ \text{trace} \left(\sum_{m=1}^{M} w_m \sigma_m^{-2} \mathbf{h}_m \mathbf{h}_m^{\mathsf{T}} \right)^{-1} \\ &\text{subject to } \mathbf{1}^{\mathsf{T}} \mathbf{w} = K, \end{aligned} \tag{8.17}$$

and

$$\begin{aligned} &\arg\max_{\mathbf{w} \in [0,1]^M} \ \lambda_{\min} \left(\sum_{m=1}^{M} w_m \sigma_m^{-2} \mathbf{h}_m \mathbf{h}_m^{\mathsf{T}} \right) \\ &\text{subject to } \mathbf{1}^{\mathsf{T}} \mathbf{w} = K, \end{aligned} \tag{8.18}$$

respectively. Both these problem formulations can be transformed into their epigraph form, leading to a semi-definite program (SDP).

Also minimizing the objective function related to the worst-case coordinate error variance in (8.14) can be expressed in the epigraph form leading to the following SDP:

$$\begin{aligned}
&\underset{t,\mathbf{w}\in[0,1]^M}{\text{minimize}} \quad t \\
&\text{subject to} \quad \mathbf{1}^\mathsf{T}\mathbf{w} = K, \\
&\begin{bmatrix} t & \mathbf{e}_j^\mathsf{T} \\ \mathbf{e}_j & \sum_{m=1}^{M} w_m \sigma_m^{-2} \mathbf{h}_m \mathbf{h}_m^\mathsf{T} \end{bmatrix} \succeq \mathbf{0}, \quad j = 0, 1, \ldots, M,
\end{aligned} \tag{8.19}$$

where $\mathbf{e}_j$ is the jth column of the identity matrix of size M.

Next, we discuss numerical experiments to illustrate the sensor placements obtained by solving the above convex programs. In Fig. 8.2, we illustrate the sparse sampler obtained by optimizing the D-optimality criterion, the A-optimality criterion and the E-optimality criterion while selecting $K = 6$ sensors in the CC problem setting. The system matrix is generated as explained in Section 8.1.2 with $M = 30$, $N = 3$, $\delta = 0.9$, and $\kappa = 0.01$. The noise variance is selected as $\sigma^2 = 1$. We observe that all three methods basically select three groups of sensors. There are three groups because we have $N = 3$ unknowns. If one measurement per group would be considered, a well-conditioned $\mathbf{H}$ matrix is obtained. The clustering into groups further boosts the energy of the measurements.

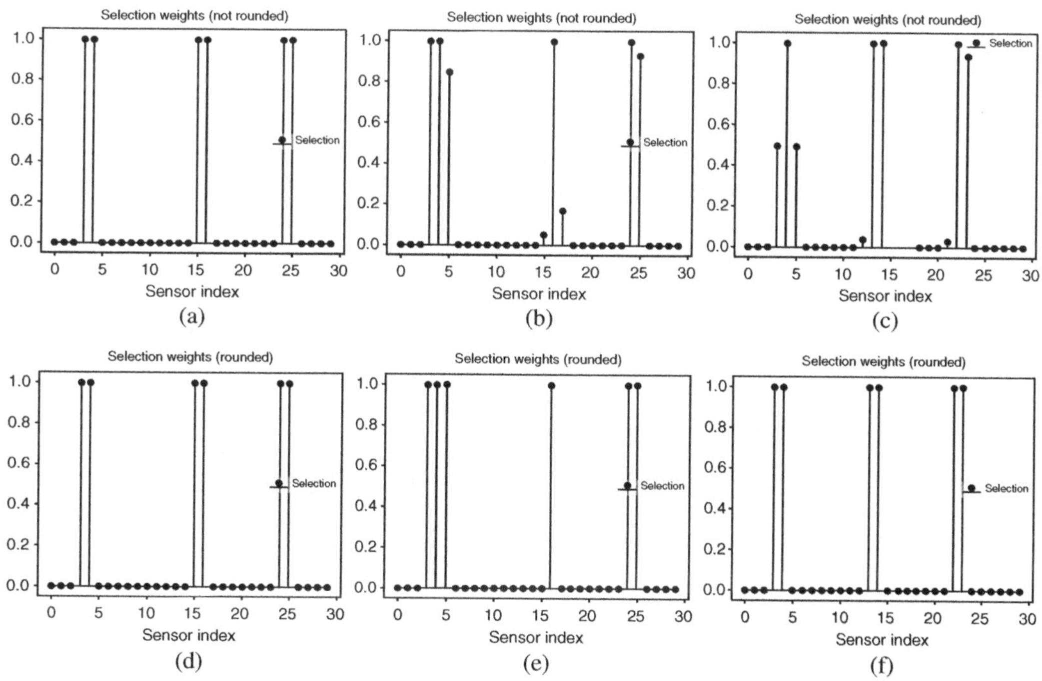

Figure 8.2 Comparison of the non-Boolean selection obtained by optimizing the D-optimality (a), A-optimality (b) and E-optimality (c) criterion while selecting $K = 6$ measurements for the setting in our running example. The recovered Boolean solution after rounding from D-optimality (d), A-optimality (e), and E-optimality (f) criteria.

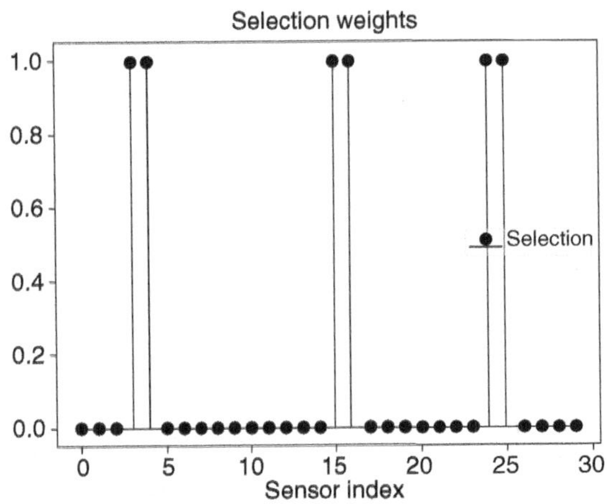

Figure 8.3 Selection obtained by greedily optimizing the D-optimality criterion while selecting $K = 6$ measurements for the setting in our running example.

The D-optimality cost function can be shown to be a submodular function in the selection variables [17]. This means that, as discussed earlier, we can greedily maximize the logdet function of the precision matrix. In Fig. 8.3, we can see that the selection pattern obtained by greedily maximizing the logdet function is similar as the rounded solutions shown in Fig. 8.2.

Although it is difficult to characterize the optimality of the convex relaxation, it is optimal for the following special case. When the parameter to be estimated is a scalar θ with $N = 1$, the error covariance is also a scalar and readily given by $E(\mathbf{w}) = \left(\sum_{m=1}^{M} w_m(h_m^2/\sigma_m^2)\right)^{-1}$. In this case, (8.5) can be optimally solved without requiring any convex relaxation. Minimizing the cost over the Boolean variable can then be optimally solved by rank ordering the signal-to-noise ratios h_m^2/σ_m^2 associated with each sensor. Specifically, the solution is obtained by setting the entries of w_m corresponding to the maximum entries of the set $\{h_m^2/\sigma_m^2, 1 \leq m \leq M\}$ to 1, and the remaining ones to zero, untill K sensors are selected (CC problem) or the performance bound is met (PC problem).

8.2.3 Correlated Observations

Now we discuss the general case with Σ being a non-diagonal matrix. The ML error covariance matrix can then be expressed as

$$\mathbf{E}(\mathbf{w}) = \left(\mathbf{H}^\top \Phi^\top(\mathbf{w})\left(\Phi(\mathbf{w})\Sigma\Phi^\top(\mathbf{w})\right)^{-1}\Phi(\mathbf{w})\mathbf{H}\right)^{-1}. \tag{8.20}$$

This matrix is now clearly a more intricate function of $\mathbf{w}$ compared to (8.15).

In what follows, we express the earlier discussed scalar functions of $\mathbf{E}(\mathbf{w})$ in a simpler form amenable to convex optimization. Firstly, we write the noise covariance matrix Σ as

$$\Sigma = \alpha \boldsymbol{I}_M + \mathbf{S}, \tag{8.21}$$

where a nonzero $\alpha \in \mathbb{R}$ is chosen such that $\mathbf{S} \in \mathbb{R}^{M \times M}$ is invertible and well conditioned. Using (8.21) in (8.20), we obtain

$$\mathbf{E}(\mathbf{w}) = \mathbf{H}^\top \Phi^\top(\mathbf{w})\left(\alpha \boldsymbol{I}_K + \Phi(\mathbf{w})\mathbf{S}\Phi^\top(\mathbf{w})\right)^{-1}\Phi(\mathbf{w})\mathbf{H}.$$

Since $\Phi^\top(\mathbf{w})\Phi(\mathbf{w}) = \text{diag}(\mathbf{w})$, we have

$$\Phi^\top(\mathbf{w})\left(a\boldsymbol{I}_K + \Phi(\mathbf{w})\mathbf{S}\Phi^\top(\mathbf{w})\right)^{-1}\Phi(\mathbf{w}) = \mathbf{S}^{-1} - \mathbf{S}^{-1}\left[\mathbf{S}^{-1} + \alpha^{-1}\,\text{diag}(\mathbf{w})\right]^{-1}\mathbf{S}^{-1}. \tag{8.22}$$

This is due to the matrix inversion lemma

$$\mathbf{C}(\mathbf{B}^{-1}+\mathbf{C}^{\top}\mathbf{A}^{-1}\mathbf{C})^{-1}\mathbf{C}^{\top} = \mathbf{A} - \mathbf{A}(\mathbf{A} + \mathbf{C}\mathbf{B}\mathbf{C}^{\top})^{-1}\mathbf{A},$$

with $\mathbf{C} = \Phi^{\top}(\mathbf{w})$, $\mathbf{B}^{-1} = \alpha \boldsymbol{I}_K$, and $\mathbf{A} = \mathbf{S}^{-1}$. Therefore, we have

$$\begin{aligned}\mathbf{E}(\mathbf{w}) &= \mathbf{H}^{\top}[\mathbf{S}^{-1} - \mathbf{S}^{-1}\left[\mathbf{S}^{-1} + \alpha^{-1}\,\mathrm{diag}(\mathbf{w})\right]^{-1}\mathbf{S}^{-1}]\mathbf{H}.\\ &= \mathbf{H}^{\top}\mathbf{S}^{-1}\mathbf{H} - \mathbf{H}^{\top}\mathbf{S}^{-1}\left[\mathbf{S}^{-1} + \alpha^{-1}\,\mathrm{diag}(\mathbf{w})\right]^{-1}\mathbf{S}^{-1}\mathbf{H}.\end{aligned} \tag{8.23}$$

Using this simpler form, for instance, the E-optimality constraint can be expressed as a linear matrix inequality (LMI) in $\mathbf{w}$, i.e., $\lambda_{\min}(\mathbf{E}(\mathbf{w})) \geq \lambda$ can be expressed as

$$\begin{bmatrix} \mathbf{S}^{-1} + \alpha^{-1}\,\mathrm{diag}(\mathbf{w}) & \mathbf{S}^{-1}\mathbf{H} \\ \mathbf{H}^{\top}\mathbf{S}^{-1} & \mathbf{H}^{\top}\mathbf{S}^{-1}\mathbf{H} - \lambda \boldsymbol{I}_N \end{bmatrix} \succeq \mathbf{0}_{M+N} \tag{8.24}$$

with $\mathbf{S}^{-1} + \alpha^{-1}\,\mathrm{diag}(\mathbf{w})$ being positive definite, which determines the choices of α. Thus, in the case of correlated observations and focusing on the PC problem, the sparse sampler can be designed by solving the following convex problem:

$$\begin{aligned} &\underset{\mathbf{w}\in[0,1]^M}{\arg\min} \quad \mathbf{1}^{\top}\mathbf{w} \\ &\text{subject to} \quad \begin{bmatrix} \mathbf{S}^{-1} + \alpha^{-1}\,\mathrm{diag}(\mathbf{w}) & \mathbf{S}^{-1}\mathbf{H} \\ \mathbf{H}^{\top}\mathbf{S}^{-1} & \mathbf{H}^{\top}\mathbf{S}^{-1}\mathbf{H} - \lambda \boldsymbol{I}_N \end{bmatrix} \succeq \mathbf{0}_{M+N}. \end{aligned} \tag{8.25}$$

This convex program can be solved using any off-the-shelf convex solvers.

We next illustrate sensor placements obtained for the two noise correlation matrices that were introduced in Section 8.1.2 [cf. (8.9) and (8.10)]. We use the same system matrix as in the uncorrelated case with $M = 30$, $N = 3$, $\delta = 0.9$, $\kappa = 0.01$, and $\sigma^2 = 1$. We set $\alpha = 10^{-4}$. While the sensor placements obtained in the uncorrelated case led to a well-conditioned system matrix, the sensor placement design with correlated observations aims to improve the conditioning taking into account the correlation across observations as measured by the error covariance matrix (8.20).

For the spatially constant correlation function, where all observations are equally correlated with correlation coefficient γ, we can see in Fig. 8.4 that as γ increases, the D-optimality cost

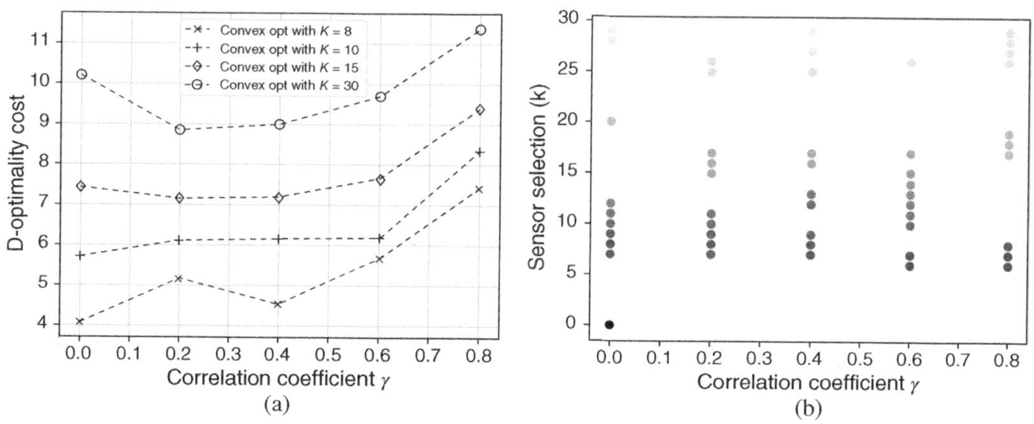

Figure 8.4 (a) Sensor placements obtained from (8.25) for different K and correlation coefficients γ when the noise has a spatially constant correlation function as in (8.10). (b) Sensor placements for different γ for $K = 10$.

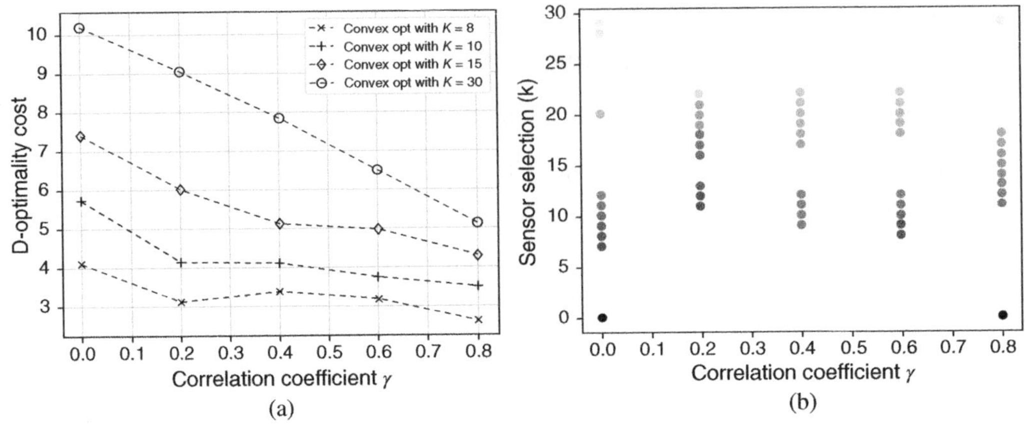

Figure 8.5 (a) Sensor placements obtained from (8.25) for different K and correlation coefficients γ when the noise has a spatially decaying correlation function as in (8.9). (b) Sensor placements for different γ for $K = 10$.

increases for a given number of sensors. This, in other words, means that if there is more correlation across sensors, fewer observations lead to better estimation quality. On the other hand, with the decaying correlation function (8.9), the effect is the opposite as seen in Fig. 8.5 with a more clustered placement.

8.3 Distributed Detection

In this section, we focus on the binary detection problem related to the model (8.2). More specifically, the two hypotheses (states) we consider are:

$$\mathcal{H}_0 : \mathbf{x} = \mathbf{n}, \tag{8.26}$$

$$\mathcal{H}_1 : \mathbf{x} = \mathbf{H}\boldsymbol{\theta} + \mathbf{n}, \tag{8.27}$$

where we will distinguish between the cases where $\boldsymbol{\theta}$ is known or not.

Irrespective of the scenario, we can relate the measurements $\mathbf{x}$ to the model

$$\mathcal{H}_0 : \mathbf{x} \sim p(\mathbf{x}|\mathcal{H}_0), \tag{8.28}$$

$$\mathcal{H}_1 : \mathbf{x} \sim p(\mathbf{x}|\mathcal{H}_1), \tag{8.29}$$

where $p(\mathbf{x}|\mathcal{H}_i)$ for $i = 0, 1$ denotes the probability density function (pdf) of $\mathbf{x}$, conditioned on the state $\mathcal{H}_i$. As we discussed in Section 8.1, we assume that $p(\mathbf{x}|\mathcal{H}_0) = \mathcal{N}(\mathbf{0}, \Sigma)$ while $p(\mathbf{x}|\mathcal{H}_1) = \mathcal{N}(\mathbf{H}\boldsymbol{\theta}, \Sigma)$, although we will often use the more general notation throughout this section.

The goal now is to solve the CC and/or PC problem in (8.5) and (8.6), respectively, for some specific detection performance measure $f(\mathbf{w})$. We next discuss some typical detection performance measures, which we approximate by more manageable functions in Section 8.3.1 and 8.3.2. If the prior hypothesis probabilities are known, i.e., in a Bayesian setting, the optimal detector minimizes the probability of error, $P_e = P(\mathcal{H}_0|\mathcal{H}_1)P(\mathcal{H}_1) + P(\mathcal{H}_1|\mathcal{H}_0)P(\mathcal{H}_0)$, where $P(\mathcal{H}_i|\mathcal{H}_j)$ is the conditional probability of deciding $\mathcal{H}_i$ when $\mathcal{H}_j$ is true and $P(\mathcal{H}_i)$ is the prior probability of the ith hypothesis. When the prior hypothesis probabilities are unknown, i.e., in a Neyman–Pearson setting, the optimal detector aims to minimize the probability of miss detection (type II error), $P_m = P(\mathcal{H}_0|\mathcal{H}_1)$, for a fixed probability of false alarm (type I error), $P_{fa} = P(\mathcal{H}_1|\mathcal{H}_0)$. The sensor selection problems for detection can then be expressed as in (8.5) and (8.6) with $f(\mathbf{w})$ replaced by either $P_e(\mathbf{w})$ (Bayesian setting) or $P_m(\mathbf{w})$ (Neyman–Pearson setting), where in the latter case we implicitly assume that $P_{fa}(\mathbf{w})$ is fixed. Note that $P_e(\mathbf{w})$, $P_m(\mathbf{w})$, and $P_{fa}(\mathbf{w})$ denote the error probabilities due to the measurement selection defined by $\mathbf{w}$.

In general, the above performance measures are not easy to optimize numerically. As a result, in the following, we present alternative measures for both known and unknown $\boldsymbol{\theta}$ cases that can be used as direct surrogates to solve the optimization problems.

8.3.1 Known $\boldsymbol{\theta}$ Parameter

Even when $\boldsymbol{\theta}$ is known, the error probabilities P_e and P_m (for a fixed P_{fa}) might not admit a known closed-form expression or their expressions might not be favorable for numerical optimization. We therefore present several weaker and simpler substitutes, which can be optimized instead of the error probabilities. These substitutes are based on the notion of distance (closeness or divergence) between the two distributions of the observations under test. They lead to tractable, if not always optimal (in terms of the error probabilities) design procedures for sparse sampling. Nevertheless, optimizing these distance measures improves the performance of any practical system.

8.3.1.1 Optimality Criteria

Let the likelihood ratio of the two hypotheses under test be defined as

$$l(\mathbf{y}) = \frac{p(\mathbf{y}|\mathcal{H}_1)}{p(\mathbf{y}|\mathcal{H}_0)}. \tag{8.30}$$

In what follows, we consider a number of distance measures that belong to the general class of Ali–Silvey distances [18], which are of the form

$$\varphi\left(\mathbb{E}_{|\mathcal{H}_i}\{\phi[l(\mathbf{y})]\}\right), \tag{8.31}$$

where $\varphi(\cdot)$ is an increasing real-valued function, $\phi[\cdot]$ is a continuous convex function on $(0, \infty)$, and $\mathbb{E}_{|\mathcal{H}_i}\{\phi[l(\mathbf{y})]\}$ indicates that $\phi[l(\mathbf{y})]$ is averaged under the pdf $p(\mathbf{y}|\mathcal{H}_i)$.

Known Prior Probabilities Under the assumption of known prior distributions for both hypotheses, a Bayesian detector can be devised. This kind of detector makes a decision based on comparing the optimal statistic to a threshold, i.e.,

$$\log l(\mathbf{y}) = \log \frac{p(\mathbf{y}|\mathcal{H}_1)}{p(\mathbf{y}|\mathcal{H}_0)} \underset{\mathcal{H}_0}{\overset{\mathcal{H}_1}{\gtrless}} \log \frac{P(\mathcal{H}_0)}{P(\mathcal{H}_1)}. \tag{8.32}$$

In the Bayesian setting, our goal is to choose the best subset of measurements that results in a prescribed Bayesian probability of error. The best achievable exponent in the Bayesian probability of error is parameterized by the Chernoff information (sometimes also referred to as the Chernoff distance) [19], and it is given by

$$C(\mathcal{H}_1||\mathcal{H}_0) = -\log \min_{0\le n\le 1} \mathbb{E}_{|\mathcal{H}_0}\{[l(\mathbf{y})]^n\}. \tag{8.33}$$

Due to the involved minimization over n, the Chernoff information in (8.33) is difficult to optimize over. Therefore, we use a special case of the Chernoff information called the Bhattacharyya distance as the optimization criterion, where the Bhattacharyya distance is obtained by fixing $n = 0.5$

in (8.33). The Bhattacharyya distance is given by

$$\mathcal{B}(\mathcal{H}_1||\mathcal{H}_0) = -\log \rho, \tag{8.34}$$

where the Bhattacharyya coefficient ρ is

$$\rho = \mathbb{E}_{|\mathcal{H}_0}\{\sqrt{l(\mathbf{y})}\}. \tag{8.35}$$

It is easy to verify from (8.35) that the Bhattacharya distance is symmetric, which means $\mathcal{B}(\mathcal{H}_1||\mathcal{H}_0) = \mathcal{B}(\mathcal{H}_0||\mathcal{H}_1)$. More importantly, the upper and lower bounds for the Bayesian probability of error can be obtained using the Bhattacharyya coefficient. The bounds are given as follows [20]

$$\frac{1}{2}\min\left(P(\mathcal{H}_0), P(\mathcal{H}_1)\right)\rho^2 \leq P_\mathrm{e} \leq \sqrt{P(\mathcal{H}_0)P(\mathcal{H}_1)}\rho. \tag{8.36}$$

Therefore, in place of the Bayesian error probability, we could maximize the Bhattacharyya distance. Furthermore, when $\int [p(\mathbf{y}|\mathcal{H}_1)]^n[p(\mathbf{y}|\mathcal{H}_0)]^{1-n}d\mathbf{y}$ is symmetric in n and the observations are independent and identically distributed, the Bhattacharyya distance is exponentially the best [21], i.e.,

$$P_\mathrm{e} \overset{\text{as.}}{=} \exp\left(-\mathcal{B}(\mathcal{H}_1||\mathcal{H}_0)\right) \text{ for } P_\mathrm{e} \to 0. \tag{8.37}$$

A similar bound as that in (8.36) can be obtained linking the Bhattacharyya distance and the more general Bayes risk

$$\mathcal{R} = \sum_{i=0}^{1}\sum_{j=0}^{1} C_{ij}\,\mathrm{Pr}\,(\mathcal{H}_i|\mathcal{H}_j)\mathrm{Pr}\,(\mathcal{H}_j). \tag{8.38}$$

The bound is given by [22]

$$\mathcal{R}_0 + \mathcal{R}_2\rho^2 \leq \mathcal{R} \leq \mathcal{R}_0 + \sqrt{\mathcal{R}_1}\rho, \tag{8.39}$$

for constants $\mathcal{R}_0, \mathcal{R}_1, \mathcal{R}_2$ dependent on the costs C_{ij} and the prior probabilities $P(\mathcal{H}_i)$, which further motivates the choice of the Bhattacharyya distance for selecting measurements in the Bayesian setting.

Note that under the independence assumption of the measurements, the likelihood $l(\mathbf{y})$ can be factorized as the product of the *local* measurement likelihoods $l_m(x) := p_m(x|\mathcal{H}_1)/p_m(x|\mathcal{H}_0)$, i.e.,

$$l(\mathbf{y}) = \frac{p(\mathbf{y}|\mathcal{H}_1)}{p(\mathbf{y}|\mathcal{H}_0)} = \prod_{m=1}^{M}[l_m(x)]^{w_m}, \tag{8.40}$$

which implies linearity in the selection variables of the Bhattacharyya distance [1], i.e.,

$$\mathcal{B}(\mathcal{H}_1||\mathcal{H}_0) = \sum_{m=1}^{M} w_m\mathcal{B}_m(\mathcal{H}_1||\mathcal{H}_0), \tag{8.41}$$

where $\mathcal{B}_m(\mathcal{H}_1||\mathcal{H}_0) := -\log \mathbb{E}_{|\mathcal{H}_0}\{\sqrt{l_m(x)}\}$ is the mth local Bhattacharya distance.

While the linearity of (8.41) eases the optimization over $\mathbf{w}$, in general, the independence on the measurements of the Bhattacharya distance allows for an offline design of the selection variable.

Unknown Prior Probabilities In the case that prior probabilities are unknown, we typically design detectors based on the Neyman–Pearson approach. In this setting, one of the error probabilities is considered fixed, e.g., P_{fa}, and the other error probability P_{m} is minimized. When this procedure is followed, the decision is again based upon the log-likelihood ratio test

$$\log l(\mathbf{y}) = \log \frac{p(\mathbf{y}|\mathcal{H}_1)}{p(\mathbf{y}|\mathcal{H}_0)} \underset{\mathcal{H}_0}{\overset{\mathcal{H}_1}{\gtrless}} \gamma, \tag{8.42}$$

where now the threshold γ is obtained by fixing P_{fa}.

For a Neyman–Pearson problem, it is known that the best achievable error exponent in the probability of error, e.g., in P_{m}, is given by the relative entropy or Kullback–Leibler (KL) divergence

$$\log P_{\mathrm{m}} \overset{\text{as.}}{=} -\mathcal{K}(\mathcal{H}_1|\mathcal{H}_0) \text{ for } P_{\mathrm{m}} \to 0. \tag{8.43}$$

where the Kullback–Leibler divergence is defined as

$$\mathcal{K}(\mathcal{H}_1|\mathcal{H}_0) := \mathbb{E}_{|\mathcal{H}_1}\{\log l(\mathbf{y})\}. \tag{8.44}$$

These properties allow to construct an upper and lower bound for P_{m}, for P_{fa} values of practical interest, i.e., $P_{\mathrm{fa}} \approx 0$ ([1]). That is,

$$\exp(-\mathcal{K}(\mathcal{H}||\mathcal{H}_0)) \leq P_{\mathrm{m}} \leq \frac{1}{1 + \frac{(\mathcal{K}(\mathcal{H}_1||\mathcal{H}_0) - \log \gamma)^2}{\upsilon^2}}, \tag{8.45}$$

where υ^2 is the variance of the log-likelihood ratio and γ is the threshold for a target P_{fa}.

Similarly to the Bhattacharyya distance, when the measurements are conditional independent on the hypothesis $\mathcal{H}$, the KL divergence is linear in $\mathbf{w}$

$$\mathcal{K}(\mathcal{H}_1||\mathcal{H}_0) = \sum_{m=1}^{M} w_m \mathcal{K}_m(\mathcal{H}_1||\mathcal{H}_0), \tag{8.46}$$

where $\mathcal{K}_m(\mathcal{H}_1||\mathcal{H}_0) := \mathbb{E}_{|\mathcal{H}_1}\{\log l_m(x)\}$ is the mth local KL divergence.

Note that due to the asymmetry of the KL divergence, it cannot be considered a proper distance. However, this property can be used if instead of minimizing the P_{m} for a given P_{fa} we want to minimize a given P_{fa} for a given P_{m}. To do so, we only need to interchange $\mathcal{H}_1$ and $\mathcal{H}_0$ in the previous expressions. When the symmetry of the metric is of importance, instead of the KL divergence, its symmetric version, the J-divergence can be employed:

$$\mathcal{D}(\mathcal{H}_1||\mathcal{H}_0) := \mathcal{K}(\mathcal{H}_1||\mathcal{H}_0) + \mathcal{K}(\mathcal{H}_0||\mathcal{H}_1). \tag{8.47}$$

As it is composed by two KL-divergence terms, this distance is also linear in $\mathbf{w}$ and can be used to find an upper and lower bound for $P_{\mathrm{e}} = 0.5(P_{\mathrm{fa}} + P_{\mathrm{m}})$ for general distributions. For arbitrary P_{e}, the upper bound is only possible to obtain when the observations are Gaussian [20].

Under the assumption of Gaussian observations, the discussed metrics between the different hypotheses under test are shown in Table 8.1. Here, $\boldsymbol{\mu}_i$ and Σ_i denote the mean vector and the covariance matrix of the ith distribution, respectively.

8.3.1.2 Sparse Sampler Design

Using the discussed surrogate measures as $f(\mathbf{w})$, a new formulation of the sparse sensing problems (8.5) and (8.6) can be stated. In what follows, we present different approaches for designing sparse samplers for detection, giving special attention to the Gaussian case for correlated observations.

Table 8.1 Summary of metrics for Gaussian probability distributions.

Metric	Expression	Setting
	Bhattacharyya	
$\mathcal{B}(\mathcal{H}_1 \| \mathcal{H}_0) :=$	$\frac{1}{8}(\boldsymbol{\mu}_1 - \boldsymbol{\mu}_0)^T \Sigma^{-1} (\boldsymbol{\mu}_1 - \boldsymbol{\mu}_0) + \frac{1}{2} \log \left(\frac{\det(\Sigma)}{\sqrt{\det(\Sigma_1)\det(\Sigma_0)}} \right)$, $\Sigma = 0.5(\Sigma_1 + \Sigma_0)$	Bayesian
	Kullback–Leibler	
$\mathcal{K}(\mathcal{H}_1 \| \mathcal{H}_0) :=$	$\frac{1}{2}\left(\mathrm{tr}\left(\Sigma_0^{-1}\Sigma_1\right) + \left(\boldsymbol{\mu}_1 - \boldsymbol{\mu}_0\right)^T \Sigma_0^{-1} \left(\boldsymbol{\mu}_1 - \boldsymbol{\mu}_0\right) - N + \log\left(\det(\Sigma_0)\right) - \log\left(\det(\Sigma_1)\right)\right)$	Neyman–Pearson
	J-Divergence	
$\mathcal{D}(\mathcal{H}_0 \| \mathcal{H}_1) :=$	$\mathcal{K}(\mathcal{H}_1 \| \mathcal{H}_0) + \mathcal{K}(\mathcal{H}_0 \| \mathcal{H}_1)$	Neyman–Pearson

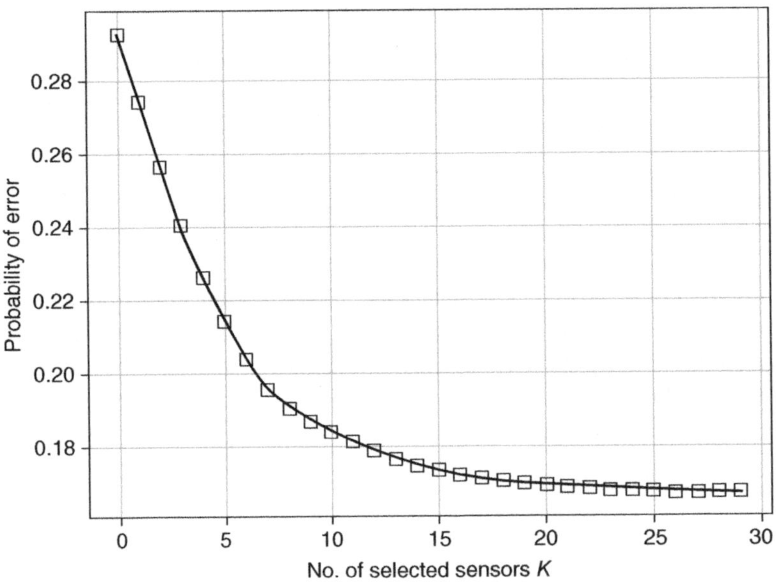

Figure 8.6 Performance of the optimal selection with conditionally independent observations in the Bayesian setting. Here, $\Sigma = \mathbf{I}$, $\boldsymbol{\theta} = \mathbf{1}$ and $P(\mathcal{H}_0) = 0.3$ and $P(\mathcal{H}_1) = 0.7$.

Conditionally Independent Observations As discussed before, when the observations are considered to be conditionally independent, the three discussed metrics are linear in $\mathbf{w}$. That is, if we collected the local distances/divergences in a vector $\mathbf{d} \in \mathbb{R}^M$, the CC and PC problems respectively simplify to

$$\underset{\mathbf{w} \in \{0,1\}^M}{\arg\max} \ \mathbf{d}^{\mathsf{T}}\mathbf{w} \quad \text{s.t.} \quad \|\mathbf{w}\|_0 = K, \tag{8.48}$$

$$\underset{\mathbf{w} \in \{0,1\}^M}{\arg\min} \ \|\mathbf{w}\|_0 \quad \text{s.t.} \quad \mathbf{d}^{\mathsf{T}}\mathbf{w} \geq \lambda. \tag{8.49}$$

Due to the linearity of the metric, it can be shown that for both problems a greedy strategy finds the optimal solution. More specifically, the largest entries in $\mathbf{d}$ are selected till we have K sensors (CC problem) or the desired performance has been reached (PC problem). As a result, the proposed selection mechanism is very attractive as its complexity is mainly that of sorting a list. An example of the performance, in terms of probability of error, of the selection for a case where $\Sigma = \mathbf{I}$ and $\boldsymbol{\theta} = \mathbf{1}$ with prior probabilities $P(\mathcal{H}_0) = 0.3$ and $P(\mathcal{H}_1) = 0.7$ is shown in Fig. 8.6.

As we will see next, for dependent observations, in general no simple mechanism exists to obtain optimal selections. However, similar to the methods developed for estimation, we discuss alternatives leveraging the convex and submodular machinery for optimizing the introduced metrics.

Dependent Observations Although thc conditional independence assumption is valid in circumstances where the noise is caused solely by the measurement acquisition, when sensors experience external noise sources or if the signal itself is stochastic in nature the independence assumption might not hold anymore. As a result, the additive property of the metrics is not valid anymore. Thus, computing the optimal sparse sampler under dependent observations requires considering nonlinear (and possibly non-convex) metrics. So, in general, we will have to solve the sparse sampler design problem [cf. (8.5) and (8.6)] using a global optimization technique [23].

Despite that there might not be a general recipe to solve problems (8.5) and (8.6) for arbitrary measurement distributions, in the particular case of correlated Gaussian observations, the presented metrics allow for a selection mechanism based on convex and submodular optimization that enables an efficient offline sampler design.

When the two distributions only differ by their means as in our binary hypothesis problem formulation [cf. (8.26) and (8.27)], from Table 8.1 it is seen that all metrics (upto a scale factor) coincide; that is

$$\mathcal{B}(\mathcal{H}_1||\mathcal{H}_0) \propto \mathcal{K}(\mathcal{H}_1||\mathcal{H}_0) \propto \mathcal{D}(\mathcal{H}_1||\mathcal{H}_0) \propto (\boldsymbol{\mu}_1 - \boldsymbol{\mu}_0)^{\mathsf{T}}\Sigma^{-1}(\boldsymbol{\mu}_1 - \boldsymbol{\mu}_0), \tag{8.50}$$

which can be interpreted as a measure of the signal-to-noise ratio, effectively defining the detection performance. By considering the selection matrix and defining $\mathbf{m} := \boldsymbol{\mu}_1 - \boldsymbol{\mu}_0$, we can consider the quantity

$$s(\mathbf{w}) := \mathbf{m}^{\mathsf{T}}\Phi^{\mathsf{T}}(\mathbf{w})\left(\Phi(\mathbf{w})\Sigma\Phi^{\mathsf{T}}(\mathbf{w})\right)^{-1}\Phi(\mathbf{w})\mathbf{m} \tag{8.51}$$

as our performance metric and solve, for instance, the PC problem

$$\underset{w\in\{0,1\}^M}{\arg\min} \quad \|\mathbf{w}\|_0 \quad \text{s.t} \quad s(\mathbf{w}) \geq \lambda. \tag{8.52}$$

Although the above problem is still hard to solve, we can consider the following convex relaxation

$$\underset{\mathbf{w}\in[0,1]^M}{\arg\min} \; \mathbf{1}^{\mathsf{T}}\mathbf{w} \tag{8.53}$$

$$\text{s.t} \quad \begin{bmatrix} \mathbf{S}^{-1} + \alpha^{-1}\,\text{diag}(\mathbf{w}) & \mathbf{S}^{-1}\mathbf{m} \\ \mathbf{m}^{\mathsf{T}}\mathbf{S}^{-1} & \lambda' \end{bmatrix} \succeq 0,$$

where $\lambda' := \lambda - \mathbf{m}^{\mathsf{T}}\mathbf{S}^{-1}\mathbf{m}$. The LMI is derived in a similar way as was done for estimation in correlated noise, i.e., by decomposing the covariance matrix Σ as $\Sigma = \alpha\mathbf{I}_M + \mathbf{S}$ with $\alpha \in \mathbb{R}$ chosen such that $\mathbf{S}$ is invertible and well conditioned, and by using the equivalence of the inequality

$$\mathbf{m}^{\mathsf{T}}\mathbf{S}^{-1}[\mathbf{S}^{-1} + \alpha^{-1}\,\text{diag}(\mathbf{w})]^{-1}\mathbf{S}^{-1}\mathbf{m} \leq \lambda' \tag{8.54}$$

with the LMI in (8.53). Note that (8.54) is derived from the constraint in (8.52) using the matrix inversion lemma and the properties of the selection matrix.

Analogously, the CC problem can be formulated using the same LMI as

$$\begin{aligned}&\underset{\mathbf{w}\in[0,1]^M,t}{\arg\min}\ t\\&\quad \text{s.t}\ \mathbf{1}^\top\mathbf{w}=K,\\&\quad\begin{bmatrix}\mathbf{S}^{-1}+\alpha^{-1}\operatorname{diag}(\mathbf{w}) & \mathbf{S}^{-1}\mathbf{m}\\ \mathbf{m}^\top\mathbf{S}^{-1} & t\end{bmatrix}\succeq 0,\end{aligned}\tag{8.55}$$

with an auxiliary variable t that has been used to formulate the epigraph form of a problem using the left-hand side of (8.54) as cost function.

As the problems (8.53) and (8.55) are convex, they can be solved efficiently using any off-the-shelf convex optimization method. Although the resulting solution is not necessarily binary, procedures such as randomized rounding can be employed to recover a feasible selection vector with good performance from the continuous solution. Although the above convex relaxations provide a way to tackle the sparse sampler design for unequal means, greedy methods can still be employed in several instances while obtaining a comparable or even better detection performance. For example, as suggested in [24], a greedy procedure that constructs a solution by greedily selecting the best solution of the original cost function $s(\mathbf{w})$ and the surrogate

$$s'(\mathbf{w}) := \begin{cases}0, & \text{if}\quad \mathbf{w}=\mathbf{0},\\ \log\det M(\mathbf{w}), & \text{otherwise},\end{cases}\tag{8.56}$$

where $M(\mathbf{w})$ denotes the matrix in the left-hand side of the LMI in (8.53) and (8.55) can be considered. As the cost function (8.56) is submodular, its greedy optimization achieves provable near-optimality guarantees. Further, because at every step it can be efficiently computed, i.e., a determinant update, this mechanism for sparse sampler design can be employed when solving the semidefinite program in (8.53) or (8.55) is not practically feasible.

An illustration of the different selections made by the convex and greedy selection approaches is shown in Fig. 8.7. In this example, we consider again $\theta = \mathbf{1}$ and ask the methods to select a

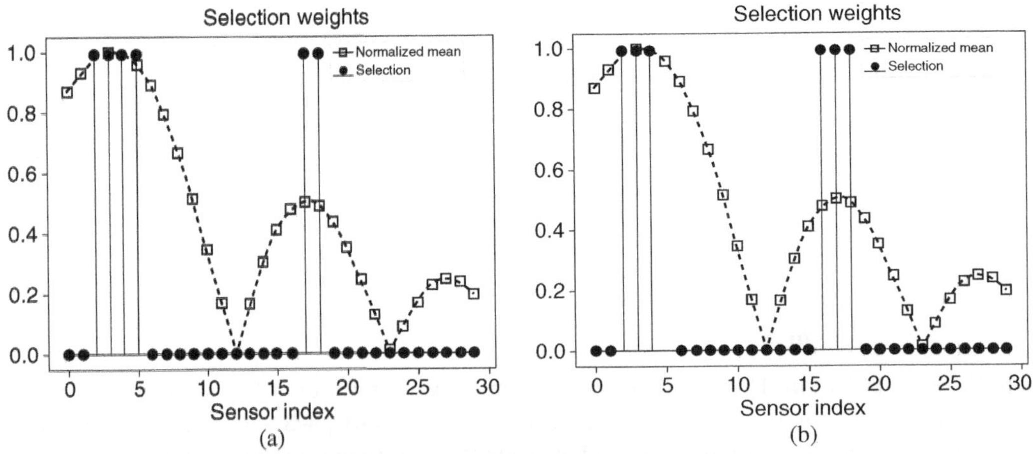

Figure 8.7 Comparison of the selection of the convex (a) and submodular (b) optimization approaches when selecting $K = 6$ measurements considering the setting of our running example under equi-correlated noise [cf. (8.10)] with $\gamma = 0.5$. In dashed lines the normalized mean is shown for illustration. Here, $\theta = \mathbf{1}$.

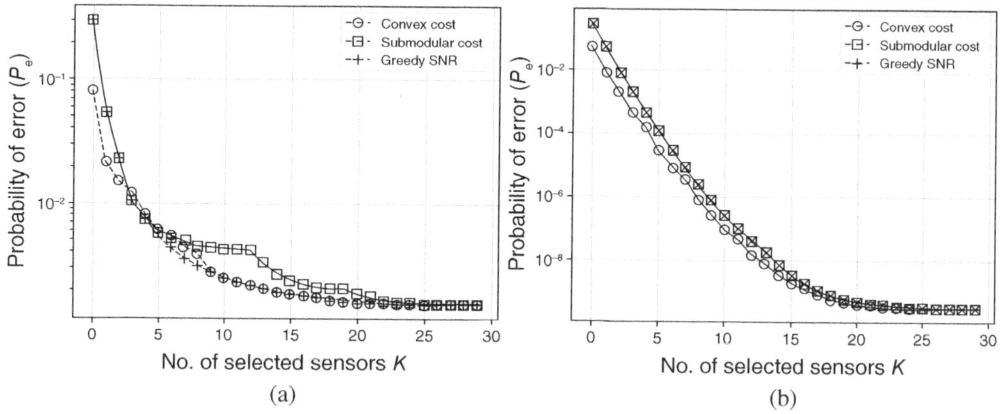

Figure 8.8 Comparison of the selection of the convex and submodular optimization approaches when selecting different number of measurements considering spatially decaying [a; cf. (8.9)] and spatially constant [b; cf. (8.10)] correlated noise. In dashed-crossed lines the performance of greedily optimizing the SNR function is shown. Here, $\theta = \mathbf{1}$ and $\gamma = 0.5$ for both types of noise.

set of cardinality $K = 6$ when the noise is equi-correlated with parameter $\gamma = 0.5$ [cf. (8.10)]. We further take $\sigma^2 = 1$. As the convex solution not necessarily provides an integer solution, here, we have selected the K entries of $\mathbf{w}$ with the highest value. To illustrate the possible impact of the mean of different measurements, the means normalized with respect to the maximum are also shown. Finally, a complete comparison for our running example, under spatially decaying or spatially constant correlated noise, is shown in Fig. 8.8. Here, we compare the methods under the same setting across different selection cardinalities K and set $\gamma = 0.5$ and $\sigma^2 = 1$ for both cases. In addition, for reference, we also include the performance of performing greedy optimization directly on the signal-to-ratio (SNR) function. Note that while the three methods achieve similar performance, the greedy methods are cheap alternatives to convex optimization which in several instances outperforms the latter even if the function is not submodular, i.e., for the Greedy SNR method. However, it is known that in certain instances this might not be the case as greedy optimization on non-submodular functions can fail, see, e.g., [24].

To illustrate the effect of the correlation pattern and strength, we consider the sensor selection problem when the noise correlation is given by respectively (8.9) and (8.10) when using the submodular selection strategy.

The result of the selection for a spatially decaying correlation function is shown in Fig. 8.9. While typically for estimation tasks the strategy selects sensors that lead to a well-conditioned measurement matrix, for detection the focus lays mostly on selecting measurements with high signal-to-noise ratio. This introduces a natural trade off between correlation and energy on the measurements. For instance, in the case of a spatially decaying noise correlation matrix, sensors seem to spread more as the correlation coefficient increases. As a result, more sensors are needed to reach the same performance as the correlation grows.

This trade off becomes less apparent when an equi-correlated noise matrix is considered. In this setting, there is no need for the sensors to have a particular arrangement to leverage correlation. This naturally leads to clear clusters based on the signal-to-noise ratio. Also, it allows for a reduced number of required sensors when the correlation coefficient increases. These effects are illustrated in Fig. 8.10.

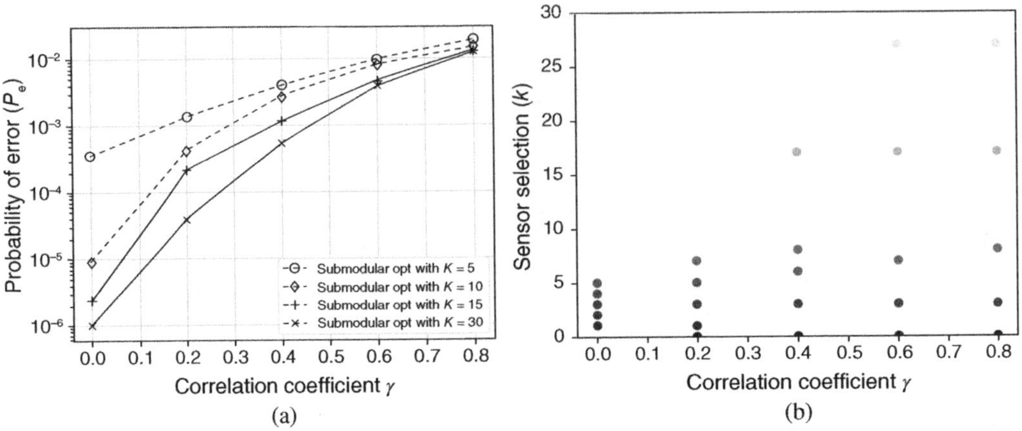

Figure 8.9 (a) Sensor selection of the submodular strategy for different numbers of sensors K and correlation coefficients γ when the noise correlation is giving by the spatially decaying correlation in (8.9). (b) Selected sensors for different γ for $K = 5$.

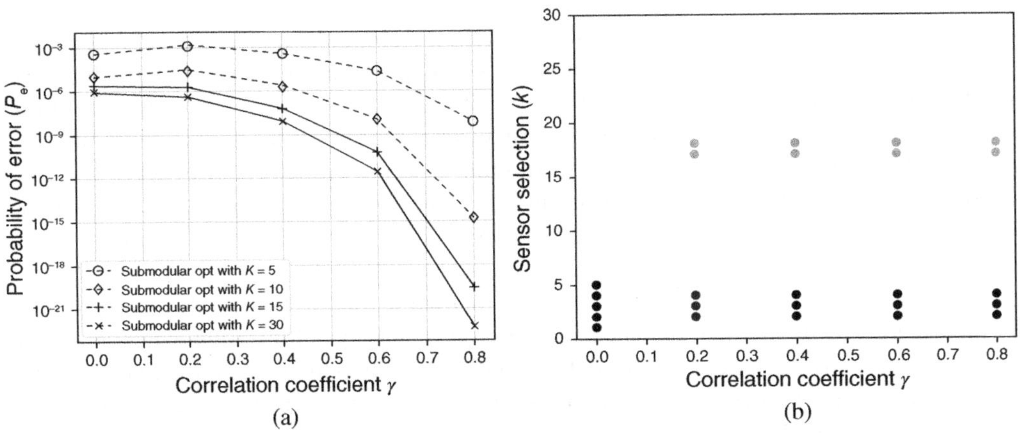

Figure 8.10 (a) Sensor selection of the submodular strategy for different numbers of sensors K and correlation coefficients γ when the noise correlation is giving by the spatially constant correlation in (8.10). (b) Selected sensors for different γ for $K = 5$.

8.3.2 Unknown θ Parameters

In many practical detection problems, complete knowledge of the signal to detect is not available. For example, only knowledge of the subspace described by our matrix $\mathbf{H}$ might be known but the precise values of $\boldsymbol{\theta}$ not. This situation can arise in direction of arrival estimation with prior knowledge of the angular sector where the signal might arrive or in the communications domain where user codes are known but not whether they are active or not.

8.3.2.1 Optimality Criteria

To tackle the design of sparse samplers for this setting, let us recall our binary hypothesis test problem

$$\mathcal{H}_0 : \mathbf{x} = \mathbf{n}, \tag{8.57}$$

$$\mathcal{H}_1 : \mathbf{x} = \mathbf{H}\boldsymbol{\theta} + \mathbf{n}. \tag{8.58}$$

while keeping in mind that in this setting $\boldsymbol{\theta}$ is an unknown parameter.

As $\boldsymbol{\theta}$ is unknown, for addressing the detection problem, we need to make use of the generalized likelihood ratio test (GRLT) which decides $\mathcal{H}_1$ if

$$L_G(\mathbf{x}) = \frac{p(\mathbf{x}; \hat{\boldsymbol{\theta}}_1)}{p(\mathbf{x}; \hat{\boldsymbol{\theta}}_0)} > \gamma, \tag{8.59}$$

where $\hat{\boldsymbol{\theta}}_i$ is the ML estimate of $\boldsymbol{\theta}$ under $\mathcal{H}_i$. In our particular case, these estimates are given, respectively, by

$$\hat{\boldsymbol{\theta}}_0 := \mathbf{0}, \tag{8.60}$$

$$\begin{aligned}\hat{\boldsymbol{\theta}}_1 = &\left(\mathbf{H}^{\mathsf{T}}\boldsymbol{\Phi}^{\mathsf{T}}(\mathbf{w})\left(\boldsymbol{\Phi}(\mathbf{w})\boldsymbol{\Sigma}\boldsymbol{\Phi}^{\mathsf{T}}(\mathbf{w})\right)^{-1}\boldsymbol{\Phi}(\mathbf{w})\mathbf{H}\right)^{-1} \\ &\times \mathbf{H}^{\mathsf{T}}\boldsymbol{\Phi}^{\mathsf{T}}(\mathbf{w})\left(\boldsymbol{\Phi}(\mathbf{w})\boldsymbol{\Sigma}\boldsymbol{\Phi}^{\mathsf{T}}(\mathbf{w})\right)^{-1}\boldsymbol{\Phi}(\mathbf{w})\mathbf{x},\end{aligned} \tag{8.61}$$

where to compute the ML estimate under $\mathcal{H}_1$ prewhitening after sampling has been performed. In this case, it can be shown that the logarithm of the GRLT is distributed as [25]

$$2\ln L_G(\mathbf{x}) \sim \begin{cases} \chi_N^2, & \text{under } \mathcal{H}_0, \\ \chi_N^2(\lambda(\mathbf{w}; \boldsymbol{\theta}),) & \text{under } \mathcal{H}_1, \end{cases} \tag{8.62}$$

where N is the dimensionality of $\boldsymbol{\theta}$ and

$$\lambda(\mathbf{w}; \boldsymbol{\theta}) = \boldsymbol{\theta}^{\mathsf{T}}\mathbf{H}^{\mathsf{T}}\boldsymbol{\Phi}^{\mathsf{T}}(\mathbf{w})\left(\boldsymbol{\Phi}(\mathbf{w})\boldsymbol{\Sigma}\boldsymbol{\Phi}^{\mathsf{T}}(\mathbf{w})\right)^{-1}\boldsymbol{\Phi}(\mathbf{w})\mathbf{H}\boldsymbol{\theta} \tag{8.63}$$

is the non-centrality parameter of the chi-squared distribution under $\mathcal{H}_1$. This leads to a detector performance defined by

$$P_{\text{fa}} = Q_{\chi_N^2}(\gamma'), \tag{8.64}$$

$$P_{\text{d}} = Q_{\chi_N^2(\lambda(\mathbf{w};\boldsymbol{\theta}))}(\gamma'), \tag{8.65}$$

where γ' is the modified threshold for the test in (8.62).

8.3.2.2 Sparse Sampler Design

Based on this performance metric, we can design a sparse sampler which maximizes P_{d} for a fixed P_{fa}. This is achieved by noticing that the P_{d} is a monotone function of $\lambda(\mathbf{w}; \boldsymbol{\theta})$. Therefore, by maximizing the non-centrality parameter (8.63), the power of the test can be maximized. However, as $\boldsymbol{\theta}$ is unknown a priori, direct maximization of $\lambda(\mathbf{w}; \boldsymbol{\theta})$ is not possible. In the following, we briefly discuss three alternatives to design sparse samplers using the noncentrality parameter as selection criterion for the designs. For more details and an extended formulation of the unknown parameter case, see, e.g., [26].

Average Design If there is some prior belief that the parameter lies within a known set Ω, a straightforward design approach can be based on maximizing the average noncentrality parameter, i.e.,

$$\max_{\mathbf{w}\in[0,1]^M} \int_{\boldsymbol{\theta}\in\Omega} \lambda(\mathbf{w}; \boldsymbol{\theta})p(\boldsymbol{\theta})d\boldsymbol{\theta}. \tag{8.66}$$

This design is equivalent to the one obtained through a Bayesian approach where a prior $p(\boldsymbol{\theta})$ is given to all the possible vectors $\boldsymbol{\theta} \in \Omega$. This approach requires an integration over the domain Ω. Note that when Ω is discretized and the integral is approximated by a summation, the previously discussed approach based on LMIs, see, e.g., (8.55), can be directly applied as $\lambda(\mathbf{w}, \boldsymbol{\theta})$ exhibits the same structure as the SNR expression in (8.51).

Worst Case (Max–Min) Design Alternatively, we could consider a worst-case design which aims to select samples that maximize the noncentrality parameter in the worst case, i.e.,

$$\max_{\mathbf{w}\in[0,1]^M} \min_{\theta\neq 0} \lambda(\mathbf{w};\theta). \tag{8.67}$$

Given the structure of the noncentrality parameter, this design problem becomes an eigenvalue maximization problem; that is

$$\max_{\mathbf{w}\in[0,1]^M} \lambda_{\min}\left(\mathbf{H}^\top\Phi^\top(\mathbf{w})\left(\Phi(\mathbf{w})\Sigma\Phi^\top(\mathbf{w})\right)^{-1}\Phi(\mathbf{w})\mathbf{H}\right), \tag{8.68}$$

which is analogous to the sparse sampler design problem using the E-optimality criterion when the ML error covariance matrix is used [cf. (8.20)].

Log-Det Design In many instances, the max–min criterion for a composite hypothesis test has an inherently pessimistic nature [27]. Therefore, a less stringent design, based on the D-optimality criterion could be to design samplers based on the problem

$$\max_{\mathbf{w}\in[0,1]^M} \log\det\left(\mathbf{H}^\top\Phi^\top(\mathbf{w})\left(\Phi(\mathbf{w})\Sigma\Phi^\top(\mathbf{w})\right)^{-1}\Phi(\mathbf{w})\mathbf{H}\right), \tag{8.69}$$

which instead of only focusing on the minimum eigenvalue, considers the energy distribution across the different directions as it aims to maximize the product of the eigenvalues. Note that, again, this problem is analogous to that of sparse samplers for estimation when the D-optimality criterion and the ML error covariance are used.

8.4 Conclusions

Distributed sensing has been investigated in this chapter. More specifically, we looked at the optimal placement of sensors to carry out a specific inference task such as estimation or detection. To simplify the presentation, a linear Gaussian model was considered and the task was to estimate the unknown variables or to detect whether there is a signal or not. The latter detection problem was studied for known and unknown signals. Even for the considered linear model, the problem statement is not easy to tackle, especially in case of correlated noise (i.e., conditionally dependent observations). Simulations on a field estimation problem show that sensors are selected that result in a well-conditioned and highly energetic measurement matrix. Furthermore, in case of correlated noise, this behavior is traded off with the correlation function. For a spatially decaying noise correlation function, this can potentially result in requiring more sensors to reach the same estimation performance as for uncorrelated noise. In case of a constant noise correlation function, sensors do not have to be close to each other to exploit the correlation, and we see that less sensors are needed to reach the same performance as in the uncorrelated noise case. For field detection a similar behavior is observed except that in this case the conditioning of the selected measurement matrix is not that important as the focus is more on selecting measurements with a high signal-to-noise ratio.

References

1 Chepuri, S.P. and Leus, G. (2016). Sparse sensing for statistical inference. *Foundations and Trends in Signal Processing* 9 (3–4): 233–368.

2 Molisch, A. and Win, M. (2004). MIMO systems with antenna selection. *IEEE Microwave Magazine* 5 (1): 46–56.

3 Blu, T., Dragotti, P.-L., Vetterli, M. et al. (2008). Sparse sampling of signal innovations. *IEEE Signal Processing Magazine* 25 (2): 31–40.

4 Eldar, Y.C. and Michaeli, T. (2009). Beyond bandlimited sampling. *IEEE Signal Processing Magazine* 26 (3): 48–68.

5 Vaidyanathan, P.P. and Pal, P. (2011). Sparse sensing with co-prime samplers and arrays. *IEEE Transactions on Signal Processing* 59 (2): 573–586.

6 Krause, A., Singh, A., and Guestrin, C. (2008). Near-optimal sensor placements in Gaussian processes: theory, efficient algorithms and empirical studies. *The Journal of Machine Learning Research* 9 (2): 235–284.

7 Joshi, S. and Boyd, S. (2009). Sensor selection via convex optimization. *IEEE Transactions on Signal Processing* 57 (2): 451–462.

8 Ranieri, J., Chebira, A., and Vetterli, M. (2014). Near-optimal sensor placement for linear inverse problems. *IEEE Transactions on Signal Processing* 62 (5): 1135–1146.

9 Chepuri, S.P. and Leus, G. (2015). Sparsity-promoting sensor selection for non-linear measurement models. *IEEE Transactions on Signal Processing* 63 (3): 684–698.

10 Liu, S., Fardad, M., Varshney, P.K., and Masazade, E. (2014). Optimal periodic sensor scheduling in networks of dynamical systems. *IEEE Transactions on Signal Processing* 62 (12): 3055–3068.

11 Bajovic, D., Sinopoli, B., and Xavier, J. (2011). Sensor selection for event detection in wireless sensor networks. *IEEE Transactions on Signal Processing* 59 (10): 4938–4953.

12 Cambanis, S. and Masry, E. (1983). Sampling designs for the detection of signals in noise. *IEEE Transactions on Information Theory* 29 (1): 83–104.

13 Yu, C.-T. and Varshney, P.K. (1997). Sampling design for Gaussian detection problems. *IEEE Transactions on Signal Processing* 45 (9): 2328–2337.

14 Coutino, M., Chepuri, S., and Leus, G. (2018). Near-optimal sparse sensing for Gaussian detection with correlated observations. *IEEE Transactions on Signal Processing* 66 (15): 4025–4039.

15 Chepuri, S.P. and Leus, G. (2016). Sparse sensing for distributed detection. *IEEE Transactions on Signal Processing* 64 (6): 1446–1460.

16 Nemhauser, G.L., Wolsey, L.A., and Fisher, M.L. (1978). An analysis of approximations for maximizing submodular set functions—I. *Mathematical Programming* 14 (1): 265–294.

17 Shamaiah, M., Banerjee, S., and Vikalo, H. (2010). Greedy sensor selection: leveraging submodularity. *49th IEEE Conference on Decision and Control (CDC)*, 2572–2577. IEEE.

18 Ali, S. and Silvey, S.D. (1966). A general class of coefficients of divergence of one distribution from another. *Journal of the Royal Statistical Society: Series B: Methodological* 28 (1): 131–142.

19 Cover, T.M. and Thomas, J.A. (1991). Information theory and statistics. *Elements of Information Theory* 1 (1): 279–335.

20 Kadota, T. and Shepp, L. (1967). On the best finite set of linear observables for discriminating two Gaussian signals. *IEEE Transactions on Information Theory* 13 (2): 278–284.

21 Kailath, T. (1967). The divergence and Bhattacharyya distance measures in signal selection. *IEEE Transactions on Communication Technology* 15 (1): 52–60.

22 Kobayashi, H. and Thomas, J.B. (1967). Distance measures and related criteria. *5th Annual Allerton Conference on Circuit and System Theory*, 491–500.

23 Horst, R. and Pardalos, P.M. (2013). *Handbook of Global Optimization*, vol. 2. Springer Science & Business Media.

24 Coutino, M., Chepuri, S.P., and Leus, G. (2018). Submodular sparse sensing for Gaussian detection with correlated observations. *IEEE Transactions on Signal Processing* 66 (15): 4025–4039.

25 Kay, S.M. (2009). *Fundamentals of Statistical Processing: Detection Theory*, vol. 2. Pearson Education India.

26 Coutino, M., Chepuri, S.P., and Leus, G. (2017). Sparse sensing for composite matched subspace detection. *2017 IEEE 7th International Workshop on Computational Advances in Multi-Sensor Adaptive Processing (CAMSAP)*, 1–5. IEEE.

27 Feder, M. and Merhav, N. (2002). Universal composite hypothesis testing: a competitive minimax approach. *IEEE Transactions on Information Theory* 48 (6): 1504–1517.

9

Sparse Sensor Arrays for Active Sensing: Models, Configurations, and Applications

Robin Rajamäki[1,2] *and Visa Koivunen*[2]

[1]*Electrical and Computer Engineering Department, University of California San Diego, La Jolla, CA, USA*
[2]*Department of Information and Communications Engineering, Aalto University, Espoo, Finland*

9.1 Introduction

In active sensing, sensors such as antennas or piezoelectric transducers probe their surroundings by emitting self-generated energy. The signals backscattered to the receivers are then processed to extract information from the physical environment – typically in the form of target detections and parameter estimates. Classical applications of active sensing include radar, sonar, wireless communications, and medical ultrasound. Active sensing also plays a key role in emerging applications such as autonomous sensing [1], robotics [2], automotive radar [3, 4], and integrated sensing and communications [5–7].

In contrast to passive (receive-only) sensing, active sensing systems have the flexibility to control the properties of the emitted signals, commonly known as *waveforms*. Waveforms can be adapted to various tasks, such as target search or tracking, for example, by shaping the waveform in the time, frequency, polarization, or angle domain. Indeed, a key benefit of an active *multi-sensor array* is the ability to focus and steer energy in desired directions by beamforming on transmit. This provides coherent combining gain and spatial selectivity, which improves signal-to-noise-ratio (SNR), enables resolving closely spaced scatterers, reduces interference in and from undesired directions, and increases reliability of communication links.

As described in Chapter 1, *sparse sensor arrays* offer several advantages over conventional uniform array geometries. These benefits include improved resolution using fewer physical sensors, the capability to identify more scatterers than physical sensors, as well as robustness to antenna system nonidealities, such as mutual coupling. In active sensing, much of this is facilitated by the effective transmit-receive virtual array known as the *sum co-array*. The sum co-array consists of the pairwise sums of transmit and receive sensor positions. Properly designed sparse arrays with N_t physical transmitters and N_r receivers can therefore achieve a uniform virtual array with on the order of N_tN_r virtual sensors – corresponding to an increase by a factor of N_t compared to a receive-only array. The benefits of a large virtual array are well-known in applications ranging from coherent imaging [8] to multiple-input multiple-output (MIMO) radar [9]. The sum co-array is analogous to the difference co-array in passive sensing. However, whereas the difference co-array emerges from the assumption of uncorrelated source signals, the sum co-array requires no such statistical assumptions.

Most works on sparse arrays in the past decade have focused on passive sensing. As a result, active sparse arrays are relatively less explored. Many concepts originally developed for passive

Sparse Arrays for Radar, Sonar, and Communications, First Edition. Edited by Moeness G. Amin.

sparse arrays can be extended to the active case – e.g., see [10] in case of the minimum-redundancy array (MRA) [11]. Fundamental differences between the two sensing modalities nevertheless give rise to unique, sometimes subtle, design issues. For example, one may or may not have the freedom to independently choose the transmit and receive sensor positions. Hence, care must be taken when defining active array redundancy and the (active) MRA configuration. If transmitters and receivers are not constrained to occupy the same positions, completely nonredundant configurations are widely known. The most notable example is perhaps the nested geometry also known as the MIMO array [9]. However, less is known about MRAs when the physical transmit and receive arrays are constrained to have the same geometry. Such configurations – regularly used in monostatic radar and medical ultrasound applications – are of interest when the area for placing the physical sensors is tightly constrained. A disadvantage of these MRA geometries is that they are computationally expensive to find. Hence, scalable low-redundancy array geometries need to be developed. A natural question is therefore whether well-known passive array configurations, such as the Nested array [12], could be adapted to active sensing? This chapter provides a positive answer by outlining a general and pragmatic framework for generating symmetric arrays suitable for both active and passive sensing.

Another pertinent question when employing sparse arrays is how to mitigate grating lobes arising from spatial undersampling. In active sensing, the flexibility of independently controlling both the transmit (Tx) and receive (Rx) beampatterns provides a means to shape the effective Tx–Rx beampattern, which is often of primary interest in sensing, especially when employing linear processing at both the transmitter and receiver. Since this effective beampattern is the product of the Tx and Rx beampatterns, a null in one beampattern can cancel a side lobe in the other. A general model for synthesizing desirable Tx-Rx beampatterns, known as image addition, was developed by Kassam *et al.* [8, 13, 14] building on prior works on coherent and incoherent imaging in optical holography [15] and radio astronomy [16], respectively. Image addition synthesizes a desired Tx–Rx beampattern by linearly combining so-called component beampatterns obtained using different Tx–Rx beamforming weight pairs. However, when the array configuration contains redundant virtual sensors, there is an opportunity to optimize the beamforming weights. Consequently, this chapter considers the problem of minimizing the number of component beampatterns, and hence the synthesis time of the desired composite beampattern, since each component may correspond to a separate transmission. We also briefly discuss the fundamental spatio-temporal trade-off between sparsity of the array geometry and the number of component weights required to synthesize any feasible Tx–Rx beampattern.

9.1.1 Goals, Scope, and Organization

The goal of this chapter is to provide an accessible introduction to active sparse array design and signal processing. We attempt to strike a balance between providing a useful resource on active sparse arrays for practitioners, as well as highlighting problems of theoretical interest to researchers in the field.

We focus on joint transmit–receive beamforming and one-dimensional array configurations with low-redundancy; largely following our two previous articles [17, 18], and the first author's thesis [19]. For brevity, several important topics in active sensing are treated only cursorily. Notable omissions include waveform design, space-time adaptive processing, and high-resolution localization algorithms. For detailed treatments on these topics, see [20–22] and references therein.

The chapter is organized as follows. Section 9.2 presents the active sensing signal model. Section 9.3 reviews sparse array configurations for active sensing, with an emphasis on the active

MRA, and scalable low-redundancy symmetric sparse arrays. Section 9.4 examines beamforming using sparse arrays. In particular, we consider the problem of synthesizing a desired joint transmit–receive beampattern using multiple component beampatterns. Section 9.5 briefly illustrates example applications of sparse active arrays. Finally, Section 9.6 concludes the chapter.

9.1.2 Notation

We denote matrices by boldface uppercase, e.g., $\boldsymbol{A}$; vectors by boldface lowercase, $\boldsymbol{a}$; and scalars by unbolded letters, A, a. The (n, m)th element of matrix $\boldsymbol{A}$ and $\boldsymbol{A}_i$ is A_{nm} and $[\boldsymbol{A}_i]_{n,m}$, respectively. Furthermore, $\boldsymbol{A}^{\mathrm{T}}$, $\boldsymbol{A}^{\mathrm{H}}$, and $\boldsymbol{A}^*$ denote transpose, Hermitian transpose, and complex conjugation. The $N \times N$ identity matrix is denoted by $\boldsymbol{I}_N$, and the standard unit vector, consisting of zeros except for the ith entry, which is unity, by $\boldsymbol{e}_i$ (dimension specified separately). Moreover, vec(·) stacks the columns of its matrix argument into a column vector, whereas diag(·) constructs a diagonal matrix of its vector argument. The ceiling function is denoted by $\lceil \cdot \rceil$. The Kronecker and Khatri–Rao (columnwise Kronecker) products are denoted by $\otimes$ and $\odot$, respectively. Sets are denoted by bold blackboard letters, such as $\mathbb{A}$. The maximum element of a finite set of real-valued scalars $\mathbb{A}$ is denoted by max $\mathbb{A}$; the subset of integers between $a, b \in \mathbb{R}$ by $[a : b] \triangleq \{c \in \mathbb{Z} \mid a \leq c \leq b; a, b \in \mathbb{R}\}$; and the sum set of $\mathbb{A}$ and $\mathbb{B}$ by $\mathbb{A} + \mathbb{B} \triangleq \{a + b \mid a \in \mathbb{A}, b \in \mathbb{B}\}$. Finally, subscripts "t," "r," and "ξ" denote "transmitter," "receiver," and "transmitter or receiver," respectively.

9.2 Active Sensing Signal Model

This section introduces the considered active sensing signal model. For simplicity of presentation, we restrict our attention to linear transmit and receive array geometries with fully digital beamforming architectures. This implies a discrete-time model, where each transmit sensor may launch a different waveform, and the output of each receive sensor is available. For extensions to planar arrays and hybrid or analog beamforming, see [19].

9.2.1 Physical Array Model

Figure 9.1 depicts the considered active sensing MIMO model consisting of co-located Tx and Rx arrays. Co-locatedness means that the distance between the Tx and Rx array is negligible w.r.t. their distance to the scatterers, i.e., both arrays observe the same scatterer angles [9]. This is in stark contrast to distributed arrays leveraging angular diversity [23]. The Tx and Rx arrays are assumed to have a one-dimensional co-linear geometry.

The Tx array illuminates a collection of K far field point scatterers, which backscatter the impinging wavefield back at the Rx array. The baseband Tx signals are modulated by a carrier

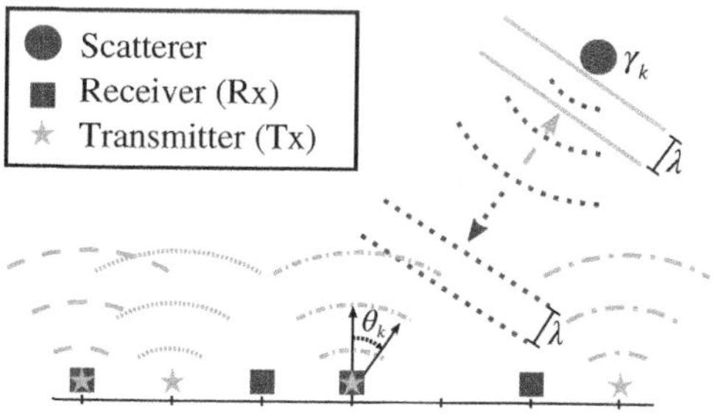

Figure 9.1 Active sensing signal model. A linear Tx array with N_{t} sensors illuminates K far field point scatterers using $N_s \leq N_{\mathrm{t}}$ linearly independent waveforms. A linear Rx array with N_{r} sensors, co-located with the Tx array, measures a superposition of the backscattered planar wavefronts.

frequency of wavelength λ before transmission. The carrier frequency is assumed much larger than the bandwidth of the baseband signal, such that a narrowband signal model approximately holds. We assume that the Tx/Rx sensors lie on a uniform grid of integer multiples of $\lambda/2$ given by set $\mathbb{D}_\xi \in \mathbb{Z}$. The number of Tx/Rx sensors is denoted by $N_\xi \triangleq |\mathbb{D}_\xi|$.

9.2.1.1 Angle-Delay-Doppler Model

The transmitted signals experience different round-trip delays depending on the distance from the sensing platform to the scatterers, and Doppler shifts due to the relative difference in radial motion between the platform and scatterers. The kth scatterer is therefore generally parametrized by scattering coefficient γ_k, azimuth angle θ_k, delay τ_k, and Doppler shift ω_k. A simple model for the continuous-time baseband Rx array output is [24]

$$\boldsymbol{y}(t) = \sum_{k=1}^{K} \gamma_k \boldsymbol{a}_\mathrm{r}(\theta_k)\boldsymbol{a}_\mathrm{t}^\mathrm{T}(\theta_k)\boldsymbol{s}(t-\tau_k)e^{j\omega_k t} + \boldsymbol{n}(t). \tag{9.1}$$

Here $\boldsymbol{a}_\xi(\theta_k) \in \mathbb{C}^{N_\xi}$ denotes the Tx/Rx array manifold vector (recall shorthand "ξ" denotes Tx or Rx), and $\boldsymbol{s}(t) \in \mathbb{C}^{N_\mathrm{t}}$ is a spatial snapshot of the transmitted waveform at time t. Term $\boldsymbol{n}(t) \in \mathbb{C}^{N_\mathrm{r}}$ models noise, interference, and clutter.

To simplify presentation, we will henceforth focus on an angle-only version of (9.1). This restriction, commonly employed in for instance MIMO radar [9, 25], is justified by the fact that the array geometry, which is of main interest here, primarily affects the sensing performance in the angular domain. For generalizations of (9.1) to frequency-selective models, see, e.g., ([26], Ch. 5).

9.2.1.2 Simplified Angle-Only Model

Suppose the K scatterers lie in the same delay-Doppler bin, and that the only unknown scatter parameters are the scattering coefficients $\boldsymbol{\gamma} \in \mathbb{C}^K$ and azimuth angles $\boldsymbol{\theta} \in [-\pi/2, \pi/2)^K$. Sufficient isolation between delay-Doppler bins can be achieved by proper waveform design and space-time processing. Setting the delay and Doppler parameters to $\tau_k = \omega_k = 0\ \forall k$, and sampling the array output at $T \in \mathbb{N}_+$ time-instances, (9.1) yields the following $N_\mathrm{t} \times T$ discrete-time model [25]

$$\boldsymbol{Y} = \sum_{k=1}^{K} \gamma_k \boldsymbol{a}_\mathrm{r}(\theta_k)\boldsymbol{a}_\mathrm{t}^\mathrm{T}(\theta_k)\boldsymbol{S}^\mathrm{T} + \boldsymbol{N} = \boldsymbol{A}_\mathrm{r}\,\mathrm{diag}(\boldsymbol{\gamma})\boldsymbol{A}_\mathrm{t}^\mathrm{T}\boldsymbol{S}^\mathrm{T} + \boldsymbol{N}. \tag{9.2}$$

Here, $\boldsymbol{A}_\xi = \boldsymbol{A}_\xi(\boldsymbol{\theta}) = [\boldsymbol{a}_\xi(\theta_1), \ldots, \boldsymbol{a}_\xi(\theta_K)] \in \mathbb{C}^{N_\xi \times K}$ is the narrowband Tx/Rx array manifold matrix modeling the relative phase shifts between the sensors induced by the planar wavefronts departing from/impinging on the Tx/Rx array. The (n, k)th entry of $\boldsymbol{A}_\xi$ depends on the nth normalized Tx/Rx sensor position $d_\xi[n] \in \mathbb{D}_\xi$, $n \in [1 : N_\xi]$, and the kth scatterer angle $\theta_k \in [-\pi/2, \pi/2)$, $k \in [1 : K]$:

$$[\boldsymbol{A}_\xi]_{n,k} = \exp(j\pi d_\xi[n]\sin\theta_k). \tag{9.3}$$

Moreover, $\boldsymbol{S} = [\boldsymbol{s}(t_1), \ldots, \boldsymbol{s}(t_T)]^\mathrm{T} \in \mathbb{C}^{T\times N_\mathrm{t}}$ is a spatio-temporal Tx *waveform matrix*, whose nth column corresponds to the waveform transmitted by the nth Tx sensor sampled at times $\{t_i\}_{i=1}^T$. For simplicity, we assume the entries of $\boldsymbol{N} \in \mathbb{C}^{N_\mathrm{r}\times T}$ are i.i.d. complex-valued zero-mean Gaussian random variables of equal variance σ^2, i.e., $N_{n,t} \sim \mathcal{CN}(0, \sigma^2)$. For more general clutter and interference models, see, e.g., ([26], Ch. 6).

Equation (9.2) can be vectorized and expressed using the Kronecker product $\otimes$ and Khatri–Rao (columnwise Kronecker) product $\odot$ as $\boldsymbol{y} \triangleq \mathrm{vec}(\boldsymbol{Y}) \in \mathbb{C}^{TN_\mathrm{r}}$:

$$\boldsymbol{y} = ((\boldsymbol{S}\boldsymbol{A}_\mathrm{t}) \odot \boldsymbol{A}_\mathrm{r})\boldsymbol{\gamma} + \boldsymbol{n} = (\boldsymbol{S} \otimes \boldsymbol{I}_{N_\mathrm{r}})(\boldsymbol{A}_\mathrm{t} \odot \boldsymbol{A}_\mathrm{r})\boldsymbol{\gamma} + \boldsymbol{n}. \tag{9.4}$$

Here, $\gamma \in \mathbb{C}^K$ is the vector of scattering coefficients, $\boldsymbol{n} \triangleq \text{vec}(\boldsymbol{N})$ is the vectorized noise matrix, and $\boldsymbol{I}_{N_r}$ is the $N_r \times N_r$ identity matrix. Equation (9.4) follows from the following Kronecker/Khatri–Rao product identities: $\text{vec}(\boldsymbol{H}\text{diag}(\boldsymbol{f})\boldsymbol{G}^\top) = (\boldsymbol{G} \odot \boldsymbol{H})\boldsymbol{f}$ and $(\boldsymbol{GH}) \odot (\boldsymbol{FD}) = (\boldsymbol{G} \otimes \boldsymbol{F})(\boldsymbol{H} \odot \boldsymbol{D})$ for matrices of appropriate dimensions – see ([27], Theorem 2) for details.

9.2.1.3 Waveform Matrix

Waveform matrix $\boldsymbol{S}$ crucially impacts the performance of the active sensing system. For example, the choice of $\boldsymbol{S}$ affects the realized transmit combining gain, the identifiability of the unknown model parameters $\boldsymbol{\theta}$ and γ in (9.4), and the fidelity to which they can be recovered in the presence of noise and interference. While Section 9.4 elaborates on the role of $\boldsymbol{S}$ in beamforming, here we only briefly discuss some basic properties of $\boldsymbol{S}$. The reader interested in waveform design is referred to the rich literature on the topic; for example, see [20, 28–30].

Waveform Rank A fundamental property of $\boldsymbol{S}$ is its rank, $N_s \triangleq \text{rank}(\boldsymbol{S})$. The canonical example of a *unit* rank system ($N_s = 1$) is single-input-multiple-output (SIMO) – also known as phased array – radar ([31], Ch. 8), which corresponds to transmitting a single appropriately phased shifted (delayed) and possibly scaled waveform from each Tx sensor. In contrast, full waveform rank ($N_s = N_t$) is usually employed in MIMO radar [9, 32], which transmits linearly independent (typically orthogonal) waveforms from each of the N_t transmitters. Typically, orthogonal waveforms are used when the radar is in a target search mode. An intermediate waveform rank ($N_s \in [2 : N_t - 1]$) usually trades off field of view for coherent transmit combining gain [33–35]. The appropriate choice of N_s is therefore task-dependent. For example, $N_s = 1$ may be desirable when focusing energy on single target for tracking purposes, whereas $N_s = N_t$ is suitable when searching for new targets without *a priori* knowledge of their directions. Increasing N_s also facilitates resolving more targets simultaneously, extending the set of achievable beampatterns, as well as illuminating broader angular intervals while mitigating interference.

Waveform rank may also be constrained by hardware costs, power consumption, or the available computational resources. For example, full waveform rank ($N_s = N_t$) can be infeasible due to the required number of expensive and power hungry radio and intermediate-frequency (RF-IF) front-ends at the transmitter. Waveform rank may also affect the hardware complexity or computational burden of the receiver. For instance, a dedicated filter is required for each transmitted waveform if matched or mismatched filtering is employed.

Practical Waveform Constraints The transmit power of the sensing system is always limited in practice. Typical power constraints include the total power constraint $\|\boldsymbol{S}\|_F^2 \leq \zeta > 0$ or the per sensor power constraint $\|\boldsymbol{S}_{:,n}\|_2^2 \leq \zeta/N_t > 0 \forall n$. Constant modulus waveforms $|S_{t,n}| = c > 0 \forall (t, n)$ are also often favored because a high peak-to-average power ratio leads to the inefficient use of amplifiers. This is especially the case in radar applications, where high Tx power is needed to achieve a desired SNR and coverage over a large surveillance volume. Another key property of the system determined by the waveform is the delay-Doppler response, which is often desired to closely approximate the ideal thumbtack function.

9.2.2 Virtual Array Model

This section introduces the canonical virtual array model known as the sum co-array [8]. The sum co-array consists of the pairwise sums of Tx–Rx sensor position pairs, and arises naturally in active sensing.

Definition 9.1 ***(Sum co-array)*** The sum co-array $\mathbb{D}_\Sigma$ is defined as

$$\mathbb{D}_\Sigma \triangleq \mathbb{D}_t + \mathbb{D}_r = \{d_t + d_r \mid d_t \in \mathbb{D}_t; d_r \in \mathbb{D}_r\},$$

where $\mathbb{D}_t$ and $\mathbb{D}_r$ denote the set of Tx and Rx sensor positions, respectively. The cardinality of the sum co-array is denoted by $N_\Sigma \triangleq |\mathbb{D}_\Sigma|$.

The sum co-array is closely related to the difference co-array, which emerges under the assumption of uncorrelated source signals and hence finds wide use in passive sensing (see Chapter 1). Recall that the difference co-array (or difference set) of array configuration $\mathbb{D}$ is the set of pairwise sensor position differences $\mathbb{D}_\Delta \triangleq \mathbb{D} - \mathbb{D} = \{d_1 - d_2 \mid d_1, d_2 \in \mathbb{D}\}$. Unlike the difference co-array, the sum co-array does not require any statistical assumptions (on the scattering coefficients) to emerge, but is rather a consequence of the co-locatedness and phase coherence of the Tx and Rx arrays.

Similarly to the physical array, a virtual steering matrix $\boldsymbol{A}_\Sigma \in \mathbb{C}^{N_\Sigma \times K}$ can be associated with the sum co-array. The (ℓ, k)th entry of this matrix is

$$[\boldsymbol{A}_\Sigma]_{\ell,k} = \exp(j\pi d_\Sigma[\ell] \sin\theta_k). \tag{9.5}$$

Matrix $\boldsymbol{A}_\Sigma$ is related to the effective Tx–Rx array steering matrix $\boldsymbol{A}_t \odot \boldsymbol{A}_r$ as

$$\boldsymbol{A}_t \odot \boldsymbol{A}_r = \Upsilon^\top \boldsymbol{A}_\Sigma, \tag{9.6}$$

where $\Upsilon \in \{0,1\}^{N_\Sigma \times N_t N_r}$ is a *redundancy pattern matrix* [17]. This binary matrix characterizes the virtual sensor multiplicities. Specifically, it forms a one-to-many map from the unique elements of the sum co-array to the corresponding physical Tx–Rx sensor pairs.

Definition 9.2 ***(Redundancy pattern)*** The (ℓ, i)th entry of the binary *redundancy pattern matrix* $\Upsilon \in \{0,1\}^{N_\Sigma \times N_t N_r}$ is

$$\Upsilon_{\ell,i} \triangleq \begin{cases} 1, & \text{if } d_t\left[\lceil i/N_r \rceil\right] + d_r\left[i - (\lceil i/N_r \rceil - 1)N_r\right] = d_\Sigma[\ell] \\ 0, & \text{otherwise.} \end{cases} \tag{9.7}$$

Here, $d_\xi[n] \in \mathbb{D}_\xi$ is the nth Tx/Rx sensor position and $d_\Sigma[\ell] \in \mathbb{D}_\Sigma$ is the ℓth sum co-array element position.

Matrix Υ satisfies by definition $\Upsilon\Upsilon^\top = \operatorname{diag}(\boldsymbol{v}_\Sigma)$, where $\boldsymbol{v}_\Sigma \in \mathbb{N}_+^{N_\Sigma}$ is a positive integer-valued vector containing the multiplicities of the unique virtual sensors. We assume without loss of generality that the virtual sensor positions $\mathbb{D}_\Sigma = \{d_\Sigma[\ell]\}_{\ell=1}^{N_\Sigma}$ and physical Tx/Rx sensor positions $\mathbb{D}_\xi = \{d_\xi[n]\}_{n=1}^{N_\xi}$ are indexed in ascending order, with their first sensors at the origin, i.e.,

$$d_\xi[1] = 0 < d_\xi[2] < \cdots < d_\xi[N_\xi] \text{ and } d_\Sigma[1] = 0 < d_\Sigma[2] < \cdots < d_\Sigma[N_\Sigma]. \tag{9.8}$$

A uniform sum co-array $\mathbb{D}_\Sigma = [0, N_\Sigma - 1]$ is also called *contiguous*. A key advantage of a contiguous co-array is that the co-array manifold $\boldsymbol{A}_\Sigma$ is a Vandermonde matrix, which guarantees the identifiability of up to $K \propto N_\Sigma$ scatterers – regardless of their configuration, provided the angles are unique.

Example 9.1 ***(Redundancy pattern of ULA)*** Consider the uniform linear array (ULA) $\mathbb{D}_t = \mathbb{D}_r = \{0, 1\}$ with $N_t = N_r = 2$ Tx/Rx sensors. The sum co-array is contiguous and given by $\mathbb{D}_\Sigma = \{0, 1, 2\}$. Consequently, by (9.7) the redundancy pattern matrix Υ and multiplicity vector $\boldsymbol{v}_\Sigma = \Upsilon\Upsilon^\top \mathbf{1}$ are

$$\Upsilon = \begin{bmatrix} 1 & 0 & 0 & 0 \\ 0 & 1 & 1 & 0 \\ 0 & 0 & 0 & 1 \end{bmatrix} \quad \text{and} \quad \boldsymbol{v}_\Sigma = \begin{bmatrix} 1 \\ 2 \\ 1 \end{bmatrix}.$$

The second row of Υ has two nonzero entries, since the second virtual sensor (corresponding to element 1) can be represented as the sum of the first Tx sensor and the second Rx sensor or vice versa, i.e., $1 = 0 + 1 = 1 + 0$.

Equations (9.5) and (9.7) enable expressing the measurement vector in (9.4) as

$$\boldsymbol{y} = \underbrace{(\boldsymbol{S} \otimes \boldsymbol{I})\Upsilon^{\mathrm{T}}}_{\triangleq \boldsymbol{C}} \boldsymbol{A}_{\Sigma}\boldsymbol{\gamma} + \boldsymbol{n}. \tag{9.9}$$

This reformulation allows re-interpreting (9.4) as a virtual array measurement model. Specifically, if the waveform rank is $N_s < N_{\Sigma}/N_{\mathrm{r}}$, then $\boldsymbol{y}$ is a *compressed* measurement of the virtual sensor outputs of the sum co-array. The compression is due to the potentially rank-deficient matrix $\boldsymbol{C} \triangleq (\boldsymbol{S} \otimes \boldsymbol{I})\Upsilon^{\mathrm{T}} \in \mathbb{C}^{TN_{\mathrm{r}} \times N_{\Sigma}}$ satisfying $\mathrm{rank}(\boldsymbol{C}) \leq \min(N_s N_{\mathrm{r}}, N_{\Sigma})$. Note that $\boldsymbol{C}$ can be tall, yet column-rank deficient, since long waveforms $T \gg N_{\mathrm{t}}$ may be employed in practice. This is typically the case when using phase-modulated radar codes and pulse compression. Matrix $\boldsymbol{C}$ also reveals that the waveform and array geometry interact via redundancy pattern matrix Υ. This interpretation and its implications remain to be fully explored in future work.

9.3 Sparse Array Configurations

This section presents examples of sparse array configurations for active sensing. Section 9.3.1 starts by categorizing active array geometries based on the number of shared sensors between the Tx and Rx arrays. We focus on geometries with fully overlapping Tx and Rx arrays (shared transceivers) that maximize the size of the sum co-array (assumed contiguous) for a given number of physical sensors. Section 9.3.2 discusses the class of MRAs, which are optimal in this regard. However, finding these array configurations typically requires solving a combinatorial optimization problem, which is impractical even when the number of physical sensors is well below a hundred. Therefore, Section 9.3.3 introduces scalable, but possibly suboptimal configurations, whose sum co-array is nevertheless contiguous and grows order-wise optimally in the number of sensors. As a bonus, these geometries are symmetric by construction, which implies that both the sum and difference co-array is contiguous, and hence the arrays are suitable for both active and passive sensing. For active sparse array design approaches beyond the sum co-array perspective adopted herein, see, e.g., Chapter 10 and references therein.

9.3.1 Categorization of Array Configurations Based on Overlap Between Tx and Rx Arrays

A meaningful definition of the MRA requires considering any constraints imposed on the admissible sensor positions, as in active sensing, one may or may not have the freedom to choose the Rx and Tx sensor positions independently. In particular, the number of distinct physical sensors (either transmitting, receiving, or transceiving), $|\mathbb{D}_{\mathrm{t}} \cup \mathbb{D}_{\mathrm{r}}|$, may be less than the total number of transmitters and receivers, $|\mathbb{D}_{\mathrm{t}}| + |\mathbb{D}_{\mathrm{t}}|$. These two quantities, together with the number of shared sensors (transceivers), $|\mathbb{D}_{\mathrm{t}} + \mathbb{D}_{\mathrm{r}}|$, satisfy

$$|\mathbb{D}_{\mathrm{t}}| + |\mathbb{D}_{\mathrm{r}}| = |\mathbb{D}_{\mathrm{t}} \cup \mathbb{D}_{\mathrm{r}}| + |\mathbb{D}_{\mathrm{t}} \cap \mathbb{D}_{\mathrm{r}}|.$$

Depending on the fraction of sensors shared by the Tx and Rx arrays, active array configurations naturally fall into one of the following three categories:

(a) *Fully overlapping* Tx and Rx arrays: $\mathbb{D}_{\mathrm{t}} = \mathbb{D}_{\mathrm{r}} = \mathbb{D}$.
(b) *Partially overlapping* Tx and Rx arrays: $\{0\} \subset \mathbb{D}_{\mathrm{t}} \cap \mathbb{D}_{\mathrm{r}} \subset \mathbb{D}_{\mathrm{t}} \cup \mathbb{D}_{\mathrm{r}}$.
(c) *Nonoverlapping* Tx and Rx arrays: $\mathbb{D}_{\mathrm{t}} \cap \mathbb{D}_{\mathrm{r}} = \{0\}$.

Figure 9.2 illustrates the three cases. We will mainly focus on (a) herein. Note that the Tx and Rx arrays have at least one common sensor (at the origin) due to the adopted ordering convention

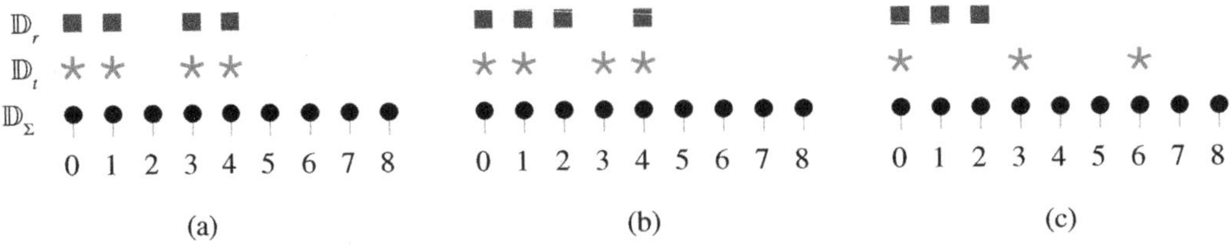

Figure 9.2 Categorization of (aligned) active array configurations based on the overlap between Tx and Rx sensors: (a) Fully overlapping, (b) partially overlapping, and (c) non-overlapping. The geometry in (a) achieves a contiguous sum co-array of exactly twice the physical Tx/Rx array aperture. The configuration in (c) attains the same co-array using fewer physical sensors at the expense of a larger physical array aperture.

in (9.8). Although the Tx and Rx arrays obviously need not share any sensors in practice, the normalized Tx–Rx array configuration satisfying (9.8) will fall into one of the above categories (a) to (c).

The total number of distinct sensors satisfies $|\mathbb{D}_t \cup \mathbb{D}_r| \in [\max(|\mathbb{D}_t|, |\mathbb{D}_r|) : |\mathbb{D}_t| + |\mathbb{D}_r| - 1]$, which in the fully overlapping and nonoverlapping cases simplifies to

$$|\mathbb{D}_t \cup \mathbb{D}_r| = \begin{cases} |\mathbb{D}|, & \text{if } \mathbb{D}_t = \mathbb{D}_r = \mathbb{D} \text{(fully overlapping)} \\ |\mathbb{D}_t| + |\mathbb{D}_r| - 1, & \text{if } \mathbb{D}_t \cap \mathbb{D}_r = \{0\} \text{(nonoverlapping).} \end{cases}$$

Fully overlapping transmitters and receivers are typically employed in pulsed systems, such as phased array radar ([31], p. 5) and medical ultrasound [36]. Nonoverlapping Tx and Rx arrays are common in applications such as full-duplex communications [37] or automotive MIMO radar [3] employing continuous wave transmission (or even a bistatic Tx–Rx array configuration). Nonoverlapping array configurations have the advantage of achieving a completely nonredundant and contiguous sum co-array, as we will see in Section 9.3.2.3. However, the physical apertures of the resulting Tx and Rx arrays can be significantly different, which may be impractical when the physical area for placing the sensors is strictly limited. In contrast, fully overlapping configurations have – by definition – the same physical Tx and Rx array aperture, which is exactly half the virtual aperture of the sum co-array, as illustrated in Fig. 9.2.

We remark that the extent to which sensors are shared between the Tx and Rx arrays is a physical property of the array. This notion of array overlap is therefore conceptually different from array partitioning or sub-array processing schemes used in, e.g., spatial smoothing [38], ESPRIT [39], multifunction radar [40], full-duplex communications [37], or joint radar and communications systems [41].

9.3.2 Minimum-Redundancy Array

A large contiguous sum co-array is desirable for achieving a high angular resolution and unambiguously resolving as many scatterers as possible. Array processing is also greatly simplified if the sum co-array is contiguous, or if it contains a large contiguous subarray. Hence, it is desirable to maximize the size of the (contiguous) co-array for a given number of physical sensors. The MRA [10, 11] is optimal in this regard, since redundancy quantifies how efficiently the sensors of the physical array are utilized toward constructing a co-array with unique contiguous virtual sensors. Originally defined for passive arrays and the difference co-array [11], the concept of redundancy was later extended to active arrays and the sum co-array [10]. See also [11, 97] for an extension to the "difference-sum" co-array, which arises in active sensing when scatterers are uncorrelated [44]. Herein, we focus on the sum co-array-based MRA for active sensing. Sections 9.3.2.1 and 9.3.2.2 formally define the concepts of redundancy and the MRA, respectively, whereas Section 9.3.2.3 briefly presents some known MRAs and discusses their structure.

9.3.2.1 Redundancy

Redundancy quantifies the repetition of the virtual sensors in the co-array. The redundancy of any array configuration with a contiguous sum co-array can be defined as ([10], Eqs. (7) and (8))

$$R(\mathbb{D}_t, \mathbb{D}_r) \triangleq \begin{cases} \frac{1}{2}|\mathbb{D}|(|\mathbb{D}|+1)/|\mathbb{D}+\mathbb{D}|, & \text{if } \mathbb{D}_t = \mathbb{D}_r = \mathbb{D} \\ |\mathbb{D}_t||\mathbb{D}_r|/|\mathbb{D}_t+\mathbb{D}_r|, & \text{otherwise.} \end{cases}$$

An array is said to be *nonredundant* if $R = 1$, and *redundant* if $R > 1$. In the case of fully overlapping Tx and Rx arrays ($\mathbb{D}_t = \mathbb{D}_r = \mathbb{D}$), the only two nonredundant array configurations are the trivial arrays $\mathbb{D} = \{0\}$ and $\mathbb{D} = \{0, 1\}$ ([18], Corollary 1). A more general definition of redundancy applicable to arrays with a noncontiguous co-array is also possible [11, 19]. In the active case, this definition can be refined further to account for the exact overlap between the Tx and Rx arrays [19]. This facilitates a more informative comparison between the redundancies of configurations with different degrees of overlap.

In the fully overlapping case, it is often of interest to evaluate the *asymptotic redundancy*. The asymptotic redundancy of a given class C of array geometries (with fully overlapping Tx/Rx sensors) is defined as

$$R_\infty(C) \triangleq \lim_{\substack{|\mathbb{D}| \to \infty \\ \mathbb{D} \in C}} R(\mathbb{D}, \mathbb{D}) = \lim_{\substack{|\mathbb{D}| \to \infty \\ \mathbb{D} \in C}} \frac{|\mathbb{D}|(|\mathbb{D}|+1)}{2(|\mathbb{D}+\mathbb{D}|)}. \tag{9.10}$$

This quantity is finite only for sparse array configuration achieving a (contiguous) sum co-array of size $|\mathbb{D}+\mathbb{D}| \propto |\mathbb{D}|^2$. Consequently, the asymptotic redundancy of the ULA is infinite, since $|\mathbb{D}+\mathbb{D}| \propto |\mathbb{D}|$. The asymptotic redundancy of any array with a contiguous sum co-array is lower bounded by $R_\infty(C_{\text{MRA}})$, where C_{MRA} is the class of MRAs, defined in Section 9.3.2.2.

9.3.2.2 Definition of MRA for Active Sensing

Moffet first proposed the MRA in the context of passive linear arrays and the difference co-array [11]. He distinguished between the *general* and *restricted* solutions, where the latter constrains the difference co-array to be contiguous. The restricted MRA thus yields the maximum number of contiguous co-array elements for a given physical aperture. Note that a general MRA can, but need not be restricted – for details see [11, 18] and ([19], pp. 108–109). Hoctor and Kassam later extended the MRA to active sensing [10]. They specifically considered the restricted MRA with a contiguous sum co-array. Passive and active MRAs correspond to extremal difference bases [45–47] and extremal additive bases [48–51], respectively. Additive bases have been studied in number theory since the early 1900s and are informally known as solutions to the *postage stamp problem* [52, 53]. The canonical version of this problem seeks a set of integers that can represent each non-negative integer smaller than some $h \in \mathbb{N}$ using the sum of at most k integers. The restricted postage stamp basis with maximal h for $k = 2$, or so-called extremal restricted two-basis, coincides with the active MRA with fully overlapping Tx and Rx sensors.

For a given number of transmitters N_t, receivers N_r, and transceivers N, the MRA can be defined as a (not necessarily unique) solution to [19]

$$\begin{aligned} &\underset{\mathbb{D}_{\xi} \subset \mathbb{N};\, h \in \mathbb{N}}{\text{maximize}}\, h \ \text{ subject to } \ \mathbb{D}_t + \mathbb{D}_r = [0 : h-1] \\ &\qquad\qquad |\mathbb{D}_t| = N_t,\ |\mathbb{D}_r| = N_r, \text{ and } |\mathbb{D}_t \cap \mathbb{D}_r| = N. \end{aligned} \tag{9.11}$$

Equation (9.11) subsumes the conventional definition of the (restricted) active MRA in the case of fully overlapping Tx and Rx arrays [10]; If $N_t = N_r = N$, then (9.11) reduces to [18]

$$\underset{\mathbb{D} \subset \mathbb{N}, h \in \mathbb{N}}{\text{maximize}}\, h \quad \text{subject to} \quad \mathbb{D}+\mathbb{D} = [0 : h-1], \quad |\mathbb{D}| = N. \tag{9.12}$$

The fully overlapping and nonoverlapping array configurations in Fig. 9.2a and c are examples of MRAs for $N_t = N_r = N = 4$ and $N_t = N_r = N + 2 = 3$, respectively. The fully overlapping MRA is redundant ($R = 1.25$), whereas the nonoverlapping array is nonredundant ($R = 1$).

The MRA can alternatively be defined for a constrained physical aperture [54, 55] or virtual aperture [19]. In the former case, the MRA is also known as the *sparse ruler* [56, 57]. This definition of the MRA is convenient when only a limited area for placing the sensors is available, or when the desired physical/virtual array is *not* one-dimensional. For example, constraining the aspect ratio of the planar or volumetric MRA [58–60] precludes degenerate linear arrays as solutions, see ([19], pp. 57–60).

9.3.2.3 Known MRAs

Nonoverlapping Tx and Rx Sensors Problem (9.11) has a simple closed-form solution in the case of nonoverlapping Tx and Rx arrays. This nonredundant array, illustrated in Fig. 9.2c, has a nested structure consisting of a dense and a sparse ULA. Specifically, the Tx and Rx sensor positions sets are

$$\mathbb{D}_t = \{0, N_r, \ldots, (N_t - 1)N_r\} \quad \text{and} \quad \mathbb{D}_r = \{0, 1, \ldots, N_r - 1\}. \tag{9.13}$$

The nested structure in (9.13) is readily verified to be a solution to (9.11); the upper bound on the size of any contiguous sum co-array $|\mathbb{D}_t + \mathbb{D}_r| \leq |\mathbb{D}_t||\mathbb{D}_r|$ is attained since $\mathbb{D}_t + \mathbb{D}_r = [0 : N_t N_r - 1]$. The definitions of $\mathbb{D}_t$ and $\mathbb{D}_r$ may naturally be reversed in (9.13) without affecting this conclusion.

Equation (9.13) is well-known [9, 61, 62], and closely related to the (passive) nested array [12], which also leverages arithmetic sequences to ensure a contiguous (difference) co-array. We note that many nonoverlapping MRA configurations exist in the two-dimensional case ([19], pp. 59–60).

Fully Overlapping Tx and Rx Sensors Finding MRAs with fully overlapping Tx and Rx sensors is generally challenging due to the combinatorial nature of (9.12). Solutions with up to $|\mathbb{D}| = 48$ physical sensors – corresponding to a contiguous sum co-array with $|\mathbb{D}_\Sigma| = 734$ virtual sensors – have been found using branch-and-bound methods in the context of additive bases [63]. The large search space of the resulting problem is tackled by employing judicious pruning strategies and concatenating solutions of a smaller-sized auxiliary combinatorial problem [51, 63]. Figure 9.3 illustrates all MRAs with $N = 11$ physical sensors (transceivers) [10]. Note that four out of six configurations are symmetric with respect to their mid-point (position 11). In fact, most known MRAs are symmetric; there exist at least one symmetric MRA for each $N \leq 48$ [51, 63]. We emphasize, however, that the MRA in (9.12) is *not* constrained to be symmetric.

9.3.3 Symmetric Sparse Array Configurations

The main drawback of the MRA is that finding it generally requires solving a combinatorial optimization problem. Consequently, generating MRAs with a large number of sensors or large aperture is usually impractical – especially when the Tx and Rx arrays are not nonoverlapping. In contrast, if a sparse array configuration allows a parametric description, then its synthesis is simplified considerably. There exist *scalable* and *order-wise optimal* array geometries that can be generated for practically any number of sensors or aperture size while achieving a co-array with on the order of $N_t N_r$ contiguous virtual sensors. This allows generating low-redundancy sparse array configurations beyond hundreds (thousands) of physical (virtual) sensors. Such geometries also upper bound the redundancy of the MRA. Together with lower bounds, the potential suboptimality with respect to the MRA can also be estimated; see [18].

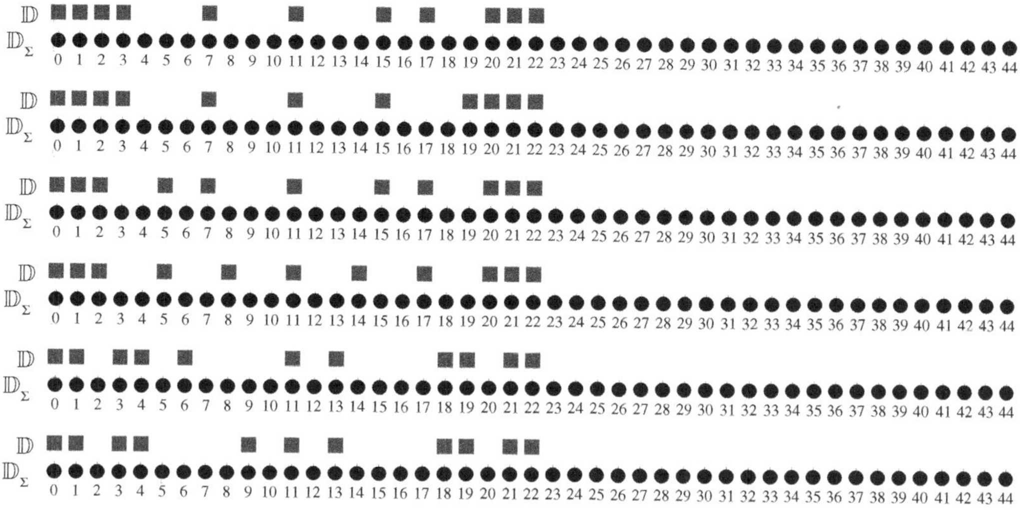

Figure 9.3 Active MRAs with $N = 11$ physical (transceiving) sensors [10]. Most of the configurations are symmetric. In fact, there exists a symmetric MRA configuration for (at least) each $N \leq 48$ [51, 63].

This section focuses on a class of *symmetric* sparse array geometries that achieve a low redundancy and contiguous sum co-array. We consider the case of fully overlapping Tx and Rx arrays. Imposing symmetry on the array geometry is not only a simplification inspired by the symmetry of most known MRAs (which are optimal but computationally intensive to find), but it also ensures equivalence between the sum and difference co-arrays. Consequently, properly designed symmetric arrays are well-suited for both active and passive sensing.

9.3.3.1 Generic Symmetric Array

Let $\mathbb{S}$ denote the set of normalized sensor positions of an array that is symmetric about $\frac{1}{2}\max \mathbb{S}$, i.e., $\mathbb{S} = \max \mathbb{S} - \mathbb{S}$. Here, $\max \mathbb{S}$ is the largest element of $\{0\} \subseteq \mathbb{S} \subset \mathbb{N}$.[1] This symmetry implies that the sum and difference co-array of $\mathbb{S}$ are identical up to a shift ([8], p. 740):

$$\mathbb{S} = \max \mathbb{S} - \mathbb{S} \implies \mathbb{S} + \mathbb{S} = \mathbb{S} - \mathbb{S} + \max \mathbb{S}. \tag{9.14}$$

Symmetric arrays are thus suited for both passive and active sensing. Symmetry also greatly simplifies array design by enabling the synthesis of arrays with a contiguous sum co-array using configurations with a contiguous difference co-array, such as the well-known Nested array [12].

Structure of Symmetric Array Any symmetric (linear) array can be described by a generator array $\{0\} \subseteq \mathbb{G} \subset \mathbb{N}$ and a non-negative offset $\ell \in \mathbb{N}$, which determines the spacing between $\mathbb{G}$ and its mirror image $\max \mathbb{G} - \mathbb{G}$. Formally, the *symmetric array* $\mathbb{S} = \mathbb{S}(\mathbb{G}, \ell)$ can be defined as[2]

$$\mathbb{S}(\mathbb{G}, \ell) \triangleq \mathbb{G} \cup (\max \mathbb{G} - \mathbb{G} + \ell). \tag{9.15}$$

We are specifically interested in $(\mathbb{G}, \ell)$ pairs yielding an $\mathbb{S}$ with a contiguous sum co-array. The desired property $\mathbb{S} + \mathbb{S} = [0 : 2\max \mathbb{S}]$ can naturally be decomposed using (9.15) into necessary and sufficient conditions involving offset ℓ and the sum/difference co-array of generator $\mathbb{G}$ ([18],

1 Consequently, position $\frac{1}{2}\max \mathbb{S}$ contains a sensor if and only if $|\mathbb{S}|$ is odd.

2 This definition is general, since symmetric arrays generated by $\mathbb{G}$ and *nonpositive* offset ℓ can equivalently be generated by $\mathbb{G}' = \max \mathbb{G} - \mathbb{G}$ and non-negative ℓ'.

Theorem 2). However, a simple sufficient condition stands out as particularly useful: If $\mathbb{G}$ has a contiguous difference co-array and ℓ is no larger than the number elements in the hole-free segment of the sum co-array of $\mathbb{G}$ (including $\{0\}$), then $\mathbb{S}$ has a contiguous sum and difference co-array (cf. [18, Theorem 2 and Corollary 2]).

Theorem 9.1 ***(Sufficient condition for contiguous co-array)*** Consider the symmetric array $\mathbb{S}$ in (9.15) with generator $\mathbb{G}$ and offset ℓ. If $\mathbb{G}-\mathbb{G} \supseteq [0 : \max \mathbb{G}]$ and $\mathbb{G}+\mathbb{G} \supseteq [0 : \ell-1]$, then $\mathbb{S}+\mathbb{S} = [0 : 2\max \mathbb{S}]$.

Proof: By (9.15), the sum co-array of the symmetric array $\mathbb{S}$ is

$$\begin{aligned}\mathbb{S}+\mathbb{S} &= \mathbb{G} \cup (\max \mathbb{G} - \mathbb{G} + \ell) + \mathbb{G} \cup (\max \mathbb{G} - \mathbb{G} + \ell)\\ &= (\mathbb{G}+\mathbb{G}) \cup (\mathbb{G}-\mathbb{G}+\max \mathbb{G}+\ell) \cup (2(\max \mathbb{G}+\ell)-(\mathbb{G}+\mathbb{G})).\end{aligned}$$

By the symmetry of the difference co-array, $\mathbb{G}-\mathbb{G} \supseteq [0 : \max \mathbb{G}] \implies \mathbb{G}-\mathbb{G} \supseteq [-\max \mathbb{G} : \max \mathbb{G}]$. Combining this with $\mathbb{G}+\mathbb{G} \supseteq [0 : \ell-1]$ implies that

$$\begin{aligned}\mathbb{S}+\mathbb{S} &\supseteq [0 : \ell-1] \cup ([-\max \mathbb{G} : \max \mathbb{G}] + \max \mathbb{G} + \ell) \cup (2(\max \mathbb{G}+\ell) - [0 : \ell-1])\\ &= [0 : \ell-1] \cup [\ell : 2\max \mathbb{G}+\ell] \cup [2\max \mathbb{G}+\ell+1 : 2(\max \mathbb{G}+\ell)]\\ &= [0 : 2(\max \mathbb{G}+\ell)].\end{aligned}$$

Noting that $\max \mathbb{G}+\ell = \max \mathbb{S}$ by (9.15) completes the proof. □

A key takeaway of Theorem 9.1 is that the design of (symmetric) array configurations with a contiguous sum co-array can be considerably simplified by leveraging geometries with a contiguous difference co-array.

Minimum-Redundancy Symmetric Array Over Generator Class $\mathcal{G}$ If $\mathbb{G}$ admits a simple parametric description, both the parameters of $\mathbb{G}$ and the offset ℓ may be optimized for minimal redundancy while imposing the sufficient conditions of Theorem 9.1 as constraints. Let $\mathcal{G}$ denote any class of array geometries with a contiguous difference co-array, i.e., $\mathbb{G}-\mathbb{G} \supseteq [0 : \max \mathbb{G}]\, \forall \mathbb{G} \in \mathcal{G}$. For a given generator class $\mathcal{G}$ and number of physical sensors N, the minimum-redundancy symmetric array over $\mathcal{G}$ is defined as any solution to

$$\underset{\mathbb{G}\in\mathcal{G},\ell\in\mathbb{N}}{\text{maximize}} \max \mathbb{G}+\ell \quad \text{subject to} \quad \begin{cases} |\mathbb{G} \cup (\max \mathbb{G}-\mathbb{G}+\ell)| = N \\ \mathbb{G}+\mathbb{G} \supseteq [0 : \ell-1]. \end{cases} \tag{9.16}$$

In particular, if (9.16) has a (not necessarily unique) solution $(\mathbb{G}^\star, \ell^\star)$ for a given $\mathcal{G}$ and N, then we denote by $\mathbb{S}(\mathbb{G}^\star, \ell^\star)$ a minimum-redundancy symmetric array for generator class $\mathcal{G}$. For general $\mathcal{G}$, (9.16) is a combinatorial problem which may be impractical to solve. However, certain choices of $\mathcal{G}$ actually admit a closed-form solution for any $N \in \mathbb{N}_+$, as we will see in Section 9.3.3.2.

9.3.3.2 Symmetric Nested Array

We now illustrate the general symmetric array design framework developed in Section 9.3.3.1. As an example of a simple yet versatile generator, we use the nested array (NA) [12], which has a contiguous difference co-array. By Theorem 9.1, the NA generator guarantees that the resulting symmetric array $\mathbb{S}$ has a contiguous sum co-array for suitable choices of offset ℓ. The convenient parametric structure of the symmetric NA also enables finding its redundancy minimizing parameters in a closed form. The symmetric NA and the NA generator are illustrated in Fig. 9.4.

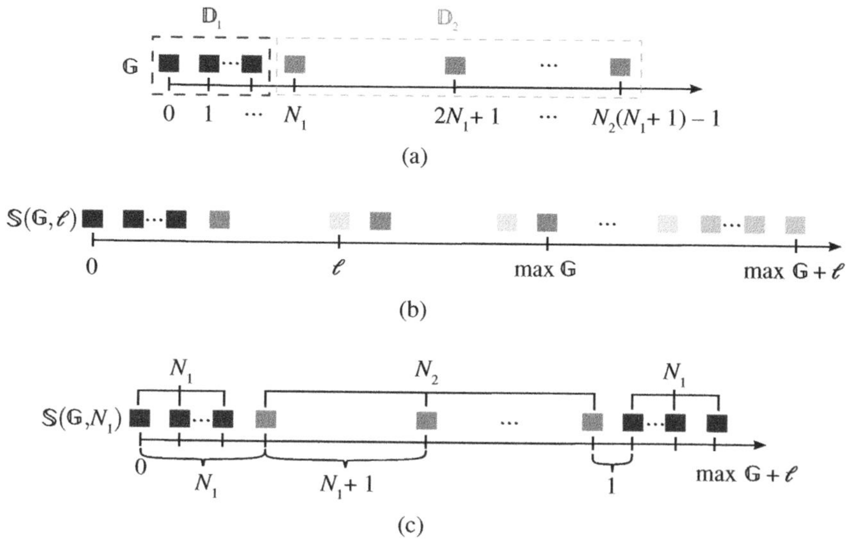

Figure 9.4 The (a) NA generator $\mathbb{G} = \mathbb{D}_{\text{NA}}$ and shift parameter ℓ define the (b) symmetric NA, which reduces to the configuration in (c), when $\ell = N_1$. The minimum-redundancy symmetric NA solving (9.16) has the structure in (c). (a) Nested array (NA), (b) Symmetric NA with offset $\ell > N_1$, and (c) Symmetric NA with offset $\ell = N_1$. Source: Adapted from [18] © 2021 IEEE.

Let $\mathcal{G}$ be the class of NAs. In this case, the generator is $\mathbb{G} = \mathbb{D}_{\text{NA}} = \mathbb{D}_{\text{NA}}(N_1, N_2)$ with non-negative parameters $N_1, N_2 \in \mathbb{N}$ and

$$\mathbb{D}_{\text{NA}} \triangleq \mathbb{D}_1 \cup (\mathbb{D}_2 + N_1), \text{ where } \begin{cases} \mathbb{D}_1 = [0{:}N_1 - 1] \\ \mathbb{D}_2 = [0{:}N_1 + 1{:}\ (N_2 - 1)(N_1 + 1)]. \end{cases} \tag{9.17}$$

By (9.15), the *symmetric* NA can therefore be written as

$$\mathbb{S}_{\text{NA}}(N_1, N_2, \ell) \triangleq \mathbb{S}(\mathbb{D}_{\text{NA}}, \ell) = \mathbb{D}_{\text{NA}} \cup (\max \mathbb{D}_{\text{NA}} - \mathbb{D}_{\text{NA}} + \ell). \tag{9.18}$$

Triplet (N_1, N_2, ℓ) controls the redundancy of $\mathbb{S}_{\text{NA}}$. For example, setting $\ell = 0$ can increase redundancy and consequently improve robustness to sensor failures ([64], Definition 2). Of special interest herein is the case $\ell = N_1$, which yields the following low-redundancy geometry:[3]

$$\begin{aligned} \mathbb{D}_{\text{CNA}}(N_1, N_2) &\triangleq \mathbb{S}_{\text{NA}}(N_1, N_2, \ell = N_1) \\ &= \mathbb{D}_1 \cup (\mathbb{D}_2 + N_1) \cup (\mathbb{D}_1 + (N_1 + 1)N_2). \end{aligned} \tag{9.19}$$

This array configuration, originally studied in the context of additive combinatorics ([48], Satz 2), is also known as the concatenated nested array (CNA) [65]. The solution to (9.16) actually has the same form as (9.19) when the elements of $\mathcal{G}$ are given by (9.17) ([18], Proposition 1). Intuitively, this is because any other offset ℓ cannot decrease redundancy. Due to its simple structure, the minimum-redundancy parameters of (9.19), and by implication (9.18), can be found in a closed form. For any given $N = 4m + k$, where $m \in \mathbb{N}, k \in \{0 : 3\}$, and the elements of $\mathcal{G}$ follow (9.17), the solution to (9.16) is $\mathbb{G}^\star = \mathbb{D}_{\text{CNA}}(N_1^\star, N_2^\star)$, $\ell^\star = N_1^\star$, where ([18], Theorem 3)

$$N_1^\star = (N - \alpha)/4, \quad N_2^\star = (N + \alpha)/2, \quad \text{and} \quad \alpha = (k + 1) \bmod 4 - 1.$$

We note that the optimal values of N_1 and N_2 are similar for the minimally redundant NA (w.r.t. the difference co-array), where $N_1^\star = N_2^\star = N/2$ for even N [12]. However, while the NA has a

3 This case is equivalent to $\ell = (N_1 + 1)k + N_1$ for any $k \in [0 : N_2]$ by reparametrization.

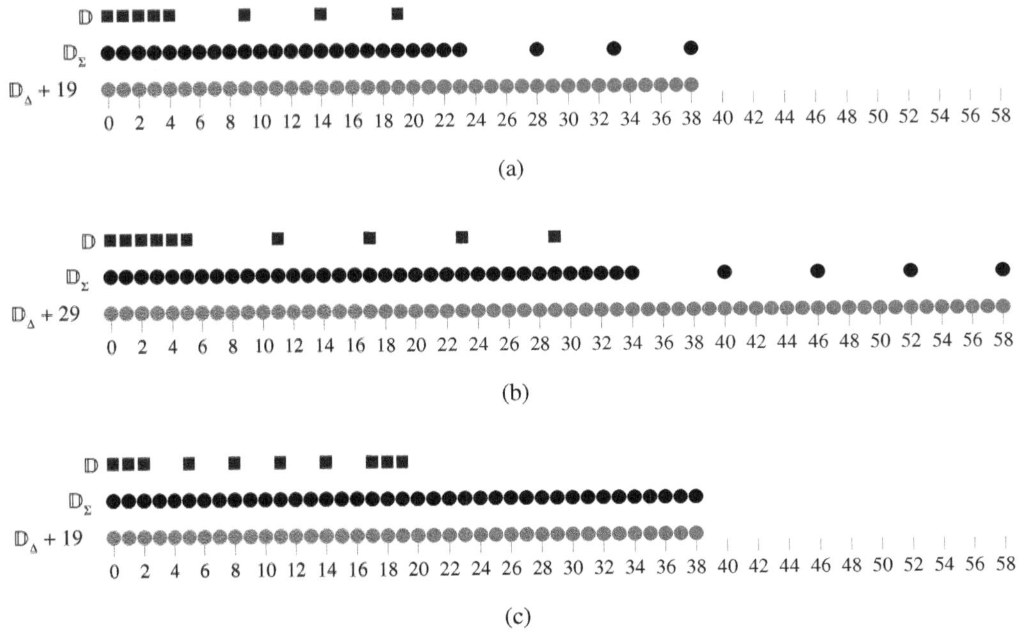

Figure 9.5 NA and symmetric NA configurations using minimum-redundancy parameters, with corresponding sum co-array $\mathbb{D}_\Sigma$ and difference co-array $\mathbb{D}_\Delta$ (shifted to non-negative half plane). The (c) symmetric NA attains a larger contiguous sum co-array than the NA, both for the same (a) physical array aperture and (b) number of physical sensors. The difference co-array of the symmetric NA is also contiguous. (a) Nested array, $N = 8$ sensors, (b) Nested array, $N = 10$ sensors, and (c) Symmetric Nested array, $N = 10$ sensors.

contiguous difference co-array, its sum co-array is noncontiguous. Indeed, Fig. 9.5 shows that the symmetric NA achieves a larger contiguous sum co-array than the NA for the same physical aperture or number of physical sensors.

The minimum-redundancy symmetric NA can be used to upper bound the asymptotic redundancy of the MRA. Specifically, denote the class of minimum-redundancy symmetric NAs by $C^\star_{\text{S-NA}} \triangleq \{\mathbb{S}^\star_{\text{NA}}(N) \mid N \in \mathbb{N}_+\}$, where $\mathbb{S}^\star_{\text{NA}}(N) \triangleq \mathbb{D}_{\text{CNA}}(N_1^\star, N_2^\star)$. The number of contiguous sum co-array elements can be shown to be $|\mathbb{S}^\star_{\text{NA}} + \mathbb{S}^\star_{\text{NA}}| = 2\max \mathbb{S}^\star_{\text{NA}} + 1 = (N^2 + 6N - 7)/4 - (\alpha - 1)^2/4 + 1$. By (9.10), the asymptotic redundancy of $C^\star_{\text{S-NA}}$ is therefore

$$R_\infty(C^\star_{\text{S-NA}}) = \lim_{N\to\infty} \frac{N(N+1)}{2((N^2 + 6N - 7)/4 - (\alpha - 1)^2/4 + 1)} = 2. \tag{9.20}$$

Consequently, the asymptotic redundancy of the MRA satisfies $R_\infty(C_{\text{MRA}}) \le R_\infty(C^\star_{\text{S-NA}}) = 2$, where C_{MRA} denotes the set of solutions to (9.12).

9.3.3.3 Other Symmetric Arrays

Beyond the NA, previously studied generator choices include the Wichmann [47] and Kløve–Mossige bases [53] originating in number theory. For array geometry design perspectives and technical details, see [18]. See also [66, 67] for symmetric arrays with a fractal structure. We remark that alternative generators in (9.15) can improve upon the redundancy of the symmetric NA. In particular, a subclass $\mathscr{K}$ of the symmetric array generated by the Kløve-Mossige basis – corresponding to a class of symmetric bases studied in the context of additive combinatorics [68] – achieves asymptotic redundancy $R_\infty(\mathscr{K}) = 23/12 < 1.9167$ ([68], Theorem, p. 177) (cf. ([18],

Proposition 2)). This lower bounds (9.20), and therefore provides a tighter upper bound on the asymptotic redundancy of the (fully overlapping) MRA. Array configurations – symmetric or asymmetric – with an even lower asymptotic redundancy, yet a contiguous sum co-array, remain to be explored.

9.4 Beamforming

This section considers *beamforming* in sensing using sparse arrays. Beamforming collectively denotes spatially directive transmission or spatial filtering upon reception, which is used to enhance (suppress) signals impinging in/from directions of interest by the principle of constructive (destructive) wave interference.

Section 9.4.1 starts by briefly reviewing Tx, Rx, and joint Tx–Rx beamforming. We focus on the joint case, which is commonly encountered in imaging applications. Section 9.4.2 considers an extension to Tx–Rx beamforming called *image addition*, where carefully chosen Tx and Rx beamforming weight pairs are used to synthesize a desired Tx–Rx beampattern by linearly combining multiple "component images." This can be interpreted as assigning desired complex values to the beamforming weights of the sum co-array. We briefly discuss methods for finding the physical beamforming weights synthesizing a desired virtual array weighting, and describe the resulting trade-off between array sparsity and the number of component images, Q. A small value of Q is often preferred to minimize the beampattern synthesis time.

9.4.1 Rx, Tx, and Joint Tx–Rx Beamforming

In transmit beamforming, the spatial self-interference pattern of the radiated wavefield, or so-called transmit beampattern, is controlled by the cross-correlation properties of the transmitted waveforms. In receive beamforming, the outputs of the Rx sensors are scaled and phase shifted prior to being coherently combined. The receive beampattern can be flexibly shaped and steered post-data-acquisition when the beamforming architecture of the receiver is fully digital. In contrast, the transmit beampattern can typically change only between transmissions, where a transmission corresponds to the duration of the waveform or the maximum round-trip time that should be unambiguously resolved. The *effective* Tx–Rx beampattern equals the product of the Tx and Rx beampatters, and hence achieves a narrower main beam or lower side lobe levels than the individual Tx and Rx beampatterns, as Fig. 9.6 illustrates. A properly designed Tx–Rx beampattern also suppresses grating lobes present in either the Tx or Rx beampattern due to spatial aliasing.

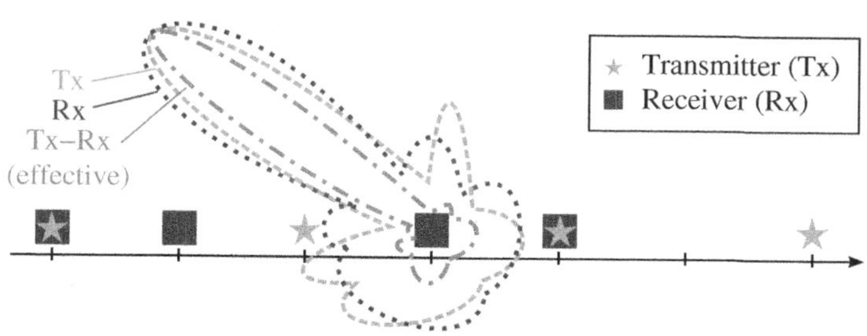

Figure 9.6 The effective Tx–Rx beampattern of an active array equals the product of the Tx and Rx beampatterns. Hence, it typically achieves a narrower main lobe and lower side lobe levels than either the Tx or Rx beampattern, which may suffer from grating lobes in case the physical array geometry is sparse. Source: Adapted from [17] © 2020 IEEE.

9.4.1.1 Receive Beamforming

The Rx beampattern magnitude, i.e., the angular magnitude response of the Rx array employing beamforming weight vector $\boldsymbol{w}_{\mathrm{r}} \in \mathbb{C}^{N_{\mathrm{r}}}$ is

$$B_{\mathrm{r}}(\theta) = |H_{\mathrm{r}}(\theta)|^2 = |\boldsymbol{w}_{\mathrm{r}}^{\mathrm{H}}\boldsymbol{a}_{\mathrm{r}}(\theta)|^2,$$

where $H_{\mathrm{r}}(\theta) \in \mathbb{C}$ is the *complex-valued Rx beampattern* or angular response,

$$H_{\mathrm{r}}(\theta) \triangleq \boldsymbol{w}_{\mathrm{r}}^{\mathrm{H}}\boldsymbol{a}_{\mathrm{r}}(\theta). \tag{9.21}$$

A typical choice for the Rx beamforming vector $\boldsymbol{w}_{\mathrm{r}}$ is the spatial matched filter $\boldsymbol{w}_{\mathrm{r}} = c\boldsymbol{a}_{\mathrm{r}}(\phi)$, where $c \in \mathbb{C}\backslash\{0\}$. The matched filter maximizes the SNR of the beamformer output in case of a single scatterer in direction ϕ embedded in zero-mean white noise (as a simple consequence of the Cauchy–Schwartz inequality). Other choices of $\boldsymbol{w}_{\mathrm{r}}$ can strike a different balance between main lobe width and side lobe levels, or adapt to the scattering environment [69]. The widely used MVDR (minimum variance distortionless response) beamformer is a prime example of such a design.

The beamformed output of the Rx array gives rise to a multiple-input single-output (MISO) model. Specifically, the received signal matrix $\boldsymbol{Y}$ in (9.2) after Rx beamforming reduces to a T-dimensional *temporal* (row) vector

$$\boldsymbol{w}_{\mathrm{r}}^{\mathrm{H}}\boldsymbol{Y} = \sum_{k=1}^{K}\gamma_k H_{\mathrm{r}}(\theta_k)\boldsymbol{a}_{\mathrm{t}}^{\mathrm{T}}(\theta_k)\boldsymbol{S}^{\mathrm{T}} + \boldsymbol{w}_{\mathrm{r}}^{\mathrm{H}}\boldsymbol{N}.$$

Rx beamforming is a key component in wireless communications, radar and passive sensing-based direction finding systems, in part due to its low computational complexity (once an appropriate $\boldsymbol{w}_{\mathrm{r}}$ is found). In practice, Rx beamforming may also be employed out of necessity due to hardware constraints imposed by a hybrid or an analog beamforming architecture.

9.4.1.2 Transmit Beamforming

The angular power distribution of the wavefield radiated by the Tx array, or Tx beampattern magnitude, is given by [70]

$$B_{\mathrm{t}}(\theta) = \boldsymbol{a}_{\mathrm{t}}^{\mathrm{H}}(\theta)\boldsymbol{S}^{\mathrm{H}}\boldsymbol{S}\boldsymbol{a}_{\mathrm{t}}(\theta).$$

Clearly, $B_{\mathrm{t}}(\theta)$ depends on the cross-correlation properties of the transmitted waveforms via positive semidefinite matrix $\boldsymbol{R}_{\mathrm{t}} \triangleq \boldsymbol{S}^{\mathrm{H}}\boldsymbol{S}$. For example, $\boldsymbol{R}_{\mathrm{t}} = \boldsymbol{a}_{\mathrm{t}}(\phi)\boldsymbol{a}_{\mathrm{t}}^{\mathrm{H}}(\phi)$ and $\boldsymbol{R}_{\mathrm{t}} = \boldsymbol{I}$ correspond to highly directive and omnidirectional transmission, respectively. The former case is also known as the phased array or spatial-matched Tx filter (in direction ϕ), whereas the latter is typically referred to as the orthogonal MIMO model. Figure 9.7 illustrates these two cases for a ULA and an MRA transmitter with $N_{\mathrm{t}} = 9$ sensors. The sparse array achieves a narrower main lobe due to its larger aperture, but its sidelobe levels are higher due to the spatial undersampling. Note that the waveform rank, $N_s = \mathrm{rank}(\boldsymbol{S}) = \mathrm{rank}(\boldsymbol{R}_{\mathrm{t}})$, differs in the phased array ($N_s = 1$) and orthogonal MIMO ($N_s = N_{\mathrm{t}}$) cases. Increasing N_s generally extends the set of achievable Tx beampatterns – for example, see [71].

For brevity, the remainder of the chapter focuses on the case of unit waveform rank $N_s = 1$, i.e., the same waveform is launched from each sensor with an appropriate phase shift. In this case, the *complex-valued Tx beampattern* $H_{\mathrm{t}}(\theta) \in \mathbb{C}$ can be defined analogously to the Rx case in (9.22) as

$$H_{\mathrm{t}}(\theta) \triangleq \boldsymbol{w}_{\mathrm{t}}^{\mathrm{H}}\boldsymbol{a}_{\mathrm{t}}(\theta), \tag{9.22}$$

where $\boldsymbol{w}_{\mathrm{t}} \triangleq \boldsymbol{S}^{\mathrm{H}}\boldsymbol{w}_s \in \mathbb{C}^{N_{\mathrm{t}}}$ is a Tx beamforming weight vector and $\boldsymbol{w}_s \in \mathbb{C}^{T}$ is the corresponding temporal waveform combining vector. Since $N_s = 1$, $|H_{\mathrm{t}}(\theta)|^2 \propto B_{\mathrm{t}}(\theta)$ holds for any $\boldsymbol{w}_s$ not in the null

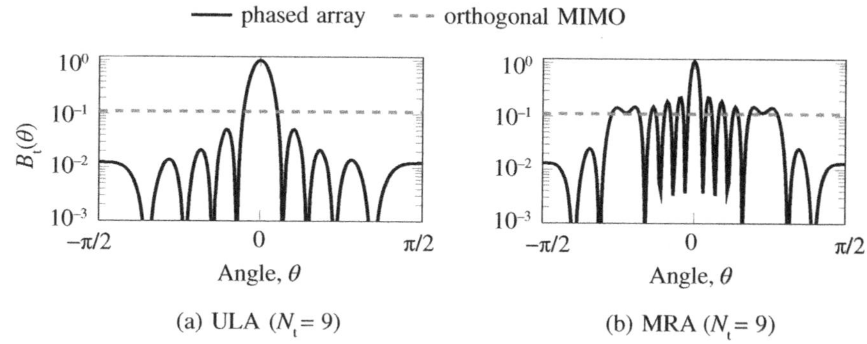

Figure 9.7 Transmit beampattern (magnitude squared) of phased array (steered to boresight) and orthogonal MIMO transmission. For the same Tx power, the phased array yields higher beamforming gain in the look direction, whereas orthogonal waveforms illuminate a wider field-of-view. (a) ULA ($N_t = 9$) and (b) MRA ($N_t = 9$).

space of $\boldsymbol{S}^{\mathrm{H}}$. To see this, write $\boldsymbol{S} = \boldsymbol{u}\boldsymbol{v}^{\mathrm{H}}$, where $\boldsymbol{u} \in \mathbb{C}^T$, $\boldsymbol{v} \in \mathbb{C}^{N_t}$ are nonzero vectors and $\|\boldsymbol{u}\|_2 = 1$. Consequently, $|H_t(\theta)|^2 = |\boldsymbol{w}_s^{\mathrm{H}}\boldsymbol{u}|^2|\boldsymbol{v}^{\mathrm{H}}\boldsymbol{a}_t(\theta)|^2 = |\boldsymbol{w}_s^{\mathrm{H}}\boldsymbol{u}|^2 B_t(\theta)$. A reasonable choice for $\boldsymbol{w}_s$ is the temporal matched filter $\boldsymbol{w}_s = c\boldsymbol{u}$, for some $c \in \mathbb{C}\backslash\{0\}$. We remark that if $N_s \geq 2$, the interpretation of (9.22) changes. In this case, $H_t(\theta)$ is a partly *virtual* Tx beampattern defined by both waveform matrix $\boldsymbol{S}$, which affects the radiated wavefield, and combining vector $\boldsymbol{w}_s$, which does not. For example, in the case of orthogonal waveforms (implying full waveform rank, $N_s = N_t$), the transmit power is radiated omnidirectionally. Hence all beamforming is performed entirely synthetically post-reception.

The received signal after Tx beamforming constitutes a SIMO model, which takes the form of an N_r-dimensional *spatial* vector

$$\boldsymbol{Y}\boldsymbol{w}_s^* = \sum_{k=1}^{K} \gamma_k \boldsymbol{a}_r(\theta_k) H_t(\theta_k) + \boldsymbol{N}\boldsymbol{w}_s^*.$$

9.4.1.3 Joint Transmit–Receive Beamforming

By (9.21) and (9.22), the *complex-valued Tx–Rx beampattern* is defined as

$$H_{tr}(\theta) \triangleq H_t(\theta)H_r(\theta) = (\boldsymbol{w}_t \otimes \boldsymbol{w}_r)^{\mathrm{H}}(\boldsymbol{a}_t(\theta) \otimes \boldsymbol{a}_r(\theta)). \tag{9.23}$$

By (9.2), the output after joint Tx–Rx beamforming can thus be written as

$$y \triangleq \boldsymbol{w}_r^{\mathrm{H}}\boldsymbol{Y}\boldsymbol{w}_s^* = \sum_{k=1}^{K} \gamma_k H_{tr}(\theta_k) + \tilde{n}, \tag{9.24}$$

where $\tilde{n} \triangleq \boldsymbol{w}_r^{\mathrm{H}}\boldsymbol{N}\boldsymbol{w}_s^* = (\boldsymbol{w}_s \otimes \boldsymbol{w}_r)^{\mathrm{H}}\boldsymbol{n} \sim \mathcal{CN}(0, \sigma^2\|\boldsymbol{w}_s\|_2^2\|\boldsymbol{w}_r\|_2^2)$ is a zero-mean complex-valued normally distributed noise term by the assumption in (9.2).

Equation (9.23) can be interpreted as the effective *point spread function* (PSF) of the Tx–Rx array. The PSF is the spatial impulse response of a linear imaging system ([72], p. 20), and hence characterizes the resolution and interference suppression capability of the joint Tx–Rx beamformer. There are many well-known approaches for realizing a desired $H_{tr}(\theta)$ by jointly choosing the Tx and Rx beampatterns $H_t(\theta)$ and $H_r(\theta)$, including co-prime beamforming [73], and various adaptive methods based on, for instance, maximizing target detection performance or signal-to-interference-plus-noise-ratio (SINR) [28, 29, 74, 75]. Related Tx–Rx beamformer design problems are also encountered in MIMO communications [76], as well as joint communications and sensing applications [77].

Sum Co-array Interpretation By (9.6), the sum co-array offers a convenient interpretation of (9.23). In particular, $H_{\mathrm{tr}}(\theta) = \boldsymbol{w}_{\Sigma}^{\mathrm{H}}\boldsymbol{a}_{\Sigma}(\theta)$, where $\boldsymbol{a}_{\Sigma}(\theta) \in \mathbb{C}^{N_{\Sigma}}$ is a sum co-array manifold vector following (9.5), and $\boldsymbol{w}_{\Sigma} = \Upsilon(\boldsymbol{w}_{\mathrm{t}} \otimes \boldsymbol{w}_{\mathrm{r}})$ can be interpreted as a *sum co-array beamforming weight vector*. Vector $\boldsymbol{w}_{\Sigma}$ depends on both the physical Tx and Rx beamforming weights, as well as the redundancy pattern $\Upsilon \in \{0, 1\}^{N_{\Sigma} \times N_{\mathrm{r}} N_{\mathrm{t}}}$ of the joint Tx–Rx array geometry. Indeed, choosing unit physical beamforming weights, $\boldsymbol{w}_{\xi} = \mathbf{1}_{N_{\xi}}$, yields sum co-array weights corresponding to the multiplicities of each virtual sensor, $\boldsymbol{w}_{\Sigma} = \boldsymbol{v}_{\Sigma}$.

Figure 9.8 shows an example of the beamforming weights and beampattern (magnitude) of both the physical array and sum co-array in case of a ULA and MRA of equivalent aperture. Both arrays have fully overlapping Tx and Rx sensors. The ULA has $N = 9$ physical sensors, whereas the MRA has $N = 6$. The Tx–Rx beampattern and beamforming weights of the co-array are a Fourier transform pair: the former is the product of the Tx and Rx beampatterns, and the latter is the convolution of the Tx and Rx beamforming weights. Recovering the Tx and Rx beamformers from the Tx–Rx beampattern can also be viewed as a polynomial factorization problem [78, 79].

9.4.2 Image Addition

The beamformer output (9.24) can be written using (9.23) as $y = \boldsymbol{w}_{\mathrm{tr}}^{\mathrm{H}}(\boldsymbol{A}_{\mathrm{t}} \odot \boldsymbol{A}_{\mathrm{r}})\boldsymbol{\gamma} + \tilde{n}$, where $\boldsymbol{w}_{\mathrm{tr}} \in \mathbb{C}^{N_{\mathrm{t}} N_{\mathrm{r}}}$ denotes the effective Tx–Rx beamforming weight vector

$$\boldsymbol{w}_{\mathrm{tr}} = \boldsymbol{w}_{\mathrm{t}} \otimes \boldsymbol{w}_{\mathrm{r}} = \mathrm{vec}\left(\boldsymbol{w}_{\mathrm{r}}\boldsymbol{w}_{\mathrm{t}}^{\mathrm{T}}\right) \tag{9.25}$$

Hence, $\boldsymbol{w}_{\mathrm{tr}}$ is constrained to vectors in $\mathbb{C}^{N_{\mathrm{t}} N_{\mathrm{r}}}$ with Kronecker structure. The idea of *image addition* [8] is to replace rank-1 matrix $\boldsymbol{w}_{\mathrm{r}}\boldsymbol{w}_{\mathrm{t}}^{\mathrm{T}}$ in (9.25) by a synthesized *Tx–Rx beamforming weight matrix*

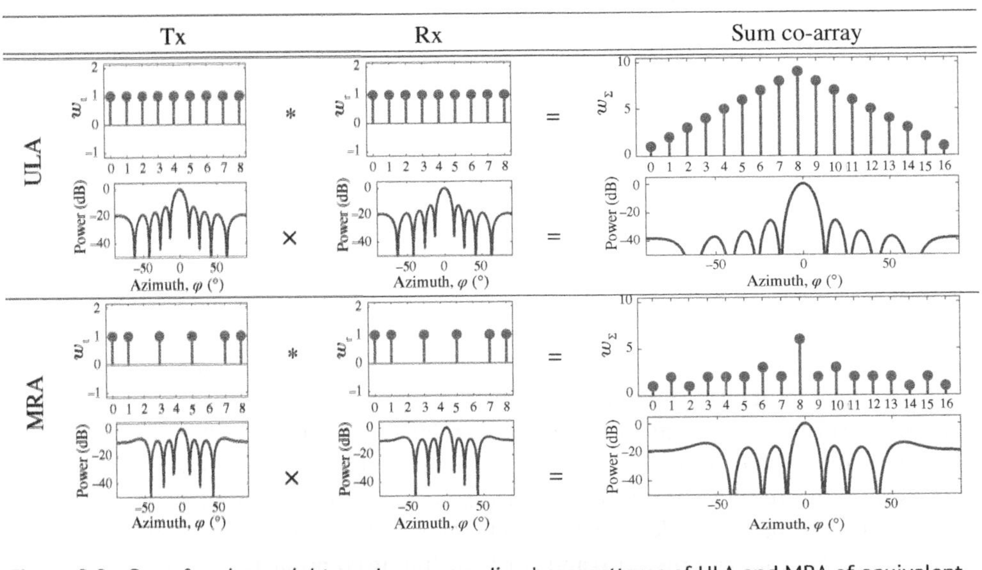

Figure 9.8 Beamforming weights and corresponding beampatterns of ULA and MRA of equivalent aperture. The Tx–Rx (product) beampattern corresponds to a co-array weighting, which is the convolution of the Tx and Rx beamforming weights. Source: Adapted from [19] © 2021 R. Rajamäki.

$\boldsymbol{W}$ of arbitrary rank. Specifically, let $\boldsymbol{W}_\xi \triangleq [\boldsymbol{w}_\xi[1], \boldsymbol{w}_\xi[2], \dots, \boldsymbol{w}_\xi[Q]] \in \mathbb{C}^{N_\xi \times Q}$ denote a Tx/Rx beamforming weight matrix with $Q \leq \min(N_\mathrm{t}, N_\mathrm{r})$ linearly independent columns. Matrix $\boldsymbol{W} \in \mathbb{C}^{N_\mathrm{r}N_\mathrm{t}}$ may then be defined as

$$\boldsymbol{W} \triangleq \boldsymbol{W}_\mathrm{r}\boldsymbol{W}_\mathrm{t}^\mathrm{T} = \sum_{q=1}^{Q} \boldsymbol{w}_\mathrm{r}[q]\boldsymbol{w}_\mathrm{t}^\mathrm{T}[q]. \tag{9.26}$$

We call $Q = \mathrm{rank}(\boldsymbol{W})$ the number of (linearly independent) *component images.*

Synthetic Tx–Rx beamforming vector $\mathrm{vec}(\boldsymbol{W})$ is realized by linearly combining the beamformed outputs of different Tx–Rx beamforming weight pairs. A crucial assumption is that the scattering scene is stationary over the acquisition period in the sense that $\boldsymbol{\gamma}$ is fixed and independent of q. Indeed, only waveform matrix $\boldsymbol{S}$ and noise matrix $\boldsymbol{N}$ in (9.2) may vary with q. Denoting the Tx–Rx beamformer output of the qth component image by $y[q] \triangleq \boldsymbol{w}_\mathrm{r}^\mathrm{H}[q]\boldsymbol{Y}[q]\boldsymbol{w}_s^*[q]$, the joint Tx–Rx beamformer output after image addition is defined as

$$\overline{y} \triangleq \sum_{q=1}^{Q} y[q] = \mathrm{vec}^\mathrm{H}(\boldsymbol{W})(\boldsymbol{A}_\mathrm{t} \odot \boldsymbol{A}_\mathrm{r})\boldsymbol{\gamma} + \overline{n}, \tag{9.27}$$

where $\overline{n} = \sum_{q=1}^{Q} \boldsymbol{w}_\mathrm{r}[q]\boldsymbol{N}[q]\boldsymbol{w}_s^*[q]$ is a noise term. Consequently, the synthesized Tx–Rx beam pattern (or synthesized PSF) becomes

$$\overline{H}_\mathrm{tr}(\theta) = \sum_{q=1}^{Q} H_\mathrm{tr}^{(q)}(\theta) = \mathrm{vec}^\mathrm{H}(\boldsymbol{W})(\boldsymbol{a}_\mathrm{t}(\theta) \otimes \boldsymbol{a}_\mathrm{r}(\theta)), \tag{9.28}$$

where $H_\mathrm{tr}^{(q)}(\theta) = H_\mathrm{t}^{(q)}(\theta)H_\mathrm{r}^{(q)}(\theta)$; $H_\mathrm{t}^{(q)}(\theta) = \boldsymbol{w}_\mathrm{t}^\mathrm{H}[q]\boldsymbol{a}_\mathrm{t}(\theta)$; and $H_\mathrm{r}^{(q)}(\theta) = \boldsymbol{w}_\mathrm{r}^\mathrm{H}[q]\boldsymbol{a}_\mathrm{r}(\theta)$ denote qth component (complex-valued) Tx–Rx, Tx, and Rx beampatterns, respectively.

Sum Co-array Interpretation

The synthesized Tx–Rx beampattern in (9.28) can be rewritten using sum co-array manifold vector $\boldsymbol{a}_\Sigma(\theta) \in \mathbb{C}^{N_\Sigma}$ as

$$\overline{H}_\mathrm{tr}(\theta) = \boldsymbol{w}_\Sigma^\mathrm{H}\boldsymbol{a}_\Sigma(\theta), \tag{9.29}$$

where $\boldsymbol{w}_\Sigma \in \mathbb{C}^{N_\Sigma}$ is the (synthesized) sum co-array beamforming weight vector

$$\boldsymbol{w}_\Sigma = \Upsilon\mathrm{vec}(\boldsymbol{W}). \tag{9.30}$$

Figure 9.9 demonstrates image addition in case of the ULA and MRA of equal aperture in Fig. 9.8. Both array configurations achieve the desired Tx–Rx beampattern due to having equivalent sum co-arrays. The MRA requires $Q = 2$ component images as it has fewer physical sensors than the ULA. Next, we briefly discuss how to select $\boldsymbol{W}$, and ultimately, $\boldsymbol{W}_\mathrm{t}$ and $\boldsymbol{W}_\mathrm{r}$ in (9.26).

9.4.2.1 Joint Optimization of Tx and Rx Beamformers

Designing beamforming weight matrix $\boldsymbol{W}$ (given an array geometry) can be cast as the following problem: find $\boldsymbol{W}$ synthesizing a desired Tx–Rx beampattern $\overline{H}_\mathrm{tr}(\phi)$ in (9.28) for $\phi \in [-\pi/2, \pi/2)$.[4] This task, which can be viewed as an FIR (finite impulse response) filter design problem, may be simplified by either discretizing the angular space or, equivalently, by satisfying (9.30) for a desired co-array weight vector $\boldsymbol{w}_\Sigma \in \mathbb{C}^{N_\Sigma}$, provided the desired beampattern $\overline{H}_\mathrm{tr}(\phi)$ is realizable. A key observation is that the system of equations in (9.30) is *underdetermined* when the array

4 The more challenging problems of only constraining the desired beampattern *magnitude* response $|\overline{H}_\mathrm{tr}(\phi)|^2$, or allowing for a variable array geometry, are not considered herein.

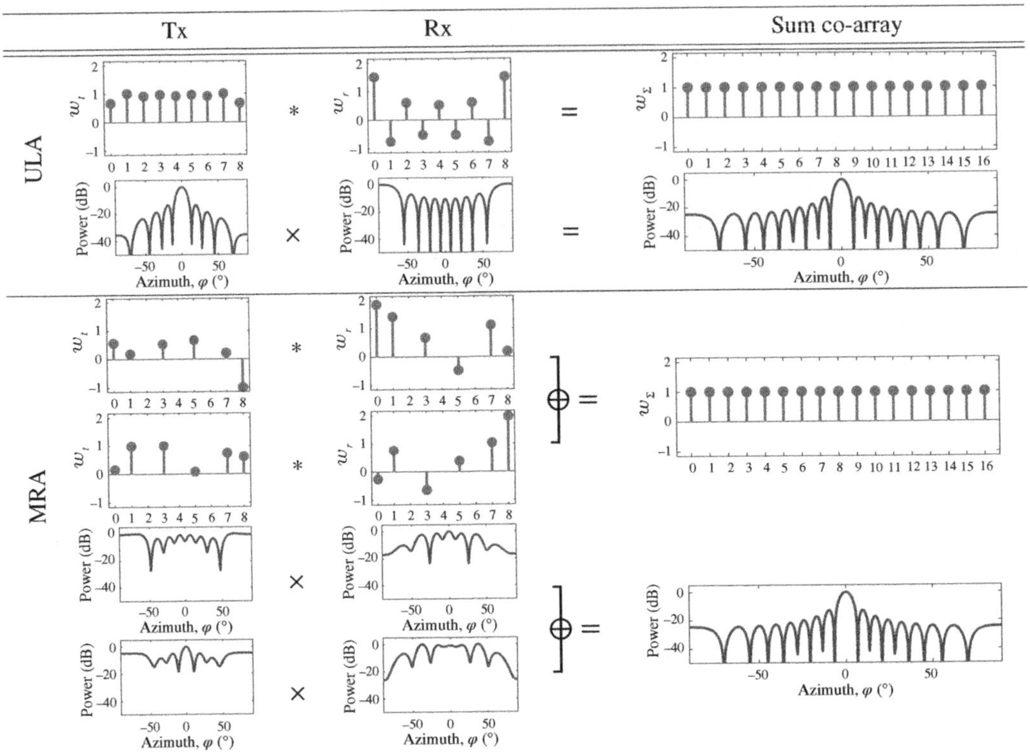

Figure 9.9 Tx–Rx beampattern synthesis using image addition. Both the ULA and MRA achieve the same effective beampattern due to having equivalent sum co-arrays. The MRA requires an additional component image due to having fewer physical sensors than the ULA.

configuration is *redundant*. That is, if $N_\Sigma < N_r N_t$, then infinitely many $\boldsymbol{W}$ satisfy $\boldsymbol{w}_\Sigma = \Upsilon \text{vec}(\boldsymbol{W})$, since $\Upsilon \in \{0,1\}^{N_\Sigma \times N_t N_r}$ has a nontrivial null space. Therefore, the question arises: how should $\boldsymbol{W}$ be chosen?

One alternative is the least-squares (LS) solution $\text{vec}(\boldsymbol{W}) = \Upsilon^\text{T}(\Upsilon\Upsilon^\text{T})^{-1}\boldsymbol{w}_\Sigma$, which has the intuitive interpretation of distributing the entries of $\boldsymbol{w}_\Sigma$ *equally* among the associated Tx–Rx sensor pairs [14, 80]. However, the LS solution does not generally yield a low-rank $\boldsymbol{W}$, which implies that unnecessarily many transmissions or linearly independent Tx waveforms may be required for achieving a desired co-array weighting. Hence, one may instead solve a low-rank matrix recovery problem, where $\text{rank}(\boldsymbol{W})$ is minimized subject to (9.30). In practice, this problem may be approximately solved by, e.g., relaxing the rank objective to the nuclear norm, or by employing alternating minimization on the rank-revealing form $\boldsymbol{W} = \boldsymbol{W}_r \boldsymbol{W}_t^\text{T}$ in (9.26) combined with bisection over the rank (inner dimension) Q ([19], pp. 85–86). Rather than satisfying $\boldsymbol{w}_\Sigma = \Upsilon \text{vec}(\boldsymbol{W})$ with equality, an approximation error tolerance can also be allowed in practice [19].

Factorizing W Given $\boldsymbol{W}$, there is generally no unique choice for physical beamforming weight matrices $\boldsymbol{W}_t$ and $\boldsymbol{W}_r$ in (9.26). However, a convenient factorization of $\boldsymbol{W}$ is given by its singular value decomposition (SVD): $\boldsymbol{W} = \boldsymbol{U}\Sigma\boldsymbol{V}^\text{H}$, where $\Sigma \in \mathbb{R}_+^{Q \times Q}$ is the diagonal matrix of positive singular values, and $\boldsymbol{U} \in \mathbb{C}^{N_r \times Q}$, $\boldsymbol{V} \in \mathbb{C}^{N_t \times Q}$ are the corresponding left-unitary singular vector matrices.

Using the SVD, we may then set $\boldsymbol{W}_\text{t} = \boldsymbol{V}^*$ and $\boldsymbol{W}_\text{r} = \boldsymbol{U}\Sigma$, for example. The factorization may be more involved in practice due to constraints imposed on $\boldsymbol{W}_\xi$, such as unit-modulus structure arising from the use of phase shifters [17, 81] or phase coded waveforms.

Spatio-Temporal Trade-Offs A fundamental question regarding sparse arrays employing beamforming is as follows: Given *any* desired co-array weight vector $\boldsymbol{w}_\Sigma \in \mathbb{C}^{N_\Sigma}$, how many component images $Q = \text{rank}(\boldsymbol{W})$ are required to satisfy $\boldsymbol{w}_\Sigma = \Upsilon\text{vec}(\boldsymbol{W})$ exactly? Intuition suggests that the sparser the array, the more component images are necessary to synthesize arbitrary co-array weightings. Indeed, the ULA with fully overlapping Tx and Rx sensors achieves *any* given $\boldsymbol{w}_\Sigma$ when $1 \leq Q \leq 2$ ([19], pp. 86–88), whereas $Q \propto N$ is necessary for sparse arrays with $N_\text{t} \propto N_\text{r} \propto N$ physical sensors and $N_\Sigma \propto N^2$ virtual sensors ([17], Proposition 1). Naturally, *some* $\boldsymbol{w}_\Sigma$ may be achieved either exactly or approximately using a smaller value of Q. For example, $Q = 1$ suffices to realize any $\boldsymbol{w}_\Sigma$ of the form $\boldsymbol{w}_\Sigma = \Upsilon(\boldsymbol{t} \otimes \boldsymbol{r})$, where $\boldsymbol{t} \in \mathbb{C}^{N_\text{t}}$ and $\boldsymbol{r} \in \mathbb{C}^{N_\text{r}}$ (i.e., set $\boldsymbol{W} = \boldsymbol{r}\boldsymbol{t}^\top$). A direction for future work is investigating practically useful Tx–Rx beampatterns that can be achieved using sparse arrays employing $Q = \mathcal{O}(1)$ component images.

9.5 Applications

We conclude the chapter by highlighting some classical, as well as timely and emerging practical applications of sparse arrays in active sensing.

9.5.1 Imaging

Imaging is a key task in applications including radar, sonar, and diagnostic medical ultrasound. An image typically consist of discrete pixels corresponding to a beamformed output such as (9.27). Consequently, the beampattern synthesis approach of Section 9.4 is well-suited for imaging using sparse arrays. Beamforming-based imaging allows extracting useful information from complex scattering environment containing continuous or distributed scatterers, potentially located in the near-field of the array. For example, in medical ultrasound imaging, the ultrasonic probe is in direct contact with the skin and hence in close-proximity to internal organs and other scatterers with intricate shapes. Despite real-world challenges, both linear and planar sparse arrays geometries employing beamforming have been successfully applied in practice [82–87]. Sparse array imaging techniques can also offset beampattern degradation due to constraints on the beamforming architecture, such as unit-modulus beamforming weights or coarse phase shifter quantization [17].

9.5.2 MIMO Radar

MIMO radar systems frequently employ sparse arrays, such as the nested geometry in (9.13), to improve angular resolution and target identifiability [9]. Topical applications of MIMO radar include autonomous sensing [1] and, in particular, automotive radar [3, 4, 88]. Sparse arrays are attractive in such applications since they enable inexpensive, compact, and mobile system architectures. Moreover, as achieving high angular resolution given a limited sensor budget is often critical, it may be necessary to extend the physical aperture to a degree where the co-array is no longer contiguous. In such cases, array interpolation techniques can help reap the benefits of a large contiguous (interpolated) co-array [89–92].

9.5.3 Wireless Communications

An application of sparse arrays in wireless communications is channel estimation. Especially at high frequencies, communication channels typically exhibit spatial sparsity with only a few strong line-of-sight components [93]. Estimating some of the channel parameters, such as directions of departure (arrival) at the transmitter (receiver), may thereby benefit from the improved resolution or identifiability offered by sparse arrays. At sub-millimeter wavelengths, hybrid or fully analog beamforming architectures may have to be employed for reasons of cost and power consumption; emphasizing the importance of beamformer design also in channel estimation [94]. Sparse array configurations and sampling techniques can be used in conjunction with such beamforming architectures to further save system resources in, e.g., channel subspace estimation [95] or direction finding in general [96–98]. Sensing is also envisioned to play an increasingly important role in 6G and beyond communications [99]. Such emerging joint communication and sensing systems [5–7] present a plethora of yet largely unexplored opportunities for sparse arrays.

9.6 Conclusions

This chapter gave an overview of active sensing using sparse arrays, with a focus on low-redundancy one-dimensional array configurations and angular-domain transmit–receive (Tx–Rx) beamforming. A simple but general signal model was described highlighting the ubiquity of the sum co-array in active sensing. A review was provided of the active MRA for various degrees of shared sensors between the Tx and Rx arrays. In the case when all Tx and Rx sensors are shared, symmetric array configurations were introduced that are scalable, meaning that they can be generated for any number of sensors N. In contrast, the corresponding MRA is computationally challenging to find and hence still unknown for $N \geq 49$. Interestingly, however, there exist symmetric MRAs for each $N \leq 48$, suggesting the possibility of an optimal class of symmetric arrays. Properly designed symmetric arrays have both contiguous sum and difference co-arrays, which makes them directly applicable to both active and passive sensing. Moreover, extending the set of achievable Tx–Rx beampatterns by image addition was considered. Synthesizing a desired Tx–Rx beampattern can be interpreted as adjusting the virtual beamforming weights of the sum co-array. When the array configuration contains redundant virtual sensors, the physical beamforming weights may be optimized to minimize the beampattern synthesis (acquisition) time. This suggests a spatio-temporal trade-off between array sparsity and the number of physical beamforming weight vectors, or "component images," required in the synthesis. Indeed, the sparser the array, the more component images are necessary to achieve arbitrary co-array beamforming weights. Finally, selected timely and emerging active sensing applications were discussed. Sparse arrays offer cost-efficiency and improved performance in many use cases.

Acknowledgment

The work of R.R. was supported by the Ulla Tuominen foundation, and the work of V.K. by the Nokia Foundation's visiting professor grant. Both R.R. and V.K. were also supported by the Finnish Defense Research Agency.

References

1 Hügler, P., Roos, F., Schartel, M. et al. (2018). Radar taking off: new capabilities for UAVs. *IEEE Microwave Magazine* 19 (7): 43–53.

2 Schouten, G. and Steckel, J. (2019). Principles of biological echolocation applied to radar sensing: applying biomimetic sensors to achieve autonomous navigation. *IEEE Signal Processing Magazine* 36 (4): 98–111.

3 Patole, S.M., Torlak, M., Wang, D., and Ali, M. (2017). Automotive radars: a review of signal processing techniques. *IEEE Signal Processing Magazine* 34 (2): 22–35.

4 Sun, S., Petropulu, A.P., and Poor, H.V. (2020). MIMO radar for advanced driver-assistance systems and autonomous driving: advantages and challenges. *IEEE Signal Processing Magazine* 37 (4): 98–117.

5 Mishra, K.V., Shankar, M.R.B., Koivunen, V. et al. (2019). Toward millimeter-wave joint radar communications: a signal processing perspective. *IEEE Signal Processing Magazine* 36 (5): 100–114.

6 Ma, D., Shlezinger, N., Huang, T. et al. (2020). Joint radar-communication strategies for autonomous vehicles: combining two key automotive technologies. *IEEE Signal Processing Magazine* 37 (4): 85–97.

7 Ahmadipour, M., Kobayashi, M., Wigger, M., and Caire, G. (2022). An information-theoretic approach to joint sensing and communication. *IEEE Transactions on Information Theory* 1. https://doi.org/10.1109/TIT.2022.3176139.

8 Hoctor, R.T. and Kassam, S.A. (1990). The unifying role of the coarray in aperture synthesis for coherent and incoherent imaging. *Proceedings of the IEEE* 78 (4): 735–752.

9 Li, J. and Stoica, P. (2007). MIMO radar with colocated antennas. *IEEE Signal Processing Magazine* 24 (5): 106–114.

10 Hoctor, R.T. and Kassam, S.A. (1996). Array redundancy for active line arrays. *IEEE Transactions on Image Processing* 5 (7): 1179–1183.

11 Moffet, A. (1968). Minimum-redundancy linear arrays. *IEEE Transactions on Antennas and Propagation* 16 (2): 172–175.

12 Pal, P. and Vaidyanathan, P.P. (2010). Nested arrays: a novel approach to array processing with enhanced degrees of freedom. *IEEE Transactions on Signal Processing* 58 (8): 4167–4181.

13 Luthra, A.K. and Kassam, S.A. (1983). Side-lobe reduction in the ring array pattern for synthetic aperture imaging of coherent sources. *The Journal of the Acoustical Society of America* 74 (3): 840–846.

14 Kozick, R.J. and Kassam, S.A. (1991). Linear imaging with sensor arrays on convex polygonal boundaries. *IEEE Transactions on Systems, Man, and Cybernetics* 21 (5): 1155–1166.

15 Gabor, D., Stroke, G.W., Restrick, R.C. et al. (1965). Optical image synthesis (complex amplitude addition and subtraction) by hollographic Fourier transformation. *Physics Letters* 18 (2): 116–118.

16 Wild, J.P. (1965). A new method of image formation with annular apertures and an application in radio astronomy. *Proceedings of the Royal Society of London. Series A: Mathematical and Physical Sciences* 286 (1407): 499–509.

17 Rajamäki, R., Chepuri, S.P., and Koivunen, V. (2020). Hybrid beamforming for active sensing using sparse arrays. *IEEE Transactions on Signal Processing* 68: 6402–6417.

18 Rajamäki, R. and Koivunen, V. (2021). Sparse symmetric linear arrays with low redundancy and a contiguous sum co-array. *IEEE Transactions on Signal Processing* 69: 1697–1712.

19 Rajamäki, R. (2021). Sparse sensor arrays for active sensing – array configurations and signal processing. PhD thesis. Aalto University. https://aaltodoc.aalto.fi/handle/123456789/109102 (accessed 13 September 2023).

20 Levanon, N. and Mozeson, E. (2004). *Radar Signals.* Wiley. ISBN: 9780471663089.

21 Guerci, J.R. (2015). *Space-Time Adaptive Processing for Radar*, 2e. Artech House. ISBN: 978-1-60807-820-2.

22 Stoica, P. and Moses, R.L. (2005). *Spectral Analysis of Signals.* Upper Saddle River, NJ: Pearson Prentice Hall.

23 Fishler, E., Haimovich, A., Blum, R.S. et al. (2006). Spatial diversity in radars – models and detection performance. *IEEE Transactions on Signal Processing* 54 (3): 823–838.

24 Friedlander, B. (2012). On signal models for MIMO radar. *IEEE Transactions on Aerospace and Electronic Systems* 48 (4): 3655–3660.

25 Bekkerman, I. and Tabrikian, J. (2006). Target detection and localization using MIMO radars and sonars. *IEEE Transactions on Signal Processing* 54 (10): 3873–3883.

26 Li, J. and Stoica, P. (2009). *MIMO Radar Signal Processing.* Wiley.

27 Liu, S. and Trenkler, G. (2008). Hadamard, Khatri-Rao, Kronecker and other matrix products. *International Journal of Information and Systems Sciences* 4 (1): 160–177.

28 Bell, M.R. (1993). Information theory and radar waveform design. *IEEE Transactions on Information Theory* 39 (5): 1578–1597.

29 Pillai, U., Li, K.Y., Selesnick, I., and Himed, B. (2011). *Waveform Diversity: Theory & Applications.* McGraw-Hill Professional. ISBN: 0071622896.

30 He, H., Li, J., and Stoica, P. (2012). *Waveform Design for Active Sensing Systems: A Computational Approach.* Cambridge University Press.

31 Skolnik, M. (1981). *Introduction to Radar Systems*, 2e. McGraw-Hill Book Company.

32 Bliss, D.W. and Forsythe, K.W. (2003). Multiple-input multiple-output (MIMO) radar and imaging: degrees of freedom and resolution. *37th Asilomar Conference on Signals, Systems and Computers*, Volume 1, 54–59, 2003.

33 Aittomäki, T. and Koivunen, V. (2007). Signal covariance matrix optimization for transmit beamforming in MIMO radars. *41st Asilomar Conference on Signals, Systems and Computers*, 182–186.

34 Fuhrmann, D.R., Browning, J.P., and Rangaswamy, M. (2010). Signaling strategies for the hybrid MIMO phased-array radar. *IEEE Journal of Selected Topics in Signal Processing* 4 (1): 66–78.

35 Hassanien, A. and Vorobyov, S.A. (2010). Phased-MIMO radar: a tradeoff between phased-array and MIMO radars. *IEEE Transactions on Signal Processing* 58 (6): 3137–3151.

36 Macovski, A. (1979). Ultrasonic imaging using arrays. *Proceedings of the IEEE* 67 (4): 484–495.

37 Everett, E., Shepard, C., Zhong, L., and Sabharwal, A. (2016). SoftNull: many-antenna full-duplex wireless via digital beamforming. *IEEE Transactions on Wireless Communications* 15 (12): 8077–8092.

38 Evans, J.E., Sun, D.F., and Johnson, J.R. (1982). Application of Advanced Signal Processing Techniques to Angle of Arrival Estimation in ATC Navigation and Surveillance Systems. *Technical report No. 582*, Massachusetts Inst of Tech Lexington Lincoln Lab.

39 Roy, R. and Kailath, T. (1989). ESPRIT – estimation of signal parameters via rotational invariance techniques. *IEEE Transactions on Acoustics, Speech, and Signal Processing* 37 (7): 984–995.

40 Stailey, J.E. and Hondl, K.D. (2016). Multifunction phased array radar for aircraft and weather surveillance. *Proceedings of the IEEE* 104 (3): 649–659.

41 Liu, F., Masouros, C., Li, A. et al. (2018). MU-MIMO communications with MIMO radar: from co-existence to joint transmission. *IEEE Transactions on Wireless Communications* 17 (4): 2755–2770.

42 Chen, C.-Y. and Vaidyanathan, P.P. (2008). Minimum redundancy MIMO radars. *IEEE International Symposium on Circuits and Systems (ISCAS)*, 45–48, May 2008.

43 Weng, C.C. and Vaidyanathan, P.P. (2011). Nonuniform sparse array design for active sensing. *45th Asilomar Conference on Signals, Systems and Computers*, 1062–1066, November 2011.

44 BouDaher, E., Ahmad, F., and Amin, M.G. (2015). Sparsity-based direction finding of coherent and uncorrelated targets using active nonuniform arrays. *IEEE Signal Processing Letters* 22 (10): 1628–1632.

45 Rédei, L. and Rényi, A. (1949). On the representation of $1, 2, \ldots, N$ by differences. *Matematicheskii Sbornik* 24 (66): 385–389.

46 Leech, J. (1956). On the representation of $1, 2, \ldots, n$ by differences. *Journal of the London Mathematical Society* s1–31 (2): 160–169.

47 Wichmann, B. (1963). A note on restricted difference bases. *Journal of the London Mathematical Society* s1–38 (1): 465–466.

48 Rohrbach, H. (1937). Ein Beitrag zur additiven Zahlentheorie. *Mathematische Zeitschrift* 42 (1): 1–30.

49 Challis, M.F. (1993). Two new techniques for computing extremal h-bases Ak. *The Computer Journal* 36 (2): 117–126.

50 Robinson, J.P. (2009). Some postage stamp 2-bases. *Journal of Integer Sequences* 12 (2): 3.

51 Kohonen, J. (2014). A meet-in-the-middle algorithm for finding extremal restricted additive 2-bases. *Journal of Integer Sequences* 17 (6): 1–11.

52 Lunnon, W.F. (1969). A postage stamp problem. *The Computer Journal* 12 (4): 377–380.

53 Mossige, S. (1981). Algorithms for computing the h-range of the postage stamp problem. *Mathematics of Computation* 36 (154): 575–582.

54 Blanton, K.A. and McClellan, J.H. (1991). New search algorithm for minimum redundancy linear arrays. *International Conference on Acoustics, Speech, and Signal Processing (ICASSP)*, Volume 2, 1361–1364, April 1991.

55 Schwartau, F., Schröder, Y., Wolf, L., and Schoebel, J. (2021). Large minimum redundancy linear arrays: systematic search of perfect and optimal rulers exploiting parallel processing. *IEEE Open Journal of Antennas and Propagation* 2: 79–85.

56 Shakeri, S., Ariananda, D.D., and Leus, G. (2012). Direction of arrival estimation using sparse ruler array design. *IEEE 13th International Workshop on Signal Processing Advances in Wireless Communications (SPAWC)*, 525–529, June 2012.

57 Romero, D., Ariananda, D.D., Tian, Z., and Leus, G. (2016). Compressive covariance sensing: structure-based compressive sensing beyond sparsity. *IEEE Signal Processing Magazine* 33 (1): 78–93.

58 Pumphrey, H.C. (1993). Design of sparse arrays in one, two, and three dimensions. *The Journal of the Acoustical Society of America* 93 (3): 1620–1628.

59 Meurisse, Y. and Delmas, J.P. (2001). Bounds for sparse planar and volume arrays. *IEEE Transactions on Information Theory* 47 (1): 464–468.

60 Kohonen, J., Koivunen, V., and Rajamäki, R. (2018). Planar additive bases for rectangles. *Journal of Integer Sequences* 21 (9): 1–25.

61 Lockwood, G.R., Li, P.-C., O'Donnell, M., and Foster, F.S. (1996). Optimizing the radiation pattern of sparse periodic linear arrays. *IEEE Transactions on Ultrasonics, Ferroelectrics, and Frequency Control* 43 (1): 7–14.

62 Cohen, R. and Eldar, Y.C. (2018). Optimized sparse array design based on the sum coarray. *IEEE International Conference on Acoustics, Speech and Signal Processing (ICASSP)*, 3340–3343.

63 Kohonen, J. (2015). Early pruning in the restricted postage stamp problem. *arXiv preprint arXiv:1503.03416*.

64 Liu, C.-L. and Vaidyanathan, P.P. (2018). Optimizing minimum redundancy arrays for robustness. *52nd Asilomar Conference on Signals, Systems, and Computers*, 79–83.

65 Rajamäki, R. and Koivunen, V. (2017). Sparse linear nested array for active sensing. *25th European Signal Processing Conference (EUSIPCO)*, Kos, Greece, August 2017.

66 Liu, C.-L. and Vaidyanathan, P.P. (2017). Maximally economic sparse arrays and Cantor arrays. *7th IEEE International Workshop on Computational Advances in Multi-Sensor Adaptive Processing, CAMSAP*, 1–5, December 2017.

67 Cohen, R. and Eldar, Y.C. (2020). Sparse array design via fractal geometries. *IEEE Transactions on Signal Processing* 68: 4797–4812.

68 Kløve, T. (1981). A class of symmetric 2-bases. *Mathematica Scandinavica* 47 (2): 177–180.

69 Van Veen, B.D. and Buckley, K.M. (1988). Beamforming: a versatile approach to spatial filtering. *IEEE ASSP Magazine* 5 (2): 4–24.

70 Fuhrmann, D.R. and San Antonio, G. (2008). Transmit beamforming for MIMO radar systems using signal cross-correlation. *IEEE Transactions on Aerospace and Electronic Systems* 44 (1): 171–186.

71 Stoica, P., Li, J., and Xie, Y. (2007). On probing signal design for MIMO radar. *IEEE Transactions on Signal Processing* 55 (8): 4151–4161.

72 Goodman, J.W. (1996). *Introduction to Fourier Optics*, 2e. McGraw Hill. ISBN: 0-07-024254-2.

73 Vaidyanathan, P.P. and Pal, P. (2011). Sparse sensing with co-prime samplers and arrays. *IEEE Transactions on Signal Processing* 59 (2): 573–586.

74 Chen, C.-Y. and Vaidyanathan, P.P. (2009). MIMO radar waveform optimization with prior information of the extended target and clutter. *IEEE Transactions on Signal Processing* 57 (9): 3533–3544.

75 Liu, J., Li, H., and Himed, B. (2014). Joint optimization of transmit and receive beamforming in active arrays. *IEEE Signal Processing Letters* 21 (1): 39–42.

76 Palomar, D.P., Cioffi, J.M., and Lagunas, M.A. (2003). Joint Tx-Rx beamforming design for multicarrier MIMO channels: a unified framework for convex optimization. *IEEE Transactions on Signal Processing* 51 (9): 2381–2401.

77 Tsinos, C.G., Arora, A., Chatzinotas, S., and Ottersten, B. (2021). Joint transmit waveform and receive filter design for dual-function radar-communication systems. *IEEE Journal of Selected Topics in Signal Processing* 15 (6): 1378–1392.

78 Mitra, S.K., Mondal, K., Tchobanou, M.K., and Dolecek, G.J. (2010). General polynomial factorization-based design of sparse periodic linear arrays. *IEEE Transactions on Ultrasonics, Ferroelectrics, and Frequency Control* 57 (9): 1952–1966.

79 Martin, C.J., Martinez-Graullera, O., Godoy, G., and Ullate, L.G. (2010). Coarray synthesis based on polynomial decomposition. *IEEE Transactions on Image Processing* 19 (4): 1102–1107.

80 Kozick, R.J. and Kassam, S.A. (1992). Coarray synthesis with circular and elliptical boundary arrays. *IEEE Transactions on Image Processing* 1 (3): 391–405.

81 Rajamäki, R., Chepuri, S.P., and Koivunen, V. (2019). Analog beamforming for active imaging using sparse arrays. *53rd Asilomar Conference on Signals, Systems and Computers*, 1202–1206. Pacific Grove, CA: Asilomar, November 2019.

82 Kozick, R.J. and Kassam, S.A. (1993). Synthetic aperture pulse-echo imaging with rectangular boundary arrays. *IEEE Transactions on Image Processing* 2 (1): 68–79.

83 Lockwood, G.R., Talman, J.R., and Brunke, S.S. (1998). Real-time 3-D ultrasound imaging using sparse synthetic aperture beamforming. *IEEE Transactions on Ultrasonics, Ferroelectrics, and Frequency Control* 45 (4): 980–988.

84 Ahmad, F., Frazer, G.J., Kassam, S.A., and Amin, M.G. (2004). Design and implementation of near-field, wideband synthetic aperture beamformers. *IEEE Transactions on Aerospace and Electronic Systems* 40 (1): 206–220.

85 Cohen, R. and Eldar, Y.C. (2018). Sparse convolutional beamforming for ultrasound imaging. *IEEE Transactions on Ultrasonics, Ferroelectrics, and Frequency Control* 65 (12): 2390–2406.

86 Cohen, R., Fingerhut, N., Varray, F. et al. (2021). Sparse convolutional beamforming for 3-D ultrafast ultrasound imaging. *IEEE Transactions on Ultrasonics, Ferroelectrics, and Frequency Control* 68 (7): 2444–2459.

87 Lehtonen, O. (2021). Medical ultrasound imaging using sparse arrays. Master's thesis. Aalto University.

88 Engels, F., Heidenreich, P., Wintermantel, M. et al. (2021). Automotive radar signal processing: research directions and practical challenges. *IEEE Journal of Selected Topics in Signal Processing* 15 (4): 865–878.

89 Qiao, H. and Pal, P. (2017). Unified analysis of co-array interpolation for direction-of-arrival estimation. *IEEE International Conference on Acoustics, Speech and Signal Processing (ICASSP)*, 3056–3060, March 2017.

90 Sun, S. and Petropulu, A.P. (2020). A sparse linear array approach in automotive radars using matrix completion. *IEEE International Conference on Acoustics, Speech and Signal Processing (ICASSP)*, 8614–8618.

91 Sun, S. and Zhang, Y.D. (2021). 4D automotive radar sensing for autonomous vehicles: a sparsity-oriented approach. *IEEE Journal of Selected Topics in Signal Processing* 15 (4): 879–891.

92 Sarangi, P., Hücümenoğlu, M.C., and Pal, P. (2022). Single-snapshot nested virtual array completion: necessary and sufficient conditions. *IEEE Signal Processing Letters* 29: 2113–2117.

93 Ayach, O.E., Rajagopal, S., Abu-Surra, S. et al. (2014). Spatially sparse precoding in millimeter wave MIMO systems. *IEEE Transactions on Wireless Communications* 13 (3): 1499–1513.

94 Shahsavari, S., Sarangi, P., and Pal, P. (2022). Beamspace ESPRIT for mmWave channel sensing: performance analysis and beamformer design. *Frontiers in Signal Processing* 1: 820617. https://doi.org/10.3389/frsip.2021.820617.

95 Haghighatshoar, S. and Caire, G. (2017). Massive MIMO channel subspace estimation from low-dimensional projections. *IEEE Transactions on Signal Processing* 65 (2): 303–318.

96 Ibrahim, M., Ramireddy, V., Lavrenko, A. et al. (2017). Design and analysis of compressive antenna arrays for direction of arrival estimation. *Signal Processing* 138: 35–47.

97 Guo, M., Zhang, Y.D., and Chen, T. (2018). DOA estimation using compressed sparse array. *IEEE Transactions on Signal Processing* 66 (15): 4133–4146.

98 Koochakzadeh, A. and Pal, P. (2020). Compressed arrays and hybrid channel sensing: a Cramér-Rao bound based analysis. *IEEE Signal Processing Letters* 27: 1395–1399.

99 Saad, W., Bennis, M., and Chen, M. (2020). A vision of 6G wireless systems: applications, trends, technologies, and open research problems. *IEEE Network* 34 (3): 134–142.

10

Sparse MIMO Array Transceiver Design in Dynamic Environment

Xiangrong Wang[1], *Weitong Zhai*[1], *and Xianghua Wang*[2]

[1] *School of Electronic and Information Engineering, Beihang University, Beijing, China*
[2] *School of Artificial Intelligence, Beijing University of Posts and Telecommunications, Beijing, China*

Multi-input multi-output (MIMO) array radar is a new type of radar system that adopts waveform diversity technology at the transmitter. In essence, different transmit antennas can transmit differed orthogonal or partially correlated signals. In this regard, MIMO radar offers more degrees of freedom (DoFs) compared with the phased array counterparts. This provides opportunities for improved radar functions, such as target detection, beamforming and direction of arrival (DOA) estimation. In Chapters 5–7 of this book, sparse receiver array design was delineated, which enhances the radar performance by fully utilizing the spatial DoFs and reduce hardware cost. Recently, with the increased demands of cost saving and improved performance, the optimization of sparse transceiver for MIMO radar has become particularly important. In this chapter, we examine the joint sparse optimization of MIMO radar transceiver for different task requirements. In Section 10.1, we briefly introduce the principle of MIMO array transceiver and review the state of the art in sparse array design. In Section 10.2, the sparse MIMO array transceiver design for enhanced adaptive beamforming is presented under the assumption of known environmental conditions (such as, the DOAs of the targets). We examine the active sparse array design enabling the maximum signal to interference plus noise ratio (MaxSINR) beamforming at the MIMO radar receiver through successive convex approximation (SCA) incorporating the two-dimensional group sparsity promoting regularization. In Section 10.3, the cognitive-driven optimization of sparse array transceiver for MIMO radar beamforming is introduced, which further eliminates the prerequisite of prior information. We propose a cognitive-driven MIMO array design where both the beamforming weights and the transceiver configuration are adaptively and concurrently optimized in dynamic operating environment via a "perception–action" cycle. In Section 10.4, we study the MIMO array transceiver thinning for enhanced multi-signal DOA estimation. Performance enhancement of DOA estimation is achieved by reconfiguring the MIMO transceiver through an antenna selection strategy leveraging the minimum Cramer–Rao Bound (CRB) criterion. The work presented in Section 10.4 amounts to performing environmental sensing followed by the unstructured sparse MIMO transceiver design. Finally, concluding remarks are provided in Section 10.5.

Sparse Arrays for Radar, Sonar, and Communications, First Edition. Edited by Moeness G. Amin.

10.1 Review of MIMO Arrays and Sparse Arrays

The concept of "MIMO" was initially applied in wireless communication systems. The transmitter and receiver of the communication system employ multiple antennas, which can effectively improve the transmission capacity. Capitalizing on the idea of MIMO communications, MIMO radar adopts a transmit waveform diversity scheme to improve the DoFs of the radar system. Compared with phased arrays, waveform diversity endows MIMO radar with the following specific advantages:

- Flexible transmit waveforms: Traditional phased arrays realize transmit beamforming by controlling the phase of each transmit array element. Typically, only a single transmit beam can be formed in a single radar pulse, and the shape of such beam is not easy to control. However, each antenna of the transmit array of MIMO radar can transmit different signals, realizing more DoFs for transmit beamforming. This has an advantage of not only searching targets with wide beams, but also tracking different targets with multiple narrow beams. In addition, facing the increasingly complex working environment, the transmit waveforms in MIMO radar can be designed to utilize prior information and adhere to various requirements placed on the operating radio frequency (RF) environment and bandwidth.
- Improved target parameter estimation: Benefiting from the additional DoFs brought by waveform diversity, MIMO radar can separate the echo signals of different transmission paths at the receiver. This is due to the underlying large virtual aperture, which yields higher spatial angular resolution compared with its phased array counterparts of only receiving DoFs.
- Improved interference and clutter suppression: Benefiting from the additional DoFs obtained by waveform diversity, MIMO radar can effectively suppress clutter and interference, in particular, for the multipath clutter generated by low altitude/ultra-low altitude targets.

We note that the traditional phased array adopts the working mode of single narrow beam scanning. For example, all antennas of phased array transmit the same waveform and the beam direction is controlled by applying different phases to the phase center. Usually, when performing the task of multi-target tracking, only single beam switching is used, which will limit time-on-target and thus hinder high-precision estimation and tracking. For MIMO radar, due to the flexibility in beam design, it can acquire the target information by forming omni-directional or specific beampatterns if so desired. In essence, a multi-beam beampattern can be easily designed. In so doing, the multi-beam beampattern, compared with the traditional narrow beam switching mode of the phased array, can observe and detect multiple targets at the same time, allowing an extended dwell time on each target, thereby improving the accuracy of estimation and tracking.

Regarding the MIMO radar the transmit waveform, that is, the selection of transmit signal, the traditional radar mostly adopts fixed waveforms (such as linear frequency modulation signal, two-phase code, Barker code, polyphase code, and so on). The waveform is selected according to some factors, such as the mainlobe width, sidelobe level, the shape of three-dimensional ambiguity function. In this case, adaptive processing is only carried out at the receive end. With the increasingly complex and diverse scenes, the working mode of fixed transmit waveform applied by traditional radar cannot obtain satisfactory results. Therefore, according to the different tasks and working scenes, the radar needs to adjust adaptively at the transmit end through dynamic perception. Combined with the concept of cognitive radar, the overall performance of radar can be greatly improved by continuously online waveform optimization.

Depending on the distance between the antennas of the transmit array and the receive array, MIMO radars can be divided into two categories: distributed MIMO radars and colocated MIMO

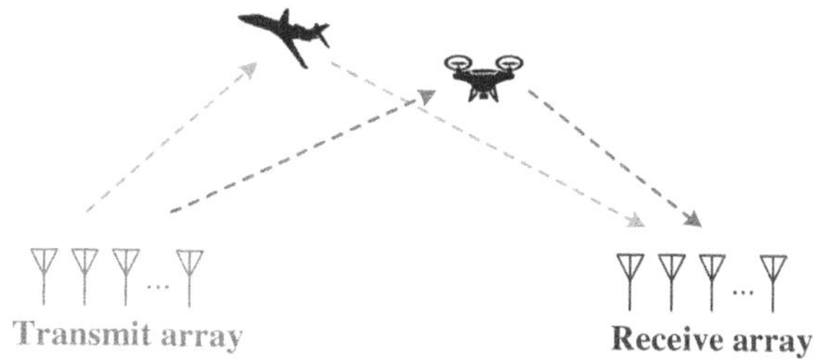

Figure 10.1 Schematic diagram of distributed MIMO Radar.

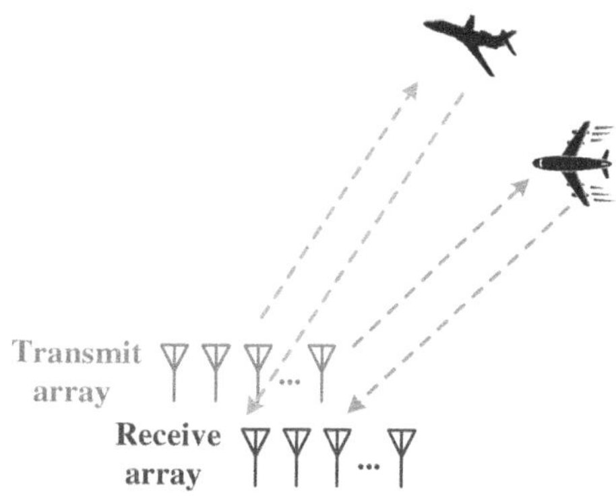

Figure 10.2 Schematic diagram of colocated MIMO radar.

radars. The schematic diagrams of these two kinds of MIMO radars are presented in Figs. 10.1 and 10.2, respectively. The distributed MIMO radar, different from the conventional multistatic radar system, enables each transmit antenna to detect the radar target from different directions by utilizing spatial diversity and increasing the inter-element spacing of the transmit array. The received echo signal can be considered as the superposition of independent fading signals. Combining these echo signals from different independent observation paths can obtain an approximately constant radar cross-section (RCS) of the target, so as to suppress the RCS flicker of the target and obtain a large spatial diversity gain. In colocated MIMO radar, the inter-element spacing is relatively small. Typically, the transmit array and receive array are closely spaced or share the same antennas. Different from phased array radar system, the colocated MIMO array enables each transmit antenna to carry arbitrary waveform signals. In terms of transmit signals, phased array radar can be regarded as a special case of colocated MIMO radar.

In this chapter, we consider the colocated MIMO array. The transmit waveform of the colocated MIMO array can be divided into two categories: orthogonal and partially correlated. When the transmit waveforms are orthogonal, the transmit beampattern is omni-directional. In this case, the virtual aperture expansion can be realized by signal separation at the receive end, so as to obtain high resolution. When the transmit waveforms are partially correlated, the transmit beampattern can be optimized to any desired shape and direction. In this case, transmit beampatterns can be designed so that the transmit energy can be reasonably allocated to the target area of interest.

We consider a uniform linear array (ULA) as an example. Assuming that the transmitter of the MIMO radar is an ULA composed of M antennas and the receiver is another ULA composed of N antennas. The transmit signal in direction θ of MIMO array can be expressed as,

$$y(t) = \mathbf{a}_T^T(\theta)\mathbf{s}(t), \tag{10.1}$$

where $(.)^T$ represents transpose, $\mathbf{a}_T(\theta)$ is the transmit steering vector and $\mathbf{s}(t)$ is the vector composed of the signal transmitted by each antenna. Then, the transmit beampattern is given by,

$$\begin{aligned} B(\theta) &= E\{y(t)y^*(t)\}, \\ &= E\{\mathbf{a}_T^T(\theta)\mathbf{s}(t)\mathbf{s}^H(t)\mathbf{a}_T^*(\theta)\}, \\ &= \mathbf{a}_T^T(\theta)E\{\mathbf{s}(t)\mathbf{s}^H(t)\}\mathbf{a}_T^*(\theta), \end{aligned} \tag{10.2}$$

where $E\{.\}$ means taking expectation. We define the autocorrelation matrix of the transmit waveform as,

$$\mathbf{R}_s = E\{\mathbf{s}(t)\mathbf{s}^H(t)\}, \tag{10.3}$$

where $(.)^H$ represents Conjugate transpose. When the transmit signals are orthogonal, $\mathbf{R}_s$ is an identity matrix, that is $\mathbf{R}_s = \mathbf{I}$. Accordingly,

$$B(\theta) = \mathbf{a}_T^T(\theta)\mathbf{I}\mathbf{a}_T^*(\theta) = M, \tag{10.4}$$

where $(.)^*$ represents taking conjugation. Equation (10.4) indicates that the radiation power in any direction is the same (which defines the omnidirectional detection situation).

When the the transmit signals are partially correlated, the matrix $\mathbf{R}_s$ is no longer an identity matrix. We can synthesize arbitrarily shaped transmit beampatterns by optimizing the matrix $\mathbf{R}_s$. A special case is the multi-carrier waveform, that is, the transmit signal of each antenna can be regarded as a linear combination of a set of orthogonal signals. In this case, the transmit signal vector can be expressed as,

$$\mathbf{s}(t) = \mathbf{W}\mathbf{s}_o(t), \tag{10.5}$$

where $\mathbf{W}$ is the weighting matrix and $\mathbf{s}_o(t)$ is the vector of M' orthogonal signals. Substituting Eq. (10.5) into Eq. (10.2), the transmit beampattern becomes,

$$\begin{aligned} B(\theta) &= E\{\mathbf{a}_T^T(\theta)\mathbf{s}(t)\mathbf{s}^H(t)\mathbf{a}_T^*(\theta)\}, \\ &= \mathbf{a}_T^T(\theta)\mathbf{W}E\{\mathbf{s}_o(t)\mathbf{s}_o^*(t)\}\mathbf{W}^H\mathbf{a}_T^*(\theta), \\ &= \mathbf{a}_T^T(\theta)\mathbf{W}\mathbf{W}^H\mathbf{a}_T^*(\theta), \\ &= \mathbf{a}_T^T(\theta)\mathbf{R}_W\mathbf{a}_T^*(\theta), \end{aligned} \tag{10.6}$$

where $\mathbf{R}_W = \mathbf{W}\mathbf{W}^H$ is the weight autocorrelation matrix. According to Eq. (10.6), the optimization problem of the transmit beamforming can be transformed into the optimization of the beamforming weight matrix $\mathbf{W}$ or the autocorrelation matrix $\mathbf{R}_W$.

At the receive end, the received signal vector can be expressed as,

$$\begin{aligned} \mathbf{x}(t) &= \eta\mathbf{a}_R(\theta)x(t), \\ &= \eta\mathbf{a}_R(\theta)\mathbf{a}_T^T(\theta)\mathbf{s}(t), \end{aligned} \tag{10.7}$$

where η represents the target RCS and the two-way path loss which is assumed to follow the Swerling 0 model, $\mathbf{a}_R(\theta)$ is the receive steering vector. Applying the multi-carrier model, the received signal vector can be formed into a matrix upon matched filtering,

$$\begin{aligned} \mathbf{X} &= \int_T \mathbf{x}(t)\mathbf{s}_o^H(t)dt, \\ &= \eta\mathbf{a}_R(\theta)\mathbf{a}_T^T(\theta)\mathbf{W}. \end{aligned} \tag{10.8}$$

When $\mathbf{W} = \mathbf{I}$, the transmit signals are orthogonal. In this case, $\mathbf{X}$ can be regarded as the received signal of $M \times N$ virtual antennas derived from the transmit and receive arrays. When $\mathbf{W} \neq \mathbf{I}$, the

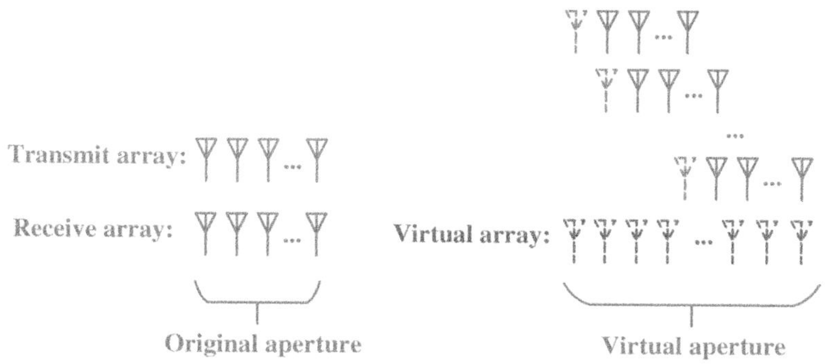

Figure 10.3 Virtual array and virtual aperture of MIMO array.

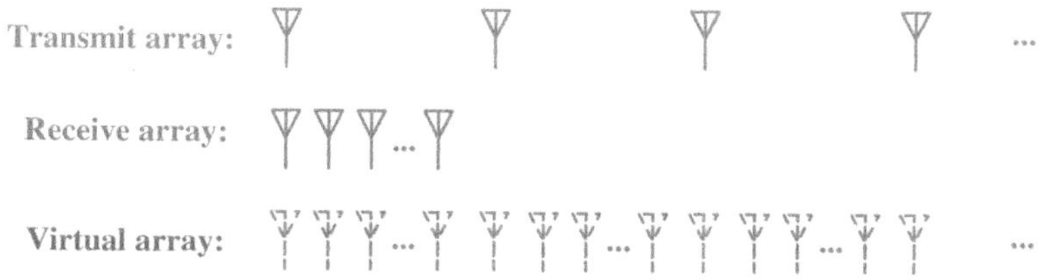

Figure 10.4 A special MIMO virtual array: sum-coarray.

transmit signals are correlated, and $\mathbf{X}$ becomes the received signal of $M' \times N$ virtual antennas derived from the virtual transmit array (whose steering vector is $\overline{\mathbf{a}}(\theta) = \mathbf{W}^T\mathbf{a}_T(\theta)$) and the receive array. Equation (10.8) indicates mathematically that the MIMO array can provide more spatial DoFs than just $M + N$.

Figure 10.3 depicts the schematic diagram of the virtual antenna. As it can be seen from the figure, the transmit array and receive array can generate a virtual array with larger aperture and more elements, which is referred to as the sum-coarray. Figure 10.4 presents another schematic diagram of a sum-coarray, for which we assume that the inter-element spacing of the receive array is d and the inter-element spacing of the transmit array is Nd. In this case, the transceiver can generate a virtual ULA consisting of MN virtual antennas with an inter-element spacing of d.

The demand for higher angular resolution in radar and communications applications has spurred the use of larger arrays. Yet, increasing the number of antenna elements leads to an increase in hardware cost and complexity, as well as higher power consumption. Therefore, various strategies have been developed to exploit the redundancy that is inherent to fully populated arrays in order to reduce the number of RF front-ends while preserving the desirable array performance. Thus, array reconfigurability through antenna selection was exploited to mitigate interference and maximize the signal to interference plus noise ratio (SINR) [1–3]. Sparse array design, aided by emerging fast antenna switching technologies, can lower the overall system cost by reducing the number of expensive front-end processing channels.

Benefiting from the larger aperture generated by the virtual arrays of MIMO radar, higher estimation accuracy and better beamforming performance can be achieved. If the sparse configuration of the MIMO transceiver is optimized, the performance of MIMO radar can be further enhanced. As mentioned above, the conventional MIMO radar has a sparse transceiver array. The inter-element spacing of the receive array is half wavelength, whereas that of the transmit array is multiple wavelengths. Hence, the MIMO sum coarray is a compact ULA with a large virtual aperture that enables high spatial resolution [4–6]. This configuration, however, may not render an optimum beamforming in terms of maximizing the output SINR (MaxSINR) for a given environment [7, 8]. Sparse MIMO array beamforming design can seek to optimize the transmit sensor locations and the corresponding waveform correlation matrix [9–11]. The optimal design of sparse MIMO transceiver can be divided into two types: structured and unstructured sparse arrays.

The environment-independent, structured sparse array design seeks to increase the number of the spatial autocorrelation lags and maximize the contiguous segment of the coarray aperture for a limited number of sensors. The main task, therein, is to enable DOA estimation involving more sources than the physical sensors [12–16]. For beamforming applications, the environment-independent design criteria typically strives to achieve desirable beampattern characteristics such as broad main lobe and minimum sidelobe levels and frequency invariant beampattern for wideband design [10, 17–21]. However, the environment-dependent, unstructured sparse receive beamformer, which strives to achieve MaxSINR, can potentially provide sparse configurations that improve target detection and estimation accuracy for the operating environment [1, 22–29].

In Section 10.2, the unstructured sparse MIMO array transceiver design for MaxSINR with known environmental information is discussed. Nonuniform sparse arrays have emerged as an effective technology which can be deployed in various active and passive sensing modalities, such as acoustics and RF applications, including GPS [2]. Fundamentally, sparse array design seeks optimum system performance, under both noise and interference, while being cognizant of the limitations on cost and aperture. Within the RF beamforming paradigm, maximizing the SINR amounts to concurrently optimizing the array configuration and beamforming weights. Equivalently, the problem can be cast as continuously selecting the best antenna locations to deliver MaxSINR under dynamic environment. In lieu of constantly placing antennas to new positions, a more practical and feasible approach is to have a uniform full array and then switch among antennas with a fixed number of front-end chains, yielding the MaxSINR sparse beamformer for a given environment.

In Section 10.3, we consider the cognitive-driven optimization of sparse MIMO array transceiver for adaptive beamforming in dynamic environment. When the environment information is unavailable, we need to learn the environmental information automatically from the received data. MIMO radars enable superior capabilities compared with standard phased arrays radar. Cognitive radar continuously interacts with the environment and update the radar parameters through the acquired knowledge [30, 31]. In MIMO radar, a "perception–action" cognition approach typically entails adaptive optimizations of the transmit waveforms and power allocation [32–34]. In addition to these parameters, array reconfiguration (hardware) can significantly improve system performance beyond that achieved with fixed antenna positions [2, 35].

For the environment-dependent design in Section 10.2, we need to estimate the DOAs of all signals in the radar field of view, including the desired signal and interferences. DOA estimation using antenna arrays enjoys a myriad of applications and has consequently commanded a significant research effort. While the DOA estimation performance is ultimately given by the specific algorithm employed, it is nonetheless lower-bounded by the CRB, which is itself intimately related to the array configuration as well as being a function of DOA [36]. The CRB has been derived in terms of array element positions for 2D and 3D arrays [37–40] and its dependence on the array configuration has been studied, see for instance [41–44]. Furthermore, the array design to optimize the DOA performance has been examined in [45–47], but the design is done offline and the resulting array is fixed. In [48], the focus was placed on DOA estimation performance and the problem of selecting the smallest number of antennas for a desired CRB was considered albeit for a ULA. In [49], antenna selection to minimize the CRB was done for arbitrary arrays with both isotropic and directional configurations. Although directional arrays are adapted to the DOA, which ensures superior performance compared to isotropic arrays, solving the optimization problem is computationally intensive, precluding an online implementation. In all the aforementioned works, the passive working mode is considered and the accuracy of DOA estimation depends on the receive array configuration. In Section 1.4, we investigate the problem of sparse MIMO array transceiver design for enhanced accuracy of multi-source DOA estimation.

In Chapter 7, we have discussed the deep learning solution of MaxSINR beamformer for spectrum sensing in dynamic environments [50, 51]. While in this chapter, we deal with iterative solutions of the sparse array design optimization problems.

10.2 Sparse MIMO Transceiver Design for MaxSINR with Known Environmental Information

Sparse array design aided by emerging fast sensor switching technologies can lower the overall system overhead by reducing the number of expensive transceiver chains. In this section, we assume a perfect knowledge of the environment and examine the active sparse array design enabling the MaxSINR beamforming at the MIMO radar receiver. The proposed approach entails an entwined design, i.e., jointly selecting the optimum transmit and receive sensor locations for accomplishing MaxSINR receive beamforming. Specifically, we consider a colocated MIMO radar platform with orthogonal transmit waveforms, and examine antenna selections at the transmit and receive arrays. The optimum active sparse array transceiver design problem is formulated as SCA alongside the two-dimensional group sparsity promoting regularization. Several examples are provided to demonstrate the effectiveness of the proposed approach in utilizing the given transmit/receive array aperture and DoFs for achieving MaxSINR beamforming.

In this section, we consider MaxSINR receive beamformer design for MIMO radar. The beamformer design, in this case, jointly selects the optimum transmit and receive sensor locations for implementing an efficient receive beamformer. We assume that the transmit sensors emit orthogonal waveforms. The MIMO radar implementing sensor-based orthogonal transmit waveforms does not benefit from coherent transmit processing gain achieved when using directional beamforming. In this context, the proposed approach is fundamentally different from the parallel sparse array MIMO beamforming designs, which primarily rely on optimizing the transmit sensor locations and the corresponding correlation matrix of the transmit waveform sequence [52–55]. The proposed approaches, therein, maximize the transmitted signal power towards the perspective target locations while minimizing the cross-correlation of the target returns to achieve efficient receive beamforming characteristics. Moreover, existing design schemes essentially pursue an uncoupled transmit/receive design which is in contrast to the approach considered in this chapter which incorporates sparse receiver design in conjunction with transmit array optimization.

The proposed transceiver beamforming design problem, in essence, amounts to configuring the transmit/receive array with the corresponding optimum receiver beamforming weights. To this end, we pursue a data dependent approach that jointly optimizes the beamforming sensor positions and weights. The optimization only requires the knowledge of the perspective desired source locations and does not assume any knowledge of the interfering sources. We optimally select M_t from M transmitters and N_r from N receivers. Such selection is a binary optimization problem, and is NP-hard. In order to avoid extensive computations associated with enumeration of all possible array configurations, we apply convex relaxation. The design problem is posed as

SCA with two-dimensional reweighted mixed $l_{1,2}$-norm penalization to jointly invoke sparsity in the transmit and receive dimensions.

10.2.1 Problem Formulation

We consider a MIMO radar with M transmitters and N receivers illuminating the scene through omni-directional beampattern. This is typically achieved through emissions of orthogonal waveforms across the transmit array. Specifically, we pursue a co-located MIMO arrangement with the transmit and receive arrays in close vicinity. As such, a far-field source is presented by equal angles of departure and arrival. Consider K target sources arriving from $\{\theta_{s,k}\}$. The received baseband data $\mathbf{x}(n) \in \mathbb{C}^{MN\times1}$ after matched filtering at the N element uniformly spaced receiver array at time instant n is given by,

$$\mathbf{x}(n) = \sum_{k=1}^{K} s_k(n)\mathbf{b}(\theta_{s,k}) + \sum_{l=1}^{L_c} c_l(n)\mathbf{b}(\theta_{j,l}) + \sum_{q=1}^{Q} \mathbf{i}_q(n) + \mathbf{v}(n), \tag{10.9}$$

where, $s_k(n) \in \mathbb{C}$ is the kth reflected target signal. The extended steering vector $\mathbf{b}(\theta)$ of the virtual array is $\mathbf{b}(\theta) = \mathbf{a}_T(\theta) \otimes \mathbf{a}_R(\theta)$. In the case of uniform transmit and receive linear arrays with a respective inter-element spacing of d_t and d_r, the transmit and receive steering vectors are given by,

$$\mathbf{a}_T(\theta) = [1 \;\; e^{j2\pi(d_t/\lambda)\cos\theta} \ldots \;\; e^{j2\pi(M-1)(d_t/\lambda)\cos\theta}]^T, \tag{10.10}$$

and

$$\mathbf{a}_R(\theta) = [1 \;\; e^{j2\pi(d_r/\lambda)\cos\theta} \ldots \;\; e^{j2\pi(N-1)(d_r/\lambda)\cos\theta}]^T. \tag{10.11}$$

The variance of additive Gaussian noise $\mathbf{v}(n) \in \mathbb{C}^{MN\times1}$ is σ_v^2 at the receiver output. There are L_c interferences $c_l(n)$ mimicking target reflected signal and Q narrowband interferences $\mathbf{i}_q(n) \in \mathbb{C}^{MN\times1}$. The latter is defined as the Kronecker product (which is represented by $\otimes$) of the receiver steering vector $\mathbf{a}_R(\theta_{i,q})$ and the matched filtering output of the interference $\mathbf{j}_q(n)$ such that $\mathbf{i}_q(n) = \mathbf{j}_q(n) \otimes \mathbf{a}_R(\theta_{i,q})$. The received data vector $\mathbf{x}(n)$ is then linearly combined to maximize the output SINR. The output signal $y(n)$ of the optimum beamformer for MaxSINR is given by [56],

$$y(n) = \mathbf{w}^H\mathbf{x}(n), \tag{10.12}$$

where $\mathbf{w}$ is the beamformer weight at the receiver. The optimal solution $\mathbf{w}_o$ can be obtained by solving the optimization problem that seeks to minimize the interference power at the receiver output while preserving the desired signal. The constraint minimization problem can be cast as,

$$\begin{aligned} \underset{\mathbf{w}\in\mathbb{C}^{MN}}{\text{minimize}} \quad & \mathbf{w}^H\mathbf{R}_{\mathbf{x}}\mathbf{w}, \\ \text{s.t.} \quad & \mathbf{w}^H\mathbf{R}_s\mathbf{w} = 1, \end{aligned} \tag{10.13}$$

where the source correlation matrix is $\mathbf{R}_s = \sum_{k=1}^{K} \sigma_{s,k}^2 \, \mathbf{b}(\theta_{s,k})\mathbf{b}^H(\theta_{s,k})$, with $\sigma_{s,k}^2 = E\{s_k(n)s_k^H(n)\}$ denoting the average received power from the kth target return. The data correlation matrix, $\mathbf{R}_{\mathbf{x}} \approx (1/T)\mathbf{x}(n)\mathbf{x}(n)^H$, is directly estimated from the T received data snapshots. The solution to the optimum weights in (10.13) is given by $\mathbf{w}_o = \mathscr{P}\{\mathbf{R}_{\mathbf{x}}^{-1}\mathbf{R}_s\}$, with the operator $\mathscr{P}\{.\}$ representing the principal eigenvector of the input matrix. This optimum solution yields the MaxSINR, SINR_o, given by [56],

$$\text{SINR}_o = \Lambda_{\max}\{\mathbf{R}_{\mathbf{x}}^{-1}\mathbf{R}_s\}, \tag{10.14}$$

which is the MaxSINR given by the maximum eigenvalue ($\Lambda_{\max}$) of the product of the two matrices, the inverse of interference plus noise correlation matrix and the desired source correlation matrix. It is clear that the resulting solution holds irrespective of the array configuration, whether the array is uniform or sparse. For the latter, the performance of MaxSINR beamformer is intrinsically tied to the array configuration. The sparse optimization of the above formulation is explained in Section 10.2.2.

10.2.2 Sparse Array Transceiver Design

The separate sensor selection problem for a joint transmit and receive design is a combinatorial optimization problem and cannot be solved in polynomial time. We formulate the sparse array design problem by exploiting the structure of the received signal model and solve it by applying sequential convex approximation. To exploit the sparse structure of the joint transmit and receive sensor selection, we introduce a two-dimensional $l_{1,2}$-mixed norm regularization to recover group sparse solutions. One dimension pertains to the transmitter sparsity, whereas the other sparsity pattern is associated with the receiver side. Moreover, this two-dimensional sparsity pattern is entwined and coupled, meaning that when one transmitter is discarded, all N receiving data pertaining to this transmitter is zero. Similarly, the receiver is discarded only when its received data corresponding to all the M transmitters is zero. The structure of the optimal sparse beamforming weight vector $\mathbf{w} \in \mathbb{C}^{MN}$ is elucidated in (10.15), where $\mathbf{w}(i,j)$ denotes the entries of $\mathbf{w}$ indexed from i to j. Also, ✓ denotes a sensor location activated, and $\times$ denotes a sensor not activated.

$$\overbrace{\underbrace{\begin{pmatrix}\checkmark\\ \times\\ \times\\ \checkmark\\ \vdots\\ \times\end{pmatrix}}_{(\text{Tx 1 active})}}^{\mathbf{w}(1,N)}\;\overbrace{\underbrace{\begin{pmatrix}\checkmark\\ \times\\ \times\\ \checkmark\\ \vdots\\ \times\end{pmatrix}}_{(\text{Tx 2 active})}}^{\mathbf{w}(N+1,2N)}\;\overbrace{\underbrace{\begin{pmatrix}\times\\ \times\\ \times\\ \times\\ \vdots\\ \times\end{pmatrix}}_{(\text{Tx 3 inactive})}}^{\mathbf{w}(2N+1,3N)}\;\cdots\;\overbrace{\underbrace{\begin{pmatrix}\checkmark\\ \times\\ \times\\ \checkmark\\ \vdots\\ \times\end{pmatrix}}_{(\text{Tx } M \text{ active})}}^{\mathbf{w}(N(M-1)+1,MN)}. \tag{10.15}$$

In (10.15), each column vector denotes the receive beamformer weights for a fixed transmit location. It is noted that the optimal sparse beamformer, corresponding to the active transmit and receive locations, follows a group sparse structure along the transmit and receive dimensions (the consecutive N sensors are discarded vertically or the consecutive M sensors are discarded horizontally). It is evident that the missing transmit sensor at position 3, for example, translates to the sparsity along all the corresponding N entries of $\mathbf{w}$ (N consecutive $\times$ vertically in (10.15)). Similarly, the group sparsity is also invoked across the received signals corresponding to all transmitters (M consecutive $\times$ horizontally in (10.15)).

10.2.2.1 Group Sparse Solutions via SCA

The problem in (10.13) can equivalently be rewritten by swapping the objective and constraint functions as follows,

$$\begin{aligned} \underset{\mathbf{w}\in\mathbb{C}^{MN}}{\text{minimize}} \quad & \mathbf{w}^H\overline{\mathbf{R}}_s\mathbf{w} \\ \text{s.t.} \quad & \mathbf{w}^H\mathbf{R}_x\mathbf{w} \le 1, \end{aligned} \tag{10.16}$$

where $\overline{\mathbf{R}}_s = -\mathbf{R}_s$. The sparse MIMO configuration of uniformly spaced receivers and transmitters with a respective inter-element spacing of d_r and $d_t = Nd_r$ is employed for data collection. The covariance matrix $\mathbf{R_x}$ of the full receive virtual array can be then obtained by the matrix completion method proposed in [25].

The beamforming weight vectors are generally complex valued, whereas the quadratic functions are real. This observation allows expressing the problem with only real variables which is typically accomplished by replacing the correlation matrix $\overline{\mathbf{R}}_s$ by $\tilde{\mathbf{R}}_s$ and concatenating the beamforming weight vector accordingly [57],

$$\tilde{\mathbf{R}}_s = \begin{bmatrix} \text{real}(\overline{\mathbf{R}}_s) & -\text{imag}(\overline{\mathbf{R}}_s) \\ \text{imag}(\overline{\mathbf{R}}_s) & \text{real}(\overline{\mathbf{R}}_s) \end{bmatrix},$$

$$\tilde{\mathbf{w}} = \begin{bmatrix} \text{real}(\mathbf{w}) \\ \text{imag}(\mathbf{w}) \end{bmatrix}. \tag{10.17}$$

Similarly, the received data correlation matrix $\mathbf{R_x}$ is replaced by $\tilde{\mathbf{R}}_\mathbf{x}$. The problem in (10.16) then becomes,

$$\begin{aligned} \underset{\tilde{\mathbf{w}} \in \mathbb{R}^{2MN}}{\text{minimize}} \quad & \tilde{\mathbf{w}}' \tilde{\mathbf{R}}_s \tilde{\mathbf{w}}, \\ \text{s.t.} \quad & \tilde{\mathbf{w}}' \tilde{\mathbf{R}}_\mathbf{x} \tilde{\mathbf{w}} \leq 1, \end{aligned} \tag{10.18}$$

where $'$ denotes transpose operation. After expressing the constraint in terms of real variables, we convexify the objective function by utilizing the first order approximation iteratively,

$$\begin{aligned} \underset{\tilde{\mathbf{w}} \in \mathbb{R}^{2MN}}{\text{minimize}} \quad & \mathbf{m}^{(k)'} \tilde{\mathbf{w}} + b^{(k)}, \\ \text{s.t.} \quad & \tilde{\mathbf{w}}' \tilde{\mathbf{R}}_\mathbf{x} \tilde{\mathbf{w}} \leq 1, \end{aligned} \tag{10.19}$$

where, $\mathbf{m}$ and b, updated at the $k+1$th iteration, are given by $\mathbf{m}^{(k+1)} = 2\tilde{\mathbf{R}}_s \tilde{\mathbf{w}}^{(k)}$, $b^{(k+1)} = -\tilde{\mathbf{w}}^{(k)'} \tilde{\mathbf{R}}_s \tilde{\mathbf{w}}^{(k)}$, respectively. Finally, to invoke sparsity in the beamforming weight vector, the re-weighted mixed $l_{1,2}$-norm is adopted primarily for promoting group sparsity.

$$\underset{\tilde{\mathbf{w}},\mathbf{c},\mathbf{r}}{\text{minimize}} \quad \mathbf{m}^{(k)'} \tilde{\mathbf{w}} + b^{(k)} + \alpha(\mathbf{p}'\mathbf{c}) + \beta(\mathbf{q}'\mathbf{r}) \tag{10.20}$$

$$\text{s.t.} \quad \tilde{\mathbf{w}}' \tilde{\mathbf{R}}_\mathbf{x} \tilde{\mathbf{w}} \leq 1, \tag{10.20a}$$

$$||\mathbf{P}_i \odot \tilde{\mathbf{w}}||_2 \leq c_i, \tag{10.20b}$$

$$0 \leq c_i \leq 1, \quad i = 1, \dots, M \tag{10.20c}$$

$$||\mathbf{Q}_j \odot \tilde{\mathbf{w}}||_2 \leq r_j, \tag{10.20d}$$

$$0 \leq r_j \leq 1, \quad j = 1, \dots, N \tag{10.20e}$$

$$\mathbf{1}'_M \mathbf{c} = M_t, \tag{10.20f}$$

$$\mathbf{1}'_N \mathbf{r} = N_r, \tag{10.20g}$$

and

$$\mathbf{P}_i = [\overbrace{0 \ldots 0 \ldots 0}^{\substack{N \text{ elements} \\ \text{1st group}}} \ldots \overbrace{1 \ldots 1 \ldots 1}^{\substack{N \text{ elements} \\ i\text{th group}}} \ldots \overbrace{1 \ldots 1 \ldots 1}^{\substack{N \text{ elements} \\ (M+i)\text{th group}}} \ldots \overbrace{0 \ldots 0 \ldots 0}^{\substack{N \text{ elements} \\ 2M\text{th group}}}]', \tag{10.21}$$

Algorithm 10.1: SCA-sparse transmit/receive beamformer design

Input: $M, N, M_t, N_r, \alpha = \beta, \alpha_0 = \beta_0$, target direction θ_s, auto-correlation matrix $\mathbf{R}_s$ and $\mathbf{R}_x$
Output: Optimal weight $\mathbf{w}$, location and number of receivers and transmitters.
Initialization:
Initialize $\mathbf{p}$, $\mathbf{q}$, $\mathbf{m}$ as all one's vectors. Set $\alpha_0 = 1$ and $\beta_0 = 1$. Initialize the tradeoff parameters α and β according to the sparsity requirement, such as $\alpha = \beta = 0.5$.
while $||\tilde{\mathbf{w}}^{(k+1)} - \tilde{\mathbf{w}}^{(k)}||_2 \geq 10^{-5}$ **do**
a) Convert $\mathbf{w}$, $\mathbf{R}_s$ and $\mathbf{R}_x$ to the real domain to get $\tilde{\mathbf{w}}$, $\tilde{\mathbf{R}}_s$ and $\tilde{\mathbf{R}}_x$ according to Eq. (10.17).
b) Update $\tilde{\mathbf{w}}^{(k+1)}$, $\mathbf{c}^{(k+1)}$, $\mathbf{r}^{(k+1)}$ according to Eq. (10.20).
c) Update $\mathbf{p}^{(k+1)}$ and $\mathbf{q}^{(k+1)}$ according to (10.23) and (10.24).
d) Convert $\tilde{\mathbf{w}}^{(k+1)}$ back into the complex solution $\mathbf{w}^{(k+1)}$ by $\mathbf{w}^{(k+1)}(i) = \tilde{\mathbf{w}}^{(k+1)}(i) + j\tilde{\mathbf{w}}^{(k+1)}(i + MN)$.
end while
return $\mathbf{w}$

$$\mathbf{Q}_j = [\overbrace{\underbrace{0\ldots0}_{(j-1)\,0s}\ 1 0\ldots0}^{N \text{ elements of the 1st group}}\ldots\overbrace{\underbrace{0\ldots0}_{(j-1)\,0s}\ 1 0\ldots0}^{N \text{ elements of the } 2M\text{th group}}]'. \tag{10.22}$$

Here, $\odot$ means the element-wise product, $\mathbf{c} \in [0, 1]^M$ and $\mathbf{r} \in [0, 1]^N$ are two auxiliary selection vectors taking values between 0 and 1, $\mathbf{P}_i \in \{0, 1\}^{2MN}$ in (10.20b) is the transmission selection vector, which is used to select the real and imaginary parts of the N weights corresponding to the ith transmitter, as shown in Eq. (10.21). Equations (10.20c) and (10.20f) indicate that we can select M_t transmitters at most. Similarly, $\mathbf{Q}_j$ in (10.20d) is the receiver selection vector, as shown in (10.22). Equations (10.20e) and (10.20g) indicate that we can select N_r receivers at most. The two parameters α and β are used to control the sparsity of the transmitters and the receivers, $\mathbf{p}$ and $\mathbf{q}$ are the reweighting coefficient vectors for the transmitters and receivers respectively and the detailed update method is given as follows.

10.2.2.2 Reweighting Update

A common method of updating the reweighting coefficient is to take the reciprocal of $|\mathbf{w}|$ [58]. However, this cannot control the number of elements to be selected. Thus, similar to [59], in order to make the number of selected antennas controllable, we update the weights using the following formula,

$$p_i^{(k)} = \frac{1 - c_i^{(k)}}{1 - e^{-\beta_0 c_i^{(k)}} + \epsilon} - \left(\frac{1}{\epsilon}\right)\left(c_i^{(k)}\right)^{\alpha_0} \tag{10.23}$$

$$q_i^{(k)} = \frac{1 - r_i^{(k)}}{1 - e^{-\beta_0 r_i^{(k)}} + \epsilon} - \left(\frac{1}{\epsilon}\right)\left(r_i^{(k)}\right)^{\alpha_0} \tag{10.24}$$

Here, α_0 and β_0 are two parameters that control the shape of the curve, and the parameter ϵ avoids the unwanted explosive case. The essence of this re-weight update method is to take a large positive penalty for a entry close to zero and a small negative reward for the entry close to 1. As a result, the entries of the two selection vectors $\mathbf{c}$ and $\mathbf{r}$ tend to be either 0 or 1. Through iterative regression, we can select M_t transmitters and N_r receivers. In addition, by controlling the values of α and β, we can obtain different sparsities of transmitters and receivers [59, 60]. The proposed algorithm for joint transmit/receive beamformer design is elaborated further in Algorithm 1.

10.2.3 Simulation

In this section, we demonstrate the effectiveness of our proposed sparse MIMO radar from the perspective of output SINR. We compare the performance of the optimal MIMO array transceiver configured according to the proposed algorithm with randomly configured MIMO arrays. In practice, the interference may be caused by co-existence in the same bandwidth or being deliberately positioned at some angles transmitting the same waveform as targets of interest. We consider both kinds of interferences.

10.2.3.1 Example 1

In this example, we fix the direction of the interferences, and change the arrival angle of the target from 0° to 90°. We consider a full uniform linear MIMO array consisting of eight transmitters ($M = 8$) and eight receivers ($N = 8$). We select $M_t = 4$ transmitters and $N_r = 4$ receivers. For the receiver, we set the minimum spacing of the sensors to $\lambda/2$, while for the transmitter, we set the minimum spacing of the sensors to $N\lambda/2$. Suppose there are two interferences with an interference to noise-ratio (INR) of 13 dB and arrival angles of $\theta_q = [40°, 70°]$. The SNR of the desired signal is fixed at 20 dB. We simulate co-existing interferences and deliberate interferences. The output SINR versus target angle is shown in Fig. 10.5. In this figure, case 1 corresponds to two deliberate interferences, whereas case 2 represents two co-existing interferences. It can be seen that, in both cases, the optimal sparse MIMO arrays exhibit better performance than randomly selected sparse arrays. Figure 10.6, depicts the configuration of the optimal MIMO array for a target at 80°.

10.2.3.2 Example 2

In this example, we consider the scenario where the interferences are spatially close to the target which causes a great adverse impact on the array performance. We change the angle of the target

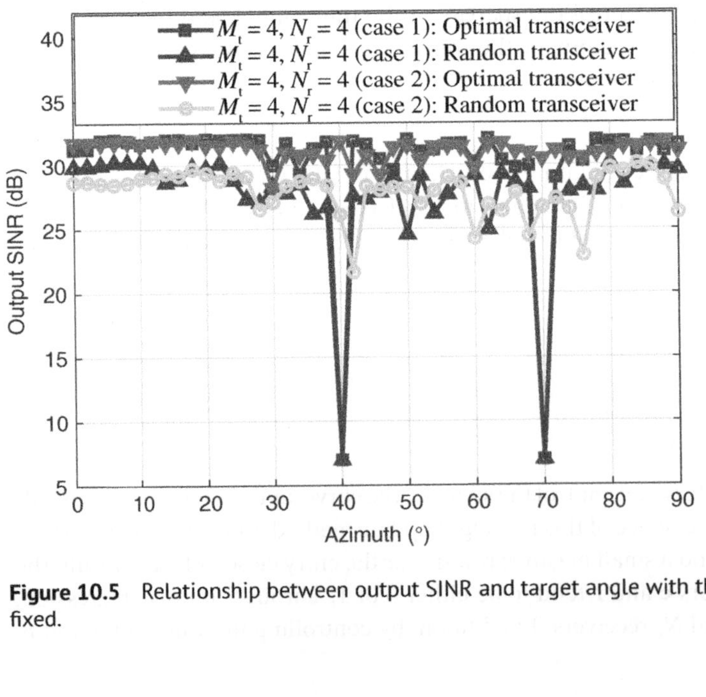

Figure 10.5 Relationship between output SINR and target angle with the directions of interferences being fixed.

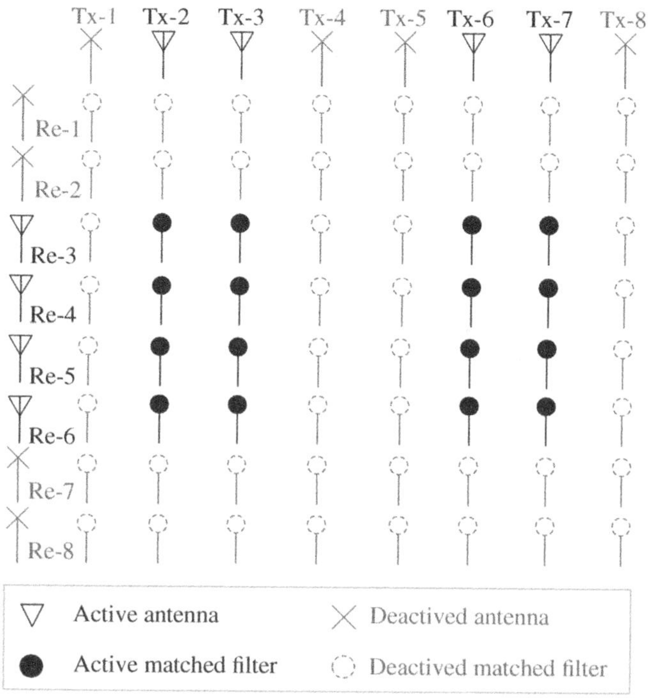

Figure 10.6 Optimal MIMO transceiver configuration when the target is from 80°.

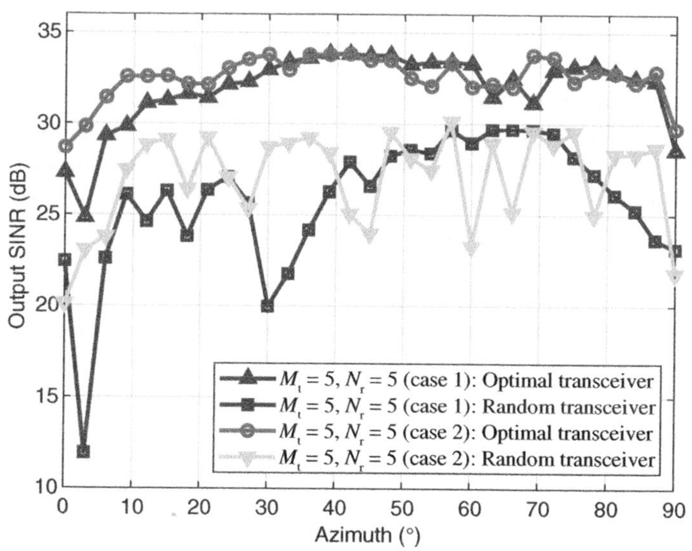

Figure 10.7 Relationship between the output SINR and the target angle when the interferences are closed to the target.

from 0° to 90°. For each target angle, two interferences are generated from the proximity of ±5° away from the target. We consider a full uniform linear MIMO array consisting of 20 transmitters ($M = 20$) and 20 receivers ($N = 20$). We select five transmitters and five receivers to compose the sparse MIMO array. The other simulation parameters remain the same as those in example 1. Again, we plot the output SINR versus target angle, as shown in Fig. 10.7. In case 1, the two interferences are deliberate, whereas in case 2, both interferences are co-existing. It can be observed that when the interferences are spatially close to the target, the proposed optimal MIMO sparse array exhibits more superiority than randomly configured sparse MIMO arrays.

10.3 Cognitive-Driven Optimization of Sparse Transceiver for Adaptive Beamforming

In Section 10.2, the optimum sparse MIMO array transceiver design is performed assuming that the environmental information is completely known. This may pose challenging problems for practical implementation, especially in a dynamic operating environment. In such environment, a cognitive MIMO radar would be capable of adjusting system parameters adaptively by sensing and learning. Beamforming performance of MIMO radar is guided by both beamforming weight coefficients and the transceiver configuration. In this section, we examine a cognitive-driven MIMO array design where both the beamforming weights and the transceiver configuration are adaptively and concurrently optimized under different environmental conditions. The perception–action cycle involves data collection of full virtual array, covariance reconstruction and joint design of the transmit and receive arrays by antenna selections. The optimal transceiver array design is realized by promoting two-dimensional group sparsity via iteratively minimizing reweighted mixed $l_{2,1}$-norm, with constraints imposed on transceiver antenna spacing for proper transmit/receive isolation. Simulations are provided to demonstrate the "perception–action" capability of the cognitive sparse MIMO array in achieving enhanced beamforming and anti-jamming in dynamic target and interference environment.

MIMO radars enable superior capabilities compared with standard phased arrays radar. Cognitive radar continuously interacts with the environment and update the radar parameters through the acquired knowledge [30, 31]. In MIMO radar, "perception–action" cognition approach typically entails adaptive optimizations of the transmit waveforms and power allocation [32–34]. In addition to these parameters, array reconfiguration (hardware) can significantly improve system performance beyond that achieved with fixed antenna positions [2, 35]. Motivated by this fact, we propose a cognitive-driven sparse MIMO array design method that incorporates sparse transceiver array optimization. The schematic system diagram is shown in Fig. 10.8, where a fully-connected switching network is employed for cognitive array reconfiguration. The task is to continuously and cognitively select different subsets of antennas from a large array to deliver the best performance under time-varying environment.

Sparse array design, aided by emerging fast antenna switching technologies, can lower the overall system cost by reducing the number of expensive front-end processing channels. The conventional MIMO radar has a sparse transceiver array. The inter-element spacing of the receive array is half wavelength, whereas that of the transmit array is multiple wavelengths. Hence, the MIMO sum coarray is a compact ULA with a large virtual aperture that enables high spatial resolution [4–6]. This configuration, however, may not render an optimum beamforming in terms of maximizing the output signal-to-interference-plus-noise ratio (MaxSINR) [7, 8]. Sparse MIMO array beamforming

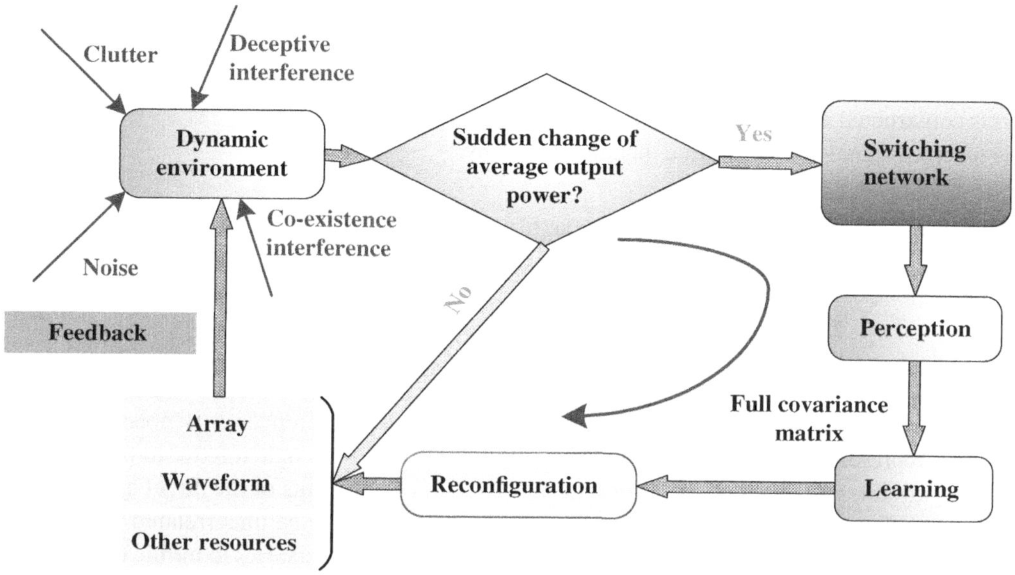

Figure 10.8 Schematic diagram of cognitive MIMO Radar.

design can achieve optimization of the transmit sensor locations and the corresponding waveform correlation matrix for synthesizing a desired beampattern [9–11]. This approach, however, essentially pursues an decoupled transmit/receive design. In this chapter, we pursue the coupled design approach and seek optimum sparse transceiver, configuring the MaxSINR beamformer at the MIMO radar receiver. We assume a colocated MIMO radar platform with orthogonal transmit waveforms. The adaptive beamforming is implemented on the virtual array after matched filtering. In the proposed design, the transceiver array configuration and beamforming weights are concurrently optimized to seek the best output SINR performance. This requires a cognitive operation where the present sensed system information is utilized to determine the next system parameters.

In the first step of the proposed cognitive beamforming, the MIMO radar senses the environment by sequentially switching to different sets of antennas for data collection, based on which a full covariance matrix of the large virtual array is constructed. In the second step of learning, a constrained design optimization problem is formulated and solved to simultaneously provide the configurations of both the transmit and receive arrays associated with MaxSINR. The radar will then act by switching on the selected transmit and receive antennas, and applying the resultant optimum beamforming weights after matched filtering. It is noted that cycle is triggered each time the output SINR degrades from a certain threshold performance, thereby making antenna selection a cognitive operation. A reweighted mixed $l_{2,1}$-norm minimization is employed to promote a two-dimensional group sparsity for MIMO transceiver design. Moreover, for practical consideration, the optimum cognitive MIMO radar is designed under antenna isolation constraints to control the minimum spacing between transmitter and receiver antennas.

10.3.1 Full Covariance Construction

In the first step, the MIMO radar senses the RF environment by sequentially switching to different sets of sparse arrays. This action enables data collection and a follow-on construction of the full covariance matrix corresponding to the virtual transceiver after matched filtering. Consider a cognitive MIMO radar with a total of M antennas. Later, only M_t transmit antennas and N_r (in this section, we assume that $M_t = N_r$) receive antennas will be optimally selected from the M available antennas. The full covariance matrix corresponding to co-located M-antenna transmit and receive arrays is constructed as follows.

The received signal at the nth time instant is given by,

$$\mathbf{x}(n) = \eta_0(n)\mathbf{a}_R(\theta_0)\mathbf{a}_T^T(\theta_0)\mathbf{s}(n) + \sum_{l1=1}^{L_1} j_{l_1}(n)\mathbf{a}_R(\theta_{j,l_1}) + \sum_{l2=1}^{L_2} \eta_{l_2}(n)\mathbf{a}_R(\theta_{j,l_2})\mathbf{a}_T^T(\theta_{j,l_2})\mathbf{s}(n) + \mathbf{v}(n), \tag{10.25}$$

where $\mathbf{s}(n) \in \mathbb{C}^{M\times 1}$ represents the M orthogonal transmit signals, L_1 represents the number of co-existing interferences, L_2 represents the number of deceptive interferences, θ_0 denotes the target angle, $j_{l_1}(n)$ represents the l_1th co-existing interference whose arrival angle relative to the receiver is θ_{j,l_1}. The variables $\eta_0(n)$ and $\eta_{l_2}(n)$ represent the reflection coefficients of the target and the l_2th deceptive interference, or spoofer, respectively, which are assumed to be uncorrelated and follow complex Gaussian distribution. Deceptive interference is used by adversaries to mimic target echo to interfere with target detection and thus exhibits the same waveform with radar transmit signal. The arrival angle of the l_2th deceptive interference is θ_{j,l_2}, $\mathbf{v}(n) \in \mathbb{C}^{M\times 1}$ is additive white Gaussian noise. Also, $\mathbf{a}_T(\theta)$ and $\mathbf{a}_R(\theta)$ represent the steering vectors of the transmit and receive arrays which are defined in the same way as (10.10) and (10.11).

After matched filtering, we obtain the data matrix $\mathbf{Y}(m) \in \mathbb{C}^{M\times M}$ for the mth radar pulse.

$$\mathbf{Y}(m) = \eta_0(m)\mathbf{a}_R(\theta_0)\mathbf{a}_T^T(\theta_0)\mathbf{R} + \sum_{l1=1}^{L_1} \mathbf{a}_R(\theta_{j,l_1})\mathbf{i}_{l_1}^T(m) + \sum_{l2=1}^{L_2} \eta_{l_2}(m)\mathbf{a}_R(\theta_{j,l_2})\mathbf{a}_T^T(\theta_{j,l_2})\mathbf{R} + \mathbf{V}'(m), \tag{10.26}$$

where $\mathbf{R} = \sigma_s^2\mathbf{I}$ is the auto-correlation matrix of M orthogonal transmit signals with equal power of σ_s^2, $\mathbf{i}_{l_1}(m) = \sum_n j_{l_1}(n)\mathbf{s}^*(n)$ is the output of matched filtering of the l_1th interference. Since the interference is not related to the transmit waveforms, $\mathbf{i}_{l_1}(m)$ still follows Gaussian distribution. We assume that the reflection coefficients obey the Swerling II target model, i.e., they change from one pulse repetition period to another. Vectorizing the data matrix $\mathbf{Y}(m)$, we generate the data vector received by the virtual array,

$$\begin{aligned}\mathbf{y}(m) &= \text{vec}(\mathbf{Y}(\mathbf{m})) \\ &= \eta_0(m)\sigma_s^2\mathbf{b}(\theta_0) + \sum_{l1=1}^{L_1} \mathbf{a}_R(\theta_{j,l_1}) \otimes \mathbf{i}_{l_1}(m) \\ &\quad + \sum_{l2=1}^{L_2} \eta_{l_2}(m)\sigma_s^2\mathbf{b}(\theta_{j,l_2}) + \mathbf{v}'(m),\end{aligned} \tag{10.27}$$

where the defination of $\mathbf{b}(\theta_i)$ is given in (10.9). The covariance matrix of the full virtual array is given by,

$$\mathbf{R}_x = E\left\{\mathbf{y}(m)\mathbf{y}^H(m)\right\} \approx \frac{1}{T}\sum_{m=1}^{T} \mathbf{y}(m)\mathbf{y}^H(m) \tag{10.28}$$

where a total of T pulses is assumed for sensing. From Eqs. (10.26), (10.27), and (10.28), we can observe that in order to obtain the covariance matrix of the full virtual array, we need to find the data matrix $\mathbf{Y}(m)$.

As the data matrix $\mathbf{Y}$ corresponds to the full virtual array, it is a challenging problem to construct $\mathbf{Y}$ using a sparse array with a small number of antennas. In the sequel, we propose a time-multiplexing method to sequentially switch on different sets of antennas for data collection and followed full covariance construction. Matrix $\mathbf{Y}$ is a Hankel matrix, that is, the elements along each anti sub-diagonal are equal. Thereby, we only need to estimate the elements on the first row and the last column to infer all elements of the full matrix.

Suppose we select M_t transmitters and M_t receivers from the M-antenna array, respectively. In this case, the M-antenna array is divided into K subarrays and each subarray comprises M_t consecutive antennas as shown in Fig. 10.9; thus $M = KM_t$. Accordingly, we divide the matrix $\mathbf{Y}$ into K^2 sub-matrices with the same size of $M_t \times M_t$ as follows,

$$\mathbf{Y} = \begin{bmatrix} \mathbf{Y}_{11} & \mathbf{Y}_{12} & \cdots & \mathbf{Y}_{1K} \\ \mathbf{Y}_{21} & \mathbf{Y}_{22} & \cdots & \mathbf{Y}_{2K} \\ \cdots & \cdots & \cdots & \cdots \\ \mathbf{Y}_{(K-1)1} & \mathbf{Y}_{(K-1)2} & \cdots & \mathbf{Y}_{(K-1)K} \\ \mathbf{Y}_{K1} & \mathbf{Y}_{K2} & \cdots & \mathbf{Y}_{KK} \end{bmatrix} \tag{10.29}$$

Utilizing the Hankel matrix structure, if the $2K - 1$ submatrices on the first row and the Kth column are estimated, we can then recover the entire matrix $\mathbf{Y}$, and subsequently obtain the covariance matrix using Eqs. (10.27) and (10.28). The $2K - 1$ submatrices can be estimated via multiplexing to different sets of antennas according to the following procedure within one radar pulse.

- For submatrix $\mathbf{Y}_{1i} (2 \leq i \leq K)$, we use the first M_t antennas as the receiving array, and use the ith group containing from $[(i-1)M_t + 1]$th to (iM_t)th antennas to transmit M_t orthogonal waveforms. After matched filtering at the receiver, we can estimate matrix $\mathbf{Y}_{1i}$, and the other matrices in the first row by sequentially multiplexing to next transmit array.
- For submatrix $\mathbf{Y}_{jK} (2 \leq j \leq K-1)$, we use the $[(K-1)M_t + 1]$th to (KM_t)th antennas as the transmit array, and switch the receive array sequentially from $(j-1)M_t + 1$ to jM_t. After matched filtering at the receiver, we can estimate matrix $\mathbf{Y}_{jK}$, and the other matrices in the Kth column.
- For submatrix $\mathbf{Y}_{11}$, we take the 1st to M_tth antennas as the receiving array and the $M_t + 1$th to $(2M_t)$th antennas as the transmitting array. The set of orthogonal transmit waveforms are then phase rotated by $e^{j(2\pi/\lambda)(M_t d)\cos\theta_0}$, which is equivalent to the selection of the transmit array composed of the 1st to M_tth antennas. After matched filtering at the receiver, we can estimate matrix $\mathbf{Y}_{11}$. The matrix $\mathbf{Y}_{KK}$ can be estimated in the same way.

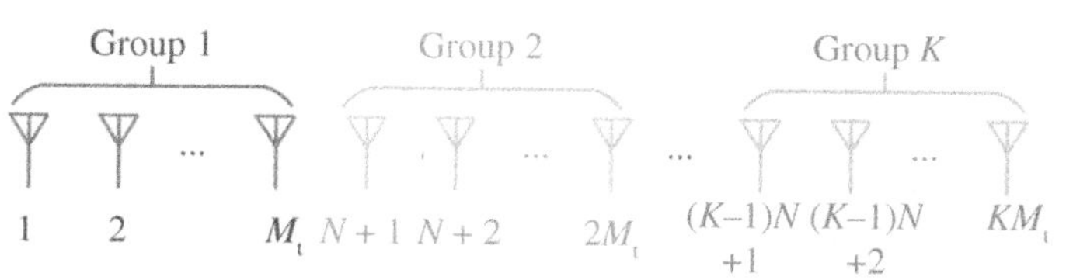

Figure 10.9 Division of the large M-antenna uniform linear array into K groups.

10.3.2 Optimal Transceiver Design

After obtaining the covariance matrix of the full virtual array, the next step for cognitive MIMO radar is to determine the optimal transceiver, including both the beamforming weights and array configurations.

10.3.2.1 Beamforming for MIMO Radar

When the M antennas are used for both the transmit and receive arrays, as the case in Section 10.3.1, the output of the MIMO receiver is given by,

$$z(m) = \mathbf{w}^H\mathbf{y}(m), \tag{10.30}$$

where beamforming weight vector $\mathbf{w} \in \mathbb{C}^{M^2\times 1}$ is applied to the virtual array at the receiver after matched filtering. The optimal of MaxSINR beamformer is referred to as Capon beamfomrer, which can be obtained by minimizing the noise and interference power without weakening the desired signal [29, 56, 61]. That is,

$$\begin{aligned} \underset{\mathbf{w}\in\mathbb{C}^{M^2\times 1}}{\text{minimize}} \quad & \mathbf{w}^H\mathbf{R}_\mathbf{x}\mathbf{w}, \\ \text{s.t.} \quad & \mathbf{w}^H\mathbf{b}(\theta_0) = 1, \end{aligned} \tag{10.31}$$

where s.t. represents "subject to." The solution of the above constrained optimization problem is given by,

$$\mathbf{w}_\text{o} = [\mathbf{b}^H(\theta_0)\mathbf{R}_x^{-1}\mathbf{b}(\theta_0)]^{-1}\mathbf{R}_x^{-1}\mathbf{b}(\theta_0). \tag{10.32}$$

10.3.2.2 Sparse Transceiver Design

For MIMO radar, beamforming in the receiver is implemented on the M^2-sensor virtual array after matched filtering, as shown in Fig. 10.10. There are M virtual sensors associated with each transmit or receive physical sensor. If a sensor is not activated, it implies that none of the M virtual sensors associated with this sensor is selected. When we leverage the sparsity of beamforming weight vector to design sparse arrays, it suggests that the M consecutive weights vertically or horizontally are all zeros. For example, in Fig. 10.10, the first sensor is not activated for waveform transmission, so none of the virtual sensors in the first column are selected. Similarly, the second sensor is not activated for receiving signals, so all the virtual sensors in the first row are discarded.

It can be inferred from the above analysis that a two-dimensional group sparsity promoting method can be employed to design the sparse transceiver configuration of MIMO radar. Suppose we choose M_t out of the M sensors as the transmit array and another M_t different sensors as the receiving array. When a sensor is activated, there is at least one nonzero weight coefficient related to the corresponding M virtual sensors. On the contrary, when a sensor is not activated, the weights of the corresponding M virtual sensors all assume zero values. Thus, the sensor selection scheme needs to warrant that the weights of the entire row or column of virtual sensors in Fig. 10.9 are either all zero or have at least one nonzero entry.

The beamforming weight vectors are generally complex valued, whereas the quadratic function in Eq. (10.31) are real. This observation allows expressing the problem with only real variables, which is typically accomplished by converting the problem from the complex domain to the real

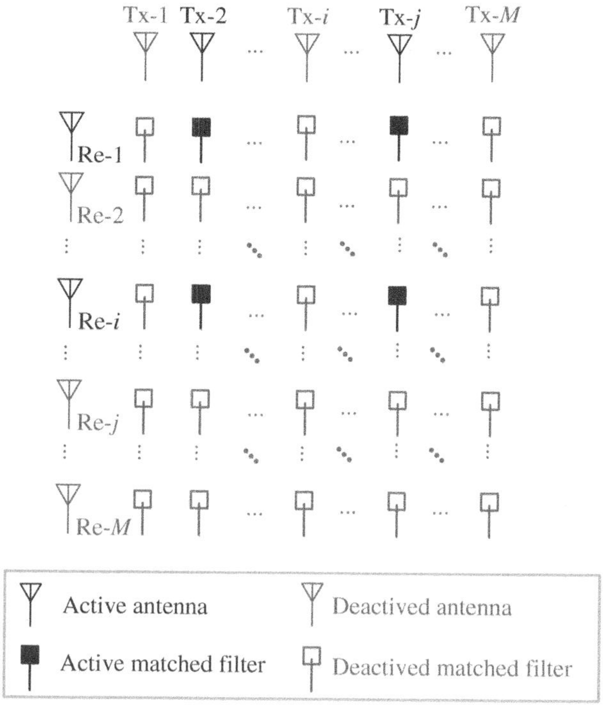

Figure 10.10 Schematic diagram of virtual sensor for cognitive MIMO radar and the illustration of two-dimensional group sparsity.

domain and concatenating the vectors accordingly,

$$\tilde{\mathbf{R}}_x = \begin{bmatrix} \text{real}(\mathbf{R}_x) & -\text{imag}(\mathbf{R}_x) \\ \text{imag}(\mathbf{R}_x) & \text{real}(\mathbf{R}_x) \end{bmatrix}, \tilde{\mathbf{w}} = \begin{bmatrix} \text{real}(\mathbf{w}) \\ \text{imag}(\mathbf{w}) \end{bmatrix},$$
$$\tilde{\mathbf{b}}(\theta_0) = \begin{bmatrix} \text{real}(\mathbf{b}(\theta_0)) \\ \text{imag}(\mathbf{b}(\theta_0)) \end{bmatrix}, \tag{10.33}$$

where $\tilde{\mathbf{R}}_x \in \mathbb{R}^{2M^2 \times 2M^2}$, $\tilde{\mathbf{w}} \in \mathbb{R}^{2M^2 \times 1}$ and $\tilde{\mathbf{b}}(\theta_0) \in \mathbb{R}^{2M^2 \times 1}$ are the real domain vectors of $\mathbf{R}_x$, $\mathbf{w}$, and $\mathbf{b}(\theta_0)$, respectively. Therefore, the sensor selection problem of sparse array construction can be described as the following optimization,

$$\underset{\tilde{\mathbf{w}},\mathbf{c},\mathbf{r}}{\text{minimize}} \quad \tilde{\mathbf{w}}\tilde{\mathbf{R}}_x\tilde{\mathbf{w}} \tag{10.34}$$

$$\text{s.t.} \quad \tilde{\mathbf{w}}^H\tilde{\mathbf{b}}(\theta_0) = 1, \tag{10.34a}$$

$$||\mathbf{P}_i \odot \tilde{\mathbf{w}}||_2 = c_i, \quad i = 1, \ldots, M \tag{10.34b}$$

$$||\mathbf{c}||_0 = M_t, \tag{10.34c}$$

$$||\mathbf{Q}_j \odot \tilde{\mathbf{w}}||_2 = r_j, \quad j = 1, \ldots, M \tag{10.34d}$$

$$||\mathbf{r}||_0 = M_t, \tag{10.34e}$$

$$\mathbf{c}^H\mathbf{r} = 0, \tag{10.34f}$$

where $\odot$ denotes the element-wise product, $\mathbf{P}_i$ and $\mathbf{Q}_j$ are the transmit selection matrix and the receive selection matrix respectively which are defined in (10.21) and (10.22). Constraints (10.34c) and (10.34e) indicate that M_t sensors are selected for each transmitting and receiving array, and $\|.\|_0$ denotes the l_0 norm. Constraint (10.34f) implies that the transmitting array and receiving array do not share any sensors.

10.3.2.3 Reweighted $l_{2,1}$-norm

The l_0-norm constraints in (10.34c) and (10.34e) are nonconvex, which renders the above optimization problem difficult to solve. We utilize the l_1-norm to approximate the l_0-norm [62]. To further promote sparsity, an iterative reweighted l_1-norm was proposed in [58]. Based on this idea, we employ an iterative reweighted mixed $l_{2,1}$-norm to promote two-dimensional group sparsity, which can be described as follows,

$$\underset{\tilde{\mathbf{w}},\mathbf{c},\mathbf{r}}{\text{minimize}} \quad \tilde{\mathbf{w}}\tilde{\mathbf{R}}_x\tilde{\mathbf{w}} + \alpha(\mathbf{p}^H\mathbf{c}) + \beta(\mathbf{q}^H\mathbf{r}) \tag{10.35}$$

$$\text{s.t.} \quad \tilde{\mathbf{w}}^H\tilde{\mathbf{b}}(\theta_0) = 1, \tag{10.35a}$$

$$\|\mathbf{P}_i \odot \tilde{\mathbf{w}}\|_2 \leq c_i, \tag{10.35b}$$

$$0 \leq c_i \leq 1, \quad i = 1, \dots, M \tag{10.35c}$$

$$\|\mathbf{Q}_j \odot \tilde{\mathbf{w}}\|_2 \leq r_j, \tag{10.35d}$$

$$0 \leq r_j \leq 1, \quad j = 1, \dots, M \tag{10.35e}$$

$$\mathbf{1}_M^H\mathbf{c} = M_t, \quad \mathbf{1}_M^H\mathbf{r} = M_t, \tag{10.35f}$$

$$\mathbf{c}^H\mathbf{r} = 0, \tag{10.35g}$$

where α and β are two parameters to balance between the array sparsity and output SINR of the transceiver. A common method of updating the reweighting coefficient is to take the reciprocal of the absolute weight value. However, this fails to control the number of elements to be selected. Thus, similar to Section 10.2.2.2, in order to control the number of selected antennas, we update the reweighting coefficient $\mathbf{p}$ and $\mathbf{q}$ in the same way as (10.23) and (10.24) for each iteration.

In practical implementation, a close distance between the transmit and receiving antennas will cause power leakage and unwanted coupling effect. To achieve high isolation between the transmit and receive channels, which is always desirable, the selected transmit and receive antennas should be sufficiently spatially separated. Therefore, for the transceiver array design, the two positions neighboring the selected transmitting antenna should be vacant and not hosting any receive antenna. To satisfy this requirement, we replace the constraint in (10.35h) with the following new constraints,

$$\begin{aligned} &c_i + r_{i-1} + r_i + r_{i+1} \leq 1, \quad i = 2, \dots, M-1, \\ &c_1 + r_1 + r_2 \leq 1, \\ &c_M + r_M + r_{M-1} \leq 1. \end{aligned} \tag{10.36}$$

10.3.3 Optimized Transceiver Reconfiguration

We obtain the optimal transceiver configuration in the current environment through sensing and learning. In the action phase, and according to the two vectors **c** and **r** from Eq. (10.35), we switch on the selected M_t antennas indicated by **c** to transmit the M_t orthogonal waveforms, and switch on the selected M_t receiving antennas indicated by **r** to receive data. After matched filtering, we perform beamforming with the weights **w** obtained from Eq. (10.35). Consequently, a new transceiver array will be reconfigured to adapt to the new environment for performance improvement. When the environment changes, the output SINR decreases, instigating the beginning of another "perception–action" cycle.

10.3.4 Simulations

In this Section, we demonstrate the effectiveness of the proposed cognitive-driven sparse transceiver design for MIMO radar in different scenarios.

10.3.4.1 Example 1

In this example, we simulate the work flow of the proposed cognitive MIMO radar and evaluate its performance. We assume that there are a total of 18 antennas with an inter-element spacing of $\lambda/2$, from which $M_t = 4$ antennas are selected for transmitting and another $M_t = 4$ antennas for receiving. In the first part of the simulation, that is, before time t_1, the target is at angle 65° with the signal-to-noise ratio (SNR) is 20 dB. There are two deceptive interferences with an interference-to-noise ratio (INR) of 15 dB located at angle 50° and 60°, respectively. At a certain moment t_1, the environment suddenly changes, and another interference close to the target at 63° is switched on. We can observe that the output SINR decreases abruptly starting from time instant t_1, as shown in Fig. 10.11. This triggers the cognitive mechanism of the MIMO radar, which re-initiates sensing and then learning from time t_1 to time t_2. It reconfigures the optimal transceiver arrays responding to the environmental change at time t_2. Accordingly, the output

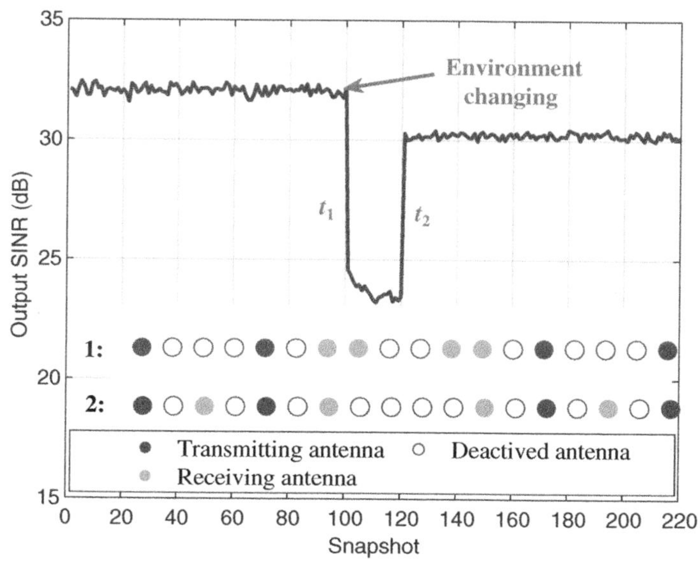

Figure 10.11 Response of cognitive MIMO radar to environmental changes.

SINR is now maximized under the new environment. The two optimum transceiver configurations 1 and 2 before and after the environmental change are shown in Fig. 10.11. Evidently, the selection of transmit antennas remains the same, while the configuration of receiving array has significantly changed. In this experiment, the output SINR of the new environment is slightly lower than that of the initial environment, which is due to the additional interference closed to the expected signal.

10.3.4.2 Example 2

In this example, we compare the performance of the optimal transceiver with that of conventional transceiver that comprises antennas indexed by 1, 5, 9, and 13 for transmitting and antennas indexed by 15, 16, 17, and 18 for receiving. Two deceptive interferences are impinging on the array from 60° and 70°, respectively, with an INR of 15 dB. The scanning angle is changing from 0° to 90° and the SNR is set as 20 dB. Other parameters remain the same as in example 1. We configure an optimal transceiver for each scanning angle. The curve of output SINR versus scanning angle is plotted in Fig. 10.12, where we can observe that when the interferences are widely separated from the target, the superiority of the proposed optimal transceiver is not obvious. On the other hand, when the interferences move closer to the target, the performance of the optimal transceiver exhibits noticeable improvement compared to that of conventional MIMO array.

10.3.4.3 Example 3

From example 2, we can argue that the optimal cognitive MIMO radar behaves well when the interferences are close to the target, which is difficult to deal with in practice. Thus, in this example, we focus on this specific scenario. For each scanning angle, we investigate two cases, where the incident angles of two deceptive interference are ±5° relative to the target in case 1 and then ±3° in case 2. Again, we plot the curve of output SINR versus the scanning angle, as shown in Fig. 10.13.

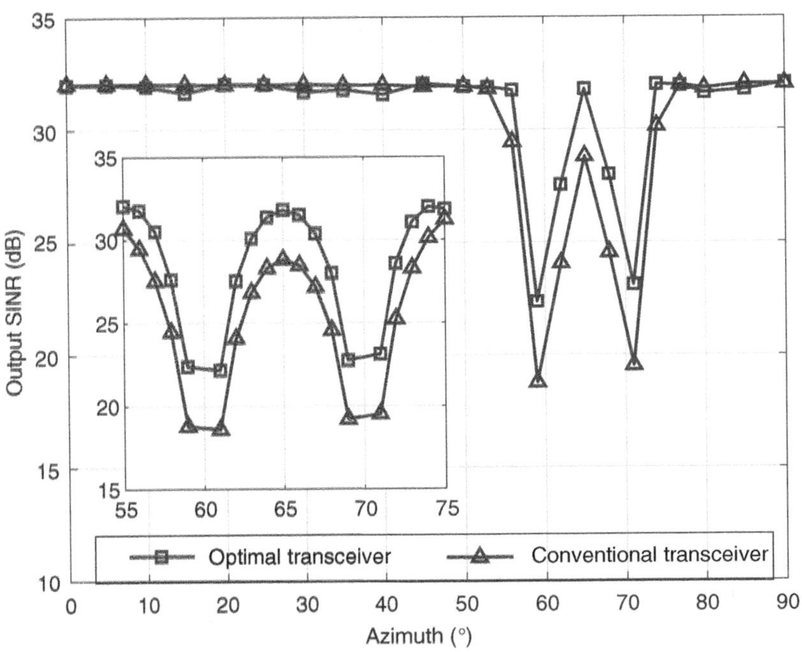

Figure 10.12 Relationship between output SINR and target angle with the directions of interferences being fixed.

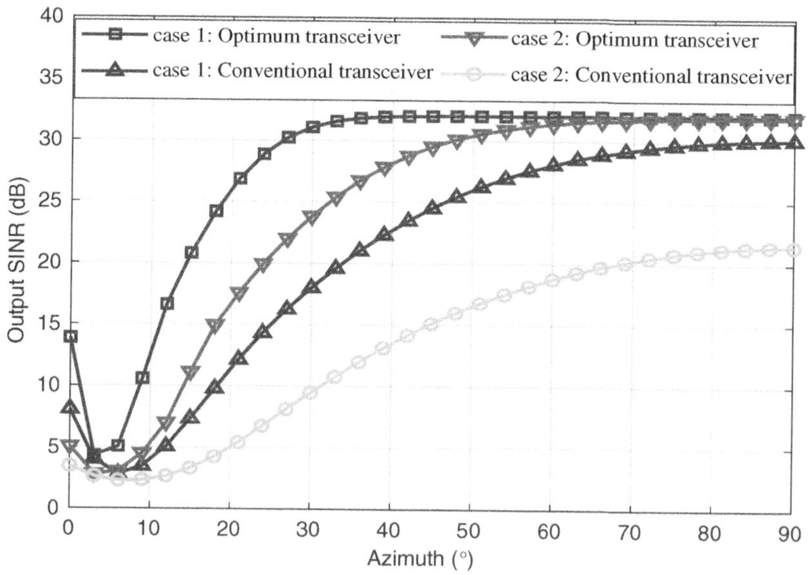

Figure 10.13 Relationship between the output SINR and the target angle when the interferences are close to the target.

It can be observed that the optimal transceiver array selected by the optimal cognitive MIMO radar clearly improves the performance. Moreover, the closer the interference is relative to the target, the higher the improvement. Additionally, there is an interesting phenomenon as seen from Fig. 10.13, that the linear MIMO transceiver behaves worse in the endfire direction compared with the broadside direction. The reason is that the beampatterns of linear arrays are broadened in the endfire direction, thus limiting their spatial resolution.

10.3.4.4 Example 4

In this example, we examine the effect of the number of candidate antennas and the antenna spacing on the output performance of our optimal sparse transceivers. The output SINR versus the scanning angle is depicted in Fig. 10.14. Again, the linear array exhibits degraded performance when steering in the endfire direction. This phenomenon can be ameliorated by increasing the total array aperture. We can see that when the sensor spacing is $\lambda/2$, the performance of the optimal sparse transceiver improves with increased number of candidate antennas. For a fixed number of antennas, increasing the inter-element spacing improves the performance of the optimal sparse transceiver at lower scanning angle. However, the performance becomes unstable when scanning to broadside direction.

10.4 Sparse MIMO Transceiver Design for Multi-source DOA Estimation

Sections 10.2 and 10.3 discuss the optimum sparse MIMO array transceiver design in terms of MaxSINR. In this section, we proceed to examine another important function of MIMO arrays, that is DOA estimation. As mentioned in Section 10.1, the sum-coarray of a co-located orthogonal MIMO array is a large virtual array with MN virtual antennas, and thus the number of estimated

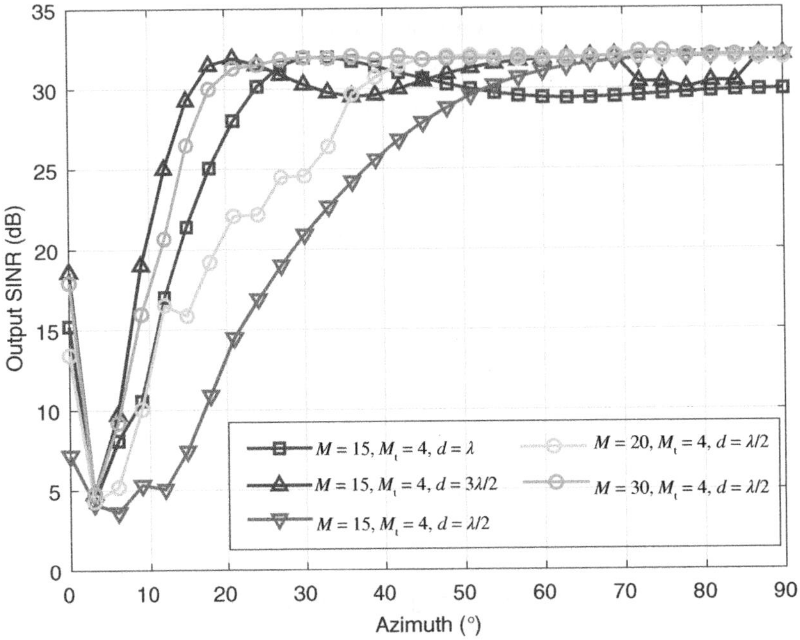

Figure 10.14 The influence of the total number of sensors and the antenna spacing on the output SINR of optimal sparse transceiver array.

sources can be dramatically increased to MN. Usually, both transmit and receive arrays are uniform, while the inter-element spacing of transmit array is set as $d_{\rm t} = Nd_{\rm r}$ with $d_{\rm r}$ denoting the inter-element spacing of the receive array. Such design of MIMO array transceiver can generate an MN-element linear uniform virtual array, in turn enabling various kinds of DOA estimation algorithms. However, this MIMO array configuration does not necessarily lend itself to high accuracy. As shown in [49], the accuracy of DOA estimation depends on the receive array configuration. In this section, we examine the problem of sparse MIMO array transceiver design for enhanced multi-source DOA estimation.

10.4.1 Cramer–Rao Bound of Multi-source DOA Estimation

Consider a collocated MIMO radar system where the transmitter and receiver are closely separated compared to the distance to the far-field sources. The transmitter comprises a ULA of M antennas and the receiver a ULA of N antennas. Let $d_{\rm t}$ and $d_{\rm r}$ denote the inter-element spacing of the transmit and receive arrays respectively. The MIMO radar transmits a set of M linearly independent signals, $\mathbf{s}(t) = \{s_1(t), \dots, s_M(t)\}$, with a covariance matrix satisfying the orthogonality condition, that is,

$$\mathbf{R}_{\rm s} = \int_0^{T_{\rm p}} \mathbf{s}(t)\mathbf{s}^H(t) = \mathbf{I}, \tag{10.37}$$

where $T_{\rm p}$ is the pulse duration. The superscript "H" represents conjugate transpose.

Assume that K far-field narrow-band sources are located in directions $\boldsymbol{\theta} = [\theta_1, \dots, \theta_K]$. The received data vector can be written as,

$$\mathbf{x}(t, \tau) = \mathbf{A}_{\rm r}\text{diag}[\boldsymbol{\beta}(\tau)]\mathbf{A}_{\rm t}^T\mathbf{s}(t) + \mathbf{v}(t, \tau), \tag{10.38}$$

where t denotes the fast-time index (i.e., time within the radar pulse) and τ represents the slow-time index (i.e., pulse number). $\boldsymbol{\beta}(\tau) = [\beta_1(\tau), \ldots, \beta_K(\tau)]^T$ is the reflection coefficient which is rapidly changing, and may not satisfy the Swirling II model. The noise vector $\mathbf{v}(t,\tau) = [v_1(t,\tau), \ldots, v_N(t,\tau)]^T$ is modeled as a zero-mean, complex white Gaussian process with power σ_n^2. The transmitting array manifold matrix $\mathbf{A}_t$ is given by $\mathbf{A}_t(\theta) = [\mathbf{a}_T(\theta_1), \ldots, \mathbf{a}_T(\theta_K)]$, where the transmitting steering vector is defined by (10.10). Similarly, the receive array manifold $\mathbf{A}_r$ is denoted as $\mathbf{A}_r(\theta) = [\mathbf{a}_R(\theta_1), \ldots, \mathbf{a}_R(\theta_K)]$, and the receive steering vector is defined by (10.11).

Performing the matched-filtering to the received signal vector by the transmitted orthogonal waveforms, we obtain

$$\begin{aligned}\mathbf{X}(\tau) &= \int_0^{T_p} \mathbf{x}(t,\tau)\mathbf{s}^H(t)dt \\ &= \mathbf{A}_r(\theta)\mathrm{diag}[\boldsymbol{\beta}(\tau)]\mathbf{A}_t^T(\theta) + \mathbf{N}(\tau),\end{aligned} \tag{10.39}$$

where the additive noise is $\mathbf{N}(\tau) = \int_0^{T_p} \mathbf{v}(t,\tau)\mathbf{s}^H(t)dt$. The elements of the matrix $\mathbf{N}(\tau)$ are independent and identically, zero-mean Gaussian with variance σ_n^2.

Applying vectorization to Eq. (10.39), we obtain

$$\mathbf{y}(\tau) = \mathrm{vec}(\mathbf{X}(\tau)) = \mathbf{A}\boldsymbol{\beta}(\tau) + \mathbf{n}(\tau), \tag{10.40}$$

where $\mathbf{A} = [\mathbf{b}(\theta_1), \ldots, \mathbf{b}(\theta_K)]$ with the vector $\mathbf{b}(\theta_k)$ is defined in (10.9), $\mathbf{n}(\tau) = \mathrm{vec}\{\mathbf{N}(\tau)\}$ is the extended additive Gaussian noise term with zero mean and covariance $\sigma_n^2\mathbf{I}$. As mentioned above, when $d_t = Nd_r$, $\mathbf{b}(\theta_k)$ is the steering vector of a ULA with MN virtual antennas uniformly spaced by d_r.

Based on the signal model in Eq. (10.40), the CRB of multi-source DOA estimation is [63],

$$\mathrm{CRB} = \frac{\sigma_n^2}{2J}\left\{\mathrm{Re}[\{\mathbf{D}^H(\mathbf{I} - \mathbf{A}(\mathbf{A}^H\mathbf{A})^{-1}\mathbf{A}^H)\mathbf{D}\} \odot \mathbf{T}^T]\right\}^{-1}, \tag{10.41}$$

where J is the total number of snapshots and $\mathrm{Re}[\cdot]$ denotes the real part of a complex entry. The first-order derivative of $\mathbf{a}_i$ with respect to $\omega_i = \frac{2\pi}{\lambda}d_r\sin(\theta_i)$ is denoted as,

$$\mathbf{d}_i = \frac{\partial \mathbf{a}_i}{\partial \omega_i} = [0, je^{j\omega_i}, \ldots, j(MN-1)e^{j(MN-1)\omega_i}], \tag{10.42}$$

and $\mathbf{D} = [\mathbf{d}_1, \ldots, \mathbf{d}_K]$. The matrix $\mathbf{T}$ is the source covariance matrix, i.e.,

$$\mathbf{T} = E\{\boldsymbol{\beta}(\tau)\boldsymbol{\beta}^H(\tau)\}. \tag{10.43}$$

When sources are uncorrelated, the matrix $\mathbf{T}$ becomes diagonal with the power P_i, $i = 1, \ldots, K$ along the diagonal. Accordingly, the CRB is also a diagonal matrix with each diagonal element representing the DOA estimation accuracy of the corresponding detection target.

10.4.2 Sparse MIMO Array Transceiver Design in the Metric of CRB

We take the maximum value of CRBs of all estimated DOAs as the objective function for array thinning, i.e.,

$$\begin{aligned} f &= \max\{[\mathrm{CRB}]_{ii,1\le ii\le K}\}, \\ &= \max_{1\le i\le K}\left\{\frac{\sigma_n^2}{2N}\{P_i\mathbf{d}_i^H(\mathbf{I} - \mathbf{A}(\mathbf{A}^H\mathbf{A})^{-1}\mathbf{A}^H)\mathbf{d}_i\}^{-1}\right\}. \end{aligned} \tag{10.44}$$

Define the selection vector $\mathbf{z} \in \{0,1\}^{N_t N_r}$, where "1" denotes the corresponding virtual antenna is selected and "0" discarded. According to the properties of MIMO transceiver, "z" exhibits the characteristics of group sparsity. From the perspective of antenna selection, Eq. (10.44) can be rewritten as,

$$f(\mathbf{z}) = \max_{1 \le i \le K} \{P_i^{-1}[\mathbf{d}_i^H \text{diag}(\mathbf{z})\mathbf{d}_i - \mathbf{d}_i^H \text{diag}(\mathbf{z})\mathbf{A}(\mathbf{A}^H \text{ diag}(\mathbf{z})\mathbf{A})^{-1}\mathbf{A}^H \text{ diag}(\mathbf{z})\mathbf{d}_i]^{-1}\}. \tag{10.45}$$

The optimization problem is then formulated as,

$$\begin{aligned} \min \quad & f(\mathbf{z}), \\ \text{s.t.} \quad & \mathbf{z} \in \{0,1\}^{N_t N_r}, \\ & (1/N_2)\mathbf{p}_i^T \mathbf{z} \in \{0,1\}, i = 1, \dots, N_t, \\ & (1/N_1)\mathbf{q}_j^T \mathbf{z} \in \{0,1\}, j = 1, \dots, N_r, \\ & \|\mathbf{Pz}\|_0 = N_1, \\ & \|\mathbf{Qz}\|_0 = N_2. \end{aligned} \tag{10.46}$$

Note that the selection vector $\mathbf{z}$ exhibits a group sparsity structure, we define

$$\mathbf{p}_i = [\underbrace{0, \dots, 0}_{(i-1)N_r \text{ 0s}}, \underbrace{1, \dots, 1}_{N_r \text{ 1s}}, 0, \dots, 0]^T \in \mathbb{R}^{N_t N_r \times 1}, \tag{10.47}$$

$$\mathbf{q}_j = [\underbrace{0, \dots, 0, 1, 0, \dots, 0}_{\substack{\text{all elements are 0 except} \\ \text{for the } j\text{th element is 1} \\ (N_r \text{ elements})}}, \dots, \underbrace{0, \dots, 0, 1, 0, \dots, 0}_{\substack{\text{all elements are 0 except} \\ \text{for the } j\text{th element is 1} \\ (N_r \text{ elements})}}]^T \in \mathbb{R}^{N_t N_r \times 1}. \tag{10.48}$$

where $\mathbf{p}_i$ and $\mathbf{q}_j$ are the transmit and receive selection vectors of the ith transmit antenna and the jth receive antenna respectively, which are used to select all virtual matched filters related to the corresponding antenna. $\mathbf{P}$ and $\mathbf{Q}$ are matrices defined as $\mathbf{P} = [\mathbf{p}_1, \dots, \mathbf{p}_{N_t}]^T$ and $\mathbf{Q} = [\mathbf{q}_1, \dots, \mathbf{q}_{N_t}]^T$, respectively.

It is obvious that both the objective function and the cardinality constraints in the above problem are nonconvex. We will introduce some convex relaxation strategies to transform the original problem into a convex form.

As for the nonconvex objective function, we can transform it into the following equivalent linear matrix inequality (LMI) constraint optimization problem utilizing the Schur complement of a block matrix, that is,

$$\min_{\mathbf{z},\delta} \ -\delta, \tag{10.49}$$

$$\text{s.t.} \begin{bmatrix} \mathbf{A}^H \text{ diag}(\mathbf{z})\mathbf{A} & \mathbf{A}^H \text{ diag}(\mathbf{z})\mathbf{d}_i \\ \mathbf{d}_i^H \text{ diag}(\mathbf{z})\mathbf{A} & \mathbf{d}_i^H \text{diag}(\mathbf{z})\mathbf{d}_i - \frac{\delta}{P_i} \end{bmatrix} \succeq 0, 1 \le i \le K. \tag{10.49a}$$

The LMI constraint in Eq. (10.49a) implies that,

$$\begin{aligned} P_i^{-1}\{&\mathbf{d}_i^H \text{diag}(\mathbf{z})\mathbf{d}_i - \mathbf{d}_i^H \text{diag}(\mathbf{z})\mathbf{A}(\mathbf{A}^H \text{ diag}(\mathbf{z})\mathbf{A})^{-1} \\ &\mathbf{A}^H \text{ diag}(\mathbf{z})\mathbf{d}_i\}^{-1} \le \frac{1}{\delta}, i \in \{1, 2, \dots, K\} \end{aligned} \tag{10.50}$$

As for the nonconvex cardinality constraints, the binary of variable "z" can be promoted via an improved reweighted l_1 iteration method. In this case, for the mth iteration, (10.46) is transformed into,

$$\begin{aligned}
\min \quad & -\delta + \mu_1 \mathbf{w}_1^{(m)T}\mathbf{z}_1 + \mu_2 \mathbf{w}_2^{(m)T}\mathbf{z}_2 \\
\text{s.t.} \quad & 0 \le \mathbf{z} \le 1, \\
& (1/N_2)\mathbf{p}_i^T\mathbf{z} \in \{0,1\}, i = 1, \dots, N_t, \\
& (1/N_1)\mathbf{q}_j^T\mathbf{z} \in \{0,1\}, j = 1, \dots, N_r, \\
& \mathbf{z}_1 = (1/N_2)\mathbf{P}\mathbf{z},\ 0 \le \mathbf{z}_1 \le 1,\ \mathbf{1}^T\mathbf{z}_1 = N_1, \\
& \mathbf{z}_2 = (1/N_1)\mathbf{Q}\mathbf{z},\ 0 \le \mathbf{z}_2 \le 1,\ \mathbf{1}^T\mathbf{z}_2 = N_2, \\
& \begin{bmatrix} \mathbf{A}^H \operatorname{diag}(\mathbf{z})\mathbf{A} & \mathbf{A}^H \operatorname{diag}(\mathbf{z})\mathbf{d}_i \\ \mathbf{d}_i^H \operatorname{diag}(\mathbf{z})\mathbf{A} & \mathbf{d}_i^H \operatorname{diag}(\mathbf{z})\mathbf{d}_i - \frac{\delta}{P_i} \end{bmatrix} \succeq 0,\ i = 1, \dots, K.
\end{aligned} \tag{10.51}$$

where μ_1 and μ_2 are manually set trade-off parameters, $\mathbf{w}_1^{(m)}$ and $\mathbf{w}_2^{(m)}$ are reweighted coefficient vectors of the transmit antennas and receive antennas, respectively. Note that we relax the binary constraint on $\mathbf{z}$ to a box constraint $0 \le \mathbf{z} \le 1$. And then a reweighted l_1-norm is adopted for promote the boolean sparsity of the selection vector. For the mth iteration, the reweighted coefficient of the ith transmit antenna, that is the ith element of $\mathbf{w}_1$, is updated by,

$$(\mathbf{w}_1^{(m)})_i = \frac{1 - (\mathbf{z}_1^{(m-1)})_i}{1 - e^{-\beta_0 (\mathbf{z}_1^{(m-1)})_i} + \epsilon} - \left(\frac{1}{\epsilon}\right)((\mathbf{z}_1^{(m-1)})_i)^{\alpha_0}, \tag{10.52}$$

where $\mathbf{z}_1^{(m-1)}$ is calculated by the optimal solution of the $m-1$th iteration. Similarly, the reweighted coefficient of the jth receive antenna, that is the jth element of $\mathbf{w}_2$, is updated by,

$$(\mathbf{w}_2^{(m)})_j = \frac{1 - (\mathbf{z}_2^{(m-1)})_j}{1 - e^{-\beta_0 (\mathbf{z}_2^{(m-1)})_j} + \epsilon} - \left(\frac{1}{\epsilon}\right)((\mathbf{z}_2^{(m-1)})_j)^{\alpha_0}. \tag{10.53}$$

10.4.3 Simulations

In this section, we verify the DOA estimation performance of the proposed method. We consider the same transceiver model and interference environment as that of example 1 in Section 10.2.3. Similarly, we select four antennas from the full transmit array and four antennas from the full receive array to form the sparse transceiver. We fix the incident angle of the desired signal to be 80°, and then perform DOA estimation on all three signal sources (the desired signal and two interferences from 40° and 70°). By applying (10.51), the optimal structure obtained of the sparse transceiver is depicted in Fig. 10.15. As can be seen from Fig. 10.15, we select the 1st, 3rd, 6th, and 8th transmit antennas and the 2nd, 4th, 5th, and 7th receive antennas to form the optimal sparse transceiver. To further verify the performance of the optimal transceiver, we compare its estimation performance with another two sparse transceivers. We conduct 1000 Monte-Carlo simulations and calculate the mean square error (MSE) of the estimated DOAs under three different transceivers. The MSE versus input SNR curves are given in Fig. 10.16. In Fig. 10.16, for the curve marked with squares, we select the 1st, 2nd, 3rd and 4th transmit antennas and the 1st, 3rd, 5th and 7th receive antennas to form a sparse transceiver, which can generate a virtual ULA after matched filtering. For the curve marked with circles, we apply the transceiver depicted in Fig. 10.6. It can be seen from Fig. 10.16 that the estimation accuracy of the optimal sparse transceiver selected by the proposed algorithm is significantly higher than that of the other two cases. With increased SNRs of the input signals,

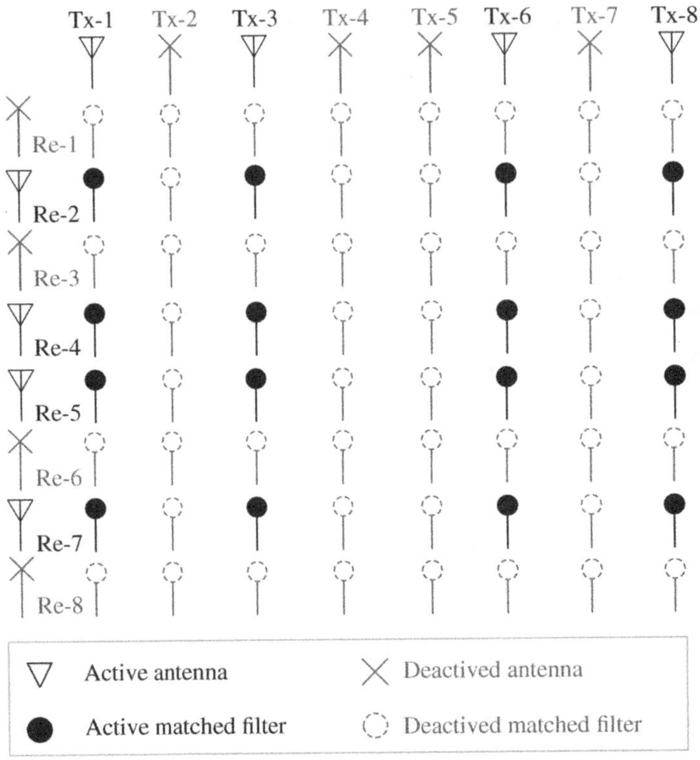

Figure 10.15 Optimal MIMO transceiver configuration selected by the proposed method.

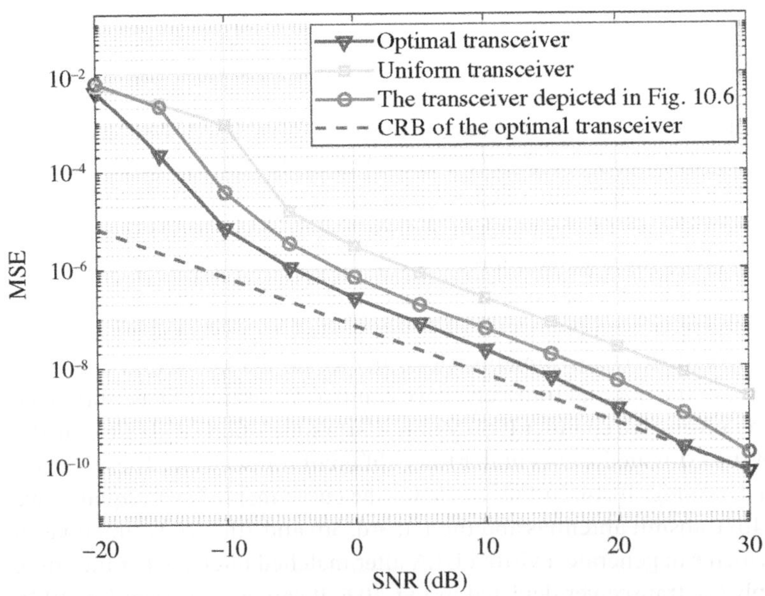

Figure 10.16 Relationship between MSE and input SNR for different transceivers.

the MSE of the optimal transceiver can gradually approach its CRB. This example shows that the optimum sparse MIMO array for MaxSINR does not necessarily preserve its optimality in the task of DOA estimation.

10.5 Conclusion

In this chapter, we examined the sparse MIMO array transceiver design for different tasks and under different criteria. The problem of sparse transceiver design for MIMO radar was considered under the assumption of prior knowledge of the source/interference environment. For a given number of transmitting and receiving antennas, a sparse MIMO array structure in terms of MaxSINR was jointly designed. It was shown by simulations that the proposed sparse MIMO transceiver selection algorithm provides superior interference suppression performance over randomly designed MIMO transceiver arrays. We proposed a novel cognitive MIMO radar which adaptively optimizes both beamforming weights and transceiver array configuration for MaxSINR beamforming in a dynamic environment. This cognitive-driven optimization framework can learn the environment automatically, thus eliminating the assumption of any prior knowledge. From the same large array, a given number of antennas were selected to construct the transmit and receive arrays jointly. The two arrays were not allowed to have overlapping or adjacent antennas for improved isolation. During perception in the underlying perception–action cycle, the data are collected by sequentially switching among different sets of antennas to construct the covariance matrix of a full virtual array. The simulations showed that the proposed cognitive-driven sparse MIMO array design can adaptively reconfigure the transceiver to maximize the beamforming performance with dynamically changing environments. In Section 10.4, we proceeded to examine another important function of MIMO arrays, namely, DOA estimation. We optimized the configuration of MIMO array transceiver from the perspective of minimizing the CRB. Through inequality scaling and first-order Taylor expansion, we transformed the original problem into an iterative convex optimization form. The simulation results showed that the transceiver optimized by this method can achieve higher DOA estimation accuracy in terms of minimum aggregate CRB. In all the optimization designs, a re-weighted mixed-norm reweighted regularization was utilized to promote two-dimensional group sparsity for sparse MIMO transceiver design.

References

1 Wang, X., Aboutanios, E., Trinkle, M., and Amin, M.G. (2014). Reconfigurable adaptive array beamforming by antenna selection. *IEEE Transactions on Signal Processing* 62 (9): 2385–2396.

2 Amin, M.G., Wang, X., Zhang, Y.D. et al. (2016). Sparse arrays and sampling for interference mitigation and DOA estimation in GNSS. *Proceedings of the IEEE* 104 (6): 1302–1317.

3 Nosrati, H., Aboutanios, E., and Smith, D. (2020). Multi-stage antenna selection for adaptive beamforming in MIMO radar. *IEEE Transactions on Signal Processing* 68: 1374–1389.

4 Liu, C. and Vaidyanathan, P.P. (2017). Correlation subspaces: generalizations and connection to difference coarrays. *IEEE Transactions on Signal Processing* 65 (19): 5006–5020.

5 Hassanien, A., Amin, M.G., Zhang, Y.D., and Ahmad, F. (2015). High-resolution single-snapshot DOA estimation in MIMO radar with colocated antennas. *2015 IEEE Radar Conference (RadarCon)*, 1134–1138.

6 Hassanien, A. and Vorobyov, S.A. (2011). Transmit energy focusing for DOA estimation in MIMO radar with colocated antennas. *IEEE Transactions on Signal Processing* 59 (6): 2669–2682.

7 Gogineni, S. and Nehorai, A. (2011). Target estimation using sparse modeling for distributed MIMO radar. *IEEE Transactions on Signal Processing* 59 (11): 5315–5325.

8 Wang, X., Hassanien, A., and Amin, M.G. (2019). Dual-function MIMO radar communications system design via sparse array optimization. *IEEE Transactions on Aerospace and Electronic Systems* 55 (3): 1213–1226.

9 Masood, M., Afify, L.H., and Al-Naffouri, T.Y. (2015). Efficient coordinated recovery of sparse channels in massive MIMO. *IEEE Transactions on Signal Processing* 63 (1): 104–118.

10 Wang, X., Tan, C.P., Wu, F., and Wang, J. (2021). Fault-tolerant attitude control for rigid spacecraft without angular velocity measurements. *IEEE Transactions on Cybernetics* 51 (3): 1216–1229.

11 Kalogerias, D.S. and Petropulu, A.P. (2014). Matrix completion in colocated MIMO radar: recoverability, bounds theoretical guarantees. *IEEE Transactions on Signal Processing* 62 (2): 309–321.

12 Moffet, A. (1968). Minimum-redundancy linear arrays. *IEEE Transactions on Antennas and Propagation* 16 (2): 172–175.

13 Pal, P. and Vaidyanathan, P.P. (2010). Nested arrays: a novel approach to array processing with enhanced degrees of freedom. *IEEE Transactions on Signal Processing* 58 (8): 4167–4181.

14 Amin, M.G., Vaidyanathan, P.P., Zhang, Y.D., and Pal, P. (2017). Editorial for coprime special issue. *Digital Signal Processing* 61 (Suppl. C): 1–2. Special Issue on Coprime Sampling and Arrays.

15 Qin, S., Zhang, Y.D., and Amin, M.G. (2015). Generalized coprime array configurations for direction-of-arrival estimation. *IEEE Transactions on Signal Processing* 63 (6): 1377–1390.

16 Ahmed, A., Zhang, Y.D., and Zhang, J. (2019). Coprime array design with minimum lag redundancy. *ICASSP 2019 –2019 IEEE International Conference on Acoustics, Speech and Signal Processing (ICASSP)*, May 2019, 4125–4129.

17 Bae, M.H., Sohn, I.H., and Park, S.B. (1991). Grating lobe reduction in ultrasonic synthetic focusing. *Electronics Letters* 27 (14): 1225–1227.

18 Lu, J. and Greenleaf, J. (1994). Study of two-dimensional array transducers for limited diffraction beams. *IEEE Transactions on Ultrasonics, Ferroelectrics, and Frequency Control* 41 (5): 724–739.

19 Doles, J.H. and Benedict, F.D. (1988). Broad-band array design using the asymptotic theory of unequally spaced arrays. *IEEE Transactions on Antennas and Propagation* 36 (1): 27–33.

20 Murino, V., Trucco, A., and Tesei, A. (1997). Beam pattern formulation and analysis for wide-band beamforming systems using sparse arrays. *Signal Processing* 56 (2): 177–183.

21 Hamza, S.A. and Amin, M.G. (2020). Sparse array beamforming design for wideband signal models. *IEEE Transactions on Aerospace and Electronic Systems* 57 (2): 1211–1226.

22 Wang, X., Amin, M.G., and Wang, X. (2018). Optimum sparse array design for multiple beamformers with common receiver. *2018 IEEE International Conference on Acoustics, Speech and Signal Processing*, April 2018, 4183–4186.

23 Wang, X., Amin, M.G., and Cao, X. (2017). Optimum adaptive beamformer design with controlled quiescent pattern by antenna selection. *2017 IEEE Radar Conference (RadarConf)*, May 2017, 0749–0754.

24 Hamza, S.A. and Amin, M.G. (2018). Optimum sparse array receive beamforming for wideband signal model. *2018 52nd Asilomar Conference on Signals, Systems, and Computers*, October 2018, 89–93.

25 Hamza, S.A. and Amin, M.G. (2019). Sparse array design utilizing matrix completion. *2019 Asilomar Conference on Signals, Systems, and Computers*, November 2019.

26 Hamza, S.A., Amin, M., and Fabrizio, G. (2018). Optimum sparse array design for maximizing signal-to-noise ratio in presence of local scatterings. *2018 IEEE International Conference on Acoustics, Speech and Signal Processing (ICASSP)*, April 2018.

27 Wang, X., Amin, M., and Cao, X. (2018). Analysis and design of optimum sparse array configurations for adaptive beamforming. *IEEE Transactions on Signal Processing* 66 (2): 340–351.

28 Hamza, S.A. and Amin, M.G. (2019). Hybrid sparse array design for under-determined models. *ICASSP 2019 –2019 IEEE International Conference on Acoustics, Speech and Signal Processing (ICASSP)*, May 2019, 4180–4184.

29 Hamza, S.A. and Amin, M.G. (2019). Hybrid sparse array beamforming design for general rank signal models. *IEEE Transactions on Signal Processing* 67 (24): 6215–6226.

30 Haykin, S. (2006). Cognitive dynamic systems. *Proceedings of the IEEE* 94 (11): 1910–1911.

31 Greco, M.S., Gini, F., Stinco, P., and Bell, K. (2018). Cognitive radars: on the road to reality: progress thus far and possibilities for the future. *IEEE Signal Processing Magazine* 35 (4): 112–125.

32 Huleihel, W., Tabrikian, J., and Shavit, R. (2013). Optimal adaptive waveform design for cognitive MIMO radar. *IEEE Transactions on Signal Processing* 61 (20): 5075–5089.

33 Bergin, J.S., Guerci, J.R., Guerci, R.M., and Rangaswamy, M. (2015). MIMO clutter discrete probing for cognitive radar. *2015 IEEE Radar Conference (RadarCon)*, 1666–1670.

34 Sharaga, N., Tabrikian, J., and Messer, H. (2015). Optimal cognitive beamforming for target tracking in MIMO radar/sonar. *IEEE Journal of Selected Topics in Signal Processing* 9 (8): 1440–1450.

35 Karamalis, P., Skentos, N., and Kanatas, A. (2004). Selecting array configurations for MIMO systems: an evolutionary computation approach. *IEEE Transactions on Wireless Communications* 3 (6): 1994–1998.

36 Stoica, P. and Nehorai, A. (1990). Performance study of conditional and unconditional direction-of-arrival estimation. *IEEE Transactions on Acoustics, Speech, and Signal Processing* 38 (10): 1783–1795.

37 Hua, Y. and Sarkar, T. (1991). A note on the Cramer-Rao bound for 2-D direction finding based on 2-D array. *IEEE Transactions on Signal Processing* 39 (5): 1215–1218.

38 Mirkin, A. and Sibul, L. (1991). Cramer-Rao bounds on angle estimation with a two-dimensional array. *IEEE Transactions on Signal Processing* 39 (2): 515–517.

39 Nielsen, R. (1994). Azimuth and elevation angle estimation with a three-dimensional array. *IEEE Journal of Oceanic Engineering* 19 (1): 84–86.

40 Baysal, U. and Moses, R. (2003). On the geometry of isotropic arrays. *IEEE Transactions on Signal Processing* 51 (6): 1469–1478.

41 Vertatschitsch, E. and Haykin, S. (1991). Impact of linear array geometry on direction-of-arrival estimation for a single source. *IEEE Transactions on Antennas and Propagation* 39 (5): 576–584.

42 Korso, M.N.E., Boyer, R., Renaux, A., and Marcos, S. (2010). Conditional and unconditional Cramer–Rao bounds for near-field source localization. *IEEE Transactions on Signal Processing* 58 (5): 2901–2907.

43 Gazzah, H. and Delmas, J.P. (2014). Crb-based design of linear antenna arrays for near-field source localization. *IEEE Transactions on Antennas and Propagation* 62 (4): 1965–1974.

44 Delmas, J.-P., Gazzah, H., and El Korso, M.N. (2019). Improved localization of near-field sources using a realistic signal propagation model and optimally-placed sensors. *Digital Signal Processing* 95: 102579.

45 Gazzah, H. and Abed-Meraim, K. (2009). Optimum ambiguity-free directional and omnidirectional planar antenna arrays for DOA estimation. *IEEE Transactions on Signal Processing* 57 (10): 3942–3953.

46 Keskin, F. and Filik, T. (2020). An optimum volumetric array design approach for both azimuth and elevation isotropic DOA estimation. *IEEE Access* 8: 183903–183912.

47 Wandale, S. and Ichige, K. (2020). On the DOA estimation performance of optimum arrays based on deep learning. *2020 IEEE 11th Sensor Array and Multichannel Signal Processing Workshop (SAM)*, 1–5.

48 Roy, V., Chepuri, S.P., and Leus, G. (2013). Sparsity-enforcing sensor selection for DOA estimation. *2013 5th IEEE International Workshop on Computational Advances in Multi-Sensor Adaptive Processing (CAMSAP)*, 340–343.

49 Wang, X., Aboutanios, E., and Amin, M.G. (2015). Adaptive array thinning for enhanced DOA estimation. *IEEE Signal Processing Letters* 22 (7): 799–803.

50 Hamza, S.A. and Amin, M.G. (2023). Sparse array design for optimum beamfoming using deep learning. *IEEE Transactions on Aerospace and Electronic Systems* 1–12. https://doi.org/10.1109/TAES.2023.3285201.

51 Hamza, S.A., Juretus, K., Amin, M.G., and Ahmad, F. (2023). Deep learning sparse array design using binary switching configurations. *ICASSP 2023 –2023 IEEE International Conference on Acoustics, Speech and Signal Processing (ICASSP)*, 1–5.

52 Roberts, W., Xu, L., Li, J., and Stoica, P. (2011). Sparse antenna array design for MIMO active sensing applications. *IEEE Transactions on Antennas and Propagation* 59 (3): 846–858.

53 Cheng, Z., Lu, Y., He, Z. et al. (2018). Joint optimization of covariance matrix and antenna position for MIMO radar transmit beampattern matching design. *2018 IEEE Radar Conference (RadarConf18)*, April 2018, 1073–1077.

54 Hamza, S.A. and Amin, M. (2020). Planar sparse array design for transmit beamforming (Conference Presentation). In: *Radar Sensor Technology XXIV*, International Society for Optics and Photonics, vol. 11408 (ed. K.I. Ranney and A.M. Raynal). SPIE. https://doi.org/10.1117/12.2557923.

55 Hamza, S.A. and Amin, M.G. (2020). Sparse array design for transmit beamforming. *2020 IEEE International Radar Conference (RADAR)*, 560–565.

56 Shahbazpanahi, S., Gershman, A.B., Luo, Z.-Q., and Wong, K.M. (2003). Robust adaptive beamforming for general-rank signal models. *IEEE Transactions on Signal Processing* 51 (9): 2257–2269.

57 Ibrahim, M.S., Konar, A., Hong, M., and Sidiropoulos, N.D. (2018). Mirror-Prox SCA algorithm for multicast beamforming and antenna selection. *2018 IEEE 19th International Workshop on Signal Processing Advances in Wireless Communications (SPAWC)*, 1–5.

58 Candes, E.J., Wakin, M.B., and Boyd, S.P. (2008). Enhancing sparsity by reweighted l1 minimization. *Journal of Fourier Analysis and Applications* 14 (5): 877–905.

59 Wang, X. and Aboutanios, E. (2020). Sparse array design for multiple switched beams using iterative antenna selection method. *Digital Signal Processing* 105: 102684. Special Issue on Optimum Sparse Arrays and Sensor Placement for Environmental Sensing.

60 Mehanna, O., Sidiropoulos, N.D., and Giannakis, G.B. (2013). Joint multicast beamforming and antenna selection. *IEEE Transactions on Signal Processing* 61 (10): 2660–2674.

61 Hamza, S.A., Amin, M.G., and Fabrizio, G. (2018). Optimum sparse array beamforming for general rank signal models. *2018 IEEE Radar Conference (RadarConf18)*, 1343–1347.

62 Kwak, N. (2008). Principal component analysis based on L1-normmaximization. *IEEE Transactions on Pattern Analysis and Machine Intelligence* 30 (9): 1672–1680.

63 Stoica, P. and Nehorai, A. (1989). Music, maximum likelihood, and Cramer-Rao bound. *IEEE Transactions on Acoustics, Speech, and Signal Processing* 37 (5): 720–741.

11

Generalized Structured Sparse Arrays for Fixed and Moving Platforms

Guodong Qin[1] and Si Qin[2]

[1] *School of Electronic Engineering, Xidian University, Xi'an, Shaanxi, China*
[2] *Microsoft China, Beijing, China*

11.1 Introduction

Sparse arrays are widely used in various applications, including communications, radar, sonar, satellite navigation, radio telescope and spectrm sensing [1–11]. Sparse arrays can estimate the direction-of-arrival (DOA) of more sources than the number of physical sensors [12–25], making them highly useful. However, the DOA estimation performance depends on the number and positions of the spatial autocorrelation lags or the virtual sensors of the coarrays. In this respect, the prototypes of classical sparse arrays are not optimal in many scenarios. Fundamentals of sparse arrays are covered in Chapter 1 of this book. This chapter aims to generalize sparse arrays and present new approaches to improve the performance of coarrays. The chapter deals with the difference coarrays in lieu of the sum coarrays associated with active sensing [26].

To achieve this goal, we introduce two ways to improve the performance of the difference coarrays: the generalized coprime array [18] and the sparse synthetic array with platform motions [27–31]. The former increases the number of degrees of freedom (DOF) of the difference coarrays by compressing the sensor spacing of the subarray and rearranging the positions of two subarrays. The latter fills the holes in the difference coarray by the virtue of array motion.

The chapter is organized as follows. Section 11.2 introduces the generalized coprime array configurations, while Section 11.3 summarizes the signal model and synthetic aperture processing for exploiting array motions. The achievable numbers of both DOF and consecutive lags after array motion for coprime and nested arrays, minimum hole array (MHA), minimum redundancy array (MRA), and sparse uniform linear array (SULA), are discussed in this section. The maximum increments in the numbers of both DOF and consecutive lags as a result of array motion are also derived. Section 11.4 introduces a dilated nested array structure to provide a hole-free difference coarray with merely a half-wavelength array motion. It is a fully filled coarray when the minimum sensor spacing does not exceed three half-wavelength units. Section 11.5 describes sparsity-based DOA estimation implementing least absolute shrinkage and selection operator (LASSO) with array motion. Section 11.6 introduces other structured arrays for fixed and moving platforms. Finally, Section 11.7 concludes the chapter.

Sparse Arrays for Radar, Sonar, and Communications, First Edition. Edited by Moeness G. Amin.

11.2 Generalized Coprime Array Configurations

11.2.1 Prototype Coprime Array and Difference Coarray

A prototype coprime array [12] is illustrated in Fig. 11.1, where M and N are coprime integers. Without loss of generality, we assume $M < N$. The unit inter-element spacing d is set to $\lambda/2$, where λ denotes the wavelength. The array sensors are positioned at

$$\mathbb{P} = \{Mnd \,|\, 0 \le n \le N-1\} \cup \{Nmd \,|\, 0 \le m \le M-1\}. \tag{11.1}$$

It is worth noting that the two subarrays share the first sensor at the 0th position, resulting in a total of $M+N-1$ sensors used in the coprime array. Additionally, the minimum inter-element spacing remains $\lambda/2$.

Denote $\mathbf{D} = [d_1, \ldots, d_{M+N-1}]^T$ as the positions of the array sensors where $d_i \in \mathbb{P}$, $i = 1, \ldots, M+N-1$, and the first sensor is assumed as the reference, i.e., $d_1 = 0$. From a pair of antennas located at the ith and kth positions in $\mathbf{D}$, the spatial correlation $\mathrm{E}[x_i(t)x_k^*(t)]$ yields the (i,k)th entry in the covariance matrix of the data vector $\mathbf{x}(t)$, $\mathbf{R}_{\mathrm{xx}}$, with lag $d_i - d_k$, where $\mathrm{E}(\cdot)$ denotes the statistical expectation operator. As such, all the available values of i and k, where $0 \le i \le M+N-1$ and $0 \le k \le M+N-1$, yield the virtual sensors of the following difference coarray:

$$\mathbb{C}_{\mathrm{P}} = \{\mathbf{z} \,|\, \mathbf{z} = \mathbf{u} - \mathbf{v}, \quad \mathbf{u} \in \mathbb{P}, \quad \mathbf{v} \in \mathbb{P}\}. \tag{11.2}$$

The difference coarray is significant because it allows for the calculation of the correlation of the received signal at all differences in set $\mathbb{C}_{\mathrm{P}}$. Applications that depend on this correlation, such as DOA estimation, can utilize the increased DOF offered by the resulting coarray structure. By using a portion or the entire set of distinct autocorrelation terms in set $\mathbb{C}_{\mathrm{P}}$ instead of the original array, an increased number of sources that can be detected by the array. The maximum number of DOF is determined by the number of unique elements in $\mathbb{L}_{\mathrm{P}}$ as defined in

$$\mathbb{L}_{\mathrm{P}} = \{l_{\mathrm{P}} \,|\, l_{\mathrm{P}} d \in \mathbb{C}_{\mathrm{P}}\}. \tag{11.3}$$

To gain a deeper understanding of difference coarrays, the self-differences of the two subarrays and their cross-differences are considered separately. The coarray is derived from the Hermitian matrix $\mathbf{R}_{\mathrm{xx}}$, and the self-difference in the coarray has positions

$$\mathbb{L}_{\mathrm{s}} = \{l_{\mathrm{s}} \,|\, l_{\mathrm{s}} = Mn\} \cup \{l_{\mathrm{s}} \,|\, l_{\mathrm{s}} = Nm\}, \tag{11.4}$$

and their corresponding mirrored positions $\mathbb{L}_{\mathrm{s}}^- = \{-l_{\mathrm{s}} \,|\, l_{\mathrm{s}} \in \mathbb{L}_{\mathrm{s}}\}$. The cross-difference has positions

$$\mathbb{L}_{\mathrm{c}} = \{l_{\mathrm{c}} \,|\, l_{\mathrm{c}} = Nm - Mn\}, \tag{11.5}$$

and their corresponding mirrored positions $\mathbb{L}_{\mathrm{c}}^- = \{-l_{\mathrm{c}} \,|\, l_{\mathrm{c}} \in \mathbb{L}_{\mathrm{c}}\}$, for $0 \le n \le N-1$ and $0 \le m \le M-1$. The full set of lags in the virtual array is given by

$$\mathbb{L}_{\mathrm{P}} = \mathbb{L}_{\mathrm{s}} \cup \mathbb{L}_{\mathrm{s}}^- \cup \mathbb{L}_{\mathrm{c}} \cup \mathbb{L}_{\mathrm{c}}^-. \tag{11.6}$$

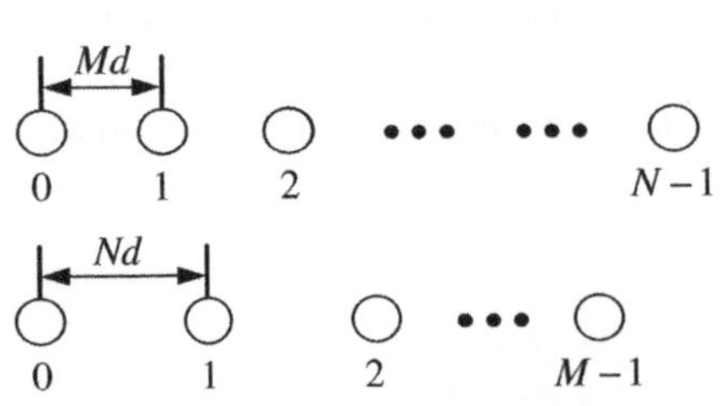

Figure 11.1 The prototype coprime array configuration.

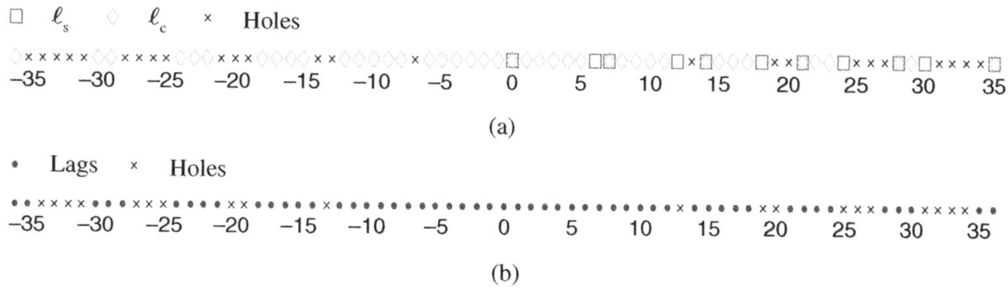

Figure 11.2 An example of prototype coprime configuration coarrays, where $M = 6$ and $N = 7$. (a) The set $\mathbb{L}_s$ and $\mathbb{L}_c$. (b) The lag positions in full set $\mathbb{L}_P$.

An example of the prototype coprime array is illustrated in Fig. 11.2, where $M = 6$ and $N = 7$. Figure 11.2a displays the self- and cross-lags described in Eqs. (11.4) and (11.5). By including the negative mirror of the above set, the full set of lags becomes symmetric, as demonstrated in Fig. 11.2b. Notice that some "holes," such as, ± 13, ± 19, ± 20, still exist in the difference coarray and are indicated by $\times$ in the figure. The total number of lags in the symmetric set provides a global upper bound for the achievable DOF.

11.2.2 Coprime Array with Compressed Inter-element Spacing

We present our main results that generalize the coarray structures through two operations, namely, *coprime array with compressed inter-element spacing (CACIS)* and *coprime array with displaced subarrays (CADiS)*. In the Section 11.2.2, we introduce the CACIS, whereas the CADiS is explored in the Section 11.2.3.

Two subarrays are examined, each with M and N sensors, where M and N are coprime. Note that, in the sequel, the condition that $M < N$ is no longer assumed. In this case, we introduce an integer compression factor p to alter the inter-element spacing of one subarray. Assume that M can be expressed as a product of two positive integers p and $\breve{M}$, i.e.,

$$M = p\breve{M}, \quad 2 \leq p \leq M. \tag{11.7}$$

It is easy to confirm that $\breve{M}$ and N are also coprime since M and N do not have common factors other than unity.

As depicted in Fig. 11.3, in the generalized coprime array, the M-element subarray has an inter-element spacing of Nd, whereas the N-element subarray has an inter-element space of $\breve{M}d = Md/p$. This means that the inter-element spacing of one subarray has been compressed by a factor of p, thus comes the term of coprime array with CACIS. Note that all arrays consist of the same $M + N - 1$ physical antenna sensors and their aperture is $(M - 1)N$, regardless of the value

Figure 11.3 The CACIS configuration.

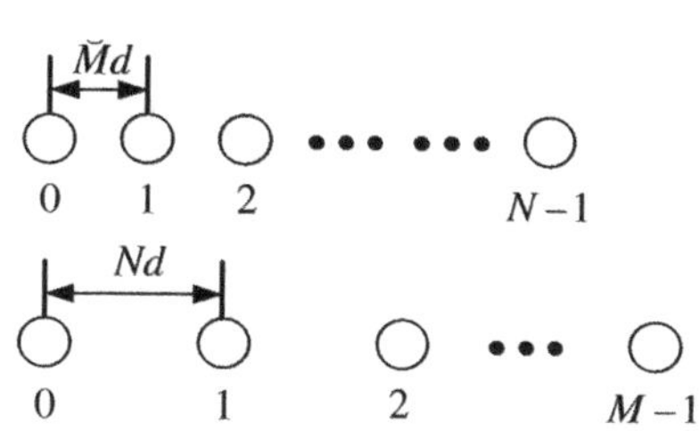

of p. It is also worth mentioning that the variation of the prototype coprime array configuration used in [32] is a special case of the CACIS configuration, which can be achieved by choosing $p = 2$.

In this array configuration, the self-lags of the two subarrays are given by the following set,[1]

$$\tilde{\mathbb{L}}_s = \{\tilde{l}_s | \tilde{l}_s = \breve{M}n\} \cup \{\tilde{l}_s | \tilde{l}_s = Nm\}, \tag{11.8}$$

and the corresponding mirrored positions $\tilde{\mathbb{L}}_s^-$, whereas the cross-lags between the two subarrays are given by

$$\tilde{\mathbb{L}}_c = \{\tilde{l}_c | \tilde{l}_c = Nm - \breve{M}n\}, \tag{11.9}$$

and the corresponding $\tilde{\mathbb{L}}_c^-$, where $0 \leq m \leq M - 1$ and $0 \leq n \leq N - 1$.

To completely exploit the DOF of the CACIS configuration, we summarize the properties of $\tilde{\mathbb{L}}_s$ and $\tilde{\mathbb{L}}_c$ in the following proposition.

Proposition 11.1 *The following facts hold for the CACIS:*

(a) There are MN distinct integers in set $\tilde{\mathbb{L}}_c$.
(b) $\tilde{\mathbb{L}}_c$ *contains all the contiguous integers in the range* $-(N-1) \leq \tilde{l}_c \leq MN - \breve{M}(N-1) - 1$.
(c) The negative values form a subset of the flipped positive values in set $\tilde{\mathbb{L}}_c$, *i.e.,* $\{\tilde{l}_c | \tilde{l}_c < 0, \tilde{l}_c \in \tilde{\mathbb{L}}_c\} \subseteqq \{-\tilde{l}_c | \tilde{l}_c > 0, \tilde{l}_c \in \tilde{\mathbb{L}}_c\}$.
(d) The self-lags form a subset of the cross-lags, i.e., $(\tilde{\mathbb{L}}_s^- \cup \tilde{\mathbb{L}}_s) \subseteqq (\tilde{\mathbb{L}}_c^- \cup \tilde{\mathbb{L}}_c)$.
(e) There are "holes" located at both positive range and negative ranges of $\tilde{\mathbb{L}}_c$. *The holes falling in the negative range are located at* $-(a\breve{M} + bN)$, *where* $a \geq 0$, $b > 0$ *are integers.*

Based on the properties (c) and (d) of Proposition 11.1, the entire lag set in the virtual array defined in Eq. (11.6) consists of $\{\tilde{l}_c | \tilde{l}_c \geq 0, \tilde{l}_c \in \tilde{\mathbb{L}}_c\} \cup \{-\tilde{l}_c | \tilde{l}_c \geq 0, \tilde{l}_c \in \tilde{\mathbb{L}}_c\}$, thus resulting in Proposition 11.2.

Proposition 11.2 *The CACIS configuration defined in Eq. (11.7) yields a virtual array such that:*

(a) It contains $2MN - (\breve{M}+1)(N-1) - 1$ *unique lags of virtual sensors;*
(b) Among the unique lags, there are $2MN - 2\breve{M}(N-1) - 1$ *consecutive integers within the range* $[-MN + \breve{M}(N-1) + 1, MN - \breve{M}(N-1) - 1]$.

The proof can be found in [18]. To illustrate the properties discussed, $M = 2\breve{M}$ is used as an example in Fig. 11.4. This configuration is equivalent to the one proposed in [32]. In this case, the virtual array consists of $3\breve{M}N + \breve{M} - N$ unique lags, with consecutive lags ranging from $[-\breve{M}N - \breve{M} + 1, \breve{M}N + \breve{M} - 1]$.

Proposition 11.2 leads us to the conclusion that, for a specific pair of M and N, smaller values of $\breve{M}$ result in more unique and consecutive coarray lags. In simpler terms, both numbers increase as the compression factor p increases. The smallest value that $\breve{M}$ can take is 1, which results in a nested array structure. This configuration provides the highest number of unique and consecutive virtual sensors.

1 $\tilde{(\cdot)}$ is used to emphasize variables corresponding to the CACIS structure.

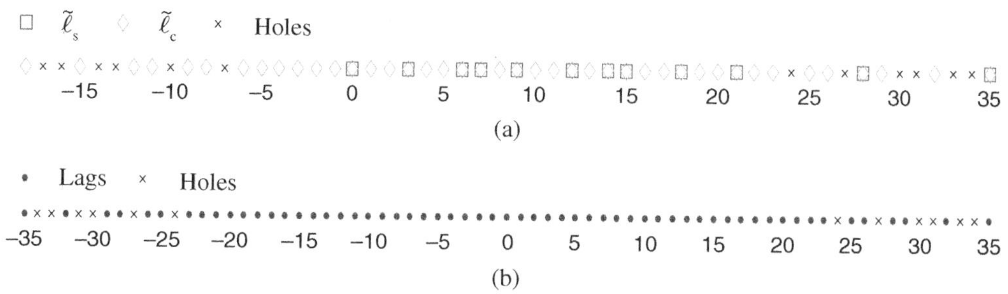

Figure 11.4 An example of CACIS configuration coarrays, where $\breve{M} = 3$, $p = 2$ and $N = 7$. (a) The set $\tilde{\mathbb{L}}_s$ and $\tilde{\mathbb{L}}_c$. (b) The full set $\tilde{\mathbb{L}}_P$.

11.2.3 Coprime Array with Displaced Subarrays

Sharing many of the same properties as MRA, the prototype coprime array and the CACIS structure offer sparse configurations in which minimum inter-element spacing remains equal to the unit spacing, typically half a wavelength, to mitigate the grating lobe problem. In addition to the issues of antenna size and mutual coupling associated with half wavelength spacing, there is a significant amount of overlap between self- and cross-lags in both the prototype coprime array and CACIS structures due to their collocated subarray placement. However, by introducing the proper displacement between the two subarrays, a new coprime array structure can lead to a larger minimum inter-element spacing and a higher number of unique lags, as well as an enlarged virtual array aperture. It should be noted, however, that the number of consecutive lags may be reduced as positive and negative lags will no longer be connected.

Consider two collinearly located uniform linear subarrays, as depicted in Fig. 11.5, where one consists of N antennas and the other with $M - 1$ antennas, resulting in a total of $M + N - 1$ sensors. This coprime array structure is referred to as a CADiS. Similar to the CACIS configuration, M and N are assumed to be coprime. The N-element subarray has an inter-element spacing of $\breve{M}d$, while the $(M - 1)$-element subarray has an inter-element spacing of Nd. The difference between the CACIS and CADiS structures is that the subarrays in the CADiS structure are collinearly placed, with the closest spacing between the two subarrays being Ld, where $L \geq \min\{\breve{M}, N\}$. It should be noted that $\breve{M} > 1$ is needed to ensure the minimum inter-element spacing is greater than unit spacing. However, the nested structure with $\breve{M} = 1$ will also be discussed later as a special case. The total number of array sensors in the CADiS structure remains $M + N - 1$, which is the same as the CACIS configuration. The minimum inter-element spacing in the CADiS is $\min\{\breve{M}, N\}d$, compared to d in the CACIS structure. Additionally, the total array aperture of the CADiS is $(MN + \breve{M}N - \breve{M} - 2N + L)d$, which is significantly larger than the $(M - 1)Nd$ of the CACIS. In practical applications, it is recommended to choose a small value for the displacement L to avoid high sidelobe peaks.

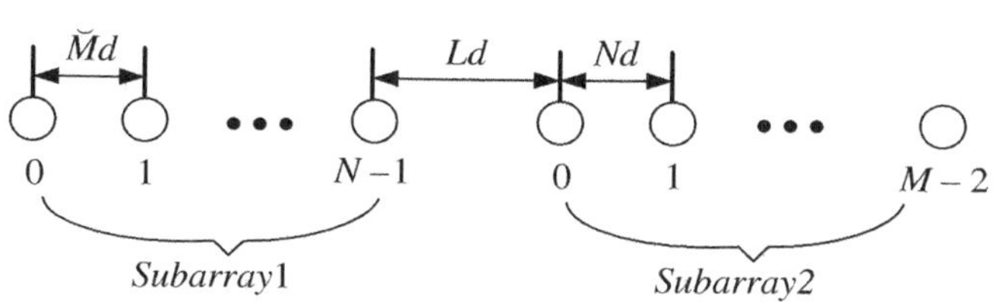

Figure 11.5 The CADiS configuration.

For the CADiS configuration, the corresponding self-lags $\bar{l}_s$ and cross-lags $\bar{l}_c$ are respectively given by[2]

$$\bar{\mathbb{L}}_s = \{\bar{l}_s | \bar{l}_s = Nm\} \cup \{\bar{l}_s | \bar{l}_s = \check{M}n\}, \tag{11.10}$$

$$\bar{\mathbb{L}}_c = \{\bar{l}_c | \bar{l}_c = \check{M}(N-1) + Nm - \check{M}n + L\}, \tag{11.11}$$

and their corresponding mirrored positions $\bar{\mathbb{L}}_s^-$ and $\bar{\mathbb{L}}_c^-$, respectively, where $0 \leq m \leq M-2$ and $0 \leq n \leq N-1$.

The following proposition reveals the properties of the resulting virtual sensors of the CADiS configuration.

Proposition 11.3 *Set $\bar{\mathbb{L}}_s$ and $\bar{\mathbb{L}}_c$ have the following properties in the CADiS configuration:*

(a) There are $(M-1)N$ distinct integers in set $\bar{\mathbb{L}}_c$.
(b) $\bar{\mathbb{L}}_c$ contains all the contiguous integers in the range $(\check{M}-1)(N-1)+L \leq \bar{l}_c \leq MN-N-1+L$.
(c) There are "holes" located at $\check{M}(N-1)-(a\check{M}+bN)+L$ in set $\bar{\mathbb{L}}_c$, where $a \geq 0$, $b > 0$ are integers.
(d) $(\bar{\mathbb{L}}_s^- \cup \bar{\mathbb{L}}_s) \nsubseteq (\bar{\mathbb{L}}_c^- \cup \bar{\mathbb{L}}_c)$.

The proof is provided in [18].

In the CACIS configuration, the negative lags form a subset of the flipped positive counterpart. Therefore, only non-negative lags in $\tilde{\mathbb{L}}_c$ are used. In the CADiS configuration, however, the negative lags do not generally overlap with the flipped positive lags because of the displacement between two subarrays, necessitating the consideration of both positive and negative lags. As such, the CADiS configuration enjoys a higher number of unique lags than the CACIS because of the utilization of negative lags. In addition, the self-lags are less likely to coincide with the cross-lags in the CADiS configuration. Consequently, the CADiS offers a larger virtual array aperture and a higher number of virtual sensors. The role of the displacement L is as follows. On the one hand, it reduces the overlaps between the self- and cross-lags. On the other hand, because $\bar{\mathbb{L}}_c$ has holes located at $\check{M}(N-1)-(a\check{M}+bN)+L$ for integers $a \geq 0$ and $b > 0$, the number of consecutive lags can be extended by choosing the value of L so that some self-lags are aligned to the cross-lag holes. As an example in Fig. 11.6, if $p=2$, $\check{M}=3$, $N=7$ and $L=\check{M}+N$, $\bar{\mathbb{L}}_s$ and $\bar{\mathbb{L}}_c$ will exhibit alignments of some holes in $\bar{\mathbb{L}}_c$ (such as 12, 14, 15, 18, and 21) with elements of $\bar{\mathbb{L}}_s$. The following proposition describes the selection of the value of L that maximizes the number of unique and consecutive lags.

Proposition 11.4 *For the CADiS configuration,*

(a) The maximum number of unique lags $2MN+2M-5$ can be achieved with $L > N(M-2)$.

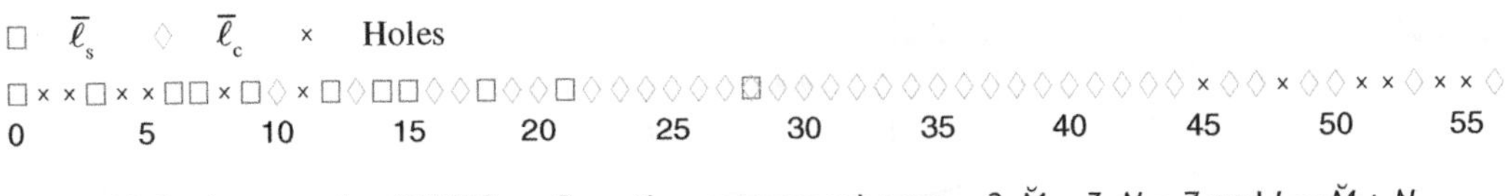

Figure 11.6 An example of CADiS configuration coarrays, where $p=2$, $\check{M}=3$, $N=7$ and $L=\check{M}+N$.

2 $\overline{(\cdot)}$ is used to emphasize variables corresponding to the CADiS structure.

(b) $L = \check{M} + N$ *is the choice that yields the largest number of consecutive lags. In this case, there are* $2MN + 2\check{M} - 1$ *unique lags, among which the range* $[(\check{M} - 1)(N - 1), MN + \check{M} - 1]$ *and its corresponding negative range* $[-MN - \check{M} + 1, -(\check{M} - 1)(N - 1)]$ *are respectively consecutive.*

The proof can be found in [18]. Based on property (b) of Proposition 11.4, it is clear that the number of unique lags increases as $\check{M}$ increases, whereas the number of the consecutive lags decreases. Particularly, for the nested array structure, i.e., $\check{M} = 1$, the positive range of consecutive lags is $[0, MN]$ and its corresponding negative range becomes $[-MN, 0]$, resulting in all unique lags to be consecutive.

For comparison, we enlist the coarray aperture, the maximum number of unique and consecutive lags for both proposed configurations in Table 11.1. As the results show, a nested structure in CACIS configurations (i.e., $\check{M} = 1$) offers the highest number of both consecutive lags and unique lags for a given coprime pair of M and N. As for CADiS, the nested structure provides the highest number only for the consecutive lags. The number of its unique lags, $2MN + 1$, on the other hand, is less than that of the CADiS structure with a large separation between the two subarrays. This means that the nested CADiS structure is most effective only when using the multiple signal classification (MUSIC) algorithm or other subspace based methods for DOA estimation, but less so when using compressive sensing (CS)-based methods. This attributed to the need to perform spatial averaging utilizing consecutive lags in [32]. It is noted that, to estimate DOAs of up to MN sources, the nested CADiS structure requires only $M + N - 1$ sensors, which are much less than the result of $2M + N - 1$ sensors as exploited in [32].

For illustrative purposes, we consider $M = 6$ and $N = 7$ with different values of the compression factor p of the two configurations, i.e., CACIS and CADiS. $L = \check{M} + N$ are considered for the CADiS configuration for the convenience of performance comparison. All configurations consist of $M + N - 1 = 12$ physical antenna sensors and the unit inter-element spacing is $d = \lambda/2$.

The virtual sensors corresponding to the CACIS and CADiS structures are respectively shown in Figs. 11.7 and 11.8. Figure 11.7a depicts the CACIS configuration example for $p = 2$, where the coprime array form a virtual array with 59 unique lags, among which 47 lags within $[-23, 23]$ are consecutive. Figure 11.7b shows for the case of $p = 3$, and the resulting virtual array has 65 unique lags, among which 59 lags within $[-29, 29]$ are consecutive. When $p = M = 6$, i.e., $\check{M} = 1$, as shown in Fig. 11.7c, the coprime array becomes the nested array structure with 71 unique lags, which are all consecutive. It is clear that both numbers of the unique and consecutive lags increase with p, and the nested array achieves the maximum number for both. For the CADiS configuration with $L = \check{M} + N$, the case of $p = 2$ is presented in Fig. 11.8a. In this case, the entire virtual array has 89

Table 11.1 Comparison of coarray aperture, number of unique and consecutive lags.

	Coarray aperture	Maximum number of unique lags	Maximum number of consecutive lags
CACIS	$(M-1)N$	$2MN - \check{M}(N-1) - N$	$2MN - 2\check{M}(N-1) - 1$
CADiS ($\check{M} > 1$)	$(N-1)\check{M} + (M-2)N + L$	$2MN + 2M - 5$	$MN - (\check{M}-1)(N-2) + 1$
(displacement L)	(arbitrary L)	$(L > N(M-2))$	$(L = \check{M} + N)$
Nested CADiS ($\check{M} = 1$)	MN	$2MN + 1$	$2MN + 1$
(displacement L)	$(L = N + 1)$	$(L = N + 1)$	$(L = N + 1)$

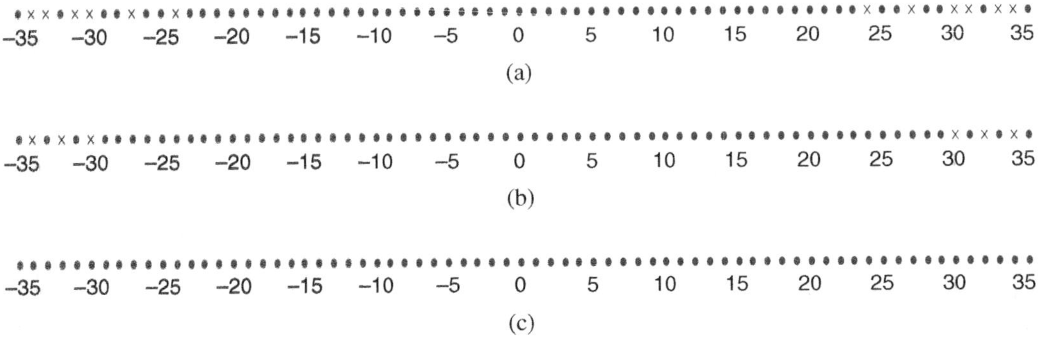

Figure 11.7 CACIS configuration coarrays, for different compression factor p ($M = 6$ and $N = 7$). (a) $p = 2$ and $\check{M} = 3$. (b) $p = 3$ and $\check{M} = 2$. (c) $p = 6$ and $\check{M} = 1$.

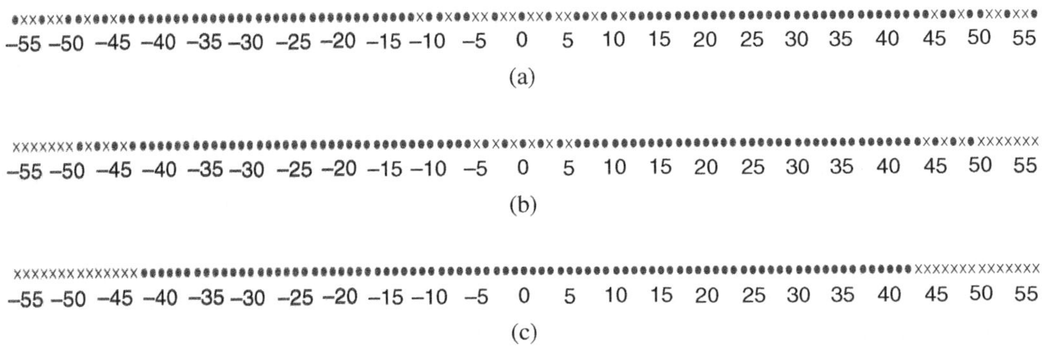

Figure 11.8 CADiS configuration coarrays with displacement $L = \check{M} + N$, corresponding the compression factor p ($M = 6$ and $N = 7$). (a) $p = 2$, $\check{M} = 3$, and $L = 10$. (b) $p = 3$, $\check{M} = 2$, and $L = 9$. (c) $p = 6$, $\check{M} = 1$, and $L = 8$.

unique lags, among which the lags within $[-44, -12]$ and $[12, 44]$ are respectively consecutive. For $p = 3$, there are 87 distinct lags, resulting in consecutive lags in $[-43, -6]$ and in $[6, 43]$, as shown in Fig. 11.8b. In Fig. 11.8c, the nested CADiS with $p = 6$ and $\check{M} = 1$ is considered as a special case. It is noted that all 85 lags in the full symmetric set of $[-42, 42]$ are consecutive.

11.3 Synthetic Structured Arrays Exploiting Array Motions

11.3.1 Array Synthetic Fundamentals

We consider a one-dimension sparse array with L_N sensors on a platform moving at a constant velocity v. The schematic is illustrated in Fig. 11.9, where the circles and rhombus represent the sensor positions of the original and the shifted arrays, respectively. The minimum inter-element spacing is denoted as $d = \lambda/2$. All the sensor positions are integer multiples of d.

The received signals from Q uncorrelated far-field narrowband sources are described as $s_q(t)$, $t = T_s, 2T_s, \dots, L_s T_s$, for $q = 1, \dots, Q$, where T_s and L_s, respectively, represent the sampling interval and the number of snapshots. The arrival angle of the qth source is denoted as θ_q. Because of the assumed short translation motion of the array, the directions of the sources with respect to the

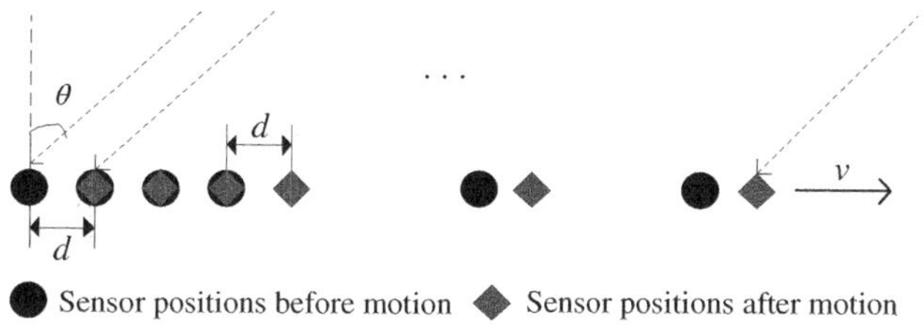

Figure 11.9 DOA estimation exploiting a moving sparse array.

sensor array can be considered unchanged during the short processing time. The output of the receive array, at time t, is expressed as

$$\begin{aligned}\mathbf{x}(t) &= \sum_{q=1}^{Q} s_q(t)\exp\left(-j2\pi vt\sin(\theta_q)/\lambda\right)\mathbf{a}(\theta_q) + \boldsymbol{\varepsilon}(t) \\ &= \mathbf{A}\mathbf{s}(t) + \boldsymbol{\varepsilon}(t),\end{aligned} \tag{11.12}$$

where $\mathbf{a}(\theta_q) = \left[1, \exp(-j2\pi\frac{d_2\sin(\theta_q)}{\lambda}), \dots, \exp(-j2\pi\frac{d_{L_N}\sin(\theta_q)}{\lambda})\right]^T$ represents the steering vector corresponding to θ_q, $q = 1, \dots, Q$, and d_l is the position of the lth array sensor, $l = 1, 2, \dots, L_N$. The first sensor is used as a reference, i.e., $d_1 = 0$. In addition, $\mathbf{s}(t) = \left[s_1(t)\exp\left(-j2\pi\frac{vt\sin(\theta_1)}{\lambda}\right), \dots, s_Q(t)\exp\left(-j2\pi\frac{vt\sin(\theta_Q)}{\lambda}\right)\right]^T$ is the signal vector. $\mathbf{A} = [\mathbf{a}(\theta_1), \dots, \mathbf{a}(\theta_Q)] \in \mathbb{C}^{L_N\times Q}$ is the array manifold matrix, and $\boldsymbol{\varepsilon}(t) \in \mathbb{C}^{L_N\times 1}$ is zero-mean additive white complex Gaussian noise vector with covariance matrix $\sigma_\varepsilon^2\mathbf{I}_{L_N}$. At time $t+\tau$, the output of the receive array becomes

$$\begin{aligned}\mathbf{x}(t+\tau) &= \sum_{q=1}^{Q} s_q(t+\tau)\exp\left(-j2\pi vt\sin(\theta_q)/\lambda\right)\cdot\exp\left(-j2\pi v\tau\sin(\theta_q)/\lambda\right)\mathbf{a}(\theta_q) + \boldsymbol{\varepsilon}(t+\tau) \\ &= \mathbf{B}\mathbf{s}(t+\tau) + \boldsymbol{\varepsilon}(t+\tau),\end{aligned} \tag{11.13}$$

where $\mathbf{B} = [\mathbf{b}(\theta_1), \dots, \mathbf{b}(\theta_Q)] \in \mathbb{C}^{L_N\times Q}$, with

$$\begin{aligned}\mathbf{b}(\theta_q) &= \exp\left(-j2\pi v\tau\sin(\theta_q)/\lambda\right)\mathbf{a}(\theta_q) \\ &= \left[\exp\left(-j2\pi v\tau\sin(\theta_q)/\lambda\right), \exp\left(-j2\pi(v\tau + d_2)\sin(\theta_q)/\lambda\right), \dots,\right. \\ &\quad \left.\exp\left(-j2\pi(v\tau + d_{L_N})\sin(\theta_q)/\lambda\right)\right]^T,\end{aligned} \tag{11.14}$$

and $\mathbf{s}(t+\tau) = \left[s_1(t+\tau)\exp\left(-j2\pi\frac{vt\sin(\theta_1)}{\lambda}\right), \dots, s_Q(t+\tau)\exp\left(-j2\pi\frac{vt\sin(\theta_Q)}{\lambda}\right)\right]^T$. The noise vector $\boldsymbol{\varepsilon}(t+\tau)$ is assumed to be uncorrelated with $\boldsymbol{\varepsilon}(t)$.

For narrowband signals with carrier frequency f, $s_q(t+\tau) = s_q(t)\exp(j2\pi f\tau)$. Accordingly, Eq. (11.13) can be rewritten as

$$\mathbf{x}(t+\tau) = \exp(j2\pi f\tau)\mathbf{B}\mathbf{s}(t) + \boldsymbol{\varepsilon}(t+\tau). \tag{11.15}$$

By choosing $v\tau = d = \lambda/2$, the steering vector at time $t+\tau$ becomes

$$\begin{aligned}\mathbf{b}(\theta_q) = &\left[\exp\left(-j\pi\sin(\theta_q)\right), \exp\left(-j\pi(1 + d_2/d)\sin(\theta_q)\right), \dots,\right. \\ &\left.\exp\left(-j\pi(1 + d_{L_N}/d)\sin(\theta_q)\right)\right]^T.\end{aligned} \tag{11.16}$$

By compensating for the phase correction factor $\exp(-j2\pi f\tau)$ using the technique described in [33], we obtain a phase synchronized received signal vector as

$$\tilde{\mathbf{x}}(t+\tau) = \mathbf{x}(t+\tau)\exp(-j2\pi f\tau) = \mathbf{Bs}(t) + \tilde{\varepsilon}(t+\tau), \tag{11.17}$$

where $\tilde{\varepsilon}(t+\tau) = \exp(-j2\pi f\tau)\varepsilon(t+\tau)$. Combining Eqs. (11.12) and (11.17) yields,

$$\mathbf{y}(t) = \begin{bmatrix} \mathbf{x}(t) \\ \tilde{\mathbf{x}}(t+\tau) \end{bmatrix} = \mathbf{A}_s\mathbf{s}(t) + \varepsilon_s(t), \tag{11.18}$$

where $\mathbf{A}_s = [\mathbf{a}_s(\theta_1), \dots, \mathbf{a}_s(\theta_Q)] \in \mathbb{C}^{2L_N \times Q}$, $\mathbf{a}_s(\theta_q) = [\mathbf{a}^T(\theta_q), \mathbf{b}^T(\theta_q)]^T \in \mathbb{C}^{2L_N \times 1}$, and $\varepsilon_s(t) = \left[\varepsilon^T(t),\ \tilde{\varepsilon}^T(t+\tau)\right]^T$.

11.3.2 The Synthetic Structured Sparse Arrays

In this section, we examine a synthetic array created by shifting a structured sparse array by half a wavelength (one unit step). This type of array is known as a coprime array and includes other sparse arrays such as nested arrays, MRA, MHA and SULA. We provide analytical expressions for both the number of DOF and consecutive lags. The former is defined as the number of non-negative unique lags in the difference coarray.

11.3.2.1 Coprime Array

In the sequel, the array at its original position is referred to as the original array, whereas that in its shifted position is termed as the shifted array. The combined structure is referred to as the synthetic array. The sensor positions of the shifted array are expressed as

$$\mathbb{P}_{c1} = \{(Mn+1)d, 0 \le n \le N-1\} \cup \{(Nm+1)d, 0 \le m \le M-1\}. \tag{11.19}$$

Combining the sensor positions of the original and the shifted arrays, we have

$$\mathbb{P}_c = \mathbb{P}_{c0} \cup \mathbb{P}_{c1}. \tag{11.20}$$

Where $\mathbb{P}_{c0} = \{Mnd \cup Nmd\}$ is the sensor positions of the original array. It is noted that some sensors corresponding to the original and the shifted arrays do overlap. This means that the number of sensors of the synthetic array becomes smaller than twice the number of sensors of the original array. Such overlapping is deemed to occur for a coprime array, because the co-array of the original coprime array contains lag-one sensor pairs.

The set formed from the difference coarray can be divided into two subsets; one is obtained from the self-lags, whereas the other is from the cross-lags. Since the self-lag positions form a subset of the cross-lag positions [34], we only consider the cross-lags in determining the number of DOF. The cross-lags between the two synthetic subarrays are given by the overall set:

$$\mathbb{S}_{ca} = \mathbb{S}_{12} \cup \mathbb{S}_{1'2'} \cup \mathbb{S}_{11'} \cup \mathbb{S}_{22'} \cup \mathbb{S}_{12'} \cup \mathbb{S}_{21'}. \tag{11.21}$$

In the above expression, subscripts 1 and 2 refer to the two subarrays in their original positions, whereas $1'$ and $2'$ refer to these subarrays in the shifted positions.

Set of Cross-Lags $\mathbb{S}_{12}$ is defined as the set of cross-lags between original subarray 1 and original subarray 2 which is given by [34]:

$$\mathbb{S}_{12} = \tilde{\mathbb{S}}_{12} \cup \tilde{\mathbb{S}}_{12}^{-} = \left\{Mk_1 - Nk_2\right\} \cup \left\{Nk_2 - Mk_1\right\}\ , \tag{11.22}$$

for $0 \le k_1 \le N-1, 0 \le k_2 \le M-1$, where $\tilde{\mathbb{S}}_{12} = \left\{Mk_1 - Nk_2\right\}$ and $\tilde{\mathbb{S}}_{12}^{-} = \left\{Nk_2 - Mk_1\right\}$.

Similarly,

$$\mathbb{S}_{1'2'} = \{Mk_1' - Nk_2'\} \cup \{Nk_2' - Mk_1'\}, \tag{11.23}$$

where $0 \le k_1^{'} \le N-1$, $0 \le k_2^{'} \le M-1$. From Eqs. (11.23) and (11.24), it is clear that $\mathbb{S}_{12}$ and $\mathbb{S}_{1'2'}$ have the same elements, i.e., $\mathbb{S}_{12} \cup \mathbb{S}_{1'2'} = \mathbb{S}_{12}$. On the other hand, $\mathbb{S}_{11'}$ is defined as the cross-lag set between subarray 1 and subarray $1^{'}$, whereas $\mathbb{S}_{22'}$ is defined as the cross-lag set between subarray 2 and subarray $2^{'}$. They are respectively given as,

$$\mathbb{S}_{11'} = \{Mk_1 - Mk_1' - 1\} \cup \{Mk_1' - Mk_1 + 1\}, \tag{11.24}$$

$$\mathbb{S}_{22'} = \{Nk_2 - Nk_2' - 1\} \cup \{Nk_2' - Nk_2 + 1\}. \tag{11.25}$$

Similarly,

$$\mathbb{S}_{12'} = \{Mk_1 - Nk_2' - 1\} \cup \{Nk_2' - Mk_1 + 1\}, \tag{11.26}$$

$$\mathbb{S}_{21'} = \{Nk_2 - Mk_1' - 1\} \cup \{Mk_1' - Nk_2 + 1\}. \tag{11.27}$$

Lemma 11.1 For sets $\mathbb{S}_{11'}$, $\mathbb{S}_{22'}$, $\mathbb{S}_{12'}$ and $\mathbb{S}_{21'}$ defined above, $\mathbb{S}_{11'} \cup \mathbb{S}_{22'} \cup \mathbb{S}_{12'} \cup \mathbb{S}_{21'} = \mathbb{S}_{12'} \cup \mathbb{S}_{21'}$. The proof can be seen in [27].

Because $\mathbb{S}_{12} \cup \mathbb{S}_{1'2'} = \mathbb{S}_{12}$ and based on Lemma 11.1, the cross-lags of the synthetic array are simplified to following set:

$$\mathbb{S}_{\rm ca} = \mathbb{S}_{12} \cup \mathbb{S}_{12'} \cup \mathbb{S}_{21'}. \tag{11.28}$$

With different combinations of the subsets, it is easy to show

$$\mathbb{S}_{12'} \cup \mathbb{S}_{21'} = \tilde{\mathbb{S}}_{12'} \cup \tilde{\mathbb{S}}_{21'} \tag{11.29}$$

where

$$\tilde{\mathbb{S}}_{12'} = \{Mk_1 - Nk_2' - 1\} \cup \{Nk_2 - Mk_1' - 1\},$$

$$\tilde{\mathbb{S}}_{21'} = \{Nk_2' - Mk_1 + 1\} \cup \{Mk_1' - Nk_2 + 1\}.$$

Then,

$$\mathbb{S}_{\rm ca} = \mathbb{S}_{12} \cup \tilde{\mathbb{S}}_{12'} \cup \tilde{\mathbb{S}}_{21'}. \tag{11.30}$$

It is noted that the new subsets $\tilde{\mathbb{S}}_{12'}$ and $\tilde{\mathbb{S}}_{21'}$ can be interpreted as the difference coarray of the original coprime array one unit step (lag) to the left and one unit step to the right. This shows that the number of unique lags increases due to motion.

We remark that if the difference coarray of the original coprime array is a filled virtual array, then there would be two additional unique lags which are created as a result of array motion, one to the left and one to the right. On the other hand, if the difference coarray of the original coprime array has holes, i.e., missing lags, the additional number of unique lags due to array motion would fill some or all of these holes. This property will be utilized to determine the number of unique lags. Because Eq. (11.30) does not depend on the specific subarray structures and sensor positions, the above conclusions are applicable to other sparse arrays.

Number of Unique Lags The number of unique lags in $\mathbb{S}_{\rm ca}$ is the sum of the number of unique lags in $\mathbb{S}_{12}$ and the number of holes in $\mathbb{S}_{12}$ which become filled due to array motion. By using a similar approach in analyzing CACIS in [18], the number of unique lags in $\mathbb{S}_{12}$ is obtained as

$$\eta_{\rm cb} = MN + M + N - 2. \tag{11.31}$$

We first identify the positions of the holes in $\mathbb{S}_{12}$ for $2 < M < N$ as

$$\mathbb{P}_{\text{hole}} = \left\{\pm(Ml_1 + Nl_2)d,\ l_1, l_2 \in \mathbb{L}\right\}, \tag{11.32}$$

where

$$\mathbb{L} = \Big\{ l_1, l_2 | 1 \leq l_1 \leq \left\lfloor N - 1 - \frac{N}{M} \right\rfloor, 1 \leq l_2 \leq \left\lfloor \left(M - \frac{2M}{N}\right) \right\rfloor, \\ |Ml_1 + Nl_2| \leq M(N-1) \Big\}, \tag{11.33}$$

and $\lfloor x \rfloor$ denotes the floor function which rounds x to the nearest integer toward zero. For the coarray shifted to the left by one unit, the condition for the holes in set $\mathbb{S}_{12}$ to be filled is:

$$Ml_1' - Nl_2' - 1 = \pm(Ml_1 + Nl_2), \tag{11.34}$$

$$Nl_2' - Ml_1' - 1 = \pm(Ml_1 + Nl_2), \tag{11.35}$$

whereas the condition for the coarray shifted to the right by one unit is given by

$$Ml_1' - Nl_2' + 1 = \pm(Ml_1 + Nl_2), \tag{11.36}$$

$$Nl_2' - Ml_1' + 1 = \pm(Ml_1 + Nl_2), \tag{11.37}$$

where $l_1, l_2 \in \mathbb{L}$, and

$$0 \leq l_1' \leq N - 1, \quad 0 \leq l_2' \leq M - 1. \tag{11.38}$$

Lemma 11.2 For an ($L_N = M + N - 1$)-element coprime array with unit inter-element spacing $d = \lambda/2$, the number of unique lags of the synthetic array after moving the original coprime array by a half wavelength is given by:

$$\eta_{\text{ca}} = \eta_{\text{cb}} + \eta_{\text{ci}} + 2, \tag{11.39}$$

where η_{ci} is the number of additional lags achieved due to array motion. It can be expressed as in Eq. (11.40),

$$\eta_{\text{ci}} = \begin{cases} 4(M-2) - 2, & N = M + 1, \\ 4(N-3) - 4, & N = 2M - 1, \\ 2[(M - 1 - \kappa_{2+})\alpha_{2+} + (N-4)\alpha_{2-}] - 2, & N = 2M + 1, \\ 2[(N - 1 - \kappa_{1+})\alpha_{1+} + (M - 1 - \kappa_{1-})\alpha_{1-}] - 4, & N = \beta M - 1, \quad \beta \geq 3, \\ 2[(M - 1 - \kappa_{2+})\alpha_{2+} + (N - 1 - \kappa_{2-})\alpha_{2-}] - 4, & N = \beta M + 1, \quad \beta \geq 3, \\ H & \text{otherwise}, \end{cases} \tag{11.40}$$

where

$$\begin{aligned} H = {} & 2[(N - 1 - \kappa_{1+})\alpha_{1+} + (M - 1 - \kappa_{1-})\alpha_{1-} + (M - 1 - \kappa_{2+})\alpha_{2+} \\ & + (N - 1 - \kappa_{2-})\alpha_{2-} - 1 - (N - 1 - \kappa_{1+})(M - 1 - \kappa_{2+}) \\ & - (M - 1 - \kappa_{1-})(N - 1 - \kappa_{2-})] - 2\max\big\{\min\{M - 1 - \kappa_{1-}, \alpha_{1+}\} \\ & \times \min\{N - 1 - \kappa_{1+}, \alpha_{1-}\}, \min\{M - 1 - \kappa_{2+}, \alpha_{2-}\} \min\{N - 1 - \kappa_{2-}, \alpha_{2+}\}\big\}, \end{aligned}$$

$$\kappa_{1+} = l_2 + l_2' + ((N - M)(l_2 + l_2') + 1)/M,$$

$$\kappa_{1-} = l_1 + l_1' - ((N - M)(l_1 + l_1') + 1)/N,$$

$$\kappa_{2+} = l_1 + l_1' + ((N - M)(l_1 + l_1') - 1)/M,$$
$$\kappa_{2-} = l_2 + l_2' - ((N - M)(l_2 + l_2') - 1)/N. \tag{11.41}$$

In addition, α_{1+} in Eq. (11.40) is the number of combinations that make $((N - M)(l_2 + l_2') + 1)$ divisible by M, and α_{1-} is the number to make $(N - M)(l_1 + l_1') + 1$ divisible by N. α_{2+} and α_{2-} are the numbers of combinations of l_1, l_1' and l_2, l_2' to make $(N - M)(l_1 + l_1') - 1$ divisible by N and $((N - M)(l_2 + l_2') - 1)$ divisible by M, respectively. The proof can be seen in [27].

Number of Consecutive Lags Note that all the holes may not be always filled due to array motion. In Figs. 11.10 and 11.11, we show two examples of an original coprime array and its corresponding synthetic array after motion. It is shown in Fig. 11.10a,b that the number of holes for both cases of $M = 4$, $N = 7$ and $M = 5$, $N = 6$ is the same. When considering the synthetic arrays, as shown in Fig. 11.11a, the difference coarray has no holes for the case of $M = 4$, $N = 7$, but Fig. 11.11b shows that one hole persists for the case of $M = 5$, $N = 6$.

It is straightforward to conclude that any two consecutive holes will be filled due to array motion. Accordingly, the first hole in the difference coarray of the synthetic array would appear at the first set of three consecutive holes in the difference coarray of the original array. Although the positions of maximum consecutive holes are given in [35] to determine the maximum frequency separation for an extended coprime array, yet the positions of consecutive holes are not analyzed, which is the purpose of Lemma 11.3.

Lemma 11.3 The position of the first positive hole in the difference coarray of the synthetic array, p_h, is expressed as

(a) For $N = M + 1, M \geq 5$,

$$p_h = 3M + N + 1;$$

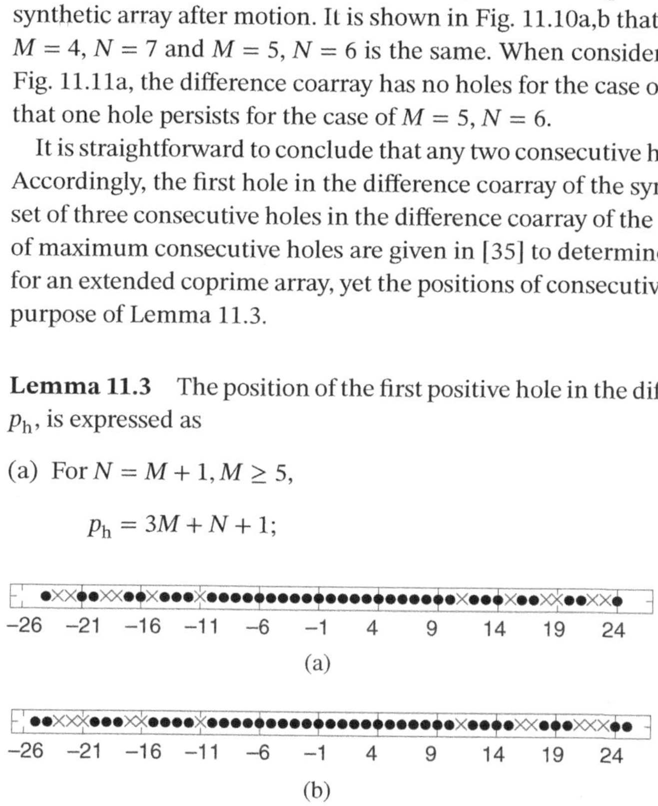

Figure 11.10 Difference coarray of original coprime array. (a) $M = 4, N = 7$; (b) $M = 5, N = 6$ (•: lags; ×: holes).

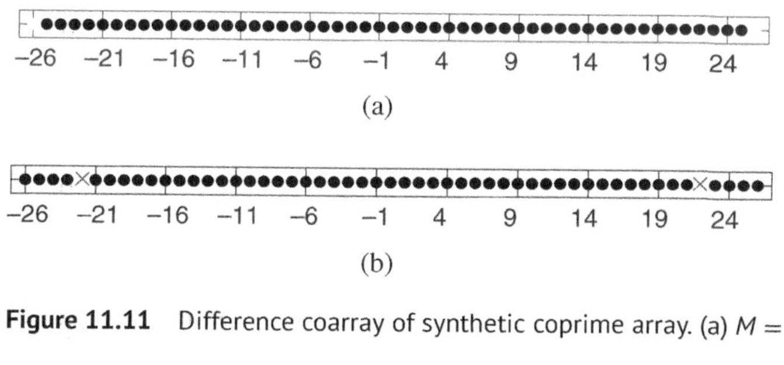

Figure 11.11 Difference coarray of synthetic coprime array. (a) $M = 4, N = 7$; (b) $M = 5, N = 6$.

(b) For $M + 1 < N < 2M - 1, M \geq 7$,

$$p_{\rm h} = \min\left[(\kappa_{\max +} + 1 - \kappa_{\rm h1+})M + (\alpha_{\rm h1+} - \alpha_{\min +} + 1)N + 1,\right.$$
$$\left.\alpha_{\rm h1-}M + (\kappa_{\max -} + 1 - \kappa_{\rm h1-})N + 1\right];$$

(c) For $N = 2M - 1$ and $M \geq 5$, or for $N = \beta M - 1, \beta \geq 3$, and $M \geq 4$,

$$p_{\rm h} = M + 3N + 1;$$

(d) For $N = \beta M + 1, \beta \geq 2$, and $M \geq 4$,

$$p_{\rm h} = (5 + 2(\beta - 2))M + N + 1;$$

(e) For $N > 2M + 1, N \neq \beta M \pm 1, \beta \geq 3$,

$$p_{\rm h} = \min\left[(\kappa_{\max} + 1 - \kappa_{\rm h1+})M + \alpha_{\rm h1+}N + 1,\right.$$
$$\left.\left(\alpha_{\rm h1-} - \alpha_{\min -} + 1\right) M + (\kappa_{\max -} + 1 - \kappa_{\rm h1-})N + 1\right].$$

(f) For other cases, no holes exist.

In the above expressions,

$$\kappa_{\rm h1+} = ((N - M)(l_2' + l_{n1}) - 1)/M + l_2' + l_{n1},$$
$$\alpha_{\rm h1+} = l_2' + l_{n1},$$
$$\kappa_{\rm h2+} = ((N - M)(l_2' + l_{n2}) - 2)/M + l_2' + l_{n2},$$
$$\alpha_{\rm h2+} = l_2' + l_{n2},$$
$$\kappa_{\rm h3+} = ((N - M)(l_2' + l_{n3}) - 3)/M + l_2' + l_{n3},$$
$$\alpha_{\rm h3+} = l_2' + l_{n3},$$
$$\kappa_{\max+} = \max(\kappa_{\rm h1+}, \kappa_{\rm h2+}, \kappa_{\rm h3+}),$$
$$\alpha_{\min+} = \min(\alpha_{\rm h1+}, \alpha_{\rm h2+}, \alpha_{\rm h3+}).$$

Similarly,

$$\kappa_{\rm h1-} = l_1' + l_{m1} - ((N - M)(l_1' + l_{m1}) + 1)/N$$
$$\alpha_{\rm h1-} = l_1' + l_{m1},$$
$$\kappa_{\rm h2-} = l_1' + l_{m2} - ((N - M)(l_1' + l_{m2}) + 2)/N$$
$$\alpha_{\rm h2-} = l_1' + l_{m2},$$
$$\kappa_{\rm h3-} = l_1' + l_{m3} - ((N - M)(l_1' + l_{m3}) + 3)/N,$$
$$\alpha_{\rm h3-} = l_1' + l_{m3},$$
$$\kappa_{\max-} = \max(\kappa_{\rm h1-}, \kappa_{\rm h2-}, \kappa_{\rm h3-}),$$
$$\alpha_{\min-} = \min(\alpha_{\rm h1-}, \alpha_{\rm h2-}, \alpha_{\rm h3-}).$$

The proof can be seen in [27].

By definitions, the number of consecutive lags is $\xi_{\rm ca} = 2p_{\rm h} - 1$, which becomes $\xi_{\rm ca} = 2M(N - 1) + 3$ if there are no holes. This is because the lags of the difference coarray, corresponding to the original array, are distributed between $-M(N - 1)$ and $M(N - 1)$. On the other hand, $\xi_{\rm cb} = 2(N + M) - 1$.

Table 11.2 Number of unique and consecutive lags for moving coprime arrays.

L_N	M, N	κ_{1+}	α_{1+}	κ_{1-}	α_{1-}	κ_{2+}	α_{2+}	κ_{2-}	α_{2-}	η_{ci}	η_{cb}	η_{ca}	NULR	ξ_{cb}	ξ_{ca}	NCLR
7	3, 5									4	21	27	1.29	15	27	1.80
8	4, 5									6	27	35	1.30	17	35	2.06
9	3, 7					1	2	2	1	8	29	39	1.35	19	39	2.05
10	5, 6									10	39	51	1.31	21	43	2.05
	4, 7									12	37	51	1.38	21	51	2.43
	3, 8	3	1	1	3					10	33	45	1.36	21	45	2.14
11	5, 7	3	2	2	3	3	4	4	3	16	45	63	1.40	23	63	2.74
12	6, 7									14	53	69	1.30	25	51	2.04
	5, 8	5	3	3	5	2	3	3	2	20	51	73	1.43	25	73	2.92
	4, 9					1	2	2	1	16	47	65	1.38	25	59	2.36
	3, 10					1	3	3	1	14	41	57	1.39	25	57	2.28

Example 11.1 Table 11.2 deals with coprime arrays and shows examples of the parameters computed from Lemmas 11.2 and 11.3 for different pairs of M and N. NULR denotes the ratio between the numbers of unique lags with and without array motion, i.e., η_{ca}/η_{cb}. NCLR denotes the ratio between the numbers of consecutive lags with and without array motion. The difference coarrays corresponding to the original and synthetic coprime arrays are, respectively, presented in Figs. 11.10 and 11.11.

From Table 11.2 and Figs. 11.10 and 11.11, we observe that the numbers of both unique and consecutive lags increase substantially due to array motion. This is largely because that all (see Fig. 11.11a) or most (see Fig. 11.11b) holes become filled. The maximum number of additional unique and consecutive lags are obtained when $2M - 1 \geq N > M + 1$ for a given number of sensors.

11.3.2.2 Other Sparse Arrays

Nested Array Similar to the coprime arrays, the difference coarray of the nested array after moving one unit step consists of the difference coarray before motion and the results of moving one unit step to the left and 1 step to the right from the original difference coarray. Because the difference coarray of two-level nested array is fully filled, the number of unique lags after the motion is expressed as

$$\eta_{na,2} = \eta_{nb,2} + 2, \tag{11.42}$$

where the expression of $\eta_{nb,2}$ is given in Table I [13].

For an optimum K-level nested array ($K = L_N - 1$), the number of unique lags before the array motion is given by $\eta_{nb,K} = L_N(L_N - 1) + 1$ [13]. The number of unique lags after the array motion

Table 11.3 Number of unique and consecutive lags for two-level nested arrays.

L_N	Optimal N_1, N_2	$\eta_{nb,2}$
Even	$N_1 = N_2 = L_N/2$	$(L_N^2 - 2)/2 + L_N$
Odd	$N_1 = (L_N - 1)/2, N_2 = (L_N + 1)/2$	$(L_N^2 - 1)/2 + L_N$

is expressed as

$$\eta_{\text{na},K} = 3L_N^2 - 13L_N + 21, \; L_N \geq 4. \tag{11.43}$$

The derivation of Eq. (11.43) is provided in [27].

MRA and MHA Because MRA and MHA do not have closed-form expressions for their array geometry and the number of achievable DOF, we analyze their DOF using examples, illustrated in Tables 11.4 and 11.5. The same sensor positions as in [21] are used. Because there may be multiple array configurations for MHA and MRA for a given number of sensors, the numbers of consecutive lags (ξ_{rb} and ξ_{hb}) are set to the maximum value.

SULA For the uniform sparse array with the inter-element spacing being three half wavelength, the sensor positions of after array motion are given by

$$\mathbb{P}_{\text{u1}} = \left\{3nd + 1, \, 0 \leq n \leq L_N - 1\right\}. \tag{11.44}$$

Because of the specific inter-element spacing, the number of unique lags in the difference array equals to $2(L_N - 1) + 1$, i.e., $\eta_{\text{sb}} = 2L_N - 1$, and the number of holes in the difference array is $4(L_N - 1)$. The number of the lags in the difference coarray of the synthetic array is expressed as

$$\eta_{\text{sa}} = 3(2L_N - 1). \tag{11.45}$$

The difference coarray after the array motion is fully filled because the number of holes between each pair of adjacent sensors is two. It is noted that, unlike other arrays, $\eta_{\text{sa}}/\eta_{\text{sb}} = 3$ is constant for the SULA.

Example 11.2 The difference coarrays of the original and synthetic arrays are respectively presented in Figs. 11.12 and 11.13 for the nested array, MRA, MHA, and SULA. From the figure, we

Table 11.4 Number of unique lags of a moving MRA.

L_N	η_{ri}	η_{rb}	η_{ra}	NULR	ξ_{rb}	ξ_{ra}	NCLR
3	2	7	9	1.29	7	9	1.29
5	2	19	21	1.11	19	21	1.11
6	2	27	29	1.07	27	29	1.07
8	2	47	49	1.04	47	49	1.04
9	2	59	61	1.03	59	61	1.03

Table 11.5 Number of unique lags of a moving MHA.

L_N	η_{hi}	η_{hb}	η_{ha}	NULR	ξ_{hb}	ξ_{ha}	NCLR
3	2	7	9	1.29	7	9	1.29
5	4	21	25	1.191	19	25	1.32
6	6	31	37	1.193	27	37	1.37
8	14	57	71	1.246	31	71	2.29
9	18	73	91	1.247	35	91	2.60

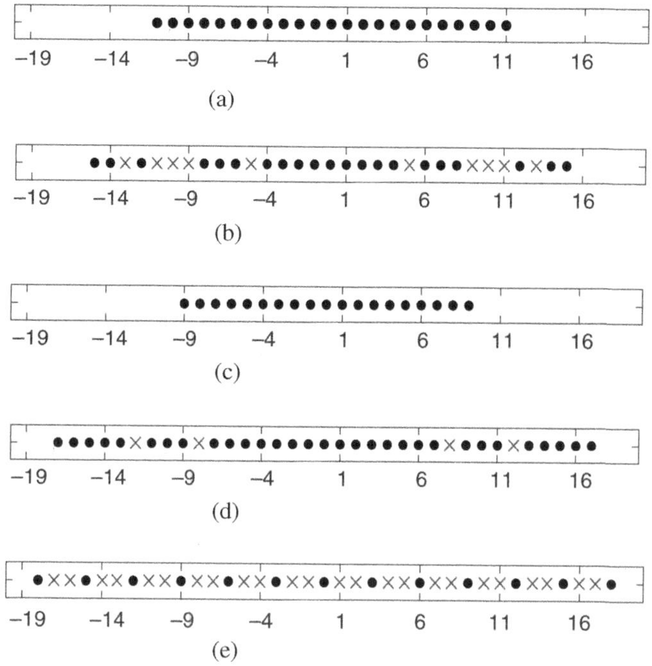

Figure 11.12 Difference coarrays of original sparse arrays. (a) Two-level nested array with $L_N = 6$; (b) Four-level nested array with $L_N = 5$; (c) MRA with $L_N = 5$; (d) MHA with $L_N = 6$; (e) SULA with $L_N = 7$.

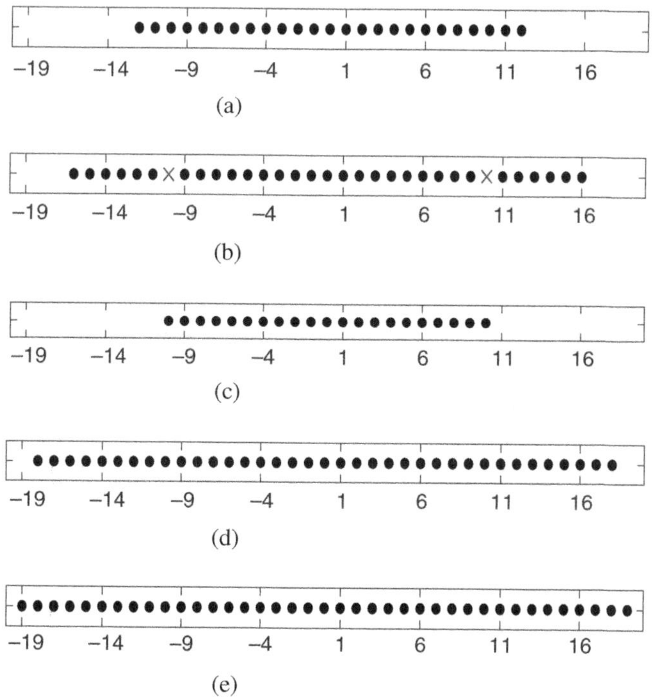

Figure 11.13 Difference coarray of synthetic sparse arrays. (a) Two-level nested array with $L_N = 6$; (b) Four-level nested array with $L_N = 5$; (c) MRA with $L_N = 5$; (d) MHA with $L_N = 6$; (e) SULA with $L_N = 7$.

observe again that the numbers of both unique and consecutive lags increase after array motion. Especially, for the SULA, there are no consecutive lags before array motion, whereas the difference coarray is fully filled after the array motion.

Remarks So far, we have analyzed the achievable DOF and the consecutive lags for different structured sparse arrays with array motion. The theoretical expressions are summarized in Table 11.6.

The following remarks are in order:

1. The difference coarrays of the synthetic two-level nested array, MRA and SULA are all fully filled. The first two original sparse arrays have fully-filled difference coarrays, while all the lags of the original SULA are disjoint.
2. The SULA achieves the highest NULR of 3, whereas the two-level nested array and MRA perform worst on the NULR and NCLR because only exactly two lags are added with the array motion, regardless of the number of sensors. The NULR and NCLR of the coprime array vary and achieve their maximum values when $2M - 1 \geq N > M + 1$ (see Table 11.2). The NULR of the K-level nested array increases as the number of sensors increases because the term $10L_N - 18/L_N^2 - L_N + 1$ monotonically decreases with L_N. The same situation happens on MHA.
3. For a given number of sensors, the K-level nested array achieves the highest number of unique lags by exploiting array motion because more holes in the original difference coarray are filled by the array motion. However, it has the smallest number of consecutive lags, which is equal to 19. MHA has the highest number of consecutive lags. The K-level nested array works better than SULA for a given number of sensors.

11.4 Structured Arrays Design for Moving Platforms

From the view of the Section 11.3, all sparse arrays can be used to form synthetic arrays with the array motion. However, these arrays are not optimal for the DOF. In this section, we introduce a new nested array for sparse array motion by expanding the sensor spacing of a conventional two-level nested array. The array configuration of the proposed dilated nested array is illustrated in Fig. 11.14. Similar to the conventional nested array, this nested array consists of two uniform linear subarrays with N_1 and N_2 sensors. However, the minimum inter-element spacing is changed to $d_r = rd$, where $r \geq 1$ is an integer. Note that, when $r = 1$, the proposed nested array degenerates to the conventional two-level nested array.

The sensor positions of the proposed nested array are expressed as

$$\mathbb{P}_{\mathrm{b}} = \{nd_r, 0 \leq n \leq N_1 - 1\} \cup \{(mN_1 + m - 1)d_r, 1 \leq m \leq N_2\}, \tag{11.46}$$

It is known that the difference coarray of the conventional two-level nested array is a fully filled array, whereas the difference coarray of the fixed position dilated nested array is not hole-free due to the sparse sensor spacing.

When the dilated nested array moves half wavelength, the sensor positions of the shifted array can be expressed as

$$\mathbb{P}_{\mathrm{a}} = \{n'd_r + d, 0 \leq n' \leq N_1 - 1\} \cup \{(m'N_1 + m' - 1)d_r + d, 1 \leq m' \leq N_2\}, \tag{11.47}$$

Table 11.6 Theoretical expressions of achievable DOF and consecutive lags for different structured sparse arrays with array motion.

	Coprime array	Two-level nested array	K-level nested array	MRA	MHA	SULA
Number of unique lags of original array	$MN + M + N - 2$	$\frac{L_N^2 - 2}{2} + L_N, L_N$ is even $\frac{L_N^2 - 1}{2} + L_N, L_N$ is odd	$L_N(L_N - 1) + 1$	See Table 11.4	See Table 11.5	$2L_N - 1$
Number of consecutive lags of original array	$2(M + N) - 1$	$\frac{L_N^2 - 2}{2} + L_N, L_N$ is even $\frac{L_N^2 - 1}{2} + L_N, L_N$ is odd	9	See Table 11.4	See Table 11.5	0
Number of unique lags of synthetic array	$MN + M + N + \eta_{ci}$	$\frac{L_N^2 + 2}{2} + L_N, L_N$ is even $\frac{L_N^2 + 3}{2} + L_N, L_N$ is odd	$3L_N^2 - 13L_N + 21$	See Table 11.4	See Table 11.5	$3(2L_N - 1)$
Number of consecutive lags of synthetic array	$2p_h - 1, p_h \neq 0$ $2M(N - 1) + 3, p_h = 0$	$\frac{L_N^2 + 2}{2} + L_N, L_N$ is even $\frac{L_N^2 + 3}{2} + L_N, L_N$ is odd	19	See Table 11.4	See Table 11.5	$3(2L_N - 1)$
NULR	$1 + \frac{\eta_{ci} + 2}{MN + M + N - 2}$	$1 + \frac{4}{L_N^2 + 2L_N - 2}, L_N$ is even $1 + \frac{4}{L_N^2 + 2L_N - 1}, L_N$ is odd	$3 - \frac{10L_N - 18}{L_N^2 - L_N + 1}$	See Table 11.4	See Table 11.5	3
NCLR	$\frac{2p_h - 1}{2(M + N) - 1}, p_h \neq 0$ $\frac{2M(N - 1) + 3}{2(M + N) - 1}, p_h = 0$	$1 + \frac{4}{L_N^2 + 2L_N - 2}, L_N$ is even $1 + \frac{4}{L_N^2 + 2L_N - 1}, L_N$ is odd	2.11	See Table 11.4	See Table 11.5	N/A

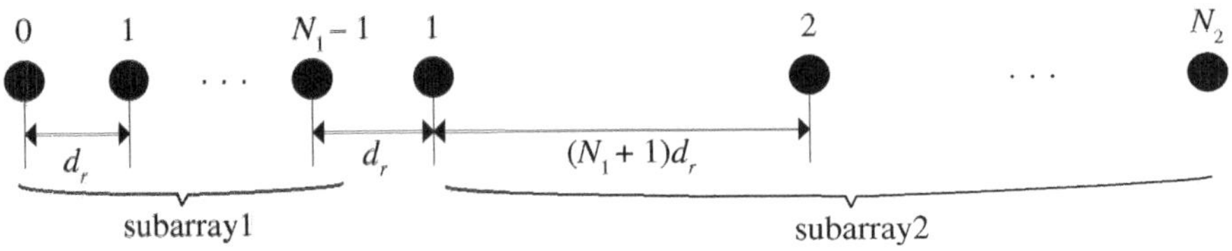

Figure 11.14 The array configuration of the dilated nested array.

The sensor positions of the synthetic dilated nested array, consisting of both original and the shifted arrays, are expressed as

$$\mathbb{P} = \mathbb{P}_b \cup \mathbb{P}_a. \tag{11.48}$$

Proposition 11.5 *The difference coarray of the synthetic dilated nested array has the following properties:*

(1) It is a fully filled array when $1 \leq r \leq 3$.

(2) There exists $2r'(N_1N_2 + N_2 - 1) + 3$ *unique lags in the coarray, where*

$$r' = \min(r, 3). \tag{11.49}$$

The proof can be seen in [28].

Because the original dilated nested array is obtained by extending the sensor spacing, the number of its unique lags is the same as that of the conventional two-level nested array. The number of the unique lags of the conventional two-level nested array is [13]

$$\xi_0 = \begin{cases} (L_N^2 - 2)/2 + L_N, & \text{for even } L_N, \\ (L_N^2 - 1)/2 + L_N, & \text{for odd } L_N, \end{cases} \tag{11.50}$$

where the optimal values of N_1 and N_2 are chosen as $N_1 = N_2 = L_N/2$ for L_N is even, and $N_1 = (L_N - 1)/2$, $N_2 = (L_N + 1)/2$ for L_N is odd [18]. From Proposition 11.5, the number of unique lags of the synthetic dilated nested array can be expressed as

$$\xi = \begin{cases} r'(L_N^2/2 + L_N - 1) + 3 - r', & L_N \text{ is even}, \\ r'(L_N^2/2 + L_N - 1/2) + 3 - r', & L_N \text{ is odd}. \end{cases} \tag{11.51}$$

Based on Eqs. (11.50) and (11.51), the number of unique lags of the synthetic dilated nested array can be simplified as

$$\xi = r'\xi_0 + 3 - r'. \tag{11.52}$$

Correspondingly, the ratio between the number of unique lags of the synthetic and that of the original dilated nested array achieves the maximum value of $\eta_{\max} = \xi/\xi_0 = 3$ when $r \geq 3$. This implies that the number of unique lags in the difference coarray of the synthetic dilated nested array through array motion is three times that of the conventional two-level nested array. Considering that the coarray corresponding to $r \geq 4$ is not hole-free and offers a short consecutive lags, $r = 3$ becomes the desirable choice to achieve hole-free coarrays.

An example is presented in Fig. 11.15. The non-negative part in difference coarrays of the synthetic dilated nested array structure with different values of r, where $L_N = 6$ is assumed, and r varies between 1 and 4. It is clear that all the difference coarrays are fully filled when $r \leq 3$. In this case, the number of consecutive lags increases as r increases, and it achieves the maximum value when $r = 3$.

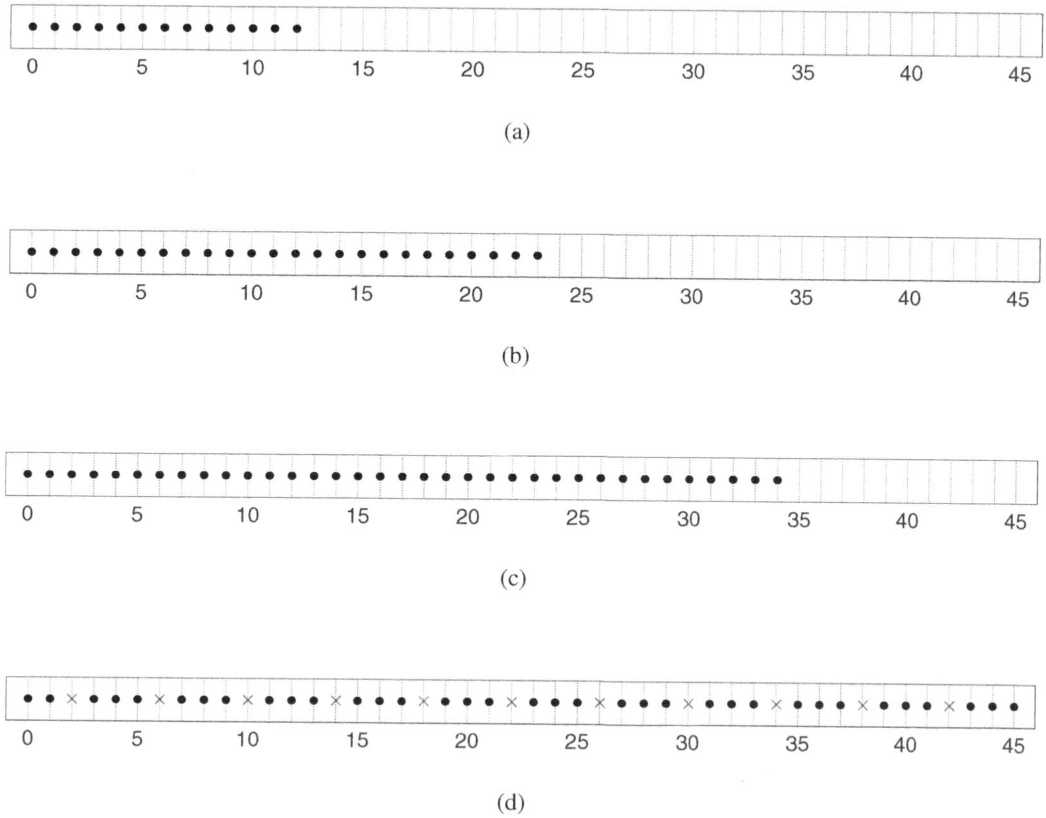

Figure 11.15 Non-negative part in the difference coarray of the synthetic dilated nested array. (a) $r = 1$; (b) $r = 2$; (c) $r = 3$; (d) $r = 4$ ($L_N = 6$, •: lags; ×: holes).

11.5 DOA Estimation Exploiting Array Motions

DOA estimation using sparse array can be processed with sparse signal reconstruction techniques [17] or coarray MUSIC [13]. Since these two techniques are discussed in the Chapter 1, only the simulations are provided here to verify the performance of the synthetic arrays.

In all simulations, the input signal-to-noise ratio (SNR) is set to 10 dB, and 1000 snapshots are used. Because each array yields a different number of DOF, different values of Q are utilized. The spatial frequencies of the Q sources, denoted as $\tilde{\theta} = d\sin(\theta)/\lambda$, are uniformly distributed between -0.49 and 0.49, which correspond to -78.5° and 78.5°, respectively. In the first simulation, [17] is used to perform sparsity-based DOA estimation, and all lags are exploited. $Q = 21$ sources are used for the coprime array with $M = 5$ and $N = 6$, whereas the value of Q is taken as 13, 18, and 17, respectively, for the four-level nested array, six-element MHA and eight-element SULA. For the MHA, the sensor positions are [0, 1, 4, 10, 15, 17].

The simulations results are shown in Figs. 11.16 and 11.17. It is clear from Fig. 11.16 that the original sparse array cannot identify the DOAs of all Q sources because of insufficient DOF. High-level sidelobes appear in Fig. 11.16d because the SULA has large inter-element spacing. On the other hand, as shown in Fig. 11.17, the proposed method correctly identify the DOAs of all Q sources due to the additional DOF achieved through array motion.

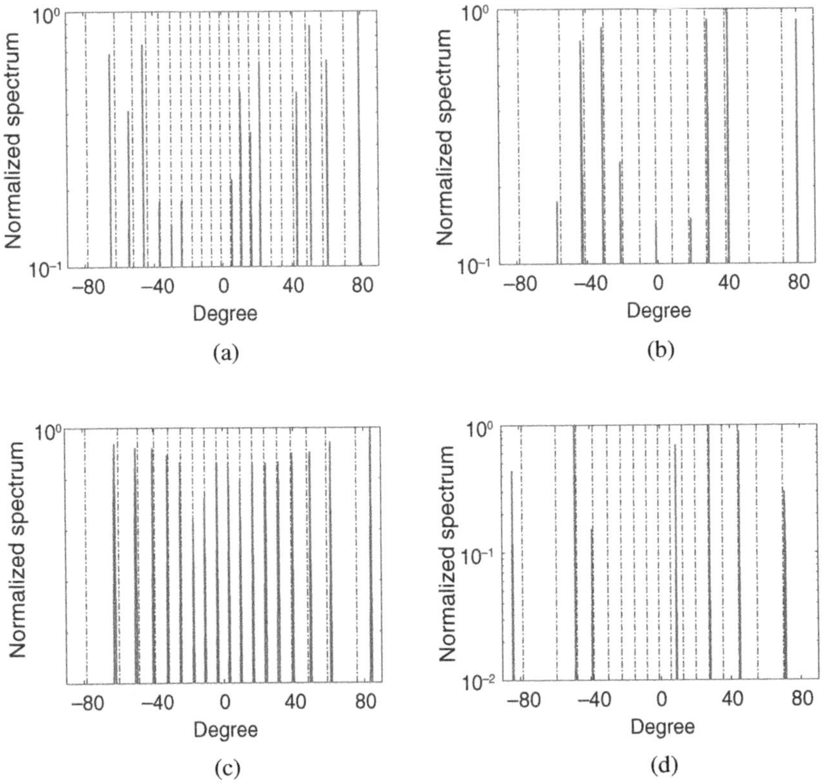

Figure 11.16 Estimated DOAs for original sparse arrays using sparsity-base technique. (a) Coprime array with $M = 5$ and $N = 6$ for 21 sources, $\eta_{cb} = 39$; (b) Four-level nested array with $L_N = 5$ for 13 sources, $\eta_{nb,K} = 21$; (c) MHA with $L_N = 6$ for 18 sources, $\eta_{hb} = 31$; (d) SULA with $L_N = 8$ for 17 sources, $\eta_{sb} = 17$.

In the second simulation example, we use the coarray MUSIC and only the consecutive lags are used. The values of Q are set to 18, 7, and 15, respectively, for the coprime array, the four-level nested array and the six-element MHA. The other parameters remain the same as in the previous simulation. The simulations results are shown in Fig. 11.18. After motion, in Fig. 11.18b,d,f, the number of consecutive lags increases, thus enabling the synthetic array to identify the DOAs of all Q sources. On the other hand, in Fig. 11.18a,c,e, the Q sources are not estimated correctly. This is because there is insufficient number of DOF.

11.6 Other Structured Arrays for Fixed and Moving Platforms

The emergence of coprime arrays has sparked interest in non-uniform arrays, resulting in numerous studies proposing various sparse configurations. Alongside CACIS and CADiS, several coprime array variants have been introduced as well. A thinned coprime array (TCA) is presented in [36] where some of the sensors of an extended coprime array are removed without affecting the aperture and the difference coarray, resulting in lower mutual coupling. Complementary coprime arrays (CCP) are derived in [37] where a dense uniform linear array (ULA) is added to an extended

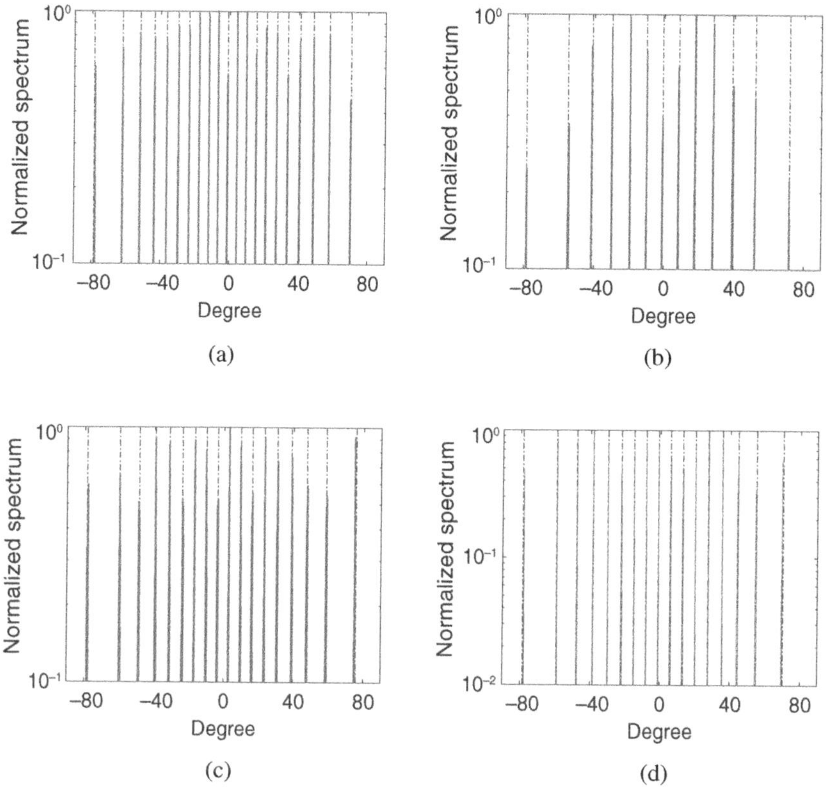

Figure 11.17 Estimated DOAs for synthetic sparse arrays using sparsity-base technique. (a) Coprime array with $M = 5$ and $N = 6$ for 21 sources, $\eta_{ca} = 51$; (b) Four-level nested array with $L_N = 5$ for 13 sources, $\eta_{na,K} = 31$; (c) MHA with $L_N = 6$ for 18 sources, $\eta_{ha} = 37$; (d) SULA with $L_N = 8$ for 17 sources, $\eta_{sa} = 45$.

coprime array to ensure contiguous difference coarray. In [38], the sensors in extended coprime arrays and CADiS are repositioned to create sliding extended coprime arrays (SECAs) and relocating extended coprime arrays (RECAs), respectively. SECAs and RECAs offer increased DOF and reduced mutual coupling compared with the original arrays. Zheng et al. [39] introduces a new array configuration based on the maximum inter-element spacing constraint (MISC) criterion. The MISC array configuration consists of three sparse ULAs plus two separate sensors which are set apart with an appropriate spacing. This configuration simultaneously increases the number of DOF and reduces the mutual coupling effects.

Most works based on sparse arrays deal with uncorrelated signals. The DOA estimation for coherent signals using moving sparse array is reported in [40]. Another important work is proposed in [41], where the sensor position errors are considered in the array synthetic process. The 2-D sparse circular array on rotating platform is provided in [42]. The DOA estimation based on the high-order cumulants is studied in [43].

Another research hotspot of moving sparse array is the array design, which aims to obtain higher number of DOF. The sparse arrays on moving platform are obtained by shifting multiple half-wavelengths in [44]. The q-dilated arrays are discussed in [45], where q is larger than 3. The improved coprime linear array configuration for moving platform is proposed in [46, 47].

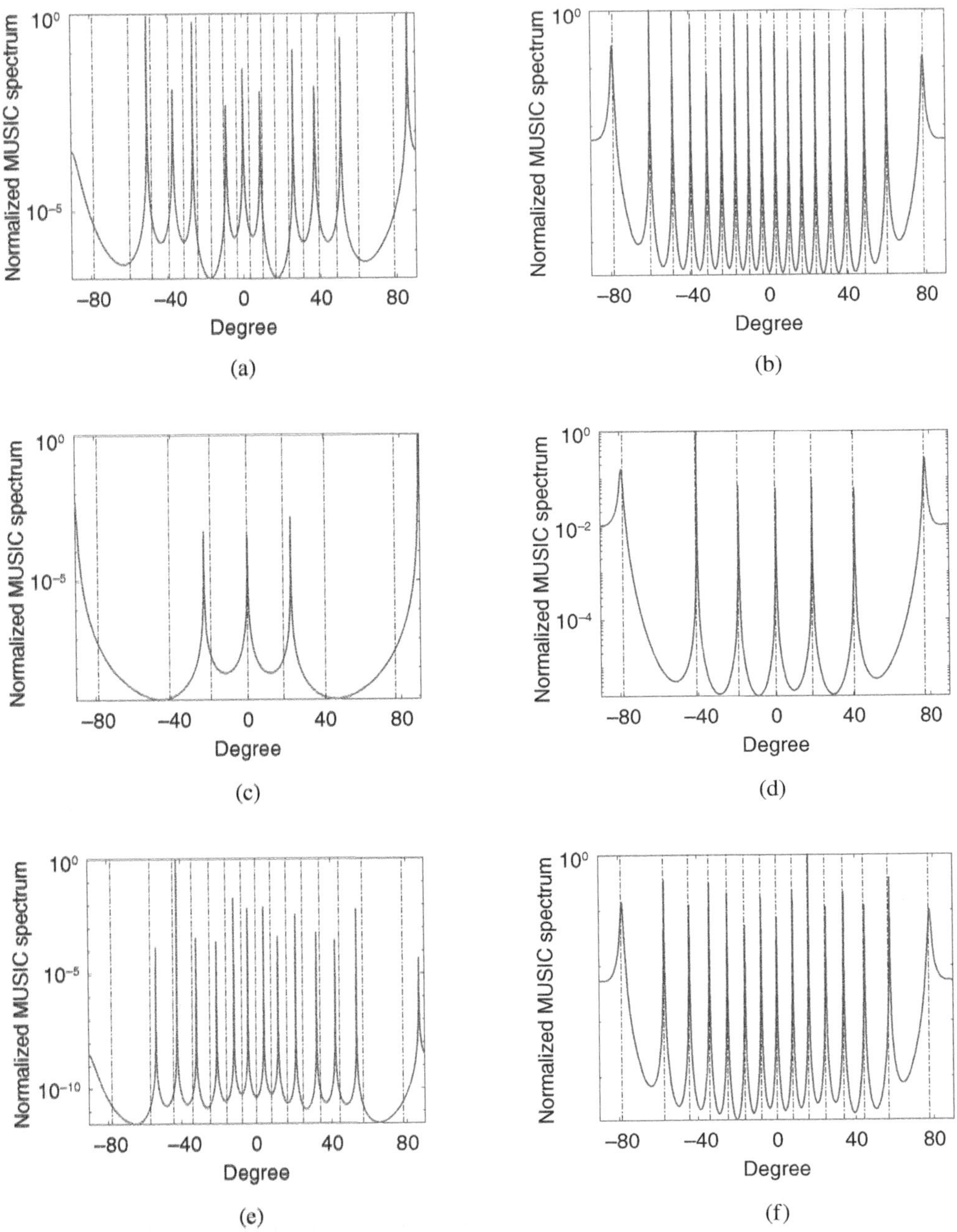

Figure 11.18 Estimated DOAs using coarray MUSIC. (a) The original coprime array with $M = 5$ and $N = 6$ for 18 sources, $\xi_{cb} = 21$; (b) The synthetic coprime array with $M = 5$ and $N = 6$ for 18 sources, $\xi_{ca} = 43$; (c) The original four-level nested array with $L_N = 5$ for 7 sources, $\xi_{nb} = 9$; (d) The synthetic four-level nested array with $L_N = 5$ for 7 sources, $\xi_{na} = 19$; (e) The original MHA with $L_N = 6$ for 15 sources, $\xi_{hb} = 27$; (f) The synthetic MHA with $L_N = 6$ for 15 sources, $\xi_{ha} = 37$. (ξ_{nb} and ξ_{na} are the number of consecutive lags for the original and synthetic four-level nested array.)

11.7 Conclusion

In this chapter, we presented the generalized structured arrays for fixed and moving platforms. The performance of the generalized arrays was evaluated using their difference coarray equivalence, and the analytical expressions of the coarray aperture, the achievable number of unique lags, and the maximum number of consecutive lags were derived for quantitative evaluation, comparison, and optimal design. The utility of these arrays was demonstrated using examples applied for DOA estimations.

References

1 Amin, M.G., Wang, X., Zhang, Y.D. et al. (2016). Sparse arrays and sampling for interference mitigation and DOA estimation in GNSS. *Proceedings of the IEEE* 104 (6): 1302–1317. https://doi.org/10.1109/JPROC.2016.2531582.

2 Yang, M., Sun, L., Yuan, X., and Chen, B. (2018). A new nested MIMO array with increased degrees of freedom and hole-free difference coarray. *IEEE Signal Processing Letters* 25 (1): 40–44. https://doi.org/10.1109/LSP.2017.2766294.

3 Gu, Y., Zhou, C., Goodman, N.A. et al. (2016). Coprime array adaptive beamforming based on compressive sensing virtual array signal. *Proceedings of the IEEE International Conference on Acoustics, Speech, and Signal Processing (ICASSP)*, 2981–2985, March 2016. https://doi.org/10.1109/ICASSP.2016.7472224.

4 Gelli, G. and Izzo, L. (1997). Minimum-redundancy linear arrays for cyclostationarity-based source location. *IEEE Transactions on Signal Processing* 45 (10): 2605–2608. https://doi.org/10.1109/78.640730.

5 Adhikari, K., Buck, J.R., and Wage, K.E. (2014). Extending coprime sensor arrays to achieve the peak side lobe height of a full uniform linear array. *EURASIP Journal on Advances in Signal Processing* 2014 (1): 148. https://doi.org/10.1186/1687-6180-2014-148.

6 Meng, H., Sun, K., Liu, Y., and Wang, X. (2011). Adaptive sparse representation for source localization with gain/phase errors. *Sensors* 11 (5): 4780–4793. https://doi.org/10.3390/s110504780.

7 Guo, M., Zhang, Y., and Chen, T. (2018). DOA estimation using compressed sparse array. *IEEE Transactions on Signal Processing* 66 (15): 4133–4146. https://doi.org/10.1109/TSP.2018.2847645.

8 Zhou, C., Gu, Y., Zhang, Y. et al. (2017). Compressive sensing-based coprime array direction-of-arrival estimation. *IET Communications* 11: 1719–1724(5).

9 Yuan, X. (2012). Coherent source direction-finding using a sparsely-distributed acoustic vector-sensor array. *IEEE Transactions on Aerospace and Electronic Systems* 48 (3): 2710–2715. https://doi.org/10.1109/TAES.2012.6237621.

10 Liao, B. and Chan, S. (2013). Direction-of-arrival estimation in subarrays-based linear sparse arrays with gain/phase uncertainties. *IEEE Transactions on Aerospace and Electronic Systems* 49 (4): 2268–2280. https://doi.org/10.1109/TAES.2013.6621815.

11 Hamza, S.A., Amin, M.G., Kirk, B., and Martone, A. (2022). Sparse array reconfigurability for source identification and angle estimation in cognitive sensing. *International Conference on Radar Systems (RADAR 2022)*, Volume 2022, 83–88. https://doi.org/10.1049/icp.2022.2296.

12 Vaidyanathan, P.P. and Pal, P. (2011). Sparse sensing with co-prime samplers and arrays. *IEEE Transactions on Signal Processing* 59: 573–586.

13 Pal, P. and Vaidyanathan, P.P. (2010). Nested arrays: a novel approach to array processing with enhanced degrees of freedom. *IEEE Transactions on Signal Processing* 58 (8): 4167–4181. https://doi.org/10.1109/TSP.2010.2049264.

14 Pal, P. and Vaidyanathan, P.P. (2012). Multiple level nested array: an efficient geometry for 2qth order cumulant based array processing. *IEEE Transactions on Signal Processing* 60 (3): 1253–1269. https://doi.org/10.1109/TSP.2011.2178410.

15 Liu, C. and Vaidyanathan, P.P. (2016). Super nested arrays: linear sparse arrays with reduced mutual coupling –Part I: Fundamentals. *IEEE Transactions on Signal Processing* 64 (15): 3997–4012. https://doi.org/10.1109/TSP.2016.2558159.

16 Liu, C. and Vaidyanathan, P.P. (2015). Remarks on the spatial smoothing step in coarray MUSIC. *IEEE Signal Processing Letters* 22 (9): 1438–1442. https://doi.org/10.1109/LSP.2015.2409153.

17 Zhang, Y., Amin, M.G., and Himed, B. (2013). Sparsity-based DOA estimation using co-prime arrays. *Proceedings of the IEEE International Conference on Acoustics, Speech, and Signal Processing*, 3967–3971, Vancouver, CA, May 2013. https://doi.org/10.1109/icassp.2013.6638403.

18 Qin, S., Zhang, Y., and Amin, M.G. (2015). Generalized coprime array configurations for direction-of-arrival estimation. *IEEE Transactions on Signal Processing* 63 (6): 1377–1390. https://doi.org/10.1109/TSP.2015.2393838.

19 Tan, Z. and Nehorai, A. (2014). Sparse direction of arrival estimation using co-prime arrays with off-grid targets. *IEEE Signal Processing Letters* 21 (1): 26–29. https://doi.org/10.1109/LSP.2013.2289740.

20 Moffet, A. (1968). Minimum-redundancy linear arrays. *IEEE Transactions on Antennas and Propagation* 16 (2): 172–175. https://doi.org/10.1109/TAP.1968.1139138.

21 Linebarger, D.A., Sudborough, I.H., and Tollis, I.G. (1993). Difference bases and sparse sensor arrays. *IEEE Transactions on Information Theory* 39 (2): 716–721. https://doi.org/10.1109/18.212309.

22 Wang, W., Ren, S., and Chen, Z. (2018). Unified coprime array with multi-period subarrays for direction-of-arrival estimation. *Digital Signal Processing* 74: 30–42. https://doi.org/10.1016/j.dsp.2017.11.015.

23 Huang, H., Liao, B., Guo, C., and Huang, J. (2018). Sparse representation based DOA estimation using a modified nested linear array. *Proceedings International Radar Conference*, 0919–0922, April 2018. https://doi.org/10.1109/RADAR.2018.8378683.

24 Zhou, C., Gu, Y., Fan, X. et al. (2018). Direction-of-arrival estimation for coprime array via virtual array interpolation. *IEEE Transactions on Signal Processing* 66 (22): 5956–5971.

25 Zhou, C., Shi, Z., Gu, Y., and Shen, X. (2013). DECOM: DOA estimation with combined MUSIC for coprime array. *Proceedings of the International Conference on Wireless Communications and Signal Processing*, 1–5, October 2013. https://doi.org/10.1109/WCSP.2013.6677080.

26 BouDaher, E., Ahmad, F., and Amin, M.G. (2015). Sparsity-based direction finding of coherent and uncorrelated targets using active nonuniform arrays. *IEEE Signal Processing Letters* 22 (10): 1628–1632. https://doi.org/10.1109/LSP.2015.2417807.

27 Qin, G., Amin, M.G., and Zhang, Y. (2019). DOA estimation exploiting sparse array motions. *IEEE Transactions on Signal Processing* 67 (11): 3013–3027. https://doi.org/10.1109/TSP.2019.2911261.

28 Qin, G., Zhang, Y., and Amin, M.G. (2019). DOA estimation exploiting moving dilated nested arrays. *IEEE Signal Processing Letters* 26 (3): 490–494. https://doi.org/10.1109/LSP.2019.2894467.

29 Qin, G. and Amin, M.G. (2020). Optimum sparse array configuration for DOA estimation on moving platforms. *Digital Signal Processing* 105: 102685. https://doi.org/10.1016/j.dsp.2020.102685.

30 Qin, G., Amin, M.G., and Zhang, Y. (2019). Analysis of coprime arrays on moving platform. submitted to *IEEE International Conference on Acoustics, Speech, and Signal Processing (ICASSP)*, Brigton, UK. ISSN 1520-6149. https://doi.org/10.1109/ICASSP.2014.6854226.

31 Qin, G. and Amin, M.G. (2021). Structured sparse array design exploiting two uniform subarrays for DOA estimation on moving platform. *Signal Processing* 180: 107872. https://doi.org/10.1016/j.sigpro.2020.107872.

32 Pal, P. and Vaidyanathan, P.P. (2011). Coprime sampling and the music algorithm. *Proceedings of the IEEE Digital Signal Processing and Signal Processing Education Meeting (DSP/SPE)*, 289–294. IEEE.

33 Stergiopoulos, S. and Sullivan, E.J. (1989). Extended towed array processing by an overlap correlator. *Journal of the Acoustical Society of America* 86: 158–171. https://doi.org/10.1121/1.398335.

34 Qin, S., Zhang, Y., Amin, M.G., and Himed, B. (2017). DOA estimation exploiting a uniform linear array with multiple co-prime frequencies. *Signal Processing* 130: 37–46.

35 BouDaher, E., Jia, Y., Ahmad, F., and Amin, M.G. (2015). Multi-frequency co-prime arrays for high-resolution direction-of-arrival estimation. *IEEE Transactions on Signal Processing* 63 (14): 3797–3808. https://doi.org/10.1109/TSP.2015.2432734.

36 Raza, A., Liu, W., and Shen, Q. (2017). Thinned coprime arrays for DOA estimation. *2017 25th European Signal Processing Conference (EUSIPCO)*, 395–399. IEEE.

37 Wang, X. and Wang, X. (2019). Hole identification and filling in *k*-times extended co-prime arrays for highly efficient doa estimation. *IEEE Transactions on Signal Processing* 67 (10): 2693–2706.

38 Zheng, W., Zhang, X., Wang, Y. et al. (2019). Extended coprime array configuration generating large-scale antenna co-array in massive MIMO system. *IEEE Transactions on Vehicular Technology* 68 (8): 7841–7853.

39 Zheng, Z., Wang, W.-Q., Kong, Y., and Zhang, Y.D. (2019). MISC array: a new sparse array design achieving increased degrees of freedom and reduced mutual coupling effect. *IEEE Transactions on Signal Processing* 67 (7): 1728–1741.

40 Li, W., Xu, X., Huang, X., and Yang, Y. (2023). Direction-of-arrival estimation for coherent signals exploiting moving coprime array. *IEEE Signal Processing Letters* 30: 304–308. https://doi.org/10.1109/LSP.2023.3261127.

41 Luo, S., Wang, Y., Wang, L., and Li, J. (2022). Direction finding for bistatic MIMO radar using a sparse moving array with sensor position errors. *IEEE Wireless Communications Letters* 11 (9): 1840–1844. https://doi.org/10.1109/LWC.2022.3184114.

42 Yadav, S. and George, N.V. (2021). Underdetermined direction-of-arrival estimation using sparse circular arrays on a rotating platform. *IEEE Signal Processing Letters* 28: 862–866. https://doi.org/10.1109/LSP.2021.3068713.

43 Wu, H., Shen, Q., Cui, W., and Liu, W. (2021). Doa estimation with nonuniform moving sampling scheme based on a moving platform. *IEEE Signal Processing Letters* 28: 1714–1718. https://doi.org/10.1109/LSP.2021.3105035.

44 Li, S. and Zhang, X. (2020). A new approach to construct virtual array with increased degrees of freedom for moving sparse arrays. *IEEE Signal Processing Letters* 27: 805–809. https://doi.org/10.1109/LSP.2020.2993956.

45 Li, S., Zhang, X., Ruan, D., and Zeng, H. (2023). Increasing the uniform degrees of freedom for moving q-dilated arrays. *IEEE Internet of Things Journal* 10 (1): 194–209. https://doi.org/10.1109/JIOT.2022.3200955.

46 Sun, L. (2023). Improved coprime array designs based on moving platform for direction of arrival estimation. *IEEE Sensors Journal* 23 (5): 5288–5298. https://doi.org/10.1109/JSEN.2023.3237751.

47 Yang, X., Wang, Y., and Chargé, P. (2021). Improved coprime linear array configuration for moving platform in DOA estimation. *IEEE Communications Letters* 25 (2): 470–473. https://doi.org/10.1109/LCOMM.2020.3029322.

12

Optimization and Learning-Based Methods for Radar Imaging with Sparse and Limited Apertures

Ammar Saleem[1], Alper Güngör[2], M. Burak Alver[3], Emre Güven[4], and Müjdat Çetin[5]

[1]*Faculty of Engineering & Natural Sciences, Sabanci University, Istanbul, Turkey*
[2]*Aselsan Research Center, Aselsan, Ankara, Turkey*
[3]*Faculty of Engineering & Natural Sciences, Maltepe University, Istanbul, Turkey*
[4]*Cognizen Arge Mühendislik ve Yazılım A.Ş., Ankara, Turkey*
[5]*Department of Electrical and Computer Eng. & Georgen Institute for Data Science, University of Rochester, Rochester, NY, USA*

12.1 Introduction

We present an overview and details of several state-of-the-art methods for synthetic aperture radar (SAR) imaging based on optimization and learning algorithms that are particularly suited to problems involving imperfect, noisy measurements from sparse, and limited observation apertures. While radar imaging conventionally relied on algorithms inspired by analytical inversion of the observation process (such as the polar format and range-Doppler algorithms) [1, 2], the emergence of scenarios involving irregular and limited apertures, including frequency band omissions, wide-angle measurements, and multi-tasking requirements motivated the development of explicit model-based methods that perform imaging through the solution of an optimization problem. Given the ill-posed nature of the imaging problems encountered in these scenarios, such optimization problems involve components that bring in regularization or prior information about the imaging problem. One such prior information that has been successfully used in the context of radar imaging is the sparsity of the scene, potentially in a transform domain. Sparsity-driven radar imaging has led to improved image quality, including increased resolution, reduced artifacts, and enhanced features in a variety of scenarios. While various optimization algorithms have been used to solve such problems, we primarily focus on state-of-the-art augmented Lagrangian-based methods such as the alternating direction method of multipliers (ADMM) that provide a flexible framework for imaging problems through variable splitting. We provide examples of SAR imaging of scenes from real measurements as well as methods that perform imaging and autofocusing simultaneously to compensate for model errors. While such methods allow the incorporation of a variety of regularizers including sparsity, explicitly constructing regularizers that capture complicated prior information from particular scene types is a challenge. In recent years, this has motivated the development of learning-based methods for radar imaging. Of particular interest are methods that continue to leverage physics-based information about the observation process, while using deep learning to construct implicit yet powerful regularizers. Interestingly, the ADMM-based framework discussed above can be used to bring in learning-based priors into image formation. We present an overview and examples of such learning-based methods for radar imaging from sparse and limited apertures, with a particular focus on Plug-and-Play (PnP) methods and convolutional neural network (CNN)-based denoisers, along with a set of results demonstrating their behavior.

Sparse Arrays for Radar, Sonar, and Communications, First Edition. Edited by Moeness G. Amin.

The rest of this chapter is organized as follows. Section 12.2 describes the sensing process and the resulting mathematical observation model for SAR imaging. That section also presents various data availability scenarios that we will consider in the rest of the chapter and that lead to limited and sparse observation apertures. Section 12.3 is dedicated to optimization-based methods for sparsity-driven SAR imaging. The section starts with an overview of relevant ideas from compressed sensing (CS) and describes how those ideas have led to the formulation of feature-enhanced SAR imaging methods. Then we turn to the main focus of the section, namely proximal algorithms, in particular ADMM, for SAR imaging in Section 12.3.3. That section develops an ADMM-based algorithm for SAR imaging and presents illustrative examples. Then, Section 12.3.4 considers the problem of model errors in SAR imaging and presents an ADMM-based approach for joint image formation and autofocusing along with illustrative examples on real SAR data. Section 12.4 changes gears and describes how state-of-the-art machine learning-based methods can be incorporated into the process of SAR image formation. The section starts with a brief discussion on the use of dictionary learning (DL)-based methods and then turns to the main focus, namely the use of deep learning-based regularizers for SAR imaging. Interestingly, the ADMM approach developed in earlier parts of the chapter provides a convenient framework for combining machine learning-based regularizers/priors and the physics-based observation model presented earlier in the chapter. That section presents a set of qualitative and quantitative results demonstrating the potential impact of learning for SAR imaging. We provide several concluding remarks at the end of the chapter.

12.2 SAR Observation Model

SAR is an imaging radar, usually mounted on a moving platform, that transmits and receives electromagnetic pulses to and from an illuminated area. The motion of the radar creates a synthetic aperture. SAR imaging is generally performed in one of two distinct modes: stripmap mode and spotlight mode. In stripmap mode, SAR imaging is performed with a fixed antenna direction with respect to the radar platform. This causes an antenna beam to pass over a strip on the ground. In spotlight mode, the antenna is continuously steered to illuminate a fixed area. High-bandwidth pulses, usually linear frequency-modulated (LFM) waveforms, are transmitted to illuminate the focused area (radius R) and the echoes are received and subsequently processed. Figure 12.1 illustrates a basic observation geometry for spotlight mode SAR.

Let $s(t)$ be the LFM pulse transmitted by the radar:

$$s(t) = \begin{cases} e^{j\omega_0 t + j\beta t^2}, & |t| \leq \frac{T_p}{2}, \\ 0, & \text{otherwise}, \end{cases} \tag{12.1}$$

where ω_0 is the carrier frequency and 2β is the FM rate. The radar transmits the real component of the transmitted signal at angle θ as shown in Fig. 12.1, then, after several assumptions [4], the returned echo signal $q_\theta(t)$ can be written as follows:

$$q_\theta(t) = \text{Re}\left\{ \int h_\theta(v) \exp(j\omega_0(t - \tau(v)) + j\beta(t - \tau(v))^2) dv \right\}. \tag{12.2}$$

$$h_\theta(v) = \iint_{x^2+y^2 \leq R^2} \delta(v - x\cos\theta - y\sin\theta) f(x,y) dx\, dy. \tag{12.3}$$

$h_\theta(v)$ is a projection of the reflectivity field $f(x,y)$ onto the corresponding θ as illustrated in Fig. 12.1 and $\tau(v)$ is the twoway propagation time of a wave from the radar platform to points in the

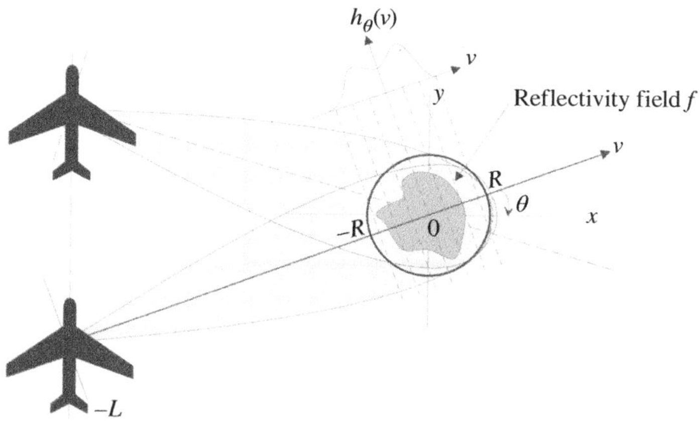

Figure 12.1 Ground plane geometry for spotlight-mode SAR sensing. The radar is mounted on an aircraft traversing along the region of interest (a circular region $-R$ to R) at a distance L while illuminating at various observation angles θ. Source: Adapted from [3].

scene at radial coordinates v and back. The received signal is subjected to various pre-processing steps, including multiplication by in-phase and quadrature components of a delayed version of the transmitted signal. Then, using the projection-slice theorem and other approximations, the pre-processed SAR data $g_\theta(t)$ can be formulated as a bandpass filtered Fourier transform (FT) of the projections $h_\theta(v)$ of the field $f(x, y)$ [4], given by:

$$g_\theta(t) = \int_{|v|\leq R} h_\theta(v) \exp(-jUv) dv \tag{12.4}$$

where $U = \frac{2}{c}(\omega_0 + 2\beta(t - \tau_0))$, τ_0 is the round-trip delay to the center of the ground patch, and c is the speed of light. We can relate $g_\theta(t)$ to $f(x, y)$ by substituting Eq. (12.3) into Eq. (12.4), as follows:

$$g_\theta(t) = \iint_{x^2+y^2\leq R^2} f(x, y) \exp\{-jU(x\cos\theta + y\sin\theta)\} dx\, dy \tag{12.5}$$

In practice, the observations at the each observation angle are noisy samples of the continuous received signal $g_\theta(t)$. This sampling in the time domain results in sampling of the spatial spectrum of the underlying reflectivity field. These samples lie in an annular region in the two-dimensional Fourier transform domain on a polar grid. A process called motion compensation (see Fig. 12.2) corrects the nonideal motion effects in the received signal. Then the noisy samples can be interpolated to a rectangular grid and concatenated into a measurement vector $\mathbf{y}$, which will be the input

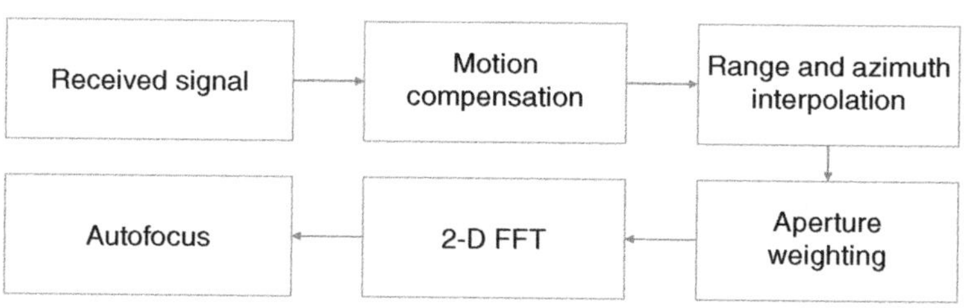

Figure 12.2 Typical processing steps for conventional SAR image formation.

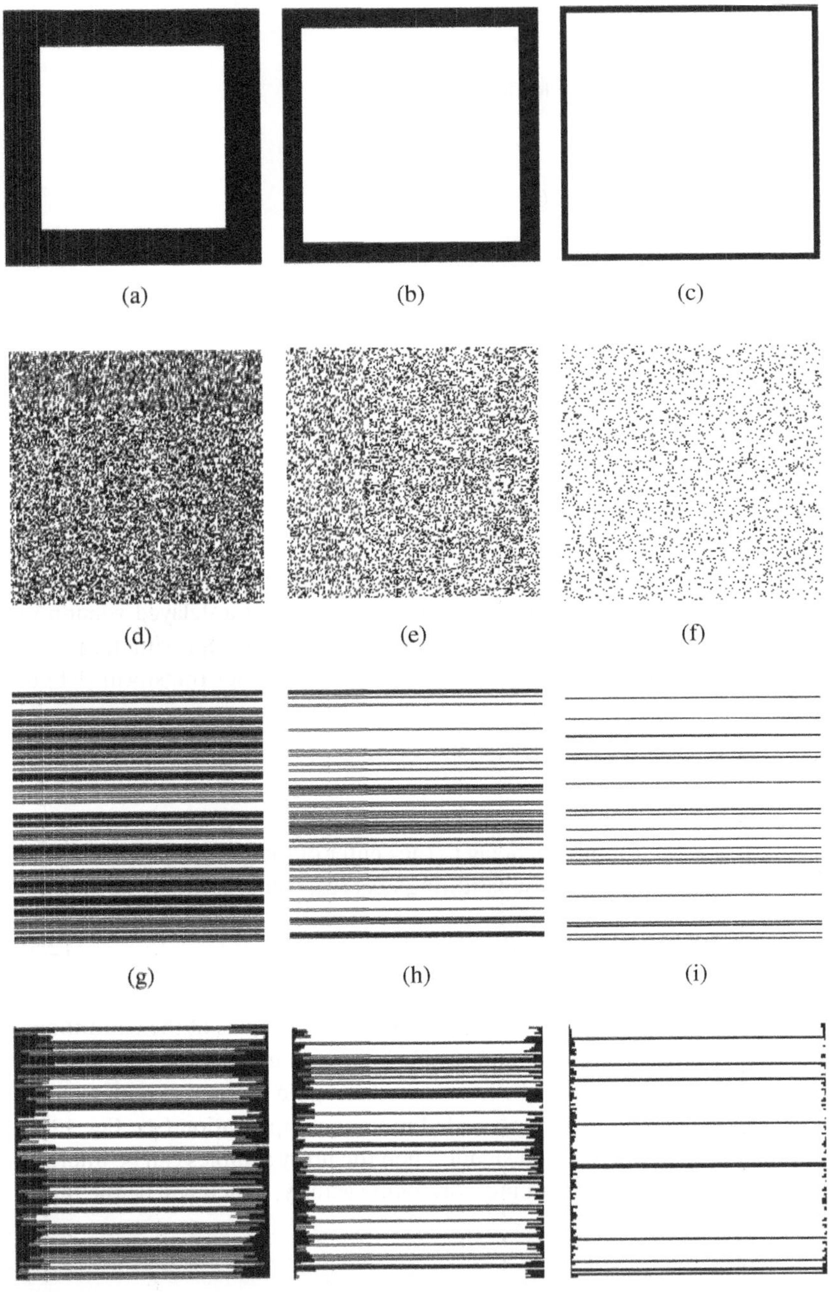

(a) (b) (c) (d) (e) (f) (g) (h) (i) (j) (k) (l)

Figure 12.3 Data selection matrix **M** configurations with white and black color depicting available and unavailable phase history samples, respectively. The rows of the matrix represent the frequency selectivity while the columns relate to look angle θ selectivity. Each row of the figure represents a particular type of selection matrix ((a–c) 2D-rectangular, (d–f) 2D-random, (g–i) 1D-linear, (j–l) 1D-random) and each column of the figure represents a particular data availability level η ((a, d, g, j) $\eta = 50\%$, (b, e, h, k) $\eta = 70\%$, (c, f, i, l) $\eta = 90\%$).

for the image formation algorithms presented in this chapter. Standard SAR image formation then performs aperture weighting followed by a 2D Fourier transformation. In addition, an optional autofocusing step can be applied after forming the image to correct the errors due to nonideal propagation media and the residual effects of nonideal motion. A typical flow of these processing steps is shown in Fig. 12.2. Further details can be found in standard texts on SAR imaging [1, 5, 6].

We use a discretized linear SAR observation model describing the relationship between the vector containing the complex-valued SAR reflectivity field and the phase history data collected in a spotlight SAR mode and pre-processed as described above. In our notation, $\mathbf{x} \in \mathbf{C}^N$ denotes the underlying image column-stacked as a vector and the matrix $\mathbf{B} \in \mathbf{C}^{M \times N}$ denotes the observation kernel that relates $\mathbf{x}$ to the vector of measurements $\mathbf{y} \in \mathbf{C}^M$:

$$\mathbf{y} = \mathbf{B}\mathbf{x} + \mathbf{n}, \tag{12.6}$$

where $\mathbf{n} \in \mathbf{C}^M$ denotes the additive noise vector, each element of which is assumed to be from a zero-mean independent identically distributed (i.i.d.) bivariate Gaussian distribution for real and imaginary parts. It is possible to work with alternative data domains. When the measurement vector $\mathbf{y}$ is in the phase history domain, as we assume here, the matrix $\mathbf{B}$ is a spatial Fourier transform operator; alternatively, if the data vector is a conventionally reconstructed image, then the observation matrix is a convolution operator associated with the point spread function that characterizes the entire chain of operations in the sensing and conventional imaging process.

The discrete forward model $\mathbf{B}$ can be expressed as the multiplication of a data selection matrix $\mathbf{M} \in \mathbf{R}^{M \times N}$ and the full Fourier transform matrix $\mathbf{U} \in \mathbf{R}^{N \times N}$, i.e., $\mathbf{B} = \mathbf{M}\mathbf{U}$. The data selection matrix $\mathbf{M}$ encodes aperture limitations and sparsity, primarily due to limited bandwidth and look angles, leading to structured sparse apertures. Another sparse-aperture scenario occurs due to missing look angles during the observation process, or frequency band omissions, leading to somewhat unstructured or even randomly distributed sparse apertures. Several nontraditional sensing scenarios, involving those with unmanned aerial vehicles (UAV) with complex motion around the scene, as well as multi-static and passive sensing can also lead to irregular and sparse apertures. Figure 12.3 shows typical data selection matrices showing the effect of data availability defined in terms of percentage $\eta = M/N * 100$.

The model in Eq. (12.5) and its discrete, noisy version in Eq. (12.6) assume a perfectly known location of the aircraft and ideal propagation media, which implies perfect knowledge of the phase of the collected measurements. However, in many cases, these assumptions may not be valid, resulting in erroneous estimation of the underlying scenes from the measurements. For this purpose, it is important to incorporate unknown phase errors.

12.3 Model-Based Imaging and the Role of Sparsity

12.3.1 Overview

In SAR image formation, the unknown, complex-valued reflectivity field is estimated from the observed phase history data, which is also complex-valued. The complex nature of the SAR sensing process captures both amplitude scaling and phase shifting of the transmitted waveform by the underlying scatterers. For most SAR scenes, the phase of the reflectivity at a certain spatial location can be modeled to be random, and uncorrelated with the phase at other locations [7].

The solution to the radar imaging problem involves solving an inverse scattering problem by reconstructing a spatial map of reflectivity from scattered electric field measurements. For 70 years of radar development, imaging techniques have been proposed to take advantage of parsimony in sparse or compressible scenes [8–11]. These techniques aim to suppress sidelobes and improve resolution of scattering locations. A key approach in many radar processing tasks involves finding sparse solutions to underdetermined linear equations, which is a topic addressed by the field of CS [12].

The primary focus of CS is to solve the inverse problem of recovering a signal $\mathbf{x} \in C^N$ from noisy linear measurements $\mathbf{y} = \mathbf{B}\mathbf{x} + \mathbf{n}$, where $\mathbf{B} \in C^{M \times N}$ is the forward operator with $M \ll N$. The solution to this ill-posed and underdetermined problem is primarily solved through regularization on the unknown scene, with two assumptions: (i) the unknown signal $\mathbf{x}$ is K-sparse in some transformation domain (i.e., has at most K nonzero entries) or is compressible with K significant coefficients, and (ii) the noise process is bounded by $\|\mathbf{n}\|_2 < \epsilon$.

CS theory provides strong results that guarantee stable solutions for a class of forward operators $\mathbf{B}$ that satisfy certain properties. One such class of operators is defined by bounding the singular values of the submatrices of $\mathbf{B}$. Specifically, consider restricted isometric constant (RIC) δ_K for a forward operator $\mathbf{B}$ be the smallest $\delta \in (0, 1)$ that satisfies

$$(1 - \delta_K)\|\mathbf{x}\|_2^2 \leq \|\mathbf{B}\mathbf{x}\|_2^2 \leq (1 + \delta_K)\|\mathbf{x}\|_2^2 \tag{12.7}$$

for all vectors $\mathbf{x}$ with at most K nonzero entries. Various works in the literature have shown that a bounded or small RIC can guarantee exact recovery of sparse signals in some transformation domain. For large problems, estimating and testing the RIC is not possible. A tractable yet conservative bound on the RIC can be obtained through the mutual coherence among the columns of $\mathbf{B}$.

Under a set of assumptions, CS offers stable recovery of noisy, compressible signals through a computationally tractable ℓ_1-regularized inverse problem:

$$\min_x \|\mathbf{x}\|_1 \text{ subject to } \|\mathbf{B}\mathbf{x} - \mathbf{y}\|_2 \leq \epsilon. \tag{12.8}$$

CS combines three essential elements: low mutual coherence among representation atoms defined by observation models and signal dictionaries, nonlinear reconstruction algorithms with low complexity, and, most importantly, sufficient conditions for provably stable reconstruction. The performance guarantees have affirmed the long-standing use of various sparse reconstruction algorithms, reinvigorated the use of pseudorandom waveforms or radar apertures with low coherence, and highlighted the usefulness of combining these two practices.

The core principles of CS, including sparse reconstruction algorithms and mutual coherence, have been prevalent in array processing literature for over 60 years. In the context of high-resolution imaging, the scattering response of an object can be accurately approximated as a sum of responses from individual reflectors, known as scattering centers. These centers provide a concise and physically meaningful description of the object. The scattering phenomenon is modeled as a superposition of responses from plates and dihedrals. In various works, several algorithms have been proposed to leverage the sparsity or compressibility of the scene reflectivity field to regularize underdetermined linear inverse problems [13].

12.3.2 Feature-Enhanced Sparse SAR Imaging

SAR imaging is an ill-posed and underdetermined problem, due to the fact that the number of measurements is much smaller than the number of unknowns. Sparsity-driven imaging can be a good candidate for a solution to this problem and can be defined as a constrained optimization problem as follows:

$$\begin{aligned} &\underset{\mathbf{x}}{\text{minimize}} \quad f(\mathbf{x}) \\ &\text{subject to } \|\mathbf{Bx} - \mathbf{y}\|_2 \leq \epsilon, \end{aligned} \tag{12.9}$$

One can also define an unconstrained version of this optimization problem using a regularization term:

$$\underset{\mathbf{x}}{\text{minimize}} \ \|\mathbf{Bx} - \mathbf{y}\|_2^2 + \lambda f(\mathbf{x}), \tag{12.10}$$

This cost function consists of two terms: the data fidelity term, which computes the difference between the data vector $\mathbf{y}$ and the linear transformation of the image vector $\mathbf{x}$ through $\mathbf{B}$, and the regularizer $f(\mathbf{x})$, which conditions the reconstructed reflectivity field according to the expected reflectivity characteristics of the scene. The formulated optimization problem is challenging to solve, due to both the random-phase nature of the reflectivities, and the presence of nonquadratic regularization (NQR) functions, which are needed for effective, robust feature enhancement. Initial work on these problems involved the development and use of quasi-Newton methods, built upon a half-quadratic-type regularization, to provide numerical solutions [3]. We will broadly refer to these SAR imaging methods as NQR techniques.

Common choices for $f(\mathbf{x})$ for NQR include ℓ_p-norm, total variation (TV) (ℓ_1-norm of spatial gradients), their combination, and ℓ_1-norm of the wavelet transform, all of which are performed on the reflectivity field. The quality of the reconstructed scene is affected by the choice of parameters λ or ϵ. The value of ϵ, which sets the radius of error in the constrained form, can be selected based on the signal-to-noise ratio (SNR) value estimated from the data. In addition, the discrepancy principle [14] can be used to evaluate the parameter selection. Other techniques that can be used for parameter section include generalized cross validation and Stein's unbiased risk estimation (SURE) [15].

12.3.2.1 Strong Scatterer Enhancement

SAR images often have sparse components in the image domain, especially when the imaged region contains strong scatterers in a weak background. CS has demonstrated that, with enough measurements and an incoherent basis for sampling, ℓ_1-norm minimization reconstruction and sparsest solution are practically equivalent. This makes the ℓ_1-norm regularizer $f(\mathbf{x}) = \|\mathbf{x}\|_1$ an effective way to obtain a sparse solution, which enhances strong scatterers or point targets in the image.

It has also been observed that, even though nonconvex, using ℓ_p-norm for $p < 1$ often produces even sparser results than using ℓ_1-norm [3, 8, 16]. This is a better approximation to solving the NP-hard problem with $p = 0$, although the strong nonconvexity may increase the likelihood of iterative methods converging to locally optimal solutions.

12.3.2.2 Region Enhancement

Most SAR object recognition algorithms use region or shape-based features. As an example, shapes of the object, shadow, and background regions are used for target classification. With conventional SAR images, segmentation of such regions is particularly difficult due to speckle. Accordingly, it is of interest to reduce variability in homogeneous regions, while preserving discontinuities at region boundaries. Such behavior has been obtained in real-valued image restoration and reconstruction problems by using constraints of the form $f(\mathbf{x}) = \|\nabla\mathbf{x}\|_1$, where ∇ is a discrete approximation to the 2D derivative operator. Using this objective function is commonly referred as total variation (TV) [17]. One could also consider replacing the ℓ_1-norm above with an ℓ_p-norm ($0 < p \leq 1$) for stronger research enhancement behavior. In the context of complex-valued SAR imaging, considering the spatially random-phase nature of the reflectivity fields, one can use the gradients of the magnitude of the reflectivity field for region enhancement, leading to $f(\mathbf{x}) = \|\nabla|\mathbf{x}|\|_1$ [3, 8].

12.3.2.3 Point Target and Region Enhancement

In addition to using a single objective function, another possibility is to use a weighted sum of multiple objective functions. Such a cost function can be expressed as:

$$\begin{aligned} &\underset{\mathbf{x}}{\text{minimize}} \quad \alpha_1 f_1(\mathbf{x}) + \alpha_2 f_2(\mathbf{x}) + \cdots + \alpha_m f_m(\mathbf{x}) \\ &\text{subject to} \qquad \|\mathbf{B}\mathbf{x} - \mathbf{y}\|_2 \leq \epsilon, \end{aligned} \tag{12.11}$$

where α_i, $i \in \{1, \ldots, m\}$, are hyper-parameters serving as the weights of each objective function.

Most SAR images consist of multiple features in a single scene, therefore, a combination of strong scatterers with uniform regions can be enforced through an optimization problem on the basis of NQR [3] as follows:

$$\begin{aligned} &\underset{\mathbf{x}}{\text{minimize}} \quad \alpha_1 \|\mathbf{x}\|_{p_1}^{p_1} + \alpha_2 \|\nabla|\mathbf{x}|\|_{p_2}^{p_2} \\ &\text{subject to} \quad \|\mathbf{B}\mathbf{x} - \mathbf{y}\|_2 \leq \epsilon, \end{aligned} \tag{12.12}$$

where p_1, p_2 are hyper-parameters acting as the sparsity-enhancing parameters of each objective function. Here, the first and second terms are aimed at enhancing point-based and region-based features, respectively, and the constraint enforces data fidelity. One could also write a regularized, unconstrained version of this problem.

12.3.2.4 Transform Domain Enhancement

Linear transforms, such as the discrete cosine or wavelet transform, are known to sparsify natural images, making them a popular choice for compressive sensing reconstructions [12, 18]. By enhancing the characteristics captured by these transforms, sparsity-driven reconstructions can effectively recover a compressed signal from limited measurements given by $f(\mathbf{x}) = \|\mathbf{W}|\mathbf{x}|\|_1$ [19].

Figure 12.4 illustrates some of the sparsity-driven feature-enhanced SAR imaging methods described in this section. As it can be observed, the performance of the recovery techniques relies heavily on the regularization model used for reconstruction.

While this section provided a brief overview of sparsity-driven imaging methods, our focus in this chapter is the solution of the types of optimization problems formulated here through proximal algorithms, which we turn to next.

12.3.3 Proximal Algorithms for SAR Imaging

Quasi-Newton methods, such as Broyden–Fletcher–Goldfarb–Shanno (BFGS) and limited-memory BFGS (L-BFGS) can be used to solve the optimization problems formulated above by approximating the inverse of the Hessian matrix using gradient information. One of the main advantages of quasi-Newton methods is their ability to converge faster than gradient descent-based methods by leveraging the curvature of the objective function. However, they can be computationally expensive for large-scale problems due to the need to store and update the Hessian matrix.

Proximal algorithms are widely used for constrained optimization problems. In contrast to quasi-Newton methods, proximal algorithms do not require the computation of the Hessian matrix or its inverse. Instead, they solve the optimization problem by applying proximal operators, which are simple and efficient functions that can be computed analytically or numerically. One of the main advantages of proximal algorithms is their ability to handle a wide range of constraints,

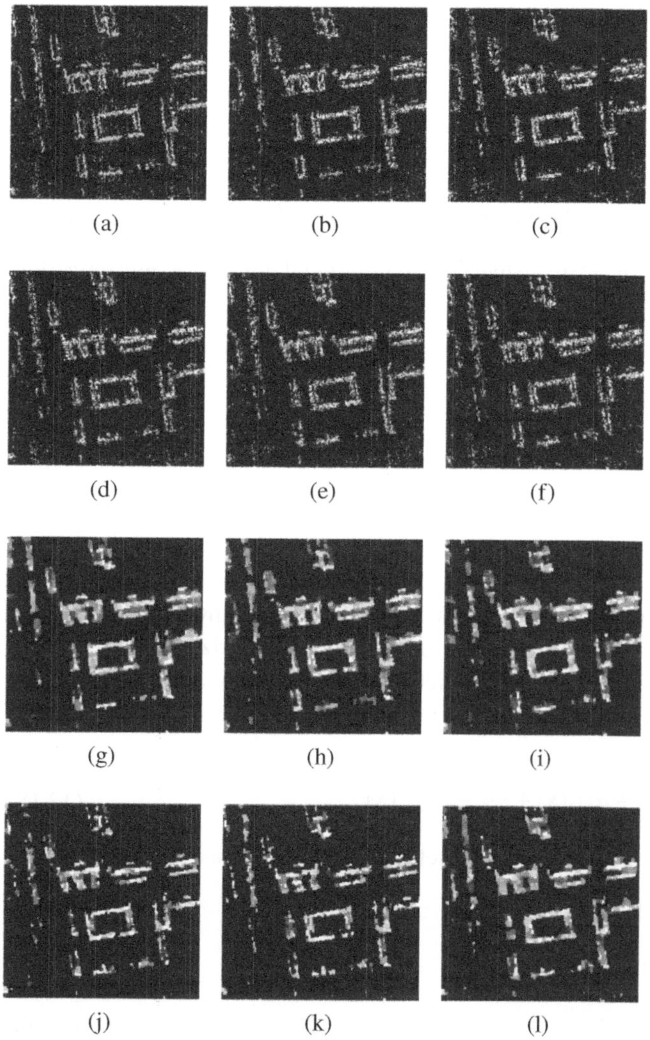

Figure 12.4 Illustration of different feature enhanced SAR imaging approaches considering three different data selection matrices at 50% data availability. (a, d, g, j) 2D-rectangular, (b, e, h, k) 2D-random, (c, f, i, l) 1D-linear. (a–c) conventional reconstruction, (d–f) strong scatter enhancement, (g–i) region enhancement using total variation, (j–l) point target and region enhancement. Source: [20]. Alver et al. 2020/from IEEE.

including sparsity, non-negativity, and TV [8, 21]. Moreover, they can be easily solved using parallel computing resources and are suitable for large-scale problems.

Specifically, ADMM based solutions hold great promise for practical applications. In Section 12.3.3.1 we provide an introduction to ADMM and in Section 12.3.3.2 we formulate ADMM-based methods for sparsity-driven SAR imaging.

12.3.3.1 Alternating Direction Method of Multipliers

ADMM was first introduced in the 1970s, with roots going back to the 1950s, and later rediscovered in 2011 [22]. ADMM incorporates ideas involving dual decomposition, method of multipliers, proximal methods, and variable splitting. It decomposes the problem into simpler sub-problems and solves each sub-problem iteratively. ADMM is particularly advantageous when the sub-problems have an analytical solution or an efficient algorithm to solve them. The algorithm is capable of solving problems of the following form:

$$\underset{\mathbf{x}}{\text{minimize}}\ g(\mathbf{x}) + h(\mathbf{z})\ \text{subject to}\ \mathbf{Gx} + \mathbf{Hz} = \mathbf{c}. \tag{12.13}$$

First, the augmented Lagrangian is formed:

$$L_\mu(\mathbf{x}, \mathbf{z}, \mathbf{d}) = g(\mathbf{x}) + h(\mathbf{z}) + \frac{\mu}{2}\mathbf{d}^T(\mathbf{Gx} + \mathbf{Hz} - \mathbf{c}) + \frac{\mu}{2}\|\mathbf{Gx} + \mathbf{Hz} - \mathbf{c}\|_2^2. \tag{12.14}$$

Then, an alternating minimization is used to solve the augmented Lagrangian minimization problem. At each iteration, the augmented Lagrangian is minimized over each variable as follows:

$$\mathbf{x}_{k+1} = \underset{\mathbf{x}}{\text{argmin}}\, L_\mu(\mathbf{x}, \mathbf{z}_k, \mathbf{d}_k), \tag{12.15}$$

$$\mathbf{z}_{k+1} = \underset{\mathbf{z}}{\text{argmin}}\, L_\mu(\mathbf{x}_{k+1}, \mathbf{z}, \mathbf{d}_k), \tag{12.16}$$

$$\mathbf{d}_{k+1} = \underset{\mathbf{d}}{\text{argmin}}\, L_\mu(\mathbf{x}_{k+1}, \mathbf{z}_{k+1}, \mathbf{d}). \tag{12.17}$$

12.3.3.2 ADMM-Based SAR Reconstruction

The optimization problems formulated in Section 12.3.2 can be solved using ADMM-based algorithms. Here we will consider the general problem in (12.11) which involves a weighted sum of multiple objective functions, enabling the use of multiple regularization constraints. We will use the term *hybrid* to refer to such cost functions. An ADMM-based solution can be obtained by first transforming the problem in (12.11) into its unconstrained counterpart as follows:

$$\underset{\mathbf{x}}{\text{minimize}}\ \iota_{E(\epsilon,\mathbf{B},\mathbf{y})}(\mathbf{Bx}) + \alpha_1 f_1(\mathbf{x}) + \alpha_2 f_2(\mathbf{x}) + \cdots + \alpha_m f_m(\mathbf{x}), \tag{12.18}$$

where $\iota_{E(\epsilon,\mathbf{B},\mathbf{y})}(\mathbf{Bx})$ is the indicator function on a feasible set $E(\epsilon, \mathbf{B}, \mathbf{y})$ defined as:

$$E(\epsilon, \mathbf{B}, \mathbf{y}) = \left\{\mathbf{x} \in \mathbf{C}^N\ :\ \|\mathbf{Bx} - \mathbf{y}\|_2 \leq \epsilon\right\} \tag{12.19}$$

and the indicator function $\iota_S(\mathbf{s})$ is defined as follows:

$$\iota_S(\mathbf{s}) = \begin{cases} 0, & \text{if}\, \mathbf{s} \in S, \\ +\infty, & \text{if}\, \mathbf{s} \notin S. \end{cases} \tag{12.20}$$

Let m be the number of separable objective functions. Let us make the following substitutions in (12.14) by dividing $\mathbf{z}$ into $m + 1$ vectors as:

$$\begin{aligned} \mathbf{G} &= \left[\mathbf{B}^\mathrm{T}\ \mathbf{I}\ \cdots\ \mathbf{I}\right]^\mathrm{T}, \quad \mathbf{H} = -\mathbf{I}, \mathbf{c} = \mathbf{0}, \\ \mathbf{z} &= \left[\left(\mathbf{z}^{(0)}\right)^\mathrm{T} \left(\mathbf{z}^{(1)}\right)^\mathrm{T}\ \cdots\ \left(\mathbf{z}^{(m)}\right)^\mathrm{T}\right]^\mathrm{T}, \end{aligned} \tag{12.21}$$

where $\mathbf{z}^{(0)}$ enforces data fidelity, and $\mathbf{z}^{(1)}, \ldots, \mathbf{z}^{(m)}$ correspond to each separable objective function. We update the objective function as

$$\begin{aligned} &\underset{\mathbf{z}}{\text{minimize}}\ \iota_{E(\epsilon,\mathbf{I},\mathbf{y})}\left(\mathbf{z}^{(0)}\right) + \alpha_1 f_1(\mathbf{z}^{(1)}) + \cdots + \alpha_m f_m(\mathbf{z}^{(m)}) \\ &\text{subject to} \qquad \mathbf{Gx} - \mathbf{z} = \mathbf{0}, \end{aligned} \tag{12.22}$$

which can be solved using ADMM.

The basic steps for solving (12.22) involve, first forming the augmented Lagrangian using (12.14) and then performing various independent iterative operations using ADMM

Algorithm 12.1: ADMM with a hybrid cost function

1. Set $k = 0$, choose $\mu > 0$, $\mathbf{z}_0^{(i)}$, $\mathbf{d}_0^{(i)}$, α_i for all i
2. **repeat**
3. $\mathbf{x}_{k+1} = (m\mathbf{I} + \mathbf{B^H B})^{-1} \left[\sum_{i=1}^{m} \mathbf{d}_k^{(i)} + \mathbf{z}_k^{(i)} + \mathbf{B}^H \left(\mathbf{d}_k^{(0)} + \mathbf{z}_k^{(0)}\right)\right]$
4. **for i = 1, …, m**
5. $\quad \mathbf{z}_{k+1}^{(i)} = \mathbf{\Psi}_{f_i \frac{\alpha_i}{\mu}} \left(\mathbf{x}_{k+1} - \mathbf{d}_k^{(i)}\right)$
6. $\quad \mathbf{d}_{k+1}^{(i)} = \mathbf{d}_k^{(i)} - \mathbf{x}_{\mathbf{k+1}} + \mathbf{z}_{k+1}^{(i)}$
7. **endfor**
8. $\mathbf{z}_{k+1}^{(0)} = \mathbf{\Psi}_{\iota_{E(\epsilon,\mathbf{I},\mathbf{y})}} \left(\mathbf{B}\mathbf{x}_{k+1} - \mathbf{d}^{(0)}\right)$
9. $\mathbf{d}_{k+1}^{(0)} = \mathbf{d}_k^{(0)} - \mathbf{B}\mathbf{x}_{\mathbf{k+1}} + \mathbf{z}_{k+1}^{(0)}$
10. $k \leftarrow k + 1$
11. **until** some stopping criterion is satisfied.

equations (12.15)–(12.17). Each operation requires solving a proximal mapping for a function $\Gamma(\mathbf{w})$ with a step size of μ, a slack input $\mathbf{v}$, and is defined as:

$$\text{prox}_\Gamma(\mathbf{v}) = \Psi_\Gamma(\mathbf{v}) = \arg\min_{\mathbf{w}} \Gamma(\mathbf{w}) + \frac{1}{2\mu} \|\mathbf{w} - \mathbf{v}\|_2^2. \tag{12.23}$$

The resulting ADMM with a hybrid cost function is summarized in Algorithm 12.1.

The inversion operation in Step 3 can be carried out using only two FFTs per iteration with details described in [8]. Therefore, in addition to leveraging proximal mapping functions, the algorithm can be executed using only one forward FFT and one inverse FFT per iteration, which is comparable to the cost of greedy methods such as Orthogonal Matching Pursuit [23].

Step 5 of the algorithm requires proximal mappings on complex-valued vectors. Proximal mappings of real-valued vectors for the functions given in Section 3.2 are defined in [21, 24]. For an input of $\mathbf{v}$ and a step size of μ, the complex extensions for the functions are derived in [8] and are summarized in Table 12.1.

Table 12.1 Proximal operators associated with the previously described objective functions [8, 21, 24, 25].

Objective function	Real-valued proximal mapping	Complex-valued proximal mapping
$\|\mathbf{x}\|_1$	$\text{soft}(\mathbf{v}, \frac{1}{\mu})$	$\text{soft}(\|\mathbf{v}\|, \frac{1}{\mu}) \exp\{j\angle\mathbf{v}\}$
$\|\mathbf{x}\|_p$	$\mathbf{w} = \mathbf{v}^{1-p}$ $\frac{1}{\mathbf{w}} \cdot \text{soft}(\mathbf{w} \cdot \mathbf{v}, \frac{1}{\mu})$	$\mathbf{w} = \|\mathbf{v}\|^{1-p}$ $\frac{1}{\mathbf{w}} \cdot \text{soft}(\mathbf{w} \cdot \|\mathbf{v}\|, \frac{1}{\mu}) \exp\{j\angle\mathbf{v}\}$
$\text{TV}(\|\mathbf{x}\|)$	$\text{ChambolleProj}(\mathbf{v})$	$\text{ChambolleProj}(\|\mathbf{v}\|) \exp\{j\angle\mathbf{v}\}$
$\|\mathbf{W}\mathbf{x}\|_1$	$\mathbf{W}^{-1}\, \text{soft}(\mathbf{W}\mathbf{v}, \frac{1}{\mu})$	$\mathbf{W}^{-1}\, \text{soft}(\|\mathbf{W}\mathbf{v}\|, \frac{1}{\mu}) \exp\{j\angle\mathbf{W}\mathbf{v}\}$
$\text{TV}(\|\mathbf{x}\|) + \alpha\|\mathbf{x}\|_1$	$\text{soft}(\text{prox}_{\text{TV}}(\mathbf{v}, \frac{1}{\mu}), \frac{\alpha}{\mu})$	$\text{soft}(\text{prox}_{\text{TV}}(\|\mathbf{v}\|, \frac{1}{\mu}), \frac{\alpha}{\mu}) \exp\{j\angle\mathbf{v}\}$

a) Implementation details for soft thresholding and Chambolle projection can be found in [21] and [24], respectively.

Finally, step 8 of the algorithm requires a projection onto ℓ_2-norm ball operation. For $\mathbf{s} = \mathbf{B}\mathbf{x}_{k+1} - \mathbf{d}_k^{(0)}$, that can be computed as follows:

$$\Psi_{\iota_{E(\epsilon,\mathbf{I},\mathbf{y})}}(\mathbf{s}) = \begin{cases} \mathbf{s}, & \text{if } \|\mathbf{s}-\mathbf{y}\|_2 \le \epsilon, \\ \mathbf{y} + \epsilon\frac{(\mathbf{s}-\mathbf{y})}{\|\mathbf{s}-\mathbf{y}\|_2}, & \text{if } \|\mathbf{s}-\mathbf{y}\|_2 > \epsilon. \end{cases} \tag{12.24}$$

12.3.3.3 Illustrative Examples

This section provides numerical demonstrations as well as visual representations of the ADMM-based approach for complex-valued SAR imaging. We first start by providing a computational time comparison in reconstructing 512 × 512 images from the TerraSAR-X dataset. Table 12.2 summarizes run-times on a single CPU for the described ADMM-based technique [8] to the feature-enhanced reconstruction method (FERM) [3] for two different undersampling types, 2D-rectangular and 2D-random sampling, as shown in Fig. 12.3, and for varying data availability levels/percentages. Note that FERM solves an unconstrained regularized optimization problem using a quasi-Newton method.

Since FERM employs an unconstrained formulation, a parameter matching procedure was employed for a fair comparison. First, FERM was run until convergence for a specific parameter set, which was followed by setting the ϵ value in the constrained formulation based on the converged computed output of FERM. As can be seen in Table 12.2, ADMM achieves similar cost function values in much less time.

The ADMM technique we presented here was also employed to reconstruct images based on real data collected by ASELSAN's SARPER™ system [26]. [8]. The reference image obtained by 2D inverse Fourier transformation of all of the available 5000 × 3500 data samples is shown in Fig. 12.5. These data have then been undersampled using a rectangular data selection matrix and image reconstruction is performed using the hybrid ADMM approach described here. Figure 12.6 shows the conventional minimum-norm image reconstructed from only 3.2% of the overall data and Fig. 12.7 shows the hybrid ADMM reconstruction from the same limited data. As visual inspection demonstrates, artifacts in the conventional image can be suppressed using hybrid ADMM, resulting in an image quality similar to the reference image, despite the extremely sparse nature of the measurement aperture.

Table 12.2 Computation times, error, and cost-ratios for FERM and hybrid ADMM (average computation times and cost function values over 200 runs) where N_{ADMM} represents the ADMM iteration count

Mask type	Data percentage (%)	t_{FERM} (s)	t_{ADMM} (s)	N_{ADMM}	$\frac{\epsilon_{\text{ADMM}}}{\epsilon_{\text{FERM}}}$	$\frac{\text{cost}(x_{\text{ADMM}})}{\text{cost}(x_{\text{FERM}})}$
Rectangular	75	10.0	2.6	34	0.994	0.90
	50	12.5	2.4	34	0.992	0.90
	25	14.4	2.2	34	0.986	0.92
	11	16.7	1.0	17	0.981	0.91
	6	17.8	1.0	16	0.994	0.74
Random	39	12.4	2.7	38	0.957	0.90
	22	14.9	2.9	42	0.922	0.94
	12	17.6	3.0	47	0.909	0.90
	6	19.7	3.1	48	0.974	0.93

Source: Adapted from [8].

Figure 12.5 Reference image from SARPER™ – airborne SAR system developed by ASELSAN. Source: [8]. H. Emre Güven et al. 2016/from IEEE.

Figure 12.6 Minimum-norm reconstruction from 3.2% of reference samples in Fourier space. Source: [8]. H. Emre Güven et al. 2016/from IEEE.

Figure 12.7 Reconstruction with the ADMM-based method from 3.2% of reference samples in Fourier space. Source: [8]. H. Emre Güven et al. 2016/from IEEE.

12.3.4 Imaging in the Presence of Model Errors

While forming SAR images, it is necessary to correct the errors originating from navigation errors or nonideal propagation media. Usually, autofocus algorithms are used to handle and correct these errors after the image formation process. However, in challenging sparse aperture and limited data scenarios, performing autofocusing as a post-processing operation after image formation may not produce satisfactory results. A more direct approach is to correct phase errors jointly while forming images, although the computational requirement remains a challenge in this case.

Several previous approaches have proposed techniques for compressive SAR imaging and autofocusing [27–32]. In this section, we focus on optimization-based joint estimation of phase errors and the underlying image. First, we briefly discuss sparsity-driven autofocus (SDA) [27] followed by an ADMM-based approach for joint estimation of phase errors and the reflectivity image from sparse-aperture data.

12.3.4.1 Sparsity-Driven Autofocus

We start by modeling the characteristics of phase errors. The errors originating from the navigation system can be modeled as one-dimensional phase errors in the azimuthal direction [27]. These errors create additive perturbations to the phase of the measurements in the Fourier domain, which can be modeled using an element-wise vector multiplication. Thus, imaging in the presence of phase errors mostly requires solving the following problem:

$$\begin{aligned} \underset{\mathbf{x},\phi}{\text{minimize}} \quad & \alpha_1 f_1(\mathbf{x}) + \cdots + \alpha_m f_m(\mathbf{x}) \\ \text{subject to} \quad & \|\mathbf{Bx} - \mathbf{y} \cdot \exp\{j\phi\}\|_2 \leq \epsilon, \end{aligned} \tag{12.25}$$

Algorithm 12.2: Sparsity-driven autofocus (SDA) [27]

1. **repeat**
2. Solve (12.26) over $\mathbf{x}$ with $\phi = \phi_k$ and set $\mathbf{x}_{k+1}$ equal to the result
3. Solve (12.26) over ϕ with $\mathbf{x} = \mathbf{x}_{k+1}$ and set ϕ_{k+1} equal to the result
4. $k \leftarrow k+1$
5. **until** some stopping criterion is satisfied.

where $\exp\{j\phi\}$ denotes the phase error term and $\cdot$ denotes element-wise multiplication. The problem (12.25) is nonconvex and finding its global optimum is an NP-hard problem.

SDA [27] solves the unconstrained version of the problem defined in (12.25). It does so by dividing the problem into two variables and using coordinate descent. The cost function is minimized in an alternating fashion over $\mathbf{x}$ and ϕ. The unconstrained form of the problem can be stated as follows:

$$\underset{\mathbf{x},\phi}{\text{minimize}}\ \alpha_1 f_1(\mathbf{x}) + \cdots + \alpha_m f_m(\mathbf{x}) + \frac{\lambda}{2}\|\mathbf{B}\mathbf{x} - \mathbf{y}\cdot\exp\{j\phi\}\|_2^2. \tag{12.26}$$

The steps of the SDA approach are summarized in Algorithm 12.2.

SDA uses a quasi Newton algorithm to solve step 2 of the algorithm. However, the unconstrained form of the ADMM described in Section 12.3.3 can also be used to solve step 2. It is also possible to use the constrained form to solve step 2 of the algorithm keeping the rest of the steps unchanged [28]. Step 3 of the algorithm solves the following problem:

$$\underset{\phi}{\text{minimize}}\ \|\mathbf{B}\mathbf{x}_{k+1} - \mathbf{y}\cdot\exp\{j\phi\}\|_2^2. \tag{12.27}$$

This problem has an analytical solution:

$$\phi_{(k+1)}(t) = -\arctan\left(\frac{\mathrm{Im}\left\{(\mathbf{B}\mathbf{x}_{(k+1)})_t^H\mathbf{y}_t\right\}}{\mathrm{Re}\left\{(\mathbf{B}\mathbf{x}_{(k+1)})_t^H\mathbf{y}_t\right\}}\right), \tag{12.28}$$

for all azimuth values t, where t denotes the azimuth index such that $\mathbf{y}_t$ denotes all elements of the vector $\mathbf{y}$ associated with the azimuth index t [25, 27].

12.3.4.2 Autofocusing with Compressive SAR Imaging Using ADMM

In this section, we discuss solving the problem in (12.25) using an ADMM-based approach, in particular the autofocusing hybrid ADMM approach introduced in [25]. This algorithm solves the constrained optimization problem given in Eq. (12.25) by using the phase error term, ϕ, as a variable in ADMM.

Let us first discuss possible advantages of using an ADMM-based approach for solving the problem in (12.26). First, the problem is not convex. Thus, the optimization technique as well as its initialization become important factors affecting convergence and the resulting reconstruction and phase error estimation performance. An early work for improving convergence relies on improving the image update step in SDA [28]. The technique introduced in [28], called AIRWALM, relies on using coordinate descent as in SDA, however solves the problem in step 2 of Algorithm 12.2 using ADMM, in constrained form. While constrained form allows for easier choice of data fidelity error bound, ϵ, using ADMM helps speed up the overall algorithm. Furthermore, an interesting idea is to perform only one iteration of ADMM, while keeping the dual variables of ADMM for fast convergence. Hence, the resulting algorithm within the coordinate-descent/ADMM framework simply includes a phase estimation step at the end of each iteration. This modification can also be seen as a model update during each iteration of ADMM. This voids any guarantees for convergence ADMM possesses. However, in practice, a consistent behavior of the algorithm was observed in [28].

Autofocusing hybrid ADMM: The approach we consider here, autofocusing hybrid ADMM [25, 33], uses the phase error, ϕ, as a variable in ADMM. First, let us refer back to Sections 12.3.3.1 and 12.3.3.2 for the necessary derivations. The following variables are defined to solve the problem (12.25) using ADMM:

$$\mathbf{z} = \begin{bmatrix} \mathbf{z}^{(\phi)} \\ \mathbf{z}^{(0)} \\ \mathbf{z}^{(1)} \\ \cdots \\ \mathbf{z}^{(m)} \end{bmatrix}, \quad \mathbf{G} = \begin{bmatrix} \mathbf{B} \\ \mathbf{I} \\ \cdots \\ \mathbf{I} \end{bmatrix}, \quad \mathbf{H} = \begin{bmatrix} 0 & -\mathbf{I} & \cdots & 0 \\ & & \cdots & \\ & & \cdots & \\ 0 & 0 & \cdots & -\mathbf{I} \end{bmatrix}, \quad \mathbf{c} = 0,$$

which divides $\mathbf{z}$ into $m+2$ sub-vectors where $\mathbf{z}^{(\phi)}$ is the phase error correction term associated with the $\mathbf{y} \cdot \exp\{j\phi\}$ operation, $\mathbf{z}^{(0)}$ is associated with the data fidelity term that accounts for the consistency between the reconstructed image and the gathered data, and $\mathbf{z}^{(1)} \ldots \mathbf{z}^{(m)}$ are associated with each of the separable objective functions as in Section 12.3.3. This setting ensures that $\mathbf{x} = \mathbf{z}^{(1)} = \cdots = \mathbf{z}^{(m)}$, $\mathbf{Bx} = \mathbf{z}^{(0)}$, and does not enforce any linear constraint on $\mathbf{z}^{(\phi)}$. The objective function is set in its final form as follows:

$$\underset{\mathbf{z}}{\text{minimize}} \;\; \iota_{E(\epsilon,\mathbf{I},\mathbf{y}\mathbf{z}^{(\phi)})}\left(\mathbf{z}^{(0)}\right) + \iota_{(\||\mathbf{z}^{(\phi)}|-1\|_2^2=0)}\left(\mathbf{z}^{(\phi)}\right) + \alpha_1 f_1(\mathbf{z}^{(1)}) + \cdots + \alpha_m f_m(\mathbf{z}^{(m)}), \tag{12.29}$$

where $\iota_{(\||\mathbf{z}^{(\phi)}|-1\|_2^2=0)}$ is the indicator function forcing magnitude of $\mathbf{z}^{(\phi)}$ to 1 and $\iota_{E(\epsilon,\mathbf{I},\mathbf{y}\mathbf{z}^{(\phi)})}\left(\mathbf{z}^{(0)}\right)$ is the indicator function associated with the inequality $\|\mathbf{Bx} - \mathbf{y}\mathbf{z}^{(\phi)}\|_2 \leq \epsilon$.

The described ADMM-based technique results in similar steps to those in Algorithm 12.1 except for the data fidelity projection steps. While the $\mathbf{x}$-update is equivalent to the update step in Algorithm 12.1 and the update steps for $\mathbf{z}^{(1)} \ldots \mathbf{z}^{(m)}$ are equivalent to (12.23), the data fidelity step is modified. The algorithm requires solving the problem below for data fidelity, instead of a projection onto the ℓ_2-norm ball:

$$\mathbf{z}_{k+1}^{(0)}, \mathbf{z}_{k+1}^{(\phi)} = \arg\min_{\mathbf{z}^{(0)},\mathbf{z}^{(\phi)}} \iota_{E(\epsilon,\mathbf{I},\mathbf{y}\mathbf{z}^{(\phi)})}\left(\mathbf{z}^{(0)}\right) + \iota_{(\||\mathbf{z}^{(\phi)}|-1\|_2^2=0)}\left(\mathbf{z}^{(\phi)}\right) + \frac{\mu}{2}\|\mathbf{Bx}_{k+1} - \mathbf{z}^{(0)} - \mathbf{d}_k^{(0)}\|_2^2. \tag{12.30}$$

As shown in [25], the solution to Eq. (12.30) can be efficiently computed. The problem can be cast back into a constrained form:

$$\begin{aligned} \arg\min_{\mathbf{z}^{(0)},\mathbf{z}^{(\phi)}} \quad & \|\mathbf{Bx}_{k+1} - \mathbf{z}^{(0)} - \mathbf{d}_k^{(0)}\|_2^2 \\ \text{subject to} \quad & \|\mathbf{z}^{(0)} - \mathbf{y} \cdot \mathbf{z}^{(\phi)}\|_2 \leq \epsilon \\ & \||\mathbf{z}^{(\phi)}| - 1\|_2^2 = 0. \end{aligned} \tag{12.31}$$

The optimal solution can then be computed in two steps as shown in [25]:

$$\mathbf{z}_{k+1}^{(\phi)}[i] = \frac{\mathbf{y}_{(i)}^H\left(\mathbf{Bx}_{k+1} - \mathbf{d}_k^{(0)}\right)_{(i)}}{\left|\mathbf{y}_{(i)}^H\left(\mathbf{Bx}_{k+1} - \mathbf{d}_k^{(0)}\right)_{(i)}\right|}, \quad \text{for all } i, \tag{12.32}$$

$$\mathbf{z}_{k+1}^{(0)} = \Psi_{\iota_{E\left(\epsilon,\mathbf{I},\mathbf{y}\cdot\mathbf{z}_{k+1}^{(\phi)}\right)}}(\mathbf{Bx}_{k+1} - \mathbf{d}_k^{(0)}). \tag{12.33}$$

Here, $\mathbf{z}_{k+1}^{(\phi)}[i]$ corresponds to the phase correction factor of a row in the phase-history domain; $\mathbf{y}_{(i)}$ and $\left(\mathbf{Bx}_{k+1} - \mathbf{d}_k^{(0)}\right)_{(i)}$ represent the corresponding rows in the phase-history domain. The resulting method is given in Algorithm 12.3.

Algorithm 12.3: Autofocusing ADMM with a hybrid cost function

1. Set $k = 0$, choose $\mu > 0$, $\mathbf{z}_0^{(i)}$, $\mathbf{d}_0^{(i)}$, α_i for all i
2. **repeat**
3. $\mathbf{q}_k = \sum_{i=1}^{m} \mathbf{d}_k^{(i)} + \mathbf{z}_k^{(i)}$
4. $\mathbf{x}_k = \frac{1}{m}\left[\mathbf{q}_k + \frac{1}{m+1}\mathbf{B}^H\left(m\left(\mathbf{d}_k^{(0)} + \mathbf{z}_k^{(0)}\right) - \mathbf{B}\mathbf{q}_k\right)\right]$
5. $\mathbf{B}\mathbf{x}_{k+1} = \frac{1}{m+1}\left[\mathbf{B}\mathbf{q}_k + \left(\mathbf{d}_k^{(0)} + \mathbf{z}_k^{(0)}\right)\right]$
6. **for** $\mathbf{i} = 1, \ldots, \mathrm{m}$
7. $\quad \mathbf{z}_{k+1}^{(i)} = \mathbf{\Psi}_{f_i \frac{\alpha_i}{\mu}}\left(\mathbf{x}_{k+1} - \mathbf{d}_k^{(i)}\right)$
8. $\quad \mathbf{d}_{k+1}^{(i)} = \mathbf{d}_k^{(i)} - \mathbf{x}_{\mathbf{k+1}} + \mathbf{z}_{k+1}^{(i)}$
9. **endfor**
10. $\mathbf{z}_{k+1}^{(\phi)}[i] = \frac{\mathbf{y}_{(i)}^H\left(\mathbf{B}\mathbf{x}_{k+1} - \mathbf{d}_k^{(0)}\right)_{(i)}}{\left|\mathbf{y}_{(i)}^H\left(\mathbf{B}\mathbf{x}_{k+1} - \mathbf{d}_k^{(0)}\right)_{(i)}\right|}$, for all i
11. $\mathbf{z}_{k+1}^{(0)} = \mathbf{\Psi}_{\iota_{E\left(\epsilon, \mathbf{I}, \mathbf{y}\cdot\mathbf{z}_{k+1}^{(\phi)}\right)}}(\mathbf{B}\mathbf{x}_{k+1} - \mathbf{d}_k^{(0)})$
12. $\mathbf{d}_{k+1}^{(0)} = \mathbf{d}_k^{(0)} - \mathbf{B}\mathbf{x}_{\mathbf{k+1}} + \mathbf{z}_{k+1}^{(0)}$
13. $k \leftarrow k + 1$
14. **until** some stopping criterion is satisfied.

Forward-looking autofocusing ADMM: Next, we discuss an accelerated phase estimation technique for the ADMM-based approach. As discussed earlier, since the overall optimization problem is nonconvex, the path that the optimization problem follows is a critical factor affecting its convergence. To accelerate phase error estimation, Algorithm 12.3 can be re-organized, without any change in terms of computational cost. To this end, we need to compute the variable $\mathbf{z}_{k+2}^{(\phi)}$, instead of $\mathbf{z}_{k+1}^{(\phi)}$ within each iteration. Here, we observe that computation of $\mathbf{z}_{k+2}^{(\phi)}$ depends on the $\mathbf{B}\mathbf{x}_{k+2} - \mathbf{d}_{k+1}^{(0)}$ term, that can be computed as follows:

$$\mathbf{B}\mathbf{x}_{k+2} - \mathbf{d}_{k+1}^{(0)} = \frac{1}{m+1}\left[\mathbf{B}\mathbf{q}_{k+1} - m\mathbf{d}_{k+1}^{(0)} + \mathbf{z}_{k+1}^{(0)}\right]. \tag{12.34}$$

Expanding $\mathbf{d}_{k+1}^{(0)}$ term, we get:

$$\mathbf{B}\mathbf{x}_{k+2} - \mathbf{d}_{k+1}^{(0)} = \frac{1}{m+1}\left[\mathbf{B}\mathbf{q}_{k+1} - m\mathbf{d}_{k}^{(0)} + m\mathbf{B}\mathbf{x}_{k+1} + (1-m)\mathbf{z}_{k+1}^{(0)}\right]. \tag{12.35}$$

For $m > 1$, the forward-looking estimation procedure requires computation of $\mathbf{z}_{k+1}^{(0)}$. This eventually requires running the phase error estimation step twice, one for estimating $\mathbf{z}_{k+1}^{(0)}$ and one for estimating $\mathbf{z}_{k+2}^{(0)}$.

However, for the case with $m = 1$, this forward-looking procedure can be implemented efficiently without any additional computational cost:

$$\mathbf{B}\mathbf{x}_{k+2} - \mathbf{d}_{k+1}^{(0)} = \frac{1}{2}\left(\mathbf{B}\mathbf{q}_{k+1} - \mathbf{d}_{k}^{(0)} + \mathbf{B}\mathbf{x}_{k+1}\right). \tag{12.36}$$

Here, $\mathbf{B}\mathbf{q}_{k+1}$ can be computed efficiently by moving step 3 of the algorithm after step 9. The resulting steps are summarized in Algorithm 12.4.

12.3.4.3 Illustrative Examples

This section provides numerical demonstrations as well as visual representations of the augmented Lagrangian method-based approach for joint autofocusing and image estimation for complex-valued SAR imaging.

Algorithm 12.4: Forward-looking autofocusing ADMM

1. Set $k = 0$, choose $\mu > 0, \mathbf{z}_0^{(i)}, \mathbf{q}_0, \mathbf{d}_0^{(i)}, \alpha_i$ for all i
2. **repeat**
3. $\mathbf{x}_k = \frac{1}{m}\left[\mathbf{q}_k + \frac{1}{m+1}\mathbf{B}^H\left(m\left(\mathbf{d}_k^{(0)} + \hat{\mathbf{z}}_k^{(0)}\right) - \mathbf{B}\mathbf{q}_k\right)\right]$
4. $\mathbf{B}\mathbf{x}_{k+1} = \frac{1}{m+1}\left[\mathbf{B}\mathbf{q}_k + \left(\mathbf{d}_k^{(0)} + \hat{\mathbf{z}}_k^{(0)}\right)\right]$
5. **for i** = 1, …, m
6. $\quad \mathbf{z}_{k+1}^{(i)} = \mathbf{\Psi}_{f_i \frac{\alpha_i}{\mu}}\left(\mathbf{x}_{k+1} - \mathbf{d}_k^{(i)}\right)$
7. $\quad \mathbf{d}_{k+1}^{(i)} = \mathbf{d}_k^{(i)} - \mathbf{x}_{\mathbf{k+1}} + \mathbf{z}_{k+1}^{(i)}$
8. **endfor**
9. $\mathbf{q}_{k+1} = \sum_{i=1}^{m} \mathbf{d}_{k+1}^{(i)} + \mathbf{z}_{k+1}^{(i)}$
10. $\mathbf{z}_{k+2}^{(\phi)}[i] = \frac{\mathbf{y}_{(i)}^H\left(\mathbf{B}\mathbf{q}_{k+1} - \mathbf{d}_k^{(0)} + \mathbf{B}\mathbf{x}_{k+1}\right)_{(i)}}{\left|\mathbf{y}_{(i)}^H\left(\mathbf{B}\mathbf{q}_{k+1} - \mathbf{d}_k^{(0)} + \mathbf{B}\mathbf{x}_{k+1}\right)_{(i)}\right|}$, for all i
11. $\hat{\mathbf{z}}_{k+1}^{(0)} = \mathbf{\Psi}_{\iota_{E\left(\epsilon, \mathbf{I}, \mathbf{y} \cdot \mathbf{z}_{k+2}^{(\phi)}\right)}}(\mathbf{B}\mathbf{x}_{k+1} - \mathbf{d}_k^{(0)})$
12. $\mathbf{d}_{k+1}^{(0)} = \mathbf{d}_k^{(0)} - \mathbf{B}\mathbf{x}_{\mathbf{k+1}} + \hat{\mathbf{z}}_{k+1}^{(0)}$
13. $k \leftarrow k + 1$
14. **until** some stopping criterion is satisfied.

First, autofocusing ADMM is compared to a coordinate-descent-based technique in terms of phase error estimation performance. To this end, we present a simulation-based study with strong scatterers using ℓ_p-norm enhancement. However, phase error estimation error is ambiguous to linear gain and offset, since an offset would only multiply the overall image with a constant phase and a linear gain would shift the image by a few pixels in one dimension circularly, which may not have an effect on the objective functions. Hence, to compare to ground-truth phase error, both estimated and ground-truth phases are unwrapped and a line is fit between them. Finally, phase error estimation error (PE) is computed as follows:

$$PE = \min_{a,b} \frac{1}{t} \|a\phi_e - \phi + b\|_2, \tag{12.37}$$

where ϕ_e is the estimated phase, ϕ is the reference phase, t denotes the number of elements; a, b are the parameters of the line.

Figure 12.8 shows phase estimation error versus iteration using a 2D-random data selection operator, leading to a random sparse aperture, for two techniques and varying p parameters for the ℓ_p-norm. Overall performance of autofocusing ADMM is better than the coordinate descent based AIRWALM approach. The ADMM-based algorithm converges faster to a better phase error estimate. Choosing a smaller p-value improves phase estimation quality as well as convergence speed.

We demonstrate the performance of forward-looking autofocusing ADMM with hybrid features on real data collected by ASELSAN's SARPER™ system [26] using a 1D random undersampling mask. The hybrid features we use here are aimed at strong scatterer enhancement using an ℓ_1-norm regularizer and region enhancement using TV. Figure 12.9 shows the reference image reconstructed using full data followed by the conventional phase gradient autofocus (PGA) algorithm [34]. Figure 12.10a shows the conventional reconstruction from undersampled data. As can be seen, the image is corrupted by undersampling artifacts and phase errors.

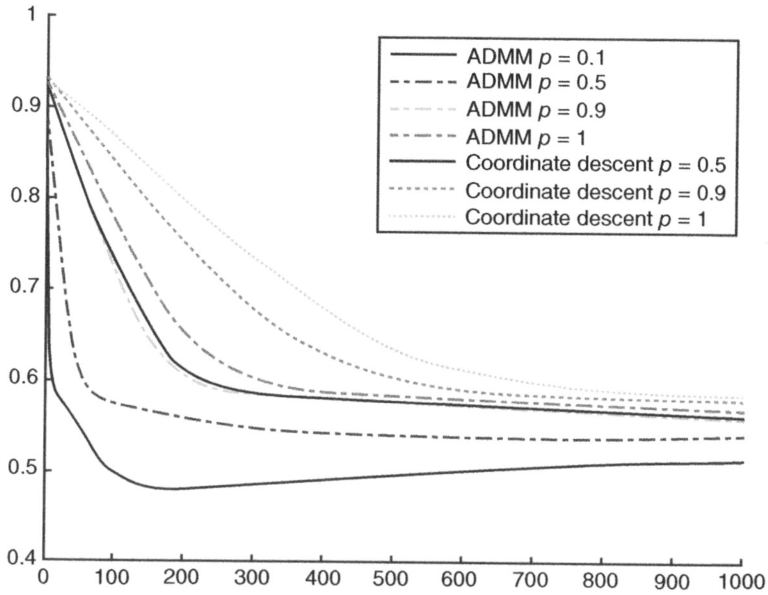

Figure 12.8 Root-mean-square error (RMSE) in phase error estimation in radians versus iteration for autofocusing ADMM and AIRWALM (coordinate descent) with different p values using a random mask for data reduction in the Fourier space. Source: [33]/IEEE.

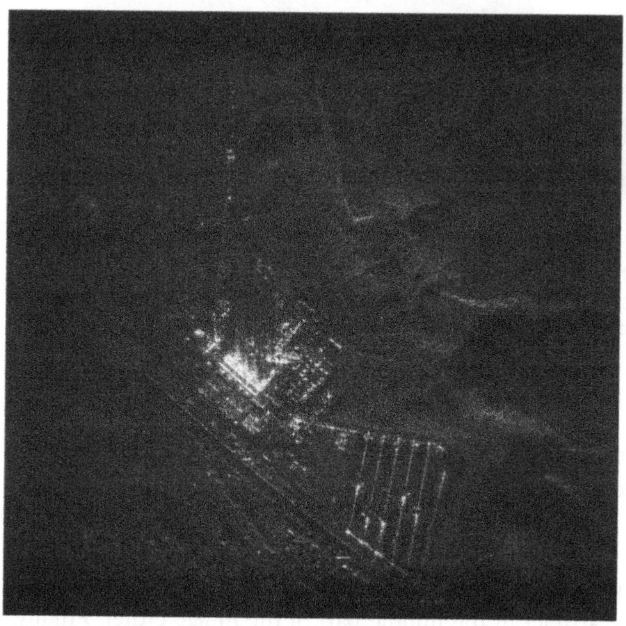

Figure 12.9 Reference image reconstructed using full-data followed by PGA.

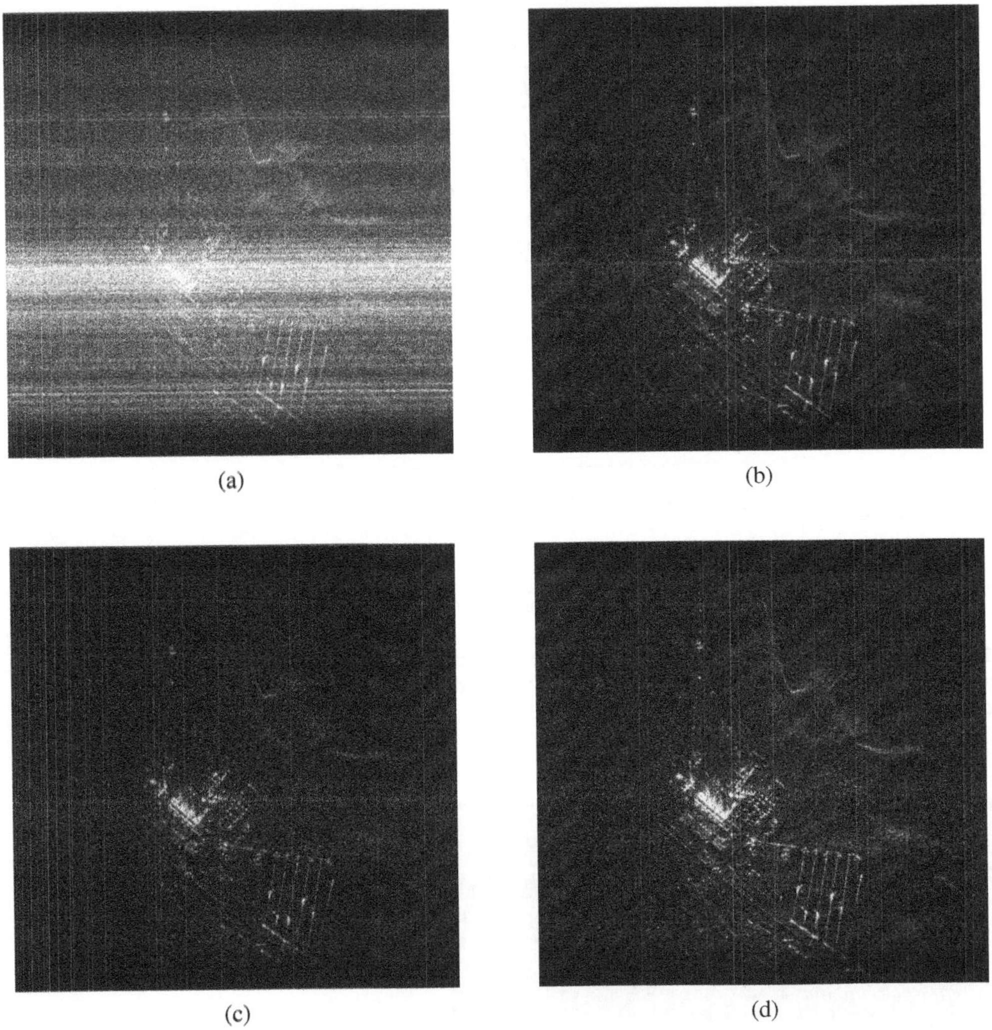

Figure 12.10 Reconstruction results from phase-corrupted data with 25% 1D-random undersampling. (a) Conventional reconstruction. (b) Compressed sensing reconstruction (with $\alpha_{TV} = 0.2$) followed by PGA [8]. (c) Sparsity-driven autofocus [27]. (d) Forward-looking autofocusing ADMM with hybrid features ($\alpha_{TV} = 0.2, \alpha_1 = 0.8$).

Figure 12.10b shows the result of reconstruction followed by autofocusing. The image is first reconstructed using the CS technique described in the Section 12.3.3.2 (disregarding the phase errors) followed by autofocusing using PGA. As can be seen, there is visible horizontal streaking artifact in the middle part of the image. Figure 12.10c shows joint autofocusing and image estimation using only an ℓ_p-norm regularizer and the SDA algorithm. It seems that this approach could not remove the streaking artifact in the middle part of the image either. Finally, Fig. 12.10d shows the results of joint autofocusing and image reconstruction using Autofocusing ADMM with hybrid features. The resulting image is a very good estimate of the reference image and does not involve the streaking artifact that was visible in the other autofocusing methods considered here.

12.4 Learning-Based SAR Imaging

12.4.1 Overview

The optimization problem in (12.10) is formulated to enforce consistency with measured data and promote known, assumed, or learned properties in $\mathbf{x}$ by introducing a prior enforcing function f. This section focuses on learning useful features for the SAR image reconstruction problem directly from a set of SAR images called a training data set. Early learning-based SAR imaging approaches involved sparse representation with learned dictionaries [35]. A more recent learning-based approach [20, 36] incorporates deep learning-based denoisers into the SAR image reconstruction process via thc PnP priors framework [37].

12.4.2 Dictionary Learning-Based SAR Image Reconstruction

Dictionary-based sparse representation of a column-stacked image $\mathbf{x}$ can be represented by $\mathbf{x} \simeq \mathbf{D}\boldsymbol{\alpha}$ such that $\|\boldsymbol{\alpha}\|_0 < K$, where K controls the sparsity of the representation. A learnable dictionary captures the most common spatial features from a set of images in an iterative DL process. The selected weighted combination of the elements of this dictionary results in the representation of an image.

SAR image reconstruction can be performed using a DL based process to solve (12.10). The process consists primarily of two steps: DL and image reconstruction. The DL step can be performed either offline, using a pre-existing training set, or with an unsupervised online process. The DL-based reconstruction can be formulated as follows:

$$\{\hat{\mathbf{x}}, \hat{\mathbf{D}}, \hat{\boldsymbol{\alpha}}\} = \underset{\mathbf{x},\mathbf{D},\alpha}{\operatorname{argmin}} \lambda \|\mathbf{y} - \mathbf{B}\mathbf{x}\|_2^2 + \sum_i \|\mathbf{E}_i|\mathbf{x}| - \mathbf{D}\alpha_i\|_2^2 + \sum_i \mu_i \|\alpha_i\|_0, \tag{12.38}$$

where $\mathbf{D}$ is the dictionary, and the reflectivity field $\mathbf{x}$ is converted into smaller patches via the patch extraction function $\mathbf{E}$. For the ith patch, α_i are the sparse coefficients to be learned, and $\mathbf{E}_i$ extracts its corresponding patch from $\mathbf{x}$. The variable α denotes the set of all α_i. For details about the solution of (12.38) and the sequence of steps, see [35].

12.4.3 Plug-and-Play Methods for SAR Image Reconstruction

PnP priors is a framework introduced by Venkatakrishnan et al. [37] which shows that, once an optimization problem in the form of (12.10) is divided into subproblems via a proximal method such as ADMM using (12.15)–(12.17), the subproblem (12.16) can be reformulated to be a Gaussian denoiser. This observation, under a few assumptions, opens up the possibility of plugging a wide range of denoising algorithms into ADMM, giving it the name PnP priors. Among such priors, off-the-shelf denoisers such as BM3D [38], nonlocal means (NLM) [39], NCSR [40], Gaussian mixture models (GMM), as well as deep learning-based CNN models [20, 36] are most commonly used. In this section, the focus will be on PnP with a CNN-based denoiser.

12.4.3.1 PnP-CNN-SAR Image Denoiser with ADMM

An off-the-shelf CNN-based denoiser is primarily structured for real-valued images. For SAR imaging, which involves the reconstruction of a complex-valued field as discussed earlier in this chapter, the regularization function f is applied on the magnitude of reflectivity field $\mathbf{x}_m := |\mathbf{x}|$. The objective function in (12.10) still involves a complex-valued reflectivity field $\mathbf{x} = \Theta\mathbf{x}_m$, where Θ is a diagonal matrix $\Theta = \text{diag}(\theta)$ containing the phases of $\mathbf{x}$ stated in a vector as $\theta = e^{j\angle\mathbf{x}}$ and

$\angle$ denotes the phase.[1] The phase is modeled as a uniformly distributed random variable and uncorrelated with the phase at other locations [7]. Applying these change of variables declarations to (12.10), and putting it into the form of (12.13), results in the following optimization problem:

$$\underset{\mathbf{x}_m,\Theta}{\text{minimize}}\ \|\mathbf{y}-\mathbf{B}\Theta\mathbf{x}_m\|_2^2+\lambda f(\mathbf{z})\ \text{subject to}\ \mathbf{x}_m-\mathbf{z}=\mathbf{0}, \tag{12.39}$$

here $g(\mathbf{x}) := \|\mathbf{y}-\mathbf{B}\Theta\mathbf{x}_m\|_2^2$ is a dual variable function, $h(\mathbf{z}) := \lambda f(\mathbf{z})$, $\mathbf{G} := \mathbf{I}$, $\mathbf{H} := -\mathbf{I}$, and $\mathbf{c} := \mathbf{0}$. The three step augmented Lagrangian minimization problem from (12.15)–(12.17) takes the following form:

$$\{\mathbf{x}_m^{(k+1)},\Theta^{(k+1)}\} = \underset{\mathbf{x}_m,\Theta}{\text{argmin}}\,||\mathbf{y}-\mathbf{B}\Theta\mathbf{x}_m||_2^2+\frac{\mu}{2}||\mathbf{x}_m-\mathbf{z}^{(k)}+\mathbf{d}^{(k)}||_2^2, \tag{12.40}$$

$$\mathbf{z}^{(k+1)} = \underset{\mathbf{z}}{\text{argmin}}\,\lambda f(\mathbf{z})+\frac{\mu}{2}||\mathbf{x}_m^{(k+1)}-\mathbf{z}+\mathbf{d}^{(k)}||_2^2, \tag{12.41}$$

$$\mathbf{d}^{(k+1)} = \mathbf{x}_m^{(k+1)}-\mathbf{z}^{(k+1)}+\mathbf{d}^{(k)}. \tag{12.42}$$

The PnP denoiser is inferred from (12.41) by rewriting the equation as follows:

$$\mathbf{z}^{(k+1)} = \underset{\mathbf{z}}{\text{argmin}}\,\frac{1}{2(\sqrt{\lambda/\mu})^2}||\mathbf{x}_m^{(k+1)}-\mathbf{z}+\mathbf{d}^{(k)}||_2^2+f(\mathbf{z}), \tag{12.43}$$

which is equivalent to denoising the image $\mathbf{z}^{(k)} = \{\mathbf{x}_m^{(k+1)}+\mathbf{d}^{(k)}\}$ with a Gaussian denoiser $\mathcal{D}_\sigma(\mathbf{z})$ with a noise level $\sigma=\sqrt{\lambda/\mu}$ [41]. Therefore, a CNN-based denoiser can also replace this minimization problem with a one-step inference with the assumption that CNN is proper and closed. A general convergence is also ensured by adhering to Lipschitz conditions.

12.4.3.2 Phase Estimation for PnP-CNN-SAR

The solution of (12.40) is further divided into two sub-problems; phase estimation and magnitude estimation. This section focuses on the phase estimation sub-problem. In this regard, let us introduce a matrix $\mathbf{X}_m = \text{diag}(\mathbf{x}_m)$ whose diagonal elements are the reflectivity magnitudes $\mathbf{x}_m$. The magnitudes of the elements of $\boldsymbol{\theta}$ denoted as $|\theta_i|$ should be 1. Then, the phase estimation problem can be expressed as:

$$\boldsymbol{\theta}^{(k+1)} = \underset{\boldsymbol{\theta}}{\text{argmin}}\,||\mathbf{y}-\mathbf{B}\mathbf{X}_m^{(k)}\boldsymbol{\theta}||_2^2+\lambda_\theta\sum_{i=1}^{N}(|\theta_i|-1)^2. \tag{12.44}$$

The solution to this minimization problem is achieved through a fixed point iterative algorithm, which can also be shown to be equivalent to a particular quasi-Newton iterative minimization algorithm:

$$\mathbf{G}\boldsymbol{\theta}^{(n+1)} = (\mathbf{B}\mathbf{X}_m^{(k)})^H\mathbf{y}+\lambda_\theta\boldsymbol{\theta}^{(n)}, \tag{12.45}$$

where

$$\mathbf{G} = (\mathbf{B}\mathbf{X}_m^{(k)})^H(\mathbf{B}\mathbf{X}_m^{(k)})+\lambda_\theta\mathbf{I}. \tag{12.46}$$

Each iteration in (12.45) involves solving a set of linear equations. This is an indication that the approach here solves the nonquadratic optimization problem in (12.44) by turning it into a series

1 The variable $\boldsymbol{\theta}$ containing the phases of the complex-valued reflectivity field here should not be confused with the same variable used for look angles in the early part of the chapter. The distinction should be clear from context.

Algorithm 12.5: Algorithm for PnP-CNN-SAR

Require: $\mathbf{y}, \mathbf{B}, \zeta, D_\sigma(x), K, T.$
$\mathbf{x}^{(0)} \leftarrow \mathbf{B}^T\mathbf{y}$ *FFT-based reconstruction*
$\mathbf{X}_m^{(0)} \leftarrow \text{diag}\{\mathbf{x}_m^{(0)}\}, \mathbf{\Theta}^{(0)} \leftarrow e^{\angle \mathbf{x}^{(0)}}, \mathbf{z}^{(0)} \leftarrow \mathbf{x}_m^{(0)}, \mathbf{d}^{(0)} \leftarrow 0$
$k \leftarrow 0, n \leftarrow 0$
repeat
$\tilde{\mathbf{x}}^{(k)} \leftarrow \mathbf{z}^{(k)} - \mathbf{d}^{(k)}, \mathbf{x}_m^{[0]} \leftarrow \mathbf{x}_m^{(k)}$
repeat
  Calculate $\mathbf{\Theta}^{(t)}$ using (12.44) *Phase Estimation*
  Solve for $\mathbf{x}_m^{(t)}$ using (12.47) *Magnitude Estimate*
  $t \leftarrow t+1$
until $\frac{\|\mathbf{x}_m^{[t+1]} - \mathbf{x}_m^{[t]}\|}{\|\mathbf{x}_m^{[t]}\|} \leq \zeta$ and $t \leq T$
$\mathbf{x}_m^{(k+1)} \leftarrow \mathbf{x}_m^{[t]}, \tilde{\mathbf{z}}^{(k)} \leftarrow \mathbf{x}_m^{(k+1)} + \mathbf{d}^{(k)}$
$\mathbf{z}^{(k+1)} \leftarrow D_\sigma(\tilde{\mathbf{z}}^{(k)})$ *Prior update*
$\mathbf{d}^{(k+1)} \leftarrow \mathbf{d}^{(k)} + \mathbf{x}_m^{(k+1)} - \mathbf{z}^{(k+1)}$
$k \leftarrow k+1$
until $\frac{\|\mathbf{d}^{(k+1)} - \mathbf{d}^{(k)}\|}{\|\mathbf{d}^{(k)}\|} \leq \zeta$ and $k \leq K$
return $\mathbf{x}_m^{(k+1)}$

of quadratic problems. This linear set of equations can efficiently be solved by the conjugate gradient algorithm with the convergence criterion $\frac{\|\theta^{(n+1)} - \theta^{(n)}\|}{\|\theta^{(n)}\|} < \zeta$. Finally for further computation, the matrix $\Theta^{(k+1)} = \text{diag}(\theta^{(k+1)})$ is stored.

12.4.3.3 Magnitude Estimation for PnP-CNN-SAR

The magnitude estimation problem in (12.40) can be expressed as follows:

$$\mathbf{x}_m^{(k+1)} = \underset{\mathbf{x}_m}{\text{argmin}} ||\mathbf{y} - \mathbf{B}\Theta^{(k+1)}\mathbf{x}_m||_2^2 + \frac{\mu}{2}||\mathbf{x}_m - \mathbf{z}^{(k)} + \mathbf{d}^{(k)}||_2^2. \tag{12.47}$$

Let $\tilde{\mathbf{x}}^{(k)} = \mathbf{z}^{(k)} - \mathbf{d}^{(k)}$, then sub-problem (12.47) has a closed form solution, i.e., taking the derivative with respect to $\mathbf{x}_m$ and equating it to zero, results in the following:

$$\left(\frac{\mu}{2}\mathbf{I} + \Theta^H\mathbf{B}^H\mathbf{B}\Theta\right)\hat{\mathbf{x}}_m = \Theta^H\mathbf{B}^H\mathbf{y} + \frac{\mu}{2}\tilde{\mathbf{x}}. \tag{12.48}$$

We solve this complex-valued linear system using the conjugate gradient algorithm. This produces a complex-valued solution while $\mathbf{x}_m$ must be real-valued, therefore, its magnitude is considered only. Note that within each iteration of ADMM, sub-problems (12.44) and (12.47) are solved iteratively until the convergence criterion is satisfied, before moving to the sub-problem (12.41).

12.4.3.4 Auxiliary Update for PnP-SAR

The final Lagrangian subproblem (12.42) is not an optimization problem, rather it is a direct update on the Lagrange multiplier $\mathbf{d}$. Algorithm 12.5 shows the summary of this method, named PnP-CNN-SAR. Note that $\mathbf{x}_m^{[t]}$ stands for the outputs of the inner loop, i.e., the updates within each iteraton, while $\mathbf{x}_m^{(k)}$ stands for the outputs of the outer loop, i.e., the overall ADMM algorithm. Typical values for K and T are 20 and 100, respectively.

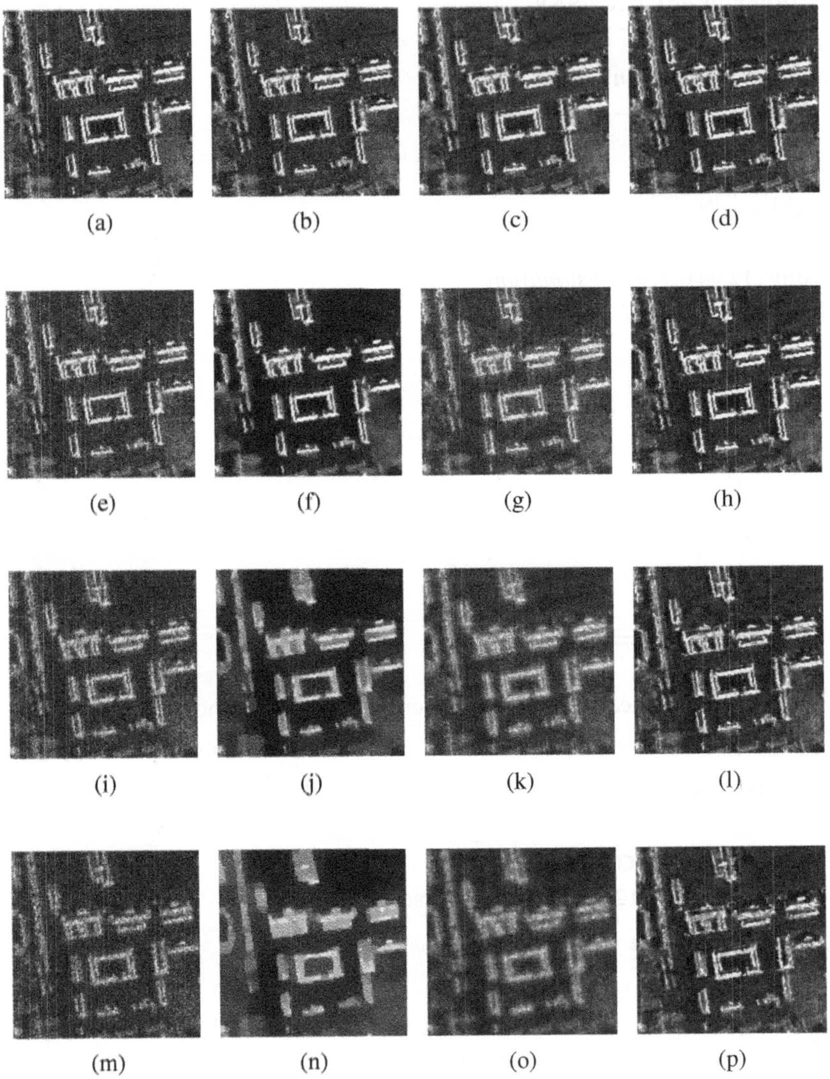

Figure 12.11 Reconstruction results for 2D-rectangular data limitation with noise $\sigma_n = 0.1\,\sigma_y$. First row: full data, second row: 90% data, third row: 70% data, fourth row: 50% data. First column: FFT-based reconstruction, second column: NQR-based reconstruction, third column: DL-based reconstruction, fourth column: PnP-CNN-SAR-based reconstruction. (a) SNR: 26.74, SSIM: 0.968. (b) SNR: 24.26, SSIM: 0.969. (c) SNR: 21.27, SSIM: 0.975. (d) SNR: 37.94, SSIM: 0.998. (e) SNR: 13.12, SSIM: 0.781. (f) SNR: 12.89, SSIM: 0.581. (g) SNR: 10.03, SSIM: 0.695. (h) SNR: 18.13, SSIM: 0.895. (i) SNR: 8.11, SSIM: 0.649. (j) SNR: 7.03, SSIM: 0.420. (k) SNR: 6.96, SSIM: 0.568. (l) SNR: 11.83, SSIM: 0.779. (m) SNR: 5.91, SSIM: 0.530. (n) SNR: 5.51, SSIM: 0.371. (o) SNR: 5.51, SSIM: 0.456. (p) SNR: 8.46, SSIM: 0.670.

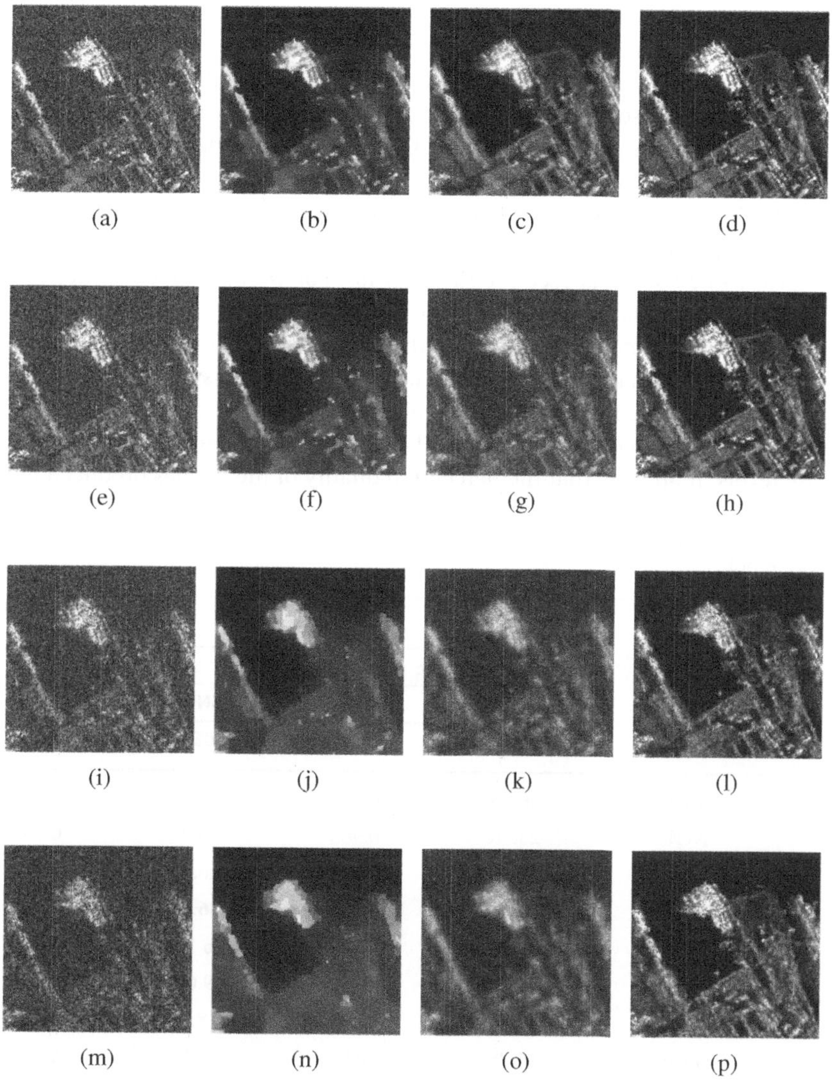

Figure 12.12 Reconstruction results for 2D-rectangular data limitation with noise $\sigma_{\mathrm{n}} = \sigma_{\mathbf{y}}$. First row: full data, second row: 90% data, third row: 70% data, fourth row: 50% data. First column: FFT-based reconstruction, second column: NQR-based reconstruction, third column: DL-based reconstruction, fourth column: PnP-CNN-SAR-based reconstruction. (a) SNR: 5.42, SSIM: 0.373. (b) SNR: 9.68, SSIM: 0.540. (c) SNR: 15.33, SSIM: 0.922. (d) SNR: 39.86, SSIM: 0.995. (e) SNR: 5.28, SSIM: 0.350. (f) SNR: 8.87, SSIM: 0.538. (g) SNR: 7.55, SSIM: 0.432. (h) SNR: 17.12, SSIM: 0.873. (i) SNR: 5.01, SSIM: 0.302. (j) SNR: 6.52, SSIM: 0.464. (k) SNR: 6.29, SSIM: 0.354. (l) SNR: 10.37, SSIM: 0.745. (m) SNR: 4.69, SSIM: 0.262. (n) SNR: 5.99, SSIM: 0.379. (o) SNR: 5.59, SSIM: 0.310. (p) SNR: 8.11, SSIM: 0.629.

12.4.4 Illustrative Examples

In this section, we compare the reconstruction of TerraSAR-X[2] images using PnP priors with CNN denoisers (PnP-CNN-SAR) [20, 36] with conventional fast Fourier transform (FFT)-based imaging, NQR with point and region enhancement [3] introduced in Section 12.3.2.3, and DL based imaging [35] introduced in Section 12.4.2. The experiments are conducted for three types of data limitations, namely, 2D-rectangular, 2D-random, and 1D-linear, four data availability levels, i.e., 50%, 70%, 90%, and 100%, and two noise levels, i.e., $\sigma_{\mathbf{n}} \in \{0.1, 1\} \times \sigma_{\mathbf{y}}$, where $\sigma_{\mathbf{y}}$ is the standard deviation of the magnitude of the phase history data. Note that this is additional noise we insert into the phase histories. Reconstruction methods are compared in terms of SNR and the structural similarity index measure (SSIM) [42] values of the reconstructed images.

For the NQR reconstruction, parameters α_1 and α_2 are optimized to obtain the highest SNR for each image in each scenario while $p = 0.9$ is used for all scenarios. For DL-based reconstruction, the dictionary and the sparse coefficients are learned as described in [35].

For PnP-CNN-SAR, a CNN-based denoiser is trained with 100% data availability and $\sigma_{\mathbf{n}} = \sigma_{\mathbf{y}}$. This network learns to map conventionally reconstructed, i.e., FFT-based, noisy images to clean images. The same network is used to reconstruct images in all scenarios.

Figures 12.11 and 12.12 show the reconstruction results for 2D-rectangular data limitation under two different noise levels. It appears that the perceptual quality of the images produced by

Table 12.3 Average SNR and SSIM values for different noise and 2D-rectangular data availability levels with the best results shown in bold

Available data level (%)	Method	Noise level			
		$0.1\sigma_y$		$1.0\sigma_y$	
		SNR (dB)	SSIM –	SNR (dB)	SSIM –
100	FFT	26.65	0.973	6.33	0.457
	NQR	23.56	0.963	9.7	0.535
	DL	22.95	0.976	16.13	0.924
	PnP	**40.03**	**0.998**	**36.82**	**0.997**
90	FFT	13.23	0.802	6.16	0.423
	NQR	11.64	0.683	9.09	0.486
	DL	12.71	0.783	8.9	0.538
	PnP	**16.78**	**0.9**	**15.85**	**0.857**
70	FFT	8.41	0.632	5.84	0.362
	NQR	7.49	0.427	7.19	0.355
	DL	8.65	0.587	7.54	0.419
	PnP	**10.75**	**0.736**	**10.34**	**0.693**
50	FFT	6.22	0.51	5.42	0.315
	NQR	5.86	0.295	6.6	0.301
	DL	6.85	0.45	6.51	0.333
	PnP	**8.11**	**0.58**	**8.09**	**0.562**

2 TerraSAR-X data were provided by the German Aerospace Center (DLR) via AO account MTH3665.

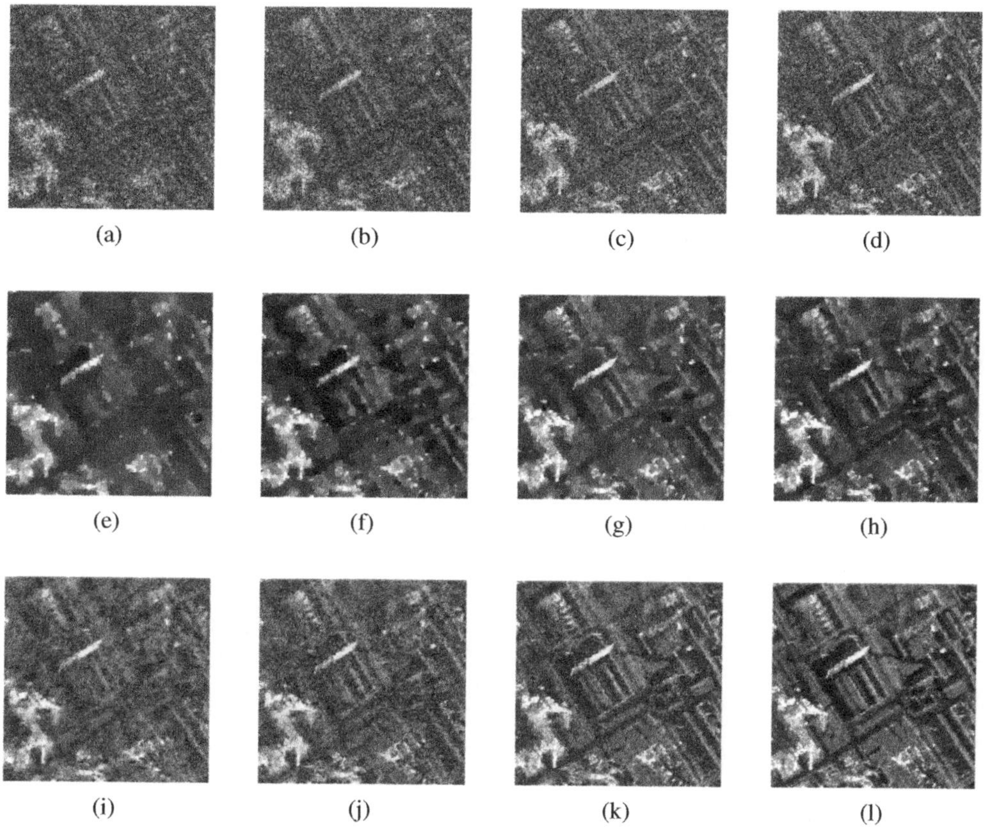

Figure 12.13 Reconstruction results for 2D-random data limitation. (a, e, i) 50% data, (b, f, j) 70% data, (c, g, k) 90% data, (d, h, l) full data. (a–d) FFT-based reconstruction, (e–h) NQR-based reconstruction, (i–l) PnP-CNN-SAR-based reconstruction.

PnP-CNN-SAR is better, and especially in the case of higher noise. While these two figures provide a visual flavor of the behavior of the learning-based SAR imaging approach, Table 12.3 shows the average SNR and SSIM values obtained over 31 test images, verifying the improved performance offered by the PnP-CNN-SAR method. The reconstruction examples for 2D-random and 1D-linear data limitations, shown in Figures 12.13 and 12.14, respectively, suggest that PnP-CNN-SAR offers similar advantages with various sparse observation apertures.

12.5 Conclusion

We presented state-of-the-art methods for sparsity-driven, optimization-based methods as well as deep learning-based methods for SAR imaging with limited and sparse apertures. These methods provide compelling alternatives to conventional SAR imaging methods, especially in scenarios involving limited and sparse observation apertures, low-quality data, and model imperfections. Both types of methods are primarily aimed to address the ill-posed nature of the SAR imaging problem in these scenarios. This requires the incorporation of explicit or implicit prior information or constraints. Sparsity-driven imaging does that simply by assuming the sparsity of the underlying reflectivity field, possibly in some transform domain, leading to various forms of feature

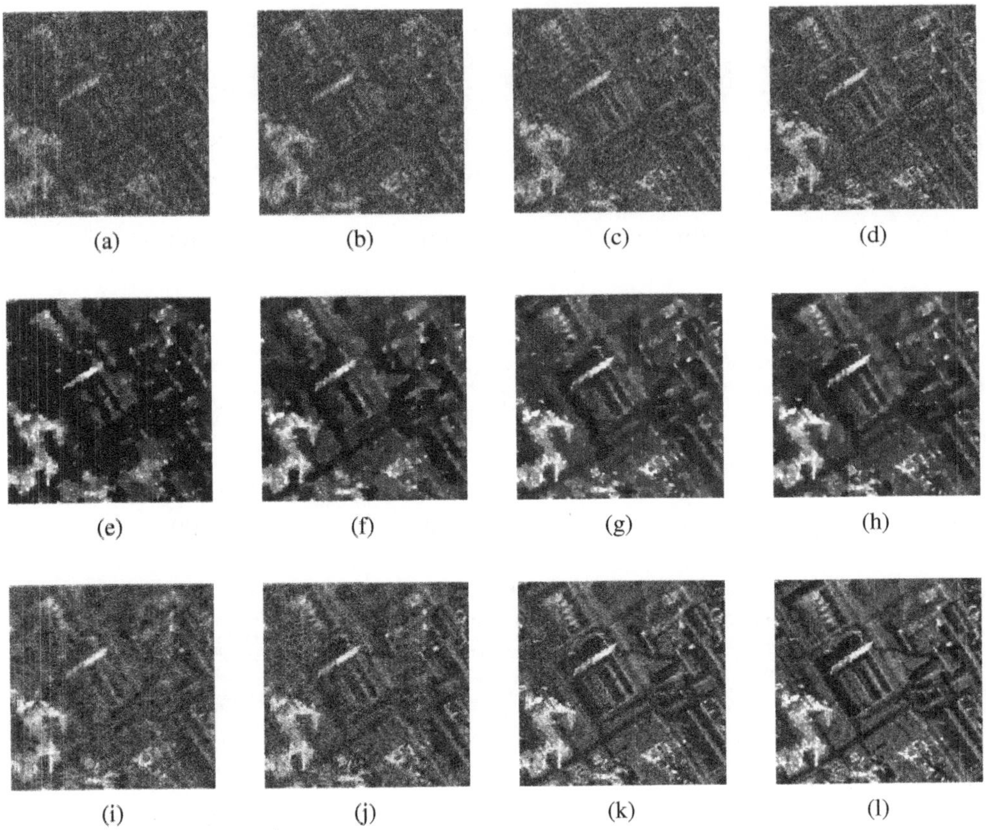

Figure 12.14 Reconstruction results for 1D-linear data limitation. (a, e, i) 50% data, (b, f, j) 70% data, (c, g, k) 90% data, (d, h, l) full data. (a–d) FFT-based reconstruction, (e–h) NQR-based reconstruction, (i–l) PnP-CNN-SAR-based reconstruction.

enhancement. This perspective has significantly benefited from theoretical work on CS, demonstrating that, under a set of assumptions, such sparse solutions can be obtained by using practical numerical algorithms. In the context of SAR imaging, sparsity has proven to be a useful asset, leading to improved resolution, reduced artifacts, and enhanced features. In this chapter, we focused on state-of-the-art proximal algorithms, in particular ADMM, to solve the optimization problems involved in sparsity-driven SAR imaging, both in the case of perfectly known observation models and in the presence of model errors. While sparsity is a useful constraint that can improve the SAR imaging process, it is too simple to capture the complex statistical behavior exhibited by reflectivity fields. Advances in deep learning provide an opportunity to learn such behavior and incorporate it as prior information into the SAR image formation process. We presented an approach developed with that perspective, namely PnP-CNN-SAR, which, while leveraging the physics-based observation model, brings in a CNN-based regularizer into computational SAR imaging within the proximal algorithms framework. The illustrative examples we presented in this chapter demonstrate the potential of such learning-based methods. In the near future, we expect to see increased interest in further development of machine learning-based methods for various radar imaging tasks. In-depth analysis of the behavior of such methods will be needed to build trust and enable their widespread use.

References

1 Jakowatz, C.V., Wahl, D.E., Eichel, P.H. et al. (1996). *Spotlight-Mode Synthetic Aperture Radar: A Signal Processing Approach*. Kluwer Academic Publishers.

2 Walker, J.L. (1980). Range-Doppler imaging of rotating objects. *IEEE Transactions on Aerospace and Electronic Systems* AES-16 (1): 23–52. https://doi.org/10.1109/TAES.1980.308875.

3 Çetin, M. and Karl, W.C. (2001). Feature-enhanced synthetic aperture radar image formation based on nonquadratic regularization. *IEEE Transactions on Image Processing* 10 (4): 623–631.

4 Munson, D.C., O'Brien, J.D., and Jenkins, W.K. (1983). A tomographic formulation of spotlight-mode synthetic aperture radar. *Proceedings of the IEEE* 71 (8): 917–925.

5 Carrara, W.G., Goodman, R.S., and Majewski, R.M. (1995). *Spotlight Synthetic Aperture Radar: Signal Processing Algorithms* (Artech House Remote Sensing Library). Artech House.

6 Soumekh, M. (1999). *Synthetic Aperture Radar Signal Processing*, vol. 7. New York: Wiley.

7 Munson, D.C. and Sanz, J.L.C. (1984). Image reconstruction from frequency-offset Fourier data. *Proceedings of the IEEE* 72 (6): 661–669.

8 Emre Güven, H., Güngör, A., and Çetin, M. (2016). An augmented Lagrangian method for complex-valued compressed SAR imaging. *IEEE Transactions on Computational Imaging* 2 (3): 235–250.

9 Xu, J.-p., Pi, Y.-n., and Cao, Z.-j. (2012). Bayesian compressive sensing in synthetic aperture radar imaging. *IET Radar, Sonar & Navigation* 6 (1): 2–8.

10 Patel, V.M., Easley, G.R., Healy, D.M., and Chellappa, R. (2010). Compressed synthetic aperture radar. *IEEE Journal of Selected Topics in Signal Processing: Special Issue on Compressive Sensing* 4 (2): 244–254.

11 Gang, X., Zhang, B., Yu, H. et al. (2022). Sparse synthetic aperture radar imaging from compressed sensing and machine learning: theories, applications, and trends. *IEEE Geoscience and Remote Sensing Magazine* 10 (4): 32–69. https://doi.org/10.1109/MGRS.2022.3218801.

12 Donoho, D.L. (2006). Compressed sensing. *IEEE Transactions on Information Theory* 52 (4): 1289–1306.

13 Potter, L.C., Ertin, E., Parker, J.T., and Cetin, M. (2010). Sparsity and compressed sensing in radar imaging. *Proceedings of the IEEE* 98 (6): 1006–1020.

14 Scherzer, O. (1993). The use of Morozov's discrepancy principle for Tikhonov regularization for solving nonlinear ill-posed problems. *Computing* 51 (1): 45–60.

15 Batu, O. and Çetin, M. (2011). Parameter selection in sparsity-driven SAR imaging. *IEEE Transactions on Aerospace and Electronic Systems* 47 (4): 3040–3050. https://doi.org/10.1109/TAES.2011.6034687.

16 Chartrand, R. (2007). Exact reconstruction of sparse signals via nonconvex minimization. *IEEE Signal Processing Letters* 14 (10): 707–710.

17 Rudin, L.I., Osher, S., and Fatemi, E. (1992). Nonlinear total variation based noise removal algorithms. *Physica D: Nonlinear Phenomena* 60 (1–4): 259–268.

18 Lustig, M., Donoho, D.L., Santos, J.M., and Pauly, J.M. (2008). Compressed sensing MRI. *IEEE Signal Processing Magazine* 25 (2): 72–82.

19 Samadi, S., Çetin, M., and Masnadi-Shirazi, M.A. (2011). Sparse representation-based synthetic aperture radar imaging. *IET Radar, Sonar & Navigation* 5 (2): 182–193.

20 Alver, M.B., Saleem, A., and Çetin, M. (2020). Plug-and-play synthetic aperture radar image formation using deep priors. *IEEE Transactions on Computational Imaging* 7: 43–57.

21 Parikh, N. and Boyd, S. (2014). Proximal algorithms. *Foundations and Trends®in Optimization* 1 (3): 127–239.

22 Boyd, S., Parikh, N., Chu, E. et al. (2011). Distributed optimization and statistical learning via the alternating direction method of multipliers. *Foundations and Trends®in Machine learning* 3 (1): 1–122.

23 Teke, O., Gurbuz, A.C., and Arikan, O. (2014). A robust compressive sensing based technique for reconstruction of sparse radar scenes. *Digital Signal Processing* 27: 23–32.

24 Chambolle, A. (2004). An algorithm for total variation minimization and applications. *Journal of Mathematical Imaging and Vision* 20: 89–97.

25 Güngör, A., Çetin, M., and Emre Güven, H. (2022). Compressive synthetic aperture radar imaging and autofocusing by augmented Lagrangian methods. *IEEE Transactions on Computational Imaging* 8: 273–285.

26 ASELSAN website for SARPER™. http://www.aselsan.com.tr/en-us/capabilities/radar-systems/; http://www.aselsan.com.tr/en-us/capabilities/radar-systems/ (Visited on 10/18/2018).

27 Önhon, N.Ö. and Çetin, M. (2012). A sparsity-driven approach for joint SAR imaging and phase error correction. *IEEE Transactions on Image Processing* 21 (4): 2075–2088.

28 Güngör, A., Çetin, M., and Güven, H.E. (2015). An augmented Lagrangian method for autofocused compressed SAR imaging. *2015 3rd International Workshop on Compressed Sensing Theory and its Applications to Radar, Sonar and Remote Sensing (CoSeRa).* June 2015, 1–5.

29 Kang, M.-S. and Kim, K.-T. (2019). Compressive sensing based SAR imaging and autofocus using improved Tikhonov regularization. *IEEE Sensors Journal* 14 (14): 5529–5540.

30 Camlica, S., Gurbuz, A.C., and Arikan, O. (2017). Autofocused spotlight SAR image reconstruction of off-grid sparse scenes. *IEEE Transactions on Aerospace and Electronic Systems* 53 (4): 1880–1892.

31 Yang, L., Zhao, L., Zhou, S., and Bi, G. (2016). Sparsity-driven SAR imaging for highly maneuvering ground target by the combination of time-frequency analysis and parametric Bayesian learning. *IEEE Journal of Selected Topics in Applied Earth Observations and Remote Sensing* 10 (4): 1443–1455.

32 Kantor, J.M. (2020). Polar format-based compressive SAR image reconstruction with integrated autofocus. *IEEE Transactions on Geoscience and Remote Sensing* 58 (5): 3458–3468.

33 Güngör, A., Çetin, M., and Güven, H.E. (2017). Autofocused compressive SAR imaging based on the alternating direction method of multipliers. *2017 IEEE Radar Conference (RadarConf).* May 2017, 1573–1576.

34 Wahl, D.E., Eichel, P.H., Ghiglia, D.C., and Jakowatz, C.V. (1994). Phase gradient autofocus-a robust tool for high resolution SAR phase correction. *IEEE Transactions on Aerospace and Electronic Systems* 30 (3): 827–835.

35 Soğanlı, A. and Çetin, M. (2014). Dictionary learning for sparsity-driven SAR image reconstruction. *2014 IEEE International Conference on Image Processing (ICIP),* 1693–1697.

36 Alver, M.B., Saleem, A., and Çetin, M. (2019). A novel plug-and-play SAR reconstruction framework using deep priors. *2019 IEEE Radar Conference (RadarConf).* IEEE, 1–6.

37 Venkatakrishnan, S.V., Bouman, C.A., and Wohlberg, B. (2013). Plug-and-play priors for model based reconstruction. *Global Conference on Signal and Information Processing (GlobalSIP),* IEEE, 945–948.

38 Dabov, K., Foi, A., Katkovnik, V., and Egiazarian, K. (2007). Image denoising by sparse 3-D transform-domain collaborative filtering. *IEEE Transactions on Image Processing* 16 (8): 2080–2095.

39 Chan, S.H., Wang, X., and Elgendy, O.A. (2016). Plug-and-play ADMM for image restoration: fixed-point convergence and applications. *IEEE Transactions on Computational Imaging* 3 (1): 84–98.

40 Dong, W., Zhang, L., Shi, G., and Li, X. (2012). Nonlocally centralized sparse representation for image restoration. *IEEE Transactions on Image Processing* 22 (4): 1620–1630.

41 Zhang, K., Zuo, W., Gu, S., and Zhang, L. (2017). Learning deep CNN denoiser prior for image restoration. *Proceedings of the IEEE Conference on Computer Vision and Pattern Recognition*, 3929–3938.

42 Wang, Z., Bovik, A.C., Sheikh, H.R., and Simoncelli, E.P. (2004). Image quality assessment: from error visibility to structural similarity. *IEEE Transactions on Image Processing* 13 (4): 600–612.

13

Sparse Arrays for Sonar

Kaushallya Adhikari[1] and Kathleen E. Wage[2]

[1]Electrical, Computer, and Biomedical Engineering Department, University of Rhode Island, Kingston, RI, USA
[2]Electrical and Computer Engineering Department, George Mason University, Fairfax, VA, USA

13.1 Introduction

Underwater remote sensing relies primarily on acoustic signals since electromagnetic waves are significantly attenuated by the medium [1]. While originally developed for depth sounding, sonar systems are now used in a wide variety of applications including navigation, underwater communications, bathymetric surveys, marine mammal studies, defense, and oceanographic research [[2], Ch. 1]. Passive sonar systems rely on listening only while active sonar systems transmit pulses and process the received reflections. Sonar systems typically use arrays of elements to send and receive acoustic signals. Arrays improve signal-to-noise ratio and facilitate localization and tracking of sources.

This chapter focuses on sparse arrays for sonar applications. In contrast to random arrays whose sensors are distributed according to a probability distribution [[3], Ch. 8], sparse arrays have sensors located on an underlying equally spaced grid. Sparse arrays are often designed by “thinning” (sub-sampling) a uniform array. For a uniform array, the sample spacing must be less than half of the wavelength of the signal to avoid spatial aliasing. The aim of sparse array processing is to provide unambiguous localization of sources using fewer sensors than a uniform array requires. The motivation for using sparse arrays is to reduce cost and simplify logistics. Cost is reduced when fewer sensors are required to span the same aperture. Fewer sensors can also reduce weight and power requirements, and make the array easier to deploy, all of which are important for autonomous systems, e.g., [4]. To guarantee good performance and prevent aliasing, sparse arrays require careful design of sensor locations and the associated signal processing algorithms. Sparse array design has a long history. The books by Steinberg [[3], Ch. 7], Johnson and Dudgeon [[5], Sec. 3.3], and Van Trees [[6], Sec. 3.9], along with the book chapter by Holm et al. [7] provide useful background on the topic. The following paragraphs briefly review the sonar literature on sparse arrays.

The classic text by Urick [[8], Ch. 3] provides a good introduction to sonar arrays that summarizes the early literature on the topic and highlights key points about processing for sparse arrays. Early research in sparse arrays focuses on multiplicative processing as an alternative to conventional additive processing. An additive processor computes a weighted sum of the sensor data to obtain its array output. A multiplicative processor computes the product of data from individual sensors or from multiple additive arrays to obtain its output. Faran and Hills [9] and later Berman and Clay [10] propose multiplicative processors that can achieve the same beampattern as an additive

Sparse Arrays for Radar, Sonar, and Communications, First Edition. Edited by Moeness G. Amin.

processor using a significantly smaller number of sensors. Urick cites a study by Linder [11] showing that a multiplicative processor using sensors spanning an aperture of 2 wavelengths achieves the same resolution as an additive processor with sensors spanning 4.5 wavelengths. Linder highlights the importance of time-averaging to reduce cross-product terms in multiplicative processors when multiple sources are present. Welsby's 1961 paper [12] presents an early sonar experiment comparing additive and multiplicative processing, concluding that higher resolution is possible with the multiplicative system. Citing key results by Brown and Rowlands [13], Faran and Hills [9], and Jacobson [14], Urick notes that multiplicative processors require higher signal-to-noise ratios and longer averaging time than additive processors. Davies and Ward [15] further explore multiplicative processing using a nested geometry consisting of a short half-wavelength array embedded within a thinned array with greater than half-wavelength spacing. They demonstrate that appropriate tapering of the nested arrays controls beampattern side lobes, enabling good performance with fewer sensors. Davies and Ward also demonstrate the sensitivity of multiplicative processors to sensor phase and amplitude errors and highlight the problem of cross terms.

Another approach to sparse array processing is to use the sensor data to estimate the spatial correlation function and its Fourier transform (spatial spectrum). An early example of this approach in sonar is the work by Nuttall et al. [16] exploring two-dimensional frequency–wavenumber spectrum estimation using a sparse linear array. Building on Nuttall et al.'s work, Bucker [17] investigates a similar technique called *cross-sensor beamforming*. He later applies that idea to develop three beamforming algorithms for a sparse vertical billboard array [18]. These early sonar papers are referenced in the influential paper by Hoctor and Kassam [19] that describes a general framework for active and passive imaging and discusses the key role of the coarray.

More recently, Pal and Vaidyanathan's [20, 21] work on nested and coprime arrays sparked renewed interest in sparse geometries and processing techniques. This led to new algorithms such as the min processor developed independently by Di Martino and Iodice [22, 23] and Liu and Buck [24, 25]. In addition to new theoretical research on sparse arrays, recent work includes analyses of underwater experimental data sets, e.g., the publications by Chavali et al. [26], Ramirez and Rogers [27], Nannuru et al. [28], and Nair et al. [29].

The goals of this chapter are to synthesize prior work on sparse sonar arrays and review the current state of the art. It includes examples of nested and coprime arrays for active and passive sonar. Since sonar systems rely on second-order statistical characterizations of incoming data to detect, localize, and track sources, the centerpiece of the chapter is a discussion of spectral analysis and correlation estimation for sparse arrays. The tutorial below presents four key sparse processing techniques and compares their performance using an open source sonar data set. The chapter assumes the reader has some background in array processing and spectral analysis. For a brief review of basic concepts and how they relate to sparse arrays, see the article by Wage [30]. Urick [[8], Ch. 3] has an excellent summary of the design tradeoffs in sparse sonar arrays that provides a good foundation for this chapter.

The chapter is organized as follows. Section 13.2.1 is a review of conventional uniform line arrays (ULAs) that defines notation and discusses beampatterns, grating lobes, and spatial aliasing. Section 13.2.2 uses an active sensing example to illustrate how sparse transmit/receive arrays can produce a spatial response equivalent to that of a filled uniform array. Building on that example, Section 13.2.3 introduces the Mills cross sparse geometry and highlights how it is applied in multibeam echo sounding for bathymetric surveys. Following the brief tutorial on active sensing, the remainder of the chapter focuses on passive sensing with sparse arrays. Section 13.3 describes four fundamental algorithms for obtaining spatial spectra from sparse array measurements. Section 13.4 illustrates the performance of these algorithms using an open source sonar data set.

The experimental data reveal some of the practical problems of implementing sparse arrays in the underwater environment. Section 13.5 concludes the chapter with a summary of the recent literature and suggestions for further reading on the topic.

13.2 Active Sonar Processing

This section uses active sonar examples to introduce fundamental concepts about sparse arrays. It concludes with a practical example of a sparse array configuration that is used in multibeam echo-sounding systems for bathymetric surveys. The coordinate system and notation defined below and used throughout the chapter are consistent with the textbook by Van Trees [6].

13.2.1 Review of Uniform Line Arrays

Figure 13.1(a) depicts the arrival of a planewave with wavelength λ at two ULAs with sensors located on the z axis. The planewave propagation angle is θ, measured from the positive z axis. The two arrays span the same aperture. The first array has nine sensors with intersensor spacing $d = \frac{\lambda}{2}$ and the second array has five sensors with spacing $d = \lambda$. Wavelength is the ratio of the propagation speed to the temporal frequency of the signal. According to the spatial sampling theorem, the spacing of a uniform array must be less than half of the smallest wavelength in the propagating signal to avoid aliasing.

An additive beamformer sums the received sensor data after time-shifting the receptions so that a signal from the desired direction adds constructively. For narrowband planewave signals, additive

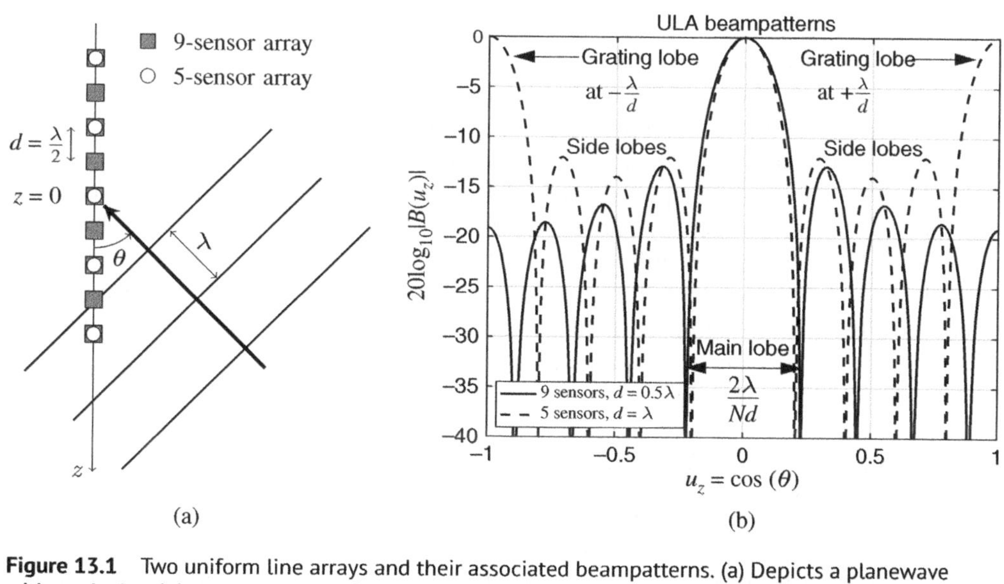

Figure 13.1 Two uniform line arrays and their associated beampatterns. (a) Depicts a planewave with angle θ arriving at two arrays. One array (square markers) has nine sensors with half-wavelength spacing and the other array (circle markers) has five sensors with one-wavelength spacing. (b) Shows beampatterns for uniform tapers and broadside steering ($\theta_s = \frac{\pi}{2}$). The two arrays have approximately equal main lobe widths since they span the same aperture. The undersampled five-sensor array beampattern has grating lobes, whereas the nine-sensor array beampattern does not.

beamforming reduces to applying a phase shift and summing, i.e., the output of an N-element array is

$$y = \sum_{n=0}^{N-1} w^*[n]x[n] = \mathbf{w}^H\mathbf{x}, \tag{13.1}$$

where $x[n]$ is the received narrowband data for the nth sensor and $w[n]$ is complex weight applied to that sensor. $\mathbf{x}$ and $\mathbf{w}$ are column vectors containing the data and weights and * and H denote the complex conjugate and Hermitian transpose, respectively. A narrowband planewave arriving at the linear array in Fig. 13.1(a) is characterized by a complex exponential replica vector $\mathbf{v}(\theta)$ defined as

$$\mathbf{v}(\theta) = \exp\left(j\frac{2\pi}{\lambda}\cos(\theta)\mathbf{z}\right) = \exp\left(j\frac{2\pi}{\lambda}u_z\mathbf{z}\right), \tag{13.2}$$

where $\mathbf{z}$ is a vector of sensor depths and $u_z = \cos(\theta)$ is the directional cosine. The complex replica represents the phase shifts between sensors as a function of the propagation angle relative to the z-axis. In conventional additive beamforming, the weight vector is a tapered version of the replica for the desired (steering) direction θ_s, i.e.,

$$\mathbf{w} = \mathbf{w}_t \odot \mathbf{v}(\theta_s), \tag{13.3}$$

where $\mathbf{w}_t$ is a vector containing the amplitude taper values and $\odot$ represents the Hadamard product. The uniform taper is the simplest, i.e., $w_t[n] = \frac{1}{N}$ for all n, though a variety of other tapers may be used, e.g., see [6, Sec. 3.1].

The beampattern is the spatial frequency response that characterizes the response to a unit-amplitude narrowband planewave input. For the additive processor, the beampattern is defined as

$$B(u_z) = \mathbf{w}^H\mathbf{v}(u_z). \tag{13.4}$$

Figure 13.1(b) shows the conventional beampatterns for the two ULAs described above when the weight vectors are computed with uniform tapers and $\theta_s = \frac{\pi}{2}$. The beampattern's main lobe width determines the array's ability to discriminate between closely spaced sources and the side lobe height determines the array's ability to detect weak sources in the presence of interfering signals at other angles. These arrays have approximately the same main lobe width since they span the same aperture. For a uniform taper the main lobe width in u_z is $\frac{2\lambda}{Nd}$ and the peak side lobe height is approximately −13 dB. The primary difference in the two beampatterns is the grating lobes (copies of the main lobe) in the beampattern of the undersampled five-sensor array. Grating lobes appear at multiples of $\frac{\lambda}{d}$ in u_z. This beampattern shows that the additive processor cannot distinguish between a planewave with angle $\theta_s = \frac{\pi}{2}$ from planewaves arriving at angles $\cos^{-1}\left(\pm\frac{\lambda}{d}\right)$.

13.2.2 Simple Active Sensing Example

This section shows how an active sonar system can use sparse aliased arrays to achieve performance similar to that of a larger system with half-wavelength sampling. In active sonar a transmitter emits a pulse and the receiver measures the pulse's reflections off targets in the environment [[8], Ch. 1-2]. This is the coherent imaging scenario described by Hoctor and Kassam [19]. Figure 13.2(a) depicts a simple active sensing scenario using nested arrays. The three-element transmit array has $\lambda/2$ spacing and the five-element receive array has $3\lambda/2$ spacing. The arrays are co-located along the x-axis and the steering angle for transmission and reception is θ (measured from the positive

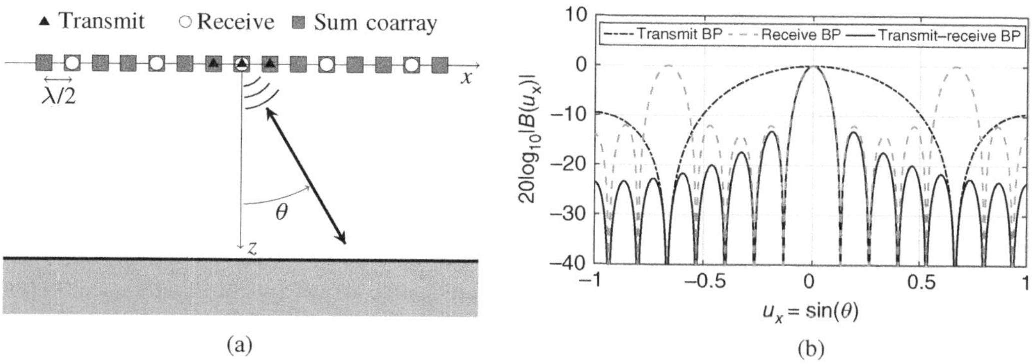

Figure 13.2 One-dimensional active sonar example using a three-element transmit array (triangles) and a five-element receive array (circles). The transmit array has $\lambda/2$ spacing while the receive array has $3\lambda/2$ spacing. This configuration produces a uniform sum coarray spanning $-7 \leq n \leq +7$. The multiplication of the transmit and receive beampatterns eliminates the ambiguity due to undersampling. The resulting two-way beampattern is equivalent to the one-way beampattern of the uniform coarray with 15 elements. (a) Transmitter, receiver, sum coarray and (b) beampatterns.

z-axis as in Section 13.2.1). The two-way beampattern is the product of the beampatterns of the transmit and receive arrays, i.e.,

$$B_{tr}(u_x) = B_t(u_x)B_r(u_x),$$

where $u_x = \sin(\theta)$ defines the angular dependence of the pattern for this geometry. Assuming reciprocity in the medium, reversing the roles of the transmitter and receiver produces the same beampattern [7]. The transmit–receive beampattern characterizes the performance of a linear processing system when targets are in the far-field of co-located arrays that have the same phase center.

Figure 13.2(b) shows the beampatterns for this simple active scenario when both the transmitter and receiver use uniform tapers. The arrays are steered to broadside, i.e., $\theta_s = 0$. Since the receive array has spacing greater than half-wavelength, $B_r(u_x)$ has grating lobes at $u_x = \pm\frac{\lambda}{d} = \pm\frac{2}{3}$. Since $B_t(u_x)$ has zeros at those locations, there are no grating lobes in the transmit–receive beampattern. Thus, the active system is not aliased and can provide unambiguous location information.

The transmit–receive beampattern is the product of two beampatterns. Since beampatterns are spatial Fourier transforms of the array weights, the product pattern corresponds to the transform of the convolution of the transmit and receive array weights. For the nested configuration in Fig. 13.2 with uniform tapers, the convolution of the weights yields

$$w_{tr}[n] = w_t[n] \star w_r[n] = \begin{cases} \frac{1}{15} & -7 \leq n \leq +7 \\ 0 & \text{otherwise} \end{cases} \tag{13.5}$$

since

$$w_t[n] = \begin{cases} \frac{1}{3} & -1 \leq n \leq +1 \\ 0 & \text{otherwise} \end{cases} \tag{13.6}$$

and

$$w_r[n] = \begin{cases} \frac{1}{5} & n = -6, -3, 0, +3, +6 \\ 0 & \text{otherwise} \end{cases}. \tag{13.7}$$

The support of the convolution is the *sum coarray* [19, 31] that defines the effective aperture of the active system. Figure 13.2(a) shows that the sum coarray for this nested active system spans

$-7 \leq n \leq +7$ with no holes. Since the coarray weighting w_{tr} is uniform, this nested active system is effectively equivalent to a uniform array with 15 sensors. As a result, the system can steer 15 orthogonal beams using only eight elements (three transmit, five receive). This example illustrates how a sparse array configuration coupled with the right processing algorithm can provide additional degrees of freedom for identifying sources.

13.2.3 Echo-Sounding and Mills Cross Array

This section describes an active sonar application where sparse arrays are commonly used. Echo sounders are active sonar systems used for bathymetric mapping and sub-bottom profiling. See Bjørnø [32] for a thorough discussion of single- and multi-beam echosounders. Multi-beam echo sounders designed for deep water rely on hull-mounted arrays, often arranged in a Mills cross configuration. The Mills cross geometry originated in radio astronomy [33], but has been applied in many other contexts, including sonar [[8], p. 62]. An active Mills cross array consists of a linear transmit array that is perpendicular to a linear receive array. In echo-sounding the transmit array is mounted along the keel (parallel to the ship's direction of travel) while the receive array is perpendicular to the keel (perpendicular to the direction of travel). See Fig. 13.3 for an illustration. When the transmit array is steered to broadside, it ensonifies a narrow strip of the bottom below the ship. The receive array can be steered to listen for echoes in a set of narrow beams spanning the ocean floor ensonified by the transmitter, allowing simultaneous measurement of depth across a wide swath with a single ping.

The Mills cross is a sparse configuration that can produce the same beampattern as a filled rectangular array with the same dimensions. To see this, consider the array of N_r transmit elements and N_r receive elements depicted in Fig. 13.4(a). As the figure indicates, the sum coarray for the Mills cross is a filled planar array with $N_t \times N_r$ elements, where $N_t = N_r = 15$ for this example. Figure 13.4(b) shows the corresponding transmit–receive beampattern for this array when both transmit and receive arrays use uniform tapers and are steered to broadside. This Mills cross configuration with 15 transmitters and 15 receivers is able to steer 225 orthogonal beams, i.e., it is equivalent to using one omnidirectional transmitter and planar receiving array containing 225 elements arranged in a square grid. A variant of the Mills cross uses sensors arranged in a "T" instead of a cross. Given the same number of transmit/receive elements, the "T" configuration achieves a filled sum coarray, similar to the standard Mills cross. Bjørnø [32] provides a thorough description of an operational deep water multibeam echo sounder using the "T" geometry.

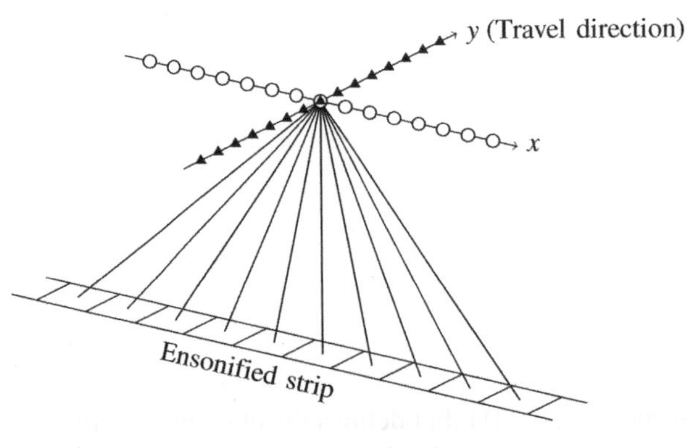

Figure 13.3 Mills cross configuration for multibeam echo-sounding. The transmit array (triangles) emits a pulse steered toward $\phi = 0$ (broadside to the transmitter) that ensonifies a narrow strip of the bottom. The receive array (circles) steers multiple beams to process the reflections from different parts of the ensonified strip.

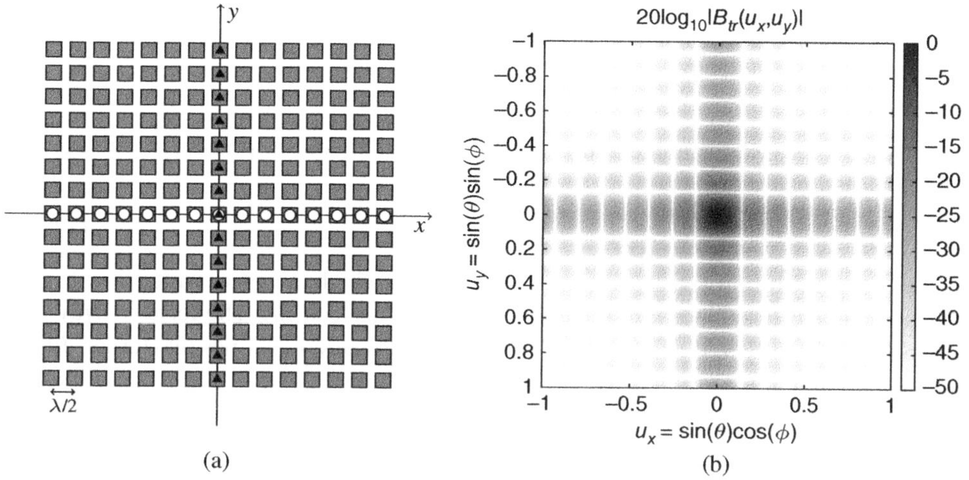

Figure 13.4 Mills cross configuration. (a) Shows a 15-element transmitter array, 15-element receiver array, and the corresponding sum coarray. (b) Shows the two-way beampattern for this active Mills cross system when the transmit and receive arrays use uniform weightings.

13.3 Passive Sonar Processing

This section is a tutorial on passive processing with sparse sonar arrays that presents four fundamental algorithms for obtaining spatial spectra from sparse ULA measurements. The coordinate system is the one introduced in Fig. 13.1(a). Since the examples only use one-dimensional arrays oriented along the z-axis, the directional cosine u_z is denoted as $u = \cos(\theta)$ for simplicity. Plots of arrays show the z-axis displayed horizontally, rather than vertically as in Fig. 13.1(a).

13.3.1 Passive ULAs and the Difference Coarray

Building on the review of ULAs in Section 13.2.1, this section defines the difference coarray used in passive sensing and emphasizes its connection to estimation of the spatial autocorrelation function (ACF). A snapshot contains the narrowband signal received by the array at a given time. For a signal consisting of one planewave and noise, the snapshot vector can be expressed as

$$\mathbf{x} = s\mathbf{v}(u_0) + \mathbf{q}, \tag{13.8}$$

where s is the signal amplitude, $u_0 = \cos(\theta_0)$ is the directional cosine of the planewave, and $\mathbf{q}$ is a vector of noise. The conventional beamformer (CBF) defined in Eq. (13.1) produces the spatial periodogram

$$|y(u)|^2 = \left|\mathbf{w}^H\mathbf{x}\right|^2 = \left|\sum_{n=0}^{N-1} w^*[n]x[n]\right|^2, \tag{13.9}$$

where $u = \cos(\theta_s)$ is the directional cosine associated with the steering direction. The periodogram is an estimate of the spatial power spectral density (PSD). Assuming the acoustic field received by the array is wide-sense stationary (WSS), the PSD estimate $\hat{P}(u)$ is equal to the spatial Fourier transform of the ACF estimate. Thus,

$$\hat{P}(u) = \mathcal{F}_s\{\hat{r}[n]\}, \tag{13.10}$$

where $\hat{r}[n]$ represents the ACF estimate and $\mathcal{F}_s$ represents spatial Fourier transform. For a given snapshot, the ACF estimate associated with the CBF is

$$\hat{r}[n] = x[n] \star x^*[-n]/D_n, \tag{13.11}$$

where $\star$ represents convolution and D_n is a normalizing factor which may vary with n.

The ACF estimate and its support are related to the difference coarray. The term *coarray* has not been defined consistently in the literature. The following discussion defines two terms: *difference coarray support* and *difference coarray weighting*. First, denote the set of all sensor locations, normalized by $\lambda/2$, in an array as A. Then, the following set of points is the difference coarray support:

$$C_d = \{n_i - n_j | n_i \in A, n_j \in A\}. \tag{13.12}$$

This definition of coarray can be traced to [34] and was popularized by Hoctor and Kassam [19]. In Chapter 1, the term "difference coarray" refers to "difference coarray support." Figure 13.5 plots a filled ULA and also depicts its difference coarray support. If a filled ULA has L sensors, its difference coarray support's range is from −(L − 1) to L − 1. There are no holes in the difference coarray support of a filled ULA. The difference coarray support indicates the lags at which the ACF can be estimated. For the array in Fig. 13.5, the ACF can be estimated from lag −6 to 6.

To define *difference coarray weighting*, consider a sensor location indicator function $i[n]$ whose value is 1 if a sensor is present at $n\lambda/2$ and 0, otherwise. The convolution

$$c[n] = i[n] \star i[-n] \tag{13.13}$$

yields the difference coarray weighting of the array while the support of the convolution is the difference coarray support defined in (13.12). The ACF of the received signal at lag n, given in (13.11), can be expressed as

$$\sum_k x^*[k]x[k+n]/D_n. \tag{13.14}$$

The difference coarray weighting function $c[n]$ indicates the number of terms in the sum of (13.14) for each lag n. Figure 13.6 illustrates the difference coarray weightings for the ULA in Fig. 13.5.

13.3.2 Difference Coarrays of Sparse Arrays

To explain how sparse arrays can have the same degrees of freedom as a filled ULA with the same aperture, this section considers difference coarrays of sparse arrays. The major motivation for using a sparse array is its ability to provide resolution that is approximately equal to the resolution of a

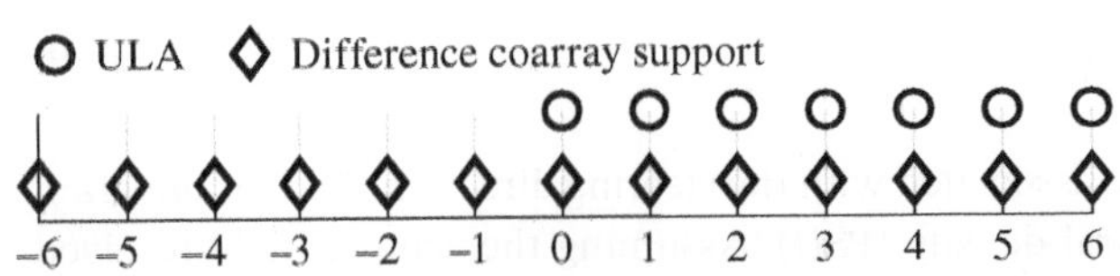

Figure 13.5 A filled ULA and its difference coarray support.

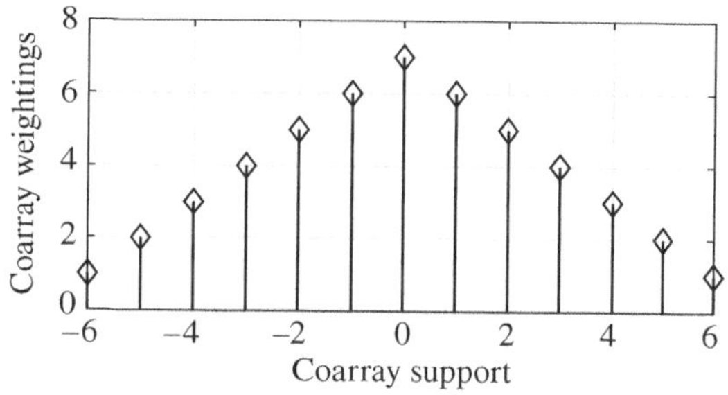

Figure 13.6 The difference coarray weightings for the ULA in Fig. 13.5.

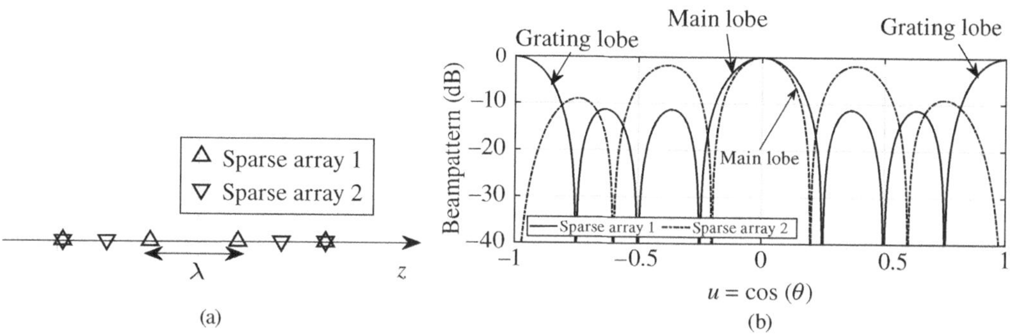

Figure 13.7 Sparse arrays and their beampatterns. (a) Two sparse arrays with four sensors each and (b) The beampatterns of the sparse arrays in (a).

filled array with the same aperture, while using fewer sensors than the filled array. Figure 13.7(a) illustrates two sparse arrays, labeled *Sparse array 1* and *Sparse array 2*. Both sparse arrays have an aperture equal to that of the seven-sensor filled array in Fig. 13.5 but each sparse array has only four sensors. Sparse array 1 is also a ULA but with an intersensor spacing of λ, which means its undersampling factor is 2. The number of major lobes in the beampattern of a sparse ULA is equal to the undersampling factor of the array. The major lobe centered at $u = 0$ is the main lobe and the remaining major lobes that have the same height as the main lobe are called grating lobes. The beampattern of Sparse array 1 is depicted in Fig. 13.7(b). The major lobe at $u = 0$ is the main lobe and the major lobes at $u = \pm 1$ are the grating lobes. The main lobe width is exactly equal to the main lobe width of the filled array in Fig. 13.7(a). Although Sparse array 1 uses only four sensors to provide the same resolution as the filled array with seven sensors, it cannot determine whether a signal is arriving from $u = 0$ or $u = \pm 1$. A sparse array can also have sensors that are nonuniformly spaced. An example of a four-sensor nonuniform sparse array is Sparse array 2 in Fig. 13.7(a). Such a nonuniform sparse array does not have grating lobes but has very high side lobes, as noted in Fig. 13.7(b). The main lobe width of Sparse array 2 is slightly less than that of Sparse array 1 in Fig. 13.7(a). However, Sparse array 2's ability to reject noise and interferers is weak due to its significantly high side lobes.

The terms difference coarray support and weightings as defined in (13.12) and (13.13) also apply to sparse arrays. Figure 13.8 plots normalized sensor locations of Sparse array 1 and Sparse array 2 from Fig. 13.7(a) and their difference coarray supports. Since these sparse arrays span from 0 to 6, their difference coarray supports span from −6 to 6. However, these sparse arrays do not have hole-free difference coarray supports. Figure 13.8 also shows an example of a sparse array (*Sparse array 3*), with the same aperture as the other two sparse arrays but a hole-free difference coarray

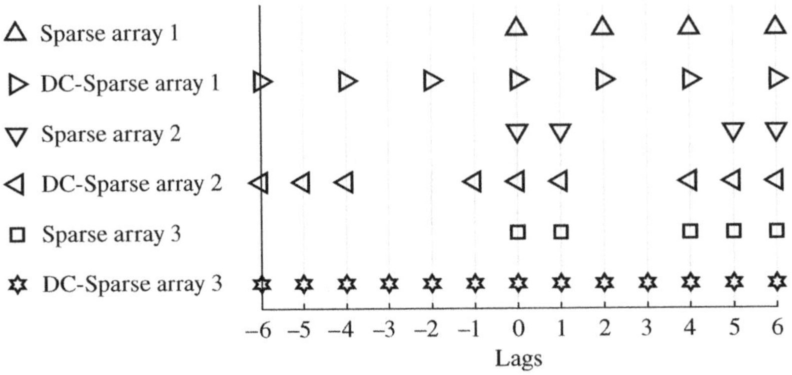

Figure 13.8 Sparse arrays and their difference coarray supports. From the top to the bottom row, the figure displays Sparse array 1 from Fig. 13.7(a) and its corresponding difference coarray support, Sparse array 2 from Fig. 13.7(a) and its corresponding difference coarray support, and Sparse array 3 with its corresponding difference coarray support.

support. The difference coarray weightings in (13.13) for the four arrays in Fig. 13.8 are illustrated in Fig. 13.9.

The difference coarray support definition in (13.12) can be generalized to cross difference coarrays. To distinguish cross difference coarrays from the difference coarrays in (13.12), we refer to C_d in (13.12) as *self* difference coarray support. Consider an array that is formed by combining two arrays, henceforth called *Subarray* a and *Subarray* b. If sets a and b contain the normalized sensor locations of Subarray a and Subarray b, respectively, two cross difference coarray supports can be defined as

$$C^{\times}_{\mathrm{d,a-b}} = \{n_i - n_j | n_i \in a, n_j \in b\}. \tag{13.15}$$

and

$$C^{\times}_{\mathrm{d,b-a}} = \{n_j - n_i | n_i \in a, n_j \in b\}. \tag{13.16}$$

Figure 13.10 illustrates a coprime array, which is a sparse array formed by interleaving two undersampled ULAs: Subarray a and Subarray b such that the undersampling factors of the two subarrays are coprime integers. The two cross difference coarray supports of the coprime array are also depicted in Fig. 13.10. A cross difference coarray support is not guaranteed to be symmetric. The figure also shows the self difference coarray support of the coprime array. In general, the self

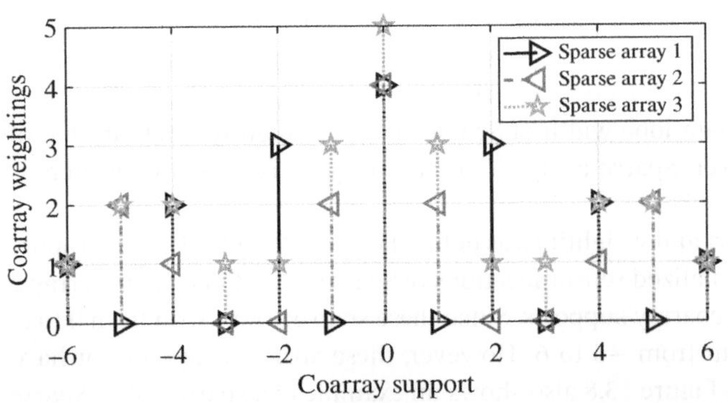

Figure 13.9 Difference coarray weightings for the three sparse arrays in Fig. 13.8.

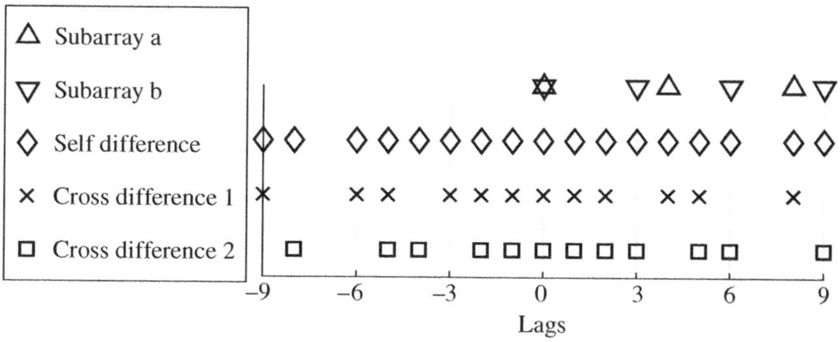

Figure 13.10 A coprime array with undersampling factors 4 and 3, the self difference coarray support, and the cross difference coarray supports.

difference coarray support contains more sensors than the cross difference coarray supports. For a coprime array the self and cross difference coarray supports may or may not be hole-free. For a nested array, which is formed by interleaving Subarray a (a filled ULA) and Subarray b (a sparse ULA whose undersampling factor is equal to the number of sensors in Subarray a or its integer multiple), all three difference coarray supports are always hole-free. An example of a nested array and its difference coarray supports are presented in Fig. 13.11.

For an array comprising subarrays with sensor locations in the sets a and b, respectively, the cross difference coarray weightings of the array are

$$c_{\mathrm{ab}}[n] = i_{\mathrm{a}}[n] \star i_{\mathrm{b}}[-n], \tag{13.17}$$

where $i_{\mathrm{a}}[n]$ and $i_{\mathrm{b}}[n]$ denote the sensor location indicator functions of Subarray a and Subarray b, respectively. The support of the convolution in (13.17) is equal to the cross difference coarray support in (13.15). Similarly, the convolution

$$c_{\mathrm{ba}}[n] = i_{\mathrm{b}}[n] \star i_{\mathrm{a}}[-n] \tag{13.18}$$

has the support equal to the cross difference coarray support in (13.16) and the convolution yields the corresponding coarray weightings, which are different from the weightings yielded by (13.17). The self and difference coarray weightings of the coprime array in Fig. 13.10 and the nested array in Fig. 13.11 are depicted in Figs. 13.12 and 13.13, respectively.

The difference coarray of an array is an important indicator of its potential to estimate ACF function and PSD. The difference coarray support indicates the lags at which the ACF can be potentially estimated whereas the difference coarray weighting indicates the number of sensor pairs available

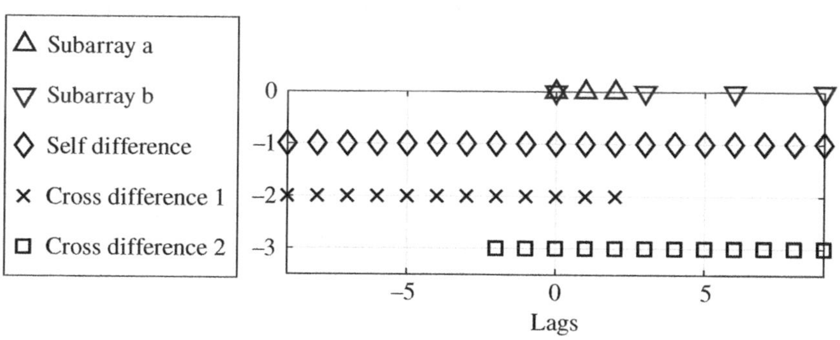

Figure 13.11 A nested array, its self difference coarray support, and cross difference coarray supports.

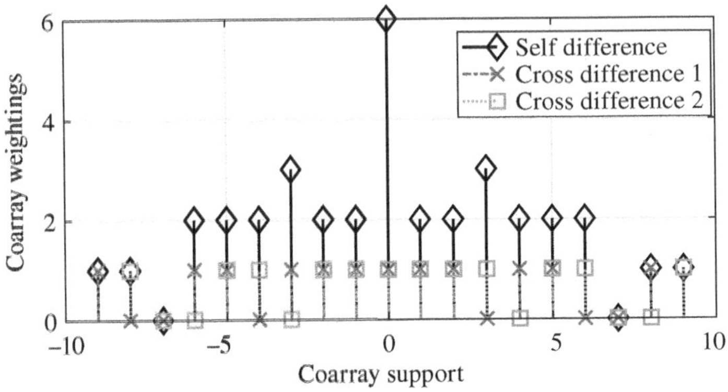

Figure 13.12 Difference coarray weightings of the coprime array in Fig. 13.10.

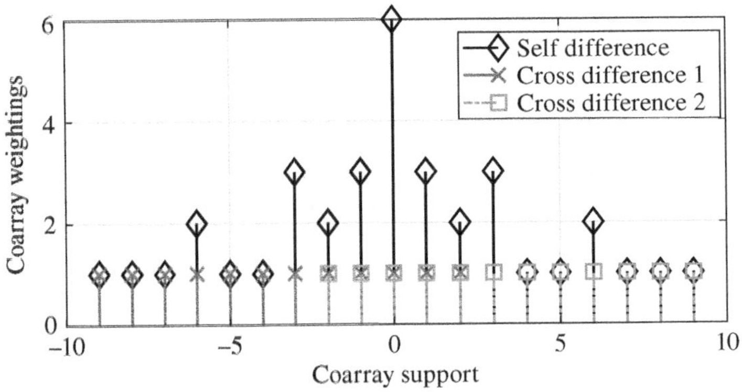

Figure 13.13 Difference coarray weightings of the nested array in Fig. 13.11.

to estimate the ACF at any given lag. A wide coarray support translates to large effective aperture of the array and narrow main lobe width of the beampattern, which means high resolution in the PSD estimation. A large difference coarray weighting translates to the availability of high number of samples for ACF estimation and thus small variance of the ACF estimate. The number of elements in a hole-free difference coarray support can be also regarded as the effective degrees of freedom of the array. For passive sensing, the effective degrees of freedom is the maximum number of source directions that can be theoretically estimated using the array. The actual number of sources that can be estimated using any sparse array, however, depends on a number of factors including SNR, number of snapshots, noise correlation, distance between the sources, and source correlation.

13.3.3 Sparse Passive Processing Algorithms

A variety of processing techniques have been employed to alleviate the aliasing artifacts in sparse sonar arrays. The four most common processors used for spatial PSD estimation are the CBF, product processor, min processor, and augmented processor. These processors implicitly or explicitly find an ACF estimate to evaluate a spatial PSD estimate. In augmented processing, the ACF estimation is explicit whereas in CBF, product processing, and min processing, the ACF estimation is implicit.

After estimating the ACF or PSD, further processing can be applied to obtain high resolution direction of arrival (DOA) estimates. Algorithms for DOA estimation include eigenanalysis-based

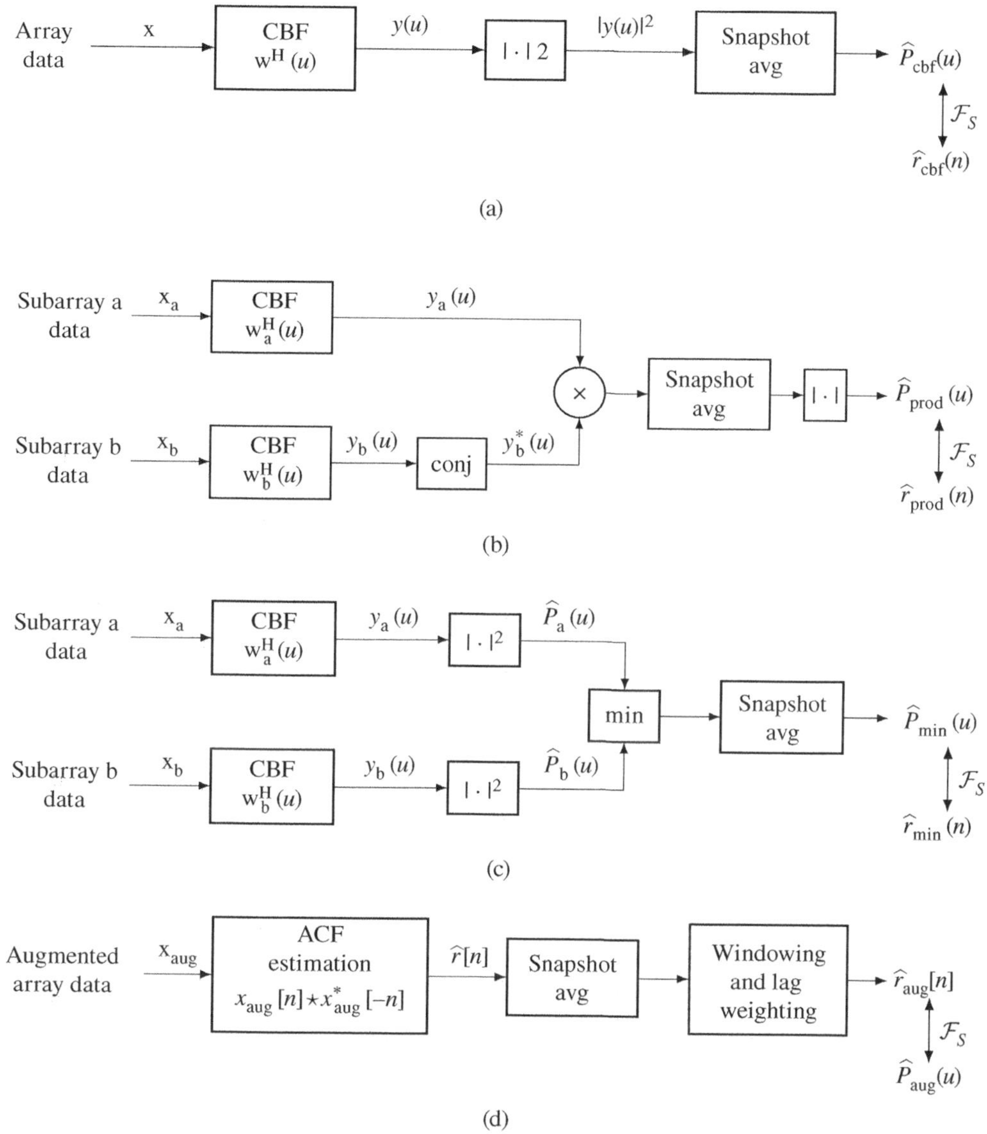

Figure 13.14 Block diagrams of the predominant processors. (a) CBF block diagram, (b) product processor block diagram, (c) min processor block diagram, and (d) augmented processor block diagram.

methods such as multiple signal classification (MUSIC) [35], min-norm method (MNM) [36], and estimation of signal parameters via rotational invariance technique (ESPRIT) [37].

Figure 13.14 shows block diagrams for the four ACF/PSD estimation methods described below.

1. *Conventional beamforming*: Figure 13.14(a) depicts the CBF, which is the standard additive processor described in Section 13.2.1. $\mathbf{x}$ contains the narrowband array data and $\mathbf{w}$ is the tapered weight vector for the sparse array (see Eqs. (13.2) and (13.3)). The squared magnitude of the CBF output, $|y(u)|^2$, is averaged over multiple snapshots to produce an estimate of the PSD which is denoted as $\hat{P}_{\text{cbf}}(u)$. The PSD estimate is the spatial Fourier transform of the ACF estimate $\hat{r}_{\text{cbf}}[n]$.

The performance of the CBF is characterized by the power beampattern of an array, $|B(u)|^2$. The power pattern is related to the self difference coarray weightings by the spatial Fourier transform. Specifically,

$$|B(u)|^2 = \mathcal{F}_{\mathsf{s}}\{c[n]\} = \sum_n c[n] \exp\left(jn\frac{2\pi d}{\lambda}u\right). \tag{13.19}$$

2. *Product processing*: In product processing, the CBF outputs of Subarray a and Subarray b are evaluated as illustrated in Fig. 13.14(b). The subarray CBF outputs are $y_{\mathsf{a}}(u) = \mathbf{w}_{\mathsf{a}}^H(u)\mathbf{x}_{\mathsf{a}}$ and $y_{\mathsf{b}}(u) = \mathbf{w}_{\mathsf{b}}^H(u)\mathbf{x}_{\mathsf{b}}$, respectively, where $\mathbf{x}_{\mathsf{a}}$ and $\mathbf{x}_{\mathsf{b}}$ denote Subarray a and Subarray b data and $\mathbf{w}_{\mathsf{a}}$ and $\mathbf{w}_{\mathsf{b}}$ denote the corresponding weight vectors. The product processor multiplies $y_{\mathsf{a}}(u)$ with the complex conjugate of $y_{\mathsf{b}}(u)$. The product is averaged over multiple snapshots and the magnitude of the average is computed to yield the PSD estimate, $\hat{P}_{\mathsf{prod}}(u)$. The performance of the product processor is characterized by the product beampattern, which is obtained by multiplying Subarray a's beampattern with the complex conjugate of Subarray b's beampattern. The product beampattern is related to the cross difference coarray weighting, i.e.,

$$B_{\mathsf{a}}(u)B_{\mathsf{b}}^*(u) = \mathcal{F}_{\mathsf{s}}\{c_{\mathsf{ab}}[n]\} = \sum_n c_{\mathsf{ab}}[n] \exp\left(jn\frac{2\pi d}{\lambda}u\right), \tag{13.20}$$

where $B_{\mathsf{a}}(u)$ and $B_{\mathsf{b}}(u)$ denote the beampatterns of Subarray a and Subarray b, respectively. Alternatively, product processing can be done by multiplying the CBF output of Subarray b with the complex conjugate of the CBF output of Subarray a. In this case, the product beampattern is given by

$$B_{\mathsf{a}}^*(u)B_{\mathsf{b}}(u) = \mathcal{F}_{\mathsf{s}}\{c_{\mathsf{ba}}[n]\} = \sum_n c_{\mathsf{ba}}[n] \exp\left(jn\frac{2\pi d}{\lambda}u\right). \tag{13.21}$$

3. *Min processing*: In min processing, the CBF outputs of Subarray a and Subarray b are first evaluated as illustrated in Fig. 13.14(c). Then, the PSD estimates corresponding to the subarray are computed as $\hat{P}_{\mathsf{a}}(u) = |y_{\mathsf{a}}(u)|^2$ and $\hat{P}_{\mathsf{b}}(u) = |y_{\mathsf{b}}(u)|^2$. The minimum processor evaluates the minimum of $\hat{P}_{\mathsf{a}}(u)$ and $\hat{P}_{\mathsf{b}}(u)$ and applies snapshot averaging to the minimum values, which results in a PSD estimate. The PSD estimate, $\hat{P}_{\mathsf{min}}(u)$, is the spatial Fourier transform of an ACF estimate denoted by $\hat{r}_{\mathsf{min}}[n]$. The performance of the min processor is characterized by the min beampattern which is given by

$$\min\left(|B_{\mathsf{a}}(u)|^2, |B_{\mathsf{b}}(u)|^2\right). \tag{13.22}$$

4. *Augmented processing*: For augmented processing, we represent the data received by a sparse array as $\mathbf{x}_{\mathsf{aug}}$, where the elements corresponding to missing sensors are 0. The augmented processor finds an initial estimate of the ACF as $\hat{r}[n] = x_{\mathsf{aug}}[n] \star x_{\mathsf{aug}}^*[-n]$. Then, $\hat{r}[n]$ is averaged over multiple snapshots. The augmented processor windows the average of $\hat{r}[n]$ to extract the hole-free section and weights the windowed function to yield the final ACF estimate. The spatial Fourier transform of the ACF estimate is evaluated to find the spatial PSD estimate, $\hat{P}_{\mathsf{aug}}(u)$.

When a nonuniform window or shading function is applied to the array data, the corresponding difference coarray weightings are obtained by replacing the sensor location indicator functions $i_{\mathsf{a}}[n]$ and $i_{\mathsf{b}}[n]$ in (13.13), (13.17), and (13.18) with the shading functions $w_{\mathsf{a}}[n]$ and $w_{\mathsf{b}}[n]$, respectively.

For post-processing using eigenanalysis-based algorithms, the ACF estimate is used to form a spatial covariance matrix estimate. The matrix estimate is used in subspace-based methods such as MUSIC, MNM, and ESPRIT which yield the pseudospectral estimate. With these algorithms, a spatial PSD estimate cannot be obtained; only signal DOAs can be estimated.

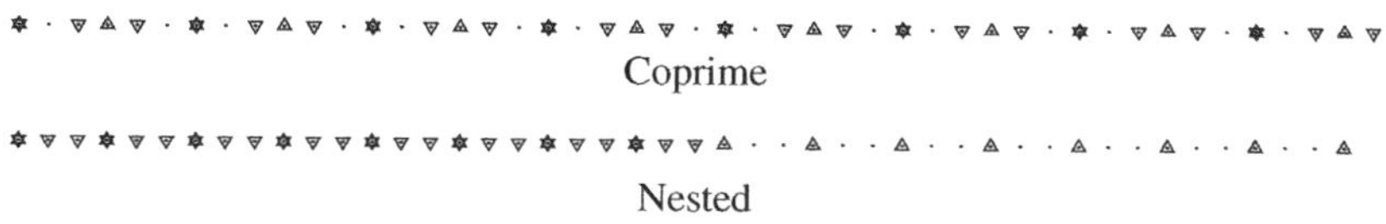

Figure 13.15 Coprime and nested arrays, each with 32 sensors. Subarray a and Subarray b sensors are denoted by ∇ and Δ, respectively.

13.3.4 Review of the Predominant Processors

This section provides a comprehensive literature survey of the four predominant sparse array processors. The following paragraph briefly describes the coprime and nested array geometries used to illustrate key findings about these processors.

In a basic linear coprime array, Subarray a has M sensors and $N\lambda/2$ spacing, while Subarray b has N sensors and $M\lambda/2$ spacing. Here, M and N are coprime integers [38]. In an extended linear coprime array, the numbers of sensors in the two subarrays are QM and QN, respectively. Figure 13.15 depicts a coprime array with $M = 3$, $N = 2$, and $Q = 8$. In a linear nested array, Subarray a has M sensors and $\lambda/2$ spacing, while Subarray b has N sensors and $M\lambda/2$ spacing [15]. In an extended nested array, the number of sensors in Subarray a is QM, where Q is an integer. Figure 13.15 depicts a nested array with $M = 3$, $N = 8$, and $Q = 8$. Multidimensional extensions to coprime and nested array geometries are defined in [21] and [39], respectively.

13.3.4.1 Conventional Beamforming

CBF has been applied to linear and planar sparse arrays. Impagliazzo et al. design a planar array on a uniform grid of $\lambda/2$ spacing, where the sensor locations are selected to minimize the side lobes [40]. They apply CBF to sonar data to provide three-dimensional acoustic images of underwater targets. Adhikari et al. [41] design linear sparse arrays maximizing the deflection coefficient for given broadband sonar data and applies CBF. When CBF is applied to a coprime or nested array, the peak side lobe height is significantly higher than that of a full array with an equal resolution. Figure 13.16 compares the CBF beampatterns of the coprime and nested arrays in Fig. 13.15 with their product and min beampatterns. The CBF, product, and min beampatterns in dB are obtained by taking *log* (with base 10) of (13.19), (13.20), and (13.22), respectively, and multiplying by 10. Unlike the peak side lobes of the product and min processors, the peak side lobe height of the CBF does not decrease with an extension of the subarrays. The CBF beampattern of a nested array has lower peak side lobe height than that of the coprime array, but the main lobe width is wider for the nested array, as evidenced by Fig. 13.16.

The spatial PSD estimate obtained using CBF on a sparse array is implicitly equal to the spatial Fourier transform of a biased estimate of the spatial ACF. One advantage of the CBF over product processing is that the CBF output is guaranteed to be non-negative definite (NND).

13.3.4.2 Product Processing

Product processing disambiguates aliasing in nested and coprime arrays. This method involves applying CBF to subarrays independently and multiplying the CBF output of one subarray with the complex conjugate of the other subarray's CBF output. Davies and Ward introduced this method for nested arrays [15]. In a nested array, the grating lobes of Subarray b are centered at $u = 2k/M$, where k is an integer while the nulls of Subarray a are centered at $u = 2k/(QM)$ as illustrated in Fig. 13.17(a). Consequently, the nulls of Subarray a remove the grating lobes of Subarray b in

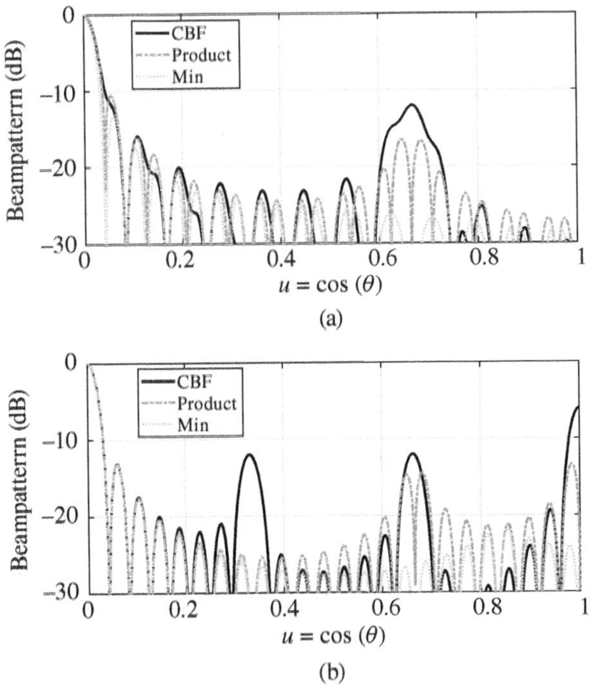

Figure 13.16 CBF, Product, and Min Beampatterns for (a) the nested array from Fig. 13.15 and (b) the coprime array from Fig. 13.15. Since the beampatterns are symmetric about $u = 0$, the plots only display $0 \leq u \leq 1$.

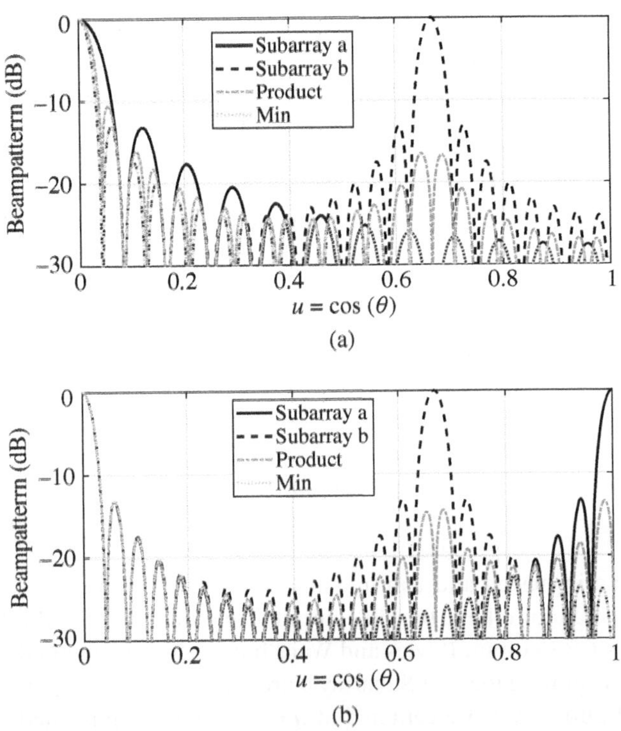

Figure 13.17 Subarray a, Subarray b, product, and Min beampatterns for (a) the nested array from Fig. 13.15 and (b) the coprime array from Fig. 13.15.

the product beampattern. The product beampattern has the same main lobe width as a full array with MN sensors. Product processing was applied to remove aliasing artifacts in coprime arrays by Vaidyanathan and Pal [21, 38]. In a coprime array, the grating lobes of Subarray a are at $u = 2k/N$ while the grating lobes of Subarray b are at $u = 2k/M$. The grating lobes' locations are unique because M and N are coprime while the main lobes overlap as illustrated in Fig. 13.17(b). Thus, multiplying the subarrays' CBF outputs reinforces the main lobes and mitigates the grating lobes. The product beampattern of a coprime array has the same main lobe width as a full array with QMN sensors.

Although product processing is effective in alleviating aliasing artifacts and providing a narrow main lobe, it has three notable disadvantages. First, both basic coprime and nested arrays have high peak side lobes. However, extending a coprime array by having Q copies of each subarray reduces the peak side lobe to the same level as a full array with an equal resolution. The value of Q required to match the peak side lobe height of a full array for different standard windows is derived in [42]. The second disadvantage of product processing is that its spatial PSD estimate is not NND [43]. However, the PSD is asymptotically unbiased as discussed in [43]. Third disadvantage of product processing is the cross terms arising from multiple signals impinging on an array. If the signals are uncorrelated, averaging the outputs of product processor can lower the cross terms at a rate of 5 dB per decade of snapshots averaged [44]. However, an underwater environment is dynamic and snapshot-deficient. As a result, enough number of snapshots might not be available for averaging and reducing the cross terms. Chavali et al. apply product processing to sonar data obtained from a coprime array in [45] and validate that the cross terms decay at a slow rate of 5 dB per 10 snapshots. Chavali and Wage have quantitatively analyzed the impact of number of snapshots on cross term mitigation, taking into account side lobe control using subarray tapers [44]. They note that while subarray tapering reduces side lobe leakage, it necessitates an increased number of snapshots for effective cross-term mitigation.

13.3.4.3 Min Processing

Min processing also removes aliasing artifacts from nested and coprime arrays [22, 25]. This method applies CBF to subarrays of a coprime or a nested array independently and it yields the final output by taking the minimum of the magnitudes of the CBF outputs. The min beampattern is also free of grating lobes because of the placements of the subarrays' grating lobes, main lobes, and nulls as described in Section 13.3.4.2. The subarray and min beampatterns for the nested and coprime arrays from Fig. 13.15 are depicted in Fig. 13.17.

One advantage of min processing over product processing is that the spatial PSD estimate of min processor output is NND. However, min processing also suffers from high side lobes and cross terms. The extension factors required in a min processor to match the peak side lobe heights of an equivalent ULA is derived for various standard windows in [46]. Chavali et al. detail the cross terms in a min processor [26]. They also compare product processing and min processing for sonar data obtained from a coprime and a nested array. For the sensor layouts for the coprime and nested arrays depicted in Fig. 13.15, [26] shows that coherent cross terms pose a significant challenge for product processing. They also demonstrate that min processing applied to the nested array is the least vulnerable to coherent cross terms among the various processor and sensor layout combinations examined in [26].

Figure 13.18 Coprime array with $N = 4$, $M = 3$, and $Q = 1$ and its self difference coarray support.

13.3.4.4 Augmented Processor

The difference coarray support of a full ULA with L sensors is always hole-free or contiguous and extends from $-(L-1)$ to $L-1$. In general, the difference coarray support of a sparse array is not contiguous. Examples of sparse arrays with contiguous difference coarray supports are minimum redundancy arrays [47] and nested arrays. Coprime arrays with coprime pairs other than $(2, 3)$ do not have contiguous difference coarray supports. However, the contiguous section of a coprime array's coarray support is, in general, longer than the coarray support of an arbitrary sparse array with an equal number of sensors and aperture. The coprime array in Fig. 13.15 extends from 0 to $46d$ and it has a contiguous difference coarray support from −46 to 46. The nested array in the figure extends from 0 to $45d$ and it has a contiguous difference coarray support from −45 to 45. To see that a coprime array's difference coarray support is not always contiguous, consider the coprime array in Fig. 13.18, where $N = 4$, $M = 3$, and $Q = 1$. While the array extends from 0 to $9d$, the contiguous difference coarray support range is only −6 to 6. Nevertheless, the contiguous section of the coarray support is still wider than that of a full array with the same number of sensors. The spatial ACF of the signals impinging on a sparse array can be estimated up to lag $\mathcal{L} - 1$ if the contiguous coarray support extends from $-\mathcal{L} + 1$ to $\mathcal{L} - 1$. Pillai et al. use the augmented processor to find the DOA estimates of narrowband plane waves [48]. They use the augmented covariance matrix in conjunction with MUSIC to estimate the DOAs of six uncorrelated sources using a four-sensor sparse array.

Solomon and Knight analyze the data received by a 32-sensor platform mounted sparse sonar array with an aperture equal to that of a 64-sensor full ULA [49]. They estimate the ACF up to lag 63. For each lag, an unbiased ACF estimate is obtained by averaging over all possible sensor pairs' data. The Fourier transform of the ACF estimate yields a PSD estimate. One drawback of the ACF and PSD estimates obtained with this approach is that they are not guaranteed to be NND.

13.3.5 Simulation Example

A clear mechanism for tapering subarrays using standard windows in product and min processors exists [42, 46]. However, there is currently no established method for tapering sparse arrays in CBF and augmented processors. Therefore, when the reduction of peak side lobe height and effective suppression of strong noise and interferers are required, the use of product and min processors is more favorable. These processors offer the advantage of tapering to achieve the desired peak side lobe height.

We present an example using simulated data to highlight the significance of array shading, subarray extension, and snapshot averaging. The scenario involves two sources with direction cosines $u_1 = -0.1$ and $u_2 = 0.04$. The SNR of the first source is 5 dB, while the SNR of the second source is 15 dB lower. Figure 13.19(a) illustrates the scanned responses of the CBF, product, min, and augmented processors using a basic coprime array ($M = 2$, $N = 3$, $Q = 1$). The scanned responses in dB for the CBF, product, min, and augmented processors are $10\,log\,\hat{P}_{\text{cbf}}(u)$, $10\,log\,\hat{P}_{\text{prod}}(u)$, $10\,log\,\hat{P}_{\text{min}}(u)$, and $10\,log\,\hat{P}_{\text{aug}}(u)$, respectively. Due to the lack of array extension, the coprime array

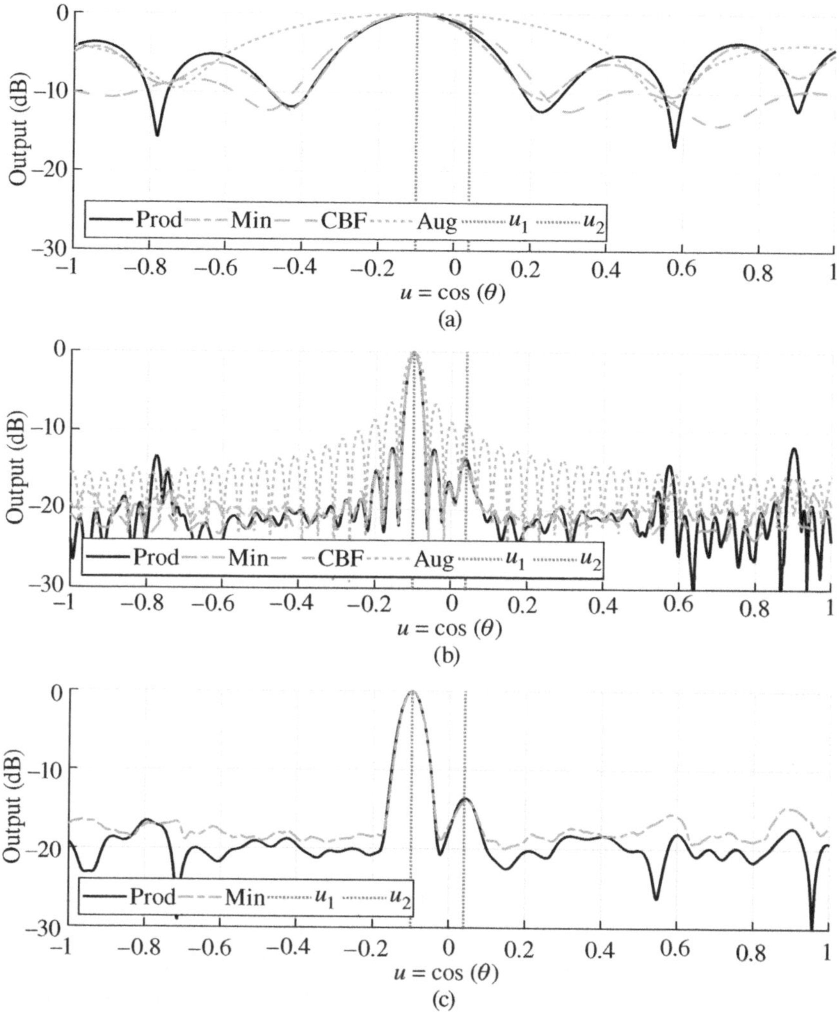

Figure 13.19 Scanned responses of different processors for a coprime array. (a) Uniform shading with $Q = 1$. (b) Uniform shading with $Q = 8.5$. (c) Hann shading with $Q = 8.5$.

exhibits high side lobes and limited resolution, preventing the resolution of the two sources by any of the processors. An extension factor of $Q = 8.5$ with uniform shading allows correct identification of the first source, but masks the second source due to high side lobes (Fig. 13.19(b)). To mitigate this, array tapering is necessary. Details on tapering for the product processor can be found in [42], while details for the min processor are provided in [25] and [46]. In this example, we employ Hann shading, which requires an extension factor of $Q = 8.5$ to match the peak side lobe height of the filled ULA with the same aperture. Consequently, the product and min processors with Hann shading (Fig. 13.19(c)) successfully resolve both sources. In all three plots in Fig. 13.19, the number of snapshots used is 50. Lowering the number of snapshots deteriorates the performance of the product processor more severely than the min processor.

13.4 Experimental Sonar Examples

To assess the performance of different processors, this section utilizes an open-source sonar data available on the SPIB website [50]. The data set consists of observations made by a 48-sensor vertical array deployed in shallow water of the Mediterranean Sea, north of Elba Island. Previous work by Chavali et al. applied product and min processing to this dataset using the arrays shown in Fig. 13.15. Their results indicated that the nested array with the min processor yielded the best performance. In addition to comparing the min and product processors, this section also includes a comparison with the CBF and augmented processors.

Figure 13.20 displays the k-ω plot obtained from approximately 301.056 seconds of data sampled at 1 kHz. The plot was obtained using CBF on a standard 48-sensor ULA. The Hann window was applied temporally, while the Chebyshev window with 30 dB attenuation was applied spatially. Figure 13.21 depicts the k-ω plots for the same dataset using both coprime and nested arrays depicted in Fig. 13.15. The left column corresponds to the coprime array, while the right column corresponds to the nested array. The panels from top to bottom in each column represent the CBF, augmented, product, and min processors, respectively. The k-ω plots of the CBF and augmented processors have higher side lobes compared to the k-ω plot of the full ULA. For both coprime and nested arrays, the augmented processor yields lower side lobes than CBF. Additionally, the k-ω plots of the product processor show higher lobes compared to the corresponding k-ω plots of the min processor for both arrays. Overall, the nested array with the min processor displays the lowest side lobes and closely matches the k-ω plot of the full ULA, which aligns with the findings in [26].

To compare product and min processors, [26] also introduced a metric called the normalized spectral error (NSE). The NSE is defined as

$$\mathrm{NSE}(\omega) = \frac{d}{2\pi}\int_{-\pi/d}^{\pi/d} \frac{(S(k,\omega) - S_{\mathrm{sparse}}(k,\omega))^2}{S^2(k,\omega)} dk, \tag{13.23}$$

where $S(k,\omega)$ represents the estimated k-ω spectrum of the full ULA and $S_{\mathrm{sparse}}(k,\omega)$ represents the estimated k-ω spectrum of the sparse processor. Figure 13.22 shows this metric for all four processors for the arrays in Fig. 13.15. For the coprime array, the augmented processor has lower NSE than the CBF. For the nested array, the augmented processor and CBF have approximately equal NSE values. For the coprime array, the product and min processors have equal NSEs at lower frequencies while at higher temporal frequencies, the product processor has a lower NSE than the augmented processor. For the nested array, the min processor has a lower NSE than the product processor, mostly at high frequencies. Among all processor–array combinations, the coprime array with the augmented processor has the smallest NSE for most frequencies, except for frequencies

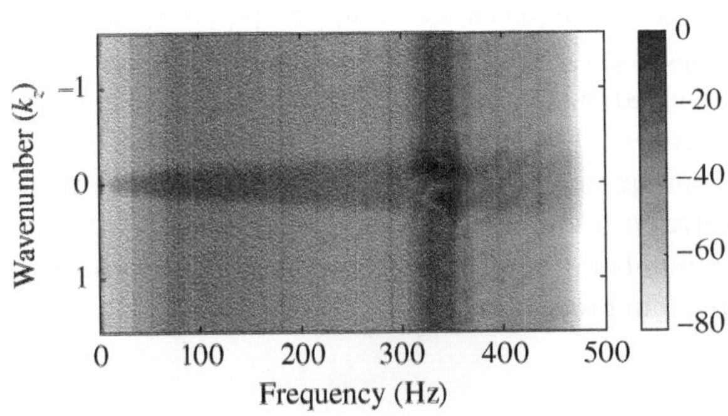

Figure 13.20 $k-\omega$ plot corresponding to the CBF for a 48-sensor ULA.

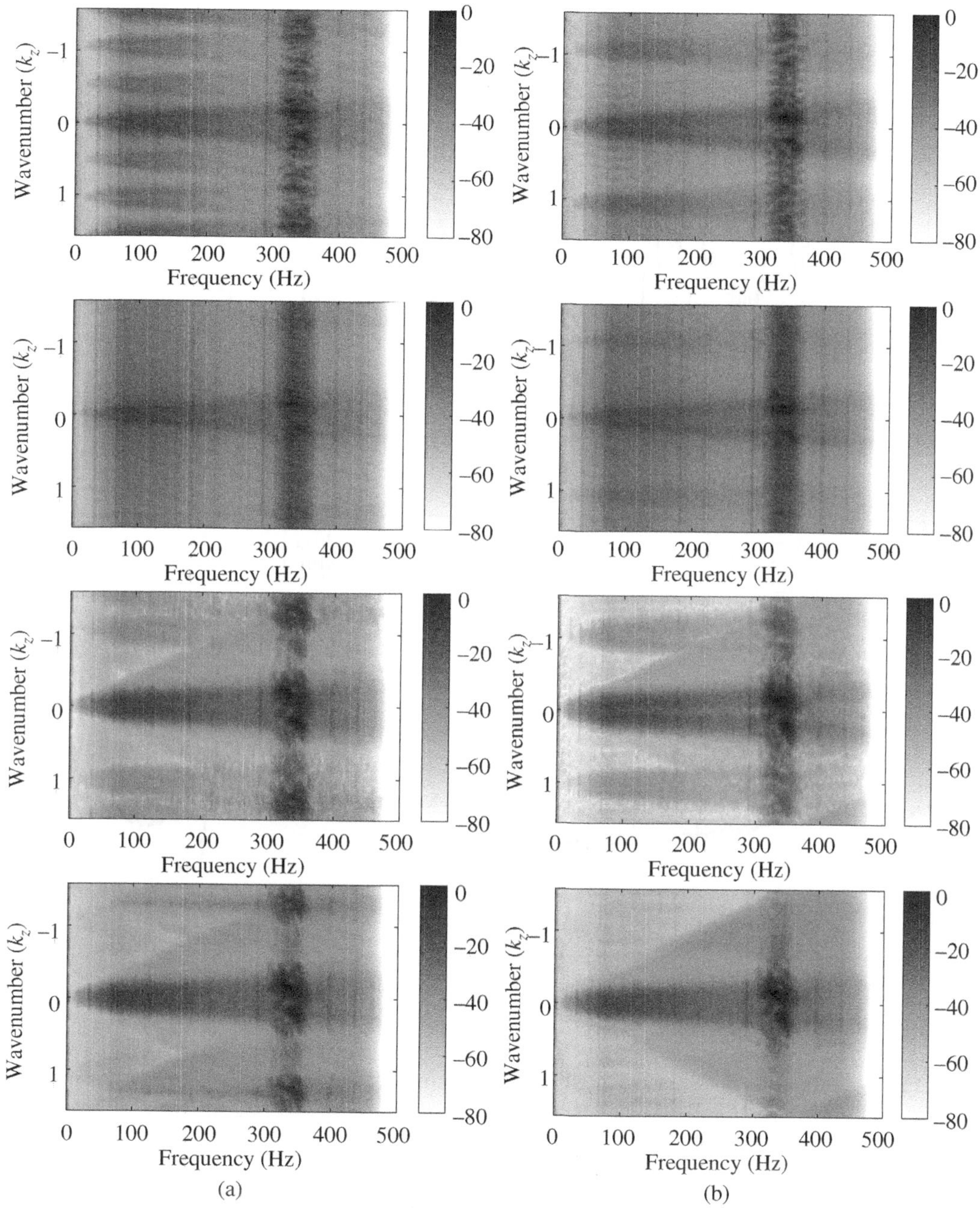

Figure 13.21 Comparison of the predominant processors for the coprime (a) and nested (b) arrays from Fig. 13.15. The $k - \omega$ plots in the figure represent the following processors, from top to bottom: CBF, augmented, product, and min processors.

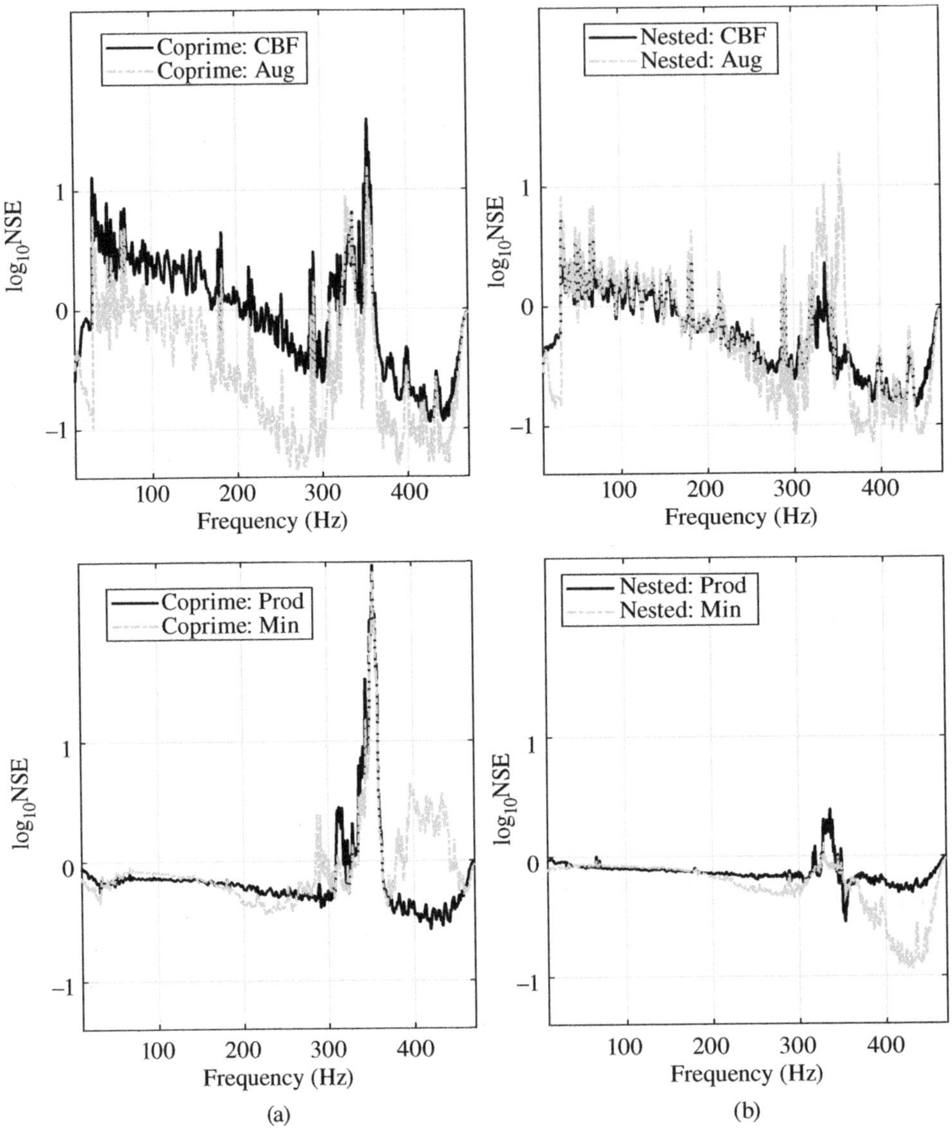

Figure 13.22 Comparison of CBF, augmented, product, and min processors for the coprime (a) and nested (b) arrays from Fig. 13.15.

close to 350 Hz. The nested array with the min processor has the smallest NSE when the frequency is close to 350 Hz.

13.5 Further Reading on Sparse Sonar

This section summarizes additional research on sparse sonar arrays and provides suggestions for further background reading. The discussion begins with theoretical analyses and new processors and then describes experimental work and application-specific designs.

Recent papers explore performance predictions for sparse arrays. Pyzdek and Culver quantify the impact of array shape perturbations on the product beampatterns of coprime arrays [51]. Bautista and Buck compare PSD estimates for coprime arrays computed using the CBF and product processors [52]. They show that the CBF processor yields a biased estimate as the aperture of the coprime array grows. The product processor produces an asymptotically unbiased estimate, which is similar to the conventional processor for a filled ULA. Bautista and Buck present an example for spatially correlated noise to illustrate the bias. Liu and Buck present a thorough analysis of the min processor's detection and estimation performance [25]. The paper derives the statistics of the min processor output for Gaussian sources and verifies those results using simulations. In another publication, Liu and Buck show that the ACF estimate obtained by evaluating the inverse spatial Fourier transform of the min PSD estimate can be used to form a Toeplitz covariance matrix estimate to use with the MUSIC algorithm [53].

Rooney and Buck [54] propose an improved PSD estimator for coprime arrays combining the product processor with Thomson's multitaper method [55]. Rooney et al. propose a multitaper version of the min processor in [56] that reduces variance. Both the product and min versions of the multitaper processors sacrifice some resolution for a reduction in variance. Rooney et al. provide a detailed analysis of the multitaper min processor and also propose *ad hoc* blends of the min and product processors that improve performance under some conditions. In another paper Rooney and Buck develop a Welch overlapping segment averaging product processor for coprime arrays that provides an alternate method of reducing variance at the expense of resolution [57].

Another method that has been applied for sparse sonar arrays is the synthetic aperture technique. Use of synthetic aperture sonar to increase the signal-to-noise ratio can be traced back to Cutrona [58]. Neudorfer *et al.* use the synthetic aperture to place data in missing sensor locations of a sparse sonar array [59]. Ramirez and Krolik use the synthetic aperture technique on a coprime array in order to make the coarray contiguous and apply spatial Fourier transform to estimate the PSD [60]. Ramirez and Rogers test this method on experimental sonar data obtained by a coprime array [27].

Two other recent publications also apply sparse array processing to underwater data sets. Nannuru et al. use sparse Bayesian learning on broadband sonar data obtained by two-dimensional coprime and nested arrays [61]. They show that the Bayesian learning method can outperform MUSIC. Nair et al. apply product processing to sonar data obtained from a twin coprime array [29]. They use cardioid processing to resolve left–right ambiguity. The paper compares the Left–Right resolved Coprime Sparse Array (LRCSA) to two other algorithms for resolving the left-right ambiguity. Results from a sea trial show good agreement between theoretical predictions and measured performance.

Sparse aliased arrays have been applied in marine mammal research. For example Thode et al. describe using a three-sensor aliased array to localize dolphins [62]. They use a frequency-domain implementation similar to an approach developed by Hinich [63]. The approach relies on averaging beamformer outputs across frequency to eliminate grating lobes. Analysis of experimental data acquired near Panama shows that Dolphin whistles have sufficient bandwidth to permit localization with the sparse array at ranges of at least 3 km. Ball and Buck implement the same broadband localization algorithm in their beamforming video recorder [64]. The recorder uses a handheld Mills cross array with a video camera mounted in the center. The beamformer location estimates are used to place a visual cue in the video recording indicating the source of vocalizations produced by marine mammals. Ball and Buck describe using the system to observe bottlenose dolphins in a lagoon.

A key advantage of sparse arrays is that they can be smaller, lighter, and easier to deploy, which is especially useful for use on autonomous underwater vehicles (AUVs). Glegg et al. describe a six-sensor sparse volumetric array mounted on an AUV [65]. They use the system to localize and track sources, e.g., a scuba diver, using a cross-correlation technique developed by Knapp and Carter [66]. Images are formed by combining the cross-correlation estimates with additive or multiplicative processing. Glegg et al. note that the multiplicative processor is similar to one of the techniques described by Berman and Clay [10]. In more recent work, researchers have investigated using coprime arrays for near-field source localization with vertical arrays in an underwater channel [67]. The vertical array is an inflatable structure that is designed to be deployed from an AUV. Li et al. [4] describes a recent laboratory and experimental study of the system.

Acknowledgements

This work was supported by the U.S. Office of Naval Research through Grants N00014-20-1-2820 (Kaushallya Adhikari) and N00014-22-1-2750 (Kathleen Wage).

References

1 Bjørnø, L. and Buckingham, M.J. (2017). General characteristics of the underwater environment. In: *Applied Underwater Acoustics*, Chapter 1, 1–84. Amsterdam, Netherlands: Elsevier. ISBN: 978-0-12-811247-2.

2 Abraham, D.A. (2019). *Underwater Acoustic Signal Processing Modeling, Detection, and Estimation*, Modern Acoustics and Signal Processing, 1e. Cham: Springer International Publishing https://doi.org/10.1007/978-3-319-92983-5.

3 Steinberg, B.D. (1976). *Principles of Aperture and Array System Design*. New York: Wiley.

4 Li, Y., Ouyang, B., Zhou, T. et al. (2022). Laboratory and field experimental study of underwater inflatable co-prime sonar array (UICSA). *Journal of Civil Engineering and Construction* 11 (1): 41–55. https://doi.org/10.32732/jcec.2022.11.1.41.

5 Johnson, D.H. and Dudgeon, D.E. (1993). *Array Signal Processing: Concepts and Techniques*. Englewood Cliffs, NJ: Prentice Hall.

6 Van Trees, H.L. (2002). *Optimum Array Processing*. New York: Wiley.

7 Holm, S., Austeng, A., Iranpour, K., and Hopperstad, J.-F. (2001). Sparse sampling in array processing. In: *Nonuniform Sampling: Theory and Practice, Information Technology: Transmission, Processing, and Storage* (ed. F. Marvasti), 787–833. Boston, MA: Springer US. ISBN: 978-1-4615-1229-5. https://doi.org/10.1007/978-1-4615-1229-519.

8 Urick, R.J. (1983). *Principles of Underwater Sound*, 3e. McGraw-Hill, Inc.

9 Faran, J.J. and Hills, R. Jr. (1952). The Application of Correlation Techniques to Acoustic Receiving Systems.

10 Berman, A. and Clay, C.S. (1957). Theory of time-averaged-product arrays. *The Journal of the Acoustical Society of America* 29 (7): 805–812. https://doi.org/10.1121/1.1909060.

11 Linder, I.W. (1961). Resolution characteristics of correlation arrays. *Journal of Research of the National Bureau of Standards, Section D: Radio Propagation* 65D (3): 245. https://doi.org/10.6028/jres.065D.032.

12 Welsby, V.G. (1961). Multiplicative receiving arrays: the angular resolution of targets in a sonar system with electronic scanning. *Journal of the British Institution of Radio Engineers* 22 (1): 5–12. https://doi.org/10.1049/jbire.1961.0077.

13 Brown, J.L. and Rowlands, R.O. (1959). Design of directional arrays. *The Journal of the Acoustical Society of America* 31 (12): 1638–1643. https://doi.org/10.1121/1.1907672.

14 Jacobson, M.J. (1957). Analysis of a multiple receiver correlation system. *The Journal of the Acoustical Society of America* 29 (12): 1342–1347. https://doi.org/10.1121/1.1908806.

15 Davies, D.E.N. and Ward, C.R. (1980). Low sidelobe patterns from thinned arrays using multiplicative processing. *IEE Proceedings F (Communications, Radar and Signal Processing)* 127 (1): 9–15. https://doi.org/10.1049/ip-f-1.1980.0002.

16 Nuttall, A.H., Carter, G.C., and Montavon, E.M. (1974). Estimation of the two-dimensional spectrum of the space-time noise field for a sparse line array. *The Journal of the Acoustical Society of America* 55 (5): 1034–1041. https://doi.org/10.1121/1.1914644.

17 Bucker, H.P. (1977). Cross-sensor beam forming with a sparse line array. *The Journal of the Acoustical Society of America* 61 (2): 494–498. https://doi.org/10.1121/1.381290.

18 Bucker, H.P. (1979). High resolution cross-sensor beamforming for a billboard array. *The Journal of the Acoustical Society of America* 65 (1): 145–147. https://doi.org/10.1121/1.382256.

19 Hoctor, R.T. and Kassam, S.A. (1990). The unifying role of the coarray in aperture synthesis for coherent and incoherent imaging. *Proceedings of the IEEE* 78 (4): 735–752. https://doi.org/10.1109/5.54811.

20 Pal, P. and Vaidyanathan, P.P. (2010). Nested arrays: a novel approach to array processing with enhanced degrees of freedom. *IEEE Transactions on Signal Processing* 58 (8): 4167–4181. https://doi.org/10.1109/TSP.2010.2049264.

21 Vaidyanathan, P.P. and Pal, P. (2011). Sparse sensing with co-prime samplers and arrays. *IEEE Transactions on Signal Processing* 59 (2): 573–586. https://doi.org/10.1109/TSP.2010.2089682.

22 Di Martino, G. and Iodice, A. (2015). Coprime synthetic aperture radar (CopSAR): a new acquisition mode for maritime surveillance. *IEEE Transactions on Geoscience and Remote Sensing* 53 (6): 3110–3123. https://doi.org/10.1109/TGRS.2014.2369035.

23 Di Martino, G. and Iodice, A. (2017). Orthogonal coprime synthetic aperture radar. *IEEE Transactions on Geoscience and Remote Sensing* 55 (1): 432–440. https://doi.org/10.1109/TGRS.2016.2608140.

24 Liu, Y. and Buck, J.R. (2015). Detecting Gaussian signals in the presence of interferers using the coprime sensor arrays with the min processor. In *2015 49th Asilomar Conference on Signals, Systems and Computers*, 370–374, November 2015. https://doi.org/10.1109/ACSSC.2015.7421150.

25 Liu, Y. and Buck, J.R. (2018). Gaussian source detection and spatial spectral estimation using a coprime sensor array with the min processor. *IEEE Transactions on Signal Processing* 66 (1): 186–199. https://doi.org/10.1109/TSP.2017.2762284.

26 Chavali, V., Wage, K.E., and Buck, J.R. (2018). Multiplicative and min processing of experimental passive sonar data from thinned arrays. *The Journal of the Acoustical Society of America* 144 (6): 3262–3274. https://doi.org/10.1121/1.5064458.

27 Ramirez, J. and Rogers, J.S. (2018). Exploiting platform motion for passive source localization with a co-prime sampled large aperture array. *The Journal of the Acoustical Society of America* 143 (6): 3946–3958. https://doi.org/10.1121/1.5042231.

28 Nannuru, S., Koochakzadeh, A., Gemba, K.L. et al. (2018). Sparse Bayesian learning for beamforming using sparse linear arrays. *The Journal of the Acoustical Society of America* 144 (5): 2719–2729. https://doi.org/10.1121/1.5066457.

29 Nair, B.M., Peter, R.J., Kumar, A., and Bahl, R. (2022). Left-Right ambiguity resolution for coprime sparse arrays. *The Journal of the Acoustical Society of America* 151 (2): 792–805. https://doi.org/10.1121/10.0009397.

30 Wage, K.E. (2018). When two wrongs make a right: combining aliased arrays to find sound sources. *Acoustics Today* 14 (3): 48–56.

31 Rajamäki, R. (2021). Sparse sensor arrays for active sensing - array configurations and signal processing. PhD thesis. Aalto University.

32 Bjørnø, L. (2017). Sonar systems. In: *Applied Underwater Acoustics*, Chapter 10 (ed. T. Neighbors and D. Bradley), 587–742. Amsterdam, Netherlands: Elsevier. ISBN: 978-0-12-811247-2.

33 Mills, B.Y. and Little, A.G. (1953). A high-resolution aerial system of a new type. *Australian Journal of Physics* 6: 272. https://doi.org/10.1071/PH530272.

34 Haubrich, R.A. (1968). Array design. *Bulletin of the Seismological Society of America* 58 (3): 977–991. https://doi.org/10.1785/BSSA0580030977.

35 Schmidt, R. (1986). Multiple emitter location and signal parameter estimation. *IEEE Transactions on Antennas and Propagation* 34 (3): 276–280. https://doi.org/10.1109/TAP.1986.1143830.

36 Kumaresan, R. and Tufts, D.W. (1983). Estimating the angles of arrival of multiple plane waves. *IEEE Transactions on Aerospace and Electronic Systems* AES-19 (1): 134–139. https://doi.org/10.1109/TAES.1983.309427.

37 Roy, R. and Kailath, T. (1989). ESPRIT-estimation of signal parameters via rotational invariance techniques. *IEEE Transactions on Acoustics, Speech, and Signal Processing* 37 (7): 984–995. https://doi.org/10.1109/29.32276.

38 Vaidyanathan, P.P. and Pal, P. (2011). Theory of sparse coprime sensing in multiple dimensions. *IEEE Transactions on Signal Processing* 59 (8): 3592–3608. https://doi.org/10.1109/TSP.2011.2135348.

39 Pal, P. and Vaidyanathan, P.P. (2012). Nested arrays in two dimensions, Part I: Geometrical considerations. *IEEE Transactions on Signal Processing* 60 (9): 4694–4705. https://doi.org/10.1109/TSP.2012.2203814.

40 Impagliazzo, J., Medeiros, M., and Kay, S. (2001). Imaging with a 2 MHz sparse broadband planar array. In *MTS/IEEE Oceans 2001. An Ocean Odyssey. Conference Proceedings (IEEE Cat. No.01CH37295)*, volume 1, 22–25, November 2001. https://doi.org/10.1109/OCEANS.2001.968660.

41 Adhikari, K., Harker, B.M., Wettergren, T.A. et al. (2021). Array shading to maximize deflection coefficient for broadband signal detection with conventional beamforming. *IEEE Access* 9: 79343–79352. https://doi.org/10.1109/ACCESS.2021.3083203.

42 Adhikari, K., Buck, J.R., and Wage, K.E. (2014). Extending coprime sensor arrays to achieve the peak side lobe height of a full uniform linear array. *EURASIP Journal on Advances in Signal Processing* 2014 (1): 1–17. https://doi.org/10.1186/1687-6180-2014-148.

43 Adhikari, K. and Buck, J.R. (2017). Spatial spectral estimation with product processing of a pair of colinear arrays. *IEEE Transactions on Signal Processing* 65 (9): 2389–2401. https://doi.org/10.1109/TSP.2017.2659642.

44 Chavali, V. and Wage, K.E. (2020). Cross term decay in multiplicative processors. *IEEE Signal Processing Letters* 27: 56–60. https://doi.org/10.1109/LSP.2019.2955815.

45 Chavali, V., Wage, K.E., and Buck, J.R. (2014). Coprime processing for the Elba Island sonar data set. In *2014 48th Asilomar Conference on Signals, Systems and Computers*, 1864–1868, November 2014. https://doi.org/10.1109/ACSSC.2014.7094791.

46 Adhikari, K. and Drozdenko, B. (2019). Design and statistical analysis of tapered coprime and nested arrays for the min processor. *IEEE Access* 7: 139601–139615. https://doi.org/10.1109/ACCESS.2019.2944109.

47 Moffet, A. (1968). Minimum-redundancy linear arrays. *IEEE Transactions on Antennas and Propagation* 16 (2): 172–175. https://doi.org/10.1109/TAP.1968.1139138.

48 Pillai, S.U., Bar-Ness, Y., and Haber, F. (1985). A new approach to array geometry for improved spatial spectrum estimation. *Proceedings of the IEEE* 73 (10): 1522–1524. https://doi.org/10.1109/PROC.1985.13324.

49 Solomon, I.S.D. and Knight, A.J. (2002). Spatial processing of signals received by platform mounted sonar. *IEEE Journal of Oceanic Engineering* 27 (1): 57–65. https://doi.org/10.1109/48.989887.

50 Circuits and Signal Processing Laboratory at Federal University of Santa Catarina (2023). Signal processing information base (SPIB). http://spib.linse.ufsc.br/ (accessed 9 November 2023).

51 Pyzdek, A. and Culver, R.L. (2013). Sensitivity of co-prime arrays to shape perturbation. *Proceedings of Meetings on Acoustics* 19 (1): 070086. https://doi.org/10.1121/1.4800922.

52 Bautista, R. and Buck, J.R. (2018). Processor dependent bias of spatial spectral estimates from coprime sensor arrays. *The Journal of the Acoustical Society of America* 143 (6): 3972–3978. https://doi.org/10.1121/1.5042411.

53 Liu, Y. and Buck, J.R. (2016). Super-resolution DOA estimation using a coprime sensor array with the min processor. In *2016 50th Asilomar Conference on Signals, Systems and Computers*, 944–948, November 2016. https://doi.org/10.1109/ACSSC.2016.7869188.

54 Rooney, I.M. and Buck, J.R. (2015). Multitapered power spectral density estimation for co-prime sensor arrays. *2015 49th Asilomar Conference on Signals, Systems and Computers*, 375–379, November 2015. https://doi.org/10.1109/ACSSC.2015.7421151.

55 Thomson, D.J. (1982). Spectrum estimation and harmonic analysis. *Proceedings of the IEEE* 70 (9): 1055–1096. https://doi.org/10.1109/PROC.1982.12433.

56 Rooney, I.M., Liu, Y., and Buck, J.R. (2018). Spatial power spectral density estimation using a multitapered coprime sensor array minimum processor. *The Journal of the Acoustical Society of America* 143 (6): 3959–3971. https://doi.org/10.1121/1.5042224.

57 Rooney, I.M. and Buck, J.R. (2019). Spatial power spectral density estimation using a Welch coprime sensor array processor. *The Journal of the Acoustical Society of America* 145 (4): 2350–2362. https://doi.org/10.1121/1.5097572.

58 Cutrona, L.J. (1975). Comparison of sonar system performance achievable using synthetic-aperture techniques with the performance achievable by more conventional means. *The Journal of the Acoustical Society of America* 58 (2): 336–348. https://doi.org/10.1121/1.380678.

59 Neudorfer, M., Kosalos, J., and Bonneau, R. (2000). Results of synthetic aperture sonar experiments. In *OCEANS 2000 MTS/IEEE Conference and Exhibition. Conference Proceedings (Cat. No.00CH37158)*, volume 1, 323–330. https://doi.org/10.1109/OCEANS.2000.881279.

60 Ramirez, J. and Krolik, J.L. (2017). Synthetic aperture processing for passive co-prime linear sensor arrays. *Digital Signal Processing* 61: 62–75. https://doi.org/10.1016/j.dsp.2016.07.021.

61 Nannuru, S., Gerstoft, P., Ping, G., and Fernandez-Grande, E. (2021). Sparse planar arrays for azimuth and elevation using experimental data. *The Journal of the Acoustical Society of America* 149 (1): 167–178. https://doi.org/10.1121/10.0002988.

62 Thode, A., Norris, T., and Barlow, J. (2000). Frequency beamforming of dolphin whistles using a sparse three-element towed array. *The Journal of the Acoustical Society of America* 107 (6): 3581–3584. https://doi.org/10.1121/1.429432.

63 Hinich, M.J. (1978). Processing spatially aliased arrays. *The Journal of the Acoustical Society of America* 64 (3): 792–794. https://doi.org/10.1121/1.382044.

64 Ball, K.R. and Buck, J.R. (2005). A beamforming video recorder for integrated observations of dolphin behavior and vocalizations (L). *The Journal of the Acoustical Society of America* 117 (3): 1005–1008. https://doi.org/10.1121/1.1831284.

65 Glegg, S.A.L., Olivieri, M.P., Coulson, R.K., and Smith, S.M. (2001). A passive sonar system based on an autonomous underwater vehicle. *IEEE Journal of Oceanic Engineering* 26 (4): 700–710. https://doi.org/10.1109/48.972113.

66 Knapp, C. and Carter, G. (1976). The generalized correlation method for estimation of time delay. *IEEE Transactions on Acoustics, Speech, and Signal Processing* 24 (4): 320–327. https://doi.org/10.1109/TASSP.1976.1162830.

67 Zhou, T., Ahmad, F., Ouyang, B., and Li, Y. (2019). Near-field source localization with an underwater inflatable sonar co-prime array. In: *Big Data: Learning, Analytics, and Applications*, vol. 10989, 26–32. SPIE. https://doi.org/10.1117/12.2520967.

14

Unconventional Array Architectures for Next Generation Wireless Communications

Nicola Anselmi[1], Sotirios Goudos[2], Giacomo Oliveri[1], Lorenzo Poli[1], Paolo Rocca[1,3], Marco Salucci[1], and Andrea Massa[4,1,5,6,7]

[1]*ELEDIA Research Center (ELEDIA@UniTN – University of Trento) DICAM – Department of Civil, Environmental, and Mechanical Engineering, Trento, Italy*
[2]*ELEDIA Research Center (ELEDIA@AUTH – Aristotle University of Thessaloniki), Thessaloniki, Greece*
[3]*ELEDIA Research Center (ELEDIA@XIDIAN – Xidian University), Xi'an, Shaanxi Province China*
[4]*ELEDIA Research Center (ELEDIA@UESTC – UESTC), School of Electronic Science and Engineering, University of Electronic Science and Technology of China, Chengdu, China*
[5]*ELEDIA Research Center (ELEDIA@TSINGHUA – Tsinghua University), Beijing, Haidian, China*
[6]*School of Electrical Engineering, Tel Aviv University, Tel Aviv, Israel*
[7]*ELEDIA Research Center (ELEDIA@UIC – University of Illinois Chicago), Chicago, IL, USA*

14.1 Introduction

The synthesis of phased array antennas with low-complex/simplified architectures plays a crucial role in the development of modern and future wireless communication systems. With the continuous growth of wireless services and applications, phased arrays have significantly evolved since their introduction, requiring multiple functionalities, larger bandwidth, higher reconfigurability, smaller size, lower weight, and cost-effective architectures [1].

Radiating elements in standard arrays are arranged on a regular grid with constant relative distances between them and equipping each element with a transmit/receive module (TRM) that enables independent control of amplitude, phase, or delay of either the transmitted or received signal with a fully-digital MIMO architecture. These types of architectures are usually referred to as *conventional* or *fullypopulated* (Fig. 14.1a). Unfortunately, the increasing performance and cost requirements of modern/future wireless communication applications makes this class of arrays often unfeasible to be used in practical scenarios. Innovative *unconventional* array architectures [2], (Fig. 14.1b,c) have thus emerged in the last years with the goal of dealing with the same performance requirements of conventional arrays but with considerably simplified and cheaper architectures.

Towards this end, array sparsening (Fig. 14.1b) has been proposed as one of the possible approaches to simplify the array architecture [3–16]. By simultaneously optimizing the inter-element distances between the array elements and the corresponding excitation weights, the number of array elements and TRMs can be significantly reduced while still achieving the user-defined radiation goals. Several approaches have been proposed in this framework, including deterministic [4, 12, 16], stochastic techniques [3, 5, 13], as well as sparseness-promoting strategies based on Compressive Sensing (CS) [6–11, 14, 15].

The possibility to simplify the array architecture while still exploiting uniformly spaced elements has been addressed, as well. More in detail, clustered arrays (Fig. 14.1c) have been widely studied

Sparse Arrays for Radar, Sonar, and Communications, First Edition. Edited by Moeness G. Amin.

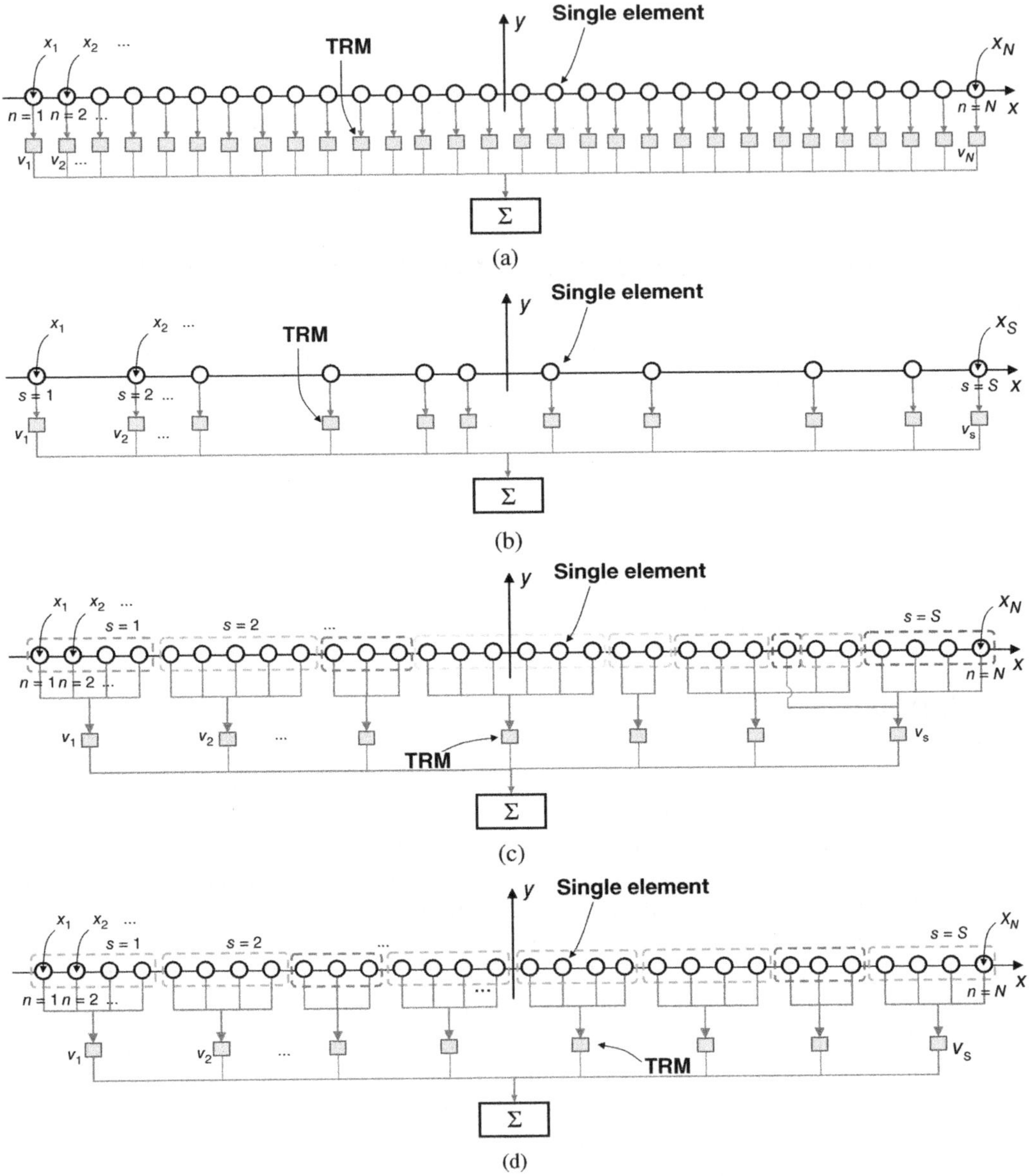

Figure 14.1 Sketch of the (a) *conventional* or *fully-populated* array architecture and (b,c,d) unconventional architectures, including examples of (b) a sparse linear array of S nonuniformly spaced elements, (c) a clustered architectures with elements grouped into S non-contiguous sub.arrays, and (d) a modular/tile array architecture where the elements are clustered into S contiguous sub-arrays of $P = 2$ pre-fixed lengths (i.e., four elements and three elements modules/tiles).

in the recent years [17–29]. These classes of array architectures recently attracted the attention of 5G/6G telecommunication and radar industry, thanks to their capability to provide maximum gain/directivity while reducing the complexity of the feeding network. This is achieved by grouping multiple elements over single TRMs. Towards this end, several techniques have been developed for the optimization of the sub-array layout, including efficient local search methods [18, 19, 23], global optimization strategies [17, 21, 22, 26], random search-based schemes [25], and CS-based approaches [24, 27].

Highly modular tiled layouts (Fig. 14.1d) have been also investigated in the last years [30–41]. Differently from clustered arrays, tiled arrays are composed by simple modules/building blocks (i.e., tiles of contiguous elements) that tessellate the antenna aperture, ensuring the complete coverage of the available area. The tiling of rectangular [30–32], hexagonal [33], and orthogonal–polygon [34] shaped apertures has been addressed with single-shape tile [30–34] and multi-shape tiles [35–41] methods.

In this framework, clustered and tiled arrays have been recently proposed for wireless communication. In particular, the problem of jointly design the single element radiator and the unconventional array architecture, while fitting 5G base station (BS) requirements, has been addressed in [26]. Moreover, capacity-driven BS array synthesis methods have been recently proposed in [28, 42], considering both clustered [28] and tiled [40, 41] architectures.

Given the focus of this Chapter, a review of several classes of methodologies recently introduced in this field will be carried out. To this end, the problem formulation regarding the design of unconventional phased array architectures will be covered, and the theoretical motivations, fundamentals, and applicative guidelines for current state-of-the-art strategies in unconventional phased array synthesis for next generation wireless communication applications will be discussed.

The Chapter is organized as follows. Section 14.2 reports the formulation of sparseness promoting techniques based on the Bayesian Compressive Sensing (BCS) method [43, 44] and Total Variation (TV) regularization [45, 46] for the design of both sparse as well as modular arrays. Afterwards, the "co-design" strategy [26] for the design of single antenna radiators and of the unconventional array strategy is described (Section 14.3) reporting an illustrative example of a 5G design. The design of unconventional BS antenna arrays aimed at maximizing the users throughput capacity in realistic 5G/6G wireless communication environments is addressed in Section 14.4. Some conclusions and final remarks will follow (Section 14.5).

14.2 Sparseness-Promoting Techniques for the Design of Unconventional Architectures

Sparseness promoting methods have been widely studied in the design of unconventional arrays starting from sparse-array designs to clustered architectures. For the sake of notation simplicity, the 1D case (linear arrays) will be illustrated in the following.

14.2.1 Sparse Array Synthesis Through Bayesian Compressive Sensing

Let us consider the design of a linear array occupying a 1D aperture $[z_{\min}, z_{\max}]$. The synthesis of a sparse arrangement of S radiating elements with a prescribed radiated pattern, $F_{\text{ref}}(\theta)$, can be formulated as follows [7]

Sparse Array Synthesis Problem – Find the minimum S value and the corresponding sparse array descriptors, namely the complex excitations vector $\mathbf{v} = \{v_s;\ s = 1, \ldots, S\}$, and the elements positions vector $\mathbf{z} = \{z_s;\ s = 1, \ldots, S\}$, that satisfy the matching constraint

$$\sum_{k=1}^{K} \left| F_{\text{ref}}(\theta_k) - \sum_{s=1}^{S} v_s e^{j2\pi z_s \sin(\theta_k)} \right|^2 \leq \varepsilon, \tag{14.1}$$

where the S element positions are assumed to belong to a user-chosen set of P arbitrary candidate locations $\overline{\mathbf{z}} = \{\overline{z}_p;\ p = 1, \ldots, P\}$ [6, 7], with $P \gg N$, N being the number of elements of the corresponding fully-populated layout, θ_k $(k = 1, \ldots, K)$ are the sampling angles, and ε is the fidelity index.

By simple manipulation [7], Eq. (14.1) can be recast into the sparse matrix form [47]

$$\mathbf{F}_{\mathrm{ref}} - \Phi\mathbf{w} = \boldsymbol{\eta}, \tag{14.2}$$

by defining the complex weights vector $\mathbf{w} = \{w_p; p = 1, ..., P\}$ where $w_p = 0$ indicates that the pth candidate location z_p is empty, while $w_p = v_s$ if $\overline{z}_p = z_s$ (i.e., the pth candidate location coincides with the s-th element position). Moreover in (14.2) $\mathbf{F}_{\mathrm{ref}} \triangleq \{F_{\mathrm{ref}}(\theta_k); k = 1, ..., K\}$ collects the $F_{\mathrm{ref}}(\theta_k)$ $(k = 1, ..., K)$ samples, $\Phi \triangleq \left\{\varphi_{kp} = e^{j2\pi z_p \sin(\theta_k)}; k = 1, ..., K,\ p = 1, ..., P\right\}$ is the problem matrix, and $\boldsymbol{\eta} \triangleq \{\eta_k; k = 1, ..., K\}$ is a zero-mean complex Gaussian vector of entries with variance σ proportional to ε [7]. It is worth noting that $\mathbf{w}$ turns out to be a sparse vector [1] since $P \gg S$. Thanks to this description, the *Sparse Antenna Synthesis Problem* can be also formulated as the following sparseness promoting problem:

Sparse Vector Synthesis Problem – Find the minimum ℓ_0-norm weight vector $\mathbf{w}$ that satisfies (14.2) where

$$\|\mathbf{w}\|_{\ell_0} \triangleq \sum_{p=1}^{P} |w_p|^0 = \sum_{s=1}^{S} |v_s|^0 = S. \tag{14.3}$$

The problem (14.3) is in principle an NP-complete one. In order to efficiently solve (14.3) the design of linear arrays has been addressed in [6, 7] by reformulating the problem in the CS framework and employing a set of BCS techniques [43, 44]. Both real-excitations and complex-excitations have been considered through a single-task (ST) [6, 43] and a multi-task (MT) [7, 44] implementation, respectively. Thanks to the probabilistic formulation and the exploitation of a fast Relevance Vector Machine solver [6, 7], BCS-based techniques have been demonstrated to be a suitable design tool for sparse array design, that allows to reach close or improved performances when compared to state-of-the-art sparse syntheses [6, 7]. In the following, the ST and MT formulations are briefly summarized in order to provide to the reader the useful insights of the BCS methods as applied to the sparse array synthesis problem, together with some illustrative examples of the proposed methodologies. It is worth noting that the illustrated formulations can be extended to planar array layouts, following the generalized BCS formulation of [8].

14.2.1.1 ST-BCS Synthesis Method

Since the ST-BCS addresses purely real-valued problems [6, 43], in order to solve (14.1) the complex-valued vectors and matrices in (14.2) are manipulated as follows

$$\widetilde{\mathbf{F}}_{\mathrm{ref}} - \tilde{\phi}\widetilde{\mathbf{w}} = \widetilde{\boldsymbol{\eta}}, \tag{14.4}$$

by defining the "extended" vectors and matrices

$$\widetilde{\mathbf{F}}_{\mathrm{ref}} = \left[\mathbf{F}_{\mathrm{ref}}, \mathcal{R}, \mathbf{F}_{\mathrm{ref}}, \mathcal{I}\right] \tag{14.5}$$

$$\tilde{\Phi} = \begin{bmatrix} \Phi_{\mathcal{R}} & -\Phi_{\mathcal{I}} \\ \Phi_{\mathcal{I}} & \Phi_{\mathcal{R}} \end{bmatrix} \tag{14.6}$$

$$\widetilde{\mathbf{w}} = \left[\mathbf{w}_{\mathcal{R}}, \mathbf{w}_{\mathcal{I}}\right] \tag{14.7}$$

$$\widetilde{\boldsymbol{\eta}} = \left[\boldsymbol{\eta}_{\mathcal{R}}, \boldsymbol{\eta}_{\mathcal{I}}\right], \tag{14.8}$$

where $\mathbf{F}_{\mathrm{ref}}, \mathcal{R} \triangleq \mathcal{R}\{\mathbf{F}_{\mathrm{ref}}\}$ and $\mathbf{F}_{\mathrm{ref}}, \mathcal{I} \triangleq \mathcal{I}\{\mathbf{F}_{\mathrm{ref}}\}$, $\Phi_{\mathcal{R}} \triangleq \mathcal{R}\{\Phi\}$ and $\Phi_{\mathcal{I}} \triangleq \mathcal{I}\{\Phi\}$, $\mathbf{w}_{\mathcal{R}} \triangleq \mathcal{R}\{\mathbf{w}\}$ and $\mathbf{w}_{\mathcal{I}} \triangleq \mathcal{I}\{\mathbf{w}\}$, $\boldsymbol{\eta}_{\mathcal{R}} \triangleq \mathcal{R}\{\boldsymbol{\eta}\}$ and $\boldsymbol{\eta}_{\mathcal{I}} \triangleq \mathcal{I}\{\boldsymbol{\eta}\}$, $\mathcal{R}\{\cdot\}$ and $\mathcal{I}\{\cdot\}$ stand for the real and the imaginary part,

1 A vector is defined to be sparse if the majority of its entries are null.

respectively, $\widetilde{\mathbf{F}}_{\text{ref}} \in \mathbb{R}^{2K}$, $\tilde{\Phi} \in \mathbb{R}^{2K\times 2P}$, $\widetilde{\mathbf{w}} \in \mathbb{R}^{2P}$, and $\widetilde{\boldsymbol{\eta}} \in \mathbb{R}^{2K}$. Accordingly, the following extended real-valued problem is formulated in the *probabilistic* framework [6]:

ST-BCS Synthesis Problem – Find the minimum ℓ_0-norm "extended" weight vector $\widetilde{\mathbf{w}}$ subject to

$$\widetilde{\mathbf{w}}^{\text{ST-BCS}} = \arg\left[\max_{\widetilde{\mathbf{w}}} \mathcal{P}\left(\widetilde{\mathbf{w}} \middle| \widetilde{\mathbf{F}}_{\text{ref}}\right)\right]. \tag{14.9}$$

The real-valued solution of (14.9) is given by [6]

$$\widetilde{\mathbf{w}}^{\text{ST-BCS}} = \frac{1}{\tilde{\sigma}^2_{\text{ST-BCS}}}\left(\frac{\tilde{\Phi}^T\tilde{\Phi}}{\tilde{\sigma}^2_{\text{ST-BCS}}} + \widetilde{\mathbf{a}}^{\text{ST-BCS}}\right)^{-1}\tilde{\Phi}^T\widetilde{\mathbf{F}}_{\text{ref}}, \tag{14.10}$$

where the superscript.^{T} indicates the transpose operator, and

- $\tilde{\sigma}^2_{\text{ST-BCS}}$ is the estimated variance of η_k $(k = 1, \ldots, K)$;
- $\widetilde{\mathbf{a}}^{\text{ST-BCS}}$ $(\widetilde{\mathbf{a}}^{\text{ST-BCS}} \in \mathbb{R}^{2P})$ is the hyperparameter vector, whose pth entry, $\tilde{a}_p^{\text{ST-BCS}}$, controls the strength of the sparseness prior over $\tilde{w}_p^{\text{ST-BCS}}$.

In order to retrieve the ST-BCS solution using (14.10), the two parameters $\widetilde{\mathbf{a}}$ and σ^2 are computed by maximizing the logarithm of the ST-BCS "marginal likelihood," $\mathcal{L}^{\text{ST-BCS}}\left(\widetilde{\mathbf{a}}, \sigma^2\right)$ [6]

$$\begin{aligned}\mathcal{L}^{\text{ST-BCS}}\left(\widetilde{\mathbf{a}}, \sigma^2\right) = -\tfrac{1}{2}\Big[&(2P)\log 2\pi + \\ &\log\left|\sigma^2 I + \tilde{\Phi}\left[\text{diag}\left(\widetilde{\mathbf{a}}\right)\right]^{-1}\tilde{\Phi}^T\right| + \\ &\widetilde{\mathbf{F}}'_{\text{ref}}\left(\sigma^2 I + \tilde{\Phi}\left[\text{diag}\left(\widetilde{\mathbf{a}}\right)\right]^{-1}\tilde{\Phi}^T\right)^{-1}\widetilde{\mathbf{F}}_{\text{ref}}\Big].\end{aligned} \tag{14.11}$$

Because of the maximization of (14.11) is not possible in an exact fashion, an iterative re-estimation of $\widetilde{\mathbf{a}}^{\text{ST-BCS}}$ and $\tilde{\sigma}^2_{\text{ST-BCS}}$ procedure [6] based on a type-II maximum likelihood approach has been used. Finally, the real and imaginary parts of the P entries of the weight vector $\mathbf{w}^{\text{ST-BCS}}$ $(\mathbf{w}_{\text{ST-BCS}} \in \mathbb{C}^P)$ are found as

$$\begin{aligned}\mathcal{R}\left\{w_p^{\text{ST-BCS}}\right\} &= \tilde{w}_p^{\text{ST-BCS}},\ p = 1, \ldots, P \\ \mathcal{I}\left\{w_p^{\text{ST-BCS}}\right\} &= \tilde{w}_p^{\text{ST-BCS}},\ p = 1, \ldots, P.\end{aligned} \tag{14.12}$$

Equation (14.10) provides a direct way to deal with complex and nonsymmetric arrays. However, such a solution bears an intrinsic limitation. The real, $\mathcal{R}\left\{w_p^{\text{ST-BCS}}\right\}$, and imaginary, $\mathcal{I}\left\{w_p^{\text{ST-BCS}}\right\}$, $(p = 1, \ldots, P)$ parts of the weights are managed as independent quantities since each $\tilde{w}_p^{\text{ST-BCS}} \in \mathbb{R}$ $(p = 1, \ldots, 2P)$ is treated as statistically independent [6]. This in turns leads to sparse BCS layouts where the array weights v_s $(s = 1, \ldots, S)$ are often either purely real or purely imaginary, neglecting that sparse complex layouts frequently exhibit non-negligible real and imaginary components at the same array locations [7]. Such a drawback yields to sub-optimal results in terms of sparsity and pattern matching [7].

14.2.1.2 MT-BCS Synthesis Method

To overcome the limitations of the ST-BCS synthesis, the MT-BCS approach [44] has been proposed and suitably customized for statistically modeling the relations between the real and imaginary parts of the array weights [7]. Towards this end, (14.2) is firstly rewritten in terms of the auxiliary weights vectors

$$\begin{cases} \widehat{\mathbf{F}}_{\mathcal{R}} - \widehat{\Phi}\mathbf{w}_{\mathcal{R}} = \widehat{\eta}_{\mathcal{R}} \\ \widehat{\mathbf{F}}_{\mathcal{I}} - \widehat{\Phi}\mathbf{w}_{\mathcal{I}} = \widehat{\eta}_{\mathcal{I}} \end{cases}, \tag{14.13}$$

where $\widehat{\eta}_{\mathcal{R}} \in \mathbb{R}^{2K}$ and $\widehat{\eta}_{\mathcal{I}} \in \mathbb{R}^{2K}$ are zero-mean complex Gaussian error vectors (with variance $\frac{\sigma^2}{2}$) such that $\widehat{\eta}_{\mathcal{R}} + \widehat{\eta}_{\mathcal{I}} = \widetilde{\eta}$, $\widehat{\Phi} \triangleq [\Phi_{\mathcal{R}}, \Phi_{\mathcal{I}}]$ is the MT observation matrix, while $\widehat{\mathbf{F}}_{\mathcal{R}} = [\mathcal{R}\{\mathbf{F}_{\mathcal{R}}\}, \mathcal{I}\{\mathbf{F}_{\mathcal{R}}\}]$ and $\widehat{\mathbf{F}}_{\mathcal{I}} = [\mathcal{R}\{\mathbf{F}_{\mathcal{I}}\}, \mathcal{I}\{\mathbf{F}_{\mathcal{I}}\}]$ ($\widehat{\mathbf{F}}_{\mathcal{R}}, \widehat{\mathbf{F}}_{\mathcal{I}} \in \mathbb{R}^{2K}$). Moreover, $\mathbf{F}_{\mathcal{R}} \in \mathbb{C}^K$ and $\mathbf{F}_{\mathcal{I}} \in \mathbb{C}^K$ satisfy the following condition

$$\mathbf{F}_{\mathcal{R}} + j\mathbf{F}_{\mathcal{I}} = \mathbf{F}_{\text{ref}}. \tag{14.14}$$

Accordingly, the real-valued MT problem can be cast in the probabilistic framework [44]

MT-BCS Synthesis Problem – Find the minimum ℓ_0-norm auxiliary weight vectors $\mathbf{w}_{\mathcal{R}}$ and $\mathbf{w}_{\mathcal{I}}$ ($\mathbf{w}_{\mathcal{R}}, \mathbf{w}_{\mathcal{I}} \in \mathbb{R}^P$) subject to

$$\begin{cases} \mathbf{w}_{\mathcal{R}}^{\text{MT-BCS}} = \arg\left[\max_{\mathbf{w}_{\mathcal{R}}} \mathcal{P}\left(\mathbf{w}_{\mathcal{R}} | \widehat{\mathbf{F}}_{\mathcal{R}}\right)\right] \\ \mathbf{w}_{\mathcal{I}}^{\text{MT-BCS}} = \arg\left[\max_{\mathbf{w}_{\mathcal{I}}} \mathcal{P}\left(\mathbf{w}_{\mathcal{I}} | \widehat{\mathbf{F}}_{\mathcal{I}}\right)\right] \end{cases}, \tag{14.15}$$

whose solution is given by

$$\begin{cases} \mathbf{w}_{\mathcal{R}}^{\text{MT-BCS}} = \left(\text{diag}\left(\widehat{\mathbf{a}}^{\text{MT-BCS}}\right) + \widehat{\Phi}^{\text{T}}\widehat{\Phi}\right)^{-1}\widehat{\Phi}^{T}\widehat{\mathbf{F}}_{\mathcal{R}}, \\ \mathbf{w}_{\mathcal{I}}^{\text{MT-BCS}} = \left(\text{diag}\left(\widehat{\mathbf{a}}^{\text{MT-BCS}}\right) + \widehat{\Phi}^{\text{T}}\widehat{\Phi}\right)^{-1}\widehat{\Phi}^{T}\widehat{\mathbf{F}}_{\mathcal{I}}, \end{cases} \tag{14.16}$$

while the corresponding estimated weight vector turns out to be [7]

$$\mathbf{w}^{\text{MT-BCS}} = \mathbf{w}_{\mathcal{R}}^{\text{MT-BCS}} + j\mathbf{w}_{\mathcal{I}}^{\text{MT-BCS}}. \tag{14.17}$$

As an illustrative example of the ST-BCS and *MT-BCS sparseness-promoting* techniques presented in this section, the synthesis of a sparse linear array, laying on the z-axis, and radiating a flat-top pattern is addressed. The reference array is a uniform $N = 14$ elements array obtained using the method presented in [48]. The angular range has been sampled uniformly in the u-space, $u_k = -1 + \frac{2k}{K}$ $k = 1, \ldots, K$, $u = \sin(\theta)$ and a uniform grid of P candidate locations has been considered $z_p = N\frac{\lambda}{2}\left(-\frac{1}{2} + \frac{p}{P}\right)$, $p = 1, \ldots, P$. Moreover the solutions are evaluated in terms of the normalized matching error

$$\xi \triangleq \frac{\int_{-1}^{1} |F_{\text{ref}}(u) - F(u)|^2 du}{\int_{-1}^{1} |F_{\text{ref}}(u)|^2 du}. \tag{14.18}$$

Figure 14.2 reports the power pattern (Fig. 14.2a) and pattern phase (Fig. 14.2b) of two BCS solutions as compared to the reference one. Similarly, (Fig. 14.2c,d) reports the corresponding amplitude and phase excitations coefficients. The BCS solutions have been obtained in order to achieve similar pattern matching accuracy, properly tailoring the noise and prior BCS control parameters (the reader is referred to [6] for the calibration of the BCS parameters for the ST-BCS approach and to [7] for the MT-BCS method). The ST-BCS solution turns out to be a nonuniform array of $S_{\text{ST-BCS}} = 22$, while the MT-BCS method allows to reduce the number of elements with respect to the reference by almost 22% (i.e., $S_{\text{MT-BCS}} = 11$). The

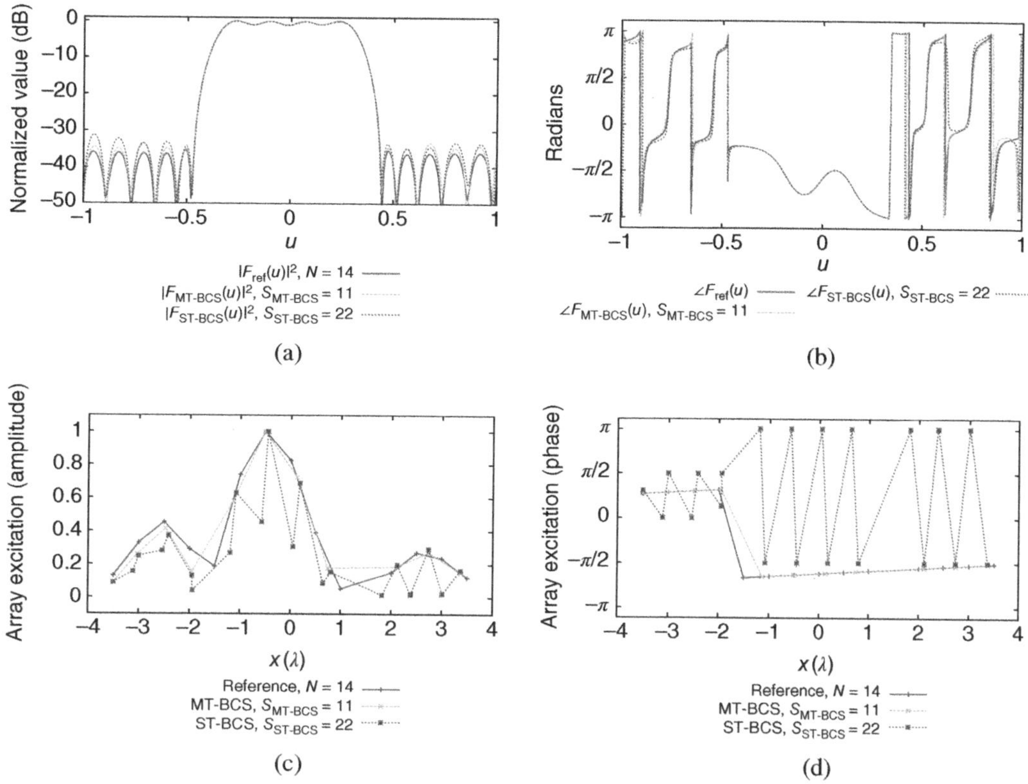

Figure 14.2 *Sparseness-Promoting Techniques (ST/MT-BCS Synthesis Method, Flat-Top Pattern, $N = 14$, $S_{ST\text{-}BCS} = 22$, $S_{MT\text{-}BCS} = 11$)* - (a) Power pattern, (b) pattern phase, (c) amplitude, and (d) phase excitation coefficients of the reference and sparse ST-BCS and MT-BCS optimized arrays.

matching with the reference pattern shows how the MT approach is more suitable to handle patterns having nonsymmetric phase components (Fig. 14.2b), as is the case of the considered flat-top beam, thanks to its higher accuracy in handling the statistical relations between the real and imaginary parts of the array weights. As expected, the lower performance of the ST technique is mainly due to the intrinsic limitation in dealing with complex array excitations, indeed unlike the MT-BCS coefficients, some ST-BCS weights are either purely real or purely imaginary [48].

14.2.2 Dictionary-Based Compressing Sensing Method

Sparseness-promoting techniques have been employed for the synthesis of other unconventional architectures, as well [24, 27]. More in detail, sub-arrayed arrays utilizing the full antenna aperture filled by elements arranged on a regular lattice but grouped and controlled at the sub-array level, have been synthesized employing dictionary-based approaches [24].

The primary concern with clustered architectures is the inevitable occurrence of quantization and grating lobes, which become increasingly problematic for wide scan angles and operational bandwidths [49]. Irregularly clustered arrays in principle enable to effectively reduce such issue, thanks to the aperiodic arrangements of the clusters phase-centers [20]. Indeed, when dealing with contiguously clustered arrays, the spacing between the clusters barycenters becomes sparse/nonuniform if irregular and/or different clusters shapes are used [20]. Furthermore, it is common to constrain the size of the synthesized sub-arrays to a limited set such that the architecture becomes modular [20, 22, 24, 27, 40]. In this case the synthesis turns out to be a selection from a pre-fixed set of sub-array lengths. In this context the synthesis of *modular clustered* arrays has been formulated in terms of a sparseness-regularized linear problem [24]. Towards this end, the clustered array is firstly described as the combination of sub-arrays selected from a "dictionary" of P elementary modules with user-defined lengths. The synthesis problem is then recast to the choice of the minimum number S of modules ($S \ll P$) and their excitations so that the clustered layout radiates a desired pattern. In such a way, the design turns out to be a sparse and linear retrieval problem where the vector of the cluster excitations of length P, having only S nonzero coefficients, is the problem unknown, while the data are the desired reference pattern, suitably sampled over a given angular range.

With reference to a linear geometry, the design problem is based on the idea that the N antennas are clustered into S sub-arrays picked from a user-defined dictionary $\mathcal{D}$ of P "candidate" modules, so that

$$\mathbf{v} = \mathcal{D}\mathbf{w}, \tag{14.19}$$

where $\mathbf{v} \triangleq \{v_n;\ n = 1, ..., N\}$ is the clustered array excitation vector, $\mathcal{D} \triangleq \{\mathbf{D}_p;\ p = 1, ..., P\}$ is the binary dictionary of clusters, $\mathbf{D}_p = \{D_{np};\ n = 1, ..., N\}$ being the pth dictionary column, whose nth entry is equal to 1 if the nth element belongs to the pth cluster, while it is equal to 0 otherwise. For visual purposes, in Fig. 14.3 an array of $N = 10$ elements is clustered into $S = 2$ subarrays, starting from a dictionary of $P = 6$ candidate clusters. In (14.19) $\mathbf{w} = \{w_p;\ p = 1, ..., P\}$ is the sub-array weighting vector, w_p ($p = 1, ..., P$) being the excitation coefficient of the pth cluster (only S entries of $\mathbf{w}$ are nonzero, $S = \|\mathbf{w}\|_{\ell_0}$; Fig. 14.3). It is worth to point out that the dictionary $\mathcal{D}$ is defined by the designer. Accordingly, the lengths of the clusters and their features are a-priori defined, such as to guarantee the physical contiguity of the elements within the clusters. Moreover, this formulation also allows the designer to enforce modularity in the final arrangement, since, by properly setting the columns of $\mathcal{D}$, the available modules lengths for the synthesis can be specified. In practical scenarios, the optimal number of clusters, S, is unknown, and since the goal is to simplify the feed

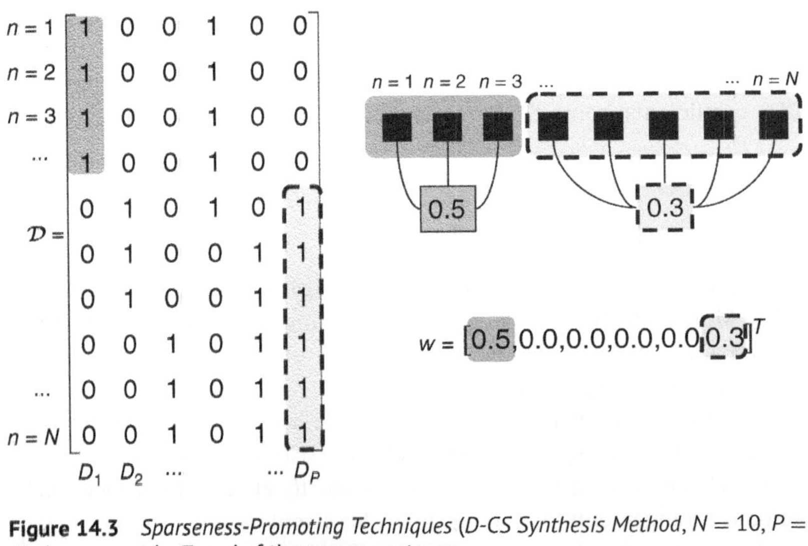

Figure 14.3 *Sparseness-Promoting Techniques (D-CS Synthesis Method, $N = 10$, $P = 6$, $S = 2$)* – Example of dictionary matrix $\mathcal{D}$ and of the sparse vector $\mathbf{w}$.

network's complexity compared to a fully populated layout, the minimum value of S is determined by searching for the sparsest $\mathbf{w}$. The arising synthesis problem can be formulated as follows:

Clustered Array Synthesis Problem– Given the user-defined dictionary $\mathcal{D}$, find the sparsest cluster excitations $\mathbf{w}^{\text{opt}}$ such that a reference far-field pattern $F_{\text{ref}}(\theta)$ is matched, as

$$\mathbf{F}_{\text{ref}} - \Phi \mathcal{D} \mathbf{w}^{\text{opt}} = \boldsymbol{\eta}. \tag{14.20}$$

It is important to note that this formulation intrinsically allows the synthesis of overlapped architectures, where some radiating elements are shared among different sub-arrays. More specifically, sub-array overlapping is enabled including in the dictionary $\mathcal{D}$ modules with common elements (i.e., different columns with nonnull entries at the same row positions). Such an option, which is not trivial to implement in other framework, allows to reduce the distance between the phase centers modules, therefore mitigating amplitude and/or phase quantization errors even if the number of TRMs are highly reduced. As a drawback, the complexity of the feeding network increases when one element must be connected to multiple sub-array ports. This approach thus enhances flexibility in the synthesis process, compared to standard clustering procedures [19], and it enables the designer to explore a wider range of trade-offs between architecture complexity and performance.

Since the unknowns vector in the problem (14.20) is a sparse one, with a linear relationship between $\mathbf{w}$ and $\mathbf{F}_{\text{ref}}$, the use of CS formulations is straightforward [11]. The number of samples K is usually chosen such that $K \ll P$, since the number of independent field samples is limited by the physical aperture of the array according to the degrees of freedom criterion [50]. Consequently, the complex kernel matrix $\Phi\mathcal{D}$ is not invertible and it is not possible to guarantee that the restricted isometry property (RIP) holds true [11]. Therefore, the convergence of simple ℓ_1-norm minimization cannot be assured [11]. Accordingly the BCS formulation turns out to be compliant with the mathematical features of the sparse problem at hand [11] since it has demonstrated satisfactory performance in tackling CS problems, even when the kernels do not conform to the RIP [6–9, 11], while alternative techniques based on basis pursuit algorithms [51–53] can be ineffective if a poor initialization is employed [52], and its accuracy is affected by the isometry constant of the problem kernel [54, 55]. Moreover, the designer usually needs an efficient synthesis procedure able to generate in fast way multiple solutions to explore the Pareto set of trade-off solutions between the number of clusters and the matching with a reference pattern. BCS method has been shown to be able to deal with high-dimensional CS problems, with solvers that easily allow to retrieve multiple trade-off solutions depending on few configuration parameters [6–8].

According to the BCS formulation, the *Dictionary Compressive Sensing* (*D-CS*) solution of the *Clustered Array Synthesis Problem* (14.20) can be found as [24]

$$\mathbf{w}^{\text{opt}} = \left[\text{diag}\left(\mathbf{a}^{\text{opt}}\right) + \mathcal{D}^{*}\Phi^{*}\Phi\mathcal{D}\right]^{-1}\mathcal{D}^{*}\Phi^{*}\mathbf{F}_{\text{ref}}, \tag{14.21}$$

where * stands for the Hermitian transpose, where in (14.21) the optimal hyperparameter vector $\mathbf{a}^{\text{opt}}$ is computed maximizing the logarithm of the marginal likelihood as

$$\begin{aligned}\mathbf{a}^{\text{opt}} = \arg\max_{\mathbf{a}} \Big\{ &\log\left(\left|I + \Phi\mathcal{D}\left[\text{diag}\left(\mathbf{a}\right)\right]^{-1}\mathcal{D}^{*}\Phi^{*}\right|\right) + \\ &\left(P + 2\upsilon_1\right)\log\left[\mathbf{F}_{\text{ref}}^{*}\left(I + \Phi\mathcal{Q}\left[\text{diag}\left(\mathbf{a}\right)\right]^{-1}\mathcal{Q}^{*}\Phi^{*}\right)\mathbf{F}_{\text{ref}} + 2\upsilon_2\right]\Big\}\end{aligned}. \tag{14.22}$$

A local search strategy has been used in [24] to solve (14.22), υ_1 and υ_2 being the control parameters of the method and I the identity matrix. Once $\mathbf{w}^{\text{opt}}$ is found (14.21), the final clustered layout can be deduced by selecting the S columns of $\mathcal{D}$ corresponding to the nonzero entries of $\mathbf{w}^{\text{opt}}$ and feeding the pth candidate cluster with the corresponding entry w_p^{opt} (the pth candidate cluster is discarded if $w_p^{\text{opt}} = 0$).

As an illustrative example, the D-CS method has been used for the synthesis of a clustered array matching a $N = 200$ Taylor pattern (SLL $= -30$ dB) through the sparsest combination of clusters grouping either two, four, or eight elements [24]. Figure 14.4a compares the synthesized

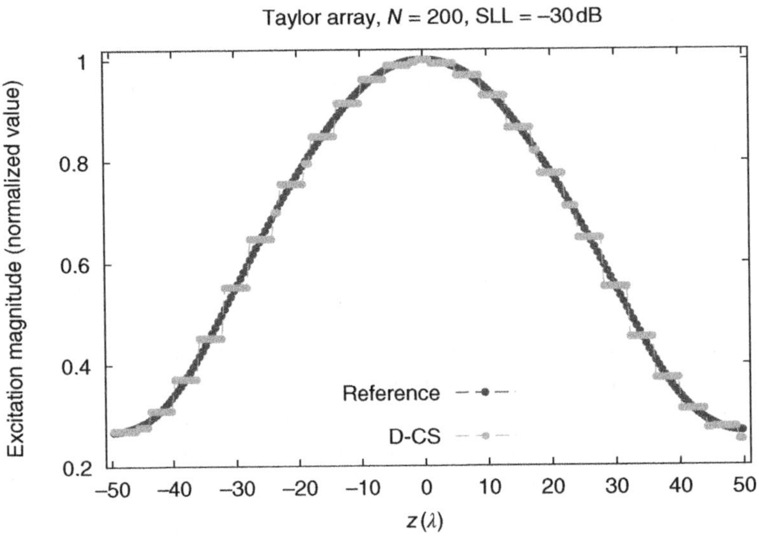

(a)

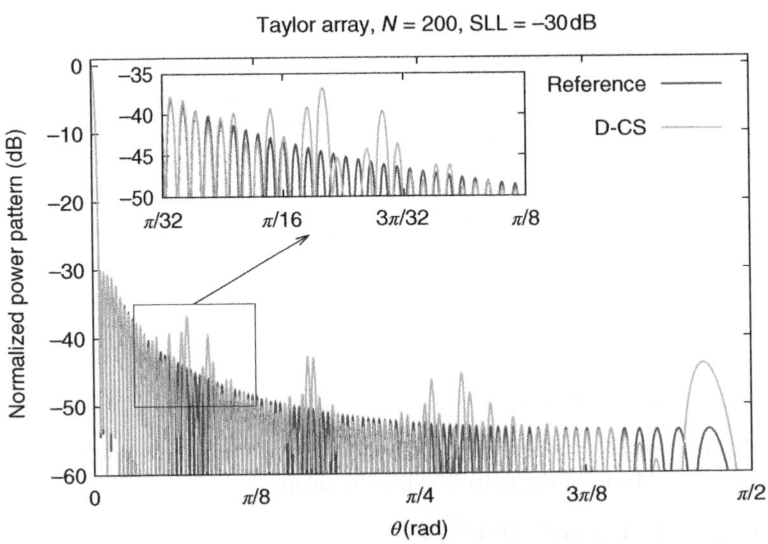

(b)

Figure 14.4 *Sparseness-Promoting Techniques (D-CS Synthesis Method, Taylor Pattern, N = 200,* SLL = −30 dB - (a) amplitude excitation coefficients and (b) normalized power patterns of the reference and sparse *D − CS* clustered arrays when considering sub-array modules of two, four, or eight elements.

D-CS equivalent excitations with respect to the reference Taylor coefficients showing how the clusters amplitudes "approximate" the behavior of the Taylor reference weighting. Moreover Fig. 14.4b compares the reference and D-CS power patterns showing how the aperiodic sparse arrangements of the three clusters modules allows to an effective sidelobe control, especially in the neighboring of the main lobe area, while considerably simplifying the resulting architecture.

14.2.3 Total-Variation Regularization Techniques

The fundamental limitation of dictionary-based unconventional array design method is related to the need for the designer to state, in advance, the candidate cluster lengths and positions. Moreover, the D-CS method may yield to configurations with overlaps between adjacent clusters, which potentially increase the complexity at the beamforming network level. While limiting the set of dictionary configurations can avoid this issue, it also requires fixing the cluster lengths and positions in advance.

In order to provide an alternative array clustering tool still exploiting the positive features of CS, while directly synthesizing contiguous clustering configurations without any constraint on the sub-array size, a reformulation of the *Clustered Array Synthesis Problem* synthesis problem in a "transformed domain" has been proposed in [27]. Such an approach stems from the observation that, the excitation coefficients of a contiguously-clustered layout have a piece-wise constant distribution (Fig. 14.5). Accordingly, the gradient of such a quantity has a naturally sparse discrete behavior with non-null entries only in correspondence with the elements between adjacent clusters. As a result, a CS-based contiguous clustering approach can be formulated

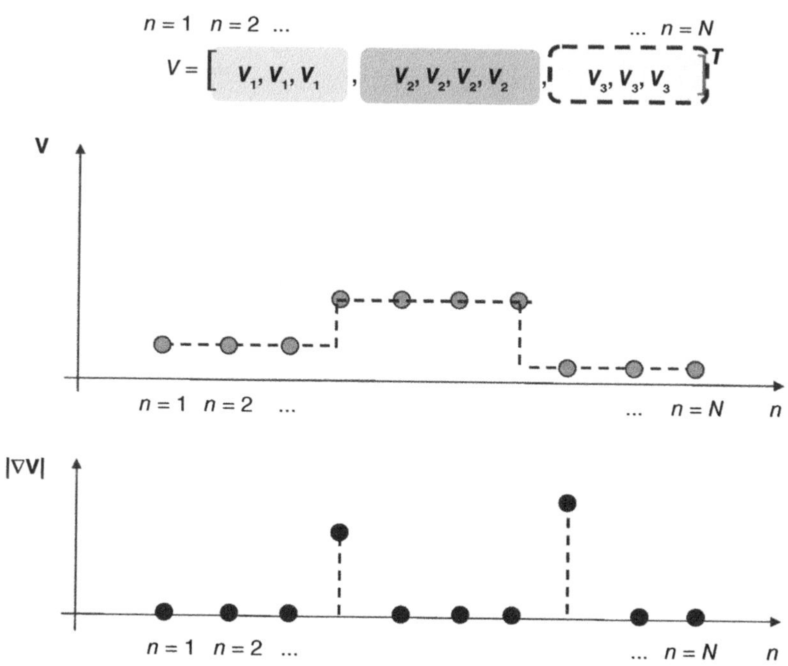

Figure 14.5 *Sparseness-Promoting Techniques (TV-CS Synthesis Method, $N = 10$, $S = 3$)* – Example of the clustered excitation vector **v** and its gradient ∇**v**.

by promoting the sparsity on the discrete gradient of the array excitations, with the aim of minimizing the number of sub-arrays in the feeding network, instead of using a user-defined dictionary of cluster combinations as in [24]. The formulation has been proposed in [27, 56] to intrinsically encode the sparsity of the discrete gradient of the unknown as a prior. By applying it to the contiguous clustering problem, the a-priori avoidance of sub-array overlapping and the elimination of the need for user-defined setup of the cluster positions and lengths, has been achieved enabling the exploration of the full design Degrees of Freedom (DoFs). Moreover, the costs/complexity of the feeding architecture is automatically minimized, since minimizing the gradient sparsity is equivalent to reduce the number of clusters. The solution of the arising problem has been achieved by exploiting the effectiveness and efficiency of TV-CS inversions [45, 46, 57–60].

From the mathematical viewpoint, the clustering problem can be recast in the CS framework by utilizing the following TV-CS formulation [45, 46, 56]

$$\min_{\mathbf{v}} \|\nabla \mathbf{v}\|_1 \text{ s.t. } F(\theta) = F_{\text{ref}}(\theta), \tag{14.23}$$

where $\|\cdot\|_1$ stands for the ℓ_1-norm. The $\min_{\mathbf{v}} \|\nabla \mathbf{v}\|_1$ objective in (14.23) is subject to a pattern matching requirement that naturally encodes the designer's need to reduce feed network costs and complexity, while maintaining appropriate control over the beam shape. In matrix form, (14.23) is re-written as follows [45, 46, 56]

$$\min_{\alpha,\mathbf{v}} \|\boldsymbol{\zeta}\|_1 \text{ s.t. } \boldsymbol{\zeta} = \nabla \mathbf{v} \text{ and } \boldsymbol{\Phi}\mathbf{v} = \mathbf{F}_{\text{ref}}. \tag{14.24}$$

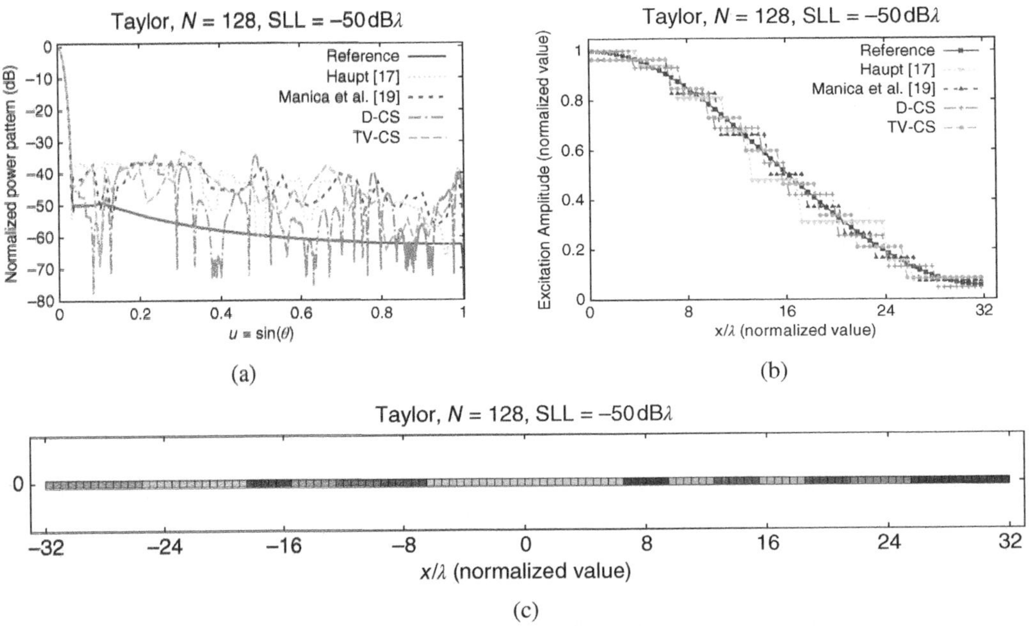

Figure 14.6 *Sparseness-Promoting Techniques (TV-CS Synthesis Method, Taylor Pattern, N = 128, SLL = –50 dB, S = 15)* – (a) Power pattern and (b) excitations amplitude of the TV-CS as compared to [17], [19] and [24]. (c) TV-CS array elements clustering configuration.

Thanks to the introduction of the auxiliary vector $\boldsymbol{\zeta} \triangleq \{\zeta_n, n = 1, ..., N\}$ [45, 46, 56] and by using the Lagrangian multiplier method [61], (14.24) can be re-written into the following *convex* equivalent expression

$$\min_{\boldsymbol{\zeta},\mathbf{v},\boldsymbol{\mu},\boldsymbol{\nu}} [\mathcal{L}(\boldsymbol{\zeta}, \mathbf{v}, \boldsymbol{\mu}, \boldsymbol{\nu})] \text{ s.t. } \boldsymbol{\zeta} = \nabla\mathbf{v} \text{ and } \boldsymbol{\Phi}\mathbf{v} = \mathbf{F}_{\text{ref}}, \tag{14.25}$$

where $\boldsymbol{\mu} \triangleq \{\mu_n, n = 1, ..., N\}$ and $\boldsymbol{\nu} \triangleq \{\nu_n, n = 1, ..., N\}$ are the Lagrangian multiplier vectors, $\mathcal{L}(\boldsymbol{\zeta}, \mathbf{v}, \boldsymbol{\mu}, \boldsymbol{\nu})$ is the Lagrangian function

$$\mathcal{L}(\boldsymbol{\zeta}, \mathbf{v}, \boldsymbol{\mu}, \boldsymbol{\nu}) \triangleq \|\boldsymbol{\zeta}\|_1 - \boldsymbol{\mu}^* (\nabla\mathbf{v} - \boldsymbol{\zeta}) - \boldsymbol{\nu}^* (\boldsymbol{\Phi}\mathbf{v} - \mathbf{F}_{\text{ref}}). \tag{14.26}$$

It is worth pointing out that, unlike D-CS [24] (14.26) does not require the user to *a-priori* specify a *dictionary* of candidate sub-arrays. As instance the synthesis of a linear array of $N = 128$ elements Taylor pattern (SLL = −50 dB) partitioned into $Q = 15$ contiguous clusters has been considered to compare the TV-CS and D-CS methods. Moreover the same solution is compared with other state-of-art methods without constraints on the subarray lengths and/or shapes [17, 19]. Figure 14.6a shows the normalized power pattern for $u \geq 0$ of the synthesized solutions together with the associated equivalent clustered amplitude excitations in Fig. 14.6b. Thanks to its pattern-matching formulation, the TV-CS solution (Fig. 14.6c) better matches the reference pattern with respect to [17, 19] (i.e., $\xi\rfloor_{\text{TV-CS}} = 2.46 \times 10^{-3}$ while $\xi\rfloor_{[\text{Haupt 2007}]} = 6.44 \times 10^{-3}$ and $\xi\rfloor_{[\text{Manica 2009}]} = 3.80 \times 10^{-3}$). Moreover, it slightly improve the D-CS solution ($\xi\rfloor_{\text{D-CS}} = 2.69 \times 10^{-3}$) where clusters of four, five, seven, and eight elements are used, thanks to the the full exploration of the design DoFs (e.g., the positions and the lengths of the array clusters) enabled by the TV-CS problem formulation.

14.3 Co-design of Unconventional Architectures and Radiating Elements

The unconventional array synthesis methods described in the previous sections considers ideal radiating sources. However, especially for next generation wireless communication, the design of the actual radiating element plays a crucial role. It is well known that 5G/6G BS antennas need to be highly flexible and high-performing, capable of supporting real-time adaptive beam reconfiguration and user tracking features, and exploiting MIMO processing. These objectives must be achieved with cost-effective architectures and antenna elements to minimize fabrication and maintenance costs. Active electronically scanned arrays (AESA) are an established technology to address 5G/6G BS antenna design, offering real-time beam control, low sidelobes, high gain, narrow beamwidth, wide steering angle, wide bandwidth, and MIMO capabilities. However, simultaneously fitting all of these requirements is very challenging, especially if inexpensive and compact arrangements are required. Conventional AESA architectures, such as fully populated and regularly clustered layouts, cannot be adopted due to their high costs or limited field-of-view, respectively. Furthermore, the peculiar features of the 5G scenario require an "atypical" approach to the design of the elementary radiator in the *AESA*, as it yields to a massive-multi-objective (MMO) design problem.

To address these challenges, an innovative *co-design* process has been proposed in [26] for the synthesis of both the elements and the array architecture for next generation 5G/6G BSs. This method simultaneously synthesizes the antenna element (embedded in the finite layout) and the array architecture taking into account real-systems electromagnetic (EM) environment. The proposed strategy is aimed at effectively handling the inter-relationships arising in a 5G/6G scenario between the single radiator and the array architecture. To this end, two fundamental challenges are addressed. Firstly, the control of the sidelobe profile of sub-arrayed layouts over wide scanning angles and bandwidths while jointly clustering excitation magnitudes and phases. Secondly, the design of the single element radiator in a *5G/6G MMO* scenario, taking into account their interactions with the surrounding elements. To overcome these challenges, the *co-design* procedure proposed in [26] combines:

- an innovative irregular clustering technique: this technique formulates the sub-arraying problem as a global optimization problem, where the degrees-of-freedom are the clustering scheme (i.e., the suitable combination of a set of admissible a-priori chosen "shapes/tiles") and the cluster excitations.
- a customized shape optimization approach to define a radiating element that satisfies the *5G/6G* electrical requirements, such as input impedance, beamwidth, isolation, and polarization, when integrated into a finite array rather than being analyzed alone or in an infinite periodic arrangement as typically done in previous literature.

In order to efficiently address this *5G/6G* design sub-problem, ad-hoc search strategies are employed in accordance with the "No-free-lunch" Theorem [62]. Specifically, a single-objective Genetic Algorithm (SO-GA) is used, which employs an innovative integer coding of the variables and specialized operators to handle irregular clustering. Additionally, the multi-objective shaped-optimization of the elementary radiator is performed using a ϵ-*MOEA* approach after an appropriate reformulation [63, 64].

14.3.1 5G Base Station Antenna Design Problem

In order to illustrate the approach proposed [26], let us consider a planar arrangement of $M \times N$ elements displaced in the (y, z)-plane and clustered in S sub-arrays ($S < M \times N$) of contiguous elements. The membership of each (m, n)th ($m = 1, \ldots, M$; $n = 1, \ldots, N$) element to the sth ($s = 1, \ldots, S$) sub-array is coded into the sub-array aggregation vector

$$\mathbf{c} \triangleq \left\{ c_{m,n} \in [1 : S];\ m = 1, \ldots, M;\ n = 1, \ldots, N \right\}. \tag{14.27}$$

Since all the elements of the sth cluster are fed with the same excitation magnitude, α_s, and phase, $\beta_s\left(\theta_0, \phi_0; f\right)$, which depends on the steering angle $\left(\theta_0, \phi_0\right)$ and the working frequency f, the far-field pattern radiated by the array is given by [2, 49]

$$F(\theta, \phi) = \sum_{m=1}^{M} \sum_{n=1}^{N} \sum_{s=1}^{S} \alpha_s e^{j\beta_s(\theta_0, \phi_0; f)} \times \delta_{sc_{m,n}} g_{m,n}(\theta, \phi; f)\, e^{j\frac{2\pi}{\lambda}\mathbf{r}_{m,n}\cdot\hat{\mathbf{r}}}, \tag{14.28}$$

where λ is the wavelength at frequency f, $\delta_{sc_{m,n}}$ is the Kronecker delta function ($\delta_{sc_{m,n}} = 1$ if $c_{m,n} = s$ and $\delta_{sc_{m,n}} = 0$ otherwise), and

$$\hat{\mathbf{r}} = \sin(\theta)\sin(\phi)\,\hat{\mathbf{y}} + \cos(\theta)\,\hat{\mathbf{z}}. \tag{14.29}$$

Moreover, $\mathbf{r}_{m,n} = y_{m,n}\hat{\mathbf{y}} + z_{m,n}\hat{\mathbf{z}}$ is the position of the the (m, n)th ($m = 1, \ldots, M$; $n = 1, \ldots, N$) array element, while $g_{m,n}(\theta, \phi; f)$ is the corresponding radiation pattern, which is a function of the H-size vector of its geometrical descriptors

$$\gamma \triangleq \left\{ \gamma_h,\ h = 1, \ldots, H \right\}. \tag{14.30}$$

Designing a BS antenna requires meeting manifold objectives and requirements related to both the array pattern (e.g., complying with the sidelobe mask for all steering angles and frequency bands of interest) and the radiation features of the array elements (e.g., impedance matching, isolation, polarization ellipticity, gain, and beamwidth) [26]. Consequently, the synthesis problem can be inherently classified as a multi-objective optimization problem when formulated as the optimal configuration of the unknown vectors:

- the sub-array vector, $\mathbf{c}$;
- the clusters amplitude coefficients $\boldsymbol{\alpha} \triangleq \{\alpha_s, s = 1, \dots, Q\}$;
- the clusters phase coefficients $\boldsymbol{\beta} \triangleq \{\beta_s(\theta_0, \phi_0; f);\ q = 1, \dots, Q;\ \theta_0 \in [\theta_0^{\min}, \theta_0^{\max}],$ $\phi_0 \in [\phi_0^{\min}, \phi_0^{\max}], f \in [f^{\min}, f^{\max}]\}$;
- the radiating element geometrical descriptors, $\boldsymbol{\gamma}$;

such that the following minimization problem is solved

$$\{\mathbf{c}, \boldsymbol{\alpha}, \boldsymbol{\beta}, \boldsymbol{\gamma}\}^{\text{opt}} = \arg \min_{\{\mathbf{c}, \boldsymbol{\alpha}, \boldsymbol{\beta}, \boldsymbol{\gamma}\}} [\Psi(\mathbf{c}, \boldsymbol{\alpha}, \boldsymbol{\beta}, \boldsymbol{\gamma})], \tag{14.31}$$

where

$$\Psi(\mathbf{c}, \boldsymbol{\alpha}, \boldsymbol{\beta}, \boldsymbol{\gamma}) \triangleq \{\Psi_j(\mathbf{c}, \boldsymbol{\alpha}, \boldsymbol{\beta}, \boldsymbol{\gamma}), j = 1, \dots, J\} \tag{14.32}$$

is the MMO cost function, comprising J terms encoding the synthesis objectives as follows [2] [26]:

- "*Array Sidelobe Mask Compliancy*" Term ($j = 1$):

$$\Psi_1(\mathbf{c}, \boldsymbol{\alpha}, \boldsymbol{\beta}, \boldsymbol{\gamma}) \triangleq \frac{1}{(f^{\max} - f^{\min})(\phi^{\max} - \phi^{\min})(\theta^{\max} - \theta^{\min})} \times \int_{f^{\min}}^{f^{\max}} \int_{\theta_0^{\min}}^{\theta_0^{\max}} \int_{\phi_0^{\min}}^{\phi_0^{\max}} \max_{\theta,\phi} \mathcal{F}\left[\text{PP}(\theta, \phi; \theta_0, \phi_0; f) - , \text{SLL}^t(\theta, \phi; \theta_0, \phi_0; f)\right] d\theta_0 d\phi_0 df \tag{14.33}$$

where $\text{PP}(\theta, \phi; \theta_0, \phi_0; f)$ and $\text{SLL}^t(\theta, \phi; \theta_0, \phi_0; f)$ are the power pattern (i.e., $\text{PP}(\theta, \phi) = |F(\theta, \phi;)|^2$) and the target sidelobe, respectively, when the array is steered towards (θ_0, ϕ_0), at frequency f, and $\mathcal{F}[\cdot]$ is the "ramp" function ($\mathcal{F}[\cdot] \triangleq [\cdot] \times \mathcal{H}[\cdot]$, $\mathcal{H}[\cdot]$ being the Heaviside function);

- "*Element Impedance Matching*" Term ($j = 2$):

$$\Psi_2(\mathbf{c}, \boldsymbol{\gamma}) \triangleq \frac{\int_{f^{\min}}^{f^{\max}} \mathcal{F}[|S_{11}(f)| - |S_{11}^t|] df}{(f^{\max} - f^{\min})}, \tag{14.34}$$

where $S_{11}(f)$ and S_{11}^t are the synthesized and the target antenna reflection coefficient at the input port at the frequency f, respectively;

- "*Element Polarization*" Term ($j = 3$):

$$\Psi_3(\mathbf{c}, \boldsymbol{\gamma}) \triangleq \frac{\int_{f^{\min}}^{f^{\max}} [\mathcal{E}(f) - \mathcal{E}^t]^2 df}{(f^{\max} - f^{\min})}, \tag{14.35}$$

where $\mathcal{E}(f)$ and $\mathcal{E}^t$ are the synthesized and the target antenna polarization ellipticity angle at the frequency f, respectively;

---

2 For the sake of notation compactness, the dependency of all the radiation quantities (i.e., pattern, gain, sidelobe, reflection coefficient, etc.) on the DoFs vector ($\mathbf{c}, \boldsymbol{\alpha}, \boldsymbol{\beta}, \boldsymbol{\gamma}$) is omitted hereinafter.

- "*Element Beamwidth*" Terms ($j = 4, 5$):

$$\begin{cases} \Psi_4(\mathbf{c}, \gamma) \triangleq \dfrac{\int_{f^{\min}}^{f^{\max}} \mathcal{F}\left[W_{\mathrm{AZ}}(f) - W_{\mathrm{AZ}}^t\right] df}{\left(f^{\max} - f^{\min}\right)} \\ \Psi_5(\mathbf{c}, \gamma) =\triangleq \dfrac{\int_{f^{\min}}^{f^{\max}} \mathcal{F}\left[W_{\mathrm{EL}}(f) - W_{\mathrm{EL}}^t\right] df}{\left(f^{\max} - f^{\min}\right)} \end{cases} \tag{14.36}$$

where

$$W_{\mathrm{AZ}}(f) = \left| \arg\min_{\phi \in [0, \phi_{\mathcal{M}}]} \left| \frac{\overline{g}(\theta_{\mathcal{M}}, \phi; f)}{\overline{g}(\theta_{\mathcal{M}}, \phi_{\mathcal{M}}; f)} - 0.5 \right| - \arg\min_{\phi \in \left[-\frac{\pi}{2}, \phi_{\mathcal{M}}\right]} \left| \frac{\overline{g}(\theta_{\mathcal{M}}, \phi; f)}{\overline{g}(\theta_{\mathcal{M}}, \phi_{\mathcal{M}}; f)} - 0.5 \right| \right| \tag{14.37}$$

$$W_{\mathrm{EL}}(f) = \left| \arg\min_{\theta \in [\theta_{\mathcal{M}}, \pi]} \left| \frac{\overline{g}(\theta, \phi_{\mathcal{M}}; f)}{\overline{g}(\theta_{\mathcal{M}}, \phi_{\mathcal{M}}; f)} - 0.5 \right| - \arg\min_{\theta \in [0, \theta_{\mathcal{M}}]} \left| \frac{\overline{g}(\theta, \phi_{\mathcal{M}}; f)}{\overline{g}(\theta_{\mathcal{M}}, \phi_{\mathcal{M}}; f)} - 0.5 \right| \right| \tag{14.38}$$

are the antenna half-power beamwidths in azimuth and elevation, while W_{AZ}^t and W_{EL}^t the corresponding target values, being $\left(\theta_{\mathcal{M}} = \frac{\pi}{2}, \phi_{\mathcal{M}} = 0\right)$ the element broadside direction, and $\overline{g}(\theta, \phi; f) = \frac{1}{M \times N} \sum_{m=1}^{M} \sum_{n=1}^{N} g_{m,n}(\theta, \phi; f)$ is the average element factor;

- "*Element Realized Gain*" Term ($j = 6$):

$$\Psi_6(\mathbf{c}, \boldsymbol{\alpha}, \boldsymbol{\beta}, \gamma) \triangleq \frac{\int_{f^{\min}}^{f^{\max}} \mathcal{F}\left[\mathcal{G}^T - \dfrac{4\pi \mu_1(f) \mu_2(f) \overline{g}(\theta_{\mathcal{M}}, \phi_{\mathcal{M}}; f)}{\int_0^{2\pi} \int_0^{\pi} \left[E(\theta, \phi; f) \sin(\theta)\right] d\theta d\phi}\right] df}{\left(f^{\max} - f^{\min}\right)}, \tag{14.39}$$

where $\mathcal{G}^t$ is the target realized gain value, while $\mu_1(f)$ and $\mu_2(f)$ are the average element efficiency and the average mismatch factor, respectively;

- "*Element Isolation*" terms ($j = 7, \ldots, J$):

$$\Psi_j(\mathbf{c}, \boldsymbol{\alpha}, \boldsymbol{\beta}, \gamma) \triangleq \frac{\int_{f^{\min}}^{f^{\max}} \mathcal{F}\left[\left|S_{i1}(f)\right| - \left|S_{i1}^t\right|\right] df}{\left(f^{\max} - f^{\min}\right)} \quad i = j - 6;\ j = 7, \ldots, J \tag{14.40}$$

where $S_{i1}(f)$ and S_{i1}^t are the synthesized and the target isolation coefficients between the reference antenna and its ith neighbor element at the frequency f, respectively.

Solving the MMO problem in (14.31) with a single step procedure is not feasible because of the unavailability of effective and efficient *MO* optimization strategies when $J \geq 4$, the complex and highly nonlinear nature of the objective functions/terms in (14.32), and the large number of DoFs, resulting in a too huge and computationally untractable search space. Accordingly, it has been split into two (still connected) simpler ones concerned with (i) the array design and (ii) the element synthesis. Namely,

- a *single-objective* (*SO*) array clustering problem where the DoFs comprise the clustering scheme $\mathbf{c}$ and the excitation vectors $\boldsymbol{\alpha}$ and $\boldsymbol{\psi}$, the *SO* cost function being defined, for a fixed radiator geometry ($\gamma = \hat{\gamma}$), as follows

$$\Psi_{\mathrm{SO}}(\mathbf{c}, \boldsymbol{\alpha}, \boldsymbol{\beta}) \triangleq \Psi_1\left(\mathbf{c}, \boldsymbol{\alpha}, \boldsymbol{\beta}, \hat{\gamma}\right); \tag{14.41}$$

- a MMO radiating element design problem where the DoFs are the geometry parameters γ, while the MMO cost function, for a fixed set of the remaining parameters ($\left\{\widehat{\mathbf{c}}, \widehat{\alpha}, \widehat{\beta}\right\}$), is given by

$$\Psi_{\text{MMO}}(\gamma) \triangleq \left\{\Psi_j\left(\widehat{\mathbf{c}}, \widehat{\alpha}, \widehat{\beta}, \gamma\right); j = 2, \ldots, J\right\}. \tag{14.42}$$

It is worth remarking that (14.41) and (14.42) are tightly coupled problems [26].

14.3.2 *Co-design* Synthesis Strategy

Because of the coupled nature of the problems formulated in Section 14.3.1, an *alternate two-step co-design* process is presented in [26] by first optimizing the cost function Φ_{SO} (14.41) when $\gamma = \widehat{\gamma}$ (Array Clustering) and then Φ_{MMO} (14.42) is minimized for a fixed DoFs configuration $\left\{\widehat{\mathbf{c}}, \widehat{\alpha}, \widehat{\beta}\right\}$ (Antenna Element Synthesis).

The alternate two-step *co-design* approach is summarized in the following iterative process:

- *Initialization* – The values of the array lattice and the aperture width are set by exploiting state-of-the-art analytic formulas [49] to avoid grating lobes within the *5G* angular steering range and bandwidth as well as to guarantee the required target directivity;
- *Antenna Element Synthesis* – A set of Pareto-optimal trial antenna shapes is determined by means of the ϵ-*MOEA*. Towards this end, the multi-objective optimization strategy is used to sample the search space by generating at each iteration U candidate configurations γ^u ($u = 1, \ldots, U$), which are ranked through the following operations:
  - *Radiator Geometry Generation* – The antenna geometry generator (AGG) operator [65], is used to infers the shape of the array element from γ^u ($u = 1, \ldots, U$), encoding/decoding the shape of the radiator starting from the vector of geometrical parameters γ embedding a combination of "parametric" and "spline"-based descriptors. To this end, the complex antenna contours have been encoded through spline curves with a small set of control points in the (y, z)-plane [26].
  - *Mutual Coupling Computation* – In order to guarantee a reliable and faithful synthesis of the radiating element, the mutual interactions among nearby radiators when arranged in a finite array layout is done through the *System Characterization* operator. This operator enables to model the elementary radiator by taking into account spacing constraints and to reliably estimate the inter-element isolation/coupling parameters by considering the *finite* size of the final arrangements, without recurring to simplified periodic models for the array structure. This guarantees to avoid further element re-optimization when the same radiator is placed within the whole array arrangement. Towards this end the radiator shape deduced from γ^u ($u = 1, \ldots, U$) is embedded in a finite small array model of $\widetilde{M} \times \tilde{N}$ elements ($\widetilde{M} \ll M, \tilde{N} \ll N$), and the coupling/isolation parameters as well as the average pattern are evaluated by feeding only the central element [26].
  - *Cost function Computation* – Concerning the optimization mechanism the original MMO is reformulated in the following multi-objective (MO) one

$$\widetilde{\Psi}_{\text{MO}}(\gamma) \triangleq \left\{\tilde{\Psi}_l\left(\widehat{\mathbf{c}}, \widehat{\alpha}, \widehat{\beta}, \gamma\right), l = 1, 2, 3\right\} \tag{14.43}$$

by aggregating, according to their physical "coherence," the components of $\Psi_{\text{MMO}}(\gamma)$ into three main cost function terms, including an *Impedances* Term

$$\tilde{\Psi}_1\left(\widehat{\mathbf{c}}, \widehat{\alpha}, \widehat{\beta}, \gamma\right) \triangleq \rho_2\Psi_2\left(\widehat{\mathbf{c}}, \gamma\right) + \textstyle\sum_{j=7}^{J} \rho_l\Psi_l\left(\widehat{\mathbf{c}}, \widehat{\alpha}, \widehat{\beta}, \gamma\right) \tag{14.44}$$

given by the weighted combination of self- and mutual-impedances terms, *a Beam Directionality* Term

$$\tilde{\Psi}_2\left(\hat{\mathbf{c}}, \hat{\boldsymbol{\alpha}}, \hat{\boldsymbol{\beta}}, \boldsymbol{\gamma}\right) \triangleq \rho_4\Psi_4\left(\hat{\mathbf{c}}, \boldsymbol{\gamma}\right) + \rho_5\Psi_5\left(\hat{\mathbf{c}}, \boldsymbol{\gamma}\right) + \rho_6\Psi_6\left(\hat{\mathbf{c}}, \hat{\boldsymbol{\alpha}}, \hat{\boldsymbol{\beta}}, \boldsymbol{\gamma}\right) \tag{14.45}$$

yielded from the combination of the beamwidth and the realized gain terms, and a *Polarization* Term

$$\tilde{\Psi}_3\left(\hat{\mathbf{c}}, \hat{\boldsymbol{\alpha}}, \hat{\boldsymbol{\beta}}, \boldsymbol{\gamma}\right) \triangleq \rho_3\Psi_3\left(\hat{\mathbf{c}}, \hat{\boldsymbol{\alpha}}, \hat{\boldsymbol{\beta}}, \boldsymbol{\gamma}\right) \tag{14.46}$$

$\rho_j, j = 2, \dots, J$, being user-defined real weighting coefficients. Since the dimension of the nonlinear MO problem is reduced to $J = 3$, the version of *ϵ-MOEA* optimization tool is applied since it proved to be effective for a global search in continuous spaces [63, 64] and to enable quick convergence while maintaining a well-distributed set of nondominated solutions during the iterative optimization process.

- "*Array Clustering*" – Starting from the array element synthesized at the convergence of the previous *ϵ-MOEA* optimization $\boldsymbol{\gamma}^{\text{opt}}$, for a given membership vector $\mathbf{c}$, the optimal values of the array excitations in both amplitude, $\boldsymbol{\alpha}$, and phase, $\boldsymbol{\beta}$, for a user-specified frequency f ($f \in \left[f^{\min}, f^{\max}\right]$) and a steering direction (θ_0, φ_0) ($\theta_0 \in \left[\theta_0^{\min}, \theta_0^{\max}\right]$, $\varphi_0 \in \left[\varphi_0^{\min}, \varphi_0^{\max}\right]$), are unique and they can be found solving a convex problem [10]. Consequently, it turns out that the DoFs of the *Array Clustering* phase are only the memberships of the array elements listed in the integer membership vector $\mathbf{c}$ and the optimization problem at hand can be classified as a *SO discrete* one, which is solved using the Integer-Coded *SO-GA* optimization by iteratively applying the operators described in [24] by minimizing the SO cost function (14.41).
- "*Array Improvement*" – The aperture/inter-element spacing are reduced by a user-defined scaling factor to minimize the array dimensions. The *co-design* loop is then repeated starting from the "*Antenna Element Synthesis*" phase until the array design is compliant with the *5G* user-defined specifications or the maximum number of iterations is reached.

As an instance of the *co-design* strategy [26], a representative unconventional antenna array design for real real-world *5G* BSs is here addressed. To this end, the design requirements feature [26] an $S_{11}^t = -10\,\text{dB}$ linear polarized radiating element, with $W_{\text{AZ}}^t = 120°$, $W_{\text{EL}}^t = 90°$, $G^t = 4\,\text{dB}$, and $S_{k1}^t = 20\,\text{dB}$, at the central frequency, f_0, for a cellular sector coverage defined by $\theta_0^{\min} = 67.5°$, $\theta_0^{\max} = 112.5°$, $\varphi_0^{\min} = -60°$, and $\varphi_0^{\max} = 60°$, and a ratio between the maximum and minimum frequencies equal to $f_r = \frac{f^{\max}}{f^{\min}} = 1.1$. The full array is composed by $N \times M = 18 \times 14$ elements located according to a rectangular lattice with spacing $d_x = 0.492\,\lambda_0$ and $d_y = 0.651\,\lambda_0$ along the x and y axis, respectively, λ_0 being the wavelength at f_0. Moreover, the *Rogers 3035* has been chosen as substrate material and a standard Teflon-filled coaxial cable has been used for feeding each elementary radiator. The "*Antenna Element Synthesis*" *ϵ-MOEA* optimization leads to the set of solution reported in the $(\tilde{\Psi}_1, \tilde{\Psi}_2)$-plane of Fig. 14.7a together with the set of Pareto optimal ones, obtained by setting $\rho_j = 1.0, j = 2, \dots, J$. Among the possible Pareto optimal trade-offs, the one reported in Fig. 14.7b has been selected, which satisfies almost all the requirements, except for a deviation from the target ellipticity at the central frequency (a more detail analysis of the solution can be found in [26]). It is worth to point out here that other trade-off solutions (e.g., with better polarization control) can be selected among the Pareto set according to the designer preferences. The solutions has been optimized in a 4×4 array environment to include the mutual

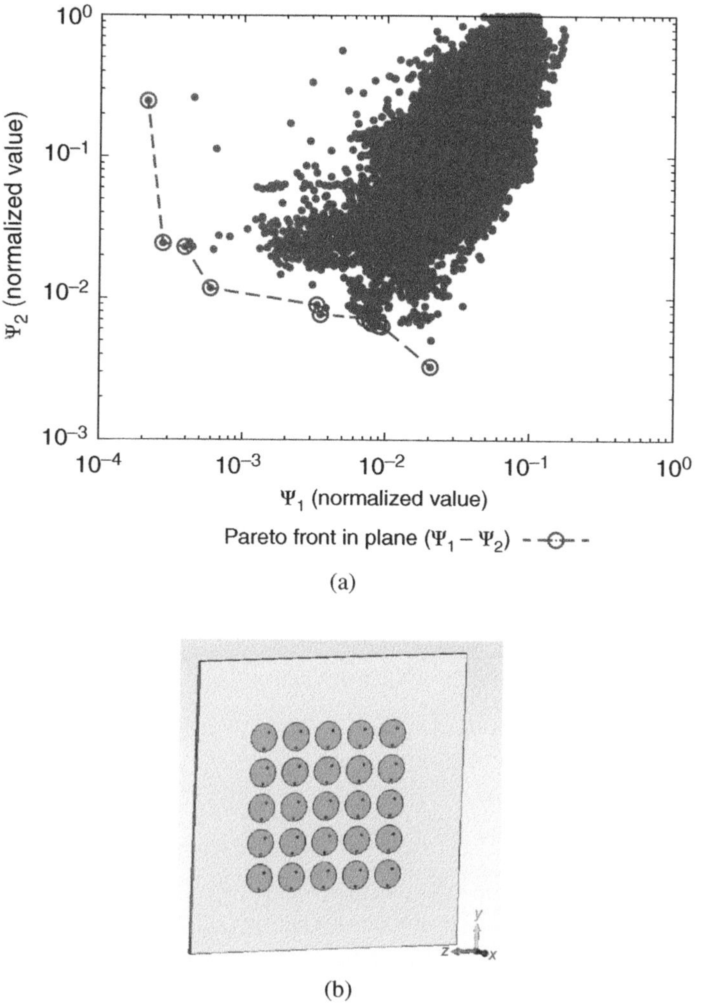

Figure 14.7 *Co-Design Strategy (Antenna Element Synthesis, $d_x = 0.492\,\lambda_0$, $d_y = 0.651\,\lambda_0$)* – (a) Plots of the of *ε-MOEA* solutions set in the $(\tilde{\Psi}_1, \tilde{\Psi}_2)$ plane and associated Pareto front, and (b) geometry of the selected Pareto-optimal trade-off solution.

coupling effects as shown in Fig. 14.7b. Dealing with the "Array Clustering" phase, an alphabet of candidate sub-arrays including square/rectangular clusters with dimensions from 1×1 up to 7×7 has been set, also considering limitations and constraints arising in the implementation of the *5G* systems. Among the *SO-GA* generated clustering schemes Fig. 14.8a reports the $S = 196$ showing the optimized irregularly clustered layouts. For illustrative purpose, the array pattern radiated when steering the main beam at the limit of the field-of-view along the principal planes, $(\theta_0, \phi_0) = (90, 60)^\circ$ and $(\theta_0, \phi_0) = (112.5, 0)$, are reported in Fig. 14.8b and c, respectively, showing that no grating lobes are present within the field of view, even at the maximum frequency.

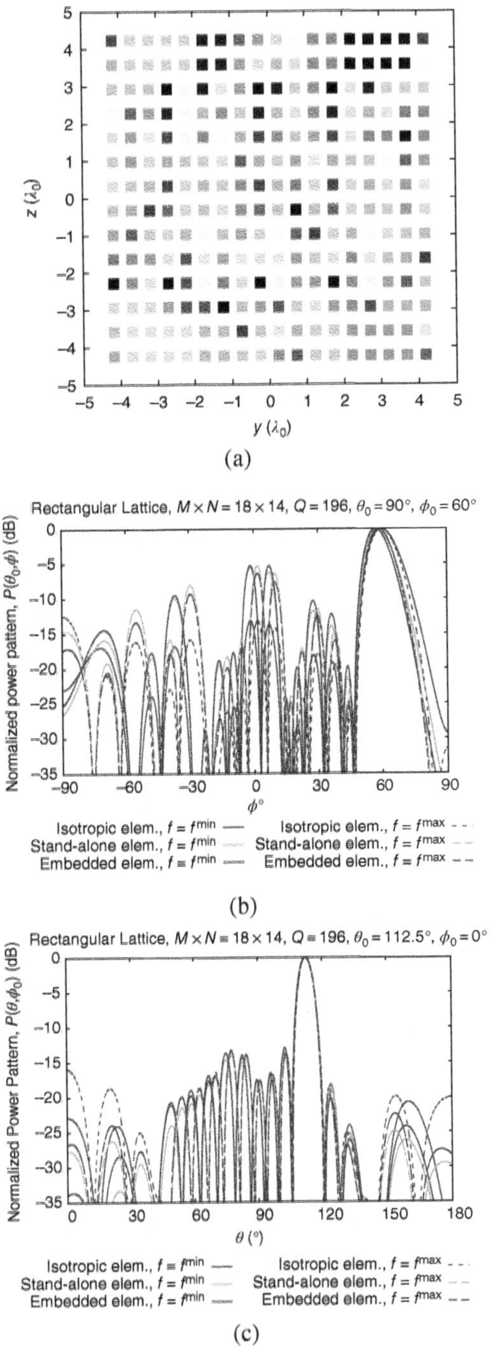

Figure 14.8 *Co-Design Strategy (Array Clustering,* $d_x = 0.492\,\lambda_0$, $d_y = 0.651\,\lambda_0$, $N \times M = 28 \times 16$, $S = 196$) – (a) Plots of the of the *SO-GA* optimized clustering layout, and (b,c) radiation pattern when steering the main beam at (b) $(\theta_0, \phi_0) = (90, 60)^\circ$, and (c) $(\theta_0, \phi_0) = (112.5, 0)^\circ$.

14.4 Capacity-Driven Synthesis of Next Generation Base Station Phased Arrays

The previously illustrated methodologies for unconventional array synthesis address "traditional" wireless performance indicators [2, 49]. However, the emergence of MIMO architectures in multi-user multi-antenna BSs has recently necessitated a re-definition of these design guidelines and objectives. In fact, most existing synthesis methods focus on optimizing the free-space radiation parameters such as directivity, half-power beamwidth, sidelobe level, and grating lobes [2, 17, 21, 49, 66–70]. This is based on the idea that the radiated power of the BS must be focused along the terminal direction to maximize the link quality and information transfer. However, in multi-user multi-antenna architectures, the signal-to-noise ratio is maximized by spatially distributing the power to constructively exploit wave scattering phenomena in the multi-path propagation environment, regardless of the arising gain, sidelobe level, or grating lobes [71, 72]. This has suggested that innovative array synthesis paradigms can be introduced to account for the propagation scenario including the radiating systems and the characteristics of the environment.

To this end, a recent general meta-paradigm for the design of unconventional antenna architectures to be used in future MIMO communication systems is described in the following. To this end, the problem of synthesizing MIMO arrays is formulated as the maximization of the quality of the communication link (i.e., the capacity). This formulation is applied to the synthesis of contiguously and irregularly clustered/tiled arrangements, which are simulated using previously validated propagation and transmitting/receiving antenna models in actual 5G setups [73]. The resulting architectural solutions are compared with those from a standard design strategy, based on the optimization of the free-space performance indexes or on the use of nonoptimized regular/periodic tilings, to point out the nonnegligible differences in terms of layouts, excitations, and radiation/communication performance.

14.4.1 Modular Array Capacity-Driven Synthesis

Let us consider a multi-user multi-antenna down-link communication scenario where a MIMO BS array consisting of a rectangular planar arrangement of $M \times N$ dual/cross-polarized (ψ being the polarization state at the TX, $\psi \in \{V, H\}$) antennas. The BS is communicating with a set of L dual-polarized (χ being the polarization state at the RX, $\chi \in \{V, H\}$) receiving antennas (i.e., the mobile-user terminals). The array elements are supposed to lie in the (y, z)-plane over a regular rectangular lattice and to be spaced by d_y and d_z along the y and z axis, respectively. To reduce the complexity and the costs of the feeding network with respect to a fully populated phased array architecture, the $M \times N$ elements are grouped into S fixed tiles ($S < M \times N$), connected at sub-array level to two TRMs, one for each polarization. Accordingly, the total number of RF chains is reduced from $2 \times (M \times N)$ to $2 \times S$, with a cost reduction proportional to $\frac{S}{M\times N}$. The membership of each (m, n)th ($m = 1, \ldots, M$; $n = 1, \ldots, N$) element to a sth ($s = 1, \ldots, Q$) sub-array is coded into the sub-array aggregation vector (14.27). Without loss of generality, the same clustering of the array elements is considered for the two polarizations (i.e., $\mathbf{c}_\psi = \mathbf{c}$, $\psi \in \{V, H\}$).

By considering a linear combination of the sub-array outputs, a tiled MIMO array generates up to $B = S$ independent beams for each ψth ($\psi \in \{V, H\}$) polarization, the radiated bth ($b = 1, \ldots, B$) ψth ($\psi \in \{V, H\}$) polarized EM field in the far-field zone being mathematically expressed as follows

$$\mathbf{F}^{(b,\psi)}(\theta, \phi) = \sum_{m=1}^{M} \sum_{n=1}^{N} \mathbf{g}_{m,n}^{\psi}(\theta, \phi) \times w_{m,n}^{(b,\psi)} e^{j\frac{2\pi}{\lambda}(y_m \sin\theta \sin\phi + z_n \cos\theta)} \quad (14.47)$$

where $\mathbf{g}^{\psi}_{m,n}(\theta,\phi)$ is the active element pattern of the (m,n)th ($m=1,\dots,M;\ n=1,\dots,N$) ψ-polarized ($\psi\in\{V,H\}$) antenna centered at $\mathbf{r}_{m,n}=\left(0,y_m,z_n\right)$, λ and (θ,ϕ) being the wavelength and the observation direction, respectively. In (14.47), the sets of $M\times N$ equivalent element excitations, $\mathbf{W}=\left\{\mathbf{w}^{(b,\psi)}\ b=1,\dots,B;\ \psi\in\{V,H\}\right\}$ being $\mathbf{w}^{(b,\psi)}=\left\{w^{(b,\psi)}_{m,n};\ m=1,\dots,M;\ n=1,\dots,N\right\}$, are function of the corresponding sets of Q sub-array complex coefficients, $\mathbf{V}=\left\{\mathbf{v}^{(b,\psi)};\ b=1,\dots,B;\ \psi\in\{V,H\}\right\}$ being $\mathbf{v}^{(b,\psi)}=\left\{v^{(b,\psi)}_{q};\ q=1,\dots,Q\right\}$, according to the following relationship

$$w^{(b,\psi)}_{m,n}=\sum_{s=1}^{S}\delta_{sc_{m,n}}v^{(b,\psi)}_{s}. \tag{14.48}$$

The L users (UEs) terminals are randomly located within the communication cell at $\{\mathbf{r}_l=\left(x_l,y_l,z_l\right);\ l=1,\dots,L\}$. Moreover, let us assume that the bth ($b=1,\dots,B$) ψ-polarized ($\psi\in\{V,H\}$) beam generated by the MIMO array is used to transmit a signal to be received by the χ-polarized ($\chi\in\{V,H\}$) antenna of the lth ($l=1,\dots,L$) UE terminal (i.e., $L=S$), also considering the hypothesis of mutual incoherence between the signals transmitted through the B beams and the noise [74]. In this scenario the signal to interference-plus-noise ratio (SINR) associated to the signal received by the χ-polarized antenna of the lth UE terminal, is defined as

$$\text{SINR}^{(\chi)}\left(\mathbf{r}_l\right)\triangleq\frac{\Pi^{(\chi)}_{\text{des}}\left(\mathbf{r}_l\right)}{\Pi^{(\chi)}_{\text{mui}}\left(\mathbf{r}_l\right)+\Pi^{(\chi)}_{\eta}\left(\mathbf{r}_l\right)}, \tag{14.49}$$

where $\Pi^{(\chi)}_{\text{des}}\left(\mathbf{r}_l\right)$ is the power associated to the desired signal, $\Pi^{(\chi)}_{\text{mui}}\left(\mathbf{r}_l\right)$ is the interference caused by the presence of the other $(L-1)$ UEs, and $\Pi^{(\chi)}_{\eta}\left(\mathbf{r}_l\right)$ is the power of the noise. More in detail, the desired power is given by

$$\Pi^{(\chi)}_{\text{des}}\left(\mathbf{r}_l\right)=\frac{\Pi_{tx}}{B}\left|\sum_{m=1}^{M}\sum_{n=1}^{N}w^{(b,\psi)}_{m,n}\mathbf{G}^{(\psi,\chi)}\left(\omega;\mathbf{r}_l,\mathbf{r}_{m,n}\right)\right|^2, \tag{14.50}$$

where Π_{tx} is the total power at the MIMO array input, which is supposed to be equally spread over the B beams, while $\mathbf{G}^{(\psi,\chi)}\left(\omega;\mathbf{r}_l,\mathbf{r}_{m,n}\right)$ is the Fourier transform of the time-domain Green's function (GF) modeling the propagation and any other EM effect between the ψ-polarized ($\psi\in\{V,H\}$) field generated by the (m,n)th ($m=1,\dots,M;\ n=1,\dots,N$) element of the MIMO array and the χ-polarized ($\chi\in\{V,H\}$) RX antenna of the lth ($l=1,\dots,L$) UE terminal [75–77]. Moreover, the multi-user interference power is equal to

$$\Pi^{(\chi)}_{\text{mui}}\left(\mathbf{r}_l\right)=\frac{\Pi_{tx}}{B}\left|\sum_{o=1,\,o\neq b}^{B}\ \sum_{m=1}^{M}\sum_{n=1}^{N}w^{(o,\psi)}_{m,n}\mathbf{G}^{(\psi,\chi)}\left(\omega;\mathbf{r}_l,\mathbf{r}_{m,n}\right)\right|^2, \tag{14.51}$$

to account for the power transmitted at the lth ($l=1,\dots,L$) UE in the position $\mathbf{r}_l$ and over the χ ($\chi\in\{V,H\}$) receiving polarization of its terminal, from the other $(B-1)$ beams generated by the MIMO array when feeding the sub-array TRMs with the coefficients $\mathbf{V}'\triangleq\{\mathbf{v}^{(o,\psi)};o=1,\dots,B;o\neq b;\ \psi\in\{V,H\}\}$. As for the noise power, it is assumed equal to $\Pi^{(\chi)}_{\eta}\left(\mathbf{r}_u\right)=\sigma^2$ for all UE terminals as well as polarization states. Substituting the above power expressions in (14.49) the SINR associated to the signal received by the χ-polarized antenna of the lth UE terminal is given by

$$\text{SINR}^{(\chi)}\left(\mathbf{r}_l\right)=\frac{\left|\sum_{m=1}^{M}\sum_{n=1}^{N}w^{(b,\psi)}_{m,n}\mathbf{G}^{(\psi,\chi)}\left(\omega;\mathbf{r}_l,\mathbf{r}_{m,n}\right)\right|^2}{\left|\sum_{o=1,\,o\neq b}^{B}\sum_{m=1}^{M}\sum_{n=1}^{N}w^{(o,\psi)}_{m,n}\mathbf{G}^{(\psi,\chi)}\left(\omega;\mathbf{r}_l,\mathbf{r}_{m,n}\right)\right|^2+\frac{B\sigma^2}{\Pi_{tx}}}. \tag{14.52}$$

From (14.52) the throughput capacity of the χ-polarization ($\chi \in \{V, H\}$) of the lth ($l = 1, ..., L$) UE [28] is derived

$$C^{(\chi)}\left(\mathbf{r}_l\right) = \log_2\left\{1 + SINR^{(\chi)}\left(\mathbf{r}_l\right)\right\}, \tag{14.53}$$

and the total *MU-MIMO sum-rate capacity* of the MIMO array in down-link is obtained as the sum of the L user throughput [73, 78]

$$C = \sum_{l=1}^{L} \sum_{\chi=V,H} C^{(\chi)}\left(\mathbf{r}_l\right). \tag{14.54}$$

The problem of synthesizing an irregular tiled array for MU-MIMO communication systems is then formulated as follows:

Capacity-Driven Modular Array Synthesis Problem -Given a MIMO array of $M \times N$ dual-polarized radiating elements centered at $\{\mathbf{r}_{m,n} = \left(0, y_m, z_n\right); m = 1, ..., M; n = 1, ..., N\}$ and i different configurations ($i = 1, ..., I$) of L UEs randomly located at the positions $\{\mathbf{r}_{l,i};\ l = 1, ..., L\}$ within the hexagonal cell, determine the optimal clustering of the array elements, $\mathbf{c}$, by using S sub-arrays chosen among a finite set of tiles/clusters shapes and the sets of sub-array complex coefficients for each ith ($i = 1, ..., I$) UEs configuration, $\widetilde{\mathbf{v}}_i^{(b,\psi)} = \left\{\widetilde{v}_{s,i}^{(b,\psi)};\ s = 1, ..., S\right\}(b = 1, ..., B;\ \psi \in \{V, H\})$ such that the average sum-rate capacity

$$\widehat{C} = \frac{1}{I}\sum_{i=1}^{I}\left\{\sum_{l=1}^{L}\sum_{\chi=V,H} C^{(\chi)}\left(\mathbf{r}_{l,i}\right)\right\} \tag{14.55}$$

is maximum and the power of the desired signal at the receiver position, $\mathbf{r}_{l,i}$ ($l = 1, ..., L$; $i = 1, ..., I$), for the χ-polarization ($\chi \in \{V, H\}$) is above a minimum threshold Π_{th}, namely

$$\Pi_{\text{des}}^{(\chi)}\left(\mathbf{r}_{l,i}\right) \geq \Pi_{\text{th}}, \tag{14.56}$$

to guarantee the connection of all UEs within the communication cell.

In the previous calculations to give statistically reliable evaluation of the down-link sum-rate capacity C, the positions of the L UEs, $\{\mathbf{r}_{l,i};\ l = 1, ..., L\}$, are varied over I different random configurations ($i = 1, ..., I$). The arising clustering problem is very challenging since the dimension of the solution space grows exponentially with the layout size [2, 19, 26]. Moreover, the solution tool must be chosen taking into account the assumptions/guidelines on the cluster type and shapes as well as the requirements in terms of geometrical or logical contiguity [2, 19, 22, 26]. Therefore, a nested random-search single-agent optimization (NRSSAO) method has been proposed in [28] to optimize the clustering configuration starting from a dictionary of cluster shapes while guaranteeing the contiguity and the modularity of the final clustered layouts. As an illustrative example, Fig. 14.9 reports the results of the NRSSAO optimization sparse modular layout where domino shaped tiles oriented either horizontally or vertically are used to assembly the BS array. More in detail, the example shows how the optimal clustering configuration changes when the capacity-driven synthesis approach (max C) is adopted, (Fig. 14.9a) with respect to optimizing traditional free-space figures of merit, such as the Directivity (max D) (Fig. 14.9b). Realistic simulation models [73] are adopted, considering a urban micro-cellular (UMi) propagation scenario in a "Madrid Grid" environment, according to the non line-of-sight International Mobile Telecommunications-Advanced (IMT-A) model ("NLOS IMT-A UMi"), where a BS comprising an array of $N \times M = 8 \times 8$ dual-polarized,

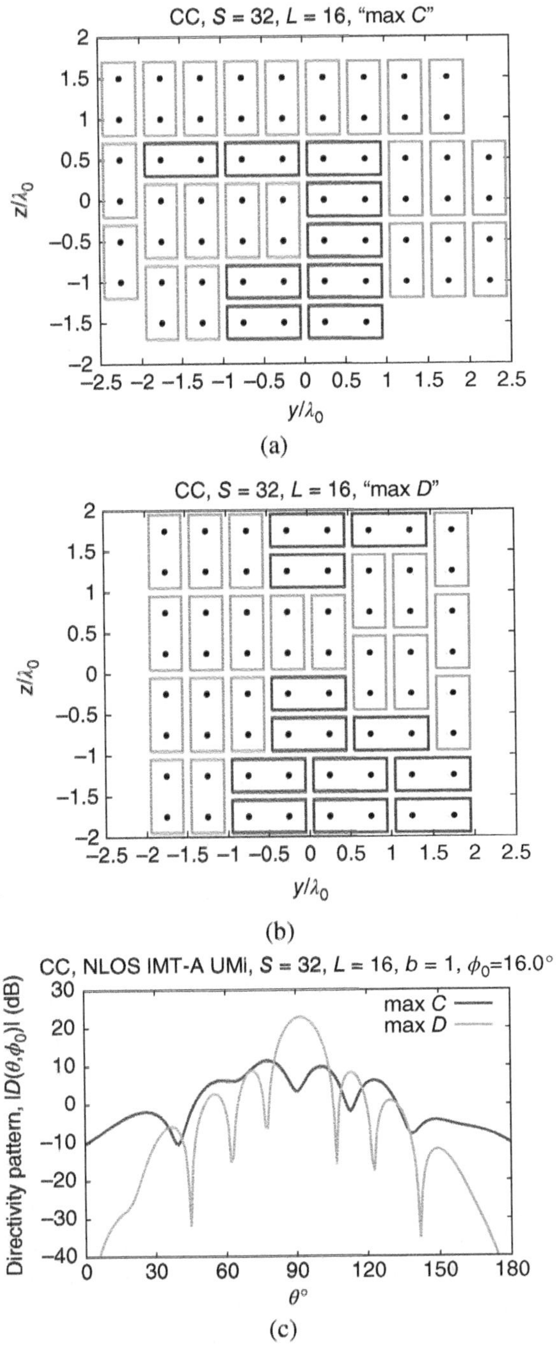

Figure 14.9 *Capacity-Driven Synthesis* (*Domino Tiles*, $d_y = d_z = 0.5\lambda$, $N \times M = 8 \times 4$, $S = 32$, $L = 16$, $I = 1$) – Clustering configurations synthesized with (a) the max C and (b) the max D methods. (c) Comparison of the directivity pattern of the $b = 1$ beam along the $\phi_0 = 16^\circ$ cut.

$\psi \in \{V, H\}$, half-wavelength spaced patch antennas resonating at $f = 2.0$ GHz, is positioned at 7 m above the ground. A Geometry-based stochastic method exploiting an approximated ray-tracing approach based on the *QuaDRiGa* implementation [79–81] has been used for the GF computation. Accordingly, the BS has been contiguously clustered into $S = 32$ domino sub-arrays, considering the same sub-array configuration for each polarization, thus halving the number of control points. By comparing the clustering configuration obtained with the capacity-driven clustering approach(Fig. 14.9a) with that from the application of the traditional max D formulation(Fig. 14.9b) one can observe that the dominoes arrangements of the max D solution follows a square aperture (Fig. 14.9b), while the maximization of $\widehat{C}$ leads to a rectangular aperture shape with a wider horizontal dimension (Fig. 14.9a) in order to allow an improved capability to exploit the multiple-scattering phenomena along the horizontal plane, where the majority of the UEs are usually located. Moreover, this is also confirmed by

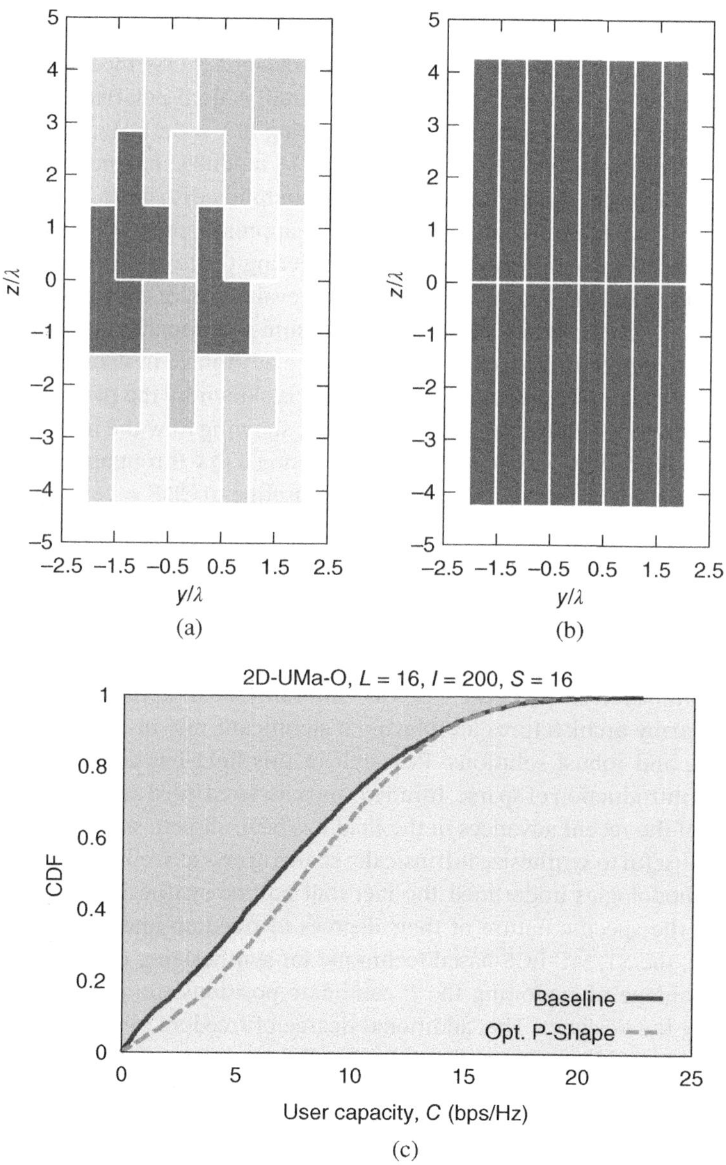

Figure 14.10 *Capacity-Driven Synthesis* (*P-shaped Tiles*, $d_y = 0.5\,\lambda$, $d_z = 0.7\,\lambda$, $N \times M = 8 \times 12$, $S = 16$, $L = 16$, $I = 200$) – Picture of (a) the optimal irregular tiling, (b) the baseline configuration, and (c) the *CDF* when adopting the optimal irregular tiling or the baseline configuration.

comparing the respective directivity patterns (e.g., the $b = 1$ beam along the $\phi_0 = 16°$ in Fig. 14.9c) showing how the max C pattern does not show a single main-lobe as in the max D approach, but instead multiple equivalent lobes with close magnitudes are generated to exploit multiple propagation paths.

Within a similar framework, a tiling methodology based on an enumerative search approach has been proposed in [40] to optimize aperiodic tilings (i.e., exact partitions of a fixed aperture shape with pre-fixed tiles) of rectangular BS array. Towards this end, either single or multiple tile shapes have been combined to guarantee the complete coverage of the aperture. As an instance of such a method, the synthesis of a $M \times N = 8 \times 12$ elements MIMO array, spaced $d_y = 0.5\,\lambda$ and $d_z = 0.7\,\lambda$, illuminating a realistic MU-MIMO communication cell site defined according to the 3GPP standards is shown [40]. The *3GPP Urban Macro (UMa)* line-of-sight (LOS) model with ISD = 500 m and the BS positioned 25 m above the ground is considered. A dual-polarized slant $\pm 45°$ ($H = +45$, $V = -45$) multi-layer stacked patch antenna, resonating at $f = 3.5$ GHz, is used as the radiating element. The goal has been to tile the array into $S = 16$ modules, having irregular P-shapes. The number of UEs has been chosen equal to $L = 16$, randomly distributed within the hexagonal cell according to $I = 200$ different configurations. The approach proposed in [40] allows to generate all the $T = 85,926$ tiling configurations exactly covering the array aperture by means of the Algorithm-X [82]. The resulting optimal tiling, after the evaluation of the whole set of tiling configurations $\mathbf{c}_t$ ($t = 1, \ldots, T$) performance, in terms average sum-rate capacity subject to the coverage condition (14.56), is reported in Fig. 14.10a. Moreover the advantage in using irregular tiled arrays with respect to the regular/baseline one (Fig. 14.10b) is shown by the plot of the Cumulative Density Function (CDF) [40] of the capacity performance, showing how the irregular P-shape tiling allows to reduce the probability of having UEs experiencing a low throughput (e.g., for $C \leq 5$ bps/Hz the CDF has been reduced from CDF = 38% of the baseline to CDF = 26% of the tiled array (Fig. 14.10c).

14.5 Final Remarks and Envisaged Trends

Phased arrays are increasingly employed in practical wireless communication applications. In this framework, unconventional array architectures are playing a significant role in driving this technology towards cost-effective and robust solutions. Research in this field has considerably grown in recent decades, with the introduction of sparse, thinned, and clustered/tiled architectures design methodologies. A review of the recent advances in the field has been shown, starting from sparseness promoting techniques useful to synthesize intrinsically sparse arrays as well as modular architectures. The reviewed methodologies underlined the fact that ad-hoc synthesis strategies must be developed according to the specific nature of their degrees of freedom and associated design challenges. More in detail, the ST/MT-BCS-based technique for synthesizing intrinsically sparse architectures has the advantage of exploiting the P candidate positions sampled within the available antenna aperture by the designer. This additional degree of freedom allows for the synthesis of irregular sparse layouts, achieving exceptional trade-offs between pattern matching accuracy and array sparsening. However, it is important to note that intrinsically sparse arrays can lead to complex beamforming networks since each element does not belong to a regular grid. Therefore, ad-hoc feeding routing becomes necessary. For this reason, clustered/tiled architectures are preferred when a simplified beamforming approach needs to be implemented. They provide a reasonable trade-off between pattern matching accuracy and the number of TRMs. Specifically, the D-CS method allows control over the size of the sub-arrays, enabling the realization of high modular architectures composed of a few types of modules/tiles. On the other hand, the TV-CS approach improves the matching with the reference pattern by relaxing the constraint on the size of the sub-arrays. Overall, the three presented sparseness-promoting methodologies enable achieving remarkable numerical efficiency, even when compared to the local search techniques employed by other sparse/clustering approaches. Synthesis times for arrays of hundreds of elements are less than one second [24, 27]. This finding reaffirms the expectations expressed in the computer science literature and further validates the effectiveness of the CS formulation for electromagnetics [11].

Recent state-of-the-art design methodologies for next generation wireless applications have been illustrated as well, to present the use of co-design strategies and capacity-oriented synthesis paradigms as effective approaches for addressing more realistic unconventional phased arrays. These methods aim at providing useful design tools that account for the "environment" (i.e., presence of EM effects and obstacles/objects) in the design phase, such as the mutual coupling effects of the single radiating element when embedded in the full array, or the multi-cell/multi-user environment of 5G/6G mobile communication sites.

Despite the effectiveness of the methods reviewed in this Chapter, many challenges and issues are still open from the methodological and practical perspectives. Indeed, future efforts are expected to deal with (i) the combination of intrinsically sparse and modular architectures in order to exploit the advantages of both types of unconventional architectures and to obtain better trade-offs between complexity and pattern performances; (ii) the use of more efficient/fast modelling tools (e.g., also based on artificial intelligence) [83–85] for the modeling of realistic EM environments, enabling the analysis of multiple and different wireless communication scenarios that can be considered during the synthesis phase, and (iii) the integration of innovative engineered materials, such as metamaterials, enabling the manipulation of EM wave propagation when combined and integrated with unconventional phased array architectures [42].

Furthermore, given the recent developments in various scientific fields, such as optical communication, photonics circuits [86], and quantum physics [87], the need for innovative phased array architectures and corresponding design methodologies in the near future is still growing.

Acknowledgments

This work benefited from the networking activities carried out within the Project "SPEED" (Grant No. 6721001) funded by National Science Foundation of China under the Chang-Jiang Visiting Professorship Program, the Project DICAM-EXC (Grant L232/2016) funded by the Italian Ministry of Education, Universities and Research (MUR) within the "Departments of Excellence 2023–2027" Program (CUP: E63C22003880001), the Project "AURORA – Smart Materials for Ubiquitous Energy Harvesting, Storage, and Delivery in Next Generation Sustainable Environments" funded by the Italian Ministry for Universities and Research within the PRIN-PNRR 2022 Program, and the Project "ICSC National Centre for HPC, Big Data and Quantum Computing (CNHPC)" funded by the European Union – NextGenerationEU within the PNRR Program (CUP: E63C22000970007). Views and opinions expressed are however those of the author(s) only and do not necessarily reflect those of the European Union or the European Research Council. Neither the European Union nor the granting authority can be held responsible for them. A. Massa wishes to thank E. Vico for her never-ending inspiration, support, guidance, and help.

References

1 Hong, W., Jiang, Z.H., Yu, C. et al. (2021). The role of millimeter-wave technologies in 5G/6G wireless communications. *IEEE Journal of Microwaves* 1 (1): 101–122.

2 Rocca, P., Oliveri, G., Mailloux, R.J., and Massa, A. (2016). Unconventional phased array architectures and design methodologies - a review. *Proceedings of the IEEE* 104 (3): 544–560.

3 Trucco, A. and Murino, V. (1999). Stochastic optimization of linear sparse arrays. *IEEE Journal of Oceanic Engineering* 24 (3): 291–299.

4 Liu, Y., Liu, Q.H., and Nie, Z. (2010). Reducing the number of elements in the synthesis of shaped-beam pattern by the forward-backward matrix pencil method. *IEEE Transactions on Antennas and Propagation* 58 (2): 604–608.

5 Goudos, S.K., Siakavara, K., Samaras, T. et al. (2011). Sparse linear array synthesis with multiple constraints using differential evolution with strategy adaptation. *IEEE Antennas and Wireless Propagation Letters* 10: 670–673.

6 Oliveri, G. and Massa, A. (2011). Bayesian compressive sampling for pattern synthesis with maximally sparse non-uniform linear arrays. *IEEE Transactions on Antennas and Propagation* 59 (2): 467–481.

7 Oliveri, G., Carlin, M., and Massa, A. (2012). Complex-weight sparse linear array synthesis by Bayesian compressive sampling. *IEEE Transactions on Antennas and Propagation* 60 (5): 2309–2326.

8 Viani, F., Oliveri, G., and Massa, A. (2013). Compressive sensing pattern matching techniques for synthesizing planar sparse arrays. *IEEE Transactions on Antennas and Propagation* 61 (9): 4577–4587.

9 Oliveri, G., Bekele, E.T., Robol, F., and Massa, A. (2014). Sparsening conformal arrays through a versatile BCS-based method. *IEEE Transactions on Antennas and Propagation* 62 (4): 1681–1689.

10 Oliveri, G. and Moriyama, T. (2015). Hybrid PSO-CP technique for the synthesis of non-uniform linear arrays with maximum directivity. *Journal of Electromagnetic Waves and Applications* 29 (1): 113–123.

11 Massa, A., Rocca, P., and Oliveri, G. (2015). Compressive sensing in electromagnetics - a review. *IEEE Antennas and Propagation Magazine* 57 (1): 224–238.

12 Shen, H., Wang, B., and Li, X. (2017). Shaped-beam pattern synthesis of sparse linear arrays using the unitary matrix pencil method. *IEEE Antennas and Wireless Propagation Letters* 16: 1098–1101.

13 Pavlenko, T., Reustle, C., Dobrev, Y. et al. (2017). Design and optimization of sparse planar antenna arrays for wireless 3-D local positioning systems. *IEEE Transactions on Antennas and Propagation* 65 (12): 7288–7297.

14 Pinchera, D., Migliore, M.D., Schettino, F. et al. (2018). An effective compressed-sensing inspired deterministic algorithm for sparse array synthesis. *IEEE Transactions on Antennas and Propagation* 66 (1): 149–159.

15 Ahmed, A. and Zhang, Y.D. (2021). Generalized non-redundant sparse array designs. *IEEE Transactions on Signal Processing* 69: 4580–4594.

16 Gu, P., He, Z., Xu, J. et al. (2021). Design of wide scanning sparse planar array using both matrix-pencil and space-mapping methods. *IEEE Antennas and Wireless Propagation Letters* 20 (2): 140–144.

17 Haupt, R.L. (2007). Optimized weighting of uniform subarrays of unequal size. *IEEE Transactions on Antennas and Propagation* 55 (4): 1207–1210.

18 Manica, L., Rocca, P., Martini, A., and Massa, A. (2008). An innovative approach based on a tree-searching algorithm for the optimal matching of independently optimum sum and difference excitations. *IEEE Transactions on Antennas and Propagation* 56 (1): 58–66.

19 Manica, L., Rocca, P., and Massa, A. (2009). Design of subarrayed linear and planar array antennas with SLL control based on an excitation matching approach. *IEEE Transactions on Antennas and Propagation* 57 (6): 1684–1691.

20 Mailloux, R.J., Santarelli, S.G., Roberts, T.M., and Luu, D. (2009). Irregular polyomino-shaped subarrays for space-based active arrays. *International Journal of Antennas and Propagation* 2009: Article ID 956524.

21 Goudos, S.K., Gotsis, K.A., Siakavara, K. et al. (2013). A multi-objective approach to subarrayed linear antenna arrays design based on memetic differential evolution. *IEEE Transactions on Antennas and Propagation* 61 (6): 3042–3052.

22 Rocca, P., Mailloux, R.J., and Toso, G. (2015). GA-based optimization of irregular sub-array layouts for wideband phased arrays design. *IEEE Antennas and Wireless Propagation Letters* 14: 131–134.

23 Yang, X., Xi, W., Sun, Y. et al. (2015). Optimization of subarray partition for large planar phased array radar based on weighted K-means clustering method. *IEEE Journal on Selected Topics in Signal Processing* 9 (8): 1460–1468.

24 Oliveri, G., Salucci, M., and Massa, A. (2016). Synthesis of modular contiguously clustered linear arrays through a sparseness-regularized solver. *IEEE Transactions on Antennas and Propagation* 64 (10): 4277–4287.

25 Avser, B., Pierro, J., and Rebeiz, G.M. (2016). Random feeding networks for reducing the number of phase shifters in limited-scan arrays. *IEEE Transactions on Antennas and Propagation* 64 (11): 4648–4658.

26 Oliveri, G., Gottardi, G., Robol, F. et al. (2017). Co-design of unconventional array architectures and antenna elements for 5G base stations. *IEEE Transactions on Antennas and Propagation* 65 (12): 6752–6767.

27 Anselmi, N., Gottardi, G., Oliveri, G., and Massa, A. (2019). A total-variation sparseness-promoting method for the synthesis of contiguously clustered linear arrays. *IEEE Transactions on Antennas and Propagation* 67 (7): 4589–4601.

28 Oliveri, G., Gottardi, G., and Massa, A. (2019). A new meta-paradigm for the synthesis of antenna arrays for future wireless communications. *IEEE Transactions on Antennas and Propagation* 67 (6): 3774–3788.

29 Oliveri, G., Gottardi, G., Anselmi, N., and Massa, A. (2022). Capacity-driven low-interference fast beam synthesis for next generation base stations. *IEEE Transactions on Antennas and Propagation* 70 (6): 4472–4484.

30 Anselmi, N., Rocca, P., and Massa, A. (2017). Irregular phased array tiling by means of analytic schemata-driven optimization. *IEEE Transactions on Antennas and Propagation* 65 (9): 4495–4510.

31 Ma, Y., Yang, S., Chen, Y. et al. (2019). Pattern synthesis of 4-D irregular antenna arrays based on maximum-entropy model. *IEEE Transactions on Antennas and Propagation* 67 (5): 3048–3057.

32 Dong, W., Xu, Z., Liu, X. et al. (2020). Irregular subarray tiling via heuristic iterative convex relaxation programming. *IEEE Transactions on Antennas and Propagation* 68 (4): 2842–2852.

33 Rocca, P., Anselmi, N., Polo, A., and Massa, A. (2020). Modular design of hexagonal phased arrays through diamond tiles. *IEEE Transactions on Antennas and Propagation* 68 (5): 3598–3612.

34 Rocca, P., Anselmi, N., Polo, A., and Massa, A. (2022). Pareto-optimal domino-tiling of orthogonal polygon phased arrays. *IEEE Transactions on Antennas and Propagation* 70 (5): 3329–3342.

35 Vigano, M.C., Toso, G., Caille, C.M., and Lager, I.E. (2009). Sunflower array antenna with adjustable density taper. *International Journal of Antennas and Propagation* 2009: Article ID 624035.

36 Karademir, S., Prokopyev, O.A., and Mailloux, R.J. (2015). Irregular polyomino tiling via integer programming with application in phased array antenna design. *Journal of Global Optimization* 65 (2): 137–173.

37 Rupakula, B., Aljuhani, A.H., and Rebeiz, G.M. (2020). Limited scan-angle phased-arrays using randomly-grouped subarrays and reduced number of phase-shifters. *IEEE Transactions on Antennas and Propagation* 68 (1): 70–80.

38 Bekers, D.J., Jacobs, S., Monni, S. et al. (2019). A Ka-band spaceborne synthetic aperture radar instrument: a modular sparse array antenna design. *IEEE Antennas and Propagation Magazine* 61 (5): 97–104.

39 Rocca, P., Anselmi, N., Polo, A., and Massa, A. (2020). An irregular two-sizes square tiling method for the design of isophoric phased arrays. *IEEE Transactions on Antennas and Propagation* 68 (6): 4437–4449.

40 Anselmi, N., Rocca, P., Feuchtinger, S. et al. (2022). Optimal capacity-driven design of aperiodic clustered phased arrays for multi-user MIMO communication systems. *IEEE Transactions on Antennas and Propagation* 70 (7): 5491–5505.

41 Ma, Y., Yang, F., Chen, K. et al. (2022). An irregular tiled array technique for massive MIMO systems. *IEEE Transactions on Wireless Communications* 21 (6): 4509–4521.

42 Oliveri, G., Salucci, M., Lombardi, R. et al. (2022). WAIM-enhanced dual-polarization sub-6 GHz wide-scan array for next generation base stations. *IEEE Transactions on Antennas and Propagation* 70 (7): 5506–5520.

43 Ji, S., Xue, Y., and Carin, L. (2008). Bayesian compressive sensing. *IEEE Transactions on Signal Processing* 56 (6): 2346–2356.

44 Ji, S., Dunson, D., and Carin, L. (2009). Multi-task compressive sampling. *IEEE Transactions on Signal Processing* 57 (1): 92–106.

45 Li, C. (2009). An efficient algorithm for total variation regularization with applications to the single pixel camera and compressive sensing. Master thesis. Rice University.

46 Li, C. (2011). Compressive sensing for 3D data processing tasks: applications, models, and algorithms. PhD thesis. Rice University.

47 Tipping, M.E. and Faul, A.C. 2003, "Fast marginal likelihood maximization for sparse Bayesian models," in Proceedings of the 9th International Workshop on Artificial Intelligence and Statistics (ed. C.M. Bishop and B. J. Frey), http://citeseer.ist.psu.edu/611465.html.

48 Isernia, T., Bucci, O.M., and Fiorentino, N. (1998). Shaped beam antenna synthesis problems: feasibility criteria and new strategies. *Journal of Electromagnetic Waves and Applications* 12: 103–137.

49 Mailloux, R.J. (2005). *Phased Array Antenna Handbook*, 2e. Norwood, MA: Artech House.

50 Bucci, O.M. and Franceschetti, G. (1989). On the degrees of freedom of scattered fields. *IEEE Transactions on Antennas and Propagation* 37: 918–926.

51 Li, G. and Burkholder, R.J. (2015). Hybrid matching pursuit for distributed through-wall radar imaging. *IEEE Transactions on Antennas and Propagation* 63 (4): 1701–1711.

52 Senyuva, R.V., Ozdemir, O., Kurt, G.K., and Anarim, E. (2016). Electromagnetic imaging of closely spaced objects using matching pursuit based approaches. *IEEE Antennas and Wireless Propagation Letters* 15: 1179–1182.

53 Chen, M.S., Liu, F.L., Du, H.M., and Wu, X.L. (2011). Compressive sensing for fast analysis of wide-angle monostatic scattering problems. *IEEE Antennas and Wireless Propagation Letters* 10: 1243–1246.

54 Wang, J. and Shim, B. (2012). On the recovery limit of sparse signals using orthogonal matching pursuit. *IEEE Transactions on Signal Processing* 60 (9): 4973–4976.

55 Chang, L.H. and Wu, J.Y. (2014). An improved RIP-based performance guarantee for sparse signal recovery via orthogonal matching pursuit. *IEEE Transactions on Information Theory* 60 (9): 5702–5715.

56 Oliveri, G., Anselmi, N., and Massa, A. (2014). Compressive sensing imaging of non-sparse 2D scatterers by a total-variation approach within the Born approximation. *IEEE Transactions on Antennas and Propagation* 62 (10): 5157–5170.

57 Goldfarb, D. and Yin, W. (2005). Second-order cone programming methods for total variation based image restoration. *SIAM Journal on Scientific Computing* 27 (2): 622–645.

58 Candes, E.J. and Tao, T. (2005). Decoding by linear programming. *IEEE Transactions on Information Theory* 51 (12): 4203–4215.

59 Bioucas-Dias, J. and Figueiredo, M. (2007). A new TwIST: two-step iterative thresholding algorithm for image restoration. *IEEE Transactions on Image Processing* 16 (12): 2992–3004.

60 Becker, S., Bobin, J., and Candès, E.J. (2011). NESTA: a fast and accurate first-order method for sparse recovery. *SIAM Journal on Imaging Sciences* 4 (1): 1–39. https://doi.org/10.1137/090756855.

61 Nash, S. and Sofer, A. (1996). *Linear and Non-Linear Programming*. Burr Ridge, IL: McGraw-Hill Higher Education.

62 Wolpert, D.H. and Macready, W.G. (1997). No free lunch theorems for optimization. *IEEE Transactions on Evolutionary Computation* 1 (1): 67–82.

63 Deb, K., Mohan, M., and Mishra, S. (2003). A Fast Multi-Objective Evolutionary Algorithm for Finding Well-Spread Pareto-Optimal Solutions. *KanGAL Report, no. 2003002*, 1-18. College of Engineering, Michigan State University.

64 Deb, K., Mohan, M., and Mishra, S. (2005). Evaluating the ε-domination based multi-objective evolutionary algorithm for a quick computation of Pareto-optimal solutions. *Evolutionary Computation* 13 (4): 501–525.

65 Lizzi, L., Viani, F., Azaro, R., and Massa, A. (2008). A PSO-driven spline-based shaping approach for ultra-wideband (UWB) antenna synthesis. *IEEE Transactions on Antennas and Propagation* 56 (8): 2613–2621.

66 Shinohara, N. (2013). Beam control technologies with a high-efficiency phased array for microwave power transmission in Japan. *Proceedings of the IEEE* 101 (6): 1448–1463.

67 Oliveri, G., Poli, L., and Massa, A. (2013). Maximum efficiency beam synthesis of radiating planar arrays for wireless power transmission. *IEEE Transactions on Antennas and Propagation* 61 (5): 2490–2499.

68 Yeap, S.B., Qing, X., and Chen, Z.N. (2015). 77-GHz dual-layer transmit-array for automotive radar applications. *IEEE Transactions on Antennas and Propagation* 63 (6): 2833–2837.

69 Xu, J., Hong, W., Zhang, H. et al. (2017). An array antenna for both long- and medium-range 77 GHz automotive radar applications. *IEEE Transactions on Antennas and Propagation* 65 (12): 7207–7216.

70 Chen, P., Hong, W., Zhang, H. et al. (2015). Virtual phase shifter array and its application on Ku band mobile satellite reception. *IEEE Transactions on Antennas and Propagation* 63 (4): 1408–1416.

71 Hong, W., Jiang, Z.H., Yu, C. et al. (2017). Multibeam antenna technologies for 5G wireless communications. *IEEE Transactions on Antennas and Propagation* 65 (12): 6231–6249.

72 Quist, B.T. and Jensen, M.A. (2009). Optimal antenna radiation characteristics for diversity and MIMO systems. *IEEE Transactions on Antennas and Propagation* 57 (11): 3474–3481.

73 Kyosti, P., Lehtomaki, J., Medbo, J., and Latva-aho, M. (2017). Map-based channel model for evaluation of 5G wireless communication systems. *IEEE Transactions on Antennas and Propagation* 65 (12): 6491–6504.

74 Carlson, C.A., Crilly, P.B., and Rutledge, J.C. (2002). *Communication Systems: An Introduction to Signals and Noise in Electrical Communication*, 4e. New York: McGraw Hill.

75 Ong, C.-J. and Tsang, L. (2003). Time domain Green's functions for lossy and dispersive multilayered media. *IEEE Microwave and Wireless Components Letters* 13 (9): 399–401.

76 Stefanski, T.P. and Reichel, B. (2017). Analytical expression for the time-domain Green's function of a discrete plane wave propagating in a 3-D FDTD grid. *IEEE Transactions on Antennas and Propagation* 65 (7): 3607–3614.

77 Farr, E.G. (2018). Characterizing antennas in the time and frequency domains. *IEEE Antennas and Propagation Magazine* 60 (1): 106–110.

78 Kaltenberger, F., Gesbert, D., Knopp, R., and Kountouris, M. (2008). Correlation and capacity of measured multi-user MIMO channels. *2008 IEEE 19th International Symposium on Personal, Indoor and Mobile Radio Communications*, Cannes, France, 1–5.

79 Jaeckel, S., Borner, K., Thiele, L., and Jungnickel, V. (2012). A geometric polarization rotation model for the 3-D spatial channel model. *IEEE Transactions on Antennas and Propagation* 60 (12): 5966–5977.

80 Jaeckel, S., Raschkowski, L., Borner, K., and Thiele, L. (2014). QuaDRiGa: a 3-D multi-cell channel model with time evolution for enabling virtual field trials. *IEEE Transactions on Antennas and Propagation* 62 (6): 3242–3256.

81 Jaeckel, S. (2017). Quasi-deterministic channel modeling and experimental validation in cooperative and massive MIMO deployment topologies. PhD dissertation. Ilmenau: TU Ilmenau.

82 Knuth, D.E. (2000). Dancing links. *Millenial Perspectives in Computer Science*, 187–214.

83 Salucci, M., Oliveri, G., Hannan, M.A., and Massa, A. (2022). System-by-Design paradigm-based synthesis of complex systems: the case of spline-contoured 3D radomes. *IEEE Antennas and Propagation Magazine* 64 (1): 72–83.

84 Massa, A. and Salucci, M. (2022). On the design of complex EM devices and systems through the System-by-Design paradigm - a framework for dealing with the computational complexity. *IEEE Transactions on Antennas and Propagation* 70 (2): 1328–1343.

85 Sotiroudis, S.P., Boursianis, A.D., Goudos, S.K., and Siakavara, K. (2022). From spatial urban site data to path loss prediction: an ensemble learning approach. *IEEE Transactions on Antennas and Propagation* 70 (7): 6101–6105.

86 Milanizadeh, M., Toso, F., Ferrari, G. et al. (2021). Coherent self-control of free-space optical beams with integrated silicon photonic meshes. *Photonics Research* 9 (11): 11002196.

87 Rocca, P., Anselmi, N., Oliveri, G. et al. (2021). Antenna array thinning through quantum Fourier transform. *IEEE Access* 9: 124313–124323.

15

MIMO Communication with Sparse Arrays

Ahmed Alkhateeb[1], Xiang Gao[2], and Elias Aboutanios[3]

[1] *School of Electrical, Computer and Energy Engineering, Arizona State University, Tempe, AZ, USA*
[2] *National Key Laboratory of Science and Technology on Communications, University of Electronic Science and Technology of China, Sichuan, China*
[3] *School of Electrical Engineering and Telecommunications, University of New South Wales, Kensington, NSW, Australia*

15.1 Introduction

Motivated by its beamforming and multiplexing gains, multiple-input multiple-output (MIMO) has become a key characteristic of current and future wireless communication systems. The need to further enhance these gains and provide more flexibility in the design of the MIMO signal processing functions motivates the use of increasingly larger arrays. However, increasing the number of antennas in the traditional way that assigns a complete radio frequency (RF) chain to each antenna brings critical challenges to the cost and power consumption. These challenges have spurred the development of new architectures for communication systems such as MIMO transceivers with low-resolution analog-to-digital converters (ADCs) Mezghani and Nossek [1], Mo et al. [2], and hybrid analog–digital architectures El Ayach et al. [3], Molisch et al. [4], Heath et al. [5], Alkhateeb et al. [6]. For the general fully–digital MIMO architecture, which allocates an RF chain per antenna, inducing array sparsity by selecting certain antenna elements or sub-arrays, can lead to a good compromise between cost, power consumption, and targeted performance metrics [7–9].

Sparse arrays differ from conventional uniform arrays in that (i) they do not have equidistant antenna spacing and (ii) the adjacent antenna spacing can be larger than half a wavelength. With the same aperture size, the sparse arrays consist of fewer antennas than the corresponding full arrays. The sparse arrays thus lead to lower spatial channel correlation between the antennas, advantageous to MIMO communications, particularly when the channels are also sparse.[1] With fewer antennas, sparse arrays require fewer transceiver chains in the hardware, resulting in a reduction of MIMO system complexity and power consumption [7–9]. In other words, we can use sparse arrays to achieve MIMO performance close to that of full arrays, while significantly reducing the complexity of MIMO systems.

In this chapter, we provide an overview of the different sparse array architectures for MIMO communication systems and their design approaches and draw insights into their potential gains in various communication scenarios. We divide these sparse array MIMO architectures into two main categories, namely, fully digital architectures with sparse arrays and hybrid analog/digital

1 Sparse channel, in contrast to rich and isotropic scattering channel, means that multipaths are sparse in a propagation environment.

Sparse Arrays for Radar, Sonar, and Communications, First Edition. Edited by Moeness G. Amin.

transceiver architectures with sparse arrays. We focus on fully digital architectures with sparse arrays in Section 15.2, whereas the hybrid analog/digital architectures are discussed in Section 15.3.

Notations: We use the following notation throughout this chapter: $\mathbf{A}$ is a matrix, $\mathbf{a}$ is a vector, a is a scalar, and $\mathcal{A}$ is a set. $|\mathbf{A}|$ is the determinant of $\mathbf{A}$, $\|\mathbf{A}\|_F$ is its Frobenius norm, whereas $\mathbf{A}^T, \mathbf{A}^H, \mathbf{A}^*$, $\mathbf{A}^{-1}, \mathbf{A}^{\dagger}$ are its transpose, Hermitian (conjugate transpose), conjugate, inverse, and pseudo-inverse respectively. $[\mathbf{A}]_{\mathcal{R},:}$ ($[\mathbf{A}]_{:,\mathcal{R}}$) are the rows (columns) of the matrix $\mathbf{A}$ with indices in the set $\mathcal{R}$, and diag($\mathbf{a}$) is a diagonal matrix with the entries of $\mathbf{a}$ on its diagonal. $\mathbf{I}$ is the identity matrix and $\mathbf{1}_N$ is the N-dimensional all-ones vector. $\mathbf{A} \circ \mathbf{B}$ is the Khatri–Rao product of $\mathbf{A}$ and $\mathbf{B}$, $\mathbf{A} \otimes \mathbf{B}$ is the Kronecker product of $\mathbf{A}$, and $\mathbf{B}$, and $\mathbf{A} \odot \mathbf{B}$ denotes the Hadamard product of $\mathbf{A}$, and $\mathbf{B}$. $\mathcal{N}(\mathbf{m}, \mathbf{R})$ is a complex Gaussian random vector with mean $\mathbf{m}$ and covariance $\mathbf{R}$. $\mathbb{E}[\cdot]$ is used to denote expectation.

15.2 Fully Digital Architectures with Sparse Arrays

When the number of antennas scales up to tens and hundreds in MIMO communications, so-called "massive MIMO," the aperture of antenna arrays becomes relatively large. For example, a 64-antenna uniform linear array (ULA) with an adjacent antenna spacing of half a wavelength may span several meters in space, e.g., 2.7 m at the 3.5 GHz frequency band. On the one hand, arrays with large aperture have relatively high spatial resolution to wireless channels, resulting in better MIMO system performance in terms of spectral efficiency [10, 11]. On the other hand, the antennas on large full arrays may not contribute equally to MIMO system performance [12]. Some antennas may be at more "important" positions, thus contributing more significantly than others. This gives us the opportunity to select the "important" antennas and discard the "trivial" ones, while still harvesting performance at sufficiently high levels. By performing such antenna selection, the conventional full arrays that are commonly used in massive MIMO systems become sparse. We call this type of architectures that still allocate an RF chain per selected antenna fully digital architectures with sparse arrays. This is because all the signal processing, such as beamforming and channel estimation, is fully done in the digital baseband domain.

Designing sparse arrays for MIMO communications can be difficult, as the array geometry should be adapted to the particular propagation environment where the system is deployed. Such adaptive designing is almost infeasible for multi-user MIMO in mobility scenarios, as the users may move from one propagation environment to another. For point-to-point MIMO, sparse arrays with fixed geometries are also hard to adapt to propagation conditions when strong reflectors or scatterers in the channels move. Instead, constructing sparse arrays through antenna selection on a full array can be a feasible approach. By setting a selection criterion, e.g., maximizing MIMO channel capacity or system rate, different subsets of antennas on the full array are selected to form sparse arrays with different geometries, which adapt to different propagation conditions. In particular, consider N_t antennas composing a full array and $L_t \ll N_t$ transceivers deployed in the system. Then, L_t out of N_t antennas are selected in, e.g., every transmission frame, forming L_t-antenna sparse arrays. The antenna selection can be realized by using an $N_t \times L_t$ RF switch in the frontend system. In this approach, instead of designing sparse arrays with fixed geometries, we need to design (i) the full antenna array, (ii) the RF switching network, and (iii) the antenna selection approach.

The radiation pattern of the full array should be designed to provide signal coverage over required distances. The array geometry then depends on the spatial properties of the propagation channels. Generally speaking, arrays spanning the horizontal direction can resolve spatial channels in azimuth angles, while arrays spanning the vertical direction can resolve elevation angles. As mentioned above, arrays with larger aperture have higher angular resolution. In the following, we discuss the other two points, namely, the switching structures in Section 15.2.1 and the antenna selection criteria and approaches in Section 15.2.2.

15.2.1 Architectures

This section provides a brief description of fully digital architectures with sparse arrays. As described in Fig. 15.1, fully digital MIMO architectures adopt a number of RF chains equal to the number of selected antennas. If N_t and L_t denote the total number of antennas and the number of RF chains, then in fully digital sparse MIMO systems, the signal transmitted by L_t RF chains is routed to L_t (out of the N_t) antennas, with each RF chain connected to only one antenna. Digital beamforming is thus performed based on the acquired channel state information (CSI). Next, we describe a multi-user MIMO system with full-digital architectures. Multi-user MIMO systems are widely adopted in 5G systems. For point-to-point MIMO systems, similar MIMO architectures and signal processing principles can be applied.

Consider a single-cell multi-user MIMO scenario, as shown in Fig. 15.1, a base station equipped with an N_t-antenna full array serves U users in the same time–frequency resource. An $N_t \times L_t$ RF switch is used to select L_t antennas out of the N_t, and connect to L_t RF chains. That is to say, we construct L_t-antenna sparse arrays with different geometries from the N_t-antenna full array. Obviously, the number of possible antenna combinations for constructing sparse arrays is $\binom{N_t}{L_t}$. When both N_t and L_t are large, antenna selection has the potential to exploit the large spatial degrees of freedom (DOFs) and outputs the *best* sparse array for the current propagation environment.

In the uplink, the CSI is estimated for all U users over all N_t base station antennas. Based on this CSI, the optimal antenna subset can be selected. In the downlink, the beamforming and data transmission are performed based on the optimal L_t-antenna sparse array. Thus, the baseband signal model can be written as

$$\mathbf{y} = \mathbf{H}^{(L_t)}\mathbf{F}\mathbf{x} + \mathbf{n}, \tag{15.1}$$

where $\mathbf{x}$ is the vector containing the data symbols for the U users, $\mathbf{F} \in \mathbb{C}^{L_t \times U}$ is the digital beamforming matrix, also known as the precoding matrix, $\mathbf{H}^{(L_t)} \in \mathbb{C}^{U \times L_t}$ is the propagation channel matrix, the

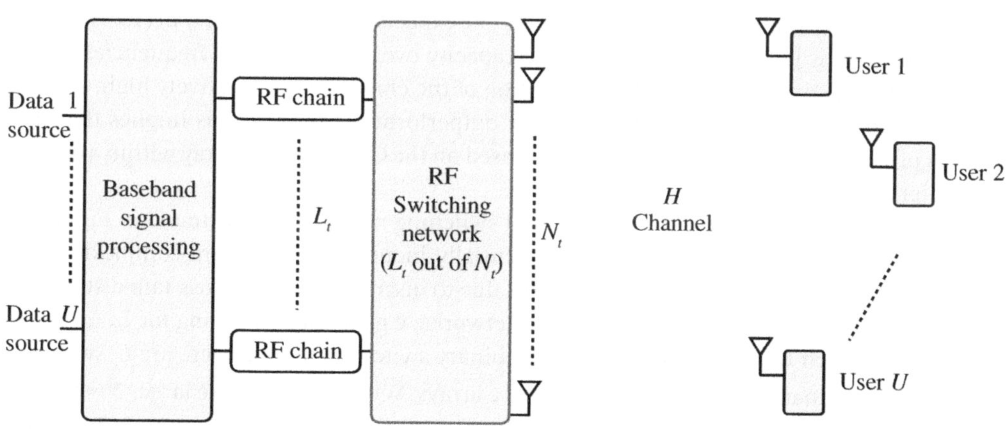

Figure 15.1 Block diagrams of the fully digital architecture with a sparse array. The RF switching network selects L_t out of the N_t antennas and connect each selected antenna with a dedicated RF chain.

superscript (L_t) indicates that L_t columns are selected from the full channel matrix $\mathbf{H}\in\mathbb{C}^{U\times N_t}$, $\mathbf{y}$ is the receive signals at the U users, and $\mathbf{n}$ represents the noise vector.

15.2.2 Design Criteria and Signal Processing Approaches

The *best* sparse array can be constructed using different criteria. One approach is to select the antenna subset that maximizes the MIMO channel capacity. This approach requires full CSI knowledge of all N_t antennas, bringing challenges on uplink channel estimation, as there are only L_t RF chains [6, 7]. Another approach is to measure the received signal strength on the N_t antennas and select the L_t with the strongest power. From the system implementation point of view, such power-based selection is much simpler than that based on full CSI. The price is, however, that the power-based selection leads to inferior MIMO performance since it only partially exploits the spatial DOFs. As N_t and L_t increase, i.e., in massive MIMO, the performance loss may become small. In the following, we discuss antenna selection based on full CSI to maximize MIMO channel capacity.

With the defined signal model in (15.1), the multi-user MIMO downlink channel capacity, achieved by dirty-paper coding (DPC) [13], is given by [14]

$$C_{\mathrm{DPC}} = \max_{\mathrm{tr}\{\mathbf{P}\}=1} \log_2 \det \left[\mathbf{I} + \rho\left(\mathbf{H}^{(L_t)}\right)^H \mathbf{P}\mathbf{H}^{(L_t)}\right], \tag{15.2}$$

where $\mathbf{P}$ is a $U \times U$ diagonal matrix for transmit power allocation to the U users, $\mathrm{tr}\{\cdot\}$ represents the trace of a matrix, and ρ is a power factor representing the ratio of the transmit power and the noise power at the receivers. The sparse array can therefore be obtained by the following optimization,

$$\left\{\mathbf{P}_{\mathrm{opt}}, \Delta_{\mathrm{opt}}\right\} = \arg \max_{\substack{\mathrm{tr}\{\mathbf{P}\}=1 \\ \Delta_i\in\{0,1\} \\ \mathrm{tr}\{\Delta\}=L_t}} \mathbb{E}\left\{\log_2 \det \left[\mathbf{I} + \rho\mathbf{P}\mathbf{H}\Delta\mathbf{H}^H\right]\right\}, \tag{15.3}$$

where Δ is an $N_t \times N_t$ diagonal matrix used for selecting L_t columns from the full channel matrix $\mathbf{H}$, and it has binary elements on the diagonal,

$$\Delta_i = \begin{cases} 1, & \text{selected} \\ 0, & \text{otherwise} \end{cases} \tag{15.4}$$

indicating whether the ith antenna is selected or not. The expectation notation $\mathbb{E}\{\cdot\}$ in (15.3) means that the optimization is performed on the average capacity over, e.g., time and frequencies. Such averaging requires the time and frequency correlation of the channel to be relatively high, so that the selected antenna combination can significantly outperform the others. This implies that the construction of sparse arrays should be performed based on the CSI of the full array within a coherence time interval and over a coherence bandwidth.

The optimization in (15.3) requires an $N_t \times L_t$ full switching network, as illustrated in Fig. 15.2. With large N_t and L_t, however, full switching potentially incurs high hardware complexity and significant loss in output signal quality, for example, due to insertion loss and cross-talk distortion. Low-complexity alternatives are partial switching networks, e.g., binary switching for $L_t = \frac{N_t}{2}$. As shown in Fig. 15.3, each RF chain is connected to a binary switch, so in total there are L_t switches and 2^{L_t} antenna combinations for constructing sparse arrays. When N_t and L_t are large, $2^{L_t} \ll \binom{N_t}{L_t}$, so the hardware complexity and signal distortion are much smaller compared to the full switch. We note that the binary switching systems are also discussed in Chapter 7 of this book, but in the context of radar applications and maximizing signal-to-interference and noise ratio [15].

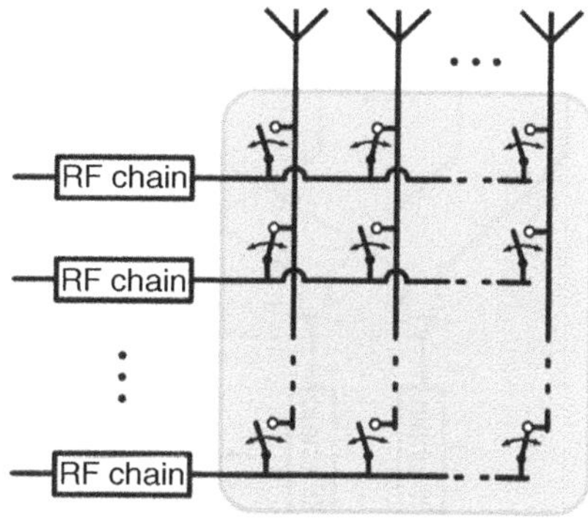

Figure 15.2 Illustration of a full switching network.

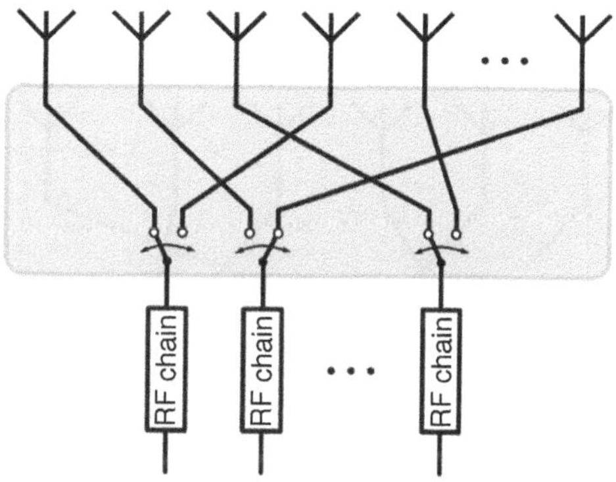

Figure 15.3 Illustration of a binary switching network.

With binary switching, the optimization problem for antenna selection becomes

$$\{\mathbf{P}_{\text{opt}}, \Delta_{\text{opt}}\} = \arg \max_{\substack{\text{tr}\{\mathbf{P}\}=1 \\ \Delta_i,\Delta_j \in \{0,1\} \\ \Delta_i+\Delta_j=1 \\ i\neq j}} \mathbb{E}\left\{\log_2 \det \left[\mathbf{I} + \rho\mathbf{P}\mathbf{H}\Delta\mathbf{H}^H\right]\right\}, \tag{15.5}$$

assuming that the ith and jth antennas, $i\neq j$, are paired. Depending on which two antennas are paired, there are different configurations of binary switching [16]. Figures 15.4–15.6 show three configurations. The first two are highly structured, with switching between adjacent antennas and switching between corresponding antennas in sub-arrays. The third uses pseudo-random assignments of the antennas to the switches.

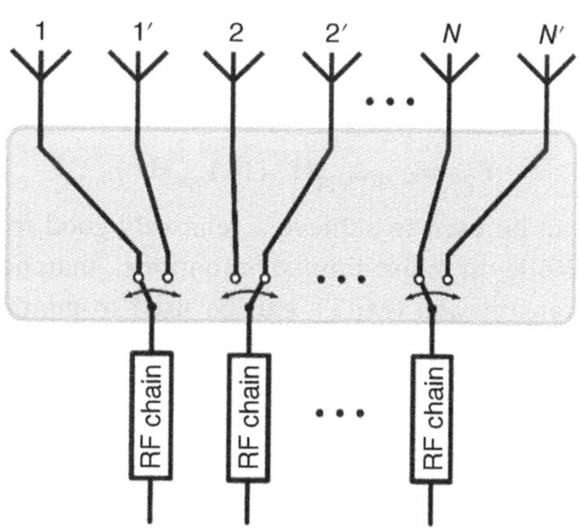

Figure 15.4 Illustration of binary switching on adjacent antennas.

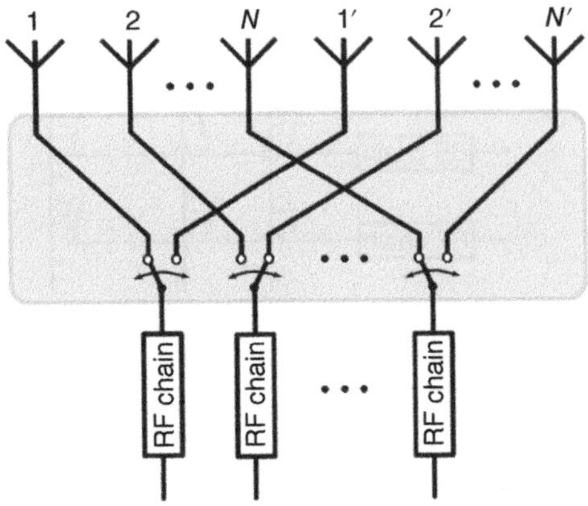

Figure 15.5 Illustration of binary switching on separate antennas.

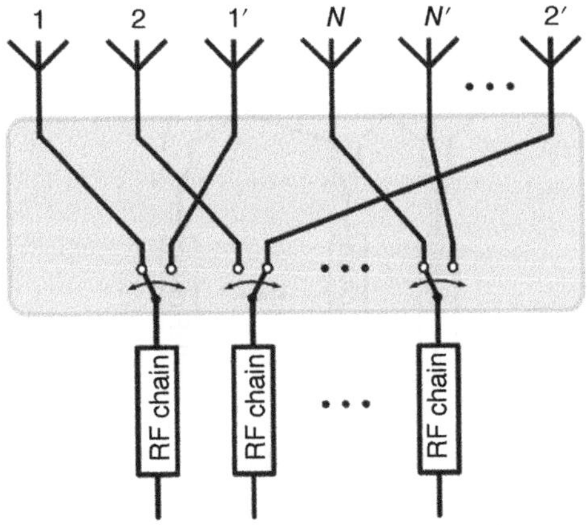

Figure 15.6 Illustration of binary switching on random antenna pairs.

The optimization problems of antenna selection based on full switching in (15.3) and binary switching in (15.5) can be solved by two steps. The first step is to assume equal power allocation among users, i.e., $P_i = \frac{1}{U}$, and find the optimal antenna subset by relaxing the constraint that each Δ_i must be a binary integer to a weaker constraint that $0 \leq \Delta_i \leq 1$. Then, (15.3) and (15.5) become convex optimization problems, solvable in polynomial time. For full switching, the L_t largest Δ_i are selected, and their indices represent the selected antennas. For binary switching, the larger one in Δ_i and Δ_j is selected. When $L_t \ll N_t$, the performance of this relaxation approaches the optimal results. The second step is to optimize the user power allocation $\mathbf{P}$ based on the constructed sparse array Δ_{opt}, which is also a convex optimization problem.

Based on the L_t-antenna sparse array, linear beamforming is performed. For instance, zero-forcing (ZF) precoding, given by

$$\mathbf{F}_{\text{ZF}} = \gamma_{\text{ZF}} \Delta_{\text{opt}} \mathbf{H}^H \left(\mathbf{H} \Delta_{\text{opt}} \mathbf{H}^H \right)^{-1}, \tag{15.6}$$

can be used to achieve a relatively good multi-user sum-rate in interference-limited scenarios.[2] While in noise-limited situations, matched-filtering (MF) beamforming, or maximum ratio transmission (MRT), can be used to guarantee a required level of receive signal-to-noise ratio (SNR)s, i.e.,

$$\mathbf{F}_{\text{MF}} = \gamma_{\text{MF}} \Delta_{\text{opt}} \mathbf{H}^H. \tag{15.7}$$

2 Interference- or noise-limited means that the performance of a multi-user MIMO system is mainly limited by inter-user interference or receive SNRs.

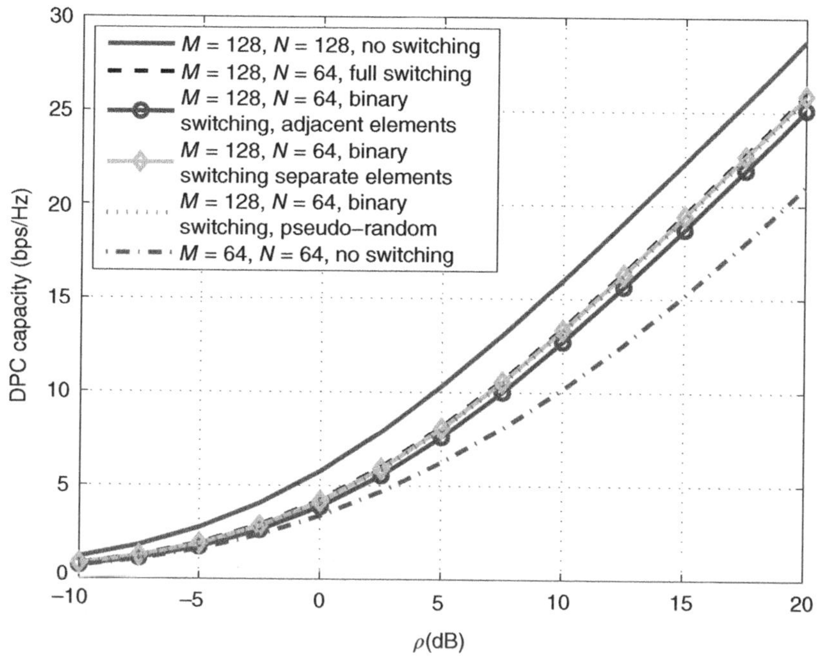

Figure 15.7 Multi-user MIMO capacities, achieved by the DPC, when an 128-antenna ULA is used to construct 64-antenna sparse arrays. The performance evaluation is based on measured channel data in a semi-urban environment, where four users are closely spaced, all with line-of-sight (LOS) propagation conditions to the base station. Source: Gao et al. [11]/IEEE.

15.2.3 Simulation Results

To illustrate the performance of the presented sparse array-based precoding approach, we plot the multi-user MIMO performance in Fig. 15.7. The base station (BS) in this system is assumed to employ an $N_t = 128$ ULA, which is used to construct an $L_t = 64$ sparse array. The base station uses this array to communicate with $U = 4$ closely located users. In Fig. 15.7, sparse array MIMO performance achieved by the full and binary switching arrays is compared. In addition, DPC capacities with 128- and 64-antenna full arrays, i.e., without switching, are also shown for comparison. We can see that by constructing sparse arrays through the 128-antenna array, MIMO performance can be significantly improved, as compared to the performance when using the 64-antenna full array. For constructing sparse arrays, there is no significant difference in using the full and binary switching approaches. The performance results in the figure indicate that constructing sparse arrays through antenna selection on a full array is feasible, and simple RF switching networks can be used to achieve relatively good performance.

15.3 Hybrid Analog–Digital Architectures with Sparse Arrays

The fully digital architectures necessitate dedicating a complete RF and baseband hardware chain for each antenna element, resulting in high cost and complexity for large arrays [17]. Thus, fully digital processing is undesirable in mmWave MIMO communication systems with large-scale arrays [18, 19]. Components such as RF up/down converters and ADCs (or digital-to-analog converters (DACs)) are both expensive and have high power consumption [20]. This led to the development of efficient and low-cost beamforming strategies for massive MIMO systems in mmWave bands. Such approaches include hybrid beamforming (or precoding) architectures [3, 20–28], beamspace signal processing techniques [29–31], lens-based analog beamforming antennas [32], and low-rate ADC methods [33].

Hybrid analog/digital architectures generally refer to the transceiver architectures where the precoding implementation is divided between the analog RF and the digital baseband domains Heath et al. [5], Alkhateeb et al. [6, 20], El Ayach et al. [3], Molisch et al. [4, 34]. The motivation for splitting the precoding implementation emerges from the interest in compromising between the low cost and power consumption of analog beamforming, which does not need to use an RF chain per antenna, and the high DOFs and flexibility of fully digital precoding in which the baseband processing has access to the signal at each antenna element. An analog precoding stage comprising analog beamformers, presents a reduced-dimensional signal to the baseband stage [35]. The analog precoding stage is generally realized as a network of phase shifters [7]. Despite being simpler than fully digital RF chains, analog phase shifters for mmWave frequencies are not easily implemented in practice [36] and can introduce a latency that may lead to performance degradation in rapidly changing channels [4]. Although the phase precision can be improved, this comes at the cost of higher power consumption [37]. Passive phase shifters, on the other hand, can alleviate the power consumption problem but they are known to incur higher losses, requiring additional amplification to maintain an acceptable output SNR [36].

Further simplification of the transmit processing chain can be made by employing switch-based sparse beamforming [4]. This strategy uses a network of switches that connect the RF chains to a subset of the available antennas, which amounts to selecting a transmit subarray. Antenna selection has been applied in other domains such as radar [38, 39] to reduce the number of RF chains while maintaining a performance that is close to the full array. It has recently been explored in the context of hybrid beamforming in various works [31, 40–43]. In [44] an online learning strategy using online convex optimization (OCO) was presented for the hybrid precoder design.

In this section, we first describe two basic and widely adopted hybrid analog/digital architectures in Section 15.3.1. We extend these basic architectures to derivatives with sparse arrays in Section 15.3.2. Finally, in Section 15.3.3, we present different design criteria and signal processing approaches for sparse hybrid analog/digital architectures.

15.3.1 Basic Hybrid Analog–Digital Architectures

Several hybrid analog–digital architectures have been proposed and investigated over the last decade Heath et al. [5], Alkhateeb et al. [6, 20], Mendez-Rial et al. [7], yet the two prevalent architectures are the fully connected array where each antenna can be connected to every RF chain, and the array of sub-arrays architecture where the antennas in a subarray have access only to one RF chain.

15.3.1.1 Fully Connected Hybrid Architecture

In the fully connected hybrid analog/digital architectures, each RF chain is connected to all the antenna elements in the array, as depicted in Fig. 15.8. If N_t, L_t, and M_t denote the number of antennas, RF chains, and number of antennas per RF chain, then, in the fully connected architecture, $N_t = M_t$. Similarly, for the receiver, let N_r, L_r, M_r represent the numbers of receive antennas, RF chains, and receive antennas per RF chains, with $N_r = M_r$. Considering a single-user

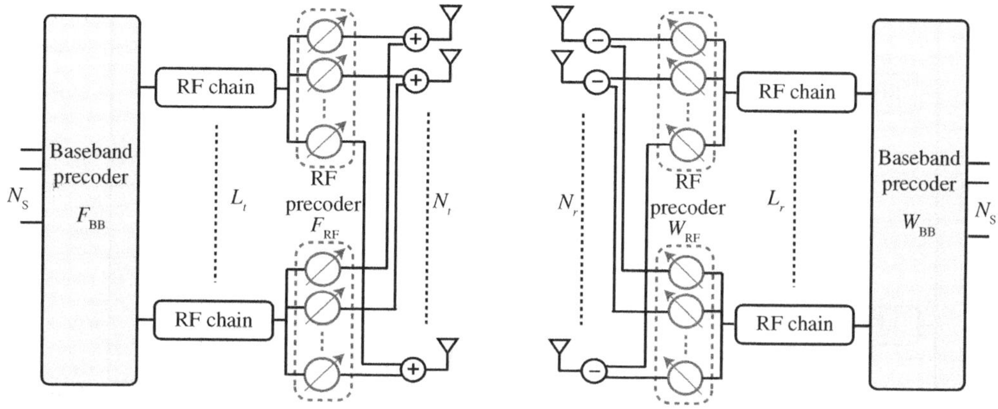

Figure 15.8 Block diagram of the fully connected hybrid analog/digital transceiver architecture. In this architecture, each RF chain is connected to all the antennas.

scenario with a hybrid analog/digital transmitter and receiver communicating via N_s data streams, the discrete-time transmitted signal $\mathbf{x}$ is

$$\mathbf{x} = \mathbf{F}\mathbf{s}, \tag{15.8}$$

where $\mathbf{s}$ is the $N_s \times 1$ vector of transmitted symbols, such that $\mathbb{E}\left[\mathbf{s}\mathbf{s}^H\right] = \frac{P_s}{N_s}\mathbf{I}_{N_s}$, and P_s is the average total transmit power. The matrix $\mathbf{F} \in \mathbb{C}^{N_t \times N_s}$ is the hybrid analog/digital precoding matrix. Here, it is important to clarify the relation between the number of streams, RF chains, and transmit antennas. Since the motivation of the hybrid analog/digital architectures is to reduce the number of RF chains, these systems usually have $N_t \geq L_t \geq N_s$ [3].

The precoder is divided between the analog and digital domains, as shown in Fig. 15.8, so that the precoding matrix $\mathbf{F}$ can be expressed as

$$\mathbf{F} = \mathbf{F}_{\mathrm{RF}}\mathbf{F}_{\mathrm{BB}}, \tag{15.9}$$

where $\mathbf{F}_{\mathrm{RF}} \in \mathbb{C}^{N_t \times L_t}$ and $\mathbf{F}_{\mathrm{BB}} \in \mathbb{C}^{L_t \times N_s}$ are the analog RF and digital baseband precoding matrices. If $\mathbf{F}_{\mathrm{RF}}$ is implemented using only analog phase shifters, then its entries have constant modulus. To impose this constraint, we normalize these entries to satisfy $\left|\left[\mathbf{F}_{\mathrm{RF}}\right]_{n,m}\right|^2 = N_t^{-1}$, where $\left|\left[\mathbf{F}_{\mathrm{RF}}\right]_{m,n}\right|$ is the magnitude of the (n, m)th element of $\mathbf{F}_{\mathrm{RF}}$. The total power constraint on the precoder is enforced by normalizing $\mathbf{F}_{\mathrm{BB}}$ such that $\|\mathbf{F}_{\mathrm{RF}}\mathbf{F}_{\mathrm{BB}}\|_F^2 = N_{\mathrm{S}}$. Now, if $\mathbf{H} \in \mathbb{C}^{N_r \times N_t}$ denotes the channel between the transmitter and receiver, then the received signal after combining can be written as

$$\mathbf{y} = \mathbf{W}_{\mathrm{RF}}^H\mathbf{W}_{\mathrm{BB}}^H\mathbf{H}\mathbf{F}_{\mathrm{RF}}\mathbf{F}_{\mathrm{BB}}\mathbf{s} + \mathbf{W}_{\mathrm{RF}}^H\mathbf{W}_{\mathrm{BB}}^H\mathbf{n}, \tag{15.10}$$

where $\mathbf{n}$ is the receive noise vector with $\mathcal{N}_{\mathbb{C}}(0, \sigma_n^2)$ entries. $\mathbf{W}_{\mathrm{RF}} \in \mathbb{C}^{N_r \times L_r}$ and $\mathbf{W}_{\mathrm{BB}} \in \mathbb{C}^{L_r \times N_s}$ represent the analog RF and digital baseband combining matrices, with $\mathbf{W}_{\mathrm{RF}}$ having the same phase shifter constraints of the RF precoding matrix.

15.3.1.2 Array of Sub-Arrays Architecture

The fully connected hybrid analog/digital architecture presented in Section 15.3.1.1 captures the full array gain. However, connecting every RF chain to every antenna increases the transceiver design and implementation complexity [45]. To reduce this complexity and increase energy efficiency, the array of sub-arrays hybrid architecture was proposed Gao et al. [46]. As illustrated in

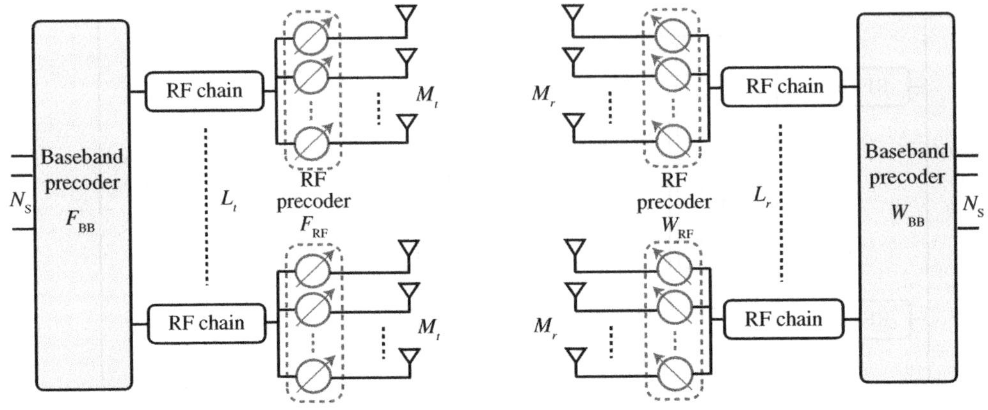

Figure 15.9 Block diagram of the array of sub-arrays hybrid analog/digital transceiver architecture. In this architecture, each RF chain is connected to a separate antenna sub-array.

Fig. 15.9, this architecture connects every RF chain to a separate subset of the antenna elements through a network of phase shifters, relaxing the need for power combiners and splitters, and reducing the number of phase shifters. Similar to the description of the fully connected architecture, for an array of subarrays hybrid architecture with L_t transmit RF chains, N_t antennas, and N_S communication streams, the transmit signal can be written exactly as in (15.8) with the precoder $\mathbf{F} = \mathbf{F}_{RF}\mathbf{F}_{BB}$. However, it is important to note that in this architecture, $N_t = L_t \times M_t$. Hence the RF precoder has the following block diagonal structure

$$\mathbf{F}_{RF} = \text{blkdiag}\left(\mathbf{f}_{RF,1}, \ldots, \mathbf{f}_{RF,L_t}\right). \tag{15.11}$$

The received signal after combining can also be written similarly to (15.10).

15.3.2 Hybrid Architectures with Sparse Arrays

Motivated by the interest in further reducing the power consumption and enhancing the achievable data rate performance, multiple hybrid architectures with *sparse array* have been proposed Gao et al. [47], Ratnam et al. [48], Payami et al. [49], Molisch et al. [4], Mendez-Rial et al. [7], Nosrati et al. [42, 44]. One main family of these architectures is switch-based hybrid beamforming. We describe two variants of these sparse hybrid architectures along with the different design criteria and signal processing approaches below.

Switch-based networks have been proposed as an alternative to phase shifters to mitigate the problems of high cost and power consumption [4, 7, 42, 44, 47–49]. In switch-based analog beamforming/combining architectures, the phase shifters in the traditional hybrid analog/digital architectures, discussed in Sections 15.3.1.1 and 15.3.1.2, are replaced with binary switches Mendez-Rial et al. [7]. Leveraging the sparse nature of massive MIMO channels in the mmWave, a switch-based approach effectively selects subsets of the available antennas in order to ensure the required performance [37]. Optimal selection of a subset of antennas from a larger array has been demonstrated to be capable of delivering a large aperture with satisfactory performance at reduced hardware cost and complexity [39, 50–56]. When the complexity of connectivity, routing, and RF multiplexing are of grave concern, the switching network can be partitioned into subsets, allowing only a few antennas per subset to be selected [7, 45, 57–59].

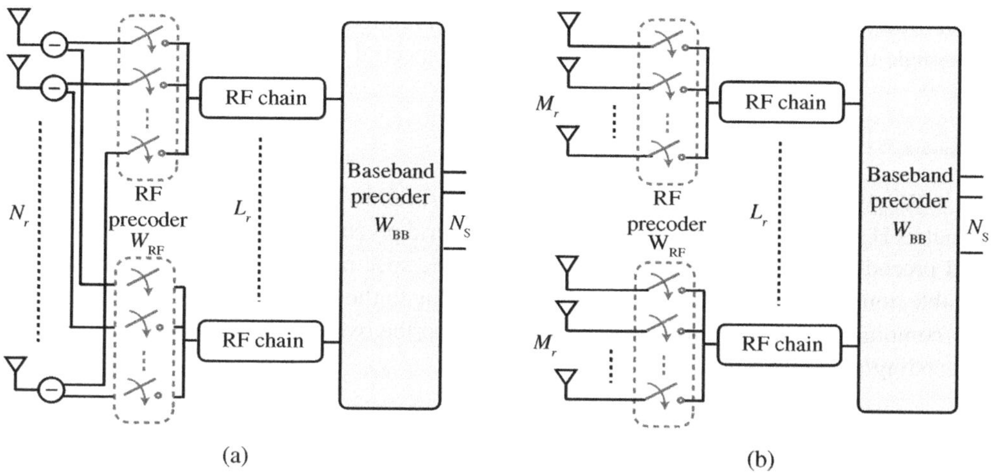

Figure 15.10 Block diagrams of two switches-based analog beamforming/combining architectures. In these architectures, switches are used to implement the analog beamforming/combining instead of the phase shifters in Figures 15.8 and 15.9. (a) Fully connected architecture. (b) Partially connected architecture.

These architectures are mainly motivated by reducing the power consumption of the RF front-end, especially at the high-frequency bands (millimeter wave/sub-terahertz). For the receiver architectures shown in Figure 15.10, the received signal after combining can be written, similar to (15.10), as

$$\mathbf{y} = \mathbf{W}_{\text{RF}}^{H}\mathbf{W}_{\text{BB}}^{H}\mathbf{H}\mathbf{F}_{\text{RF}}\mathbf{F}_{\text{BB}}\mathbf{s} + \mathbf{W}_{\text{RF}}^{H}\mathbf{W}_{\text{BB}}^{H}\mathbf{n}, \tag{15.12}$$

where, in this case, the RF precoders and combiners $\mathbf{F}_{\text{RF}}$ and $\mathbf{W}_{\text{RF}}$, are implemented using analog switches. More specifically, for the fully connected switches-based architecture in Fig. 15.10a, each receive RF chain selects a subset of the N_r antennas to be noncoherently combined before passing it through the RF chain. Therefore, each column of $\mathbf{F}_{\text{RF}}$ and $\mathbf{W}_{\text{RF}}$ will have a 1 or 0 value depending on whether or not this RF chain is connected to the corresponding antenna element. More formally, $[\mathbf{F}_{\text{RF}}]_{(\ell,m)} = 1$ if RF chain ℓ is connected to antenna n (similarly for $\mathbf{W}_{\text{RF}}$).

In the switch-based array of sub-arrays architecture in Fig. 15.10b, each antenna is routed to only one RF chain, relaxing the need for RF splitters. This reduces the power consumption while penalizing the array gain. The set of RF combiners can then be given by Mendez-Rial et al. [7]

$$\mathcal{A} = \left\{ \begin{bmatrix} \mathbf{b}_1 \\ \mathbf{0} \\ \vdots \\ \mathbf{0} \end{bmatrix} \cup \begin{bmatrix} \mathbf{0} \\ \mathbf{b}_2 \\ \vdots \\ \mathbf{0} \end{bmatrix} \cup \dots \begin{bmatrix} \mathbf{0} \\ \mathbf{0} \\ \vdots \\ \mathbf{b}_{L_r} \end{bmatrix} \in \mathbb{B}^{N_r \times 1} \,\middle|\, \mathbf{b}_\ell \in \mathbb{B}^{M_r \times 1}\ \forall \ell \right\}, \tag{15.13}$$

where $\mathbf{b}_\ell, \forall \ell$ are optimized based on the target design criteria.

15.3.3 Design Criteria and Signal Processing Approaches

In this section, we present some approaches for designing the hybrid precoding/combining matrices for the sparse array-based architectures described in Section 15.3.2.

Given the switch-based analog beamforming/combining architectures in Section 15.3.2, and considering a single-user MIMO system with N_s streams, the spectral efficiency of this system can be written as

$$R_{\text{switches}} = \log_2 \left| \mathbf{I}_{N_s} + \frac{\rho}{N_s} \mathbf{R}_n^{-1} \mathbf{H}_{\text{eff}} \mathbf{H}_{\text{eff}}^H \right|, \tag{15.14}$$

where $\mathbf{R}_n = \mathbf{W}_{\text{BB}}^H \mathbf{W}_{\text{RF}}^H \mathbf{W}_{\text{RF}} \mathbf{W}_{\text{BB}}$ is the noise covariance matrix and ρ is the SNR. The $L_r \times L_t$ effective channel matrix $\mathbf{H}_{\text{eff}} = \mathbf{W}_{\text{BB}}^H \mathbf{W}_{\text{RF}}^H \mathbf{H} \mathbf{F}_{\text{RF}} \mathbf{F}_{\text{BB}}$ includes the hybrid precoders and combiners. Designing the hybrid precoding/combining matrices to maximize the spectral efficiency in (15.14) leads to an intractable nonconvex optimization problem, mainly due to the imposed constraints on the RF precoders/combiners Mendez-Rial et al. [7]. Next, we describe two approaches for designing the hybrid precoding/combining matrices in (15.14).

15.3.3.1 Optimal Hybrid Beamforming Design Via Exhaustive Search

A widelyadopted approach for designing the hybrid precoders/combiners is to decouple the precoders and combiners design El Ayach et al. [3]. Following this approach, and temporarily assuming that the receiver can perform optimal nearest-neighbor decoding, we can write the mutual information maximization problem, focusing on the precoder design, as

$$\begin{aligned} \left\{\mathbf{F}_{\text{BB}}^{\star}, \mathbf{F}_{\text{RF}}^{\star}\right\} = \; & \arg\max \; \log_2 \left| \mathbf{I}_{N_s} + \tfrac{\rho}{N_s} \mathbf{H} \mathbf{F}_{\text{RF}} \mathbf{F}_{\text{BB}} \mathbf{F}_{\text{BB}}^H \mathbf{F}_{\text{RF}}^H \mathbf{H}^H \right|, \\ & \text{s.t.} \;\; \mathbf{F}_{\text{RF}} \in \mathcal{A}, \\ & \qquad \left\| \mathbf{F}_{\text{RF}} \mathbf{F}_{\text{BB}} \right\|_F^2 = N_S. \end{aligned} \tag{15.15}$$

where $\mathcal{A}$ is the set of candidate RF precoders, i.e., the set of all precoders that satisfy the RF hardware constraints on the switch-based analog beamforming architecture. An example of the set $\mathcal{A}$ for the architecture in Fig. 15.10b is defined in (15.13). Since the RF precoder is selected from a finite codebook, this hybrid precoding optimization problem can also be formulated as the following outer–inner problem

$$\left\{\mathbf{F}_{\text{BB}}^{\star}, \mathbf{F}_{\text{RF}}^{\star}\right\} = \underset{\mathbf{F}_{\text{RF}} \in \mathcal{A}}{\arg\max} \left\{ \begin{array}{ll} \underset{\mathbf{F}_{\text{BB}}}{\max} & \log_2 \left| \mathbf{I}_{N_s} + \tfrac{\rho}{N_s} \mathbf{H} \mathbf{F}_{\text{RF}} \mathbf{F}_{\text{BB}} \mathbf{F}_{\text{BB}}^H \mathbf{F}_{\text{RF}}^H \mathbf{H}^H \right|, \\ \text{s.t.} & \left\| \mathbf{F}_{\text{RF}} \mathbf{F}_{\text{BB}} \right\|_F^2 = N_S, \end{array} \right\} \tag{15.16}$$

where the outer maximization for the RF precoder is over the candidate set $\mathcal{A}$ and the inner optimization is for the baseband precoder. Using Proposition 1 in [60], if the singular value decomposition (SVD) decomposition of $\mathbf{H}$ is given by $\mathbf{H} = \mathbf{U}\boldsymbol{\Sigma}\mathbf{V}^H$, and if we define the SVD decomposition of the matrix $\boldsymbol{\Sigma}\mathbf{V}^H \mathbf{F}_{\text{RF}} (\mathbf{F}_{\text{RF}}^H \mathbf{F}_{\text{RF}})^{-\frac{1}{2}}$ as $\boldsymbol{\Sigma}\mathbf{V}^H \mathbf{F}_{\text{RF}} (\mathbf{F}_{\text{RF}}^H \mathbf{F}_{\text{RF}})^{-\frac{1}{2}} = \overline{\mathbf{U}}\overline{\boldsymbol{\Sigma}}\overline{\mathbf{V}}^H$, then the optimal baseband precoder that solves the inner problem is given by

$$\mathbf{F}_{\text{BB}}^{\star} = \left(\mathbf{F}_{\text{RF}}^H \mathbf{F}_{\text{RF}}\right)^{-\frac{1}{2}} \left[\overline{\mathbf{V}}\right]_{1:N_s} \Lambda, \tag{15.17}$$

where Λ is an $N_s \times N_s$ diagonal water-filling power allocation matrix, with

$$[\Lambda]_{i,i}^2 = \left(\mu - \frac{N_s}{\rho \left[\overline{\Sigma}\right]_{i,i}} \right)^+, i = 1, 2, \dots, N_s, \tag{15.18}$$

with μ satisfying

$$\sum_{i=1}^{N_s} \left(\mu - \frac{N_s}{\rho \left[\overline{\Sigma}\right]_{i,i}} \right)^+ = N_s. \tag{15.19}$$

For the RF precoder, $\mathbf{F}_{\mathrm{RF}}$ can then be optimally designed by searching over $\mathcal{A}$. If $\mathcal{A}$ is very large, sub-optimal solutions such as matching pursuit that leverage the potential sparsity of the millimeter wave channels [3, 4, 7], can be utilized to greedily search for the columns of the RF precoder $\mathbf{F}_{\mathrm{RF}}$. The hybrid combiners can be designed following a similar approach Mendez-Rial et al. [7].

15.3.3.2 Hybrid Beamformer Design Via Convex Optimization

Considering the mutual information expression in (15.14), if the optimal combiner were to be used at the receive side, the transmit mutual information simplifies to [42]

$$\mathcal{I}_{\mathrm{t}} = \log_2\left(\left|\mathbf{I}_{N_s} + \frac{\rho}{N_s}\mathbf{H}\mathbf{F}_{\mathrm{RF}}\mathbf{F}_{\mathrm{BB}}\mathbf{F}_{\mathrm{BB}}^H\mathbf{F}_{\mathrm{RF}}^H\mathbf{H}^H\right|\right). \tag{15.20}$$

This reduces the problem to the transmit side only and allows the transmit precoder to be designed. Once this is done, it can be put back into the mutual information expression to enable the receive side beamformer to be obtained. In this manner, the problem is transformed into two separate problems for the design of the precoding and combining matrices.

It is known that the unconstrained optimum precoder is $\mathbf{F}_{\mathrm{opt}} = \mathbf{V}_{N_s}\Lambda$, where $\mathbf{V}_{N_s}$ is the matrix comprising the N_s right singular vectors of the channel matrix that correspond to the largest N_s singular values. The diagonal matrix Λ expresses the power allocation to the difference streams, which may be obtained via water-filling power allocation [61]. Assume that the mmWave system and propagation channel parameters permit a hybrid precoder $\mathbf{F}_{\mathrm{RF}}\mathbf{F}_{\mathrm{BB}}$ that is sufficiently close to $\mathbf{F}_{\mathrm{opt}} = \mathbf{V}_{N_s}$ to be obtained. Then, it can be shown that

$$\mathcal{I}_{\mathrm{t}} = \log_2\left(\left|\mathbf{I}_{N_s} + \frac{\rho}{N_s}\mathbf{H}_{N_s}\mathbf{F}_{\mathrm{RF}}\mathbf{F}_{\mathrm{RF}}^H\mathbf{H}_{N_s}^H\right|\right). \tag{15.21}$$

The task at this stage is to obtain $\mathbf{F}_{\mathrm{RF}}$. To this end, $\mathcal{I}_t$ can be further approximated as [3]

$$\mathcal{I}_{\mathrm{t}} \approx \log_2\left(\left|\mathbf{I}_{N_s} + \frac{\rho}{N_s}\Sigma_{N_s}^2\right|\right) - \left(N_s - \|\mathbf{V}_{N_s}^H\mathbf{F}_{\mathrm{RF}}\|_F^2\right), \tag{15.22}$$

where Σ_{Ns} is the diagonal matrix with the largest N_s singular values of the channel matrix along the diagonal. Now the maximization of $\mathcal{I}_t$ in terms of $\mathbf{F}_{\mathrm{RF}}$ can be cast as

$$\max_{\mathbf{F}_{\mathrm{RF}}} \ \|\mathbf{V}_{N_s}^H\mathbf{F}_{\mathrm{RF}}\|_F^2,$$

$$\text{s.t.} \quad \mathbf{F}_{\mathrm{RF}} \in \{0,1\}, \tag{15.23}$$

$$\mathrm{rank}(\mathbf{H}_{N_s}\mathbf{F}_{\mathrm{RF}}) = N_s, \tag{15.24}$$

$$\|\mathbf{F}_{\mathrm{RF}}\mathbf{F}_{\mathrm{BB}}\|_F^2 = N_s. \tag{15.25}$$

This optimization problem is still nonconvex as it involves a maximization of the norm, as well as nonconvex constraints. In order to avail a solution, constraints (15.24) and (15.25) are dropped and are subsequently addressed using rank control and normalization steps [42]. The Frobenius norm maximization can then be approached by relaxing the first constraint to $0 \leq \mathbf{F}_{\mathrm{RF}}(i,j) \leq 1$ for all i and j where $\mathbf{F}_{\mathrm{RF}}(i,j)$ is the (i,j)th element of $\mathbf{F}_{\mathrm{RF}}$. While still nonconvex, the problem can be solved using a sequential convex programming strategy to give a suboptimal solution iteratively [42]. Thus optimization problem at the ℓth iteration was derived to be

$$\max_{\mathbf{F}_{\mathrm{RF}}} \ f(\mathbf{F}_{\mathrm{RF}}, \mathbf{F}_{\mathrm{RF}}^{(\ell-1)}),$$

$$\text{s.t.} \quad 0 \leq \mathbf{F}_{\mathrm{RF}}(i,j) \leq 1, \quad i,j = 1, \ldots, N_{\mathrm{t}}(k_{\mathrm{t}}). \tag{15.26}$$

After initialization, the sequential convex programming (SCP) procedure approaches a local optimum of (15.26) iteratively.

As the rank constraint was dropped, the SCP procedure devised above does not guarantee that the rank of the precoder is N_s. Therefore, in order to ensure the rank constraint is met, the SCP procedure is modified by checking the rank of the precoder at each iteration. If the rank constraint were not met, then the precoder estimate is perturbed using a Gaussian random matrix as follows

$$\mathbf{F}_{\text{RF}}^{(\ell)} \leftarrow \mathbf{F}_{\text{RF}}^{(\ell)} + \eta \mathcal{F}_{\text{RF}}, \tag{15.27}$$

where η is a scaling factor and

$$\text{vec}(\mathcal{F}_{\text{RF}}) \sim \mathcal{N}\left(\mathbf{0}, \mathbf{I}_{N_t k_t}\right), \tag{15.28}$$

where $\mathcal{N}\left(\text{vec}\left(\mathbf{F}_{\text{RF}}^{(\ell)}\right), \mathbf{I}_{N_t k_t}\right)$ the normal distribution with mean $\text{vec}\left(\mathbf{F}_{\text{RF}}^{(\ell)}\right)$ and covariance $\mathbf{I}_{N_t k_t}$.

The resulting method is presented in Algorithm 15.1 and is referred to as the SHD-NM algorithm. We also note that a second approach based on majorization theory and employing a QR decomposition with quadratic update, referred to SHD-QRQU, was also proposed in [42]. Furthermore, the work considered the case of a partially connected architecture and proposed a solution by introducing a connectivity matrix $\mathbf{G}$ and incoporating it into the SHD-NM procedure. Both SHD-QRQU and the partially connected case are not detailed here and the reader is referred to [42] for further details.

Algorithm 15.1: Switch-based hybrid design by norm maximization (SHD-NM)

Input : $\mathbf{H}, L, I$

Decompose $\mathbf{H} = \mathbf{USV}^H$

Initialize $\mathbf{F}_{\text{RF}}^{(0)}$ randomly such that its elements are uniformly distributed between 0 and 1

Set $\ell = 0, i = 0$

while $\ell \leq L$ and $i \leq I$ **do**

    Solve (15.26) to obtain $\mathbf{F}_{\text{RF}}^{(\ell)}$

    Construct $\tilde{\mathbf{H}} = \mathbf{H}_{N_s} \mathbf{F}_{\text{RF}}^{(\ell)}$

    Update $\mathbf{F}_{\text{BB}}$

    Calculate $\mathcal{I}^{(\ell)}$ based on (15.20)

    if $\mathcal{I}^{(\ell)} \geq \mathcal{I}^{(\ell-1)}$ and $\text{rank}\left(\tilde{\mathbf{H}}\right) = \text{N}_s$ **then**

        $\bar{\mathbf{F}}_{\text{RF}} \leftarrow \mathbf{F}_{\text{RF}}^{(\ell)}$

        $\ell \leftarrow \ell + 1, i \leftarrow 0$

    else

        $\mathbf{F}_{\text{RF}}^{(\ell-1)} \leftarrow \eta \mathcal{F}_{\text{RF}},$

        s.t. $\mathcal{F}_{\text{RF}} \sim \mathcal{N}\left(\text{vec}\left(\bar{\mathbf{F}}_{\text{RF}}\right), \mathbf{I}\right)$

        $i \leftarrow i + 1$

Round $\mathbf{F}_{\text{RF}}$

Output: $\mathbf{F}_{\text{RF}}, \mathbf{F}_{\text{BB}}$

Simulation Results The SHD-NM algorithm, described above, was simulated and compared to the following techniques [42]:

- UOP: Unconstrained Optimum Precoder
- SSP: Spatially Sparse Precoder [3]
- SHD-NM: Switch-based Hybrid Design by Norm Maximization (Algorithm 15.1)
- SHD-UGD: Switch-based Hybrid Design by Unified Greedy Algorithm [35]

For the purposes of the simulations, we assume a clustered channel model [3] with $N_{\text{cl}} = 8$ clusters and $N_{\text{ray}} = 20$ rays. The ray angles of departure (AoDs) and Angles of Arrival (AoAs) are drawn from a Laplacian distribution, while their complex amplitudes are complex normal, maintaining an average power of 1 in each cluster. The system comprises a switch-based beamformer with $N_t = 64$, $k_t = 4$, $N_r = 64$, and $k_4 = 4$. The transmitter and receiver comprise uniform planar arrays (UPA) with half wavelength inter-element spacing. Full details of the simulations can be found in [42].

We first present the spectral efficiency of the algorithms as a function of the SNR. The results shown in Fig. 15.11. In this scenario $k_t = N_s = 4$ which is a class of hybrid networks that has been studied frequently in literature [17, 35, 62]. The digital precoding matrix $\mathbf{F}_{\text{BB}}$ is then a square matrix, and the unitary structure of $\mathbf{F}_{\text{BB}}$ permits the joint design of $\mathbf{F}_{\text{RF}}$, $\mathbf{F}_{\text{BB}}$ to be decoupled. Looking at the figure, we see that increasing N_s restricts the ability of the hybrid network to approximate the optimal unconstrained precoder resulting in a slight degradation in performance. This is seen in the gap between the SSP and UOP curves. Also, as demonstrated in the figure, the SHD-NM algorithm outperforms the greedy method (SHD-UGD). Finally, we point out that switch-based network generally sacrifices some performance compared to the unconstrained and phase shift based hybrid structures in order to achieve the hardware cost and power savings.

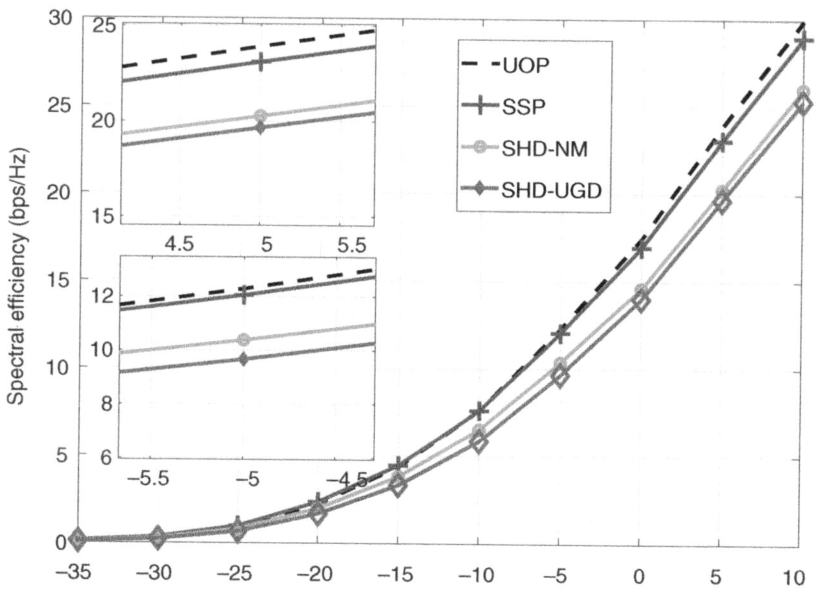

Figure 15.11 Spectral efficiency attained by different hybrid design methods for $N_t = 64$ and $N_r = 16$ transmit and receive antennas, respectively. The mmWave channel comprises $N_{\text{cl}} = 8$ clusters and $N_{\text{ray}} = 10$ rays in each cluster. The number of RF chains and data streams are $k_t = 4$ and $N_s = 4$, respectively.

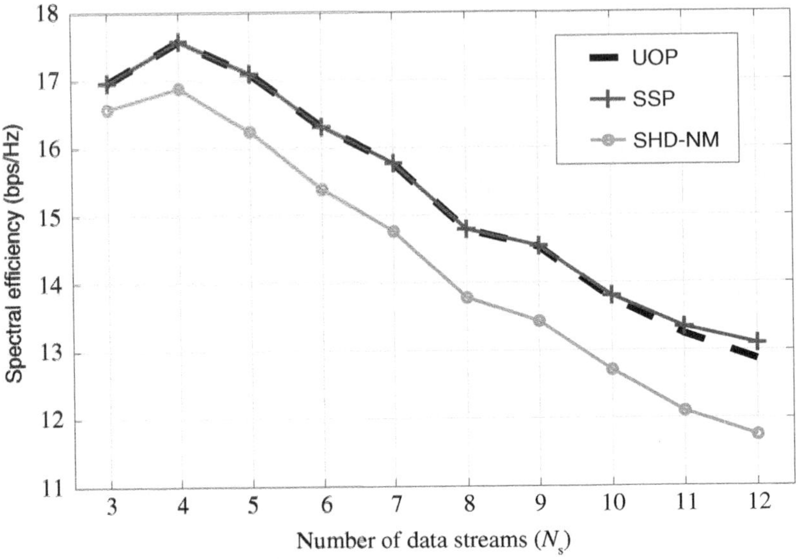

Figure 15.12 Spectral efficiency attained by various hybrid design methods for $N_t = 64$ and $N_r = 16$ transmit and receive antennas, respectively. The mmWave clustered channel comprises $N_{cl} = 8$ clusters and $N_{ray} = 10$ rays in each cluster. The number of RF chains is $k_t = 12$ and the number of data streams (N_s varies over the range $[3, \ldots, 12]$).

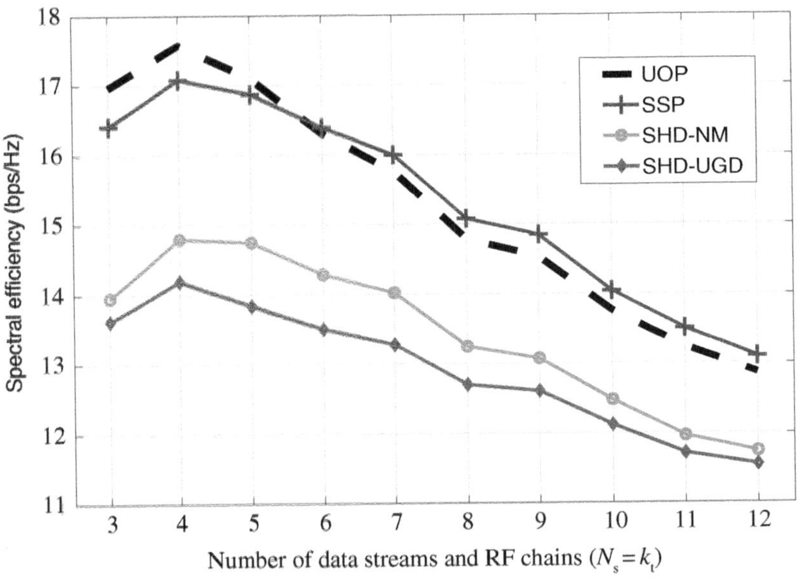

Figure 15.13 Spectral efficiency exhibited by various hybrid design methods for $N_t = 64$ and $N_r = 16$ transmit and receive antennas respectively. The mmWave channel comprises $N_{cl} = 8$ clusters and $N_{ray} = 10$ rays in each cluster. A variable number of data streams, $N_s = 1, \ldots, 12$, are communicated by an equal number of RF chains, $N_s = k_t$.

Fixing the SNR, we present the spectral efficiency as a function of the number of data streams in Fig. 15.12. In this example, a 64 × 16 UPA having 12 RF chains at both the transmit and receive sides ($k_t = k_r = 12$) is used. The SNR is set to 0 dB and N_s is varied over the range [3, … , 12]. Firstly, notice that the spectral efficiency decreases with increasing N_s. The SHD-NM algorithm shows a reasonable performance as it tracks the optimum precoder, especially for $N_s \leq 9$. For larger N_s, the feasible solution space shrinks as the digital beamformer consumes more of the DOFs provided by the virtual eigen-channels, thus diminishing the capability of SHD-NM deteriorates.

Finally, the performance in a scenario with varying N_s is obtained for a hybrid network with $k_t = N_s$ and the results shown in Fig. 15.13. Again here a UPA of size 64 × 16 UPA is used and the SNR is set to 0 dB. the simulations are run for $k_t = k_r = N_s = 3, \dots, 12$. In this case, the number of network structures that afford unequal power allocations for the SSP algorithm increases significantly, thus leading to SSP outperforming the UOP as N_s increases. The SHD-NM algorithm, on the other hand, exhibits better performance than SHD-UGD.

15.4 Conclusion and Future Directions

Adopting sparse arrays has the potential of reducing the cost and power consumption of large-scale MIMO systems. Further, when efficiently designed and constructed, the sparse array-based MIMO systems could approach the performance of traditional full-array fully digital architectures. This chapter reviewed possible architectures for MIMO systems with sparse arrays with both fully digital and hybrid analog/digital architectures. It also discussed different criteria for designing these arrays, formulated the key signal processing problems, and described efficient ways for constructing these arrays and devising their functions. Realizing the full potential of sparse arrays for MIMO communication systems still requires investigating several key problems. For example, it is important to explore the use of machine learning to dynamically adapt these sparse arrays based on the specific site and deployment. Towards this direction, we note that Chapter 6 of this book considers deep learning based design of hybrid beamforming with sparse arrays. It is also important to develop approaches that further reduce the antenna selection complexity, especially for systems with very large arrays such as reconfigurable intelligent surfaces. All that highlights a promising area for future research and advances in MIMO communication systems.

References

1 Mezghani, A. and Nossek, J.A. (2007). On ultra-wideband MIMO systems with 1-bit quantized outputs: performance analysis and input optimization. *2007 IEEE International Symposium on Information Theory*, 1286–1289. https://doi.org/10.1109/ISIT.2007.4557400.

2 Mo, J., Alkhateeb, A., Abu-Surra, S., and Heath, R.W. (2017). Hybrid architectures with few-bit ADC receivers: achievable rates and energy-rate tradeoffs. *IEEE Transactions on Wireless Communications* 16 (4): 2274–2287. https://doi.org/10.1109/TWC.2017.2661749.

3 El Ayach, O., Rajagopal, S., Abu-Surra, S. et al. (2014). Spatially sparse precoding in millimeter wave MIMO systems. *IEEE Transactions on Wireless Communications* 13 (3): 1499–1513. https://doi.org/10.1109/twc.2014.011714.130846.

4 Molisch, A.F., Ratnam, V.V., Han, S. et al. (2017). Hybrid beamforming for massive MIMO: a survey. *IEEE Communications Magazine* 55 (9): 134–141. https://doi.org/10.1109/MCOM.2017.1600400.

5 Heath, R.W., Gonzalez-Prelcic, N., Rangan, S. et al. (2016). An overview of signal processing techniques for millimeter wave MIMO systems. *IEEE Journal of Selected Topics in Signal Processing* 10 (3): 436–453. https://doi.org/10.1109/JSTSP.2016.2523924.

6 Alkhateeb, A., Mo, J., Gonzalez-Prelcic, N., and Heath, R.W. (2014). MIMO precoding and combining solutions for millimeter-wave systems. *IEEE Communications Magazine* 52 (12): 122–131. https://doi.org/10.1109/MCOM.2014.6979963.

7 Mendez-Rial, R., Rusu, C., Gonzalez-Prelcic, N. et al. (2016). Hybrid MIMO architectures for millimeter wave communications: phase shifters or switches? *IEEE Access* 4: 247–267. https://doi.org/10.1109/ACCESS.2015.2514261.

8 Park, S., Alkhateeb, A., and Heath, R.W. (2017). Dynamic subarrays for hybrid precoding in wideband mmWave MIMO systems. *IEEE Transactions on Wireless Communications* 16 (5): 2907–2920. https://doi.org/10.1109/TWC.2017.2671869.

9 Teng, Y., Zhao, Y., Wei, M. et al. (2022). Sparse hybrid precoding for power minimization with an adaptive antenna structure in massive MIMO systems. *IEEE Transactions on Wireless Communications* 21 (7): 5279–5292. https://doi.org/10.1109/TWC.2021.3139097.

10 Gao, X. and Liu, X. (2019). Design of rectangular antenna array structure for massive MIMO. *Proceedings of the International Conference on Wireless Communications and Signal Processing (WCSP)*. https://doi.org/10.1109/WCSP.2019.8928094.

11 Gao, X., Edfors, O., Rusek, F., and Tufvesson, F. (2015). Massive MIMO performance evaluation based on measured propagation data. *IEEE Transactions on Wireless Communications* 14 (7): 3899–3911. https://doi.org/10.1109/TWC.2015.2414413.

12 Gao, X., Edfors, O., Tufvesson, F., and Larsson, E.G. (2015). Massive MIMO in real propagation environments: do all antennas contribute equally? *IEEE Transactions on Communications* 63 (11): 3917–3928. https://doi.org/10.1109/TCOMM.2015.2462350.

13 Costa, M. (1983). Writing on dirty paper (corresp.). *IEEE Transactions on Information Theory* 29 (3): 439–441. https://doi.org/10.1109/TIT.1983.1056659.

14 Vishwanath, S., Jindal, N., and Goldsmith, A. (2003). Duality, achievable rates, and sum-rate capacity of Gaussian MIMO broadcast channels. *IEEE Transactions on Information Theory* 49 (10): 2658–2668. https://doi.org/10.1109/TIT.2003.817421.

15 Hamza, S.A., Juretus, K., Amin, M.G., and Ahmad, F. (2023). Deep learning sparse array design using binary switching configurations. *ICASSP 2023 - 2023 IEEE International Conference on Acoustics, Speech and Signal Processing (ICASSP)*, Rhodes Island, Greece, 1–5. https://doi.org/10.1109/ICASSP49357.2023.10094662.

16 Gao, X., Edfors, O., Tufvesson, F., and Larsson, E.G. (2015). Multi-switch for antenna selection in massive MIMO. *Proceedings of the IEEE Global Communications Conference (GLOBECOM)*. https://doi.org/10.1109/GLOCOM.2015.7417765.

17 Rusu, C., Mendez-Rial, R., Gonzalez-Prelcic, N., and Heath, R.W. (2016). Low complexity hybrid precoding strategies for millimeter wave communication systems. *IEEE Transactions on Wireless Communications* 15 (12): 8380–8393. https://doi.org/10.1109/twc.2016.2614495.

18 Bogale, T.E. and Le, L.B. (2014). Beamforming for multiuser massive MIMO systems: digital versus hybrid analog-digital. *2014 IEEE Global Communications Conference*. IEEE, December 2014. https://doi.org/10.1109/glocom.2014.7037444.

19 Sohrabi, F. and Yu, W. Hybrid digital and analog beamforming design for large-scale MIMO systems. In *2015 IEEE International Conference on Acoustics, Speech and Signal Processing (ICASSP)*. IEEE, April 2015. https://doi.org/10.1109/icassp.2015.7178507.

20 Alkhateeb, A., El Ayach, O., Leus, G., and Heath, R.W. (2014). Channel estimation and hybrid precoding for millimeter wave cellular systems. *IEEE Journal of Selected Topics in Signal Processing* 8 (5): 831–846. https://doi.org/10.1109/JSTSP.2014.2334278.

21 Gholam, F., Via, J., and Santamaria, I. (2011). Beamforming design for simplified analog antenna combining architectures. *IEEE Transactions on Vehicular Technology* 60 (5): 2373–2378. https://doi.org/10.1109/tvt.2011.2142205.

22 Han, S., Chih-Lin, I., Xu, Z., and Rowell, C. (2015). Large-scale antenna systems with hybrid analog and digital beamforming for millimeter wave 5G. *IEEE Communications Magazine* 53 (1): 186–194. https://doi.org/10.1109/mcom.2015.7010533.

23 Nsenga, J., Bourdoux, A., and Horlin, F. (2010). Mixed analog/digital beamforming for 60 GHz MIMO frequency selective channels. *2010 IEEE International Conference on Communications*. IEEE, May 2010. https://doi.org/10.1109/icc.2010.5502689.

24 Pi, Z. (2012). Optimal transmitter beamforming with per-antenna power constraints. *2012 IEEE International Conference on Communications (ICC)*. IEEE, June 2012. https://doi.org/10.1109/icc.2012.6364537.

25 Roh, W., Seol, J.-Y., Park, J. et al. (2014). Millimeter-wave beamforming as an enabling technology for 5G cellular communications: theoretical feasibility and prototype results. *IEEE Communications Magazine* 52 (2): 106–113. https://doi.org/10.1109/mcom.2014.6736750.

26 Sayeed, A. and Behdad, N. (2010). Continuous aperture phased MIMO: basic theory and applications. *2010 48th Annual Allerton Conference on Communication, Control, and Computing (Allerton)*. IEEE, September 2010. https://doi.org/10.1109/allerton.2010.5707050.

27 Venkateswaran, V. and van der Veen, A.-J. (2010). Analog beamforming in MIMO communications with phase shift networks and online channel estimation. *IEEE Transactions on Signal Processing* 58 (8): 4131–4143. https://doi.org/10.1109/tsp.2010.2048321.

28 Zhang, X., Molisch, A.F., and Kung, S.-Y. (2005). Variable-phase-shift-based RF-baseband codesign for MIMO antenna selection. *IEEE Transactions on Signal Processing* 53 (11): 4091–4103. https://doi.org/10.1109/tsp.2005.857024.

29 Brady, J., Behdad, N., and Sayeed, A.M. (2013). Beamspace MIMO for millimeter-wave communications: system architecture, modeling, analysis, and measurements. *IEEE Transactions on Antennas and Propagation* 61 (7): 3814–3827. https://doi.org/10.1109/tap.2013.2254442.

30 Sayeed, A.M. (2002). Deconstructing multiantenna fading channels. *IEEE Transactions on Signal Processing* 50 (10): 2563–2579. https://doi.org/10.1109/tsp.2002.803324.

31 Wang, X. and Aboutanios, E. (2020). Sparse array design for multiple switched beams using iterative antenna selection method. *Digital Signal Processing* 102684. https://doi.org/10.1016/j.dsp.2020.102684.

32 Gao, X., Dai, L., Zhou, S., et al. (2018). Beamspace channel estimation for wideband millimeter-wave MIMO with lens antenna array. In *2018 IEEE International Conference on Communications (ICC)*. IEEE, May 2018. https://doi.org/10.1109/icc.2018.8422131.

33 Mo, J. and Heath, R.W. (2014). High SNR capacity of millimeter wave MIMO systems with one-bit quantization. *2014 Information Theory and Applications Workshop (ITA)*. IEEE, February 2014. https://doi.org/10.1109/ita.2014.6804238.

34 Chalise, B.K., Amin, M.G., Martone, A. et al. (2022). Optimum hybrid MVDR beamformer with sparse signal recovery approach. *2022 IEEE Radar Conference (RadarConf22)*, New York City, NY, USA, 1–6, https://doi.org/10.1109/RadarConf2248738.2022.9764295.

35 Jiang, Y., Feng, Y., and Varanasi, M.K. (2018). Hybrid beamforming for massive MIMO: a unified solution for both phase shifter and switch networks. *2018 10th International Conference on Wireless Communications and Signal Processing (WCSP)*. IEEE, October 2018. https://doi.org/10.1109/wcsp.2018.8555527.

36 Poon, A.S.Y. and Taghivand, M. (2012). Supporting and enabling circuits for antenna arrays in wireless communications. *Proceedings of the IEEE* 100 (7): 2207–2218. https://doi.org/10.1109/jproc.2012.2186949.

37 Ahmed, I., Khammari, H., Shahid, A. et al. (2018). A survey on hybrid beamforming techniques in 5G: architecture and system model perspectives. *IEEE Communications Surveys & Tutorials* 20 (4): 3060–3097. https://doi.org/10.1109/comst.2018.2843719.

38 Nosrati, H., Aboutanios, E., and Smith, D. (2020). Multi-stage antenna selection for adaptive beamforming in MIMO radar. *IEEE Transactions on Signal Processing* 68: 1374–1389. https://doi.org/10.1109/TSP.2020.2973544.

39 Wang, X., Aboutanios, E., Trinkle, M., and Amin, M.G. (2014). Reconfigurable adaptive array beamforming by antenna selection. *IEEE Transactions on Signal Processing* 62 (9): 2385–2396. https://doi.org/10.1109/tsp.2014.2312332.

40 Hamza, S.A. and Amin, M.G. (2019). Hybrid sparse array beamforming design for general rank signal models. *IEEE Transactions on Signal Processing* 67 (24): 6215–6226. https://doi.org/10.1109/TSP.2019.2952052.

41 Nosrati, H., Aboutanios, E., Smith, D., and Wang, X. (2019). Switch-based hybrid precoding in mmWave massive MIMO systems. *2019 27th European Signal Processing Conference (EUSIPCO)*. IEEE, September 2019. https://doi.org/10.23919/eusipco.2019.8902688.

42 Nosrati, H., Aboutanios, E., Wang, X., and Smith, D. (2022). Switch-based hybrid beamforming for massive MIMO communications in mmWave bands. *Signal Processing* 200: 108659. https://doi.org/10.1016/j.sigpro.2022.108659.

43 Rajamäki, R., Chepuri, S.P., and Koivunen, V. (2020). Hybrid beamforming for active sensing using sparse arrays. *IEEE Transactions on Signal Processing* 68: 6402–6417. https://doi.org/10.1109/TSP.2020.3032657.

44 Nosrati, H., Aboutanios, E., and Smith, D. (2021). Online switch-based hybrid beamforming for massive MIMO systems. *2020 28th European Signal Processing Conference (EUSIPCO)*. IEEE, January 2021. https://doi.org/10.23919/eusipco47968.2020.9287431.

45 Nosrati, H., Aboutanios, E., and Smith, D.B. (2018). Spatial array thinning for interference cancellation under connectivity constraints. *2018 IEEE International Conference on Acoustics, Speech and Signal Processing (ICASSP)*. IEEE, Apr 2018. https://doi.org/10.1109/icassp.2018.8462389.

46 Gao, X., Dai, L., Han, S. et al. (2016). Energy-efficient hybrid analog and digital precoding for MmWave MIMO systems with large antenna arrays. *IEEE Journal on Selected Areas in Communications* 34 (4): 998–1009. https://doi.org/10.1109/JSAC.2016.2549418.

47 Gao, Y., Vinck, H., and Kaiser, T. (2018). Massive MIMO antenna selection: switching architectures, capacity bounds, and optimal antenna selection algorithms. *IEEE Transactions on Signal Processing* 66 (5): 1346–1360. https://doi.org/10.1109/tsp.2017.2786220.

48 Ratnam, V.V., Molisch, A.F., Bursalioglu, O.Y., and Papadopoulos, H.C. (2018). Hybrid beamforming with selection for multiuser massive MIMO systems. *IEEE Transactions on Signal Processing* 66 (15): 4105–4120. https://doi.org/10.1109/tsp.2018.2838557.

49 Payami, S., Ghoraishi, M., Dianati, M., and Sellathurai, M. (2018). Hybrid beamforming with a reduced number of phase shifters for massive MIMO systems. *IEEE Transactions on Vehicular Technology* 67 (6): 4843–4851. https://doi.org/10.1109/tvt.2018.2807921.

50 Amin, M.G., Wang, X., Zhang, Y.D. et al. (2016). Sparse arrays and sampling for interference mitigation and DOA estimation in GNSS. *Proceedings of the IEEE* 104 (6): 1302–1317. https://doi.org/10.1109/jproc.2016.2531582.

51 Gharavi-Alkhansari, M. and Gershman, A.B. (2004). Fast antenna subset selection in MIMO systems. *IEEE Transactions on Signal Processing* 52 (2): 339–347. https://doi.org/10.1109/tsp.2003.821099.

52 Molisch, A.F. and Win, M.Z. (2004). MIMO systems with antenna selection. *IEEE Microwave Magazine* 5 (1): 46–56. https://doi.org/10.1109/mmw.2004.1284943.

53 Nosrati, H., Aboutanios, E., and Smith, D.B. (2017). Receiver-transmitter pair selection in MIMO phased array radar. *2017 IEEE International Conference on Acoustics, Speech and Signal Processing (ICASSP)*. IEEE, March 2017. https://doi.org/10.1109/icassp.2017.7952748.

54 Nosrati, H., Aboutanios, E., and Smith, D.B. (2017). Array spatial thinning for interference mitigation by semidefinite programming. *2017 25th European Signal Processing Conference (EUSIPCO)*. IEEE, August 2017. https://doi.org/10.23919/eusipco.2017.8081606.

55 Wang, X., Greco, M.S., and Gini, F. (2021). Adaptive sparse array beamformer design by regularized complementary antenna switching. *IEEE Transactions on Signal Processing* 69: 2302–2315. https://doi.org/10.1109/TSP.2021.3064183.

56 Wang, X., Amin, M., and Cao, X. (2018). Analysis and design of optimum sparse array configurations for adaptive beamforming. *IEEE Transactions on Signal Processing* 66 (2): 340–351. https://doi.org/10.1109/TSP.2017.2760279.

57 El Ayach, O., Heath, R.W., Rajagopal, S., and Pi, Z. (2013). Multimode precoding in millimeter wave MIMO transmitters with multiple antenna sub-arrays. *2013 IEEE Global Communications Conference (GLOBECOM)*. IEEE, December 2013. https://doi.org/10.1109/glocom.2013.6831611.

58 Ratnam, V.V., Molisch, A.F., and Papadopoulos, H.C. (2016). MIMO systems with restricted pre/post-coding—capacity analysis based on coupled doubly correlated Wishart matrices. *IEEE Transactions on Wireless Communications* 15 (12): 8537–8550. https://doi.org/10.1109/twc.2016.2616337.

59 Wang, X., Tan, C.P., Wang, Y., and Zhang, Z. (2021). Active fault tolerant control based on adaptive interval observer for uncertain systems with sensor faults. *International Journal of Robust and Nonlinear Control* 31 (8): 2857–2881.

60 Alkhateeb, A. and Heath, R.W. (2016). Frequency selective hybrid precoding for limited feedback millimeter wave systems. *IEEE Transactions on Communications* 64 (5): 1801–1818.

61 Palomar, D.P. and Fonollosa, J.R. (2005). Practical algorithms for a family of waterfilling solutions. *IEEE Transactions on Signal Processing* 53 (2): 686–695. https://doi.org/10.1109/tsp.2004.840816.

62 Sohrabi, F. and Yu, W. (2016). Hybrid digital and analog beamforming design for large-scale antenna arrays. *IEEE Journal of Selected Topics in Signal Processing* 10 (3): 501–513. https://doi.org/10.1109/jstsp.2016.2520912.

Index

Sparse Arrays for Radar, Sonar, and Communications, First Edition. Edited by Moeness G. Amin.

Printed and bound by CPI Group (UK) Ltd, Croydon, CR0 4YY
17/09/2026
14964313-0006